THE ORIGINS OF THE WORLD

THE ORIGINS OF THE WORLD

The Invention of Nature in the 19th Century

Edited by Laura Bossi

Musée d'Orsay/The Montreal Museum of Fine Arts
Muséum national d'histoire naturelle/Gallimard

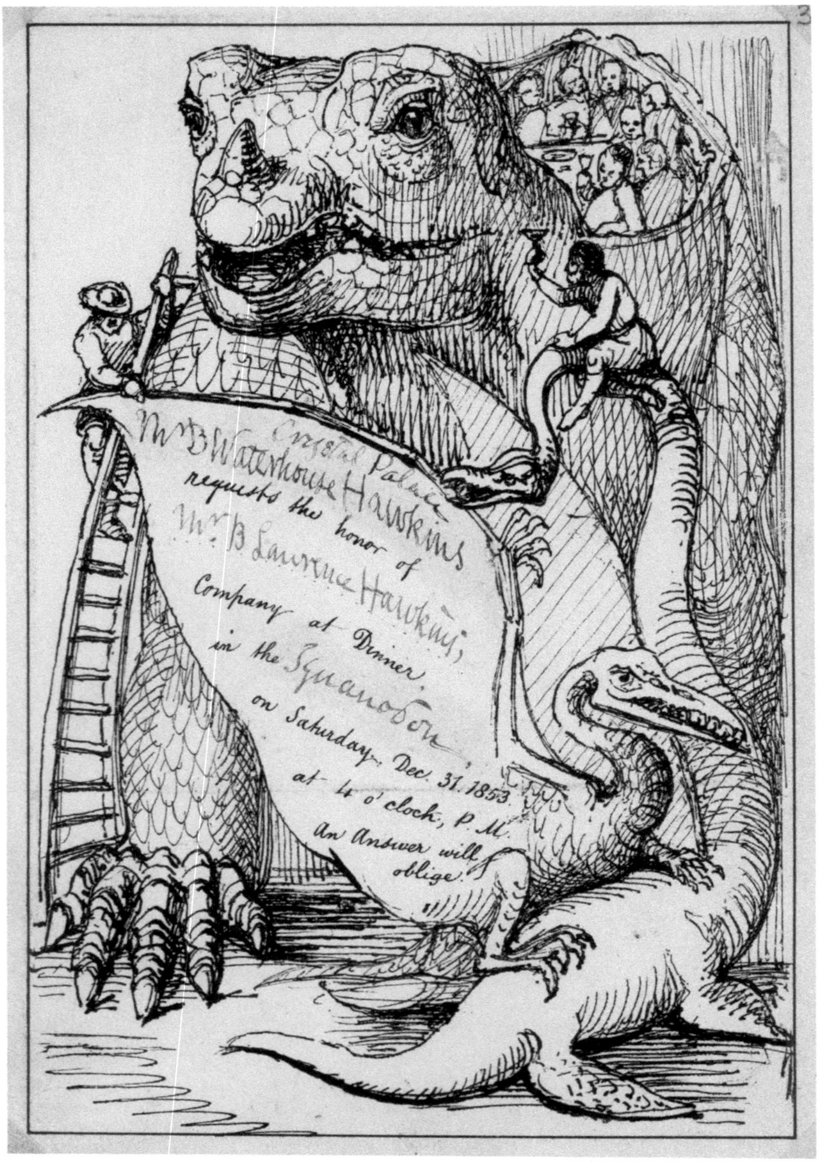

1
Benjamin Waterhouse Hawkins,
Invitation to the Iguanodon
Dinner at the Crystal Palace, 1853
Pencil and ink
London, The Natural History Museum

This catalogue was published in conjunction with the exhibition *Les origines du monde. L'invention de la nature au XIXe siècle/ The Origins of the World: The Invention of Nature in the 19th Century*

Paris, Musée d'Orsay, January – May 2021

Montreal, The Montreal Museum of Fine Arts, June 15 – October 12, 2021

The exhibition was organized by the Musée d'Orsay and Musée de l'Orangerie, Paris, and the Montreal Museum of Fine Arts.

EPMO

ÉTABLISSEMENT PUBLIC DES MUSÉES D'ORSAY ET DE L'ORANGERIE

MONTREAL MUSEUM OF FINE ARTS

The exhibition was produced with the outstanding partnership of the Muséum national d'histoire naturelle, represented by Professor Guillaume Lecointre.

MUSÉUM NAL HIST NATURELLE

With the generous support in Montreal of

Hydro Québec

The project was developed under the direction of Laurence des Cars, President, Musée d'Orsay and Musée de l'Orangerie, and Nathalie Bondil, Vice-Chair, Canada Council for the Arts.

CHIEF CURATOR
Laura Bossi,
Neurologist and science historian, Paris

EXHIBITION CURATOR, PARIS
Élise Dubreuil,
Curator of Decorative Arts, Musée d'Orsay
with the collaboration of
Lionel Britten,
Head of Documentation, Musée d'Orsay
Marie-Liesse Boquien, Jérôme Legrand and Clémentine Lemire,
documentary research officers, Musée d'Orsay

EXHIBITION CURATORS, MONTREAL
Jennifer Laurent,
Curator of Modern and Contemporary Decorative Arts, The Montreal Museum of Fine Arts

Hilliard T. Goldfarb,
Senior Curator – Collections and Curator of Old Masters, The Montreal Museum of Fine Arts

SCIENTIFIC ADVISORS
Pietro Corsi,
Emeritus Professor of the History of Science, University of Oxford and EHESS, Paris

Dario Gamboni,
Professor of Art History, Université de Genève

Guillaume Lecointre,
Teacher-researcher in Systematics and Evolution and professor, Muséum national d'histoire naturelle, Paris

EXHIBITION ORGANIZERS

MUSÉE D'ORSAY

Laurence des Cars
President, Musée d'Orsay and Musée de l'Orangerie

Francis Steinbock
Director, Administration

Virginie Donzeaud
Deputy Director, Administration

Sylvie Patry
Director, Curatorial Affairs and Collections

Hélène Flon
Director, Exhibitions

Élise Bauduin
Exhibition Manager

Guillaume Blanc
Director, Audiences

Luc Bouniol-Laffont
Director, Cultural Programming and Auditoriums

Aurélie Cauchy-Laure
Director, Development and International Relations

Hélène Charbonnier
Director, Digital Media

Amélie Hardivillier
Director, Communications

Exhibition design
Martin Michel

Lighting design
François Austerlitz

Graphic design
Costanza Matteucci and Sylvain Reymondon

THE MONTREAL MUSEUM OF FINE ARTS

Pierre Bourgie
President, Board of Trustees

Stéphane Aquin
Director

Yves Théoret
Deputy Director

Danielle Champagne
Director General, MMFA Foundation

Pascale Chassé
Director, Communications

Mary-Dailey Desmarais
Chief Curator

Mélanie Deveault
Director, Education and Wellness

Carolina Calle Sandoval
Head, Exhibition Administration

Sylvie Ouellet
Project Coordinator

France Trinque
Research Assistant

Richard Gagnier
Head, Conservation

Exhibition design
Exhibitions Production, MMFA with
Kanva, Montreal

To all of the collectors and institutions whose generous support made this exhibition possible, including those who have preferred to remain anonymous, we extend our deepest gratitude:

Austria

Linz
LENTOS Kunstmuseum Linz
Oberösterreichische Landesmuseum

Vienna
Albertina
Leopold Museum
Naturhistorisches Museum

Belgium

Brussels
Institut royal des sciences naturelles de Belgique
Musées royaux d'Art et d'Histoire, musée du Cinquantenaire

Canada

Montreal
McGill University Library, Rare Books and Special Collections
Canadian Centre for Architecture
McGill University, Redpath Museum

Ottawa
National Gallery of Canada

Czech Republic

Prague
Galerie Zdeněk Sklenář

France

Besançon
Musée des Beaux-Arts et d'Archéologie

Dijon
Musée des Beaux-Arts

La Roche-sur-Yon
Musée de la Roche-sur-Yon

Le Havre
Musée d'Art moderne André-Malraux
Muséum d'histoire naturelle

Le Puy-en-Velay
Musée Crozatier

Les Eyzies
Musée national de Préhistoire

Montbard
Musée Buffon

Montbéliard
Musées de Montbéliard

Nancy
Musée de l'École de Nancy

Paris
Beaux-Arts de Paris
Bibliothèque Mazarine
Bibliothèque nationale de France
Centre Pompidou, musée national d'Art moderne/Centre de création industrielle
Musée Carnavalet – Histoire de Paris
Musée des Arts décoratifs
Musée du Louvre
Musée Marmottan Monet
Musée national Gustave-Moreau
Musée Rodin
Muséum national d'histoire naturelle
Petit Palais, musée des Beaux-Arts de la Ville de Paris

Reims
Musée des Beaux-Arts

Rennes
Musée des Beaux-Arts

Sens
Musées de Sens

Germany

Berlin
Berlin-Brandenburgische Akademie der Wissenschaften
Berlinische Galerie – Landesmuseum für Moderne Kunst, Fotografie und Architektur
Die Stiftung Preußische Schlösser und Gärten Berlin-Brandenburg
Staatliche Museen zu Berlin, Nationalgalerie

Frankfurt
Städel Museum

Jena
Friedrich-Schiller-Universität, Ernst-Haeckel-Haus

Munich
Städtische Galerie im Lenbachhaus und Kunstbau München

Stuttgart
Staatsgalerie Stuttgart

Israel

Jerusalem
Israel Museum

Italy

Naples
Museo e Real Bosco di Capodimonte

Rome
Galleria Borghese

Venice
Fondazione Musei Civici di Venezia – Ca' Rezzonico, Museo del Settecento veneziano

Vicenza
Intesa Sanpaolo Collection, Gallerie d'Italia – Palazzo Leoni Montanari

Monaco

Institut océanographique, Fondation Albert Ier, Prince de Monaco – Musée océanographique de Monaco

Netherlands

The Hague
Kunstmuseum Den Haag

Otterlo
Kröller-Müller Museum

Norway

Oslo
Munch Museet

Sweden

Stockholm
Nationalmuseum
The Hilma af Klint Foundation

Switzerland

Basel
Anatomisches Museum Basel
Kunstmuseum Basel

Bern
Kunstmuseum Bern

Zug
Kunsthaus Zug

United Kingdom

Cambridge
Cambridge University Library
Sedgwick Museum of Earth Sciences, University of Cambridge

Cardiff
Amgueddfa Cymru – National Museum Wales

London
National Maritime Museum
Tate
Natural History Museum
Royal Collection Trust / HM Queen Elizabeth II

Manchester
The Whitworth, University of Manchester

Oxford
Ashmolean Museum, University of Oxford

United States

Dallas
Dallas Museum of Art

Indianapolis
Indianapolis Museum of Art at Newfields

Los Angeles
Los Angeles County Museum of Art

Milwaukee
Layton Art Collection, Inc. at the Milwaukee Art Museum

New Haven
Yale Center for British Art
Yale University Art Gallery

New York City
The Metropolitan Museum of Art

Reading
Reading Public Museum

Richmond
Virginia Museum of Fine Arts

St. Louis
Saint Louis Art Museum

Washington, D.C.
National Gallery of Art

Private collections

Collection de Bueil & Ract-Madoux
Collection of Ramy Boutros
The Jack Daulton Collection

Our thanks go as well to all those who contributed to the production of the exhibition and this catalogue:
Sébastien Allard, Karin Althaus, Christine Argot, Thomas Bach, Gabrielle Baglione, Maria Balshaw, Sylvain Bellenger, Gabriella Belli, Evelyn Benesch, Giovanna Bertazzoni, Claire Berthommier, Bernard Blistène, Christina Brandt, David Brown, Janet Browne, Jean-Gabriel de Bueil, Joe Cain, Robert Calcagno, Patrick de Carolis, Guillaume Cerutti, Cécilie Champy, Austin Chinn, Mariolina Cilurzo, Jean Clair (Gérard Régnier), Cédric Crémière, Jacques Cuisin, Servane Dargnies-de Vitry, Cécile Debray, Lucile Desmoulins, Jean-Marc Drouin, Blaise Ducos, Jannic Durand, Guillaume Faroult, Alex Farquharson, Cécile Forest, Giles Forster, Matthias Frehner, Ginny Fremlin-Key, William (Ned) Friedman, Olivier Gabet, Thomas Gaehtgens, Guillaume Gaillard, Donatien Grau, Claude Bernard Haim, Doede Hardeman, Andréa Hart, Stefanie Heckmann, Josef Helfenstein, Sten Olaf Henrichsen, Annegret Hoberg, Nathalie Houzé, Sophie Jugie, Tim Knox, Valia Lamprou, Christine Lancestremère, Gabriele Langer, Bruno Latour, Brigitte Léal, Laurent Le Bon, Christophe Leribault, Adrian Lister, Lord Lloyd Webber, Olivier Lorquin, Silvia Lucantoni, Jean-Luc Martinez, Sarah Miller, James Moore, Dominique Morel, Giuseppe Olmi, Édouard Papet, Patrizia Piscitello, Alain Prochiantz, Chloë Raymond, Sofia Rinaldi, Martin Rudwick, Lord Sainsbury, Xavier Salmon, Andrea Schlieker, Katharina Schmidt, Stéphane Schmitt, Sabine Schulze, Silvia Sebastiani, Desmond Shawe-Taylor, Sabine Sobodka, Werner Spies, Nicolas Surlapierre, Benno Tempel, Valérie Thomas, Irène Tschopp, Alfred Weidinger, Nina Zimmer.

AT THE MUSÉUM NATIONAL D'HISTOIRE NATURELLE

Bruno David,
President
Emmanuel Skoulios,
Deputy Executive Director, Resources
Jérôme Gestin,
Assistant Deputy Executive Director, Resources
Michel Guiraud,
Deputy Executive Director, Collections
Gildas Illien,
Director, Libraries and Documentation
Joëlle Garcia,
Head of Knowledge Dissemination and Mediation, Libraries and Documentation Department
Alice Lemaire,
Head of Conservation, Restoration and Digitization, Libraries and Documentation Department
Stéphanie Decobert,
Director, Communication
Agnès Parent,
Director, Audiences
Lucile Desmoulins,
Head, Consumer Publishing
With the close cooperation of all the department heads and their teams.

AT THE MUSÉE D'ORSAY

William Bartoletti, Stéphane Bayard, Catherine Bony, Myriam Bru, Manuel Caria, Carolina Carpinschi, Renaud Cesson, Augustin Chaunu, Jeanne Chevrier, Michaël Chkroun, Silvia Cristini, Bruno Dapaz, Armelle Dobrenel, Rodolphe Doucet, Béatrice Féna, Didier Frémond, Françoise Fur, Thomas Galifot, Miléna Girardin, Axelle Glapa, Fabrice Golec, Patrick Gomas, Philippe Gomas, Fiona Gomez, Gary Granchon, Laureen Grant, Marie-France Gripois, Marion Guillaud, Félix Guion-Firmin, Pascal Haffner et le détachement de sapeurs-pompiers du musée d'Orsay, les chefs de centrale, Sonia Hamza et Erwan Rivière ainsi que toutes les équipes de l'accueil et de la surveillance, Laurence Imbert, Gabrielle Lacombe, Sandrine Lambert, Cyrille Lebrun, Ludovic Le Goff, Emmanuel Leruyet, Matthieu Leverrier, Agnès Marconnet, Nathalie Mengelle, Odile Michel, Isabelle Morin Loutrel, Laurent Morizet, Jean Naudin, Romuald Picard, Patrick Porcher, Rémy Rappart, Scarlett Reliquet, Denis Rollé et les installateurs du musée d'Orsay, Patrice Schmidt, Rachel Scrivo, Élodie Tamburrini, Barbara Topalov, Catherine Tudoret.

AT THE MONTREAL MUSEUM OF FINE ARTS

Louisa Blair, Marc Chauveau, Mélisande Chauveau, Violette Chauveau, Giles Forster, Jean-François Gauvin, Gilles Genty, Anouk Gingras, Ann Marie Holland, Henry House, Anthony Wayne Howell, Erell Hubert, Louis-Xavier Joseph, Hans Larsson, Ken Leach, Annie Lussier, Lauren Williams, Janet Young.

The curators extend special thanks to all the members of the Montreal Museum of Fine Arts team for their commitment and for the support they provided throughout this project.

The exhibition *The Origins of the World: The Invention of Nature in the 19th Century* is presented in Montreal by Hydro-Québec. The Museum would like to acknowledge the invaluable contributions of its official sponsors, Air Canada and Denalt Paints, the Angel Circle of the Montreal Museum of Fine Arts and its media partners Bell, *La Presse* and *Montreal Gazette*.

The exhibition benefited from Heritage Canada's generous Canada Travelling Exhibition Indemnification Program. The Museum is grateful to the Ministère de la Culture et des Communications and to the Conseil des arts de Montréal for their ongoing support. The Museum's International Exhibition Program receives funding from the Exhibition Fund of the Montreal Museum of Fine Arts Foundation and from the Paul G. Desmarais Fund.

The Museum would also like to thank its Volunteer Guides for their unflagging support, as well as the members and various individuals, corporations and foundations that magnanimously contribute, especially the Fondation de la Chenelière, headed by Michel de la Chenelière, and Arte Musica, headed by Pierre Bourgie.

The Museum sincerely thanks all those whose generous help, encouragement and support made this exhibition and scholarly publication possible.

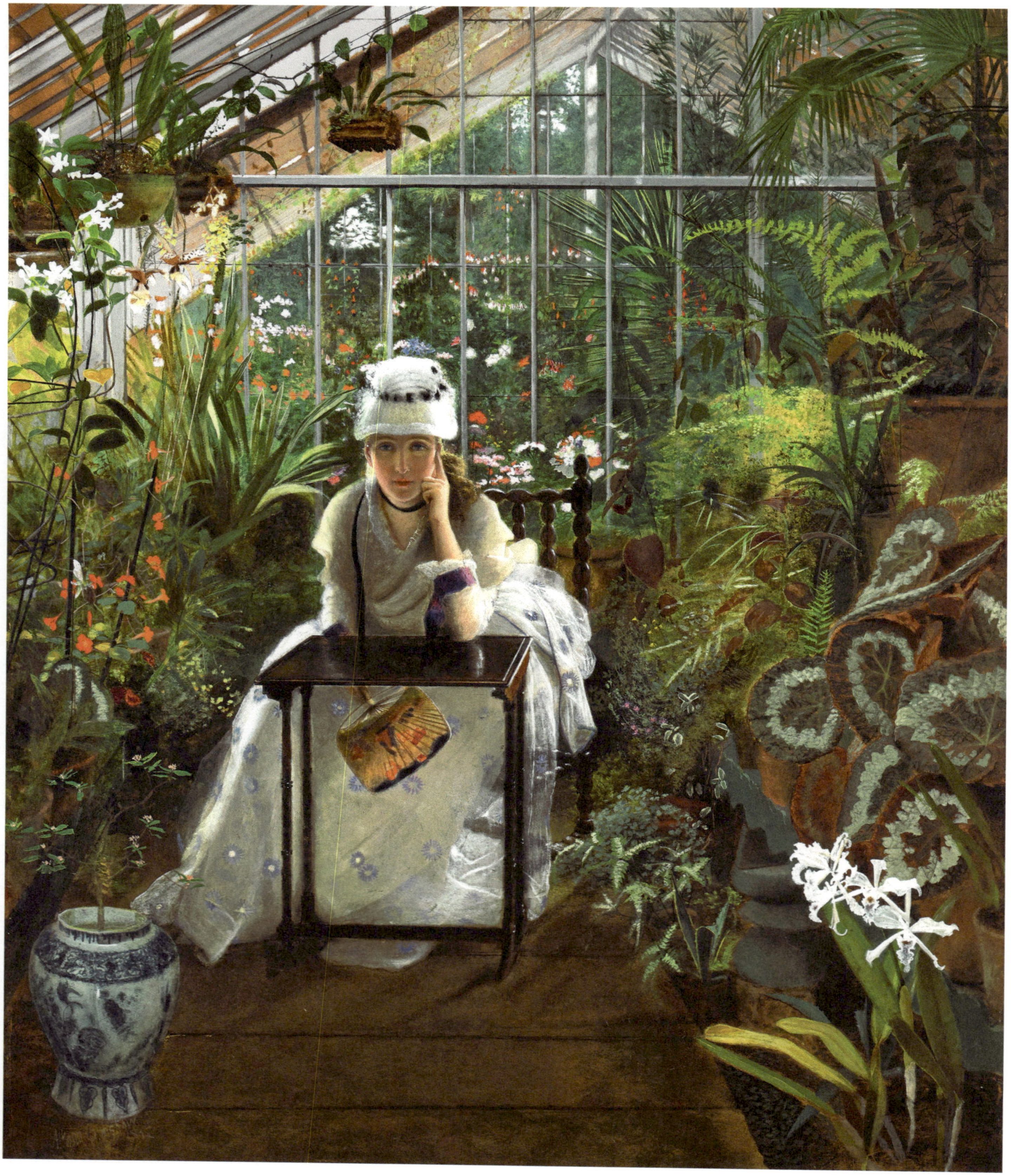

2
John Atkinson Grimshaw,
Il Penseroso, 1875
Oil on cardboard, 59.7 × 49.5 cm
Private collection

The Origins of the World: The Invention of Nature in the 19th Century would not have been possible without the contribution of the many people we wish to thank here.

We are grateful for the collaboration with the Montreal Museum of Fine Arts, with its former Director General and Chief Curator, Nathalie Bondil, who programmed the exhibition and oversaw its development with enthusiasm and conviction. The Montreal Museum of Fine Arts will host the exhibition's second showing in summer 2021. Special thanks go to the Muséum national d'histoire naturelle, a major exhibition partner and outstanding lender, and to its president, Bruno David, and all of its teams for supporting this project with exceptional generosity. Without the vital commitment of these institutions and individuals, the exhibition would have been inconceivable.

Laura Bossi, a neurologist and science historian, accepted the challenge of dealing with the exhibition's complex subject. As chief curator, she did a remarkable job working in close cooperation with the other curators: Élise Dubreuil, Curator of Decorative Arts at the Musée d'Orsay for the Paris showing, and at the Montreal Museum of Fine Arts for the Canadian showing, Hilliard T. Goldfarb, Senior Curator – Collections and Curator of Old Masters, and Jennifer Laurent, Curator of Modern and Contemporary Decorative Arts. In Paris, the exhibition was prepared by Lionel Britten, Head of Documentation at the Musée d'Orsay, and Marie-Liesse Boquien, Jérôme Legrand and Clémentine Lemire, documentary research officers. Hélène Flon, Director of Exhibitions, and Élise Bauduin, Exhibition Manager, at the Musée d'Orsay, handled the production and coordination, and Martin Michel created the exhibition design. The exhibition also benefited from the expertise and insights of top-flight scientific advisors: Pietro Corsi, Emeritus Professor of the History of Science at Oxford University and EHESS, Paris, Dario Gamboni, Professor of Art History at Université de Genève, and Guillaume Lecointre, Teacher-researcher in Systematics and Evolution, and professor at the Muséum national d'histoire naturelle. To all of them we offer our sincere gratitude for their commitment and magnificent work.

We are indebted as well to the many institutional and private collection lenders who, despite a complicated public health situation and its attendant logistical challenges, graciously agreed to entrust us with their works.

LAURENCE DES CARS
President, Musée d'Orsay and Musée de l'Orangerie

NINA ATHANASSOGLOU-KALLMYER
Emerita Professor of Art History, University of Delaware

GABRIELLE BAGLIONE
Curator, Graphic Arts Collection, Muséum d'histoire naturelle du Havre

CLAUDE BLANCKAERT
Emeritus Research Director, CNRS (Centre Alexandre-Koyré, Paris), Honorary Associate, Muséum national d'histoire naturelle, Paris

LAURA BOSSI
Neurologist and science historian, Paris

MARIE-NOËLLE BOURGUET
Emerita Professor of Modern History, Université Paris Diderot

FAE BRAUER
Professor of Art and Visual Culture and Director of Research, University of East London; Honorary Professor of Art History and Cultural Theory, UNSW Art & Design, Sydney

DANIELLE CHAPERON
Professor, Université de Lausanne

PHILIPPE COMAR
Visual artist, exhibition designer, exhibition curator and writer, Paris

PIETRO CORSI
Emeritus Professor of the History of Science, University of Oxford and EHESS, Paris

ÉLISE DUBREUIL
Curator of Decorative Arts, Musée d'Orsay, Paris

SABINE DU CREST
Lecturer in Art History, Université Bordeaux-Montaigne

DARIO GAMBONI
Professor of Art History, Université de Genève

JOËLLE GARCIA
Head of Knowledge Dissemination and Mediation, Libraries and Documentation Department, Muséum national d'histoire naturelle, Paris

JACQUELINE GOY
Scientific Officer, Institut océanographique – Fondation Albert Ier, prince de Monaco

LAURENT GRASSO
Artist

PASCALE HEURTEL
Chief Curator and Director of Libraries, Assistant to the Chancellor for Heritage, Information and Scientific and Technological Culture, Conservatoire national des arts et métiers, Paris

THIERRY HOQUET
Philosopher and professor, Université Paris Nanterre

BARBARA LARSON
Professor of Art History, University of West Florida, Pensacola

GUILLAUME LECOINTRE
Teacher-researcher in Systematics and Evolution and professor, Muséum national d'histoire naturelle, Paris

ALICE LEMAIRE
Head of Conservation, Restoration and Digitization, Libraries and Documentation Department, Muséum national d'histoire naturelle, Paris

MICHELLE LENOIR
Honorary Director of Libraries and Honorary Associate, Muséum national d'histoire naturelle, Paris

CAMILLE LORENZI
Professor of History and Geography, Lycée Arago, Villeneuve-Saint-Georges

VIRGINIE MARIS
Research Director, CNRS, Centre d'écologie fonctionnelle et évolutive, Montpellier

ELLINOR MICHEL
Scientific Associate, Department of Life Sciences, Natural History Museum, London

ARNAULD PIERRE
Professor of Art History, Sorbonne Université, Paris

NATHALIE RICHARD
Professor of Contemporary History, TEMOS Laboratory, Le Mans Université

BERTRAND TILLIER
Professor, Université Paris 1 Panthéon-Sorbonne

MARCOS UZAL
Film critic, editor-in-chief of *Cahiers du cinéma*, Paris

JULIA VOSS
Exhibition curator, Professor of Art History, Leuphana University, Lüneburg

NICOLAS WANLIN
Professor of Literature, École polytechnique (IP Paris)

PIERRE WAT
Professor of Contemporary Art History, Université Paris 1 Panthéon-Sorbonne

MARK P. WITTON
Research Associate, University of Portsmouth

CONTENTS

HISTORY OF ART, HISTORY OF IDEAS

LAURENCE DES CARS
President, Musée d'Orsay and Musée de l'Orangerie
NATHALIE BONDIL
Vice-Chair, Canada Council for the Arts

It is rare that an exhibition resonates so intensely with the spirit of its day. Indeed, what better time to examine the place of humankind in nature than the complex period we are living through, which confronts us with the burning topicality of the question?

The Origins of the World: The Invention of Nature in the 19th Century delves into the interrelation of arts and sciences in the very century in which arose the issues we continue to grapple with today. In many respects, the long nineteenth century that stretched to the outbreak of the First World War was a pivotal moment, essential to understanding our present day.

Set against the backdrop of the Industrial Revolution and the rise of colonial empires, this critical period saw the cataloguing of nature take shape at the same time as modern science, fuelled by the discoveries of countless scientific expeditions, was consolidating its substance. As knowledge progressed in every field, science organized and systematized itself; and in the century of the ordering of the world came the invention of the modern museum, replacing the eclectic cabinet of curiosities.

This was also the era in which the still-porous borders between arts and sciences allowed exchanges and back-and-forths between these two spheres of knowledge. Fed by abounding exploration, the imaginary of science and nature was still very powerful, with each milestone reached in scientific progress prompting a parallel in the arts. The exhibition reveals the close and often little-known ties that connect the development of these spheres. The nineteenth was a century of fruitful interactions, before art and science were definitively separated in the twentieth.

Above all, the nineteenth century saw the emergence of many questions of which we today are the direct heirs. It was a paradoxical time, in that humans' increasingly sophisticated knowledge of nature was coupled with an ever more predatory relationship, which was already apparent to contemporary thinkers. Writing the entry titled "Man" for the *Nouveau dictionnaire d'histoire naturelle*, published in 1817, Lamarck observed, "It seems that man is destined to exterminate himself after having rendered the earth uninhabitable." Thus, it is no surprise that the nineteenth century saw both the beginning of the massive destruction of nature and the birth of ecology.

Central to this pivotal century, of course, was the great scientific upheaval caused by the 1859 publication of Charles Darwin's *On the Origin of Species*, which radically reversed humans' long-understood place in the world, depriving them forever of the comfortable Cartesian status of "masters and possessors of nature."

Dispossessed of transcendent standing, Darwin's humans now occupied a relative position within the infinite diversity of living things, in a world understood today as an ecosystem. The history of modernity is closely linked to this narcissistic wound inflicted by the Darwinian theory: in brutally stripping humans of transcendence, it painfully questioned their place on Earth, as much in the existential dread of their own demise as in the obsession with their origins. Where did humans now fit within nature, vis-à-vis the animal world, and their own animality? All questions inherited from the nineteenth century but no less relevant today in the face of another mass extinction of species and acute awareness of our fragility. And the dual pitfall—beatification vs. predation of nature—to which we still succumb illustrates how deeply bound to that era we remain. The nineteenth century stands as the century of the genesis of scientific modernity in all respects, and while it provides keys for understanding our present-day world, it still has a great deal to tell us.

Few fine arts museums have tackled this important subject. Yet to our minds it certainly belongs there, perhaps today more than ever. True to our conviction that a fine arts museum must reflect the questions of its time, and in line with exhibitions such as *Black Models: From Géricault to Matisse*, presented at the Musée d'Orsay in spring 2019, we wanted this exhibition and the accompanying catalogue to be a true forum, open to widely diverse thinkers and disciplines invited to engage in dialogue about complex questions.

An opportunity to remind us all that the history of art, far from being an isolated discipline, is very often the mirror of the history of ideas, of sensitivities and of representations. A polyphony echoing the museum itself, in both the intentions and the execution of the exhibition. Exhibiting works of art and scientific objects side by side is always a complex but passionate gamble, with all the challenges that such couplings pose in terms of presentation and mediation, in particular for young audiences and school groups.

Opening minds to the history of ideas and to current issues, but rereading them from the long perspective of knowledge: for us, that too is the mission of a great national museum. An invitation to humility but also, and especially, to curiosity.

WHY DARWIN?

BRUNO DAVID AND
GUILLAUME LECOINTRE
Muséum national d'histoire naturelle

The nineteenth century was extraordinarily rich in scientific advancement. Over the course of those hundred years, entire fields of knowledge were opened up by the likes of Urbain Le Verrier and his discovery of Neptune, Louis Pasteur in medicine, Georges Cuvier in comparative anatomy and the geologist Charles Lyell, a close friend and admirer of Charles Darwin, whose thinking inspired the latter 1800s and still holds sway. Darwin made an enormous impact on science. His theoretical work was a program of research that has borne fruit for more than a hundred and fifty years. Enriched by confirmations and amendments, his theory has been reinforced and now fuels sciences as diverse as cultural anthropology, medicine, biochemistry and algorithmics, among others. Yet it may be surprising to learn that it underlies the latest cancer therapies, or that it explains the transmission of musical repertories in ethnomusicology, or that it elucidates our empathy towards other species. And that's without counting the topics he addressed in relation to the founding questions of disciplines as yet unnamed in his day, such as ethology, ecology and phylogenetics (in fact, the last two were named by one of his contemporaries, Ernst Haeckel). It was as a scientist that Darwin proposed a natural origin of species, including the human species, and a natural origin of our behaviours, but, unintentionally, his impact proved to be cultural, spanning the spheres of religion and the imagination in a single wave. Arts like literature, painting, sculpture and architecture came to echo his influence in speaking of nature and the way we fit into it. Darwin is the pivot point of his century. This is a heavy honour to bear, for few scientific authors have been as copiously invoked, as credited with inspiration, as fiercely contested or instrumentalized as Charles Darwin, often by those who have never read his work. Still today, he is quoted in scholarly biology articles, often to lend them authority. Still today, Darwin rumours and clichés abound, and even writers who have read his work may have done so through vision-distorting glasses. Opticians offer a variety of models on the ideological rack.

A CHOICE OF GLASSES

There is the myopic model worn by contemporary moralizers, who choose to see Darwin as a racist or, like American creationists, as the instigator of a "morally, culturally and politically destructive" wicked science. There is the old-fashioned model favoured in certain human and social science circles that see in Darwin only the root of the techno-progressive evolutionism of the very end of the nineteenth century or of the reductionist biology of the 1970s, with its overemphasis on the power of genes, known as sociobiology. These vagaries of history explain why, in the early twenty-first century, Darwin still smacks of brimstone in some French university departments. A long-time favourite is the corrective model of the great ornithologist Ernst Mayr, called "species realism," which is so misleading that the wearer loses sight of what *On the Origin of Species* is actually talking about; in fact, some lazy science historians claim it contains nothing about the origin of species. This is still the model of choice for teaching evolution in French high schools, which explains in part why evolution is still so badly understood. There is also a type of distorting glasses that anachronistically ascribe to Darwin the excesses of neo-Darwinism from the 1960s to the 1990s: instead of rightly pointing to naive reductionism and genetic determinism, they blame Darwin. Another choice is the old banally reductive Spencerian model, dear to the neo-liberal economics circles that equate natural selection with competition. And, lastly, there are the white-framed lenses of the righters-of-wrong who see Darwin as a mere repeater of ideas formulated before him by authors he fails to cite, glasses ignorant of the very nature of theoretical work in science. Darwin was dealing with a divisive subject and he knew it: he hesitated, provoking three no doubt unhoped-for effects in the year following the publication of *On the Origin of Species*.

DARWIN'S THREE EFFECTS

First, there was the *eagerly awaited* Darwin. Science had matured: Lamarck was discussed favourably, Chambers had tried his hand, Lyell held authoritative sway, Darwin was organizing his ideas in his 1837 notes and 1842 draft, and Wallace was not far behind. It was these circumstances, as well as his friends, that pushed Darwin to hastily publish *On the Origin of Species* in November 1859. The book was an immediate success, and a second edition had to be printed in early 1860.

There soon followed the *appropriated* Darwin. Beginning in 1860, Herbert Spencer cherry-picked ideas from *The Origin* to develop a competitionist, anti-statist evolutionary philosophy applied to liberal Victorian society, for which the French journalist Émile Gautier regrettably coined the term Social Darwinism in 1880, even though Darwin had rejected such thinking in his 1876 autobiography. Later, Peter Kropotkin in 1889 and Élisée Reclus in 1905 would see the natural origin of mutual aid in an expanded concept of natural selection with more complex consequences, as discussed by Darwin in 1871.

And there was the *contested* Darwin. The Darwinian gesture, which was to offer a natural explanation of the origin of species and of humans, signalled scientific independence in the discourse on the origins of everything in existence, independence from the religious authorities who, until then, had governed what was acceptable or not (think of Buffon's squabble with theologians about Earth's age in 1751). Many were irked by this show of autonomy, and in 1860 the Bishop of Oxford, Samuel Wilberforce, unleashed a polemic against the Darwinians. It bears noting that in many countries Darwin's theoretical and factual legacy is still contested: the battleground is no longer in laboratories or churches but in schools.

Aside from science, there was a fourth, slower-paced effect: that of Darwin as the inspiration for literary works, paintings, architecture and the arts in general. This book deals with that effect. Darwin's influence in this sphere did not land like a punch; rather, it spread slowly and powerfully, like an unstoppable tidal wave. His vision would change the way Western culture considered nature and humans. The purpose of this publication is to show how that came about. But first, we shall attempt to characterize what constitutes the power of the Darwinian gesture.

WHY DOES DARWIN SEEM SCANDALOUS, AND WHY IS HE TRULY PROFOUND?

Why such controversy? Because where theology saw proof of the existence of the Great Clockmaker, Darwin saw a natural origin of the correspondence between biological forms and functions. This is an old story but the debate persists: today's American neo-creationism, which adheres to the concept of Intelligent Design, revives the original principle of God as the prime mover of a perfect design that keeps the world in proper order. Conversely, Darwin, in the same gesture and with openly comparatist reasoning, proposed a natural origin of species and humankind, of their behaviours and social relationships. He began by proposing a non-theological origin of similarity: initially of similarity within species, with the principle of natural selection, then of similarity between species, with the principle of theoretical genealogy.

SIMILARITY WITHIN SPECIES

In reading *On the Origin of Species*, it is important to see the similarities that came to define species as a linguistic convenience serving to express the diversity of the living world in a structured manner. For Darwin, species do not exist. There are only individuals whose differences fluctuate randomly. Horticulturalists and breeders seize on these variations to develop genealogical lineages to produce forms unknown in nature. Together, the individuals have the capacity to multiply. Whatever the species, in every generation and without pause, many more individuals than the environment can accommodate are born. Frog spawn is composed of hundreds of eggs; the spawn of a single teleost, or bony fish, contains hundreds of thousands of eggs (the record is 47 million). Then there is chance: the individuals differ in each of their structures and each of their functions, at every scale of observation. Every biological individual is unique, hence the population is variable. Consequently, each anatomical entity, such as the eyes, manifests diverse performances within a population. And most traits that cause the performance are heritable. Then there are limits, or constraints: the individuals that see less well are less able to escape their predators or detect their prey. On average, their ability to transmit their traits is less effective. Over time, their heritable variations, and the related poorer performances, are excluded from the genealogical lineage. The deficient eyes eventually disappear; the eyes that continue to perform well are maintained. This "pruning" results in a regularity, a recognizable, efficient structure. Chance proposes, constraint disposes. With this simple principle, Darwin offers a straightforward explanation of the origin of similarity and the natural reason for which bodily organs seem to be tailored to their functions. Unlike William Paley's natural theology, which taught that the conformity between organs and their efficient operation proved that they had been built by a divine clockmaker, Darwin offers a natural origin of the correspondence, albeit relative, between form and function. Ultimately, all things that multiply, vary and transmit within a species are sooner or later bound to encounter limits in this finite world. Such is the overarching principle that the living world, to which we belong, teaches us. It is called natural selection, and today we know that it concerns all the cells of our bodies, our microbiota, and all the populations of all the species, with their behaviours and their cultures. The regularity of the structure of beings and the performance of their functions are the result of chance, of course, but not just chance: the limits imposed by their environments have structured them, at the cost of a gigantic invisible cemetery. All species, irrespective of type, exist only because they have been unsparing in individual deaths.

DARWIN DOES NOT EXPLAIN WHY THINGS CHANGE, BUT WHY THINGS DO NOT CHANGE

The explanation of why things change lies at the chemical and physical level, not the biological level. Accordingly, Darwin started from the disorder of the spontaneous variation of living beings (not from an order of transcendent origin intrinsic to the living world) and proposed a principle that explains how a regularity results from random fluctuations through the combination of the properties of variation, proliferation and transmission, which are all subject to constraints. Natural selection does not at first explain how a species changes, but it explains why it does not change even though its individual constituents do. That is what Ernst Mayr and his pedagogical heirs failed to take into account, with the result that at school we were taught that natural selection explains how species change. But the subtitle of *On the Origin of Species* contains neither the word "evolution" nor "transformation" nor "transmutation." It contains the word "preservation": *The Preservation of Favoured Races in the Struggle for Life*. It is through "pruning," through a permanent sifting out of forms that diverge too far from environmental constraints, that the average form is preserved and, for the purposes of our language, is called "species." Thus, Darwin adopted a *nominalist* vision, incompatible with the philosophical foundations of monotheisms. He completely reversed the causal principle: in biology, order or regularity is not a cause but something that must be explained. Specifically, given that the alteration of matter and entropy rule at the physical-chemical level, how, at a more integrated level, can similarity be produced *in the short term*? True, Maupertuis had already intuited this in 1754, but he did not infer an actual scientific theory from it. Shockingly, in the 1960s, the metaphor of the "genetic program" made order in biology causal again, and current efforts to eliminate this epistemological treachery are no surprise. The principle of natural selection is not *exclusively* conservative, of course: if the environment changes, as it sooner or later inevitably does, then the sifting-out process changes and the average form shifts. Natural selection also explains how species change *over the long term*.

SIMILARITY BETWEEN SPECIES

When Darwin theoretically conceived a general genealogy of the entire living world, he was answering the question of shared attributes. Any child can see that dogs and cats, too, have eyes and a mouth. What accounts for attributes shared by individuals incapable of breeding together? The substantiated scientific answer is common ancestry. Long ago, the ancestors of cats and the ancestors of dogs procreated together, and for many generations they passed on certain shared traits while other traits changed (without which dogs and cats would be similar still today). In such a family tree, the species necessarily becomes a linguistic convention. If becoming a species over the long term means seeing one's form change, or one's genealogical flow split into daughter branches, what becomes of the species? What becomes of its limits? The question cannot be resolved in essentialism, unless one disregards the diachronic nature of species, as did Ernst Mayr, who practised a "realism of species," favouring a here-and-now definition grounded more in empirical and pragmatic relevance than in theory. Darwin, no doubt unintentionally, produced a profoundly anti-essentialist scientific theory, which is why political systems heavily influenced by monotheistic religions curtail its teaching in schools.

In conclusion, without confining biological variations to transcendent concepts, Darwin explains at once the natural origin of similarity within species (why do individuals so closely resemble each other?), the similarity between species (why do I share traits with a dog or a cow?), the origin of correspondence between form and function in nature, and the origins of humans and of their behaviours. These considerations explain why Darwin was pivotal in a Western-centric century where the sciences were taken seriously, so much so that few areas of the world's cultural sphere were untouched by the question of the natural origin of what makes us.

NATURE UNVEILED TO SCIENCE

LAURA BOSSI

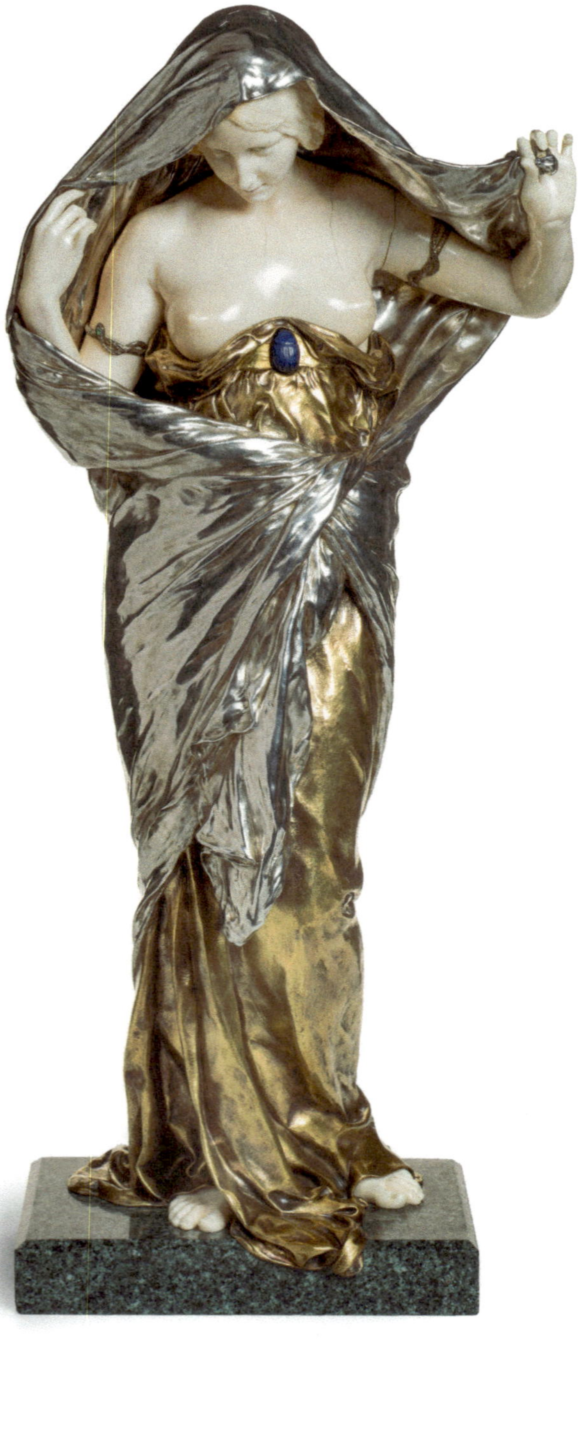

3
Louis Ernest Barrias,
Nature Unveiling Herself Before Science, 1903
Ivory, gold- and silver-plated bronze, lapis lazuli, marble, h. 58 cm
Paris, Musée des Arts décoratifs

It has never been so imperative that we think about nature, as it is currently the subject of increasingly anxious interrogation.[1] Climate change has put the human habitat to the test, revealing it to be a fragile refuge. The growing scarcity of many animal species, forecast by Rachel Carson in her 1962 book *Silent Spring*, has raised awareness of the depletion of the living world that scientists now describe as a "sixth extinction."[2] The fact that human responsibility is involved in such transformations of the environment—to the point that it has prompted the announcement of a new epoch, the Anthropocene[3]—makes us question their compatibility with our humanity, and even our survival.[4] The political scene has also seen the wider presence of environmental movements and the demand that species and ecosystems be protected in international law.[5] Numerous environmentalists, along with the critics of environmentalism, share the conviction that human beings exist outside the natural world, which they artificialize.[6] Humanity's destruction of nature is also a central theme in the work of many contemporary artists who, at times in an ingenuous or militant manner, attempt to examine our relationship with plants and animals, or imagine an "anthropized" future.

In contrast, some philosophers and scientists would find it necessary to radically reconsider our view of humankind and nature, and to reject the divide between nature and culture, nature and society, human history and the history of life and the Earth. With his "Gaia hypothesis," James Lovelock, the most fervent advocate of a vision of the Earth as a single organism, "a living being," revived myths reconnecting with the pantheism that stemmed from Romantic holism.[7]

Nevertheless, the pandemic has shown a darker side of nature. Demiurges that we are, prey to a hubris we need to control in order to "protect" an endangered natural world, we have awakened to the consciousness of being vulnerable, threatened by the possibility of a premature death. And we have remembered the ancient personifications of nature as an unpredictable woman who "manifests herself" in natural disasters—famines, storms, volcanic eruptions, earthquakes, droughts, epidemics and more.[8] Prometheus begat Pandora, who brought every evil to humankind, leaving it with hope alone.[9]

In order to "think" about our relationship with nature, it behooves us to take a detour to the nineteenth century, to the time when humanity's relationship with the natural world dramatically changed. With its great exploratory land and, later, oceanographic expeditions, the mapping of the world extended to every continent; railroads and the telegraph shattered distances and transformed transportation and communications; electricity and coal accelerated the industrial revolution that spread from Britain to other European countries and then to the United States and Japan; and urbanization wrought enormous changes in landscapes.

It was a time when science was considered the highest form of truth available to humankind. During the nineteenth century were born the fields of biology (through the work of people such as Lamarck, Treviranus . . .), paleontology (Cuvier, Blainville . . .), organic chemistry (Wöhler, Liebig . . .), physiology (Johannes Müller, Claude Bernard . . .), cell biology (Schleiden, Schwann, Virchow . . .), geology (Cuvier, Lyell . . .), bacteriology (Pasteur, Koch . . .), anthropology (Broca . . .) and ecology (Humboldt, Haeckel, Marsh, Thoreau . . .).

However, with the publication of Charles Darwin's *On the Origin of Species* in 1859 and *The Descent of Man* in 1871, it was the theory of evolution that profoundly changed our conception of humankind, its origins, its place within the natural world, its relationships with animals and its own animal nature. In his 1866 work *Generelle Morphologie der Organismen* [General morphology of organisms], the German Darwinist Ernst Haeckel postulated the "recapitulation" of the evolution of species (phylogeny) during the development of the embryo (ontogeny), and formulated the first evolutionary classification of living beings. Human beings were henceforward positioned in a genealogical tree that underscored their ape-like ancestry and kinship with the animal world. The mechanism of natural

1 See William Cronon, ed., *Uncommon Ground: Rethinking the Human Place in Nature* (New York: W. W. Norton, 1996); Philippe Descola, ed., *Les Natures en question* (Paris: Odile Jacob, 2018).

2 Richard E. Leakey and Roger Lewin, *The Sixth Extinction: Patterns of Life and the Future of Humankind* (New York: Anchor Books, 1995); Elizabeth Kolbert, *The Sixth Extinction: An Unnatural History* (New York: Henry Holt and Company, 2014).

3 Paul J. Crutzen and Eugene F. Stoermer, "The 'Anthropocene,'" *IGBP Newsletter*, no. 41 (May 2000), p. 17–18; Christophe Bonneuil and Jean-Baptiste Fressoz, *The Shock of the Anthropocene: The Earth, History and Us*, trans. David Fernbach (London: Verso, 2016).

4 Barbara Ward and René Dubos, *Only One Earth: The Care and Maintenance of a Small Planet. An Unofficial Report Commissioned by the Secretary-General of the United Nations Conference on the Human Environment* (New York: W. W. Norton, 1972).

5 Jean-Paul Deléage, *Histoire de l'écologie. Une science de l'homme et de la nature* (Paris: La Découverte, 1991).

6 Catherine and Raphaël Larrère, *Du bon usage de la nature. Pour une philosophie de l'environnement* (Paris: Flammarion, 2009).

7 James E. Lovelock, *Gaia: A New Look at Life on Earth* (Oxford: Oxford University Press, 1979); Bruno Latour, *Face à Gaïa. Huit conférences sur le nouveau régime climatique* (Paris: La Découverte, 2015).

8 Carolyn Merchant, *Autonomous Nature: Problems of Prediction and Control from Ancient Times to the Scientific Revolution* (New York and London: Routledge, 2016).

9 Pierre Hadot, *Le voile d'Isis. Essai sur l'histoire de l'idée de nature* (Paris: Gallimard, "Folio Essais" series, 2008).

selection offered a non-teleological explanation of life that at the time would be argued by the supporters of one idea of progress, and that continued to trouble the proponents of a harmonious cosmos, like those of a benevolent creation. Evolutionism also became—in the words of Haeckel—"the key to every riddle," a world view and the basis of a rational doctrine guiding all human activities. It revolutionized psychology (influencing Freud,[10] Ferenczi, Piaget and Baldwin), ethology (Lorenz), criminal anthropology (Lombroso) and the social sciences.

In this exhibition, we present a "long nineteenth century," in which Darwin and evolutionism were central to a direction followed until the interruption of World War I. We compare, from a European perspective, the key milestones in the way of scientific discoveries with their parallels in the visual arts.

Initially regarded as a divine creation intended to be kept and cultivated by Man, as of the Renaissance nature became the subject of curiosity, and then scientific study. In the first half of the nineteenth century, the discovery of the world's diversity, with the dramatic rise in the number of known species, and the comprehension of the Earth's incredibly great age became the premises of evolutionism for scientists; for artists, they engendered the proliferation of botanical and animal painting, examples of which can be seen in the work of Audubon or Gould, and, in the canvases of Carus, Church or Turner, the rise of the landscape to the status of a major art. In the second half of that century, the search for a "natural basis for beauty"—which Darwin found in sexual selection, and Haeckel in "nature's art forms"—was given expression by artists through a renewed interest in life's simplest forms, in the world of the infinitely small, in the ocean depths and in floral and animal motifs. In the light of evolutionism, the troublesome linkage of our human and animal natures inspired a revival of ape iconography (as in works by Von Max, Fremiet or Kubin) and new ways of depicting age-old human-animal hybrid figures like centaurs, minotaurs and mermaids (as in those by Böcklin, Carriès, Rodin, Moreau or Watts). Artworks by Redon, Klimt, Munch, Kupka or Kubin show Art Nouveau and Symbolism's fascination with the origins of life, ontogeny and phylogeny: unicellular forms, embryos and sea creatures insinuate their way into primordial worlds or the mysteries of motherhood.

Nonetheless, after encountering evolutionism a number of artists, such as Kandinsky, Kupka, Hilma af Klint or Mondrian, rejected the naturalization of human beings and scientism. They sought a secular immortality in occultism and spiritualism, or in the theosophy and anthroposophy that accompanied the birth of abstract art.

With its carnage, World War I spelled the end of that "long nineteenth century." The self-inflicted death of the old Europe, it inaugurated a period in which the scientistic drift of Darwinian theories—a purely "zoological" vision of humankind, social Darwinism, eugenics—would be incorporated into totalitarian ideologies that wished to create a "new man"[11] by moulding und culling from the mass of people. The use of technology would extend from the production of manufactured goods on a global scale to the making of weapons of limitless power.

At the same time, biological and genetic research has definitively confirmed our kinship with all living creatures, as well as brought our fates closer together within the same ecosystem. However, the "continuism" that places human beings within a long animal genealogy has not revived the ancient concept of a world "full" of living forms in which each is necessary to the harmony of the whole. On the contrary, the idea was established of an *Übermensch* who would leave his animal beginnings far behind him and would rise to the rank of a god-man on earth, or who would recreate humankind and the world.

There are two aspects to nature. It can be a generous mother who creates a multitude of admirable forms. It can also be a cruel mother who shows no concern for her children. It has its laws, but those laws are not human. It kills and destroys, just as it continuously generates new forms of life.

In the face of new utopias—that of a "transhumanism" dreaming of creating superhumans using technology, bioengineering and artificial intelligence, or an anti-humanist environmentalism that would repudiate civilization in order to defend an idealized natural world—it is wisdom that we must hope for. Will we be able to rethink our relationship with nature, our cradle? Will we be able to preserve its diversity, as well as, possibly, once again experience the wonderment its beauty aroused among the artists and the poets of the past?

10 Laura Bossi, "Fantaisies phylogénétiques," in *Freud. Du regard à l'écoute*, exh. cat., ed. Jean Clair (Paris: Musée d'art et d'histoire du Judaïsme/Gallimard, 2018), p. 38–49.

11 Laura Bossi, "The New Man: Degeneration and Regeneration," in *The 1930s: The Making of "The New Man,"* exh. cat., ed. Jean Clair (Ottawa: National Gallery of Canada; Paris: Gallimard, 2008), p. 36–57.

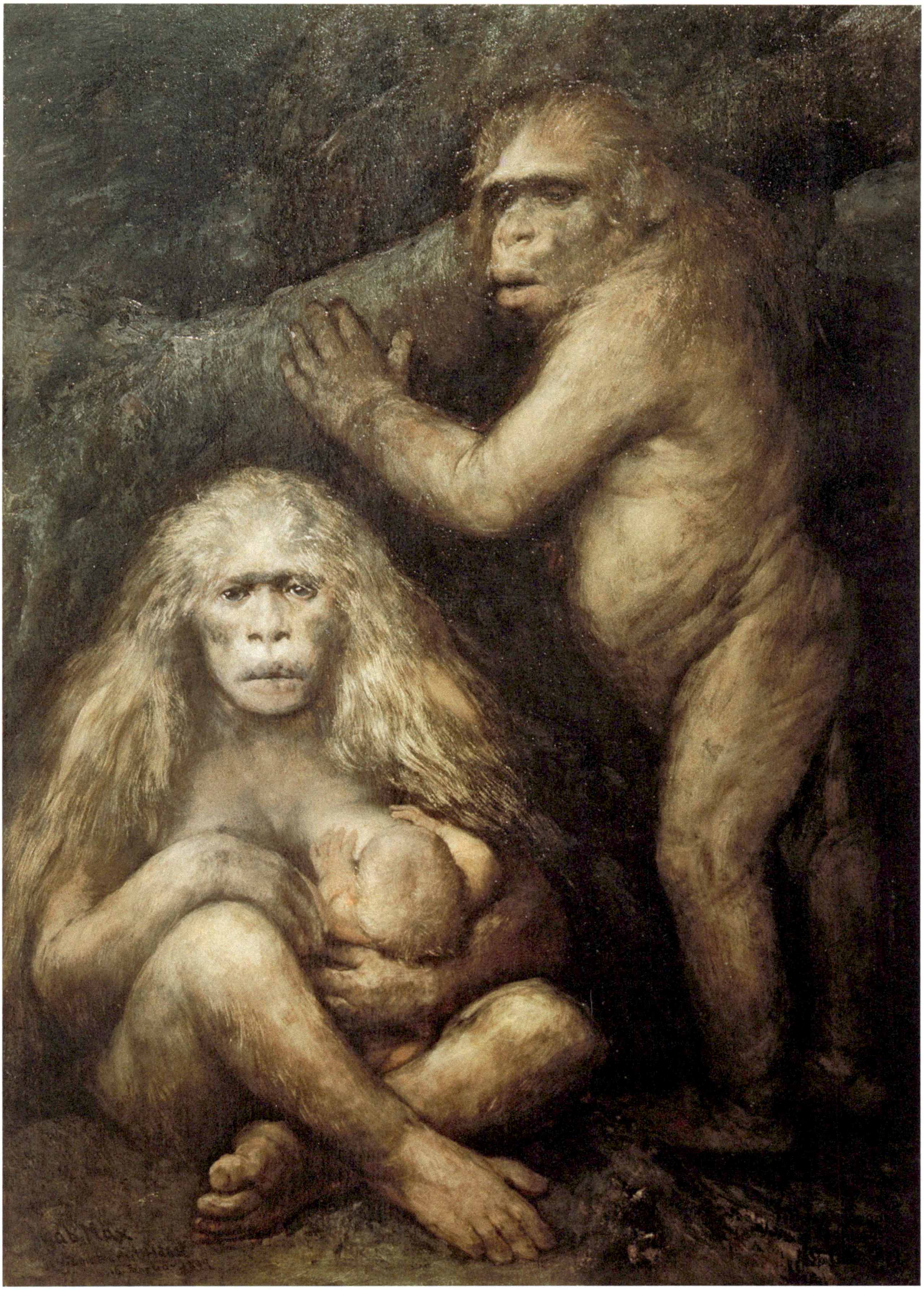

4
Gabriel von Max,
Pithecanthropus alalus, 1894
Oil on canvas, 99 × 68.5 cm
Jena, Friedrich-Schiller-Universität,
Ernst-Haeckel-Haus

“And Adam gave names to all cattle,
and to the fowl of the air,
and to every beast of the field; but for Adam
there was not found an help meet for him.”[1]

(Genesis 2:20)

PROLOGUE IN THE EARTHLY PARADISE

LAURA BOSSI

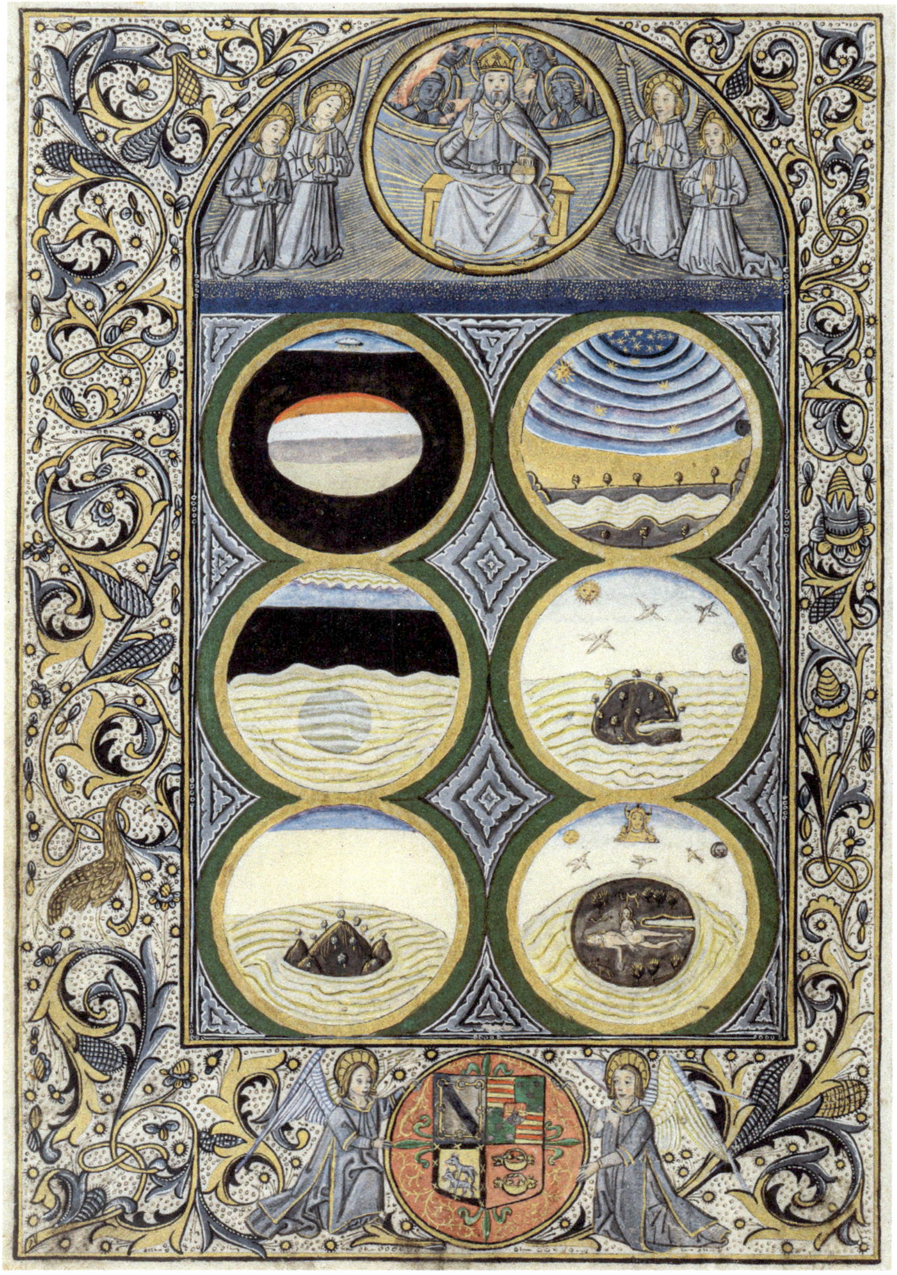

5
Nicolas Jenson, "The Creation of the World,"
in *Biblia Latina*, Venice, 1476
Book-length manuscript,
illuminated on vellum, 29.5 × 21 cm
Paris, Bibliothèque nationale de France

According to the biblical account that has shaped Western culture,[2] God created the world; it is neither eternal, as in the view of the atomist or Aristotelian tradition,[3] nor evil, as the Gnostics had it, but rather "good" and a place of harmony.

The origin of man was to be found in a garden, that of Eden, given for him to cultivate and keep.[4]

The Bible juxtaposes two Genesis narratives. The initial one describes the six stages of Creation in as many "days" (5) with, on the first, that of light; then the heavens; followed by the earth, seas and plants; the sun, moon and stars; and the birds and marine creatures. On the sixth day, God creates "every beast of the earth," then man and woman "in [His] image, after [His] likeness," and gives them dominion "over every living thing" and "over all the earth" (Genesis 1:24–31).

In the second account, God, who is both artist and gardener, forms Adam (*'adam*, meaning man) with the dust of the ground (*'adamah*)[5] and puts him in a garden that he himself has planted with "every tree that is pleasant to the sight, and good for food; the tree of life also in the midst of the garden, and the tree of knowledge of good and evil" (Genesis 2:9). In order that Adam not be alone, God fashions all the beasts of the field and the birds of the sky, and takes them to him, the only creature with the gift of language, so that he can give every animal its own name, according to its species (6, 7). However, for Adam, among them "there was not found an help meet for him" (Genesis 2:20). God therefore creates a woman out of one of Adam's ribs, and Adam welcomes her as "bone of my bones, and flesh of my flesh" (Genesis 2:23). Later, after the Fall, Adam will call her "Eve," because she will be the mother of all the living. Putting an end to God's plan, the serpent, "more subtil than any beast of the field" (Genesis 3:1), will entice the woman to eat the fruit of the tree of knowledge, and she in turn will have Adam do the same. The two will be driven out of the earthly paradise, and will come to know good and evil, as well as to suffer the punishment of labour, the pain of childbirth, and death.

Those two cryptic, contradictory narratives, separated by several centuries, tell us in poetic fashion that out of dust man is made in the image of God. The apex of creation, but also a solitary creature, different from all others through possession of the power of speech, he lives in a world of symbols, vainly seeking his like among the animals. Such stories also relate that woman is an independent being, the equal of man, but forms "one flesh" with him. Finally, they provide an explanation for the existence of evil in the natural world by ascribing it to the original sin of the first parents.[6]

Many artists have depicted that Edenic state of nature with scenes showing the creation of the animals (10), Adam (123, P. 185) and Eve (8), the naming of the animals and the Fall,[7] as well as the expulsion from paradise. Moreover, writers from Dante to Milton have dealt with it in their poetry.

In the New Testament, the theme of the Garden of Eden would be reprised—indeed, it could be said "recapitulated"—in the figure of Christ the gardener. When the risen Christ manifests himself to Mary Magdalene on Easter morning,[8] she does not recognize him immediately, for he appears to her in the guise of the keeper of the Garden of Gethsemane. Depictions of the episode can seem strange to us: Albrecht Dürer provides Christ with a straw hat and spade; Bronzino, in a painting held in the Louvre, shows his statuesque Christ leaning on the same tool; and, in the striking version by Abraham Janssens and Jan Wildens (about 1620) in the collection of Dunkirk's Museum of Fine Arts, he is also surrounded by garden vegetables (9). An explanation for that iconography can be found in the epistles of Saint Paul: through his incarnation and resurrection, Christ is the "new Adam"[9] who will bring life and return humankind to the paradise that was lost.

A third biblical garden appears in the very mineral form of the heavenly Jerusalem, glittering with precious stones, that is described in the Revelation of

1 This and all further biblical citations are from the King James version, available online: https://quod.lib.umich.edu/k/kjv/browse.html.

2 Certain fundamentalists refer to Genesis as a font of scientific truths. Yet Genesis is a narrative that belongs to the realm of myth, intended to provide answers to individuals' questions concerning their identity and future. Science is not a narrative, but a collective search for knowledge driven by doubt. There is no reason for the supposed conflict between science and religion as they are two different domains that should not infringe upon each other. See Stephen Jay Gould, *Rocks of Ages: Science and Religion in the Fullness of Life* (New York: Ballantine Books, 1999).

3 Lucretius, *On the Nature of Things*, trans. W.H.D. Rouse (Cambridge, MA: Harvard University Press, 1992).

4 "And the LORD God took the man, and put him into the garden of Eden to dress it and to keep it" (Genesis 2:15).

5 The Latin words for man (*homo*) and earth, ground (*humus*) are also related.

6 The doctrine of original sin was developed in opposition to Gnosticism by Saint Augustine at the beginning of his serving as a bishop, about 396.

7 See, for example, Stephanie Moser, *Ancestral Images: The Iconography of Human Origins* (Ithaca: Cornell University Press, 1998).

8 This is the *Noli me tangere* episode John 20:11–18 and Mark 16:9–11.

9 See Romans 5:12–21, and 1 Corinthians 15:22: "For as in Adam all die, even so in Christ shall all be made alive."

10 Laura Bossi, "Du paradis et des jardins zoologiques," in *Histoire naturelle de l'âme* (Paris: Presses universitaires de France, 2003), p. 56–71.

11 See Ronald King, *The Quest for Paradise: A History of the World's Gardens* (New York: Mayflower Books, 1979).

12 The sixteenth century saw a flowering of botanical gardens in Europe: in Padua (where the first chair for studying simples [medicinal plants] was established in 1533 at the university supported by the Venetian Republic and, in 1545, such a garden, or *orto*), Pisa (1546), Bologna (1568), Paris (1576), Leiden (1577), Leipzig (1580), Montpellier (1589), and so on. See Monique Mosser and Georges Teyssot, eds., *The Architecture of Western Gardens: A Design History from the Renaissance to the Present Day* (Cambridge, MA: MIT Press, 1991).

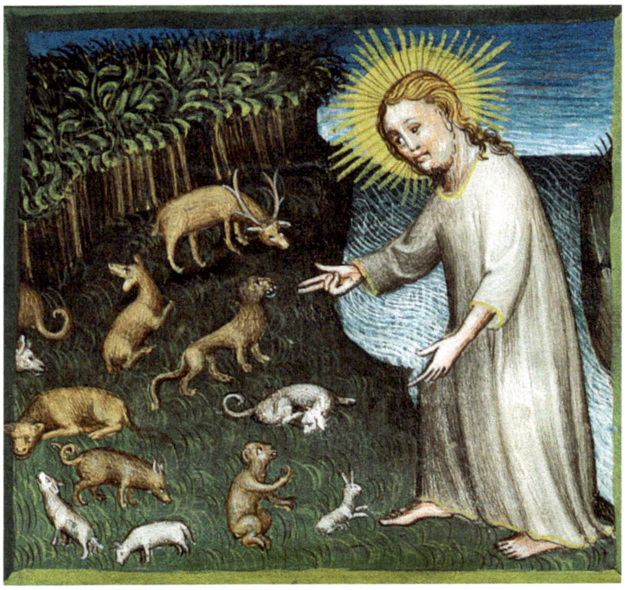

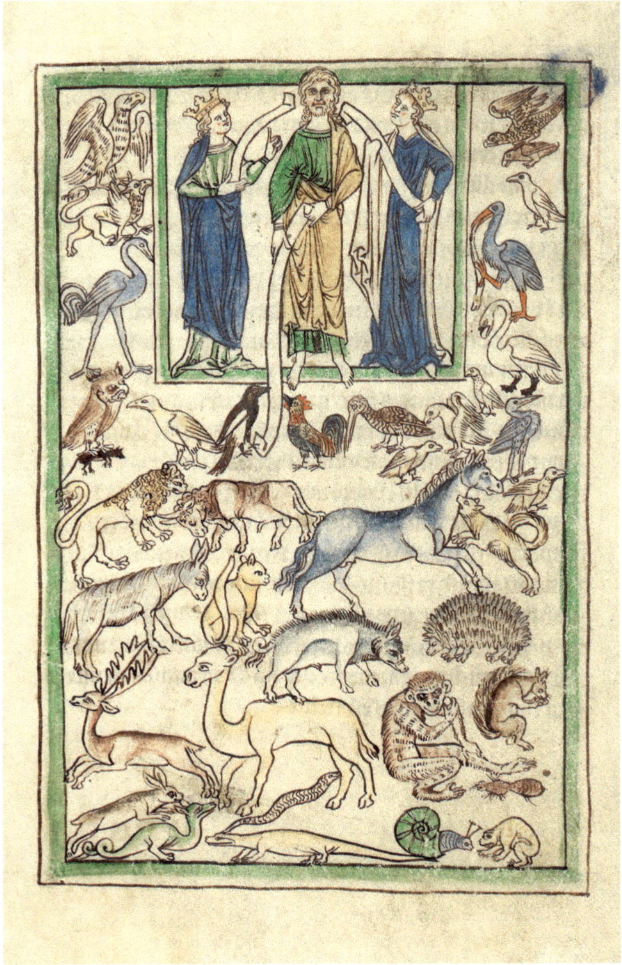

Saint John. However, it is a garden for city dwellers, where the tree of life—the sole instance of vegetation—has the slightly incongruous effect of the plants in tubs placed in new urban developments full of glass-and-steel buildings to give them the illusion of life.[10]

Those narratives and images have moulded the Western imagination of nature and are very enlightening in respect to our relationship with other creatures. The plant world is seen as an enclosed garden adorned with flowers, vegetables and beautiful grounds. In it nature is domesticated and pleasing to humans. As for the animal world, it is mainly domesticated animals, companion animals, those that are hunted or the exotic specimens gracing parks and royal menageries that are depicted. "Wild" animals such as lions or leopards make an appearance to call to mind the prophecy of Isaiah heralding the paradise in which all of creation would live in harmony. A monkey, sometimes biting into an apple—the caricature of a human being and the image of a sinful soul—can practically always be found, while animal pests such as insects, snakes or snails are only exceptionally present (7).

Should we wish to have a picture of the earthly paradise of the Middle Ages, the square cloisters of Cistercian monasteries provide us with one. They are enclosed gardens, whose four corners are reminders of the four rivers of Paradise, and the wells of flowing waters at their centres of the Fountain of Life. Saint Bernard said it explicitly: the cloister is paradise.[11] Looking at the etymology of the word "garden" reveals that it can be traced back to the Old English *geard*, meaning fence or enclosure.

In the fifteenth century, many paintings depicted paradise as an enclosed garden with its fountain; examples include Dirk Bouts's panel in the collection of the Palais Beaux-Arts Lille and the Van Eyck altarpiece in Ghent. The latter features fifty different meticulously painted plants, but not a single animal, with the exception of the Lamb, an allegory of Christ.

Under the spell of the classical tradition, the Renaissance would combine the notion of the earthly paradise with the pagan concepts of the Elysian Fields and Arcadia. The earthly paradise would often be represented in the form of parks and open, verdant gardens. The botanical gardens that sprung up first in Italy and then throughout Europe in the sixteenth century were most likely evocations of such a paradise.[12]

6
Petrus Comestor, *Bible historiale*, leaflet 6, first quarter of 15th century
Book-length manuscript, illuminated on parchment, 35 × 27 × 9 cm
Paris, Bibliothèque Mazarine

7 FIG.
Anonymous, *Adam Naming the Animals*, about 1250–1260
Pencil, ink, gouache and wash on parchment, 21 × 15.7 cm
Los Angeles, John Paul Getty Museum

From the seventeenth century until the present day, the West has endeavoured to recover or reconstruct the Garden of Eden in an ever-wider world:[13] explorers searched for it in the Old World and later in the New World.[14] Colonizers wanted to recreate it by clearing wilderness lands; naturalists, right up to the great classifier Carl Linnaeus, considered themselves new Adams naming species. Gardens, greenhouses and menageries seemed an effort to reunite the scattered parts of the Garden of Eden.

The Genesis narrative also lived on in the field of the earth sciences. While the history of the origins of the world and humankind had long been conceived in the West as being very brief, the idea of a single, unidirectional and finite history of the Earth, expressed in the "hexameral" (six-day) myth of Creation, could, in contrast to other eternalist or cyclical cosmogonies, be a model for conceptualizing an infinitely longer history of the Earth during the development of geology in the nineteenth century.[15] Count de Buffon, in his 1778 work *The Epochs of Nature*, had already outlined a vast panorama of the Earth's history divided into seven eras that mirrored the seven days of the biblical narrative, but featured a longer history before the appearance of humankind in its seventh and final period.

Chapters 6 to 9 of Genesis tell the story of the Great Flood. In that case the sin is concupiscence (the "sons of God" lust after women)[16] and God, "grieved," will "destroy all flesh, wherein is the breath of life, from under heaven," with the exception of Noah, a "just . . . and perfect" man who, along with his family, is spared. In the ark he builds according to God's instructions, Noah harbours couples—a male and a female—of "every beast . . . and all the cattle," as well as of "every creeping thing that creepeth upon the earth" to save them from the flood. When the waters recede, God promises that "while the earth remaineth, seedtime and harvest, and cold and heat, and summer and winter, and day and night shall not cease" (Genesis 8:22). He makes a covenant with Noah and hands power over "every beast of the earth . . . and green herb" to mankind. That narrative, like the one of Adam naming the animals, justifies Judeo-Christian anthropomorphism and, at the same time, underscores Noah's role in the deliverance of living things.

A great many artists have tackled the subjects of the Flood and Noah's ark in their works.[17] They range from the twelfth-century mosaics of the Cathedral of Monreale in Sicily and Saint Mark's Basilica in Venice to the thirteenth-century stained-glass windows of the Sainte-Chapelle in Paris, frescoes by Raphael and Michelangelo in Rome, and Nicolas Poussin's "tragic landscape" *The Deluge* from about 1660, as well as to depictions by scholars from the modern era such as the Jesuit Athanasius Kircher who, in his 1675 volume *Arca Noë*, tried to work out the shape, size and organization of the ark's interior (11), or the Swiss naturalist Johann Jakob Scheuchzer, whose 1731 work *Physica sacra* showed the ark surrounded by fossils, then considered "witnesses" of the Flood.[18]

In the nineteenth century, the subject would see itself enriched by the considerations of the new sciences of optics, geology and zoology, as demonstrated in the works of Joseph Mallord William Turner, inspired by Goethe's theory of colours (97, P. 141), and John Martin (84, P. 121), who inserted a conjunction of the sun, the moon and a comet in the background of his painting dealing with the subject to suggest a natural cause of the Flood.[19]

Moreover, in 1863, the Neapolitan animal painter Filippo Palizzi, a contemporary of Darwin, would show off his zoological knowledge by depicting the animals of Paris's Jardin des Plantes emerging from the ark (12).

For their part, historians have not neglected to draw parallels between the story of the Garden of Eden and other origin myths.[20]

Although the myth of the first human being, as well as of an earthly paradise,[21] is totally omitted in the Qur'ān, the Muslim paradise also resembles a garden. *The Book of Muhammad's Ladder* recounts the night journey of the Prophet, guided by the archangel Gabriel, to Heaven and Hell. It describes seven successive paradises that are gardens. The seventh, the abode of God, is a *hortus conclusus*—an enclosed

---

13 See Carolyn Merchant, "Reinventing Eden: Western Culture as a Recovery Narrative," in *Uncommon Ground: Rethinking the Human Place in Nature*, ed. William Cronon (New York: W. W. Norton, 1996), p. 132–159.

14 See Alessandro Scafi, *Alla scoperta del paradiso: un atlante del cielo sulla terra* (Palermo: Sellerio, 2011).

15 See Martin J. S. Rudwick, *Earth's Deep History: How It Was Discovered and Why It Matters* (Chicago: University of Chicago Press, 2014).

16 Certain commentators have regarded such "sons of God" (*bnei Elohim*) as being creatures belonging to the divine world, in what constitutes the theory of angelic sin. See Mathias Delcor, "Le mythe de la chute des anges et de l'origine des géants comme explication du mal dans le monde dans l'apocalyptique juive. Histoire des traditions," *Revue de l'histoire des religions*, vol. 190, no. 1 (1976), p. 3–53; Claus Westermann, *Genesis 1–11: A Continental Commentary*, trans. John J. Scullion (Minneapolis: Fortress Press, 1990).

17 See Roger Kaysel, *Arche Noah* (Baden, Switzerland: Schweizer Kindermuseum, 1992).

18 Johann Jakob Scheuchzer, *Physica sacra* (Augsburg and Ulm: Wagner, 1731–1735).

19 See, by way of comparison, Martin J. S. Rudwick, *Scenes from Deep Time: Early Pictorial Representations of the Prehistoric World* (Chicago: University of Chicago Press, 1992), p. 24; also *Visions du déluge de la Renaissance au XIXe siècle*, exh. cat., ed. Rémi Cariel and Sylvie Wuhrman (Paris: Réunion des musées nationaux, 2006).

20 See Jean Delumeau, *Une histoire du paradis*, vol. 1: *Le Jardin des délices* (Paris: Fayard, 1992).

21 See Jacqueline Chabbi, *On a perdu Adam: La création dans le Coran* (Paris: Seuil, 2019).

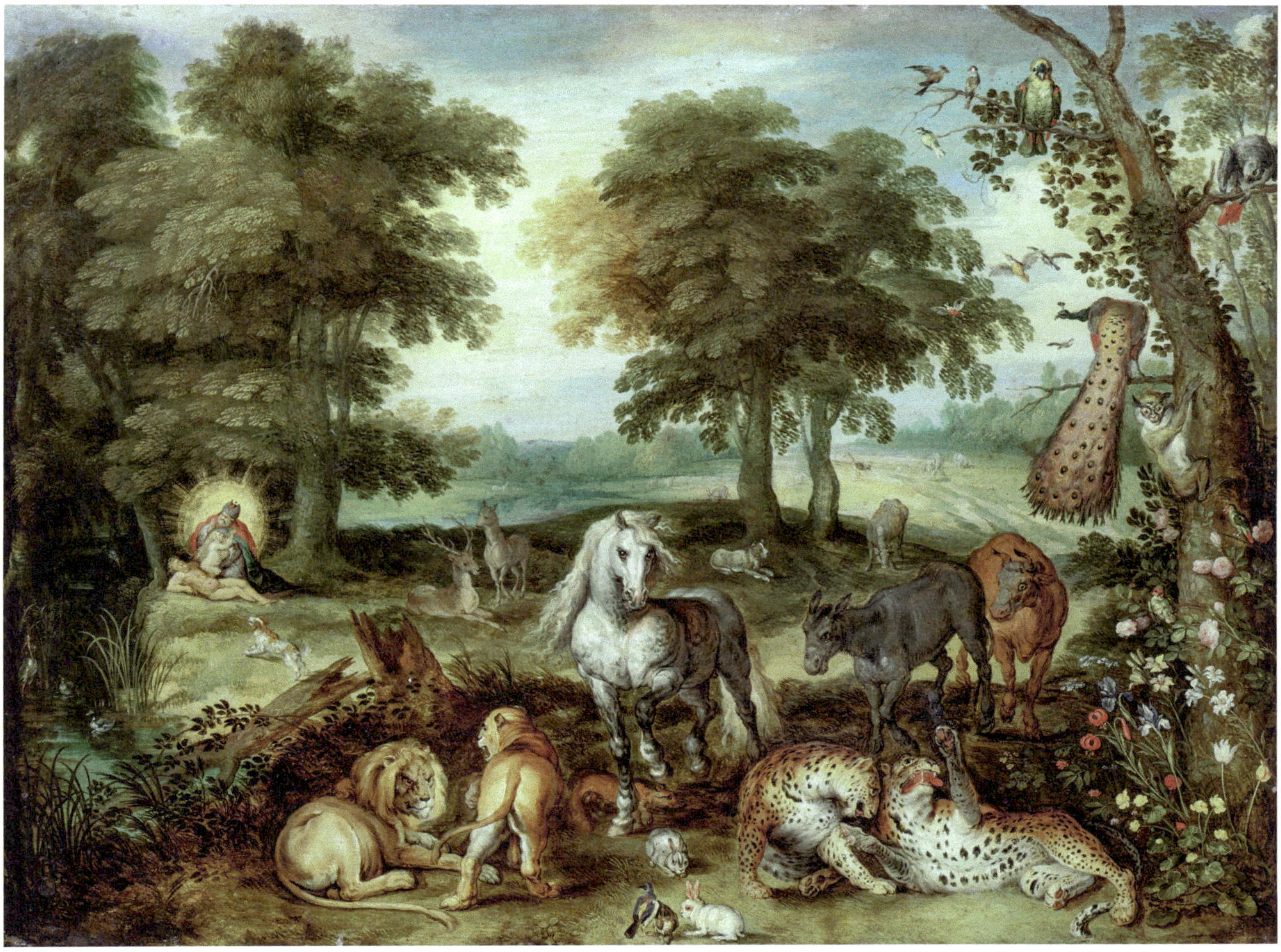

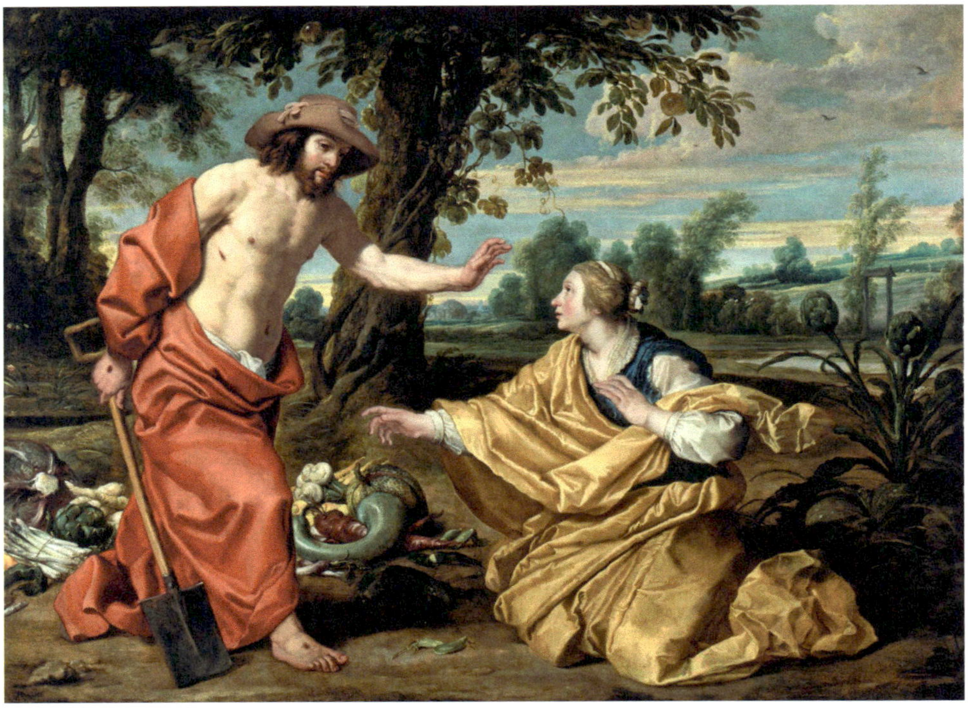

8
Jan Brueghel the Younger,
The Garden of Eden with the Creation of Eve, about 1630
Oil on copper, 48.5 × 37 cm
Besançon, Musée des Beaux-Arts et d'Archéologie

9 FIG.
Jan Wildens, Abraham Janssens,
Noli me tangere, about 1620
Oil on canvas, 155 × 206 cm
Dunkirk, Fine Arts Museum

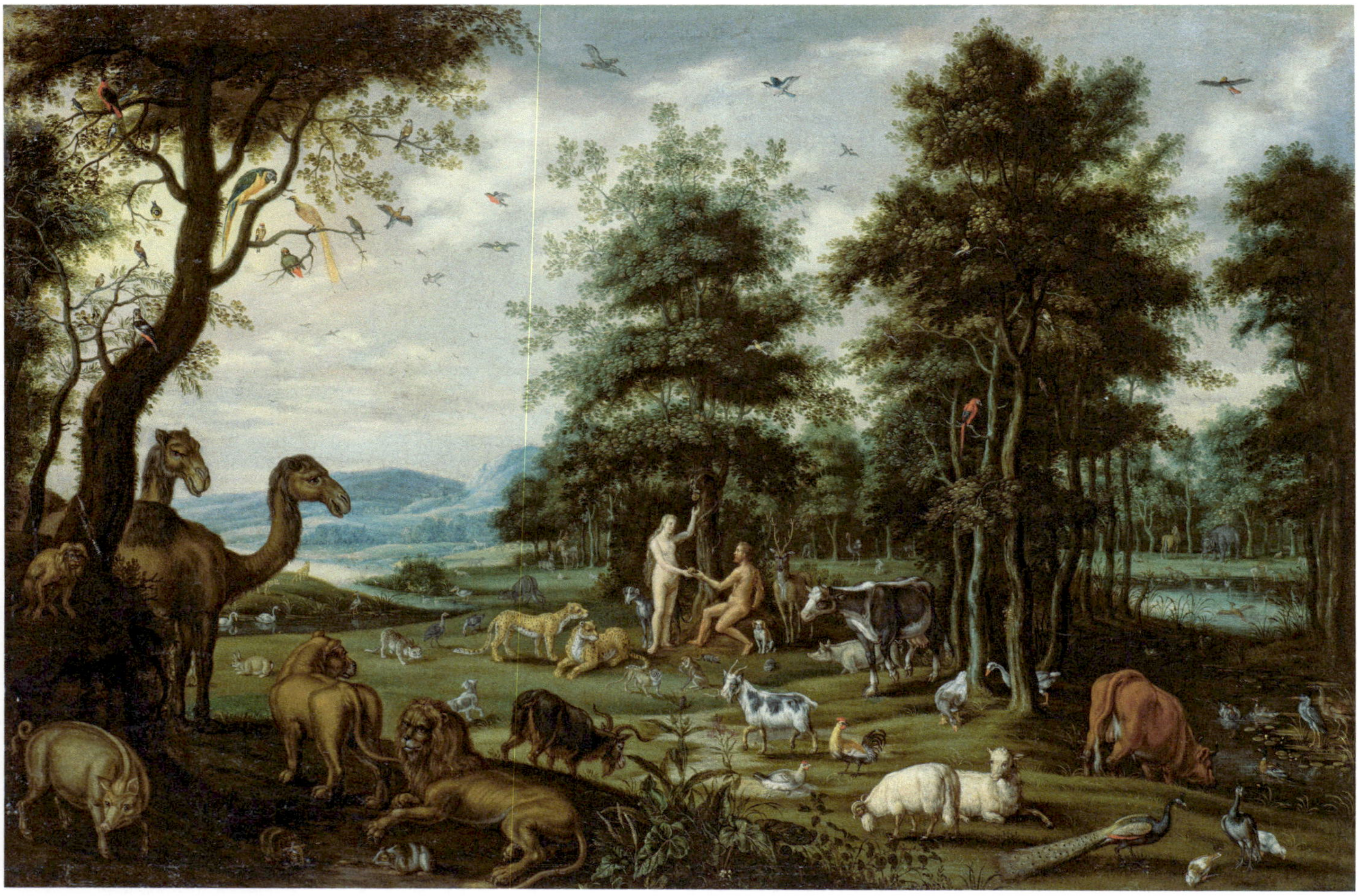

10
Isaak van Oosten,
The Earthly Paradise, 1625–1650
Oil on canvas, 58.2 × 84.8 cm
Rennes, Musée des Beaux-Arts

11
Athanasius Kircher,
"Cross-section of Noah's Ark,"
in *Arca Noë, in tres libros digesta*,
pls. 122–123, Amsterdam:
Johannes Janssonius van
Waesberge, 1675
Paris, MNHN

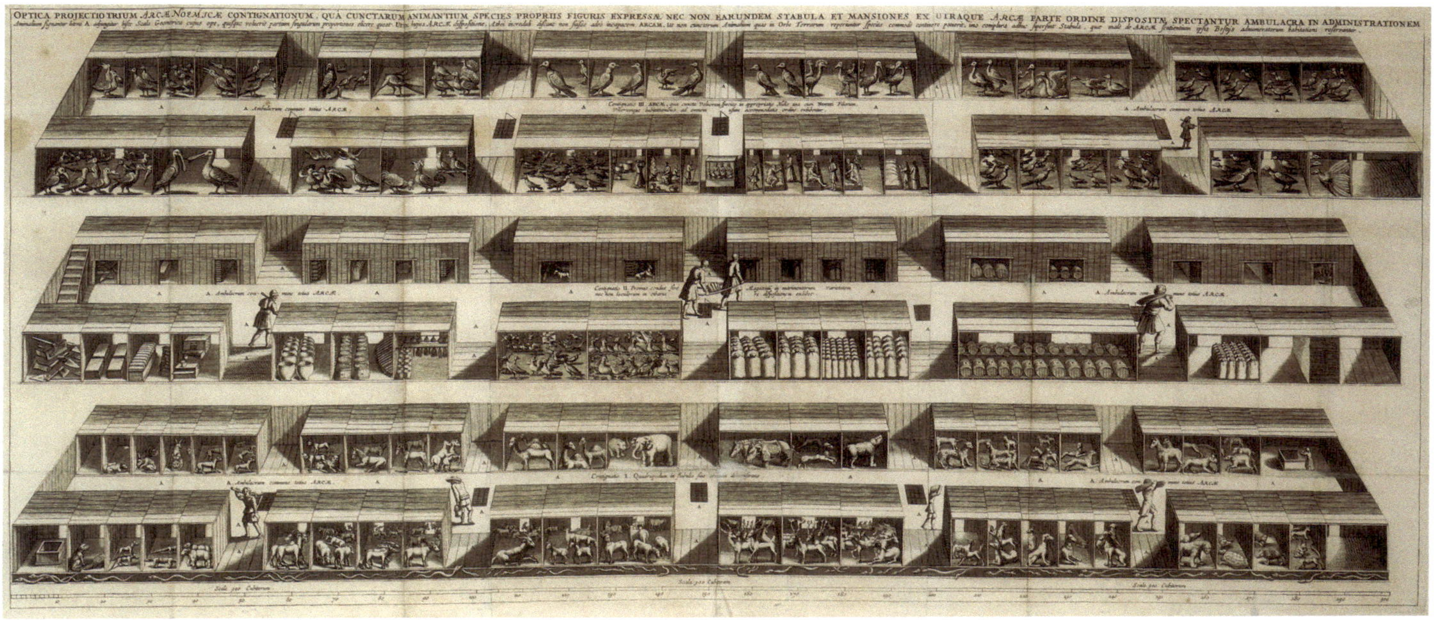

garden—filled with fountains and fruit trees, and enlivened by the melodious singing and music-making of young maidens. "Therein are rivers of water which corrupts not; and rivers of milk of which the taste changes not; and rivers of wine, a delight to those who drink, and rivers of clarified honey. And in it will they have *all kinds* of fruit, and forgiveness from their Lord."[22]

Although they lack a tree of knowledge, *The Epic of Gilgamesh* as well as Persian myths also incorporate wonderful gardens in which peace, happiness and abundance reign, and where human beings and animals live together in harmony. A joyful past within a garden is also alluded to in the Greek and Roman traditions.[23] Such a "golden age" is spoken of in *Works and Days* by Hesiod, who places it in the Islands of the Blessed, where "the grain-giving earth bears honey-sweet fruit flourishing thrice a year,"[24] and Plato makes mention of it in *Statesman*. In Latin literature, Virgil, in his "Fourth Eclogue,"[25] situates it in Arcadia, and Ovid, in *The Metamorphoses*, evokes a land unrestrainedly yielding fruit during an everlasting springtime, where, as in the paradise of the Book of Revelation, milk and honey flow spontaneously, their animal sources lost sight of, not unlike as in the minds of the children today who are completely unaware of where the bottles of milk and jars of honey in supermarkets come from.

"The Golden Age was first founded, which, without any avenger, of its own accord, without laws, practised both faith and rectitude. Punishment, and the fear *of it*, did not exist . . . The Earth itself, too, in freedom, untouched by the harrow, and wounded by no ploughshares, of its own accord produced everything; and men, contented with the food created under no compulsion, gathered the fruit of the arbute-tree, and the strawberries of the mountain, and cornels, and blackberries adhering to the prickly bramble-bushes, and acorns which had fallen from the wide-spreading tree of Jove. *Then* it was an eternal spring; and the gentle Zephyrs, with their soothing breezes, cherished the flowers produced without any seed. Soon, too, the Earth unploughed yielded crops of grain, and the land, without being renewed, was whitened with the heavy ears of corn. Then, rivers of milk, then, rivers of nectar were flowing, and the yellow honey was distilled from the green holm oak."[26]

22 *The Holy Qur'ān*, trans. Maulawi Sher 'Ali (Islamabad, Pakistan, and Tilford, United Kingdom: Islam International Publications Ltd., 2015), 47:16, available online: https://www.alislam.org/quran/Holy-Quran-English.pdf.

23 See Jean-Pierre Vernant, *Myth and Thought Among the Greeks*, trans. Janet Lloyd with Jeff Fort (New York: Zone Books, 2006).

24 Hesiod, *Works and Days*, in *The Homeric Hymns and Homerica*, trans. Hugh G. Evelyn-White (Cambridge, MA: Harvard University Press, 1914), l. 173, available online: http://www.perseus.tufts.edu/hopper/text.

25 *Virgil's Eclogues*, trans. Barbara Hughes Fowler (Chapel Hill: The University of North Carolina Press, 1997). In respect to Arcadia, see Monica Ferrando, *Il regno errante. L'Arcadia come paradigma politico* (Vicenza: Neri Pozza, 2018).

26 *The Metamorphoses of Ovid*, trans. Henry T. Riley (London: George Bell & Sons, 1893), p. 10, available online: https://www.gutenberg.org/files/21765/21765-h/files/Met_I-III.html#bookI_fableIII.

P. 32-33

12
Filippo Palizzi,
After the Deluge, 1864
Oil on canvas, 185 × 266 cm
Naples, Museo e Real Bosco di Capodimonte

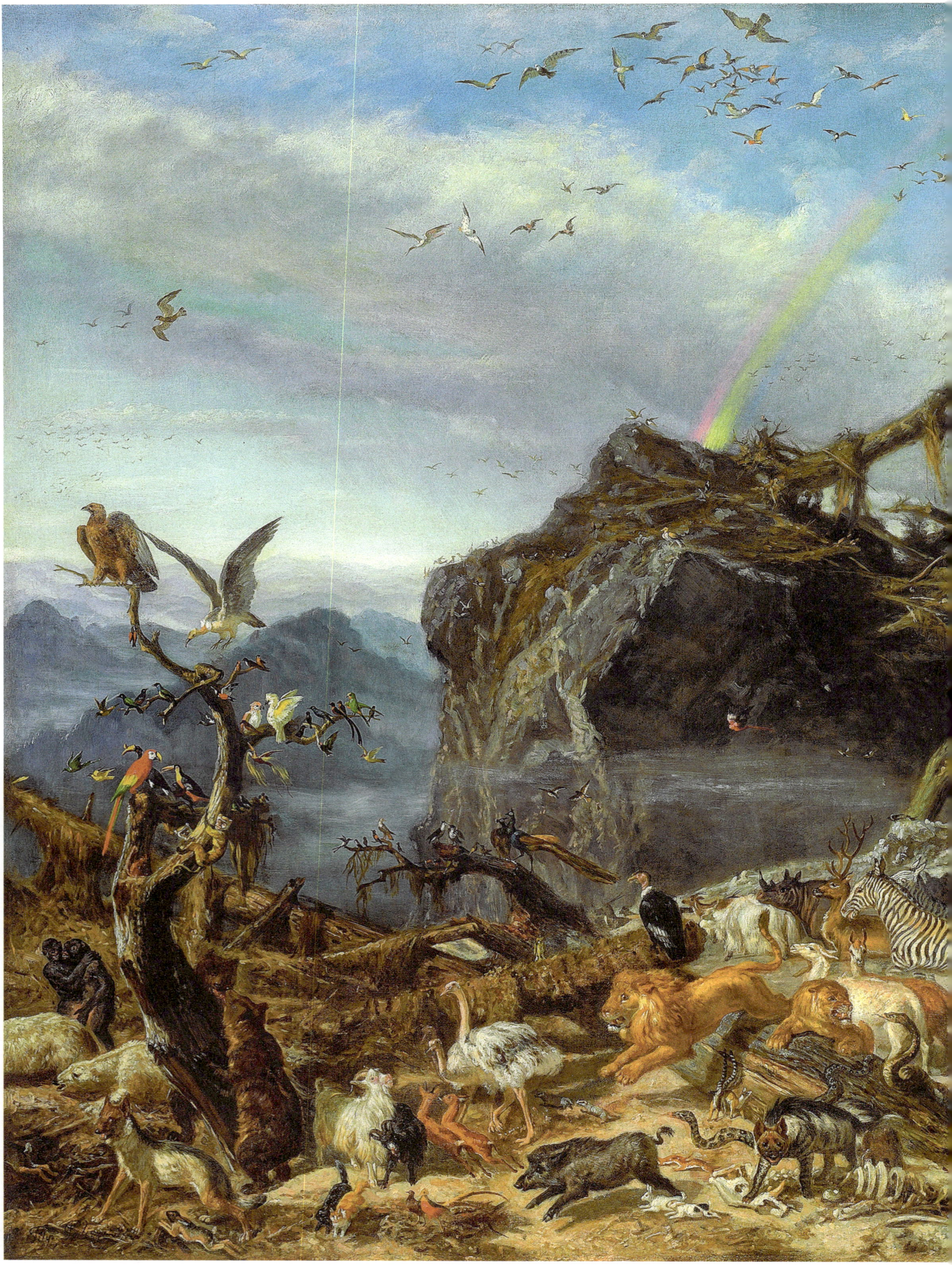

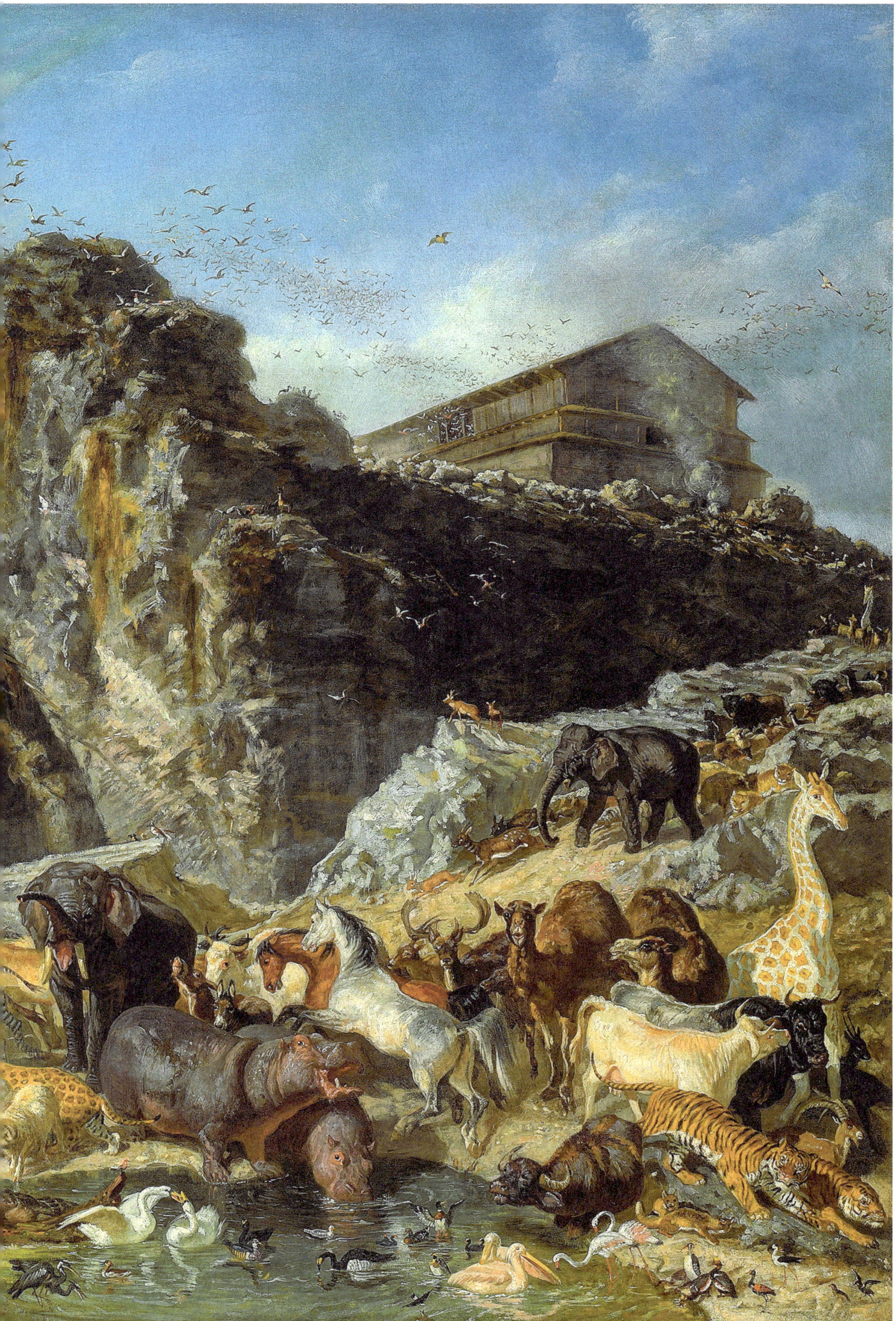

I

THE POWER OF NATURE IN EARLY MODERN EUROPE

SABINE DU CREST

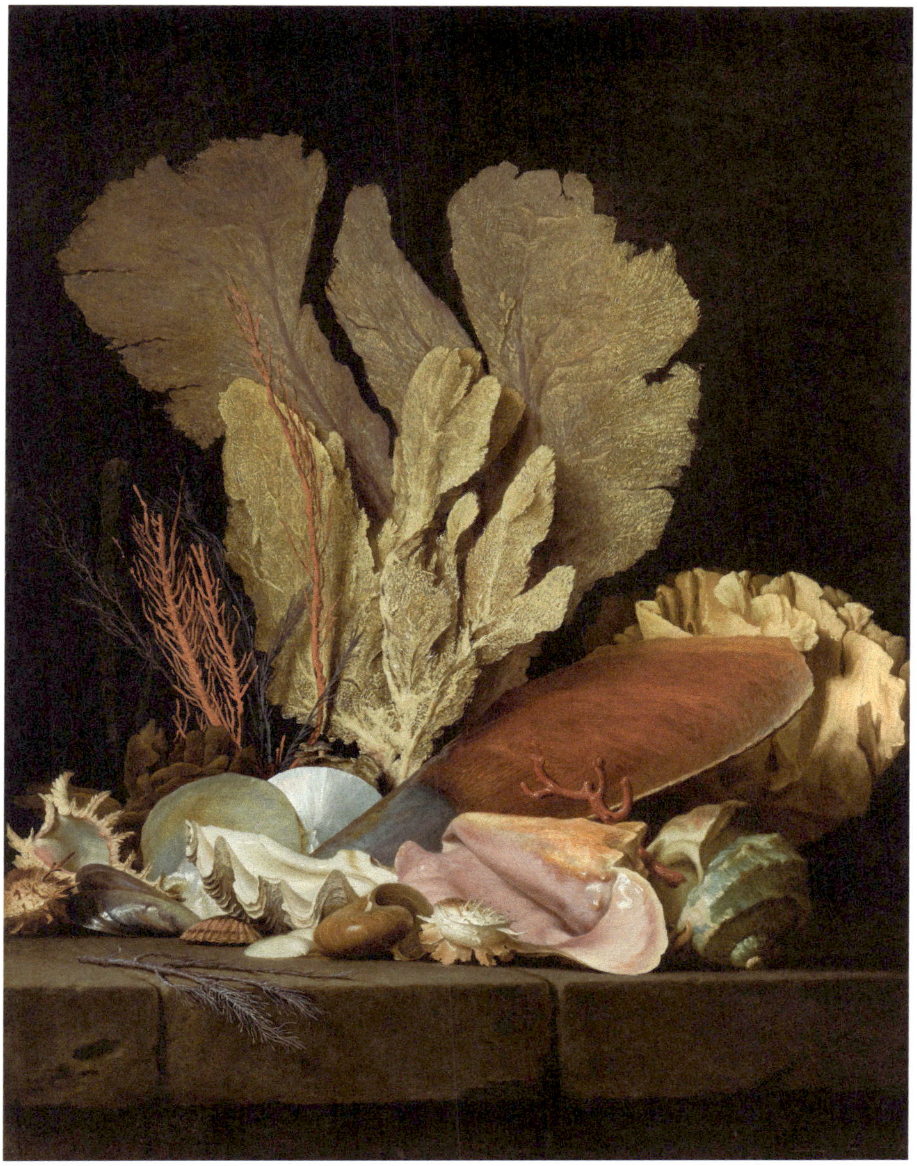

13
Anne Vallayer-Coster,
Sea Plumes, Lithophytes and Shells, 1769
Oil on canvas, 130 × 97 cm
Paris, Musée du Louvre

In a waking dream that reconstructs the New World as a maritime Garden of Eden whose treasures are being offered to Europeans, *Coral Fishing (Allegory of the Discovery of America)* by Jacopo Zucchi depicts unclothed Roman beauties, deities of a new cult arrayed in pearls and coral and surrounded by putti and fishermen, shells raised to the level of *objets d'art* and a macaw borne as a hunting trophy (14). In the foreground facing us is a monkey, also draped in jewels and seemingly coveting more, a *mise en abîme* suggesting the unrestrained determination of Europeans of the time to see, acquire, collect, transform and study these *naturalia* of the antipodes.

Between possession, fascination and imagination, the story of the powers of nature is the story of the world and of humankind's infinite quest for its origins, right down to its most distant traces in time and space, and need to unlock the secrets of the universe. A matter of science and taste, the natural world both near and far kindled the curiosity of scientists, the imagination of artists, the desire of adventurers, the interest of merchants and the will to dominate of princes. The human view of the world combined knowledge and power, especially when it came to enslaving nature, as in the Cartesian project.[1] Nature could be understood as the setting for the slow birth of science. For some, however, nature remained an eternal enigma and a place of wonder in which to decipher the presence of the divine and the emergence of signs and communication among species.[2] An entire spectrum of natural learning based on productive interactions between all orders of knowledge has found an echo, beyond the scientism of the nineteenth and twentieth centuries, in the most recent scientific discoveries, as well as in the resurgence of cabinets of curiosities in contemporary art, museums and exhibitions.

From *The Geographer* and *The Astronomer* by Vermeer to *Buffon en naturaliste* by Carmontelle, exploration of the world took various forms, as evoked in paintings. In the space of a century, a paradigm shift had occurred in Europe: the personification of a model scholar opening his eyes to nature viewed from the recesses of his study gave way to the portrait of a scientist surrounded by animals, affirming the extent of his zoological research. It is as if we were first being shown the men of science enthralled and intrigued by the hugeness of the celestial or terrestrial world now available to them, thanks to globalization and the new observational instruments, and then a luminous imagery populated by crocodiles, elephants, camels, turtles and corals, all created by the father of natural history. Between the two, the horizons of research took shape in the living world in all its forms and its direct study simply could not cease.

THEATRE OF NATURE

Jan van Kessel's cabinet paintings, in which seventeenth-century Europe is defined by a nature publicized through its curious depiction, seem to offer an apology to the natural world.[3] Of all the natural riches depicted in the paintings of flowers, fruits, butterflies, insects and marine animals appearing to have emanated from scholarly works and placed in a palace orangery framed by views of fauna, mammals, birds and reptiles, there emerges, however, through the representation in the foreground of game and drink, a question that continues to chafe about their proper use. Thence, the idea of whether the meaning, or indeed the vanity, of a world collection is essential in the modern age. The perception of nature as the place where wonders, monsters and marvels manifest themselves is firmly rooted in European history. From Pliny the Elder to Saint Augustine and then to Peiresc, the Prince of the Republic of Letters, it is through moments, or beings, or strange and disturbing events that humans confront the world and attempt to grasp it in multiple ways, where the continuous links between human beings and their environment in the form of magic, exorcism and the collection of rarities emerge.[4] All fields of knowledge, art and power become involved. In pursuit of the great secrets of nature, from Carpaccio's *St. George and the Dragon* to Salvator Rosa's disturbing nocturnal scenes, numerous paintings evoke the fear and danger emanating from natural creatures, while the shells, fruits and flowers of still lifes are odes to the beauty of the world and meditations on its fragility.

1 See Carl Havelange, *De l'œil et du monde. Une histoire du regard au seuil de la modernité* (Paris: Fayard, 1998).

2 Consider Plato's allegory of the cave and Giorgione's *Three Philosophers*. See Jean Céard, *La Nature et les prodiges. L'insolite au XVIe siècle en France* (Geneva: Droz, 1996); Stéphane Van Damme, *Seconde nature. Rematérialiser les sciences de Bacon à Tocqueville* (Paris: Les Presses du réel, 2020).

3 See Jan van Kessel, *Europe,* from the *Four Parts of the World,* Munich, Alte Pinakothek.

4 See Céard, *La Nature et les prodiges*; *Peiresc et l'Italie,* proceedings of the international colloquium (Naples, Istituto Italiano per gli studi filosofici, June 23–24, 2006), ed. Marc Fumaroli (Paris: Alain Baudry, 2009); Antoine Schnapper, *Le Géant, la licorne et la tulipe. Les cabinets de curiosités en France au XVIIe siècle* (Paris: Flammarion, "Champs" series, 2012); Ann Blair, *The Theater of Nature: Jean Bodin and Renaissance Science* (Princeton: Princeton University Press, 1997); Lorraine Daston and Katherine Park, *Wonders and the Order of Nature, 1150–1750* (New York: Zone Books, 2001).

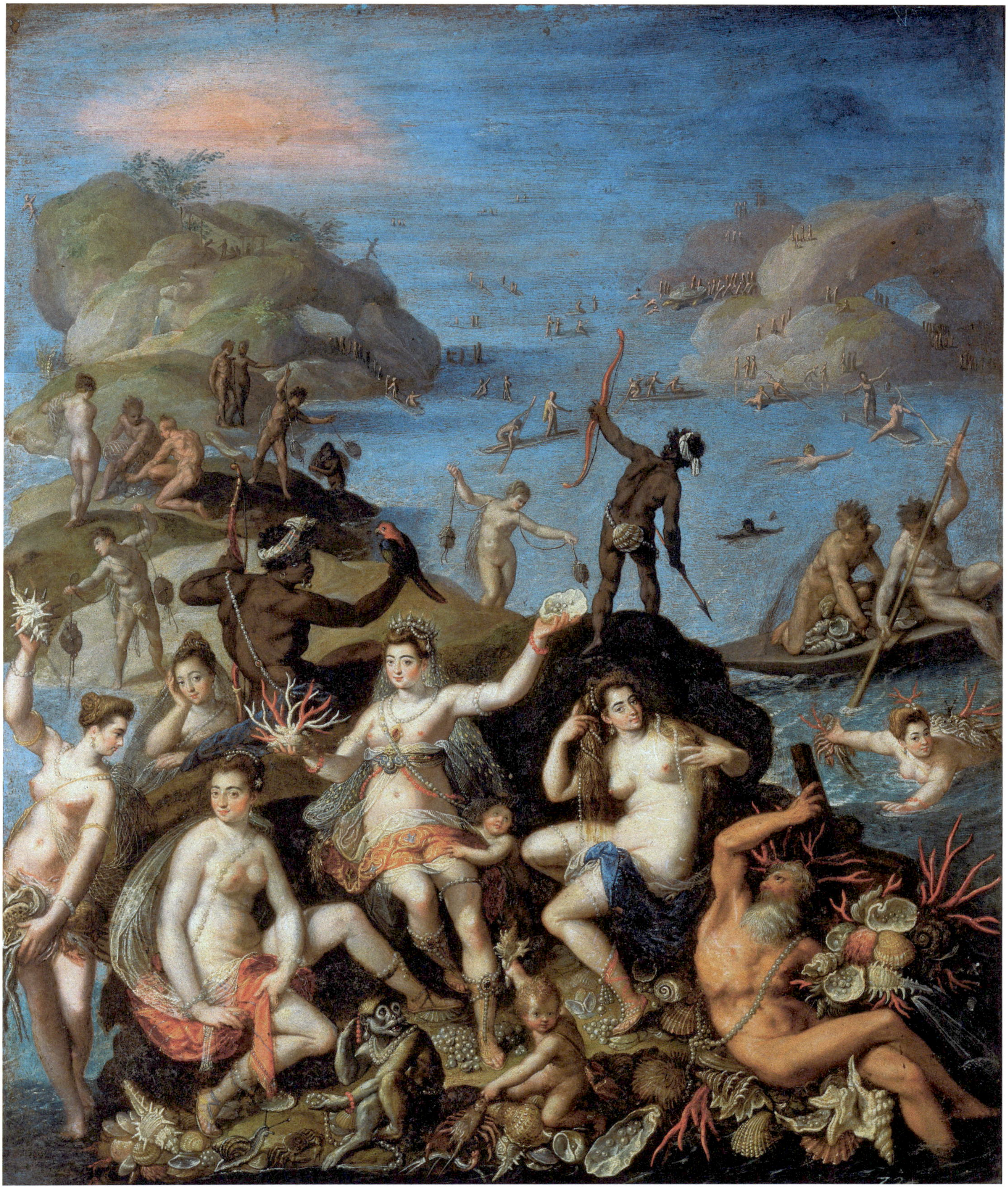

14
Jacopo Zucchi,
Coral Fishing (Allegory of the Discovery of America),
1615–1630
Oil on copper, 52 × 42.5 cm
Rome, Galleria Borghese

Taking an inventory of the world involved the precise cataloguing of nature in places of learning,[5] an undertaking in which early seventeenth-century scholars were engaged. The act of marvelling at extraordinary specimens, of which medieval church treasures and the *Wunderkammern* of the princely Renaissance courts were the theatres, gave way to comparative analysis, a principle developed by antiquarians. This began in the late sixteenth century with the cabinets of curiosities and eventually led to today's museums of natural science.[6] In these early places of knowledge, the universe was presented through its natural objects, which scholars attempted to organize using arrangements inherited from a taxonomy based on the art of memory and an awareness of a continuous chain of living organisms that made it possible to consider the microcosm as a means of elucidating the macrocosm.[7]

The accumulation of objects in cabinets of curiosities, as well as the way they were organized, was not random. The *naturalia*, one of the most important categories of materials collected in the cabinets, were arranged next to the *artificialia*—objects made by human hands—and with the *exotica*, according to a certain logic, at various angles, and with endless subsections and intersections based on each person's knowledge. The cabinet catalogues, as inventories of the known world, were themselves sources of curiosity, while being part of the staging. The cabinets, artworks and books—the theatres of curiosity—played complementary roles and had close and complex relationships. Thus was built the world view that shaped them.[8] Based on the system of analysis that prevailed in antiquarian research and religions, namely comparatism, the study of natural specimens in particular focused on the interstices and interregnums, as with corals, about which there was uncertainty for a very long time on whether to designate them as animal, mineral or plant. The categories themselves remained porous for a long period and there were multiple modes of preservation.

In Naples, at the very end of the sixteenth century, Ferrante Imperato built a space dedicated to the detailed organization of plants, minerals, birds, corals, shells and other marine animals.[9] The frontispiece of his *Natural History* in twenty-eight books is an engraved representation of his cabinet, which offers a glimpse into the way scholars thought at the time, while revealing their learned relationship with nature (15). Taking centre stage is a crocodile—the epitome of exoticism and savagery, now dried out and stuffed—hanging from the ceiling. Some of the doors of the wall cabinets have been opened for visitors, showing the spatial organization of the *naturalia* that are assembled in boxes and arranged in an order based on the knowledge drawn from the books on the shelves of the library, which occupies a whole section of the room, and also from experimental data resulting from the direct study of nature. Imperato's cabinet was a condensed version of the natural world and bore witness to the new importance acquired by reasoning rooted in physical experience. In the early seventeenth century, Sir Francis Bacon influenced the advancement of learning by promoting scientific experimentation over the darkness of ancient sources.[10]

In Bologna, in his natural history room, Ulisse Aldrovandi recreated, tamed and kept nature within reach by storing over 10,000 items, minerals and fossils in two cabinets and over 4,000 caskets, as well as dried and stuffed animals, a herbarium and a botanical garden, which was among the first in Europe. A "pinacotheca" containing watercolours and xylographs intended to publicize this room completed the system, since it was considered essential that knowledge be disseminated, and therefore all means of reproducing nature were required (16). Now aware of the need for accurate description, naturalist scholars and antiquarians alike called upon artists, who became the best conduits for the transmission of knowledge. Jacopo Ligozzi, a "universal painter," employed his talent to render nature so faithfully that the natural world portrayed in his watercolours of fish, birds and rodents seems almost hyperrealist. Filippo Napoletano, whose famous lemons and shells float on a timelessly dark background, was another such artist. While they accurately portrayed the natural world, fruit and flower paintings transfigured

5 See Giuseppe Olmi, *L'inventario del mondo. Catalogazione della natura e luoghi del sapere nella prima età moderna* (Bologna: Il Mulino, 1992); Paula Findlen, *Possessing Nature: Museums, Collecting, and Scientific Culture in Early Modern Italy* (Berkeley: University of California Press, 1994).

6 See Adalgisa Lugli, *Naturalia et mirabilia. Les cabinets de curiosités en Europe* [1983] (Paris: Adam Biro, 1998); Krzysztof Pomian, *Collectionneurs, amateurs et curieux (Paris, Venise, XVIe–XVIIIe siècle)* (Paris: Gallimard, 1987); Julius von Schlosser, *Les Cabinets d'art et de merveilles de la Renaissance tardive* [1908], preface and postface by Patricia Falguières (Paris: Macula, 2012).

7 Frances A. Yates, *The Art of Memory* (London: Routledge & Kegan Paul, 1966).

8 See *Le Théâtre de la curiosité, XVIe–XVIIe siècle*, proceedings of the international colloquium (Paris, Centre V. L. Saulnier, Université de la Sorbonne, March 8, 2007), ed. Sabine du Crest and Frank Lestringant (Paris: Presses de la Sorbonne, 2008).

9 See letter by Ferrante Imperato, 1597: "Il mio teatro di Natura non consiste in altro che in cose naturali, come dir Minerali, Animali e Piante," quoted in Olmi, *L'inventario del mondo*, p. 270, note 34.

10 See Francis Bacon, *The Advancement of Learning* (Oxford, 1605).

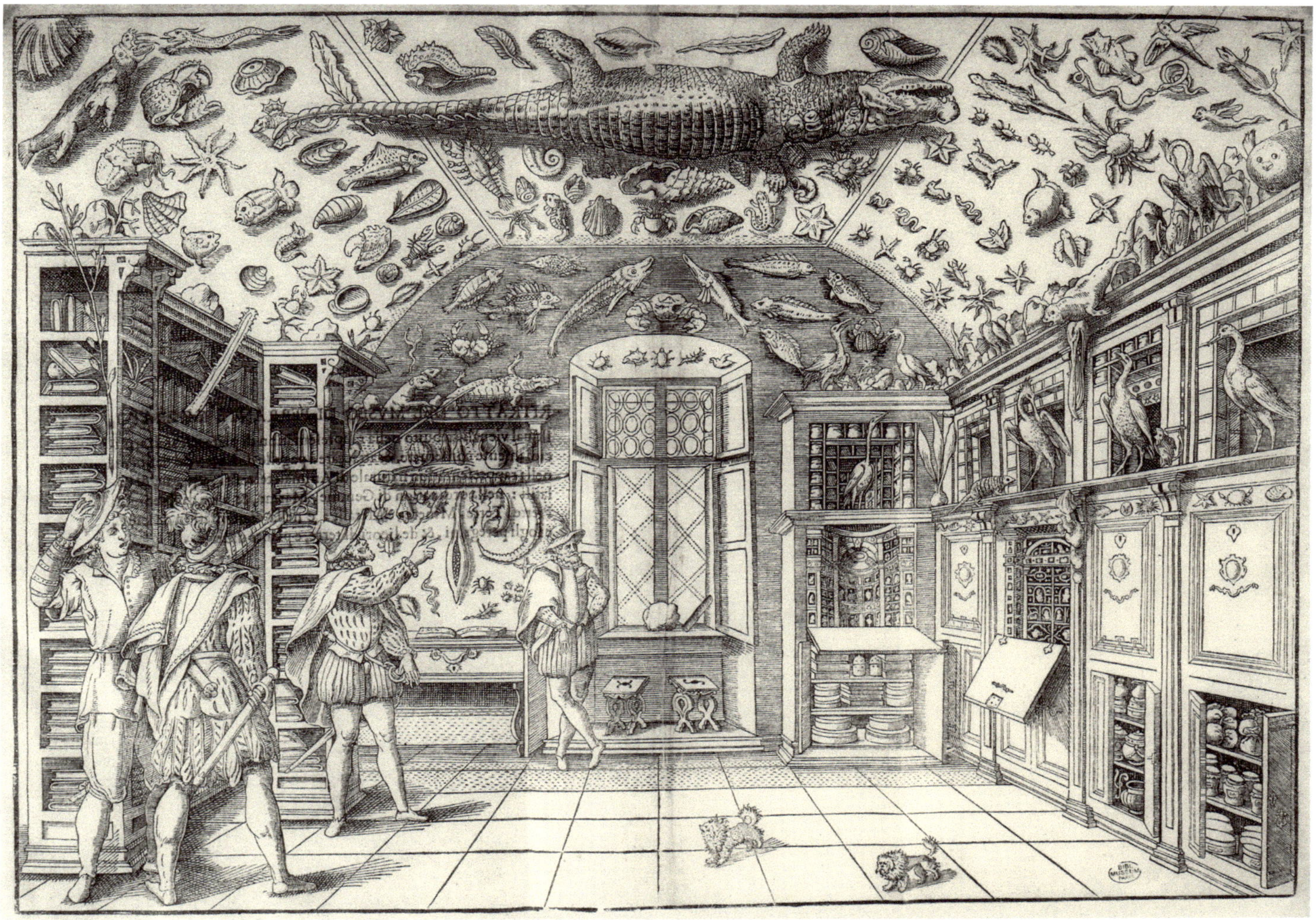

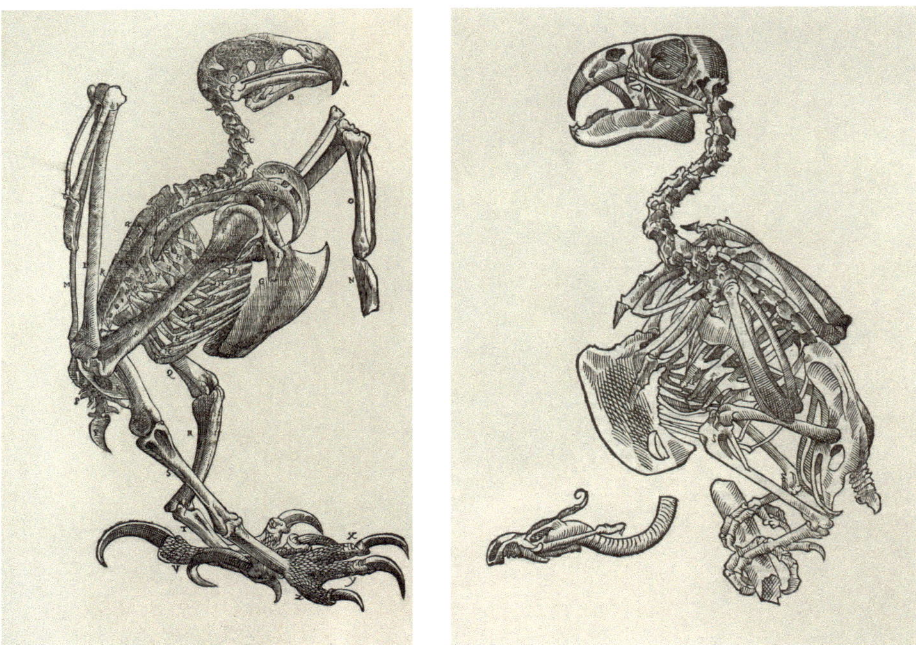

15
Ferrante Imperato,
Dell'Historia naturale di Ferrante Imperato Napolitano. Libri XXVIII,
Naples: Costantino Vitale, 1599
Paris, MNHN

16 A, B
Ulisse Aldrovandi,
"Bird Skeleton," *Ornithologiae hoc est de avibus historiae libri XII*,
Bologna, Franciscum de Franciscis, 1599–1603
Paris, MNHN

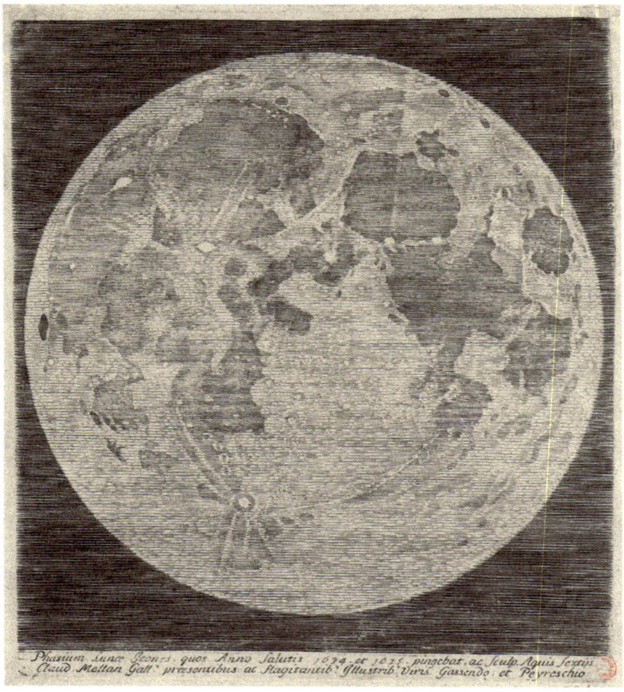

reality and encouraged meditation. Pieter Brueghel the Elder, in his bouquets, invented compositions of floral species that bloom at different times and are made to droop by gravity, yet depicted each tulip, each iris, and even the tiniest insect on a petal as faithfully as possible. Similarly, Caravaggio's *Basket of Fruit* is more a commentary on the inevitability of the passage of time and the mimetic capacities of art than a painting of everyday reality. Observational instruments, not yet perfected, were being employed throughout Europe in the early seventeenth century to study, from both near and far, insects and stars in greater depth. Peiresc used a Galilean telescope to study the phases of the Moon in the pure summer sky of Provence; these were engraved by Claude Mellan in the 1630s (17). In Rome, at the Accademia dei Lincei, named for lynx, which are known for their sharp eyesight, scientific study was connected to philosophical speculation and praise for the prince. Bees formed part of the Barberini family crest, and represented the exact intersection of antiquarian learning, poetry, emblem art, painting, architecture and science, as they were being meticulously examined with the brand new Drebbel microscope in the 1620s.[11]

This new way of depicting nature in full effervescence was applied to the exotic living world, from Albrecht Dürer's rhinoceros (28, P. 55) to the South American fruits being painted by Albert van der Eeckhout or the oriental flowers of *China Illustrata* by Athanasius Kircher, author of one of the most important cabinets of curiosities of his time, housed at the Roman College.[12] Throughout the seventeenth century, knowledge of distant worlds considerably changed not only zoology, conchyliology, horticulture, botany and the European pharmacopoeia, but also humankind's relationship with the natural world, which permeated all of society, populating its imagination with new creatures (20, 21, 22, 23). A strange little animal taken from Aldrovandi's reflections appears in the foreground of *The Rape of Helen* by the Bolognese painter Guido Reni, while the *Tenture des Indes*, a series of tapestries produced by the Gobelins from cartoons that Maurice of Nassau had given to Louis XIV, condenses Brazil into a luxuriant scenography featuring elephants, tigers and other wild animals mostly unknown until then in Europe. The *Mexican Treasure* is a major example of globalized naturalist research. This invaluable work represents a century of medicinal research that catalogues the fauna, flora and minerals of the New World, as studied by a Spanish doctor with the help of Indigenous artists. It was published in Rome by a group of Lincei academicians.[13]

There was an interest in birds, in all their forms, expressed through ornithology, garden ornamentation, interior decoration of palaces, the preservation of species through aviaries and even liturgical vestments. The naturally brilliant feathers of South American birds were used in the composition of mosaics commissioned for European religious celebrations, while birdsong was also carefully studied. The first ornithological collection, *L'Uccelliera*, a book of art and science illustrated by leading artists, was published in Rome in 1622 by Cassiano dal Pozzo, patrician of the Republic of Letters.[14] The artificially animalized interiors were set against the representation of animals on another register, creating a relationship between the inside and the outside. Jacopo

17 FIG.
Claude Mellan,
Map of the Moon: Full Moon,
1635–1636
Engraving, 20.8 × 23.3 cm
Abbeville, Musée Boucher de Perthes

11 See Sabine du Crest, "Les abeilles dans la Rome des Barberini. De la dilatation d'un insecte dans l'art," in *L'Animal au XVIIe siècle*, proceedings of the first meeting of the Centre de recherches sur le XVIIe siècle européen (Bordeaux, Université Michel-de-Montaigne, November 21, 2001), ed. Charles Mazouer (Tübingen: Gunter Narr, 2003), p. 103–118; Svetlana Alpers, *The Art of Describing* (Chicago: University of Chicago Press, 1983).

12 See *Athanasius Kircher. Il Museo del mondo*, exh. cat., ed. Eugenio Lo Sardo (Rome: De Luca, 2001); *Athanasius Kircher: The Last Man Who Knew Everything*, ed. Paula Findlen (New York: Routledge, 2004); Athanase Kircher, *China illustrata* [1667], trans. Charles D. van Tuyl (Muskogee, OK: Indian University Press, 1987), https://htext.stanford.edu/content/kircher/china/kircher.pdf.

13 See Francisco Hernández et al., *Rerum medicarum Novae Hispaniae thesaurus, seu, Plantarum animalium mineralium Mexicanorum historia* (Rome: Mascardi, 1628–1630); 2nd ed. (1651).

14 See Francesco Solinas, *"L'Uccelliera," un libro di arte e di scienza nella Roma dei primi Lincei* (Florence: Olschki, 2000).

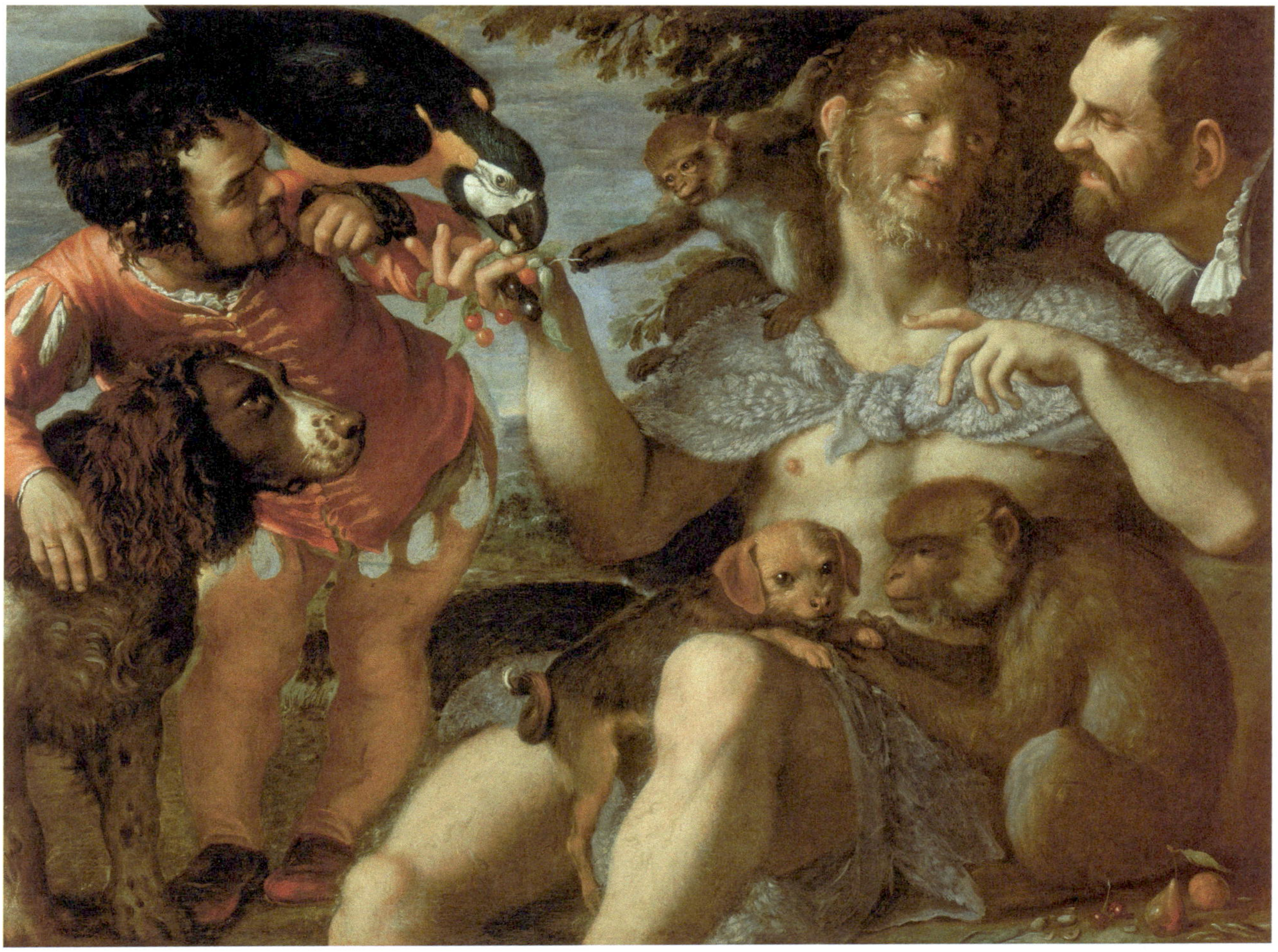

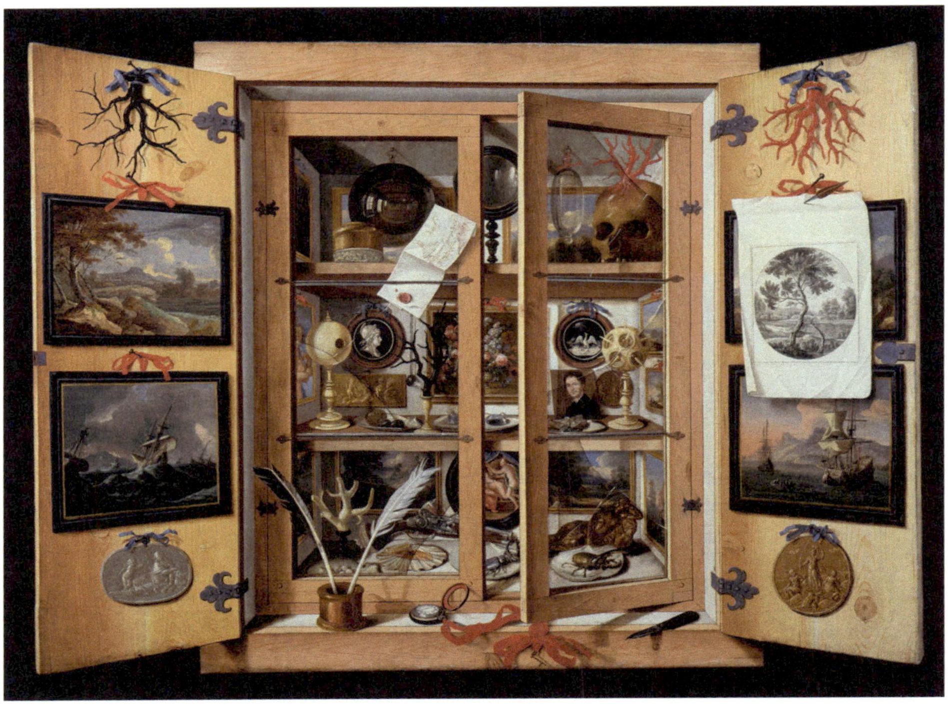

18
Agostino Carracci,
Triple Portrait: Hairy Harry, Mad Peter and Tiny Amon,
1598–1600
Oil on canvas, 101 × 133 cm
Naples, Museo di Capodimonte

19 FIG.
Andrea Domenico Remps,
Cabinet of Curiosities,
about 1675–1690
Oil on canvas, 137 × 99 cm
Florence, Museo dell'Opificio delle Pietre dure

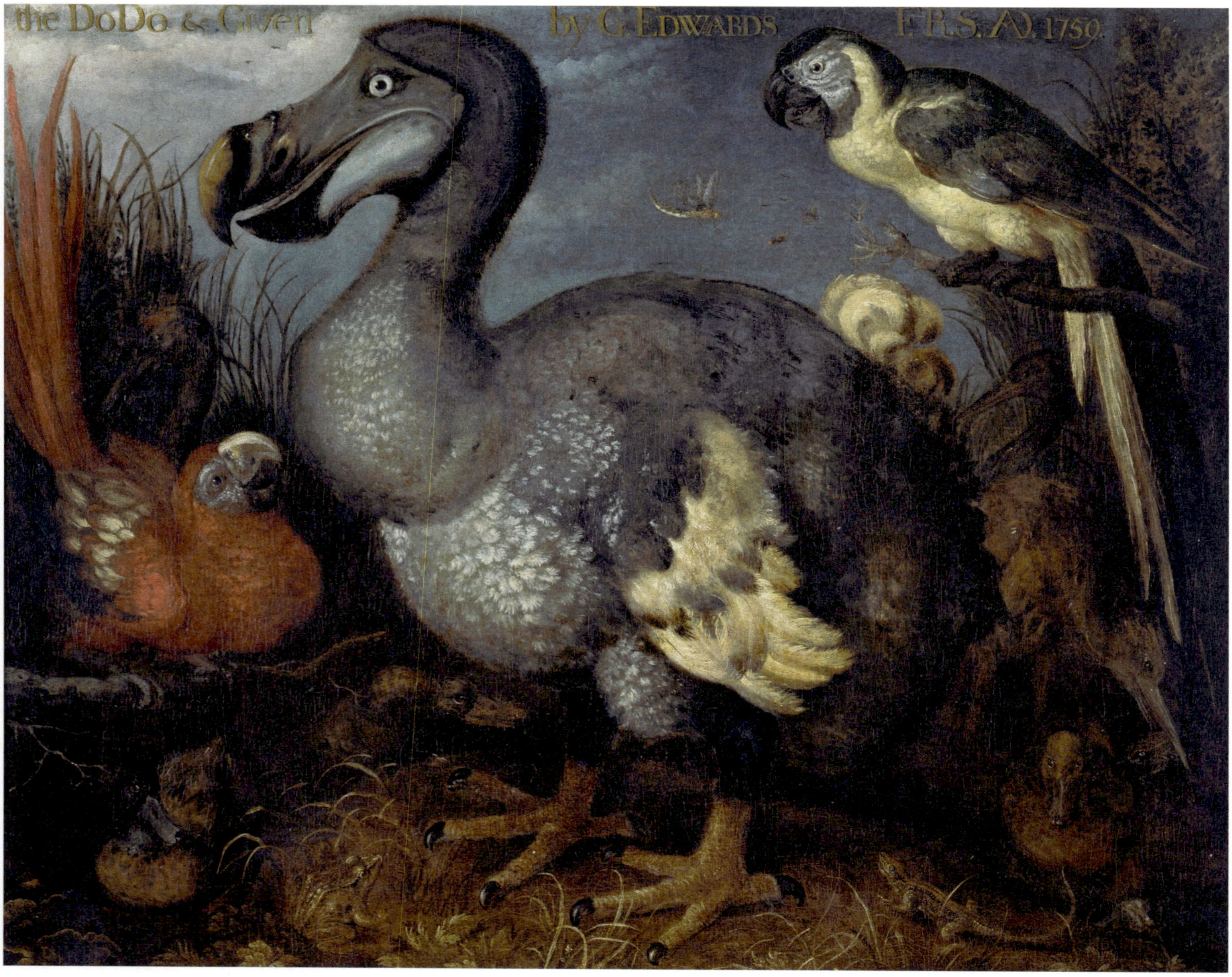

20
Roelandt Savery,
The Dodo, before 1639
Oil on canvas, 96 × 116 cm
London, The Natural
History Museum

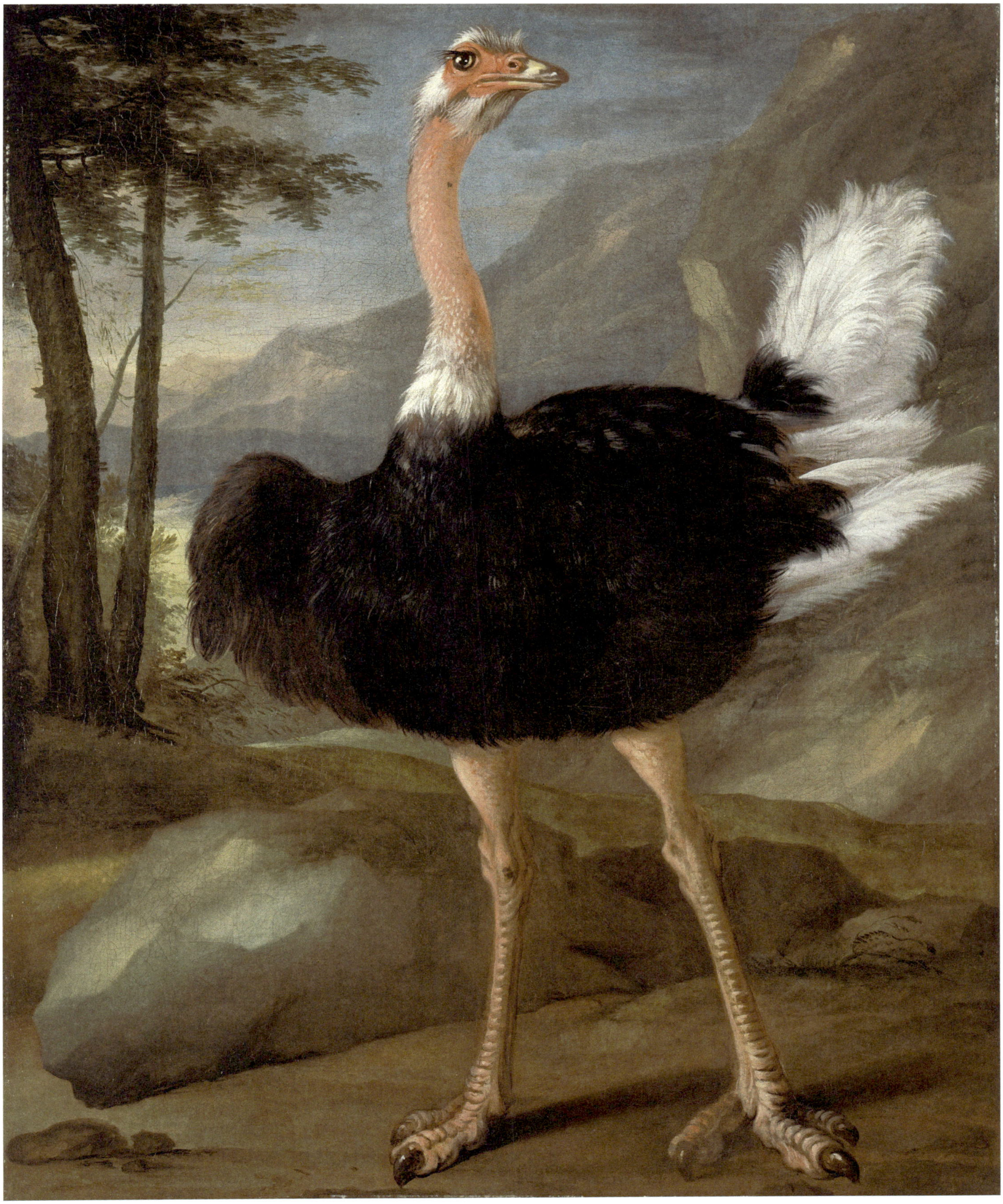

21
Nicasius Bernaerts,
Study of an Ostrich, 1665–1668
Oil on canvas, 91 × 74 cm
Paris, Musée du Louvre,
on deposit at the Musée
du château des ducs de
Wurtemberg, Montbéliard

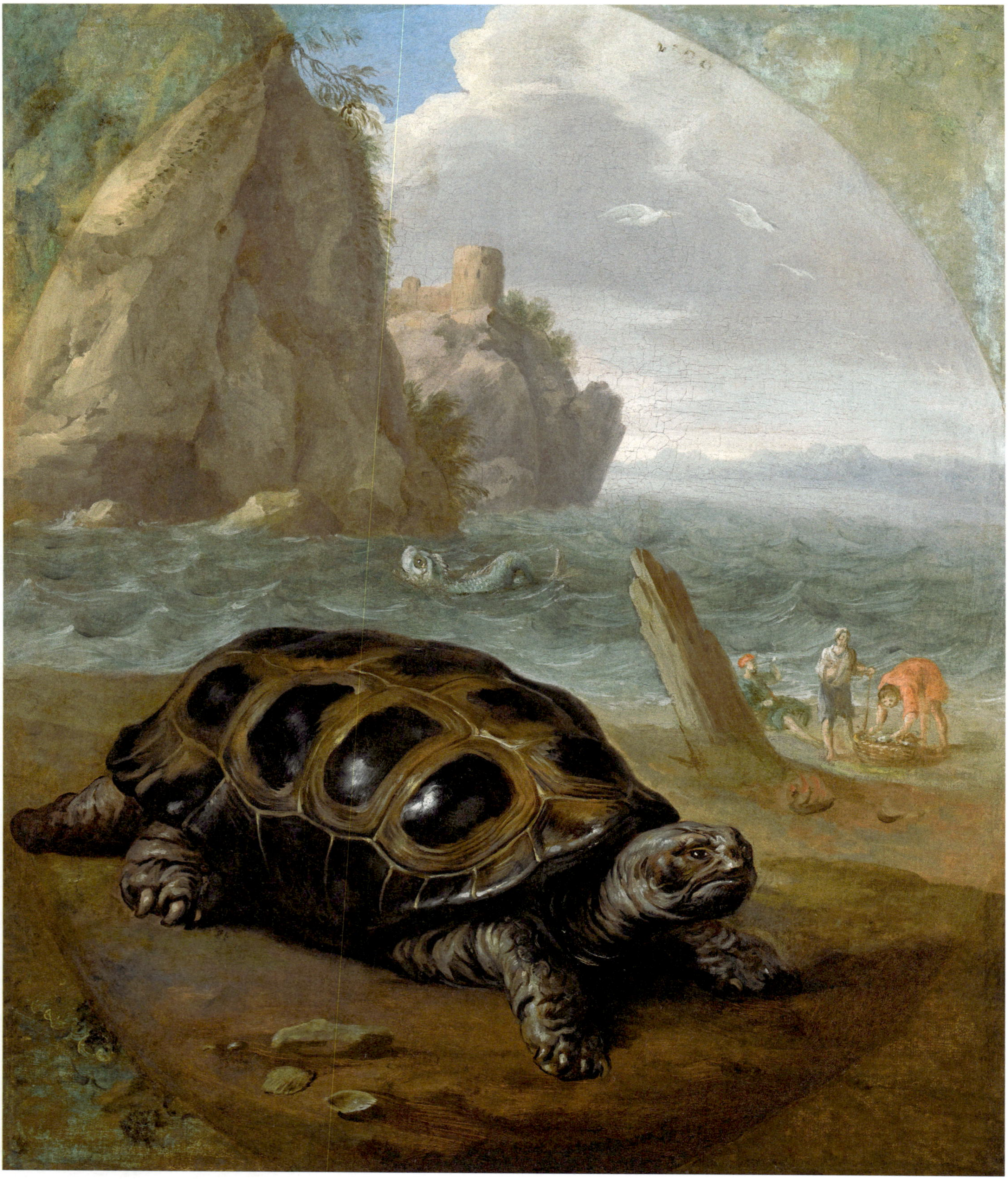

22
Nicasius Bernaerts,
Sea Turtle, 1665–1668
Oil on canvas, 84.5 × 69.5 cm
Paris, Musée du Louvre

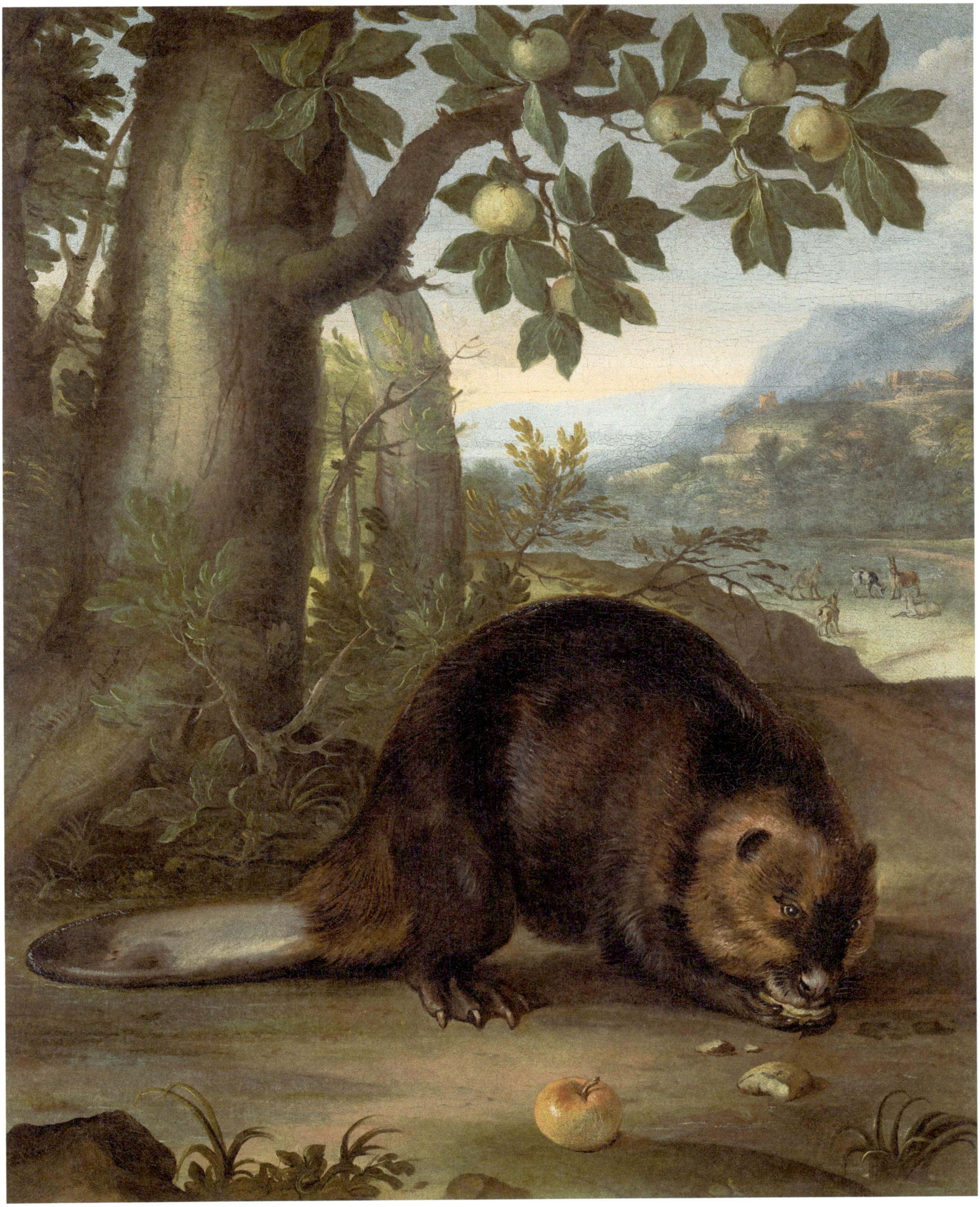

23 FIG.
Nicasius Bernaerts,
A Beaver, 1655–1668
Oil on canvas, 91 × 72.5 cm
Paris, Musée du Louvre

Zucchi was commissioned by Cardinal Ferdinando to paint aviaries of exotic birds—images designed to excite curiosity—on the vaulted ceiling of the *studiolo* in the gardens of the Villa Medici in Rome and on the walls of the Chigi Palace in Ariccia. Many interiors were designed as exteriors, and exteriors as interiors: the cool rooms of homes and gardens all over Europe, from the Villa d'Este to the park of the Palace of Versailles, recreated the natural world in an artificial form of nature, reflecting the highly intellectualized relationship of the elites with the natural world. Remarkable objects from near and far, as well as knowledge and power, blurred the boundaries between art and science: nautiluses (shells brought back from explorations of the South Pacific), once mounted on gilt bronze, became hanaps or ewers; ostrich eggs, coconuts, rhinoceros horns or gourds served as precious drinking cups; narwhal teeth placed on a carved horse's head suggested a unicorn; and bezoars adorned the *Wunderkammern* and many palaces (24). As reliquaries of a new style, these "boundary objects" were ornaments from princely collections, evoking the interweaving of the natural and the artificial, as well as the uncertain possession of the world by the men of this time.[15]

The fascination with the mystery of nature and its secrets never waned in this century. The *trompe l'oeil* of a cabinet of curiosities painted by Domenico Remps around 1680 invites the viewer to meticulously explore the objects assembled from the Medici collection, including white and black corals and landscapes (19). This cabinet, with its broken windowpanes, is a *mise en abîme* of humankind's illusory desire to endlessly explore the known world. If there was one area of natural production where scientific research was associated with an interest in the symbolic, even the sacred, and in art, it was mineralogy. This, thanks to the Medici, found a home in Florence. While Peiresc was interested in fossils and gems as natural antiquities from the earth, and sent his emissaries as far as India to find them, the Florentine princes collected hard stones from all over the world and had them examined, in particular, by the Danish scientist Niels Steensen (Nicolas Steno). They also used the stones to produce lapis lazuli and agate vases, as well as mosaic tabletops with stones of various origins and colours. These objects, created by virtuoso artists at the princes' hardstone workshop that had been founded in the late sixteenth century, were regarded as both collectibles and works of art.[16]

In this quest for the origins of humanity, and in the face of a seemingly complex world with infinite secrets to be unlocked, a unified vision of nature prevailed, connecting knowledge, power and the sacred. Being in the world, studying it and reinventing it all flowed from the same analogical process, which was based on a perception of the living as a totality, an uninterrupted network. This was true of the geognostic landscapes painted by Dürer during his journey through the Alps and of Giuseppe Arcimboldo's portrait of *Rudolph II in Vertumne*, works founded on the constant interaction between man and the world. The geological landscape of ancient times that Leonardo da Vinci painted in *The Virgin and Child with Saint Anne*, with bluish mountains in the background and pebbles in the foreground, was a pictorial expression of the Tuscan artist's ongoing quest to find a link between human history and natural history in the Arno region. It is no wonder that Paul Valéry called Leonardo "the artist of the world itself."[17] Claude Lorrain was undoubtedly the first painter to carefully study the effects of sunlight on water and buildings and then render them accordingly in his seaport paintings. However, poeticized nature long remained the place for a relationship with the divine, as in Giorgione's *The Tempest*, where nature overpowers the human figure; or in Bellini's *Saint Francis Receiving the Stigmata*, where it communes with the saint; or in Poussin's *Orpheus and Eurydice*, where it instills fear.

15 See Sabine du Crest, "Des reliquaires d'un genre nouveau," in *Cannibalismes disciplinaires. Quand l'histoire de l'art et l'anthropologie se rencontrent*, proceedings of the colloquium *Histoire de l'art et anthropologie* (Paris, Musée du quai Branly, June 21–23, 2007), ed. Anne-Christine Taylor and Thierry Dufrêne (Paris: Musée du quai Branly; Institut national d'histoire de l'art, 2009), p. 325–334; Sabine du Crest, *L'Art de vivre ensemble. Objets frontière de la Renaissance au XXI*e siècle (Rome: Gangemi, 2017).

16 Founded in 1588, the Galleria dei Lavori became the Opificio delle Pietre Dure in 1860.

17 See Paul Valéry, *Introduction à la méthode de Léonard de Vinci*: "Les actions des forces naturelles concurrentes qui font de tous les lieux de la terre autant de monuments accidentels de leur conflit," in *Carnets de Léonard de Vinci* (Paris: Gallimard, "Tel" series, 2006), p. 8.

18 See Georges-Louis Leclerc, comte de Buffon, *Histoire naturelle* (Paris: Imprimerie royale, 1749), Premier Discours, "Théorie de la Terre," p. 5. For an English translation, see James Smith Barr's *Buffon's Natural History*, vol. 1 (London: Printed for the Proprietor, 1797), https://archive.org/details/buffonsnaturalhi01buff/page/n7/mode/2up.

19 See Abbé Antoine Pluche, *Le Spectacle de la nature, ou Entretiens sur les particularités de l'histoire naturelle qui ont paru les plus propres à rendre les jeunes gens curieux et à leur former l'esprit*, vol. 4 (Paris: Vve Estienne, 1739); Pierre-Joseph Buc'hoz, *Le Jardin d'Éden. Le paradis terrestre renouvelé dans le jardin de la reine à Trianon*, 2 vols. (Paris: Chez l'auteur, 1783–1785); *Sciences et curiosités à la cour de Versailles*, exh. cat., ed. Catherine Arminjon and Béatrix Saule (Paris: Réunion des musées nationaux, 2010).

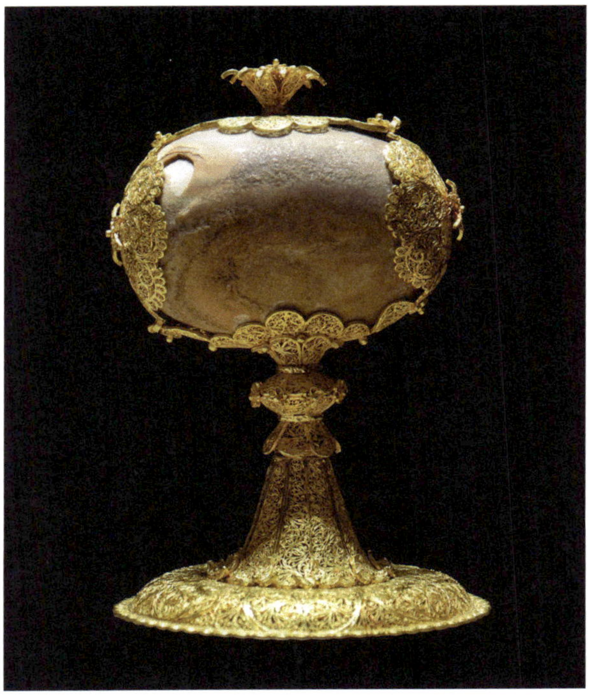

NATURAL LIGHT

The Cartesian imperative that man must become master and possessor of nature gradually took hold as specializations in specific disciplines were established and studiousness supplanted curiosity. The first official publication in 1669 of France's newly created Académie royale des sciences, the *Description anatomique d'un caméléon, d'un castor, d'un dromadaire, d'un ours et d'une gazelle* [Anatomical description of a chameleon, beaver, dromedary, bear and gazelle], established experimentation and observation as accepted methods of studying zoology, which was now freed from the necessity of referring to the ancient authorities. One of the Académie's first members, the architect and naturalist Claude Perrault, initiated this scientific revolution, which flourished in centres of learned interaction such as academies, but also at the court, in plant gardens, menageries and chemistry laboratories. In the following century, the process culminated in the classification undertaken by the Swede Carl Linnaeus, based on the discontinuity of species.

Far from being a form of princely amusement, replicas of the latest discoveries or simply fads, the scientific discourse at the Court of Versailles in the seventeenth and eighteenth centuries was at a high level, both socially and politically, overlaying technical innovation, knowledge of the world and interior decoration. Architecturally, the menagerie at Versailles was conceived as a system of observation and domination of exotic animals from faraway places. These included gazelles, guinea fowls, mongooses, Numidian cranes, macaws from tropical America, an ostrich (painted by Nicasius Bernaerts) (21), cassowaries from Indonesia (illustrated by Jean-Baptiste Oudry), an ephemeral crocodile and Louis XV's rhinoceros from Chandannagar which is still in the Muséum today. Sometimes they were diplomatic gifts, such as the Congo elephant from the king of Portugal (painted by Pieter Boel), and a tigress from the sultan of Morocco. In the salons decorated for Louis-Henri de Bourbon, Prince de Condé in Chantilly, monkeys maliciously mimicked human actions, and enlivening the dinner conversations in Louis XV's private apartments was an exotic hunt gallery featuring wild animals that had fallen victim to man. At the same time, some of these animals were also dissected. Once Count de Buffon had compiled his "well-ordered collection of all the works on nature," he began his great work, *Natural History*, which was published in thirty-six volumes from 1749 to 1789. It was based on behavioural observations and a willingness to progress from the individual to the general.[18] Increasingly, the plants and animals in the garden were associated with naturalists' cabinets, herbaria and scholarly works. Botany now encompassed science, pedagogy, civility, the court and use of the world. The spectacle of nature shaped the spirit, according to Abbé Pluche, and Marie-Antoinette recreated the Garden of Eden at the Trianon. A collection of 260 plant engravings paid tribute to the travellers, botanists, princes and financiers who made possible the extraordinary botanical collection at Trianon, from the Cape of Good Hope to French Guyana and from China to Virginia.[19] Previously unknown species were attentively acclimatized, like the strawberry plant, or the pineapple classified by the botanist Antoine-Laurent de Jussieu and painted by Oudry. Physiocrats studied the globalized management of the natural order in a century in which plants grown far away and commodities from the antipodes, such as tea, coffee, chocolate, sugar, tobacco, rice, maize, many fruits and vegetables as well as silk and

24 FIG.
Embellished Bezoar Stone,
between 1675 and 1700
Bezoar, gold filigree, 11.6 × 7.6 cm
Vienna, Kunsthistorisches
Museum

cotton, brought about lasting changes in consumption and daily life in Europe.

The owner of one of the most remarkable cabinets of natural science in the early eighteenth century, Joseph Bonnier de La Mosson, undertook to rationalize knowledge while preserving the aesthetics of its presentation. In his Parisian mansion, the wealthy financier organized nature into a row of cabinets on the ground floor where items were arranged in an order that pleased both the eye and the mind (25). The first cabinet of natural history was devoted to animals preserved in flasks, the second contained dried animals, butterflies and other insects, plants and minerals, while the third housed his herbarium, library and famous 1,000 shells, all set out in compartments lined with blue satin, a display that anticipates the watercolours by Leroy de Barde (26, 27). Upon the sale of this collection in 1745, Buffon acquired for the King's Cabinet various shells and shelves that are still in place in the Muséum today.[20] The naturalist Daubenton, who authored an article titled "Cabinet" for the *Encyclopédie* edited by Denis Diderot and Jean Le Rond d'Alembert, wrote that "a cabinet of natural history is an abridgment of all nature."[21] In his "Lettre sur le choix et l'arrangement d'un cabinet curieux" [Letter on the selection and arrangement of a cabinet of curiosities], the collector and naturalist Dezallier d'Argenville, also a contributor to the *Encyclopédie*, gave specific advice on how to organize collections in cabinets of curiosities. It was essential to compile and organize a cabinet to appreciate the progress of science in a scholarly setting that would offer a panoptic perspective on the natural world based on the principles of concentration and abbreviation, yet separated into various specialized sections. In the field of natural sciences, a subject that captivated the leading minds of that time, identifying and classifying species were major concerns, and spatial organization played a decisive role. In 1638, the Danish zoologist Ole Worm had already determined that the unicorn horn was actually a narwhal tooth; and coral, which continued to intrigue scholars, was now being viewed as a plant. In 1755, Dezallier established his authority in this domain with his work *Conchyliologie*, which was published a number of times, and with *Oryctologie*, which distinguished the study of marine animals from that of terrestrial fossils, paving the way for paleontology and geology, a term whose use was later established by the Geneva naturalist Saussure.[22]

This was also a century of technology, when useful instruments were being developed to explore the natural world, and chemistry became an essential partner in this endeavour. Lavoisier, a brilliant scientist who had fallen victim to the Reign of Terror as a tax collector, authored two works—the introduction to *Nomenclature chimique* and *Traité élémentaire de chimie*, published in 1789—that make him the father of modern chemistry. In David's full-length portrait of him at work with his wife, the new instruments from his laboratory, where they were used to study gases and water, are in the foreground in a splendid still life of glass.[23]

Like the natural elements and their nomenclature, human beings were now the object of profound study. In this grand project of unveiling and transparency, cross-sections of cadavers and plants were displayed in life-size anatomical (and botanical) wax models. A specialist in their preparation was the Italian physician and anatomist Felice Fontana, who became director of the first Museum of Physics and Natural History in Florence, which was founded by Grand Duke Pietro Leopoldo of Tuscany and opened to the public in 1775. At La Specola, as the museum came to be known, Fontana developed an encyclopedic learning pathway that would make nature's system open to all. His crowning achievement, however, was the spectacular and macabre wax cabinet, which attracted all of Europe, and in particular artists like Élisabeth Vigée Le Brun who were seeking to understand the inner secrets of facial expression.

The same passion for exploration that led some to collect plants or closely examine human entrails led others, like Lady Hamilton, to join scholarly

20 See Guillaume Glorieux, *À l'enseigne de Gersaint. Edme-François Gersaint, marchand d'art sur le pont Notre-Dame, 1694–1750* (Seyssel: Champ Vallon, 2002). See also Lafaille's cabinet in La Rochelle and Robien's in Rennes.

21 See Daubenton, in *Encyclopédie*, vol. 2, 1752, p. 489; *1740, un abrégé du monde. Savoirs et collections autour de Dezallier d'Argenville*, exh. cat., ed. Anne Lafont (Paris: Institut national d'histoire de l'art; Lyon: Fage, 2012).

22 See Antoine-Joseph Dezallier d'Argenville, *L'Histoire naturelle éclaircie dans une de ses parties principales, l'oryctologie, qui traite des terres, des pierres, des métaux, des minéraux, et autres fossiles* (Paris: De Bure, 1755) and *La Conchyliologie, ou Histoire naturelle des coquilles de mer, d'eau douce, terrestres et fossiles, avec un traité de la zoomorphose, ou représentation des animaux qui les habitent*, 3rd ed. (Paris: De Bure, 1780).

23 "In the eighteenth century, the laboratory was a special feature or component of the cabinet, a place for chemical processes and investigation." See Pascal Dubourg-Glatigny, "Laboratoire," in *1740, un abrégé du monde*, p. 151–156.

24 See Alexandre Koyré, *From the Closed World to the Infinite Universe* (Baltimore: Johns Hopkins University Press, 1957), p. 2; Marcel Gauchet, *Le Désenchantement du monde. Une histoire politique de la religion* (Paris: Gallimard, 1985); Catherine Colliot-Thélène, *Études wébériennes. Rationalités, histoires, droits* (Paris: Presses universitaires de France, 2001).

25 Koyré, ibid.

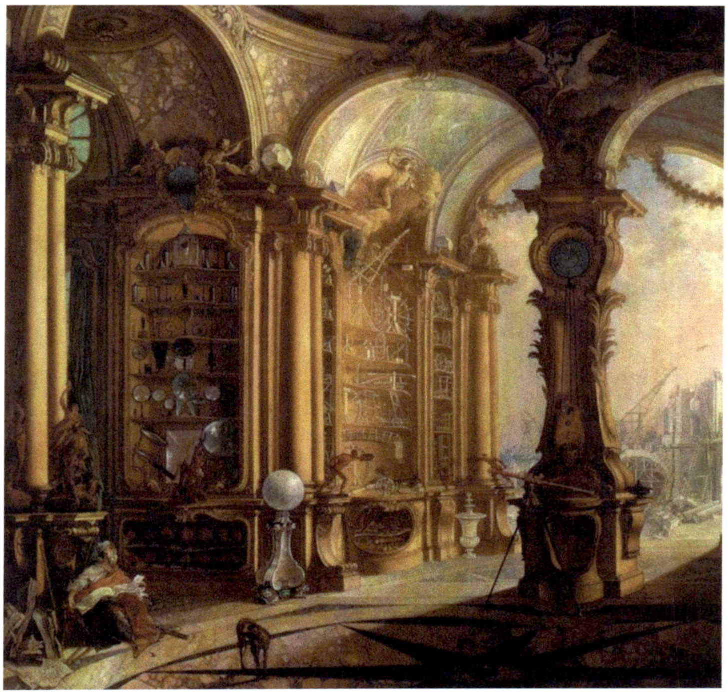

earth of Volaire. And where, in the seventeenth century, still lifes retained coded symbols, under the brushstrokes of a Chardin they could cast doubt on the very nature of things.

These centuries of natural history tell the story of cooperation from within and beyond the discipline, between humans and the world they study, through the *Wunderkammern*, as well as in books, menageries, gardens and the graphic arts. Intellectuals constantly fluctuated between enthusiasm and disenchantment. Those like Schiller who were nostalgic for a world where meaning had not been lost and where the divine seemed accessible in nature had noted the disenchantment as early as 1788. The French philosopher Alexandre Koyré wrote that, beyond the experience itself, the change of perspective and the new theoretical vision gained by the transition from the closed-off, Aristotelian world to the infinite universe of Euclidean geometry imply "the discarding by scientific thought of all considerations based on value-concepts, such as perfection, harmony, meaning and aim, and finally the utter devalorization of Being, the divorce of the world of value and the world of facts."[24] This process, in which curious knowledge and studious knowledge of the natural world have interacted, has caused man to lose "his place in the world, or more correctly perhaps . . . the very world in which he was living and about which he was thinking . . . [such that he has] had to transform and replace not only his fundamental concepts and attributes, but even the very framework of his thought."[25] For better or for worse.

excursions and search for the origins of Earth in faraway places like the slopes of Vesuvius. This spawned a fashion for scientific jewelry made from sculpted lava stone, souvenirs of the Grand Tour, and for paintings of the erupting volcano by the Chevalier Volaire (90, P. 136), while age-of-the-Earth debates engaged learned European society.

To human eyes, the sacred, the scientific, the artistic and the political all interact to some extent in the great book of nature. To study nature and represent it is also to reconstruct and reinvent it. As a field of human activity, nature becomes culture. In other words, through human intervention, nature is at work, playing a role.

Paintings that depict nature in all its forms capture these changing relationships in landscapes and compositions of flowers, fruits, shells and animals. Rather than alternating between domination and subjugation, the relationship between humans and nature in modern times—while complex and interactive—is one of recurring ambivalence and of a shifting focus between near and far, large and small. Landscapes can at times be imbued with feelings they share with the humans who blend into them. A spectacle of passion bursts forth in the raging sea of Vernet or the blazing

25 FIG.
Jacques de Lajoue,
Bonnier de La Mosson's Cabinet of Physics, 1734
Oil on canvas, 108 × 137 cm
Blessington, Russborough House & Parklands,
Beit Collection

P. 50-51

26, 27
Alexandre-Isidore Leroy de Barde
Watercolour and gouache
Birds, 1800-1828,
124.5 × 88.5 cm
A Collection of Foreign Birds,
about 1810, 126 × 90 cm
Paris, Musée du Louvre

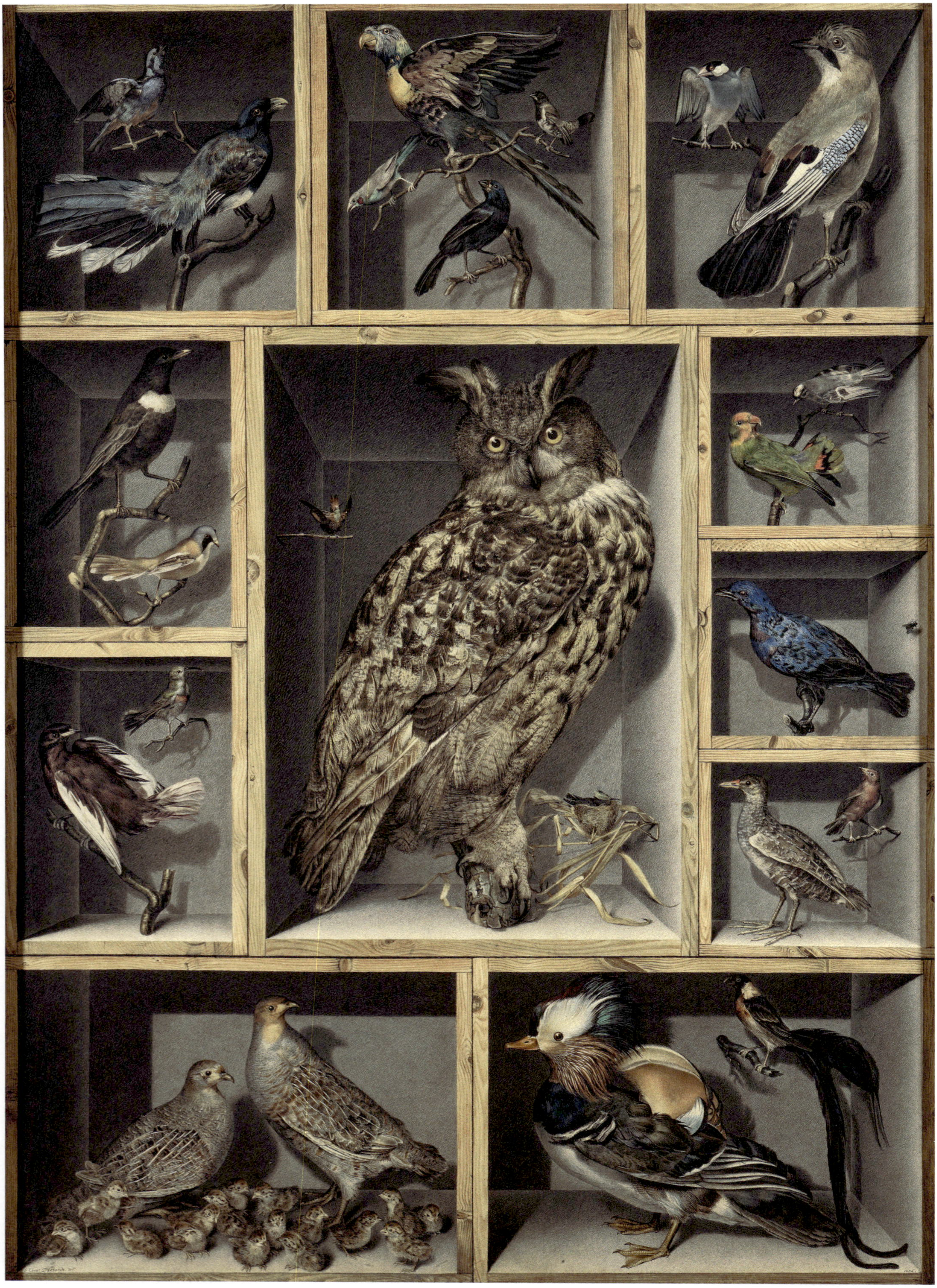

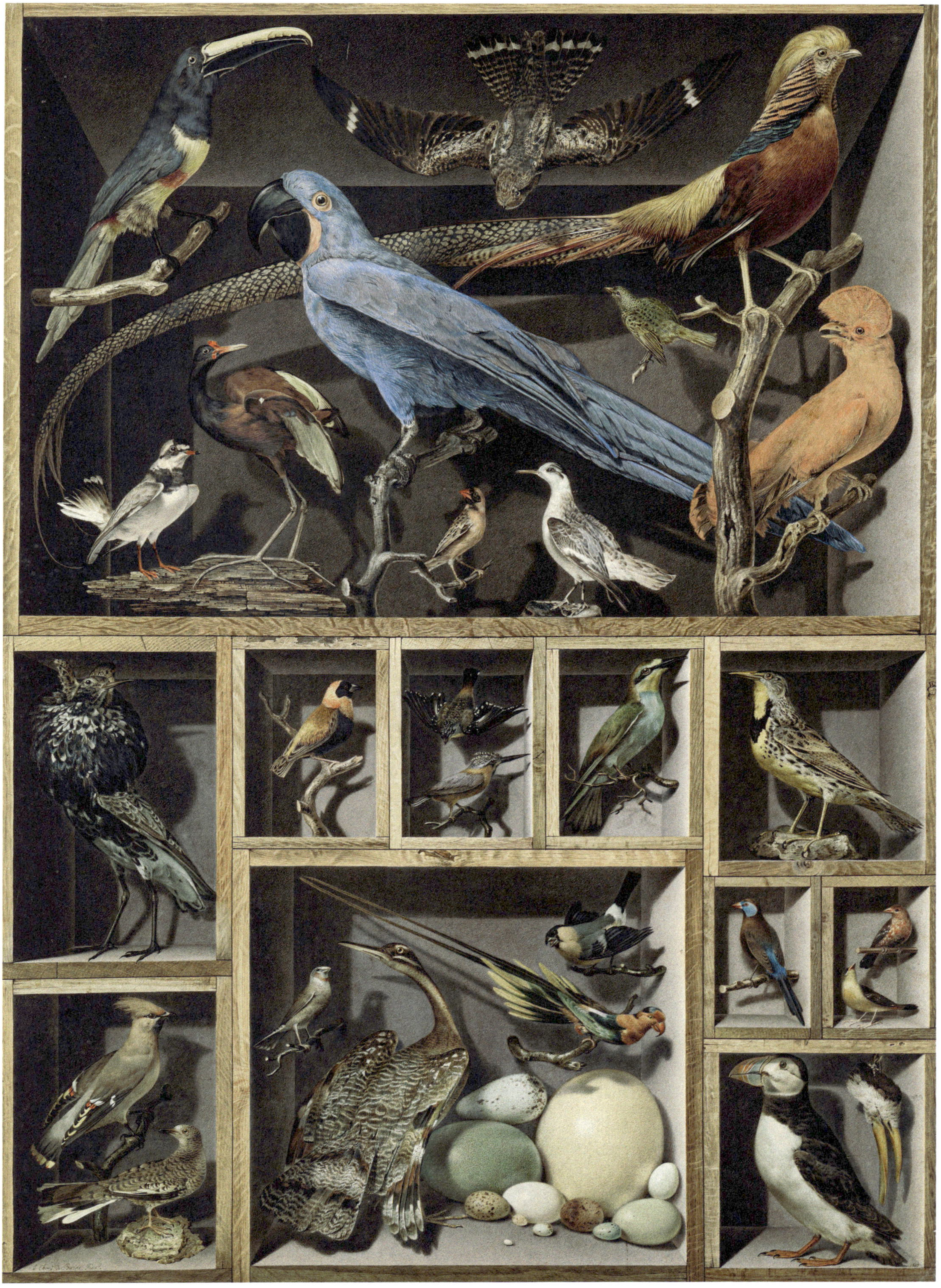

THE GRAND TOUR OF CLARA THE RHINOCEROS

LAURA BOSSI

"A rhinoceros!
I can't get over it!"
Eugène Ionesco, *Rhinoceros*[1]

With the increased opening-up of trade with the East, the eighteenth century saw the arrival in Europe of many exotic birds and animals—monkeys, camels, rhinoceroses and elephants—intended to expand the menageries of monarchs and nobles, as well as be put on public display. Some of them would become veritable celebrities, with names and sometimes biographies, and would be painted by great animal artists.

The case of Clara, a female Indian rhinoceros that was born in the region of Assam about 1738 and died in London on April 14, 1758, is one of the most exemplary.

While the Romans had rhinoceroses brought from Asia or Africa for circus games and the triumphs held to celebrate victories, following the specimen presented in the Colosseum during the Secular Games held in 248 C.E. to celebrate Rome's millennial anniversary, there were no further reports of any rhinoceroses in Europe prior to 1515,[2] when the first rhinoceros of the modern era, whose image was circulated by means of Dürer's famous engraving (28),[3] landed in Lisbon. Intended for Pope Leo X, the ship on which it was being transported sank near La Spezia, on Italy's Ligurian coast, in 1516. At the court of King Philip II of Spain, a second Indian rhinoceros, a female, lived for a longer time (from 1577 to after 1586) in the El Escorial menagerie; in Madrid she is remembered in the name of a street close to the Puerta del Sol, Calle de la Abada (*abada* being another word for rhinoceros in Spanish). A third rhinoceros arrived in London in 1684. It was sold to a businessman who charged the public a shilling to see the animal, and two shillings to sit astride it; however, it did not survive long, probably due to the inexperience of its owner. Another rhinoceros lived in captivity in London from 1739 to 1741; that creature was studied and drawn by James Parsons, who also used it as the model for a large painting held in London's Natural History Museum.[4]

Clara was the fifth rhinoceros to reach modern Europe alive, but the first to live a long time in captivity, as well as the first to gain a celebrity akin to that of Dürer's rhinoceros.

Her mother having been killed by hunters, the young orphan was adopted by a director of the Dutch East India Company, J. A. Sichterman, who had raised her as a pet and even allowed her to move around his home freely. Nonetheless, when she grew bigger he sold her to an enterprising captain, Douwe Mout van der Meer, who immediately grasped her commercial potential and arranged, at his own expense, for the animal to travel on his ship, the *Knabenhoe*, from Calcutta to Rotterdam by way of Ceylon and then around the Cape of Good Hope and the western African coast.

The six-month-long sea voyage must have been a difficult one, not so much because of the creature's size or behaviour (she apparently had a docile nature and was used to human company), but because of the huge amount of food she needed. Furthermore, as water was in short supply on the ship, Clara's hide was coated with fish oil to prevent its dehydration, which must not have given her a very pleasant smell. Following her landing safe and sound in Rotterdam on July 22, 1741, Van der Meer set her up in his hometown of Leiden, whose university was famous throughout the whole of Europe. The first images we have of her, and among the most striking, are those commissioned by the anatomist Bernhard Siegfried Albinus from Jan Wandelaar for the former's anatomical atlas, which was intended to replace the classic work by Andreas Vesalius dating from 1543. Two superb engravings show Clara, flanked by a human skeleton, from the front and the back—in one, she is grazing on grass (the main occupation of a rhinoceros, which must eat close to 70 kilograms of grass or other plants daily to stay healthy) (30). Although used as a backdrop, she is their true focus.

The engravings Van der Meer would have made shortly after to bring Clara to the attention of the general public would be of a different nature (29), akin more to advertising posters meant to whet the curiosity and prepare people's minds for an exceptional spectacle. The animal was portrayed as the biblical Behemoth described in the Book of Job (40:15—"Behold now behemoth, which I made with thee; he eateth grass as an ox"), one of the most fearsome creatures, along with the Leviathan, of land and sea.

During her stay in Leiden, cared for by her "impresario" and already offered up to the view of both scholars and the inquisitive, Clara thrived. In 1744, Van der Meer left the Dutch East India Company to devote himself entirely to his ward. Two years later, he embarked upon an amazing Grand Tour of the major cities of Europe that would last for twelve years.[5]

Set up in a horse-driven wagon, which can be seen in a painting by Pietro Longhi (33), and sometimes travelling by boat on the Rhine, Clara would wend through Germany, Austria and Switzerland, on a journey that would take her to Hanover, Berlin, Frankfurt, Breslau (now Wrocław, Poland), Vienna, Munich (possibly), Regensburg (1747), Fribourg, Dresden, Meissen, Leipzig, Kassel, Mannheim, Strasbourg, Bern, Zurich, Basel, Schaffhausen, Augsburg, Nuremberg and Würzburg, where she would be given the name *Jungfer Clara* [Miss Clara].

She caused a sensation in Berlin, where Frederick the Great, along with eminent members of his court, went to the Spittelmarkt area and braved the mud and the smell of fish oil; in Vienna, Empress Maria Theresa would see her twice, and named Van der Meer a baron of the Holy Roman Empire. Accompanied by his young son, Augustus III, Elector of Saxony and King of Poland, paid her a visit in Dresden, while in Meissen, she served as a model for Johann Joachim Kändler, the chief modeller of the town's porcelain manufactory. In Kassel, she was put up for a month in the orangery of the Landgrave of Hesse-Kassel, Frederick II, where she discovered she had a passion for oranges.

Clara next went to France, first to Reims, and then to Versailles (1749) and Paris, setting off a veritable "rhinomania" that saw allusions to the animal proliferate on clocks (31), wallpaper and fabrics, and even inspired hairstyles dubbed *à la rhinocéros*. Writing to Denis Diderot, the man of letters Friedrich Melchior Grimm noted: "All of Paris, so easily intoxicated by small things, is now enthralled by a species of animal

1 Eugène Ionesco, *Rhinoceros and Other Plays*, trans. Derek Prouse (New York: Grove Press, 1960), p. 13.

2 Leendert Cornelis Rookmaaker, "Captive Rhinoceroses in Europe from 1500 until 1810," *Bijdragen tot de Dierkunde*, vol. 43, no. 1 (January 1973), p. 39–63, available online: https://www.repository.naturalis.nl/document/547904; Leendert Cornelis Rookmaaker, *The Rhinoceros in Captivity: A List of 2439 Rhinoceroses Kept from Roman Times to 1994* (The Hague: SPB Academic Publishing, 1998); see also T. H. Clarke, *The Rhinoceros from Dürer to Stubbs, 1515–1799* (London: Sotheby's Publications, 1986).

3 Its life inspired the novel by Lawrence Norfolk, *The Pope's Rhinoceros* (London: Vintage, 1997).

4 James Parsons, "A Letter from Dr. Parsons to Martin Folkes, Esq., President of the Royal Society, Containing the Natural History of the Rhinoceros," *Philosophical Transactions of the Royal Society*, vol. 42, no. 470 (1742–1743), p. 523–541, pls. 1–3.

5 Glynis Ridley, *Clara's Grand Tour: Travels with a Rhinoceros in Eighteenth-Century Europe* (New York: Grove Press, 2004).

called the rhinoceros."[6] In his publication *Lettre sur le rhinocéros*, the priest Jean-Baptiste Ladvocat remarked that Clara's tongue was velvet-smooth and that she liked beer and wine.[7] Louis XV would have liked to keep the lovely Clara for his menagerie, but Van der Meer was not prepared to relinquish his golden goose. He asked 100,000 ecus for her—a prohibitive price that made the king abandon the idea.[8]

From Paris, Clara went to Marseille, where she boarded a ship for Naples. She arrived in Rome for the 1750 Holy Year, and lost her horn there. Leaving the Eternal City, she moved on to Bologna, Milan, Venice (where she was painted by Longhi) and Verona, then Vienna and, finally, London, where she was seen by King George II.

Information about Clara's "tour" during the years from 1752 to 1758 is less exact. She was reported to have been in Prague, in Warsaw in 1754, in Kraków and Breslau, and then in Copenhagen in 1755. The year 1758 saw her back in London, where she could be viewed at The Horse and Groom in Lambeth Market for six pence or a shilling, and where she died at the age of twenty on April 14.

It is likely more portraits of Clara were painted than of any other rhinoceros. A beautiful 1747 drawing by the physician and naturalist Petrus Camper showing her as a chubby youngster is in the collection of the University of Amsterdam. We have seen that Jan Wandelaar depicted her from the front and back in Albinus's anatomical atlas; in Augsburg, the animal painter Johann Elias Ridinger would make six drawings of her, emphasizing the folds of her skin, the three toenails on her feet, and the inexpressiveness of her eyes (rhinoceroses have very bad eyesight). Better known, however, are the paintings the Venetian noble Giovanni Grimani dei Servi commissioned from Pietro Longhi (32, 33).

In Paris, the animal painter Jean-Baptiste Oudry executed a truly life-sized portrait of Clara measuring three by four metres (34), which he exhibited at the 1749 Salon. That remarkable work was acquired by the Duke of Mecklenburg and is now held at the Staatliches Museum in Schwerin, Germany.

Count de Buffon would use Oudry's portrait of Clara in his *Histoire naturelle*, but did not indicate her name or identify her as an individual specimen;[9] she would represent the "typical" rhinoceros, rectifying once and for all the image of the animal propagated by Dürer, who had not been able to observe the animal directly himself, and had attached an additional small horn to its back. Clara would also appear in Denis Diderot and Jean Le Rond d'Alembert's *Encyclopédie*.

By turns taking on the various roles of a pet, mythical creature, biblical monster, marvel coveted by kings, fairground freak, wonder of nature worthy of a life-sized portrait by one of the greatest painters of his day, international celebrity and, finally, within the new systematization of the natural world, classic example of a species, Clara was the perfect embodiment of the transition from curiosity to studiosity that took place in the Age of Enlightenment.

6 Maurice Tourneux, ed., *Correspondance littéraire, philosophique et critique par Grimm, Diderot, Reynal, Meister, etc.* (Paris: Garnier, 1877), vol. 1, p. 272.

7 Jean-Baptiste Ladvocat, *Lettre sur le rhinocéros* (Paris: Chez Thiboust, 1749).

8 When an Indian rhinoceros was finally installed in the royal menagerie at Versailles, it would be the gift of a certain Chevalier, the governor of a distant French enclave in what would soon become British India. See Ridley, *Clara's Grand Tour*, p. 140.

9 "Le Rhinocéros," in Georges-Louis Leclerc, comte de Buffon, *Histoire naturelle, générale et particulière* (Paris: Imprimerie royale, 1754), vol. 11, pl. 7, p. 202.

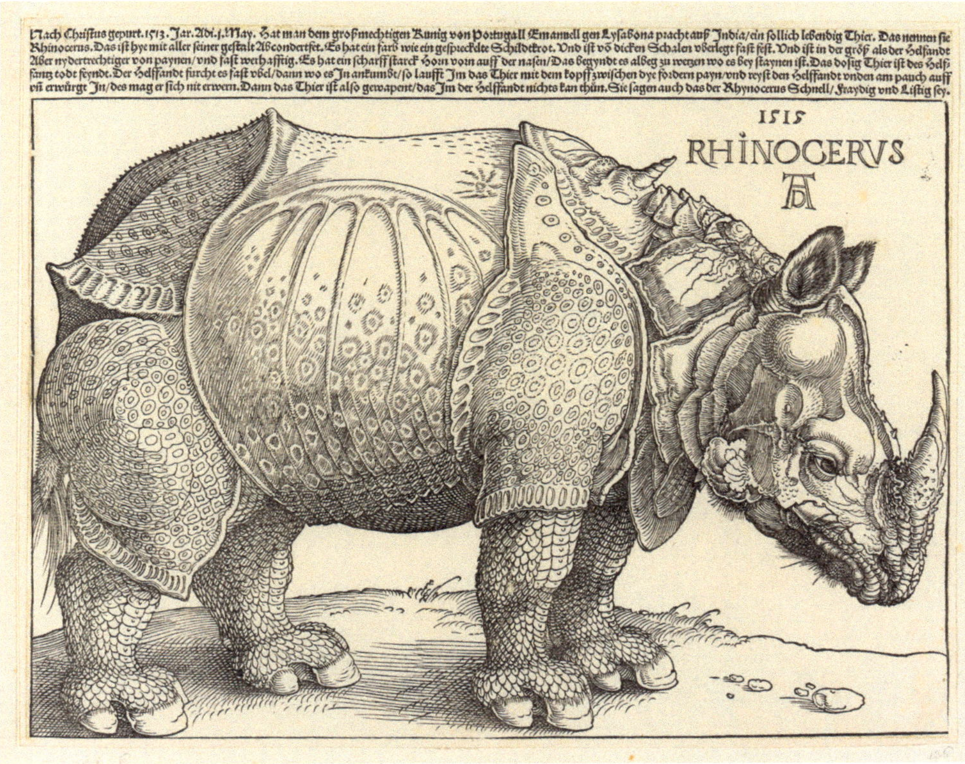

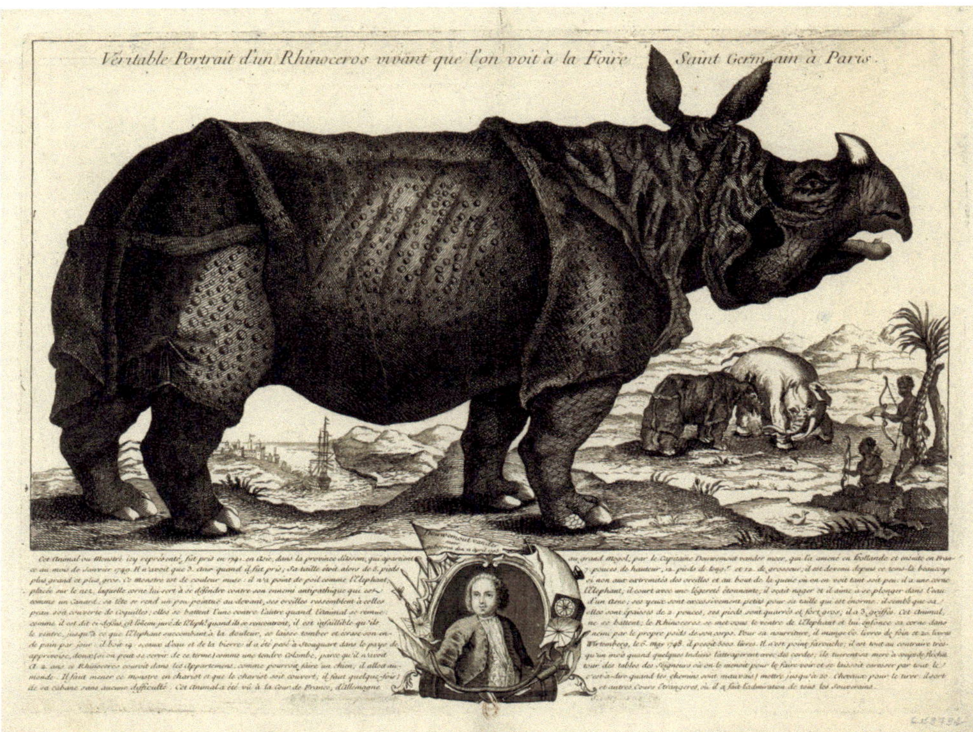

28
Albrecht Dürer,
The Rhinoceros, 1515
Engraving, woodcut,
21.6 × 30.2 cm
Paris, Musée du Louvre

29 FIG.
Étienne Charpentier (engraver),
True Likeness of a Live Rhinosceros Seen at the Saint-Germain Fair in Paris, 1749
Print, steel engraving,
30.5 × 53 cm
Paris, BnF

P. 56, 57

30
Bernhard Siegfried Albinus,
Jan Wandelaar (draughtsman),
Tabulae sceleti et musculorum corporis humani, pl. 4, Leiden:
Joan & Herman Verbeek,
1738–1747
Paris, MNHN

31
Jean-Joseph de Saint-Germain (founder),
François Viger (clockmaker),
Rhinoceros Clock, 1750
Gilt and patinated bronze with tortoiseshell and copper on the base, enamelled face,
83 × 50 × 24 cm
Paris, Musée du Louvre

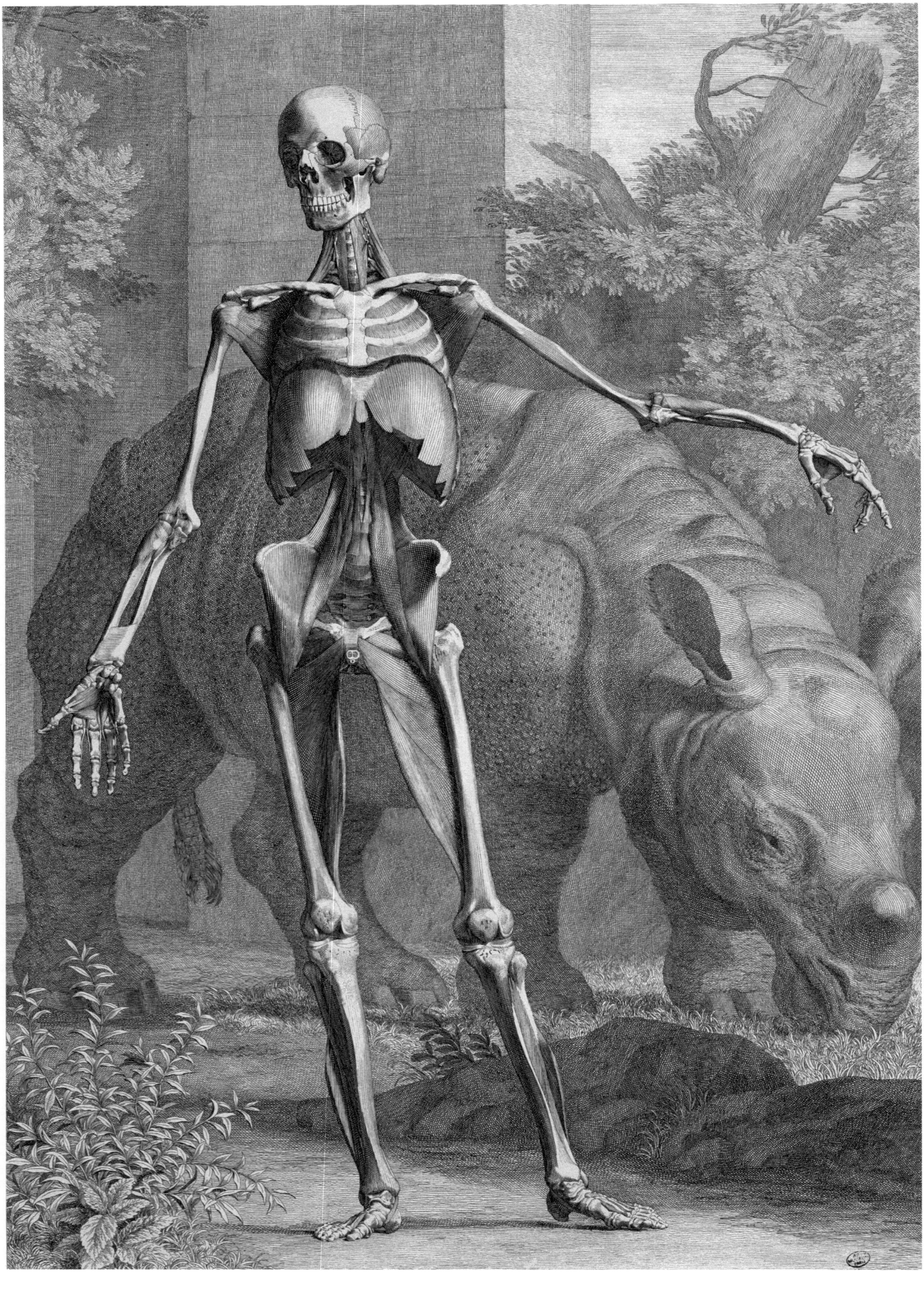

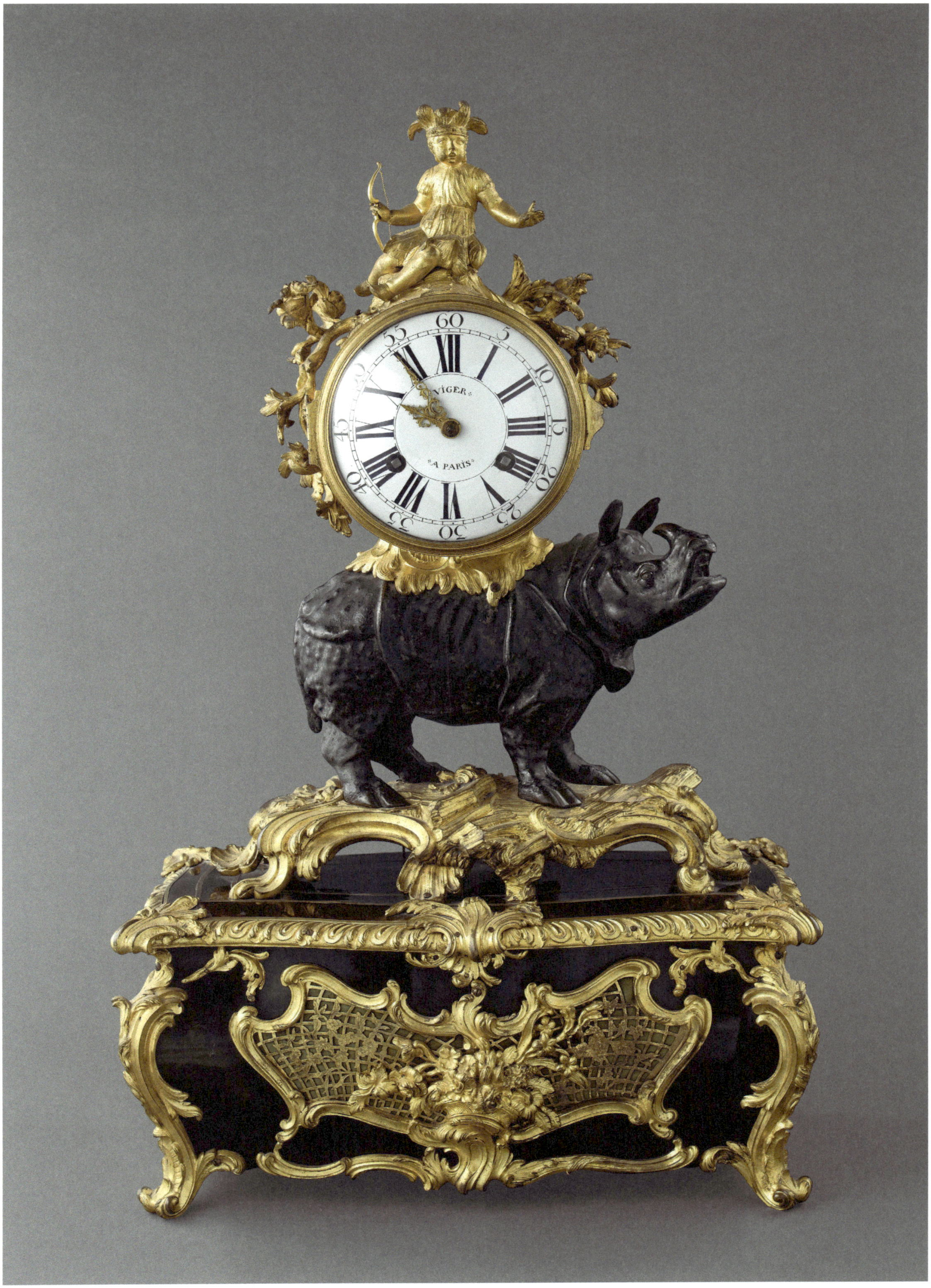
VIGER
A PARIS

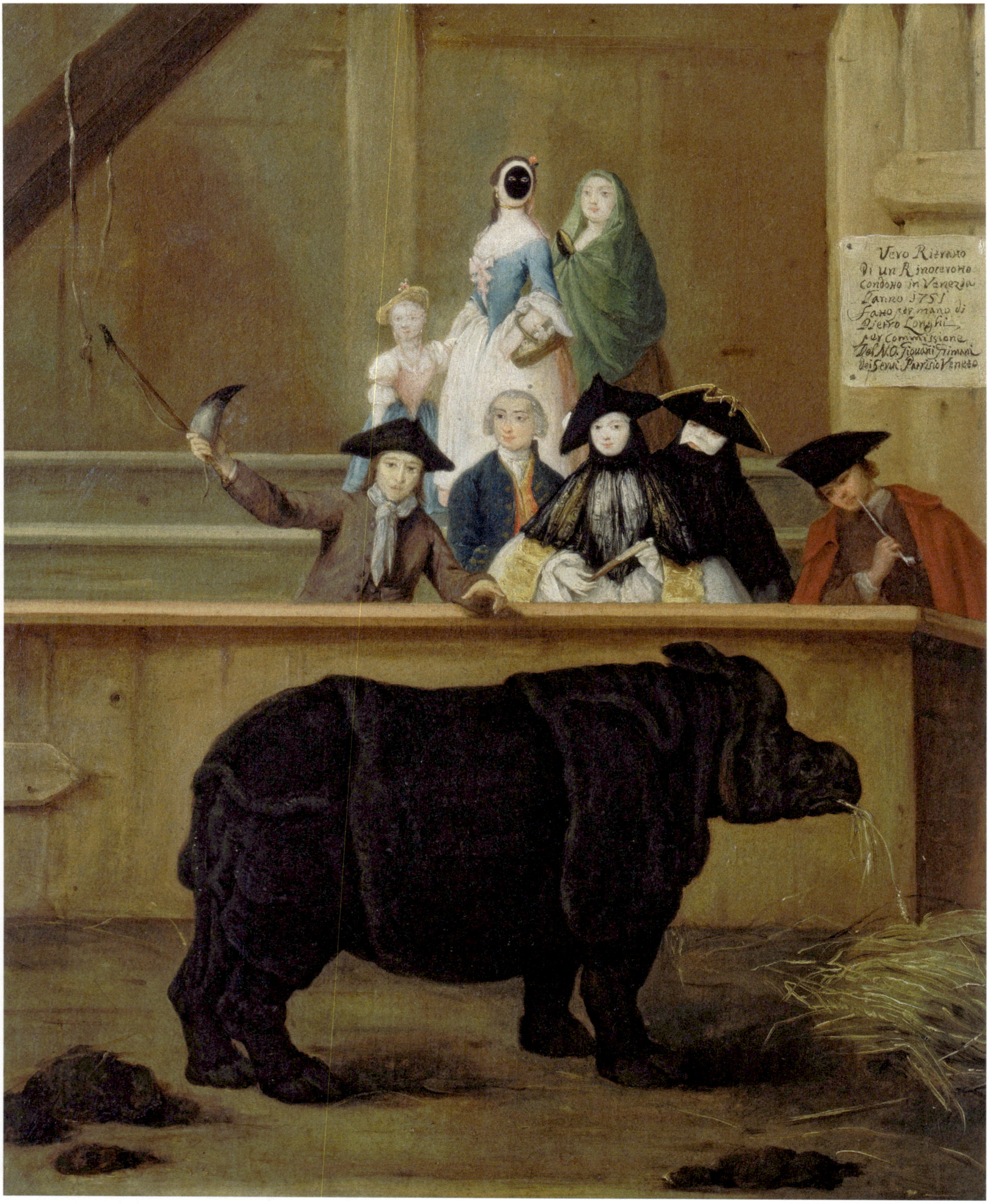

32
Pietro Longhi,
The Rhinoceros, 1751
Oil on canvas, 62 × 50 cm
Venice, Fondazione
Musei Civici di Venezia

33
Pietro Longhi
(attributed to),
The Rhinoceros, about 1751
Oil on canvas, 55.5 × 72 cm
Vicenza, Gallerie
d'Italia – Palazzo Leoni
Montanari

34 FIG.
Jean-Baptiste Oudry,
Rhinoceros, 1749
Oil on canvas, 310 × 456 cm
Schwerin, Staatliches
Museum Schwerin

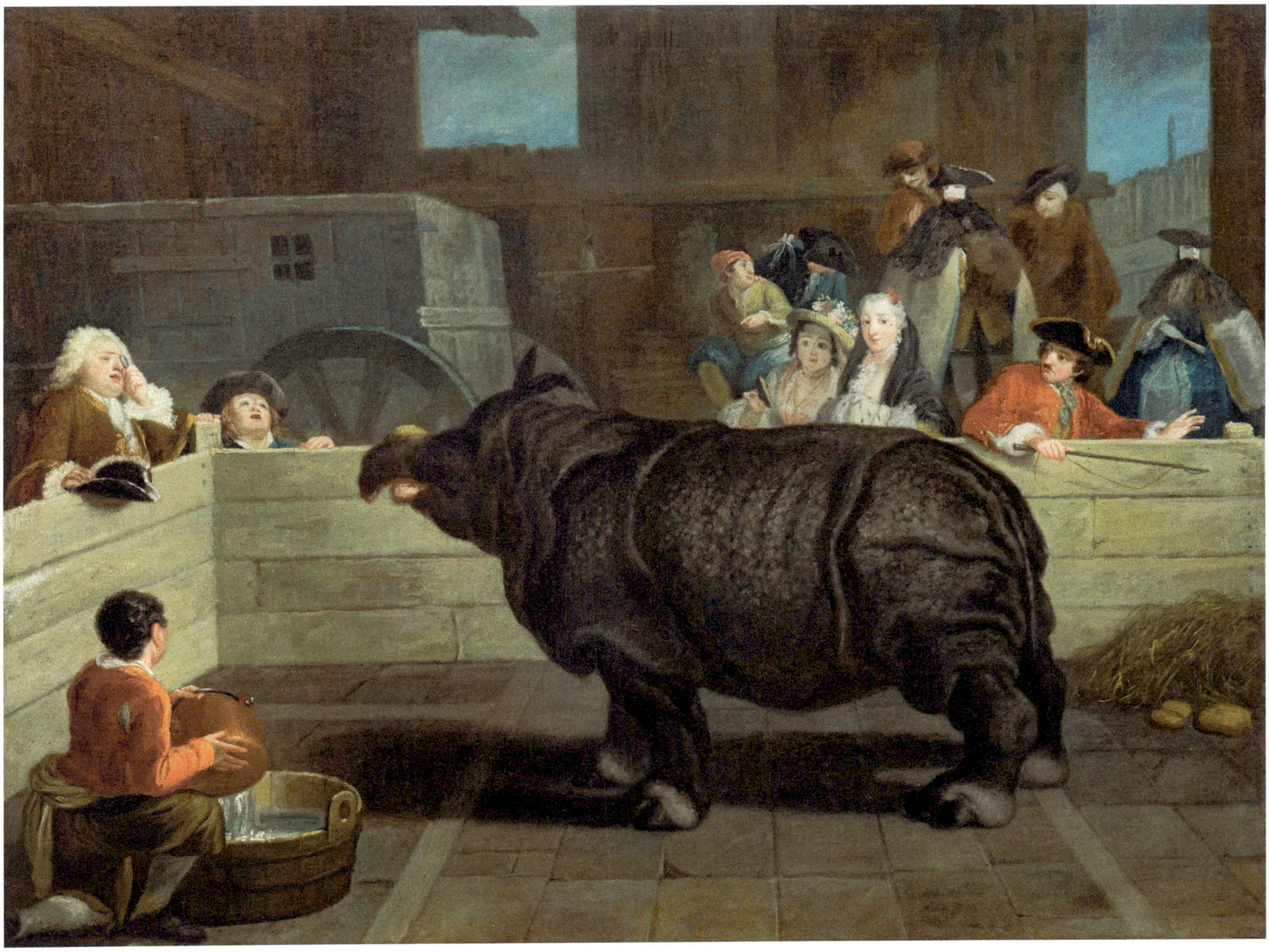

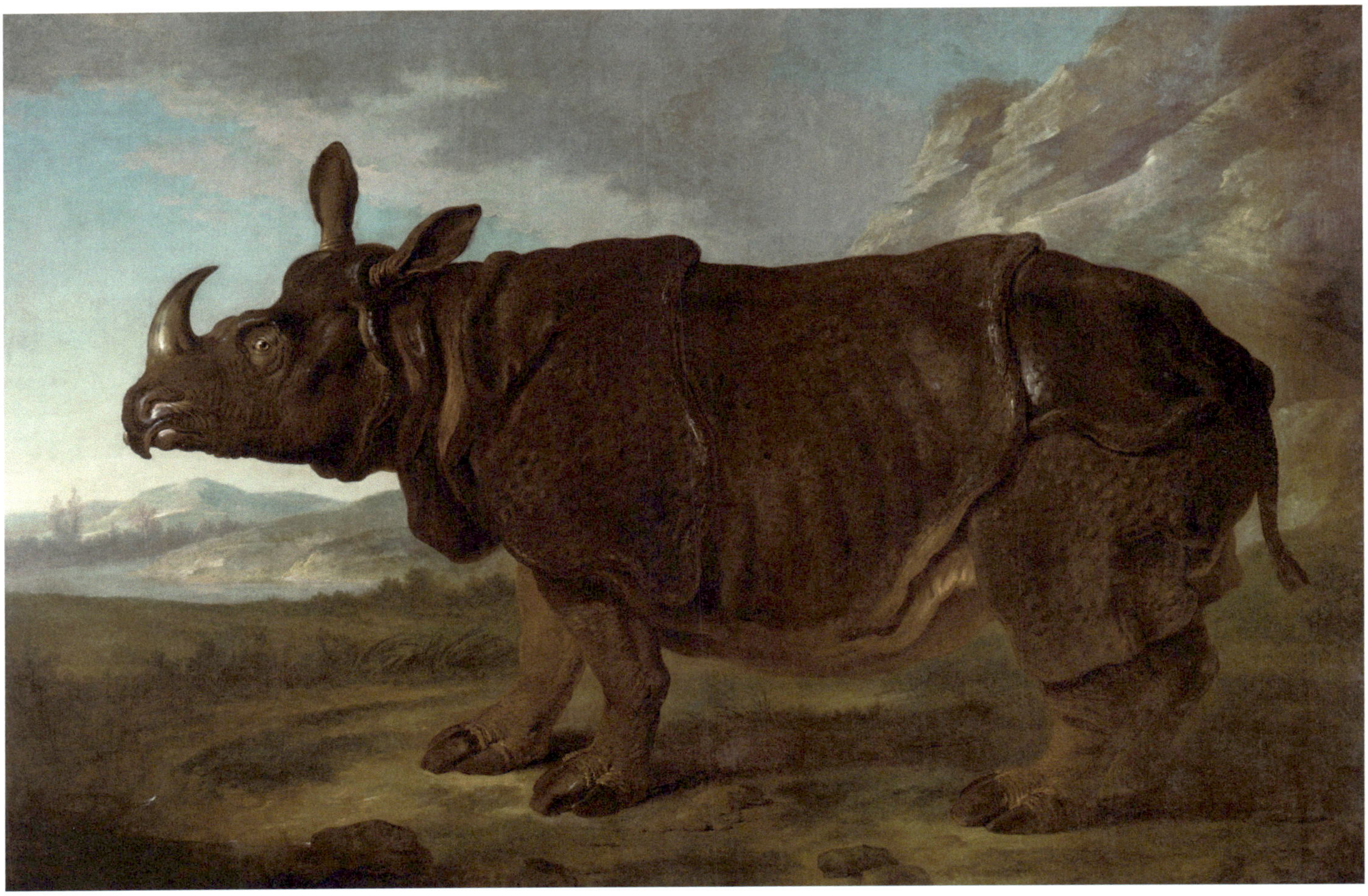

FROM AMATEURS TO SCHOLARS:

THE MUTATIONS OF A COLLECTION

PASCALE HEURTEL AND MICHELLE LENOIR

Early in the seventeenth century, three factors favoured the emergence of a new iconographic genre: floral and avian illustration. Gathered in portfolios, paintings of flowers or birds on paper or vellum,[1] some enlivened with insects and most of them captioned, were presented on their own rather than, as previously, in support of a text. These factors were the persistence, after the invention of the printing press, of the art of illumination perfected in the fifteenth century; the development of naturalistic observational drawing to illustrate scientific treatises of the Renaissance; and the inclusion of natural curiosities and living collections among the objects sought after by amateurs.

Thus it was that, in the middle of the century, while under house arrest at Château de Blois, Louis XIII's younger brother, Gaston, Duke of Orléans, a prince of great scientific and artistic culture and one of the most lavish patrons and collectors of his day, commissioned an eminent miniaturist, Nicolas Robert, to depict on vellum the flora of the vast botanical and pleasure gardens (36, 37) he continually enhanced and had entrusted to the care of his personal physician, Abel Brunyer, trained at the Faculty of Medicine in Montpellier. To this undertaking he added the indigenous and exotic birds that inhabited or migrated through his domain.

Renowned by the time of the prince's death, "all [his] books of flowers and birds" were bequeathed to his nephew Louis XIV "to be . . . placed in his Majesty's cabinet at the Louvre and used for his entertainment."[2] Convinced of their value for the fields of science and art, which contributed to the glory of the king and the grandeur of the State, the king and Jean-Baptiste Colbert, his minister of finance, decided that the work should carry on. New vellums would be executed under the same conditions. Nicolas Robert, appointed painter of miniatures to the king, and his successors, all chosen among the best miniaturists of their time, selected their models in the King's Garden in Paris and the menagerie at Versailles, working under the supervision of the Garden's teaching and research professors and, in some cases, taking part in other important state scientific enterprises, such as voyages and publications.

Louis XIV had the vellums transferred to Versailles, but after his death they were definitively returned to the Royal Library in Paris. On occasion, though, the richly bound portfolios bearing his heraldic emblems were taken to the château for presentation to distinguished court visitors along with other royal collection treasures. Meanwhile, the miniaturists continued to enrich the various series—concentrating on botanicals due to waning interest in the menagerie—under the dual control of the Keeper of the Print Cabinet and the professors of the King's Garden.

Amateur in origin yet prestigious by virtue of the rarity of its subjects, the precision of the drawings, the whiteness of the skins and the subtlety of the colours, the collection of royal vellums was and had always been a true scientific collection, one of rigour and accuracy, owing to the will of its patrons and the conditions of its production.

But, at a time when natural sciences were branching out, when scholars were becoming professionals, when the frontiers of the known world were steadily expanding, taking with them the idea and possibility of exhaustively inventorying an ever more abundant natural world, could this collection created for the pleasure of princes still serve the nation once the monarchy was no more? Answering the question in the affirmative, the constitutional authorities decided in June 1793 to transfer the precious portfolios of vellums to the Muséum d'histoire naturelle, newly established to replace the King's Cabinet and Garden.[3] However, now entirely in the hands of scholars who until then had served merely to assure its scientific accuracy, the collection would have to evolve in its manner of production, its organization and its objectives, all closely related, of course. The idea of picturing every element of the living world, as they hoped to do with the specimens, was clearly unrealistic, but it was important that the vellums remain a reference collection, a visual encyclopedia of nature's treasures. Therefore, they would be organized and completed, in the same way as the other natural history collections from the Royal Cabinet and from the noble, ecclesiastic and princely cabinets seized elsewhere in France or Europe.

The Muséum professors began by undertaking the enormous task of reclassification. First, the botanicals, most widely represented in the royal collection: under the direction of René Louiche Desfontaines, the 4,236 botanical vellums[4] were not only ordered according to the natural method described by Antoine-Laurent de Jussieu in 1789[5] but also renamed using the Linnaean nomenclature that had taken lasting hold in the eighteenth century. Next came the zoological series, under the aegis of each professor: Étienne Geoffroy Saint-Hilaire for the vertebrates, Georges Cuvier for the fish, Jean-Baptiste Lamarck for the shells and so on. Little by little, a new collection took shape: the Muséum's collection. Reorganized, it was also expanded and rebalanced.

To complete it—and, in hindsight, the magnitude of what remained to describe and depict in that era of conquest is easily understood—no single artist could suffice. Two, then three drawing masters, both for botany and for zoology, were carefully chosen; they were paid on a piecework basis for commissions suggested by the various chair holders and systematically approved by the Muséum's Assemblée des professeurs. Despite the financial difficulties of a country at war, the nascent institution was allocated the substantial budget of 3,000 livres to remunerate the naturalist artists. There was no lack of missing specimens in the collection, and no lack of commissions despite the troubled times. Lustrous flowers and exotic birds came to life under the expert brushes of brothers Pierre-Joseph and Henri-Joseph Redouté (35, 38, 39) and Pierre-François de Wailly[6] (40), recalling the finest works of Nicolas Robert, and the nineteenth century saw the addition of series of shells with incomparable pearly sheen, mammals (45, 46), insects and corals from around the globe, and even a modest series of fossils. The Ménagerie in the Jardin des Plantes, once headed by Lamarck,

1 Exceptionally fine, white, supple parchment made from the skin of stillborn calves.

2 Paris, Bibliothèque nationale de France, Manuscrits, Français, 10206, fols. 310v–311.

3 Article IV of Title III of the June 10, 1793, decree concerning the establishment of the Muséum d'histoire naturelle specifically mentions the transfer of the sixty-four portfolios of the vellum collection from the Bibliothèque nationale (heir to the royal collections that included the vellums) to the Jardin des Plantes, home of the Muséum.

4 As counted by Sébastien Gérardin, who was tasked with an initial ordering of the collection in 1804.

5 The classification principles adopted for the Jardin des Plantes were set out by Antoine-Laurent de Jussieu in *Genera plantarum* (Paris: Herissant & Barrois, 1789).

6 Long identified as Léon de Wailly, no doubt as a result of confusion with his son, the novelist Armand-François-Léon de Wailly (1804–1863).

the author of an early theory of evolution, furnished living models that joined the animals painted in the early eighteenth century in the menageries of the king and princes. As a result, the drawing masters for zoology, Nicolas Maréchal and, later, Jean-Charles Werner, represented, among others, monkeys and apes in attitudes similar to those of humans, Maréchal in scenes that were still ornamental (43), Werner in a far starker perspective that foretold the advent of scientific photography (41, 42, 44).

Amid the profusion of new theories and major discoveries that punctuate the history of natural sciences in the nineteenth century, the Muséum's collection of vellums, a survivor of another age, no doubt, but at the same time a work of science and a vector of utopia, participated in this groundswell and, in its own way, marked the changes and trends, like leaves grown on the tree of life planted by Darwin in 1859.

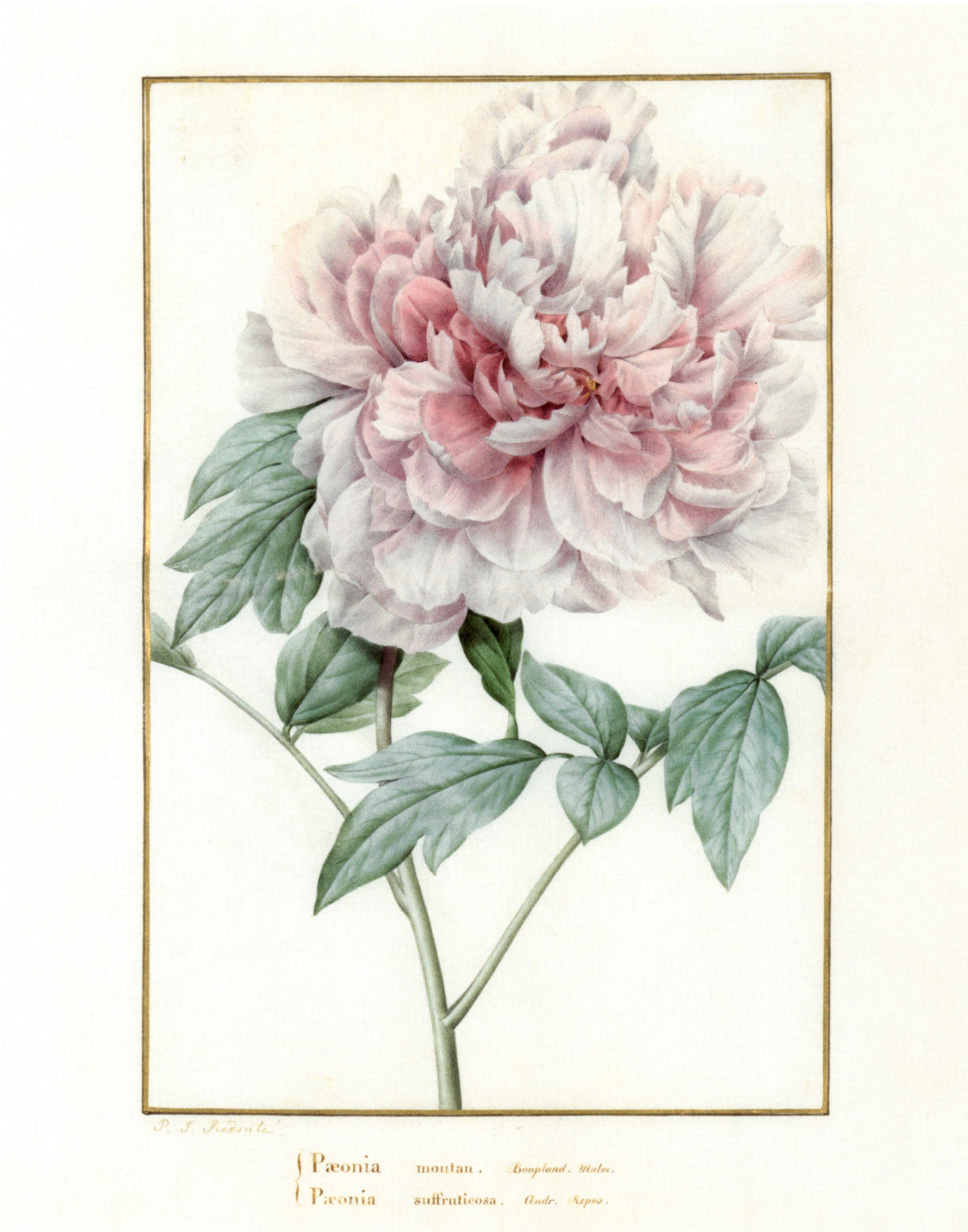

35
Pierre-Joseph Redouté,
Paeonia x suffruticosa
Andrews: *Paeonia moutan*
Bonpland, Malmaison,
late 18th century –
early 19th century
Watercolour on vellum,
46 × 33 cm
Paris, MNHN

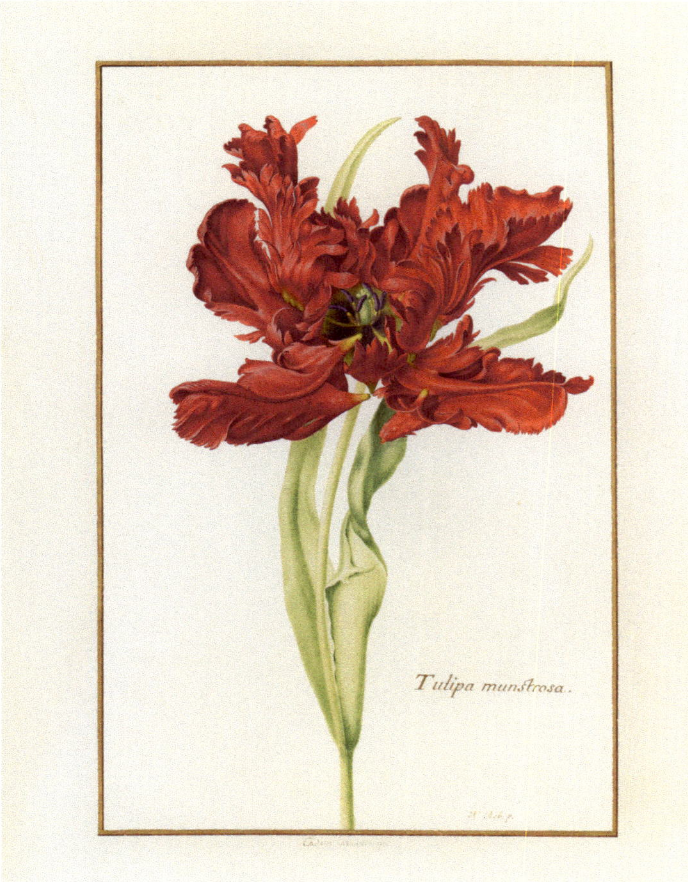

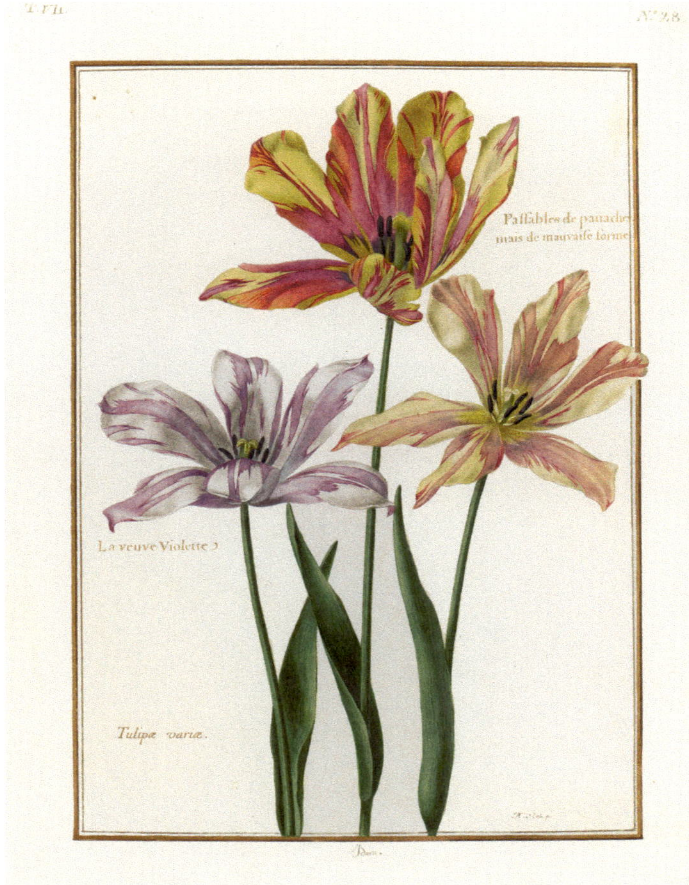

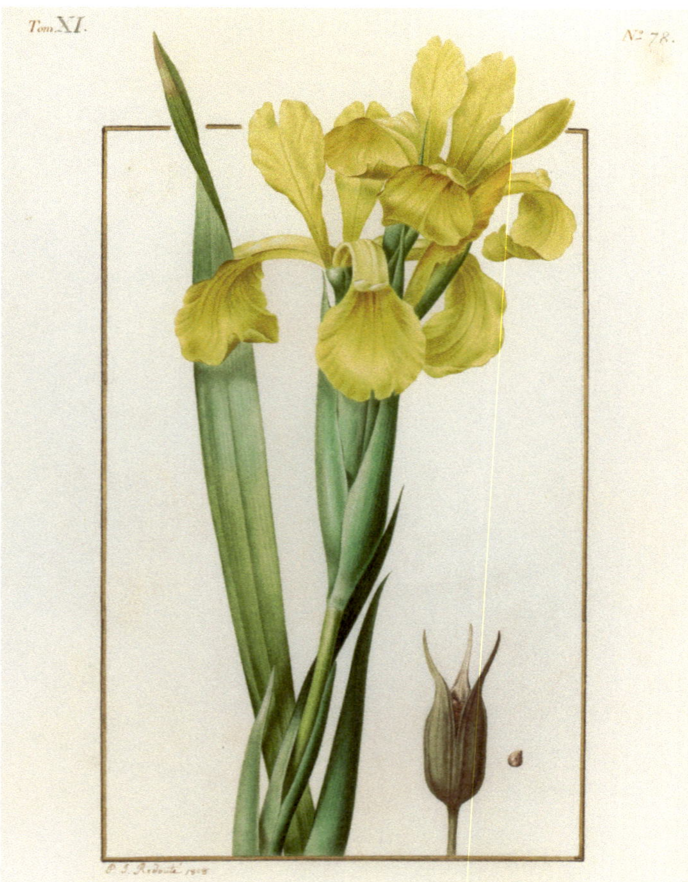

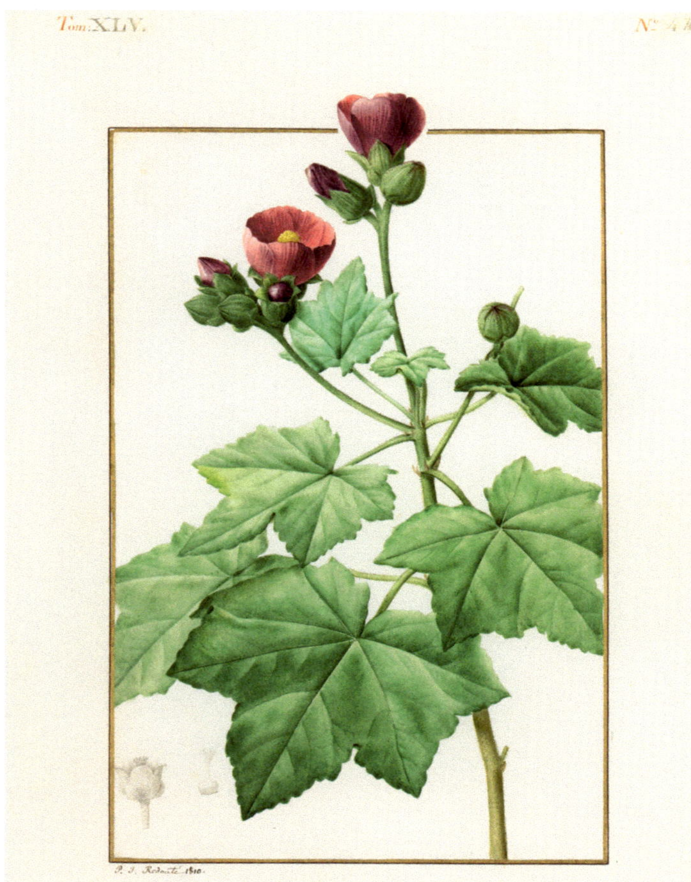

36, 37
Nicolas Robert, 17th century
Gouache on vellum, 46 × 33 cm
Tulipa munstrosa
Tulipae variae
Paris, MNHN

38, 39
Pierre-Joseph Redouté,
Watercolour on vellum, 46 × 33 cm
Iris Monnieri De Candolle, 1808
Malva umbellata Cavanilles, Mexico, 1810
Paris, MNHN

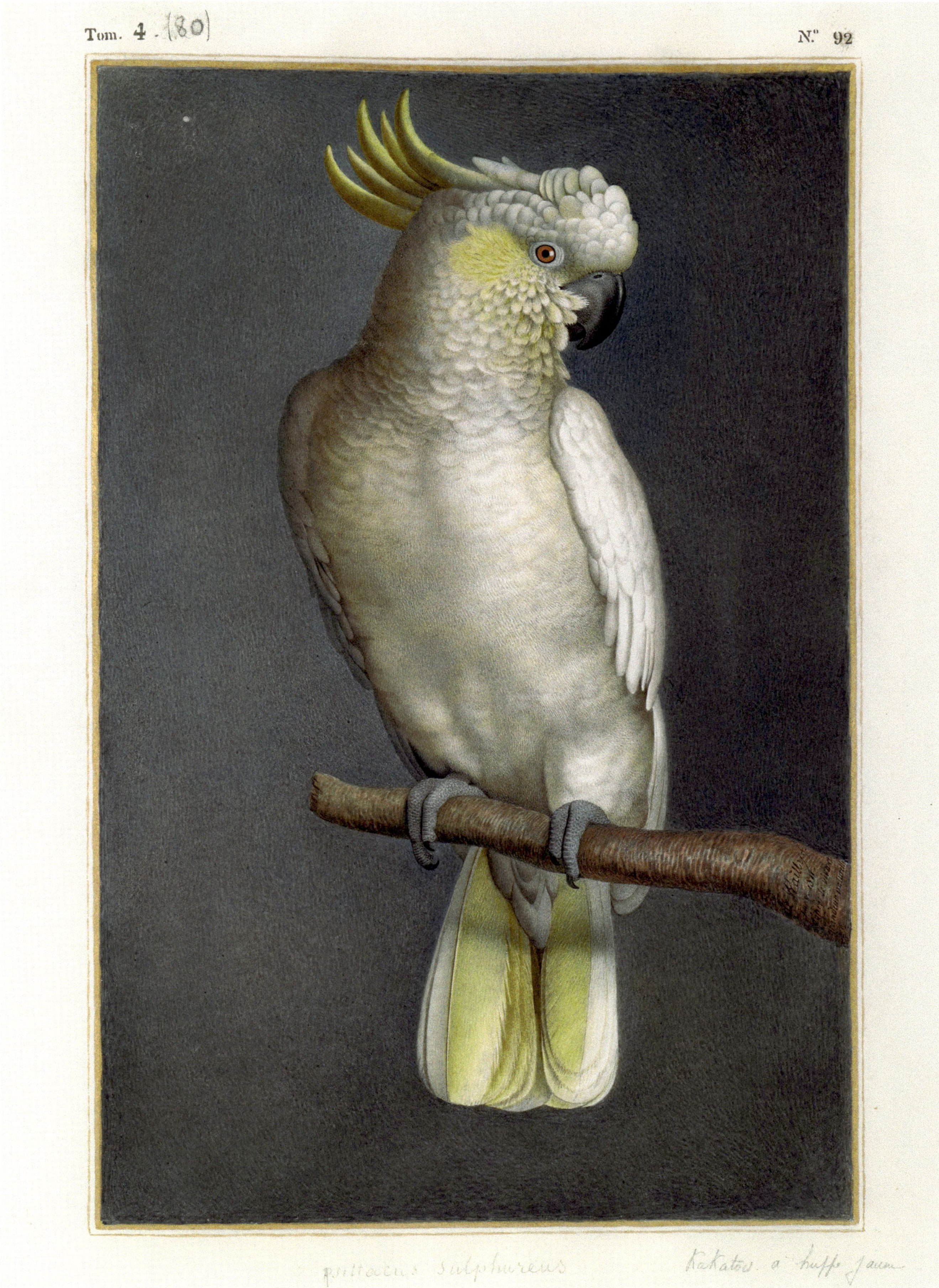

40
Pierre-François de Wailly,
Psittacus sulphureus,
Yellow-crested Cockatoo, 1811
Watercolour on vellum,
46 × 33 cm
Paris, MNHN

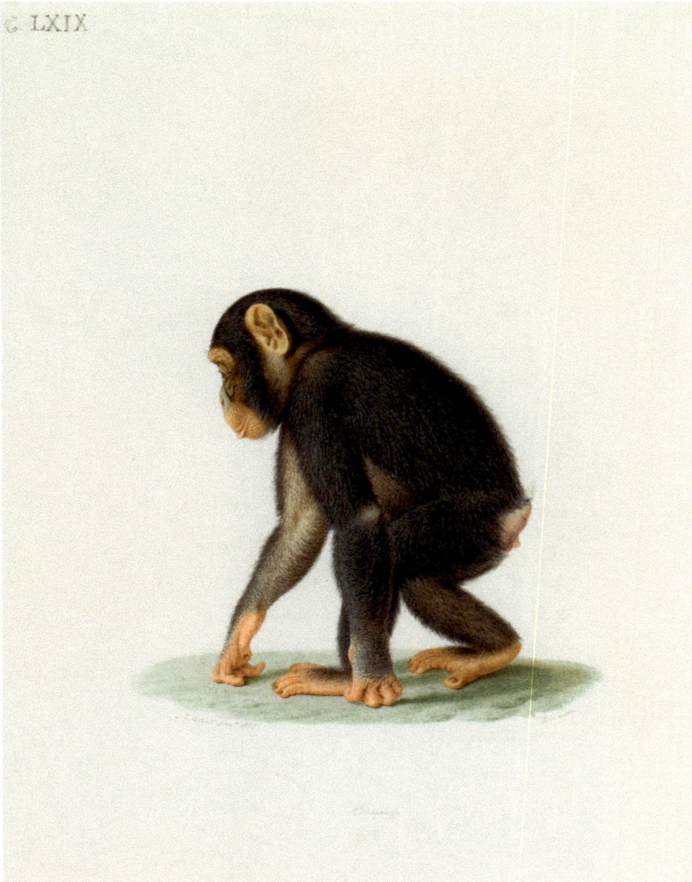

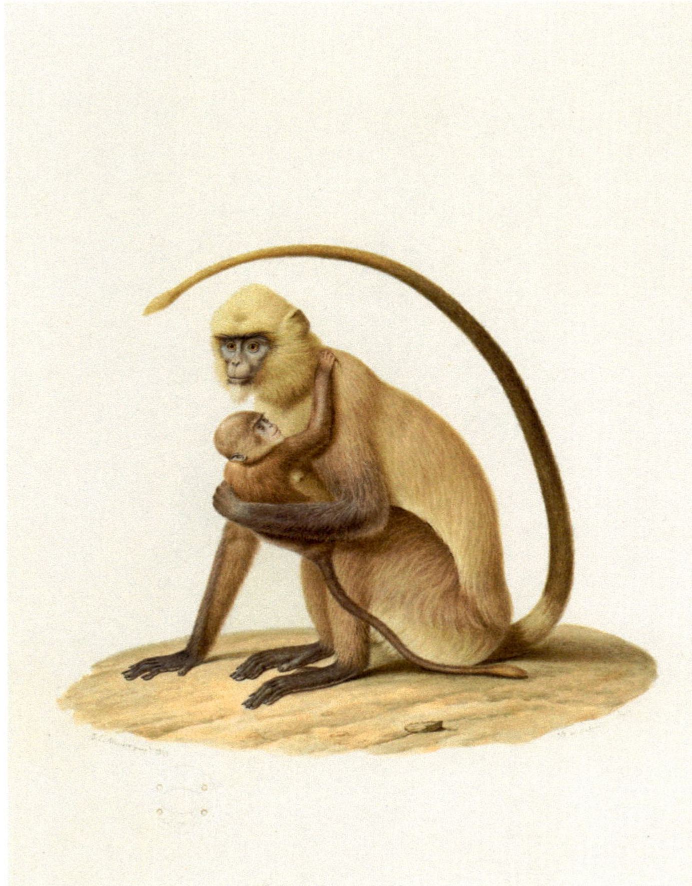

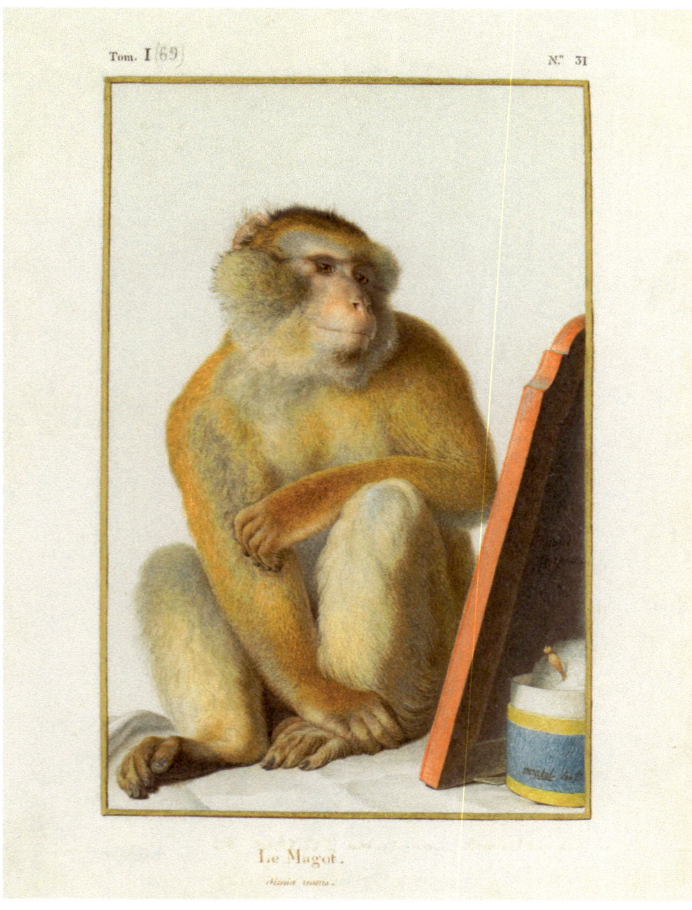

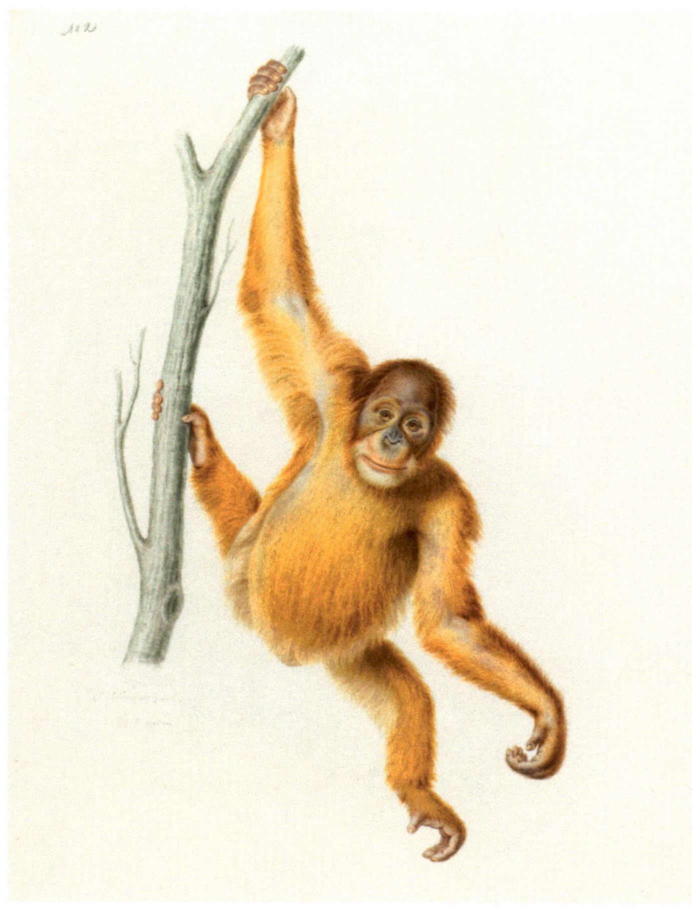

41, 42, 44
Jean-Charles Werner,
Watercolour on vellum,
46 × 33 cm

Chimpanzee, 1837
Semnopithecus dussumieri, 1842
Young Sumatran Orangutan That Died at the Ménagerie, 1836
Paris, MNHN

43
Nicolas Maréchal,
Magot. Simia inuus, 1793–1794
Watercolour on vellum, 46 × 33 cm
Paris, MNHN

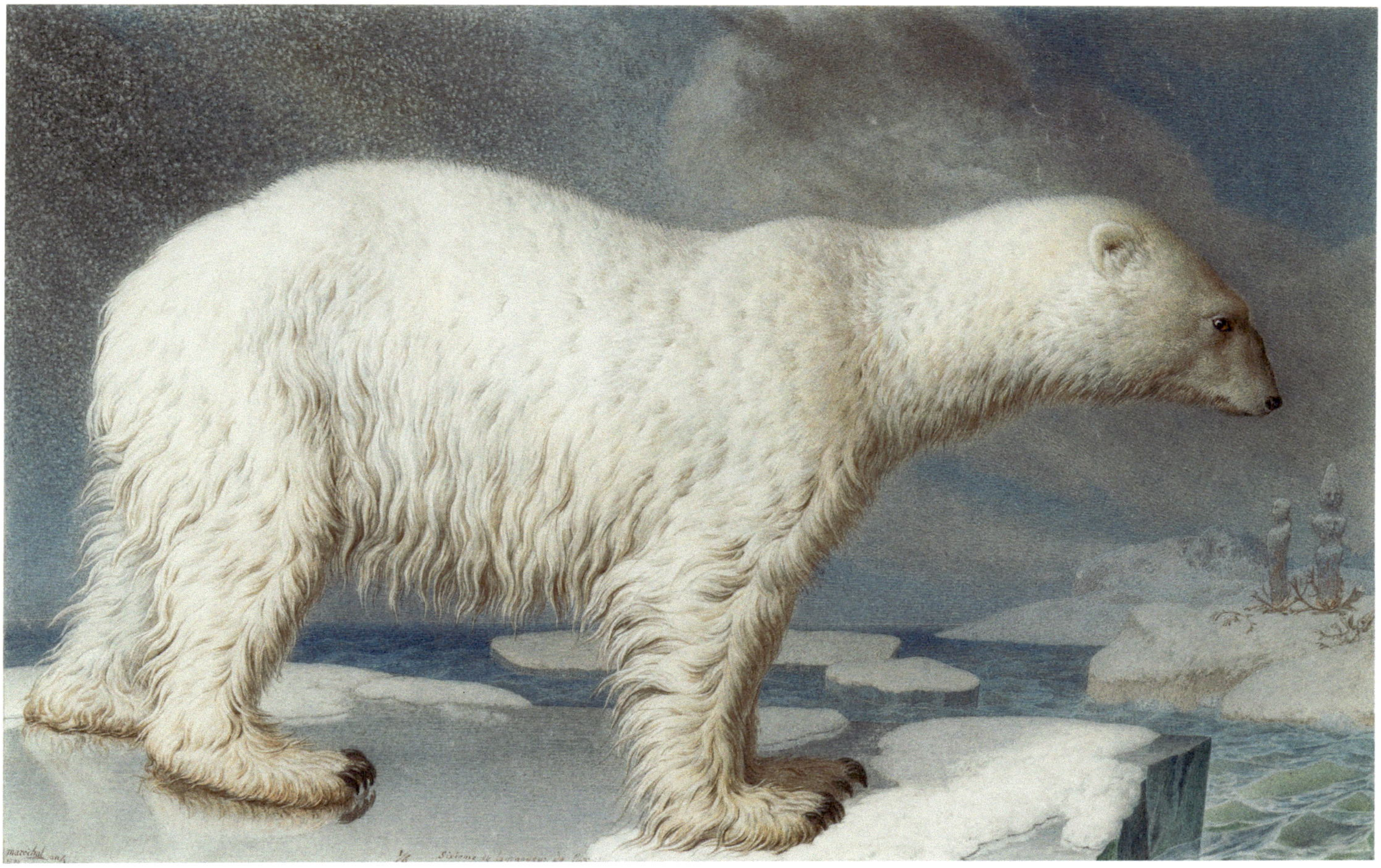

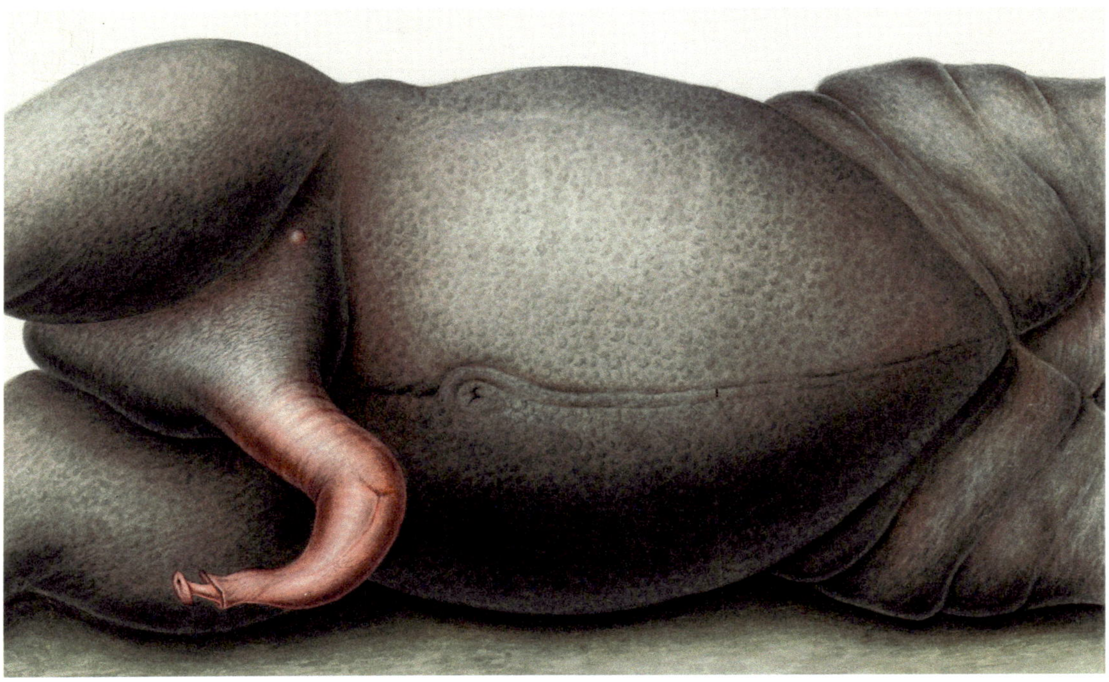

45
Nicolas Maréchal,
Polar Bear, 1796
Watercolour on vellum,
33 × 46 cm
Paris, MNHN

46
Pierre-Joseph Redouté,
Rhinoceros Underbelly with Penis,
late 18th century –
early 19th century
Watercolour on vellum,
33 × 46 cm
Paris, MNHN

II THE WORLD'S IMMENSITY, NATURE'S DIVERSITY: THE ERA OF THE GRAND VOYAGES

MARIE-NOËLLE BOURGUET

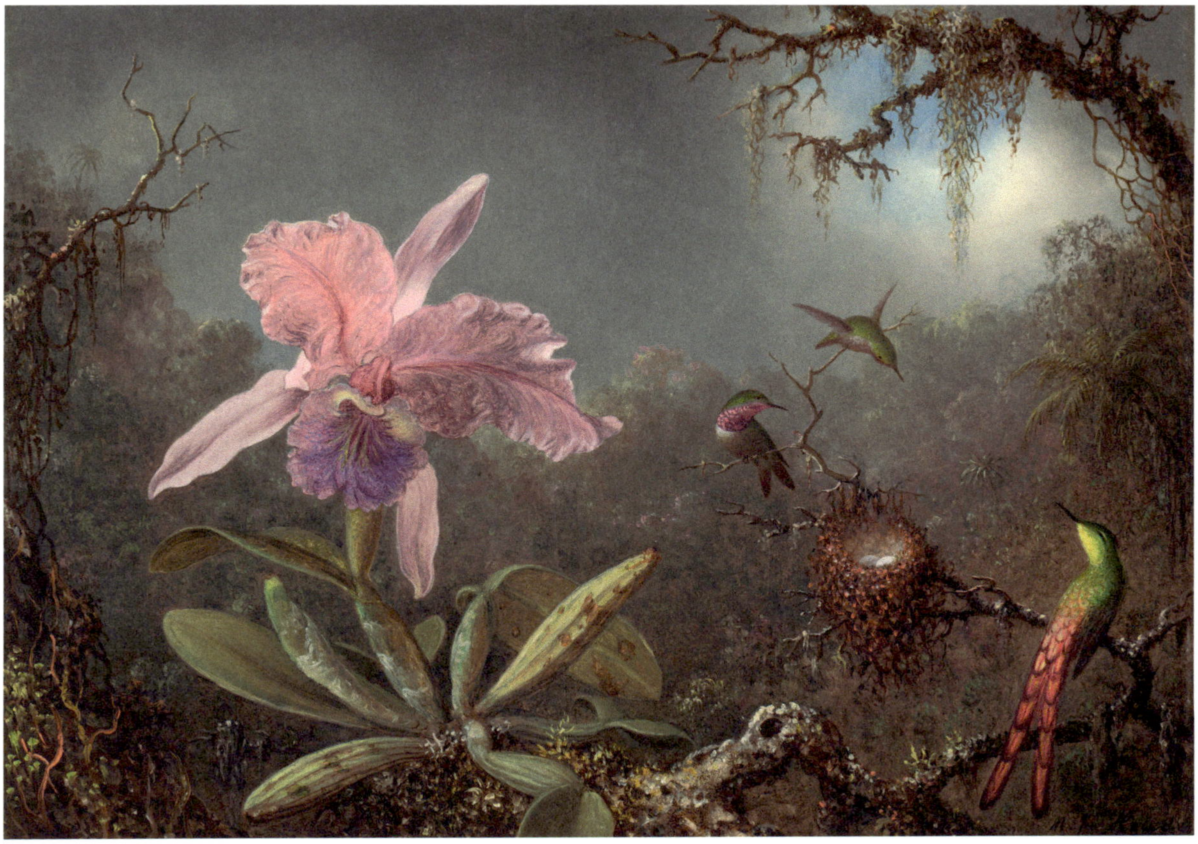

47
Martin Johnson Heade,
Cattleya Orchid and Three Brazilian Hummingbirds, 1871
Oil on wood, 34.8 × 45.6 cm
Washington, National Gallery of Art

"We are truly in a wonder-land of fertility and luxuriance; we have on all sides of us extraordinary plants, electric eels, tigers, armadillos, apes, and parrots . . . What magnificent vegetation! Cocoa-nut palms from fifty to sixty feet in height; *Poinciana pulcherrima*, with pyramidal bunches of flowers a foot high, and of a splendid bright red colour; bananas, and a host of trees with enormous leaves and sweet-smelling flowers as large as one's hand, all of which are entirely new . . . What a vast number of smaller plants, therefore, must have as yet escaped observation! How brilliant the plumage of the birds and the colours of the fishes!—even the crabs are sky-blue and gold!"[1]

When he wrote this letter to his brother Wilhelm on July 16, 1799, Alexander von Humboldt had just set foot on the Venezuelan coast, near Cumana, six weeks after taking leave of the shores of the Old World. To capture the enchantment, his pen dispensed superlatives and exclamations, in a distant echo of the words of Christopher Columbus, three centuries earlier: "Here the fishes are so different from ours that it is wonderful. There are some like cocks of the finest colors in the world, blue, yellow, red and of all colors, and others painted in a thousand ways."[2] All travellers from past centuries who had gone off to discover distant lands experienced, when faced with nature in all its diversity, particularly the lushness of the tropics, this disparity between the sumptuous spectacle unfolding before their eyes and the paucity of words to describe it. In such a state of awe and stupefaction, how could one accrue knowledge and undertake science, asked Humboldt: "Hitherto we have been running about like a couple of fools; for the first three days we could settle to nothing, as we were always leaving one object to lay hold of another. Bonpland declares he should lose his senses if this state of ecstacy were to continue."[3] It is the evocation of these practices deployed before nature's infinite variety that will be our focus here, through the lens of the voyages and expeditions of the eighteenth and nineteenth centuries.

EXPLORED SPACES

But first, let us locate these travels in time and space.[4] George Anson's circumnavigation (1740–1744) and the maritime peace achieved by the Treaty of Paris in 1763 ushered in a new era of European exploration. The loss of their American colonies set the English and French on a course towards the southern seas, in search of "*Terra Australis*," the antipodean continent whose hypothetical existence had fed the human imagination since antiquity. In the last third of the eighteenth century, an extraordinary series of expeditions—one almost every two years—headed towards the Pacific: Byron, Wallis and Carteret, Bougainville, Cook, Marion-Dufresne, Kerguelen, La Pérouse, Bligh, Marchand, Entrecasteaux, Dupetit-Thouars, Vancouver, Flinders, Baudin. The end of the century saw new actors enter the scene: Russians in the North Pacific (Aleutian Islands, Alaska), and Spanish and Creoles in the South Pacific, with Malaspina's expedition. In terms of geography, however, James Cook's second voyage (1772–1775) showed that the southern continent, if it did exist, would be found under ice, inaccessible to any and all territorial conquest. From that moment on, he and the navigators who followed him had to switch goals and explore the ocean itself, along with the plethora of islands and archipelagos scattered across it: Matthew Flinders and Nicolas Baudin circumnavigated the vast island continent that was New Holland—the future Australia. The decades that followed, with expeditions such as those of the Frenchmen Laplace and Dumont d'Urville (1839–1840), completed the exploration and charting of Oceania, the immense fifth part of the world, both maritime and insular.

In the navigators' experience, islands and continents looked like shoreline from the vessel's deck and were barely penetrated during stops and quick forays. During the same period, however, with the encouragement of political authorities and scientific institutions, geographical and naturalist exploration also reached the continental interior: Peter Simon Pallas explored Siberia on behalf of Catherine II; botanists Joseph Dombey and André Michaux were sent by Louis XVI to Chili and Peru, and to North America, respectively, in search of useful plants; Alexander von Humboldt received the King of Spain's authorization to traverse

1 Letter from Alexander von Humboldt to his brother Wilhelm, Cumana, July 16, 1799, in *Life of Alexander von Humboldt*, ed. Karl Bruhns, trans. Jane and Caroline Lassell (London: Longmans, 1873), vol. 1, p. 263 and 265.

2 Edward Everett Hale, *The Life of Christopher Columbus* (Chicago: Howe & Co., 1891), p. 80.

3 Humboldt, *Life*, p. 265.

4 Marie-Noëlle Bourguet, "The Explorer," in *Enlightenment Portraits*, ed. Michel Vovelle (Chicago and London: The University of Chicago Press, 1997), p. 257–315; Hélène Blais, *Voyages au Grand Océan. Géographies du Pacifique et colonisation, 1815–1845* (Paris: Comité des travaux historiques et scientifiques, 2005); Janet Browne, "Une science impérialiste : l'histoire naturelle britannique et les voyages d'exploration de Banks à Darwin," in *Le Muséum au premier siècle de son histoire*, ed. Claude Blanckaert, Claudine Cohen, Pietro Corsi and Jean-Louis Fischert (Paris: Muséum national d'histoire naturelle, 1997), p. 197–210.

5 Hélène Richard, "La mer sans fond ?" in *La Mer, terreur et fascination*, exh. cat., ed. Alain Corbin and Hélène Richard (Paris: Bibliothèque nationale de France/Seuil, 2004), p. 39–49. See also the essays in this catalogue, p. 86 and p. 108.

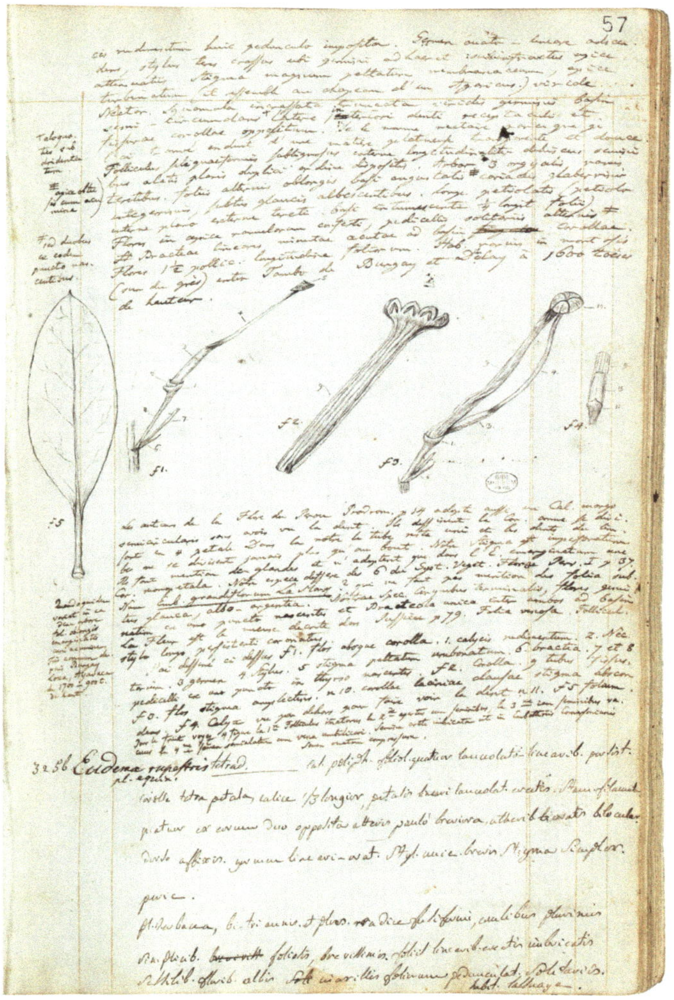

48
Aimé Goujaud, known as Bonpland, *Botanical Field Notes of the Humboldt and Bonpland Expedition*, 1799–1804
Handwritten notebook, 33 × 40 × 6 cm
Paris, MNHN

his colonies in the equinoctial Americas with Aimé Bonpland; Thomas Jefferson charged Meriwether Lewis and William Clark to explore the North American continent all the way to the Pacific; Mungo Park was sent by the African Society to discover the African interior and, some years later, René Caillié reached Timbuktu; Victor Jacquemont visited India and pushed all the way through to the centre of the Himalayas; Auguste de Saint-Hilaire, Baron Georg-Heinrich von Langsdorff and Alcide d'Orbigny penetrated the forests of Amazonia and Brazil. In contrast with the teams of sailors and scientists assembled for expeditions at sea, for these travels over land, the explorer went solo, or was accompanied by just a few men; interpreters, guides and porters were recruited on the spot. Without the option of returning, he had to adapt to the natural landscape he crossed and to the social universes of those he encountered and depended upon for his progress as well as for the transportation of his gear and his samples.

Throughout this period, a dimension of geographical space mostly escaped scientific investigation: the oceanic deeps. An object of fear and secular repulsion, the abyss warranted less scientific study as its murky depths supposedly rendered life impossible. It would take the work of Beautemps-Beaupré, who in 1791 undertook with Entrecasteaux the search for La Pérouse, before a systematic charting of the ocean's depths could be conceived; and the hydrographic campaign of HMS *Challenger* (1872–1876) before there would be proof of the presence of life at all depths of the seas and oceans, contrary to popular belief. The end of the nineteenth century would see the launch of the international hydrographic campaigns, with the exploration of the oceanic depths, a new chapter in the history of scientific expeditions.[5]

AT SEA: THE ACT OF CATALOGUING

While James Cook's *Endeavour* was anchored off the east coast of New Holland (April 28 – May 5, 1770), Joseph Banks and botanist Daniel Solander collected more than 3,600 species of plants, at least 1,400 of which were unknown in Europe; once back at sea, ensconced in their narrow, cramped cabin, the two men kept busy, one writing in his journal, the other studying the massive pile of seaweed, wood and shells heaped on the table. The scene in which painter Eduard Ender depicted Alexander von Humboldt and Aimé Bonpland in a bivouac during their sail up the Orinoco features, on land this time, a similar experience (49): amid a pile of specimens and instruments, the two scientists, armed with magnifying glasses, are examining, drawing and organizing the plants, and transcribing in their notebooks—logbook, botanical journal—the details of their samples and the measurements from the astronomical and geodetic surveys made en route (48).

These gestures, the austere routine of the voyage, which underpinned the success of the expedition, were in fact its driving force. The instructions given to the travellers before their departure are an endless refrain: what to do with a plant or a seed found among others of

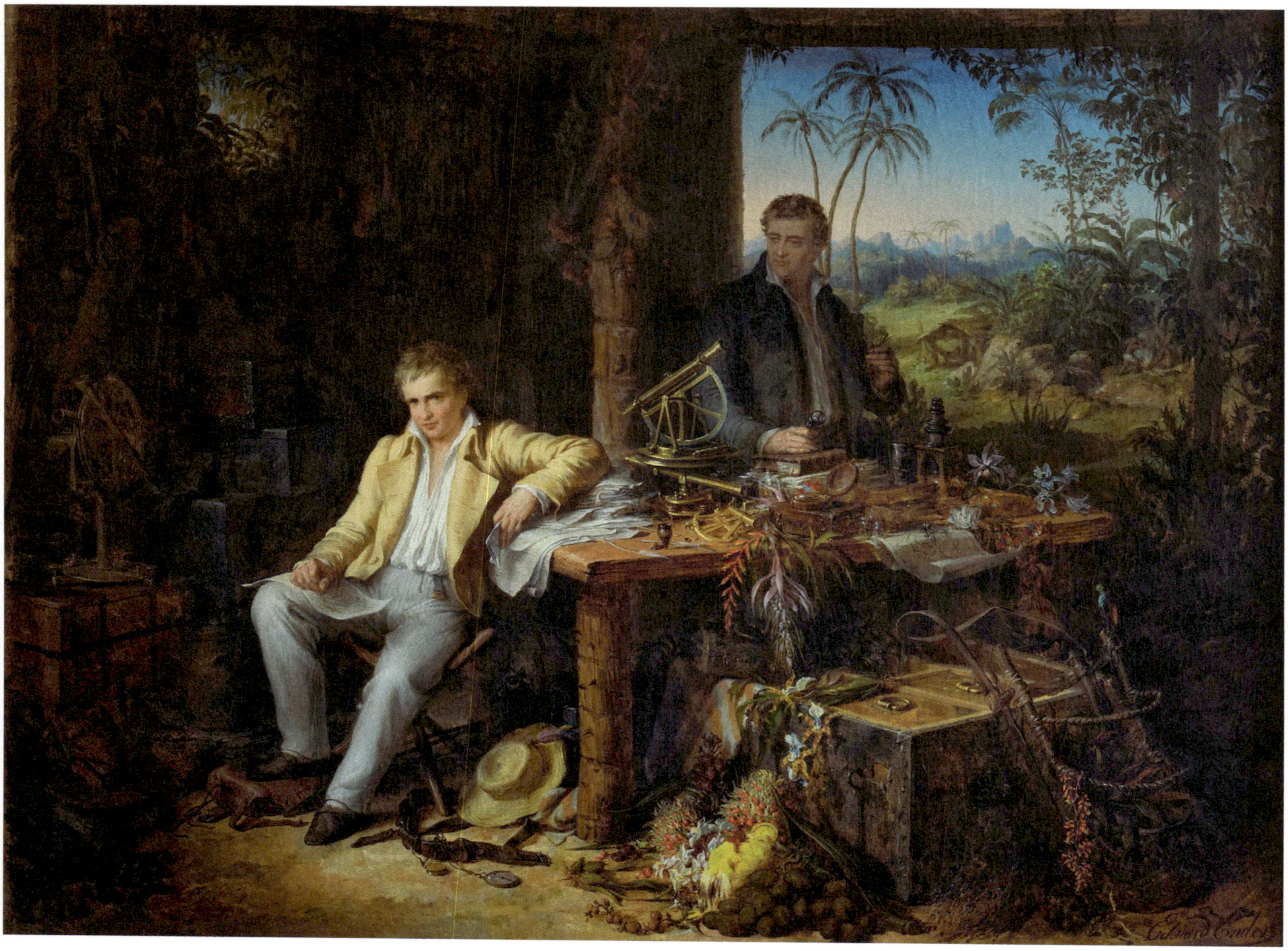

49
Eduard Ender,
Alexander Von Humboldt and Aimé Bonpland in the Jungle, about 1850
Oil on canvas, 80 × 150 cm
Berlin, Berlin-Brandenburgische Akademie der Wissenschaften

50 A, B
Aimé Goujaud, known as Bonpland, Humboldt & Bonpland herbarium, equatorial America, n.d.
44 × 28 cm
Cordia sebestena Cuba, Oreopanax
Paris, MNHN

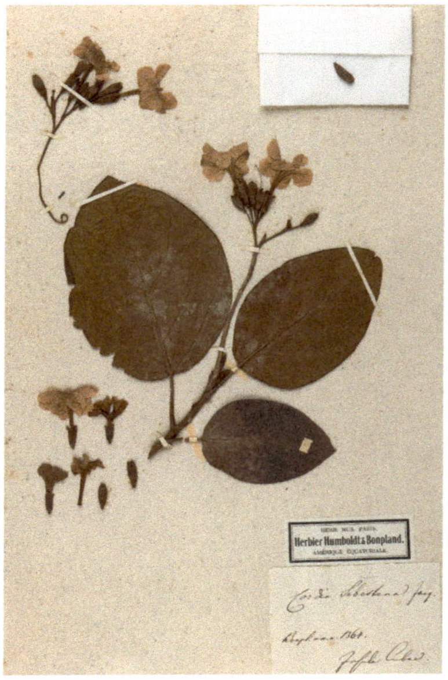

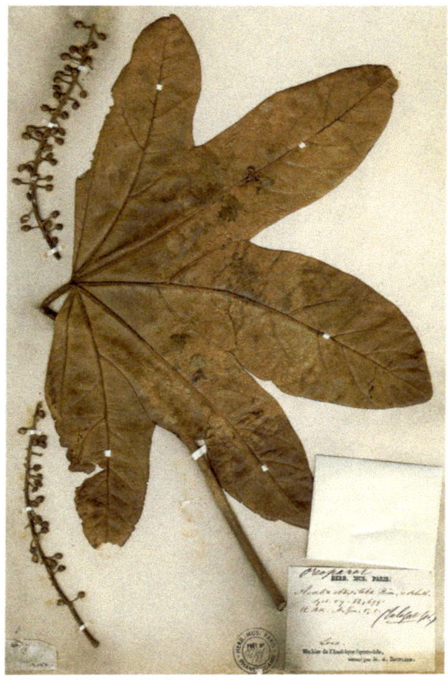

a different species; a stone without provenance; a measurement without location or date information?[6] Even the usual means of classification—herbaria, labels, inscriptions—did not always suffice. For the plants and animals that were hard to preserve, special techniques were refined: Philibert Commerson adapted the model of the dry herbarium, creating a "fish herbarium" in order to preserve dried specimens of the ichthyological collection that he amassed on the voyage with Bougainville—160 species from the Atlantic, the Brazilian coast, the shores of Madagascar and Île de France (present-day Mauritius). For fragile plants, or those collected in a condition that did not lend themselves to preservation, Humboldt and Bonpland "printed" them between two sheets of paper, capturing their general appearance. And for all of these travellers, whether it was a matter of rendering the likeness of a plant or an animal or the aspect of a landscape, the reliance on sketches and drawings was indispensable, the necessary complement to their collecting. For the grand expeditions the tendency grew to include artists among the team of scientists; these were expected to capture nature and illustrate the voyages through sketches and watercolours. For his first expedition, Cook brought onboard the naturalist-painter Sydney Parkinson; on the second voyage, landscape artist William Hodges executed the first depictions of the Pacific islands to circulate in Europe, in which he played with the effects of light and shadow.[7] While the three draughtsmen hired by Baudin would abandon the expedition while in port at Île de France, the young Charles-Alexandre Lesueur and Nicolas-Martin Petit, initially hired as assistant gunners, were soon recognized for their talent and became the official draughtsmen of the expedition: the first for his animal drawings, particularly the jellyfish he observed during the southern navigation and after his return (with François Péron, he described, drew and painted more than 120 different species—four times more than the known number up to that point) (65, 66, P. 92–93); the second for his portraits of Indigenous people, drawn with great delicacy and finesse (63, 64, P. 90–91).[8] From July 1833 to July 1834, the topographical and landscape painter Conrad Martens accompanied the expedition of HMS *Beagle* led by Robert Fitzroy all along the American coast to Patagonia (51, 52, 53).

RETURNS

Ultimately, it was upon the return that the scientific destiny of these voyages played out. Dramatically, sometimes, as in the case of the expedition to Arabia launched by the king of Denmark in 1761: when the botanist Pehr Forsskål died during the voyage, a part of his botanical and zoological collections were seized by privateers and the rest arrived in Copenhagen in a terrible state, phials and flacons having been opened at customs, specimens thrown out or prey to insects; it would take a decade before the cases were unpacked and the collections studied: some descriptions were published, but, of the dozens of species discovered by Forsskål, the science of his time would remain ignorant. Similarly, the Frenchman Dombey saw a part of his collections captured and auctioned off; the rest, laid under embargo, was split with the Spanish without giving the botanist time to label his plants; the only remnants of his collecting, aside from the herbarium, were but a few plants, derived from seeds that were brought back and planted in the King's Garden. Forgotten for almost a century in the institutions of Moscow and St. Petersburg, the rich material collected by members of Baron Langsdorff's mission to document the natural history, geology, anthropology and languages of Brazil would only be reopened in 1930.

6 Marie-Noëlle Bourguet, "La collecte du monde : voyage et histoire naturelle (fin XVIIe siècle–début XIXe siècle," in *Le Muséum au premier siècle de son histoire*, p. 163–196.

7 Bernard Smith, *Imagining the Pacific: In the Wake of the Cook Voyages* (Melbourne: Melbourne University Press, 1992), p. 111–134.

8 Jacqueline Bonnemains, Elliott Forsyth and Bernard Smith, eds., *Baudin in Australian Waters: The Artwork of the French Voyage of Discovery to the Southern Lands, 1800–1804* (Melbourne: Oxford University Press, 1988), and the essay in this catalogue, p. 86. Due to the strained relations between Baudin and the scientists onboard, the French expeditions that followed, from Freycinet on, would employ only naval officers and exclude all civilians, even for scientific tasks.

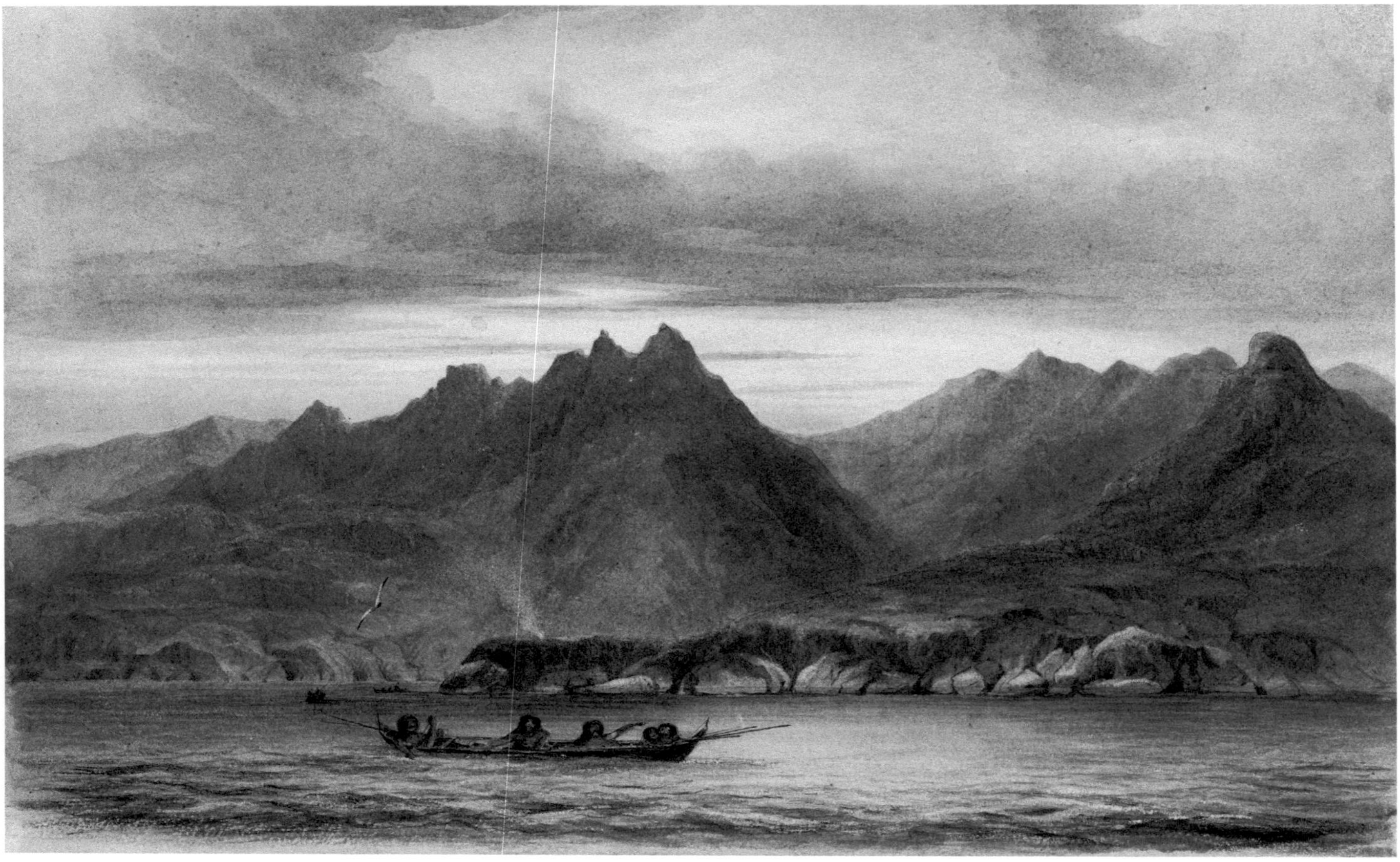

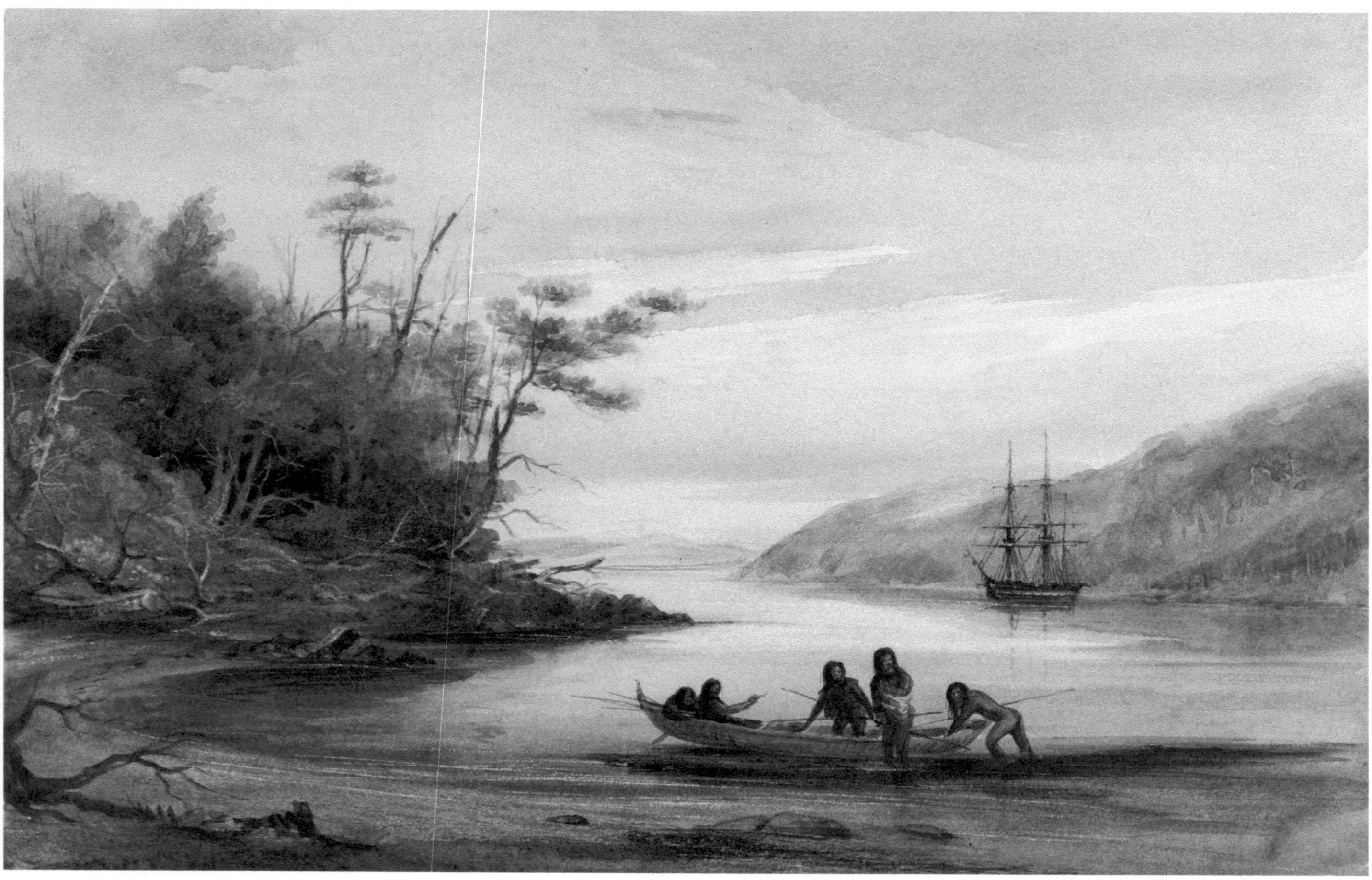

51
Conrad Martens,
Fuegians in a Canoe off Wollaston Island, 1835–1836
Graphite, watercolour, heightened with gum arabic, 21.7 × 34.7 cm
London, National Maritime Museum

52
Conrad Martens,
Fuegians and the Beagle *at Portrait Cove, Beagle Channel*, 1835–1836
Graphite, watercolour, heightened with white, 19.4 × 29.5 cm
London, National Maritime Museum

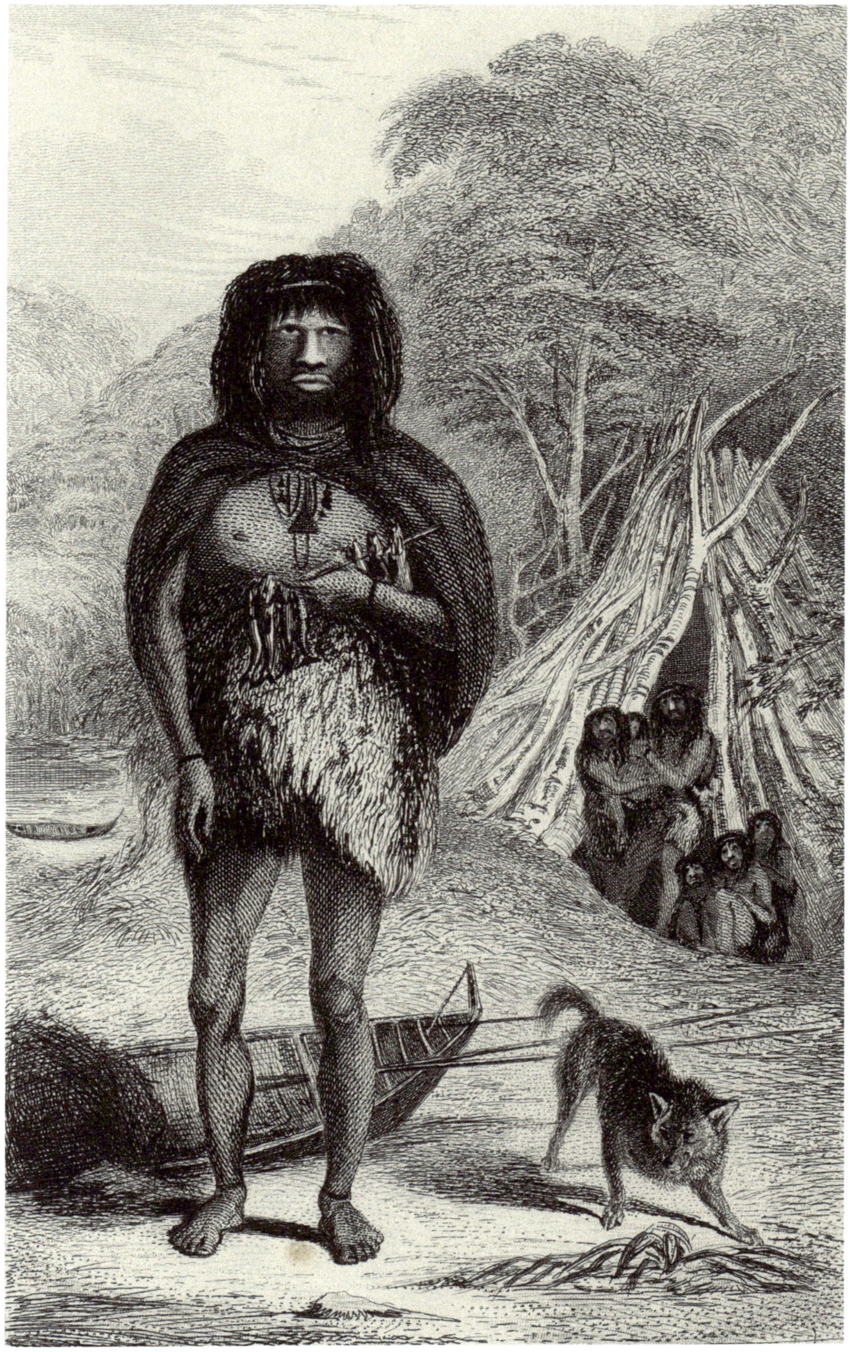

53
Robert Fitzroy,
"Fuegian," in *Narrative of the Surveying Voyages of His Majesty's Ships* Adventure *and* Beagle, vol. 2,
London: H. Colburn, 1839
Paris, MNHN

But there were also joyful, even triumphant, returns. In late summer 1804, Humboldt and Bonpland returned from their American expedition with some 60,000 samples they had botanized, corresponding to more than 6,000 plant species; they opted to gift one of their herbaria to the Muséum national d'histoire naturelle in Paris (50), and to Cuvier a number of animal specimens and fossils.[9] That same year, Antoine-Laurent de Jussieu, the director of the Muséum, enthusiastically accepted collections brought back by the survivors of the Baudin expedition: "Of all the collections that came to us from distant lands and different epochs, those carried by the ships *Le Naturaliste* and *Le Géographe* are certainly the most considerable, particularly in terms of the animal kingdom."[10] Aside from the cases of living plants and a herbarium heavy with 1,500 vegetal species, including more than 600 new ones, the wealth and novelty of the animal collections were particularly fascinating: a cassowary, live kangaroos; more than 100,000 specimens, from mammals to zoophytes, belonging to 2,500 species never before described. Already enriched by the contribution of the revolutionary confiscations and by the voyages of Commerson, Sonnerat, Dombey and Michaux, the Muséum saw its collections markedly increase with the expeditions in the early part of the nineteenth century.[11]

For the desk-bound scientists in the metropolitan centres to whom fell the task to identify, name and classify the material from every corner of the globe, this sudden increase was at once a windfall and a challenge, necessitating the search for a means of designation and classification that could be embraced by all.[12] While Linnaeus's binomial nomenclature had become widespread since the eighteenth century (Latin usage, name of genus followed by a specific qualifier), the classification system of the Swedish botanist, based on the reproductive organs (stamens and pistil), was, despite its relative ease, soon supplanted by the natural method revealed in 1789 in Antoine-Laurent de Jussieu's *Genera plantarum*. This compendium established the classification of the morphological criteria, ranked according to the principle of the subordination of characteristics: based on the observation that a plant possesses one or two cotyledons, or none, a certain number of characteristics constant within a given family follows. This method is the one Humboldt and Bonpland elected to use in the Americas. It was also the one the zoologists Cuvier and Valenciennes used in Paris, in the twenty-two volumes of their *Histoire naturelle des poissons* (1829–1850), in which more than 4,500 species were catalogued, most from the southern expeditions in the early century.

Even though these questions of classification were of great scientific import, they were not what caught the interest of the public, who were apprised of the expeditions' arrivals by the newspapers of the day. More fascinating than the herbaria or the sample cases were the plants and animals that, brought back alive, were either acclimated outdoors or placed inside hothouses, or exhibited in zoos and menageries (54). Recalling the feelings aroused in his childhood upon discovering a fan palm and a giant dragon tree in an old tower of Berlin's Botanic Garden, Humboldt underscored the importance of the sensory experience available to the inhabitant of temperate regions upon seeing exotic plants.[13] Even though these plants could only offer, in the garden or hothouse in which they were cultivated and isolated from their natural habitat, a constrained and impoverished image of the real thing, their physical presence, their realness, made them irreplaceable by pen or brush. In fact, the arrival of a plethora of new species multiplied the presence of foreign plants in the botanical gardens of Europe. To accommodate plants from the tropics, hothouses were set up in the main institutions: in Paris, after the first buildings were installed in the King's Garden by Antoine de Jussieu and Sébastien Vaillant in 1714, and by Buffon in 1788, two large steam-heated hothouses were built in the 1830s to house the "Mexican" and "Australian" plants. In 1808, when he had barely carried out the work of erudition that Humboldt expected from him for the volumes on botany for their *Voyage*, Aimé Bonpland was appointed intendant of the garden of Malmaison by Empress Josephine and placed in charge of plants cultivated in

---

9 Humboldt, "Lettre à Messieurs les professeurs, administrateurs du Muséum d'histoire naturelle," December 18, 1804, in *Lettres américaines d'Alexandre de Humboldt, 1798–1807*, ed. E.-T. Hamy (Paris: Guilmoto, 1905), p. 175.

10 Antoine-Laurent de Jussieu, "Notice sur l'expédition à la Nouvelle Hollande," *Annales du Muséum*, vol. 5 (1804), p. 7 and 10.

11 In about 1830 the total number of animal species known at the time was estimated to be almost 80,000, of which almost a third were insects, molluscs and infusoria. For plants, the number of catalogued species went from 20,000 in about 1700 to 50,000 in about 1820; and, for flowering plants alone, the total number was more than 160,000 in about 1880. See Rudolph Wagner, "The Numerical Relations of Animals," *The Edinburgh New Philosophical Journal*, vol. 15 (April–October 1833), p. 221–222; André Cailleux, "Progression du nombre d'espèces de plantes décrites de 1500 à nos jours," *Revue d'histoire des sciences et de leurs applications*, vol. 6, no. 1 (1953), p. 42–49.

12 Jean-Marc Drouin, "De Linné à Darwin : les voyageurs naturalistes," in *Éléments d'histoire des sciences*, ed. Michel Serres (Paris: Bordas, 1989), p. 321–335.

13 Alexander von Humboldt, *Cosmos. A Sketch of a Physical Description of the Universe*, vol. 2 (New York: Harper & Brothers, 1866), p. 20.

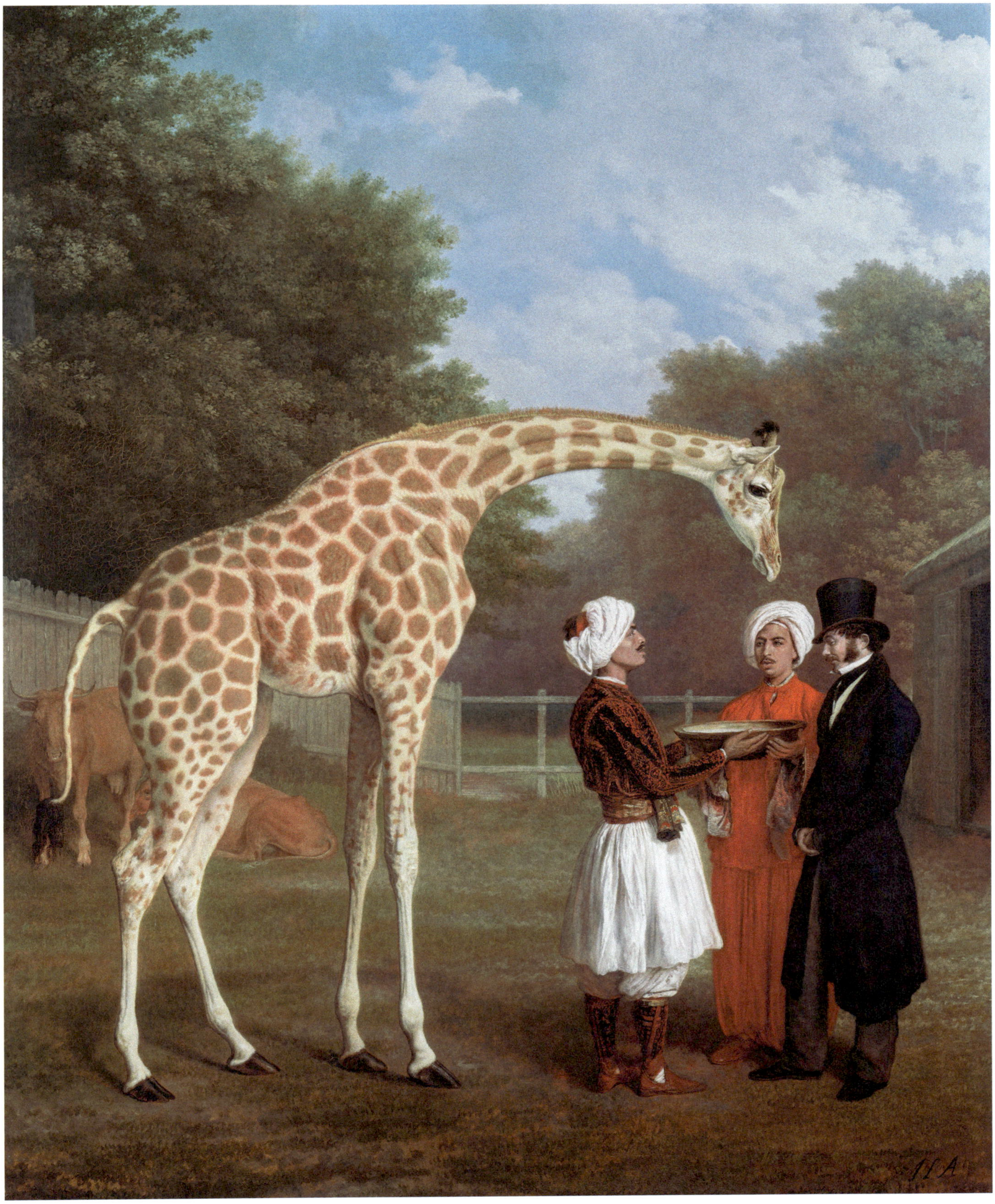

54
Jacques-Laurent Agasse,
The Nubian Giraffe, 1827
Oil on canvas, 127.3 × 101.7 cm
London, The Royal
Collection Trust

the hothouse that had been set up there. In 1848, the massive Palm House at Kew Gardens, near London, was built. In Berlin, King Frederick William III had a vast landscape garden designed for Peacock Island, Wannsee, and a palm house and a menagerie for exotic animals built by Karl Friedrich Schinkel, both highly prized by the public (55). The movement also spawned private gardens: in a notebook, Flaubert describes the gardens planted with exotic plants—"cactus, cereus from Peru, agave from America, chamaerops from Barbary, cycas from India"—"to spark the agreeable memories of a colonist or a traveller."[14] Under the Second Empire, the fashion of introducing hardwood trees of distant origins—North American and Asian—spread throughout the parks in the Lyon region and the Rhône valley, a contrast to the European conifers.

GEOGRAPHY OF PLANTS AND PHYSIOGNOMY OF THE LANDSCAPE: NATURE WRIT LARGE

Upon his arrival in the Americas, Humboldt stated that he did not wish to limit his task to that of a "descriptive botanist," preoccupied with "the discovery of new species of plants, the study of their external structure, their distinguishing characteristics, and the analogies that group them together into classes and families." Indeed, he was of the position that it is not by examining individual specimens that nature reveals itself and offers the greatest joys: "But far more thrilling than the contemplation of these marvellous objects individually is the overpowering sight of the whole mass of such magnificent vegetation, the elegant luxuriance of which is so exhilarating and, at the same time, so soothing."[15] For the traveller, it is important to grasp the global impression that nature's spectacle makes on his senses. While Humboldt shared German Romantic epistemology's holistic concept of nature, he adopted—in contrast with Schiller's speculative idealism—an unapologetically empiricist approach. By viewing nature as a whole whose elements are in constant interaction (*"Alles ist Wechselwirkung"*) and whose unity and harmony result in an interplay of multiple, opposing forces, he made it an object of study accessible to science through analysis, measurement and quantification. By means of the extraordinary panoply of instruments he brought with him to the New World, Humboldt intended to challenge the link between organic life—the world of the living—and inanimate nature—the "physics of the earth": "The study of nature . . . demands the gathering together of all the knowledge dealing with modifications of matters."[16] It is this approach that makes him systematically articulate the observation and description of plants with the study of the surrounding physical conditions—temperature, altitude, nature of the soil, etc.—as he proudly explained upon his return regarding the herbarium intended for the Muséum d'histoire naturelle in Paris: "This collection, besides having been described on the premises and containing a large number of new genera, can boast the additional advantage of exhibiting only objects whose height of growth above sea level can be indicated."[17]

Over the course of the dozen climbs he made to the highest peaks of the Andes, Humboldt paid particular attention to the stratification of the vegetation along the slopes. However, he was not the first to notice this phenomenon: Linnaeus had already imagined the zonal or altitudinal distribution of plants as a double mountain, attached to the equator by its base. Later, naturalist Jean-Louis Soulavie represented by cross-section the succession of vegetal formations along the slopes of the Vivarais mountains, from the vines and olive trees to the alpine plants. Humboldt adopted this model, pushing the analysis of the factors at play even further—altitude, temperature, air composition, pressure, humidity, light intensity, sky colour, soil chemistry, etc.—all determining factors in the life of plants, animals and, of course, humans. But he also introduced a new, very visual, dimension. Having observed that the physical factors did not so much affect the distribution of individual plants as the grouping of vegetation presenting similar morphological characteristics, he turned to the notion of physiognomy: independent of its floristic composition, vegetation's appearance must be examined in relation to the surrounding environmental conditions to which it has adapted. Thus, while they are made up of species that are taxonomically different, the tropical forests of the Americas, Africa and Asia resemble each other in their bearing and general appearance.

To summarize his research and reveal his vision of nature, Humboldt made use of representational imagery, drawing heavily on graphs and numbers: based on

14 Quoted by Stéphanie Dord-Crouslé, "Flaubert et les Manuels Roret, ou le paradoxe de la vulgarisation. L'art des jardins dans *Bouvard et Pécuchet*," in *Le Partage des savoirs, XVIIIe–XIXe siècles*, ed. Lise Andries (Lyon: Presses universitaires de Lyon, 2003), p. 93–118 (specifically p. 100–105).

15 In Alexander von Humboldt and Aimé Bonpland, *Essay on the Geography of Plants* [1807], ed. Stephen T. Jackson, trans. Sylvie Romanowski (Chicago: University of Chicago Press, 2009), p. 64; letter to his brother, July 16, 1799, in *Life*, p. 265.

16 Humboldt, *Essay*, p. 79.

17 Humboldt, "Lettre à Messieurs les professeurs, administrateurs du Muséum d'histoire naturelle," December 18, 1804, in *Lettres américaines*, p. 176.

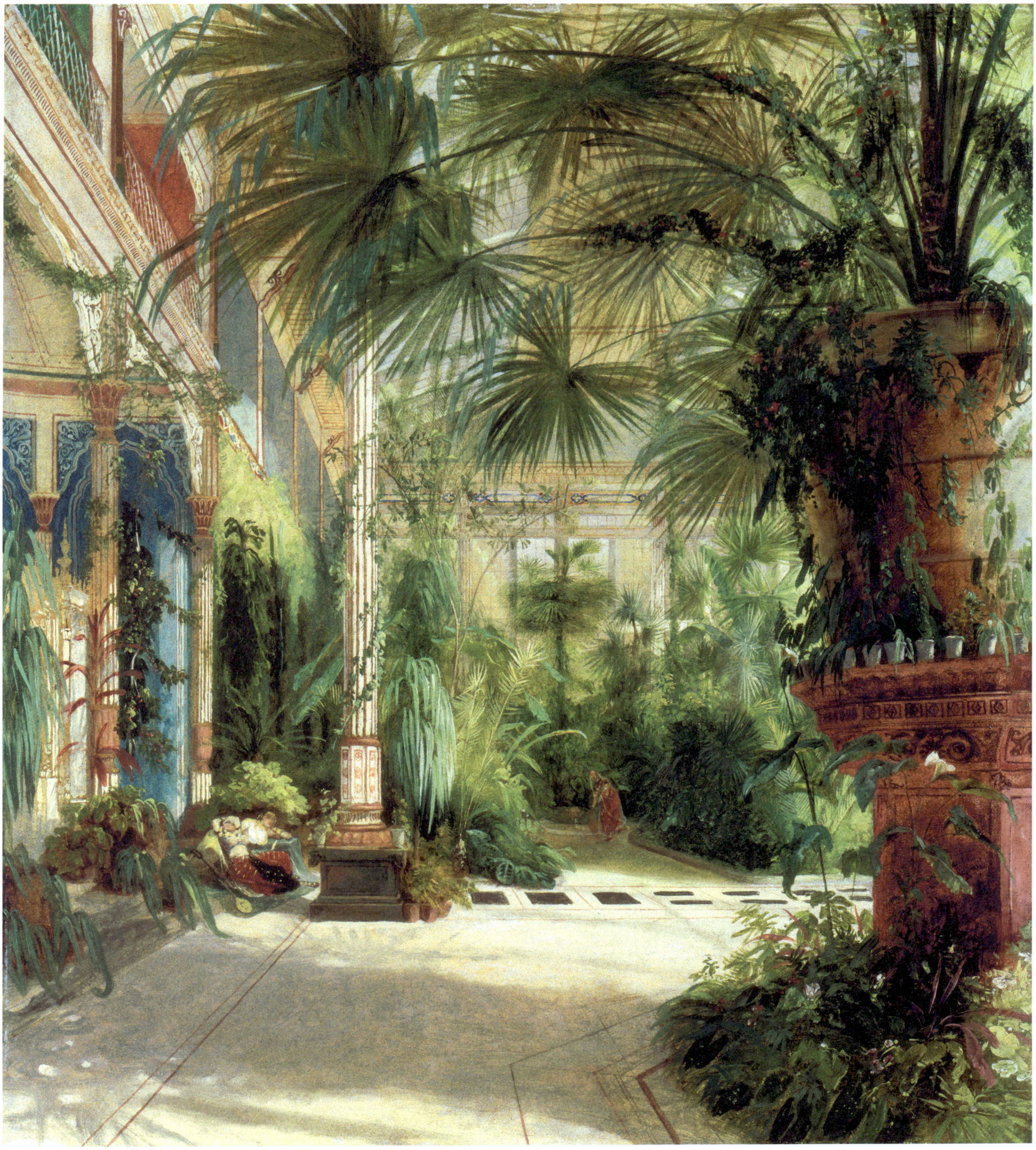

55
Carl Blechen,
The Interior of the Palm House,
1832–1833
Oil on paper mounted
on canvas, 64 × 56 cm
Berlin, Staatliche Museen
zu Berlin – Preußischer
Kulturbesitz

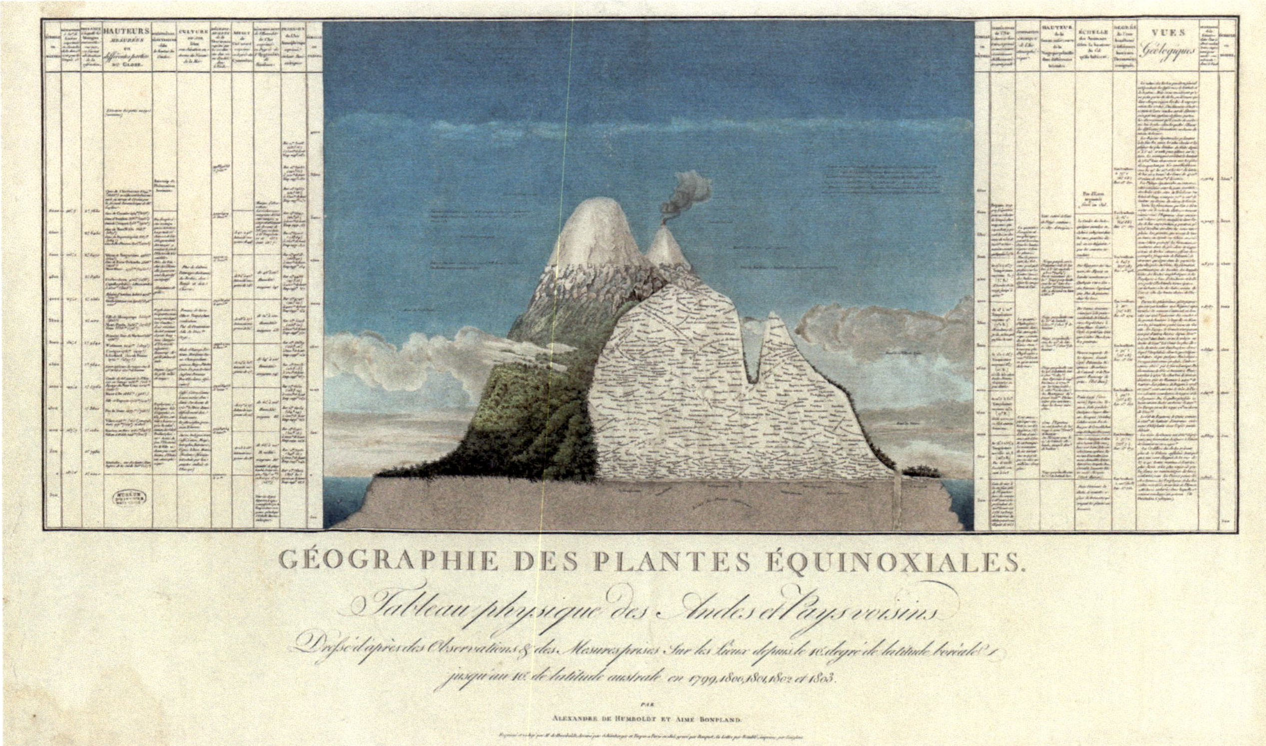

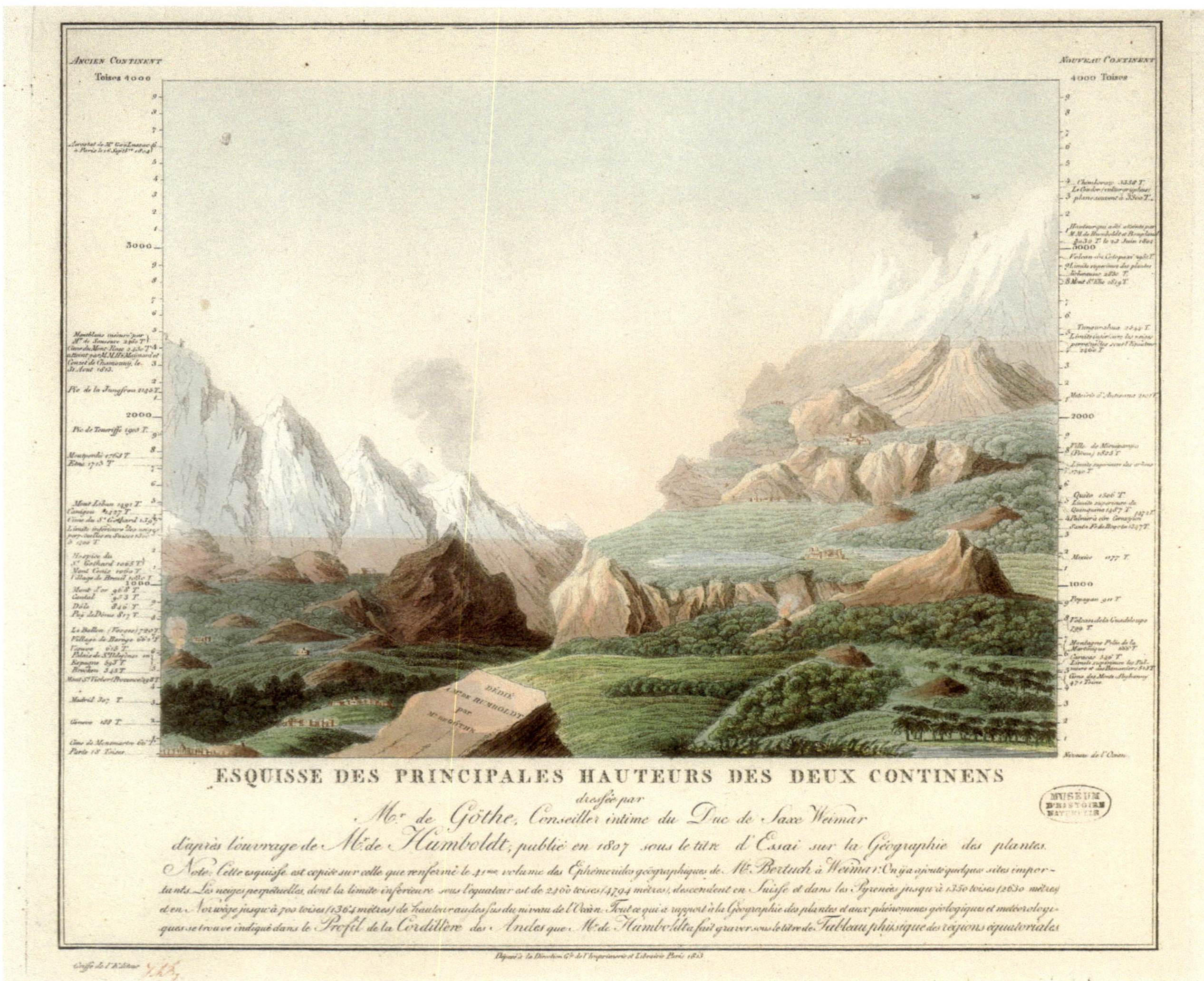

56
Aimé Goujaud, known as Bonpland, and Alexander von Humboldt, Schönberger and Turpin (draughtsmen), Bouquet and Beaublé (engravers), *Géographie des plantes équinoxiales. Tableau physique des Andes et Pays voisins*, Paris: Langlois, 1805
Printed card, 61 × 91 cm
Paris, MNHN

57
Johann Wolfgang von Goethe, *Esquisse des principales hauteurs des deux continens dressée par Mr. de Göthe*, Paris: n.p., 1813
Printed card, 47 × 55 cm
Paris, MNHN

a sketch he had made at Guayaquil after his—inconclusive—climb of Chimborazo, on June 23, 1802, he had a profile of the volcano drawn and engraved by artists upon his return to Paris (56).[18] In the centre of the image is an analytical presentation (a list of collected plants) and a colour drawing of the stratification of the landscape, from the palm tree zone until the perpetual snows. On either side, like a frame, are a series of columns containing the numerical data that Humboldt and Bonpland collected on the physical and environmental conditions and whose interaction contributed to shaping the distinct physiognomy of each of the vegetal strata: "When one ascends from sea level to the peaks of high mountains, one can see a gradual change in the appearance of the land and in the various physical phenomena in the atmosphere. The plants in the plains are gradually replaced by very different ones: woody plants decrease little by little and are replaced by herbaceous and alpine plants; higher still, one finds only grasses and cryptogams. Rocks are covered with a few lichens, even in the regions of perpetual snow."[19]

Therefore, knowledge gained confirmed the initial intuition regarding the unity of nature; even enriching the perception while revealing new delights. But how to recreate the impressions obtained in the tropics from the spectacle of nature for inhabitants of the temperate regions? Ascribing to the tradition of descriptive literature, and counting on the richness and power of language—particularly the German language—to excite the reader's imagination, Humboldt himself attempted in his *Ansichten der Natur* (1808) to suggest an anthology of the main landscapes one might encounter over the earth's surface—steppes and deserts, forests, waterfalls, volcanoes, and so on.[20] Moreover, he saw painting as the medium that could act on the imagination of the viewer to the most potent effect. In his *Essay on the Geography of Plants*, in order to help landscape painters represent the diverse landscapes, he had already developed a typology of some fifteen vegetal forms, including palm trees, fir trees, climbing plants, cacti, grasses and lichens.[21] Above all, he encouraged young artists to leave, to travel, to see for themselves the diversity of landscapes offered by nature. German artist Johann Moritz Rugendas joined Baron Langsdorff's mission at its start in Brazil, then criss-crossed Latin America, travelling through the Andes and the Argentinian Pampa: inspired by Humboldt's thoughts on the physiognomy of nature, he wanted his art to serve science and so returned from his travels with thousands of drawings and paintings (58, 59). Later in the century, American painter Frederic Edwin Church followed Humboldt's trajectory, wishing to see the same places, the Andean landscapes that the scientist had contemplated, in order to capture their unique physiognomy: by associating a profusion of detail with the vast views, his giant canvases give a sense of nature's infinity as well as its unity (91, 92, P. 136–137).

FROM NATURAL HISTORY TO THE HISTORY OF NATURE

Beyond the fascination with nature's profuse immensity that they aimed to catalogue, beyond the attempts to link the diversity of plant forms to physical and environmental conditions, the travellers and naturalists of the eighteenth and nineteenth centuries were aware of leaving one fundamental question unanswered: that of the original diversity of species on the earth's surface, which was slowly revealed by their botanical work. This question was explicitly formulated by botanist Robert Brown upon his return from his voyage of the southern seas onboard Matthew Flinders's *Investigator*: working on a herbarium of some 4,200 species from the coasts of New Holland (present-day Australia), classified according to Antoine-Laurent de Jussieu's method of natural families, he compared the composition with that of the flora of other regions in the world and found that the southern flora shared with the latter (with India and the Pacific islands in particular) less than ten percent of its species; in addition, he noted that the proportion of dicotyledonous and monocotyledonous plants was structurally very different.[22] To explain this finding, would he have to look at climate, history, even as far back as the Earth's origins? This question was also asked by Humboldt when he discovered at Tenerife a dragon tree that resembled, by its extraordinary size and its assumed antiquity, "one of the oldest inhabitants of our globe" and whose origins intrigued him: "The dracœna . . . offers a curious phenomenon with respect to the migration of plants. It has never been found in a wild state on the continent of Africa: the East Indies is its real country. By what means has this

18 Humboldt, *Géographie des plantes équinoxiales. Tableau physique des Andes et Pays voisins* (Paris: Langlois, 1805).

19 Humboldt, *Essay on the Geography of Plants*, p. 76.

20 The work was translated into French that same year by J.-B. Eyriès, *Tableaux de la nature, ou Considérations sur les déserts, sur la physionomie des végétaux, et sur les cataractes de l'Orénoque*, 2 vols. (Paris: Schoell, 1808).

21 Humboldt, *Essay on the Geography of Plants*, p. 73–74.

22 Robert Brown, *General Remarks, Geographical and Systematical, on the Botany of Terra Australis* (London: Bulmer, 1814).

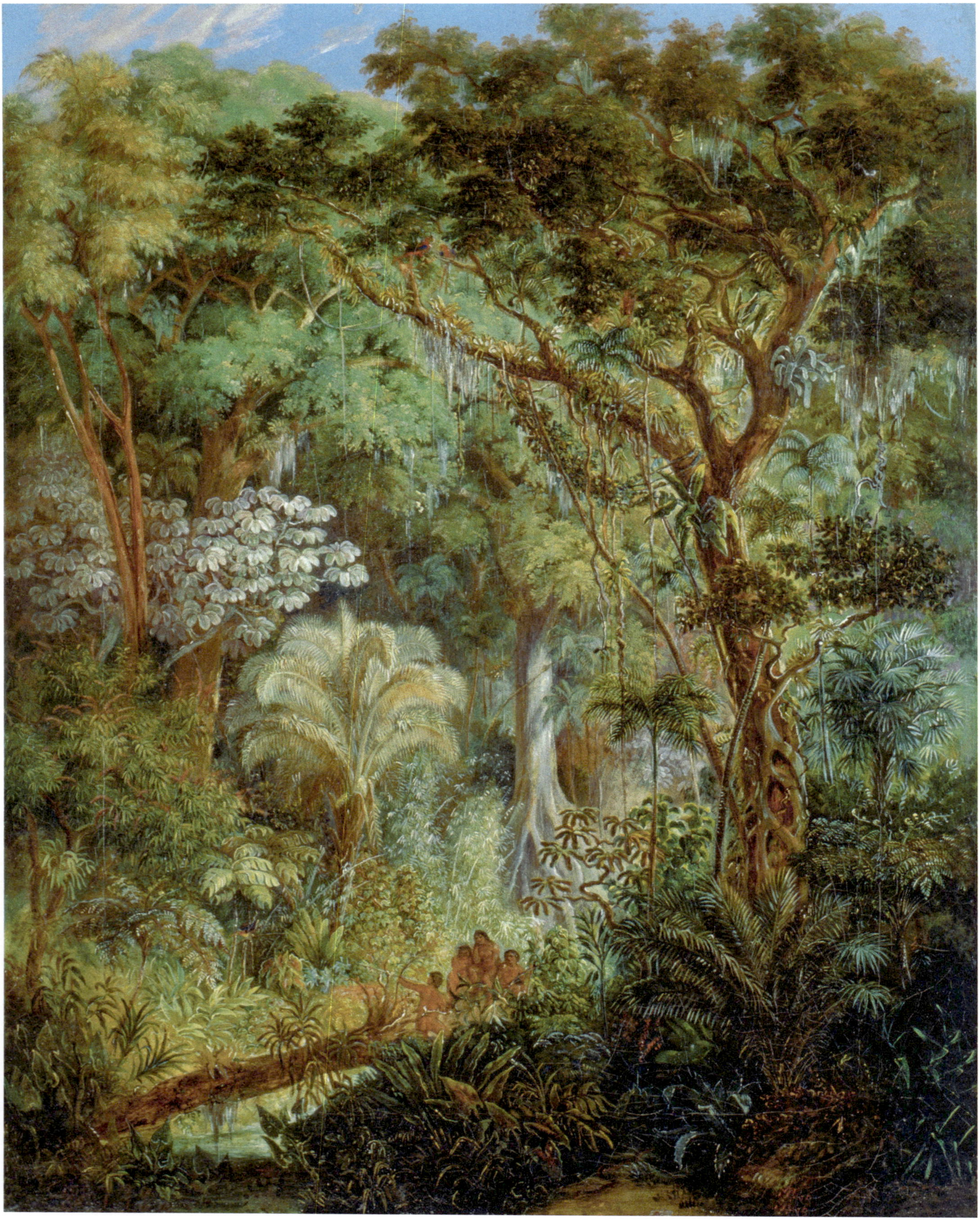

58, 59
Johann Moritz Rugendas, 1830
Oil on canvas
Indians in the Virgin Forest,
62 × 49 cm
Giant Tree in Brazil's Tropical Forest, 47 × 34 cm

Berlin, Die Stiftung Preußische Schlösser und Gärten Berlin-Brandenburg

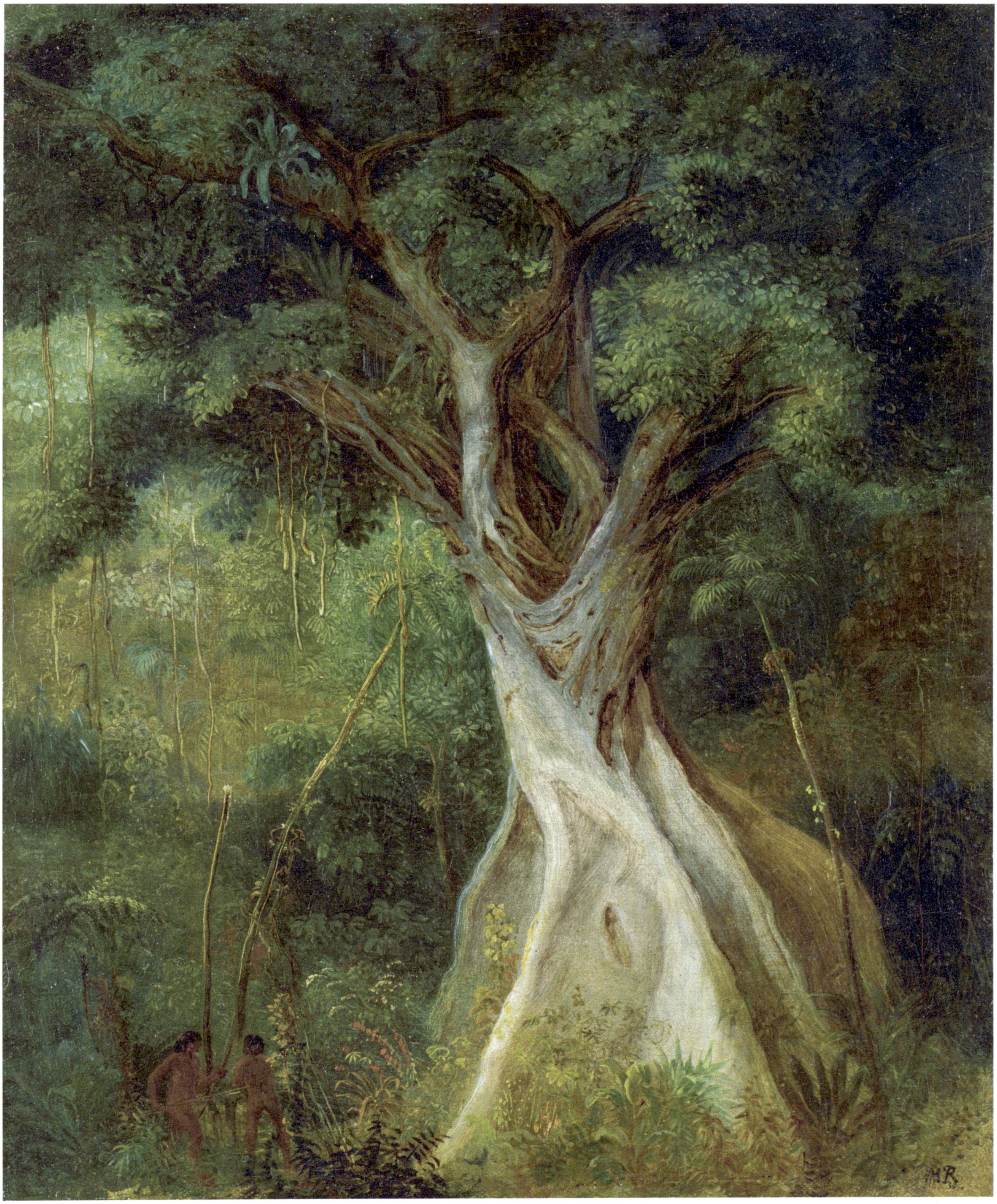

tree been transplanted to Teneriffe, where it is no way common? does its existence prove, that, at some very distant epocha, the Guanches had connections with other nations originally from Asia?"[23] For his part, regarding observations made by botanist René Louiche Desfontaines between 1783 and 1786 on the flora of North Africa, Augustin Pyramus de Candolle underscored the proximity of flora found on both coasts of the Mediterranean: "There are seas that seem to have stopped the migration of plants less than others. Such is the case for instance with the Mediterranean, whose shoreline has an almost identical vegetation: of 1,577 species observed by Mr. Desfontaines in Barbary, there are only about 300, that is to say barely 1/5, that could not be found in Europe."[24] In the early nineteenth century, the history of nature, its original diversity, its transformations, remained mostly a matter of conjecture: "In order to solve the great problem of the migration of plants, the geography of plants delves into the interior of the earth; it looks at the ancient monuments that nature has left behind in petrifications, in wood fossils, in coal strata that are the tomb of the initial plant life of our planet . . . The origin and the first homeland of the plants most useful to man and living with him since the remotest times is as impenetrable a secret as that of the land of origin of domestic animals."[25]

To fully grasp how this new notion of the history of nature could have emerged, a notion based not on some hypothetical geological accident or other catastrophe, but rather on the idea that time is long and imperceptible, it is appropriate to look at the role of the islands in the history of these voyages and in the evolution of a historical view of nature and its transformations.[26] The islands—constrained, isolated places and, in the case of the tropical islands, fascinating worlds due to the luxuriance of their flora and the strangeness of their fauna—were for the traveller-naturalists an ideal terrain for observation and reflection. In 1768, captivated by the floristic richness of the Mascarene Islands and of Madagascar, Philibert Commerson decided to abandon Bougainville's expedition in order to dedicate himself to his own exploration: thirty-four cases of herbaria, notes and specimens were sent to Paris upon his death and brought to the King's Garden. While the islands were awe-inspiring, they were also—due to their isolation, the fragility of their soil and the strong climatic variations to which they were subject—extremely fragile microcosms, and quickly degraded as soon as the Europeans settled on them. When HMS *Beagle* anchored off the shore of St. Helena in July 1836, Charles Darwin expressed his fascination with this island "situated so remote from any continent, in the midst of a great ocean, and possessing a unique Flora"; but he also well knew that it was already renowned to be a lost paradise. Since its discovery by the Portuguese in 1502, the island had undergone rapid deforestation under the hungry mouths of goats introduced by the first settlers, then, once the East India Company took control, as a result of the development of plantations; noting the comprehensive destruction of the woods, Darwin rued the fact that no catalogue of the original flora had been undertaken, which would have made it possible to measure the process in progress and to reflect on the impact of the degradation of the plant cover on the other forms of insular life—disappearance of the endemic animal species and invasion of foreign species.[27] Seen as isolated microcosms, while at the same time endowed with an abundant yet fragile nature, the islands were the first inkling of an awareness at the turn of the eighteenth and nineteenth centuries of the environmental problems and the threat that man historically imposed on nature.

But there is even more: because of certain particular characteristics of their flora and their fauna, islands and archipelagos raised questions among some early-nineteenth-century naturalists. Such was the case of geologist Leopold von Buch, an old friend of Alexander von Humboldt at the Freiberg Mining

23 Alexander von Humboldt, *Personal Narrative of Travels to the Equinoctial Regions of the New Continent, During the Years 1799–1804. By Alexander de Humboldt and Aimé Bonpland; with Maps, Plans, &c.* [1814], trans. Helen Maria Williams, 3rd ed. (London: Longman, Hurst, Rees, Orme, and Brown, 1822), vol. 1, p. 145, 146–147.

24 Augustin Pyramus de Candolle, "Géobotanique," in Frédéric Cuvier, ed., *Dictionnaire des sciences naturelles* (Strasbourg: Levrault; Paris: Le Normant, 1820), vol. 18, p. 359–422 (specifically p. 405).

25 Humboldt, *Essay n. 15*, p. 69 and 72.

26 Marie-Noëlle Bourguet, "L'île, lieu et opérateur des savoirs," in *Le Temps de l'île*, exh. cat. (Marseille: Mucem/Parenthèses, 2019), p. 78–93.

27 Charles Darwin, *Narrative of the Surveying Voyages of HMS* Adventure *and* Beagle, vol. 3: *Journal and Remarks, 1832–1836* (London: Henry Colburn, 1839), p. 581–584.

28 Leopold von Buch, *Description physique des îles Canaries* [1825], trans. from the German by Charles Boulanger (Paris: Levrault, 1836), p. 147–148.

29 Darwin, *Narrative of the Surveying Voyages of HMS* Adventure *and* Beagle, vol. 3, p. 454 and 474.

30 Charles Darwin, "An Historical Sketch of the Recent Progress of Opinion on the Origin of Species," in *On the Origin of Species by Means of Natural Selection, or The Preservation of Favoured Races in the Struggle for Life* [1859], 4th ed. (London: John Murray, 1866), p. XVIII. Darwin cites Buch's remarks that *"varieties slowly become changed into permanent species, which are no longer capable of intercrossing."* I would like to thank Pietro Corsi for drawing my attention to Darwin's reference to the work of this geologist and naturalist. Letter from Alexander von Humboldt to his brother Wilhelm, Cumana, July 16, 1799, in *Life of Alexander von Humboldt*, ed. Karl Bruhns, trans. Jane and Caroline Lassell (London: Longmans, 1873), vol. 1, p. 263 and 265.

Academy: in 1815, during an exploration of the Canaries archipelago, Buch observed that its floristic diversity was the result of its isolation, the varieties of the same species having found their way to distinct islands, where, for want of encountering and combining with other varieties, they slowly became a new species, from then on unable to intercross.[28] Twenty years later, during a stopover in the Galapagos archipelago (September 15–October 20, 1835), Charles Darwin made a similar observation about fauna: his process made it possible to retrace the way the archipelago—"a little world within itself," as he would say, as though to underscore the cognitive importance of this frame of observation—could favour the birth of another approach to nature, its diversity, its transformations. From the moment of his arrival, the naturalist was fascinated by the strange fauna that inhabited this cluster of islands, and their spectacle carried him back to antediluvian times: giant tortoises, grotesque reptiles, birds fearless of man. He undertook to observe the forms of life; he described, compared and classified them. Seeing in this group of volcanic islands, located below the equator some 600 miles west of the American coast, a distant satellite of the continent, he observed that the endemic species that inhabited it had retained the general characteristics of their American provenance. But he also noted that this fauna possessed strange, inexplicable traits. He remarked, for instance, that thirteen species or varieties of chaffinches were present, closely resembling each other except for the shape and size of their beak, which differed according to each species' lifestyle and diet. However, Darwin did not pay attention at the time to the geographical distribution of these variations: when he made this observation, he was in fact convinced that the proximity of the islands, the smallness of the archipelago and the uniformity of the climate constituted living conditions that were too homogeneous for their populations not to be too, without notable variation from one island to another. As well, he did not pick up on the statement made by the archipelago's vice governor, Nicolas Lawson, who boasted of his ability to assign each tortoise to its island of origin with just a glimpse. And if he himself recorded in his notebook regarding three species of mockingbirds captured during his long walks, that the habitat of each group corresponded to a different island, he noted the fact as a curiosity, without any sign of deeper interest. "It is the fate of every voyager, when he has just discovered what object in any place is more particularly worthy of his attention, to be hurried from it."[29]

It was only once he had returned to England—after having entrusted the ornithologist John Gould to examine the species of birds collected on the voyage, notably the chaffinches of the Galapagos; compared his collections and notes with those of the other members of the crew from HMS *Beagle*; and challenged his reflections with observations made in other places by earlier naturalists, notably Buch's remarks on the flora of the Canaries and the slow transformation of the insular varieties and species—that Darwin slowly seized on the heuristic scope of the circumstances he had observed in the archipelago.[30] Rejecting the idea of the immutability of the species, he thought about the imperceptible changes that could have taken place over the course of time, and, to explain the variations that he observed between the groups of chaffinches, resorted to the concept of "descent with modification": a variety of bird endowed by chance with a specific character (such as a strong, short beak) adapted to the conditions of one island is more likely to feed itself, and thus to survive and thrive there; conversely, another variety endowed with its own character (such as a long, narrow beak) will be better suited to survive on another island, in different living conditions. Over the course of generations, each variety will slowly colonize the insular place to which it found itself best adapted, eventually becoming a distinct species.

In 1859, the same year as the publication of the last volume of *Cosmos* and the death of Humboldt, Darwin published *Origin of Species*: twenty years after his trip to the antipodes, he turned the southern archipelago into a case study to illustrate his theory of evolution and support the theory of natural selection upon which it is based. To the infinite variety of nature—the subject at the heart of natural history, the object of cataloguing practices and the basis of the travellers' descriptions—the concept of variation would add the possibility of comprehending its evolution in time, and thus to write its history.

NICOLAS BAUDIN'S EXPEDITION TO THE AUSTRAL LANDS:

TRAVELLING ARTISTS

GABRIELLE BAGLIONE

One cannot entertain the idea of travelling to the far side of the world without at the same time providing means for recording the journey. Within the framework of a scientific expedition, the work of collecting data and specimens is methodically structured. The destination of Nicolas Baudin's "voyage of discovery to the Southern Hemisphere"[1] (1800–1804) was New Holland (present-day Australia). The voyage had three objectives: a geographical survey of the southeast coast of these little known "austral lands," the collection of natural specimens and a methodical approach to recording the customs and physical characteristics of the human inhabitants. The nineteen scientists who set sail included botanists, zoologists, mineralogists, geographers and astronomers. Some were correspondents for the newly founded Société des observateurs de l'homme, which relayed its instructions through Georges Cuvier and Joseph-Marie de Gérando.[2] Numerous natural specimens were brought back, and major contributions to geography made. Almost 3,000 drawings and manuscripts are now preserved in the Muséum d'histoire naturelle in Le Havre. Such art served the interests of a science that grew with each new discovery. Two centuries later, we are still gauging the full extent of its cultural and scientific contributions.

TWO ARTIST EXPLORERS

A seasoned navigator with a passion for the natural sciences,[3] Baudin led this expedition, which was the brainchild of Louis Antoine de Bougainville, and had the backing of the Institut de France and the approval of Napoleon Bonaparte. Baudin would bring back to France some 100,000 natural specimens along with thousands of drawings and manuscripts by artists, officers and scientists.[4]

After five months at sea, health issues prompted the three hired draughtsmen—Jacques-Gérard Milbert, Michel Garnier and Louis Lebrun—to abandon the voyage at the first port of call, the Île de France (present-day Mauritius). Two assistant gunners who had also been recruited to illustrate Baudin's personal journal were then designated as the official draughtsmen. They would prove essential to the expedition and the publication of its findings. Charles-Alexandre Lesueur was a native of Le Havre. Although he had no documents indicating artistic training, his talent as a draughtsman was immediately recognized. He was tasked with depicting animals and landscapes. Nicolas-Martin Petit, who had trained in the studio of Jacques Louis David, was to do portraits of the Indigenous people they encountered along the way. François Péron oversaw Lesueur's drawings to ensure the accuracy of his depictions. An assistant zoologist, anatomist and anthropologist, Péron became the go-to zoologist after the deaths of Stanislas Levillain and René Maugé de Cely, seasoned zoologists close to Baudin. He penned descriptions to accompany the drawings, which were extremely precise and faithfully coloured. Upon his return from the voyage, Lesueur produced watercolours on vellum with the aid of a magnifying glass; the whiteness and transparency of the skins perfectly matched the renditions of the marine invertebrates, which seemed to float on the surface of the support (65, 66).[5] Petit's undeniable talent is embodied in the strong sense of presence exuded by his models and his classical training is legible in his compositions. His models are generally depicted as bust portraits in three-quarter views. From a classical perspective, his portrait of a South African man (64), executed during a stopover at the Cape of Good Hope, leaves nothing to be desired—the model's ritual scarring nonetheless attests to his culture. A portrait of an Australian woman (63) is inscribed in a medallion, another aspect of the classical vocabulary; her hair band, her toned bust and the softness of her features reflect the influence of the image of Josephine Bonaparte—one of the key figures associated with the expedition.[6]

ORGANIZING THE COLLECTION OF SPECIMENS AND CREATING AND DISSEMINATING KNOWLEDGE

The environments covered were vast, the realities diverse and the time for reporting on them short. While France and Great Britain were enjoying a period of political peace, their competition for scientific knowledge was fierce. Geographical surveys of coastlines made it possible to draw up maps that were published in 1811. Rocks and plants were collected, studied and classified.[7] The animal specimens, which included some 2,500 new species, were described, drawn and preserved (either naturalized or preserved in fluid). Cultural artifacts were collected in New Holland, Van Diemen's Land (present-day Tasmania) and Timor.[8] The body of data was considerable.

Tasked with providing an account of the voyage that would make the results of this scientific harvest official,[9] François Péron oversaw the work of the draughtsmen.[10] His plan was to supply a considerable quantity of information in forty-one plates. This quite legitimate ambition entailed the reconstruction of observed reality, bringing together on the same plate objects seen in different locations or associating, in the form of a "family portrait" (human or animal), individuals that did not necessarily share bonds of kinship (60, 62). Péron composed portraits of representative individuals who bore traits observed on various others of their kin.

While the published images were not "instant" portraits, the figures depicted nonetheless constituted precious testimonies, providing tangible proof of what had been *observed* and giving this body of work undeniable cultural and scientific value. In Tasmania, Lesueur drew baskets made from algae—kelp—and rushes (61). The tragic history of the Tasmanians, who were decimated by illnesses brought by Western settlers and sold into slavery in other lands, particularly in the southern part of continental Australia, slowed down the traditional oral

1 For an account of the voyage, first published in French in 1807, see François Peron, *Voyage of Discovery to the Southern Hemisphere, Performed by Order of the Emperor Napoleon during the Years 1801, 1802, 1803, and 1804* (London: Richard Phillips, 1809).

2 Georges Cuvier, *Note instructive sur les recherches à faire relativement aux différences anatomiques des diverses races d'hommes* [Note on the Researches to be Carried Out Relative to Anatomical Differences between the Diverse Races of Man]; Joseph-Marie de Gérando, *The Observation of Savage Peoples*, trans. and ed. F.C.T. Moore (Berkeley: University of California Press, 1969).

3 Baudin displayed his interest in plants and their capacity for acclimatization during his expedition to Latin America from 1796 to 1798.

4 Added to this in the following years was the work carried out preliminary to the publication of the account of the voyage. This work is scattered in various institutions (Muséum d'histoire naturelle in Paris, Archives nationales de France, Archives de la marine among others) and private collections.

5 In 1806, Lesueur, who was hoping to obtain the position of painter on vellum at the Muséum, produced numerous watercolours of animals and distinguished himself by his ability to depict a great variety of species (marine specimens, mammals, birds, insects). For a discussion of the vellum paintings executed for the Muséum, see the essay in this catalogue, p. 60.

6 Living animals and plants were offered to Joséphine Bonaparte for the Château de la Malmaison, where it was hoped they would acclimatize. This is illustrated in the frontispiece of the *Voyage of Discovery to the Southern Hemisphere.*

7 The collections of natural objects are housed in the Muséum d'histoire naturelle in Paris.

8 Two hundred and six objects were collected but there is no longer any trace of them.

9 Péron oversaw the production of the first volumes of the *Voyage de découvertes aux Terres australes* and Louis de Freycinet looked after the work of publication: volume 1 of the account with plates (1807), a geographical atlas (1811), volume 2 (1816); second edition (1824).

10 Petit died in the fall of 1804, worn out, like most of the scientists, by the rigours of the expedition. His portraits were taken over by a draughtsman who remained in Paris; the features are cruder and the gaze much less present.

transmission of manufacturing skills. The drawings are now making it possible for artists to recover this knowledge; in this regard, they have become very valuable for a culture that was subjected to terrible violence. Apart from its cultural interest, the Baudin expedition can also be seen as important from an ecological standpoint as well. A large number of the marine animals described are accompanied by the geographical coordinates of the places where they were found; Péron also took readings of the temperature and the phosphorescence intensity of seawater. When compared with recent data, this information can be significant for the study of the evolution of these environments.

This body of drawings attests to exploration motivated by the ambition to make an inventory of "the world" and to classify it. Artistic and scientific talent were placed at the service of a growing body of knowledge. Today, as in 1804, the essence of this collection lies in the construction and dissemination of knowledge. It enables us to better know environments and people and illustrates the desire of a society that is seeking, through science, to know—and to measure—the immensity and diversity of the world.

60
Charles-Alexandre Lesueur,
Macropus fuliginosus (Desmarest, 1817) Kangaroo,
between 1804 and 1810
Watercolour, gouache, graphite, gum arabic on vellum, 25.7 × 40.3 cm
Le Havre, Muséum d'histoire naturelle

61 FIG.
Charles-Alexandre Lesueur,
Van Diemen's Land - Weapons, Vessels, Ornaments, 1807
Watercolour, India ink and pencil, 27 × 21.6 cm
Le Havre, Muséum d'histoire naturelle

62
Charles-Alexandre Lesueur,
Terrapene carolina (Linnaeus, 1758) Turtle,
between 1816 and 1837
Watercolour and gouache, 46.7 × 31.1 cm
Le Havre, Muséum d'histoire naturelle

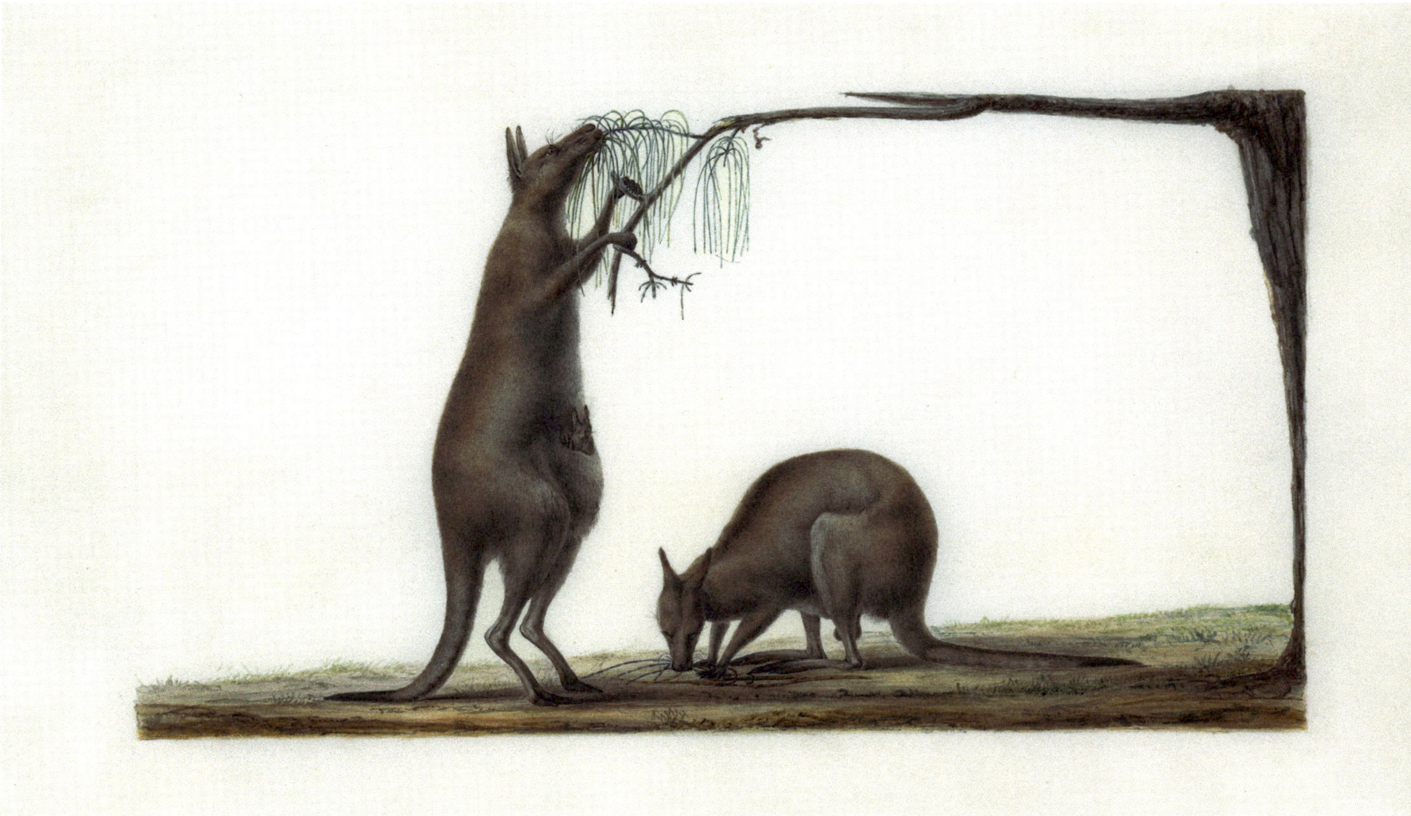

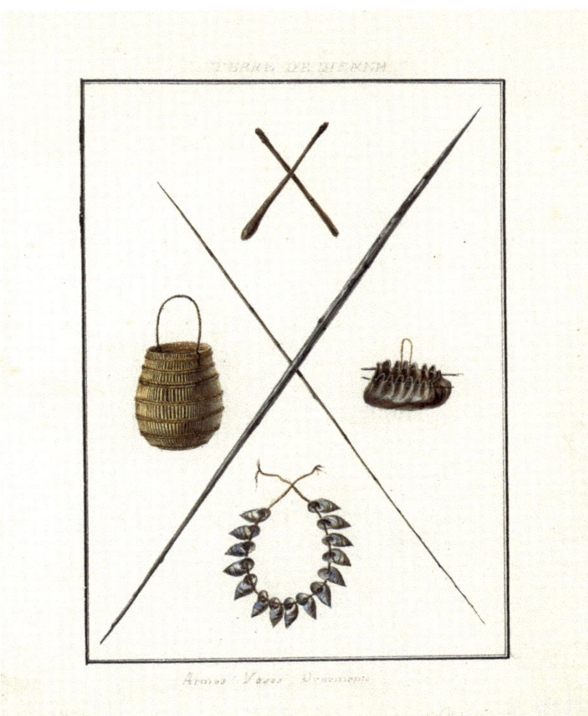
TERRE DE DIEMEN
Armes Vases Ornemens

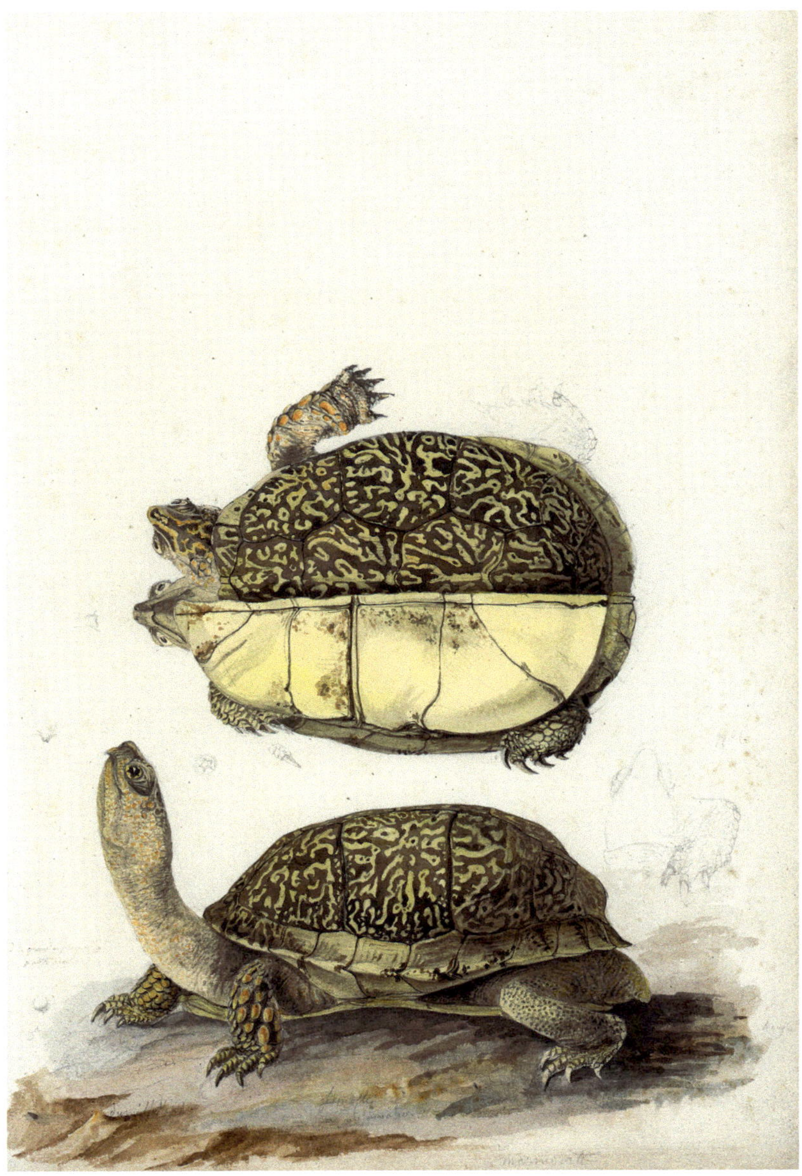

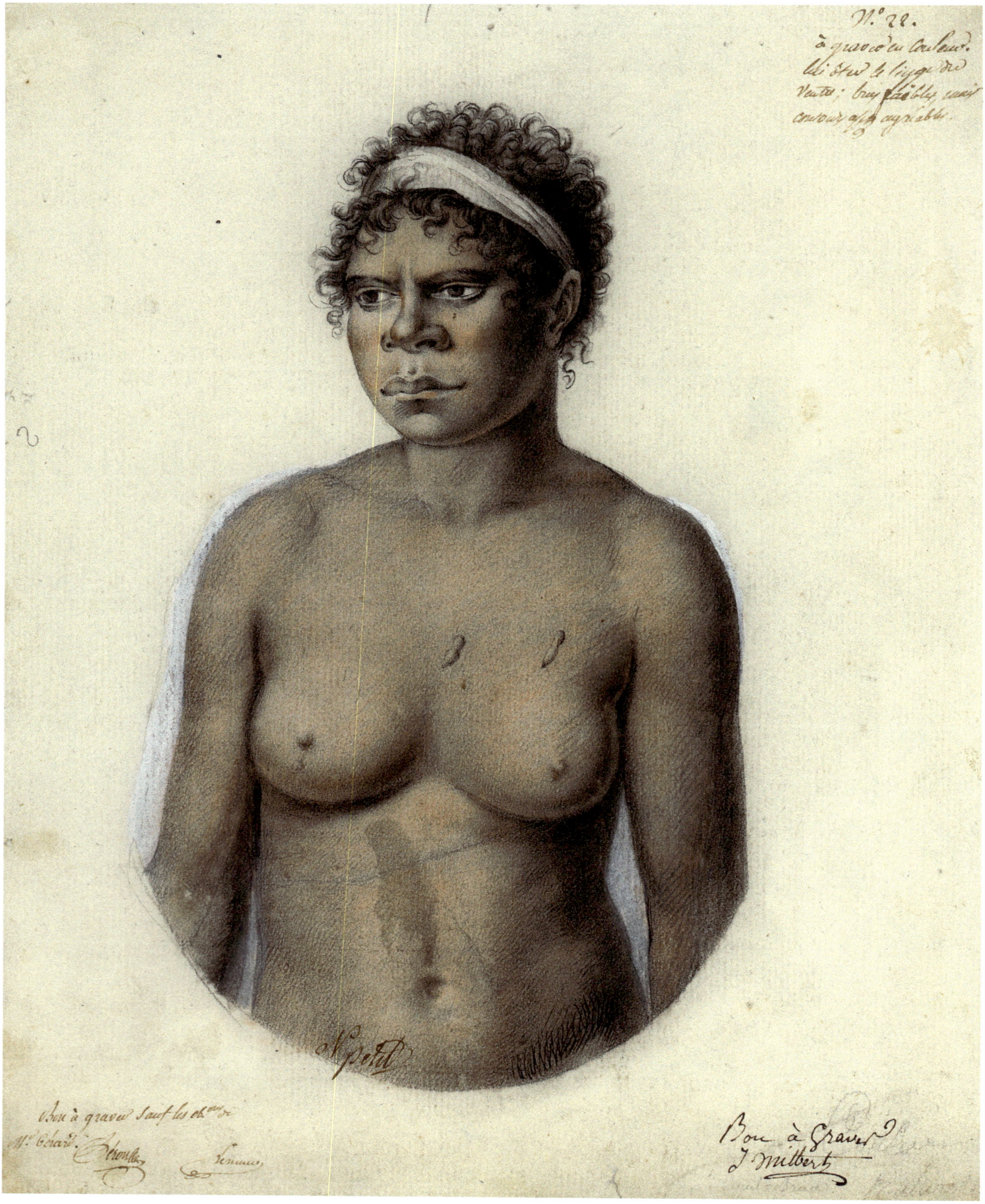

63 FIG.
Nicolas-Martin Petit,
Portrait of a Woman of New Holland (Australia), 1802–1804
Black chalk, red chalk, white highlights, brown ink and India ink, 29.7 × 24.7 cm
Le Havre, Muséum d'histoire naturelle

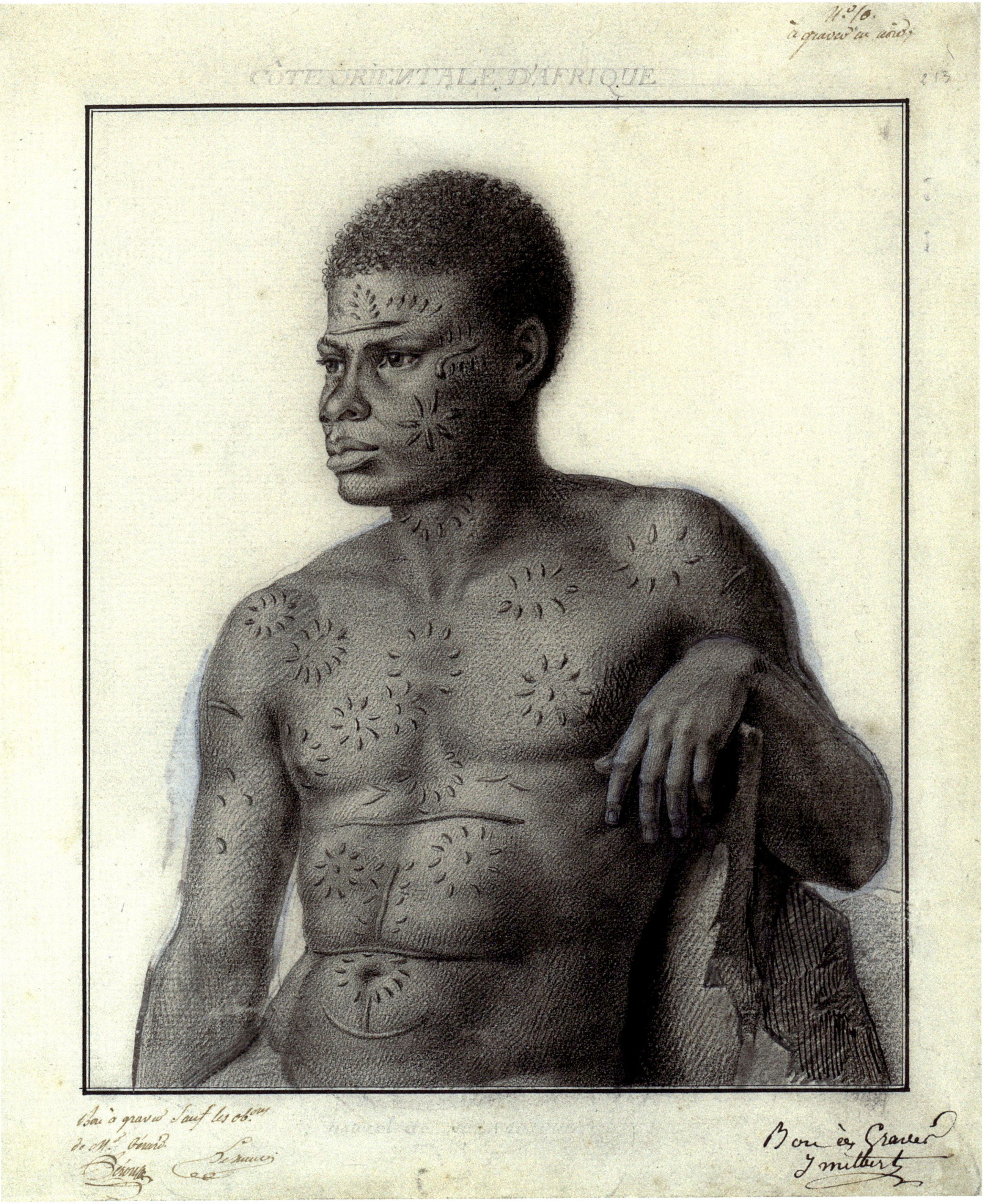

64 FIG.
Nicolas-Martin Petit,
Portrait of a Man of Coastal Mozambique Bearing Scarification Marks, 1803
Pencil, black chalk, brown ink, India ink and white highlights, 27.8 × 22.5 cm
Le Havre, Muséum d'histoire naturelle

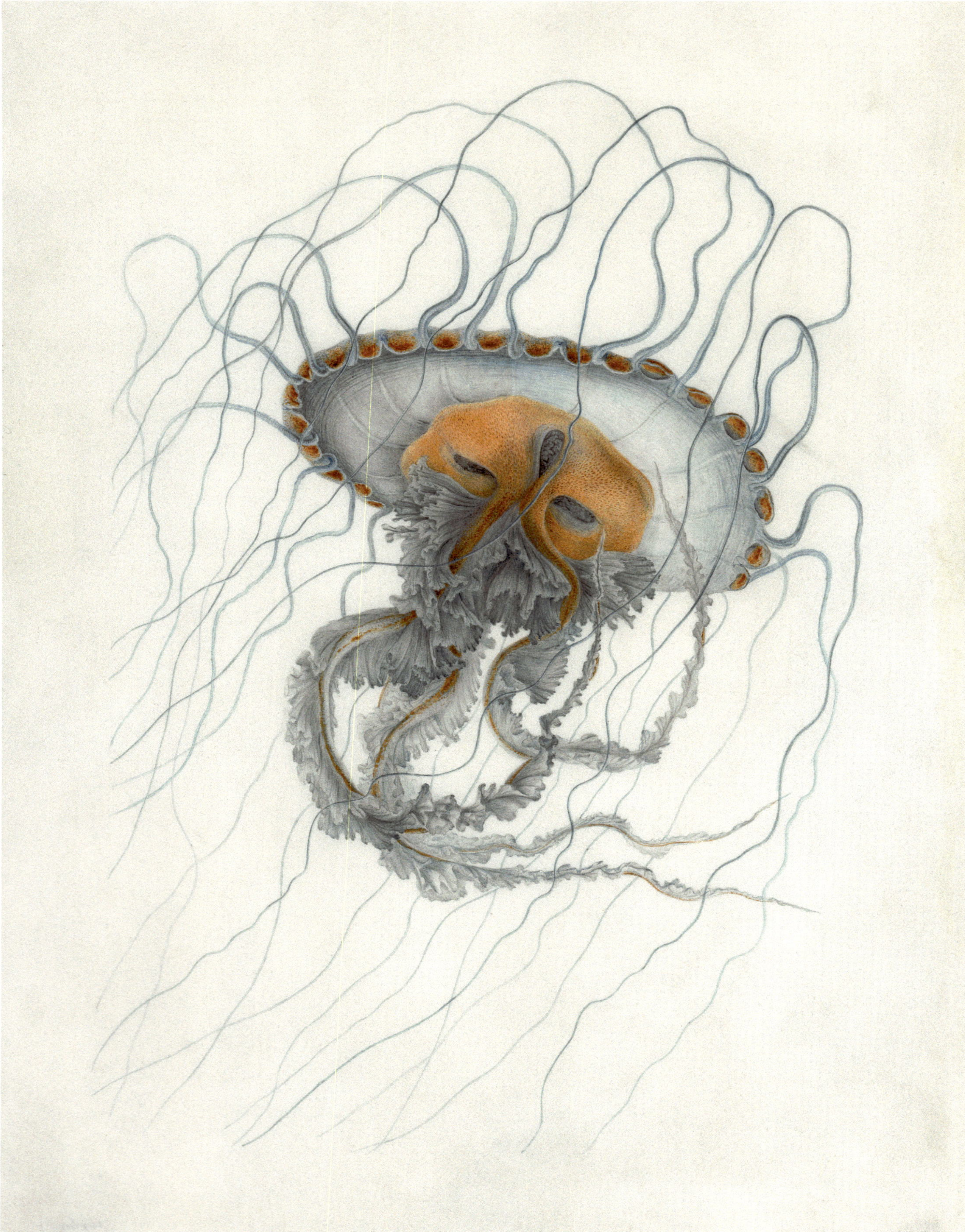

65
Charles-Alexandre Lesueur,
Chrysaora Lesueur Jellyfish,
between 1804 and 1810
Watercolour and pencil on vellum,
43.2 × 27.9 cm
Le Havre, Muséum d'histoire naturelle

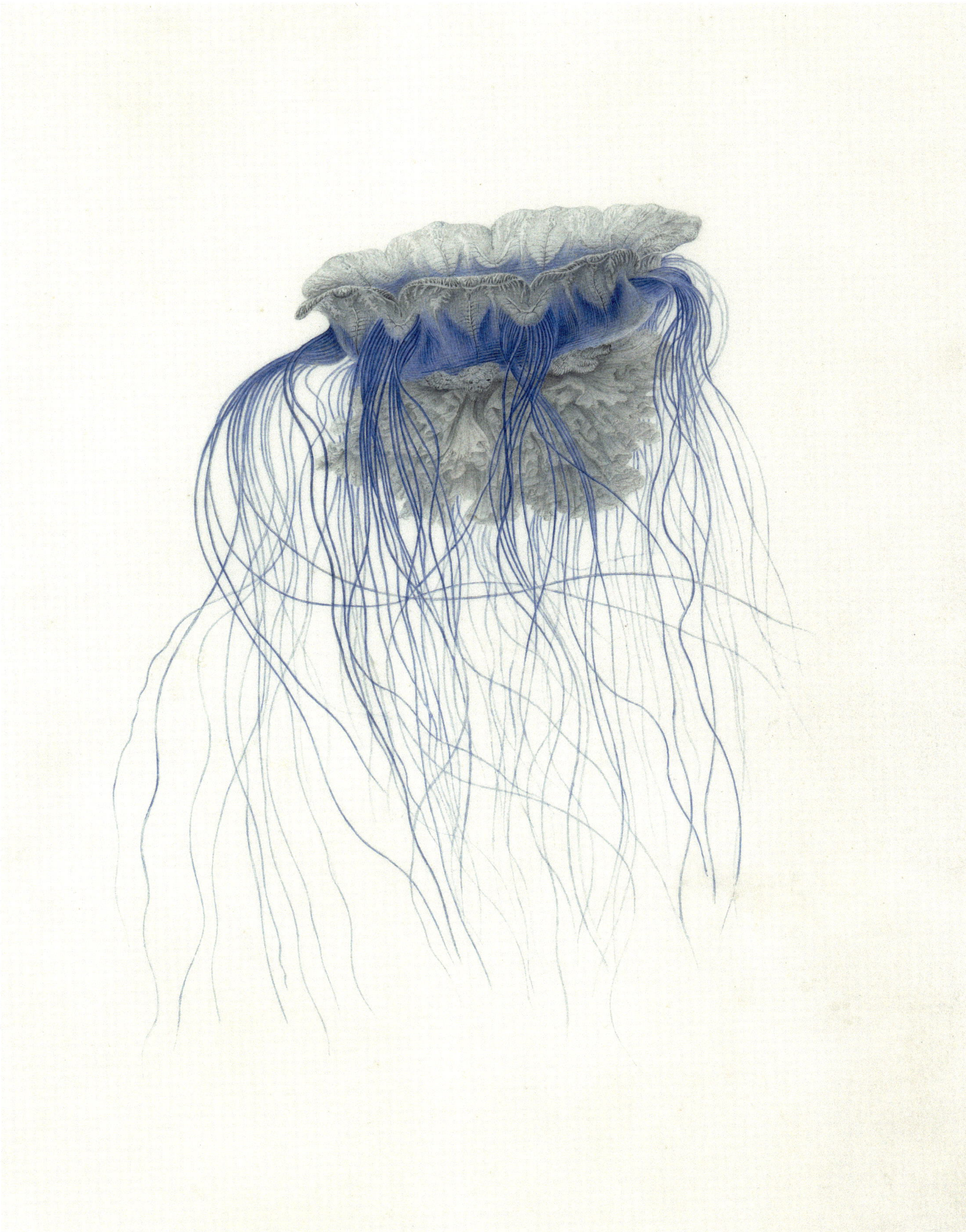

66
Charles-Alexandre Lesueur,
Cyanea lamarckii Jellyfish,
between 1804 and 1810
Watercolour on vellum, 44.1 × 28.5 cm
Le Havre, Muséum d'histoire naturelle

AQUARIUM MANIA

CAMILLE LORENZI

"Fashion, which is so often about frivolous things of no consequence, is from time to time fortunate enough to extend to things worthy of it. The vogue for aquariums seems to us to be one trend that merits encouragement."[1] This excerpt from the *Journal des débats politiques et littéraires* of 1865 is a good illustration of the phenomenon that came to be known as aquarium mania.[2] A novel laboratory instrument newly invented in England and subsequently introduced in France, the aquarium soon generated a level of interest that reached far beyond the scientific community to which it had been initially confined. Its dissemination even set off a kind of craze. This "aquarium mania" lasted only a decade but its brevity did not make it any less intense.

The aquarium was first created to observe and study aquatic life. It established itself in England in the early 1850s and spread to France in the 1850s and 1860s, thanks mainly to the efforts of Jeanne Villepreux-Power, a specialist in marine biology, who is considered the first inventor of the aquarium. She developed a water tank that made it possible to study marine specimens when brought on land. The results of her observations are recorded in a book illustrated with watercolour plates: *Observations et expériences physiques sur plusieurs animaux marins et terrestres*, published in 1860. The construction of public aquaria would similarly serve in the scientific observation of species: the first public aquarium opened its doors in London in 1853. Six years later, in 1859, the Aquarial Gardens were inaugurated in Boston and, in 1860, a Vienna Salon offered visitors an opportunity to see a vast aquarium. In France, the first establishment of this kind in Paris opened in the Jardin d'acclimatation in October 1861.

Moving from its initial status as a scientific curiosity, the aquarium quickly became a decorative object much sought after by enthusiasts, who used it to embellish their homes. Examples like the aquarium of Ferdinand Barbedienne are authentic pieces of luxury furniture that bring together the period's artistic influences (68). Admired for their picturesque qualities, the fish in Barbedienne's aquarium became rare knick-knacks much prized by the exponents of the bourgeoisie. They were assigned a place amid collections of objects that became, at the time, the exclusive preserve of the well-to-do.[3] Public aquaria, for their part, would become new popular attractions (67).

In fact, the aquarium often constituted a striking spectacle, and the acquisition of one was described as an opportunity to contemplate the canvas of marine life at one's ease. The fondness for aquaria resided in the specimens' immediate natural beauty, which stemmed from the grace of their movements and their bright colours. An article from *L'Illustration* of 1857 depicts a feverish scene: "You see them, belly up, fins spread out like sails, cutting one another off, getting out of one another's way, swimming around one another in vague, fleeting spirals, contracting and expanding like [so many versions of] Proteus conjured up by some invisible Caliban."[4]

But the aquarium mainly made it possible to show marine life as it had never been pictured before: viewers could see, directly before their eyes, scenes that in normal times could be found only at great depths, in the abysses of a world to which humans had never had any real access.

The director of the Jardin d'acclimatation stressed the novel sensations available to first-time viewers of aquaria, saying, "Doesn't what you saw there seem somewhat like the surprise that Virgil said we would have were the Earth to split open and reveal its infernal chasms and other things unknown even to the gods themselves? . . . Did you not feel that the watery abyss had opened up before your eyes and shown you things that Nature had so far kept from you?"[5]

1 *Journal des débats politiques et littéraires*, December 19, 1865, p. 3.

2 On this enthusiasm for the marvels of Nature, see Éric Baratay and Élisabeth Hardouin-Fugier, *Zoos. Histoire des jardins zoologiques en Occident, XVIe–XXe siècle* (Paris: La Découverte, 1998), p. 35–36; Bernd Brunner, *The Ocean at Home: An Illustrated History of the Aquarium* (New York: Princeton Architectural Press, 2005), p. 17–20; Camille Lorenzi, "L'Aquarium en France au XIXe siècle, 1855–1870. Diffusion sociale et culturelle d'un objet scientifique," master's thesis 1, Paris 1 Panthéon-Sorbonne, thesis director Sylvain Venayre, 2008.

3 On the material culture of the bourgeoisie, see for example Manuel Charpy, "L'ordre des choses. Sur quelques traits de la culture matérielle bourgeoise parisienne, 1830–1914," *Revue d'histoire du XIXe siècle*, no. 34 (2007), p. 105–128.

4 "Culture familière des plantes aquatiques," *L'Illustration*, November 7, 1857, p. 310.

5 Étienne Rufz de Lavison, "Sur l'aquarium du Jardin d'acclimatation," *Bulletin de la Société zoologique d'acclimatation* (Paris: Société d'acclimatation, 1863), p. XLVII–XLVIII.

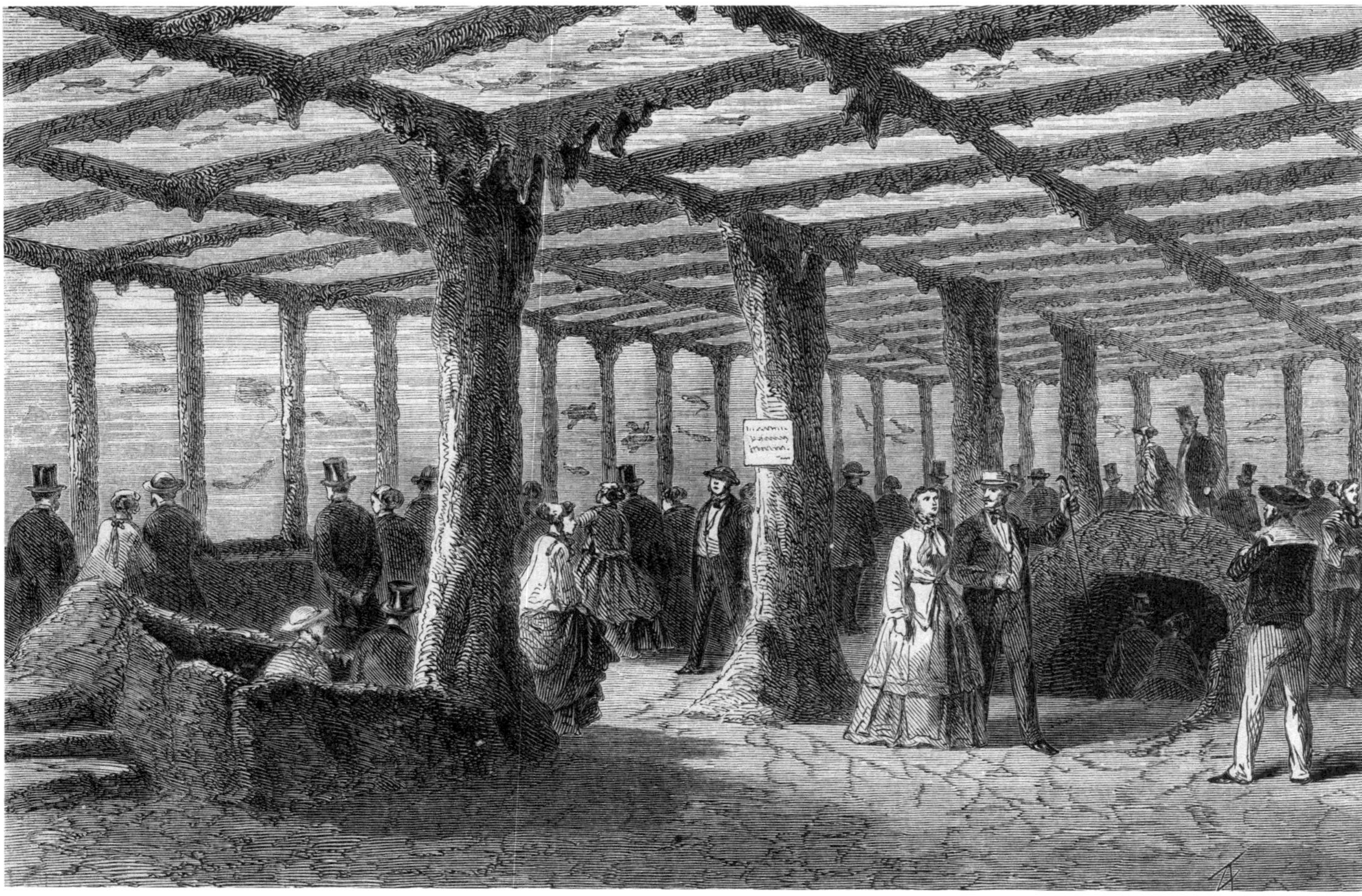

67 FIG.
Jules Gaildrau, "Saltwater Aquarium," in *L'Exposition universelle de 1867 illustrée*, Paris: 1867

68
Ferdinand Barbedienne (founder), Édouard Lièvre (draughtsman), *Aquarium and Pedestal*, about 1875
Glass, silver-plated bronze, 143 × 75 × 62 cm
Collection of Ramy Boutros

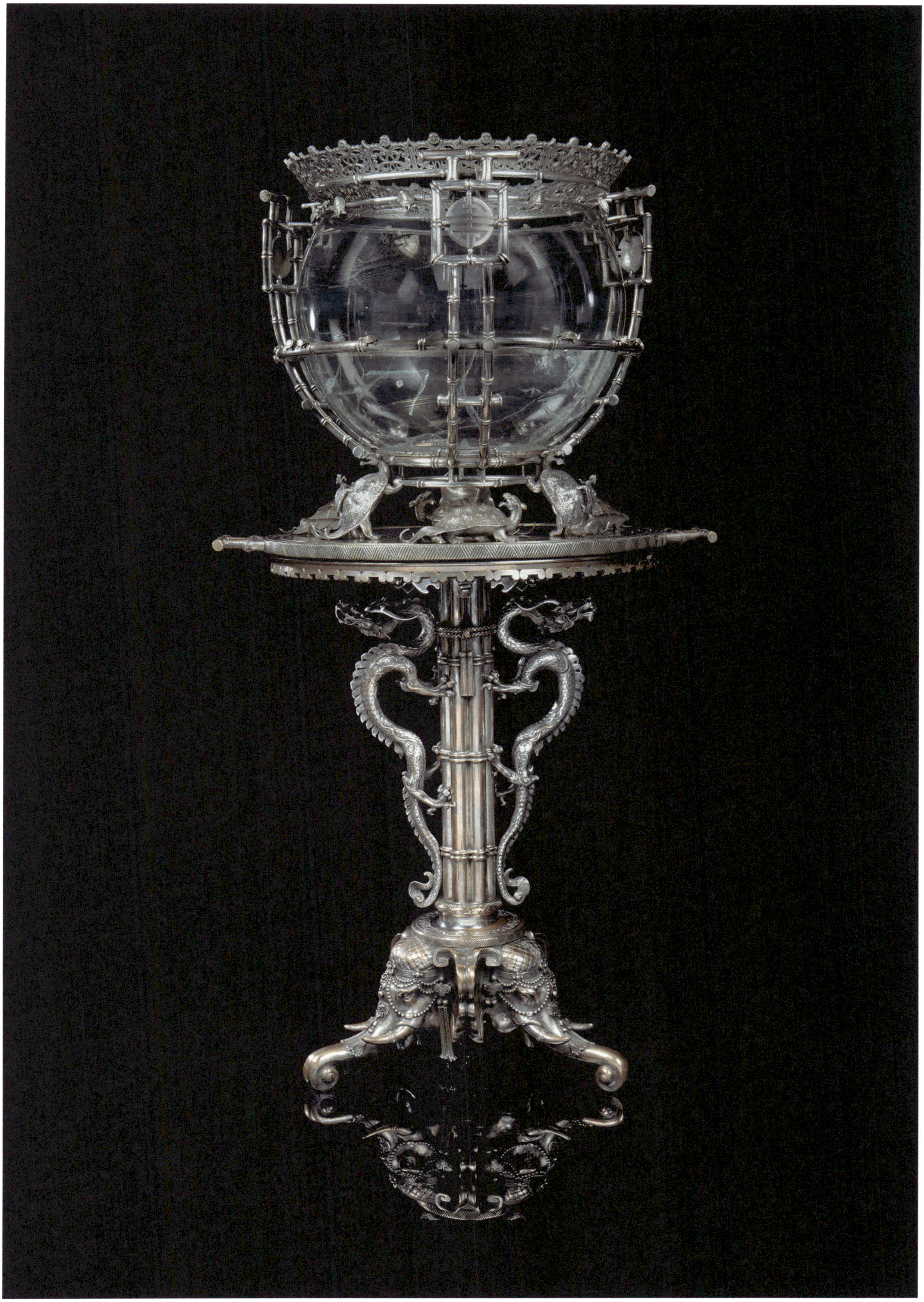

TWENTY THOUSAND LEAGUES UNDER THE SEA:

THE FIRST OCEANOGRAPHIC TREATISE

JACQUELINE GOY

Twenty Thousand Leagues Under the Sea is the novel of the oceans. Published as a serial between 1869 and 1870 in the *Magasin d'éducation et de récréation*, Jules Hetzel's literary review, it was part of the extraordinary impulse to disseminate science that existed throughout the nineteenth century (69 A-E); Jules Verne took a leading role by creating a setting for the replication of biotopes, thanks to a combination of discoveries in marine life.

Humans have always been drawn to underwater life, and fishermen were the first to recognize its riches. Marine wildlife, which only became accessible thanks to fishing gear (nets and lines), was reproduced during the Hellenistic period in fish mosaics, illustrating the beginnings of marine zoology, which could be described as gastronomic. In fact, Ned Land, the harpooner in Jules Verne's novel, classifies fish according to whether or not they can be eaten!

Following the publication of Charles Darwin's writings in 1859, the question of the mystery of the origins of life re-emerged. Jules Verne set his novel within this movement, with the intention of writing "the real book of the sea" because it is in the seas' depths that "these huge specimens of life from another age" are hidden. And he warns: "You will travel in a land of marvels" as "it was through the sea that the globe as it were began." In the novel, the narrator, Pierre Aronnax, a professor at the Muséum d'histoire naturelle in Paris, assumes the role of a scholar, questioning the origins of life, while his domestic, Conseil, classifies all species. As for Captain Nemo, whom he endows with all the virtues of social justice and with encyclopedic knowledge, he entrusts him with explaining the mechanics of his machinery and in particular the physical aspect of seawater.

While it may not be the most famous of Jules Verne's novels, it is the one whose universality is undisputed by virtue of all the authors mentioned. He pays tribute, in particular, to George Sand,[1] but also to Jules Michelet, to whom he makes numerous references and, in the passage on the battle with the octopus, to Victor Hugo.[2]

Jules Verne was fascinated by the fauna of the deep seas, about which he provides precise details. This is not a matter of mere speculation but of observations borrowed from the scholars of the time, although science historians, in particular Eric Mills, consider the pioneer of oceanographic narrative to be the Scottish naturalist Wyville Thompson, the author of *The Depths of the Sea*, published in 1873, that is to say after *Twenty Thousand Leagues Under the Sea*.[3] Jules Verne makes use of these observations to reveal the mysterious populations of the ocean's depths, drawing on the authors in Captain Nemo's library. They are all there, from the earliest navigators to his contemporaries, some of whom were the very first to dive equipped with a diving suit. Among the scholars from antiquity, Aristotle is key, then Linnaeus, who leads the series of renowned authors, followed by Buffon, considered to be the father of science, Cuvier, and Lacépède for his classification of fish, and many others.

Jules Verne was also interested in the findings of the long circumnavigation voyages that were undertaken applying a new scientific approach, voyages initiated by Bougainville, Cook, La Pérouse, d'Entrecasteaux and, in particular, Dumont d'Urville, who was awarded national honours. Indeed, at Vanikoro, he traced the wreckage of La Pérouse's ships and the tin box containing instructions annotated by King Louis XVI. In fact, Jules Verne embellished a little; these instructions had been placed in the Archives nationales, in Paris.[4] The instructions in the famous box entrusted to La Pérouse were written by the botanist André Thouin.[5] This is a good example of an aspect of Jules Verne's literary style, starting out with two separate facts and interchanging the details. But noticeably missing from these pages are Nicolas Baudin and his *Voyage of Discovery to the Southern Hemisphere*.[6] Jules Verne may have deliberately concealed the main source of his novel, as much was borrowed from the account of this journey. The long ropes of salps that Baudin describes, magnificent jellyfish, invigorating sea cucumber jam, Timor's sacred crocodiles, the Malay language, the difference between sea lions and seals, all these passages were taken from François Péron's account of the voyage, written in 1807.

In invoking the sea's luminescence, its most beautiful spectacle, Jules Verne raises the question of its phosphorescence, whose marvellous powers he recalls in these pages, citing the animals that participate in this pageant: salps, piddocks, jellyfish and so on (69 B, C). What we call, even today, phosphorus of the sea did not escape Homer in his long description of this sea of the Greeks on which "when forth ower the sea there appeareth to seamen the gleam of blazing fire."[7] Twenty-five centuries later, we learn that "this light felt as if alive."

In his novel, Jules Verne mentions another group of authors who distinguished themselves as of 1857, during the laying of telegraph cables, which then revealed life in the deep seas. These highly technical operations were as much an opportunity for scholars as for the understanding of deep-sea geology and the related fauna being studied by Alphonse Milne-Edwards.

An entire chapter of *Twenty Thousand Leagues Under the Sea* was devoted to the Gulf Stream, "a true lake in the middle of the Atlantic." While Benjamin Franklin was the first to observe the time differences between voyages heading from Europe to America and those returning, and to comprehend that a current drove the ships from west to east, it was Matthew Fontaine Maury upon whom Jules Verne relied to provide all the guidance on its hydrodynamics.

Throughout the book, repetition provides another way of allowing the reader to become acquainted, little by little, with this environment. In this way, Jules Verne points out on several occasions that "vegetable life was disappearing quicker than animal life." He also explains that unicellular algae, in particular trichodesmium, colour the waters of the Red Sea. Corals are described based on "the theory of Mr Darwin."[8] The colouring of Mediterranean red coral is mentioned by the "poetic

1 George Sand, in a letter sent to Jules Verne in June 1865, in Adolphe Brisson, "M. Jules Verne," *Le Temps*, December 1897, p. 2.

2 See the description of the battle between Gilliatt and the octopus by Victor Hugo in *Toilers of the Sea*, trans. W. Moy Thomas (London and Toronto: J. M. Dent & Sons Ltd; New York: E. P. Dutton & Co., 1911), part 2, "Malicious Gilliatt," p. 297–299.

3 Eric Mills, *Biological Oceanography: An Early History, 1870–1960* (Ithaca: Cornell University Press, 1989).

4 Archives nationales, Mar. 3 JJ 386.

5 Catherine Gaziello, *L'Expédition de Lapérouse, 1785–1788* (Paris: Comité des travaux historiques et scientifiques, 1984), p. 64.

6 See the essay in this catalogue, p. 86.

7 Homer, *The Iliad*, trans. A. T. Murray (Cambridge, MA: Harvard University Press; London: William Heinemann, Ltd., 1924), book 19, lines 376–380.

8 Jules Verne, *Twenty Thousand Leagues Under the Sea*, trans. William Butcher (Oxford: Oxford University Press, 1998), p. 124.

names of *blood flower* and *foam of blood*, the names the trade gives to its finest products,"[9] in reference to the blood of the Gorgon Medusa, which, in Greek myth, turns into coral. Echinoderms provide an opportunity to mention the earliest harvests carried out at over 700 metres, by James Ross, in Antarctica. Crustaceans are treated like sea monsters, perhaps to evoke the *Cosmographia*, by Sebastian Münster. Molluscs are described according to their decorative aspects, as it was *de rigueur* for every cabinet of curiosity to include some, just as Nemo's salon did. Unaided, they illustrate History's impact on Science. Indeed, when, after the French Revolution, these cabinets were confiscated, leading to their being placed in the Muséum, all these shells enriched the collections that Lamarck and Lacépède took upon themselves to put in order.

Jules Verne devotes a chapter to the ineluctable octopus (69 D), reviving the battle described by Victor Hugo.[10] As for fish, they swim freely in front of the portholes, representing the largest of the seas' populations (69 E). In fact, they animate the setting, with their freedom providing a kind of contrast to the protagonists' confinement. All of Jules Verne's talent as a playwright resurfaces in these pages and, if the enumeration of the species seems tedious, it adds a certain poetry with its infinity of lists.[11]

What is most important, however, is to retain several principal ideas: the sea covers seventy percent of the earth's surface, the waters are in constant movement, the wind affects the water's surface, light rays illuminate the uppermost one hundred metres where most of the algae can be found, the ocean's depths are dark and life is omnipresent (71, 72). The sea is the kingdom of the fish, and through their behaviour patterns and interactions, Jules Verne is able to introduce all of the other organisms in the marine ecosystems.

Herein lies his genius, inventing a voyage under the sea, following an itinerary where the routes of the great navigators overlap, a genuine Odyssey of modern times. In explaining evolution, *Twenty Thousand Leagues Under the Sea*, a novel with a strong philosophical connotation, had the broadest reach since Darwin's theories on the origin of the species and natural selection, theories that marked the end of the century by revealing the importance of nature, which launched the Art Nouveau movement. While the marine environment was one of Émile Gallé's favourite settings (201, 202, 203, 204, P. 292–293),[12] Jules Verne appears to be a precursor of this movement.

All the same, this visionary took a compelling approach towards change, and one should note this intuitive feature, which arises when he foresees the pollution caused by algae, this "decaying grass [that] poisoned the air," and the seas "crowded with jellyfish," and changes to the Gulf Stream that would have repercussions on "the climate of Europe [that] will undergo changes of an unforeseeable nature." Jules Verne understood well the fragile balance that governs the oceans and, early on, sent a message to future generations, encouraging them to preserve the state of the environment. This work, which he began to write in 1868, was a fundamental stage in the history of science, on an important topic in the knowledge of the ocean. Added to this is the element of the fantastic through the recreation of the marine environment and the creatures that inhabit it, as the scenes of violence between sperm whales and baleen whales are very real, the waters that produce the ocean currents transport ships more efficiently than seven-league boots, and the creatures that glitter are much more magical than all the magic wands one encounters in fairy tales. More than one hundred and fifty years later, this tale, which has preserved its extraordinary evocative powers, reveals itself to be a remarkable achievement.

---

9 Ibid., p. 157.

10 Hugo, *Toilers of the Sea,* p. 297–299.

11 See Umberto Eco, *The Infinity of Lists,* trans. Alistair McEwen (New York: Rizzoli, 2009).

12 See the essay in this catalogue, p. 288.

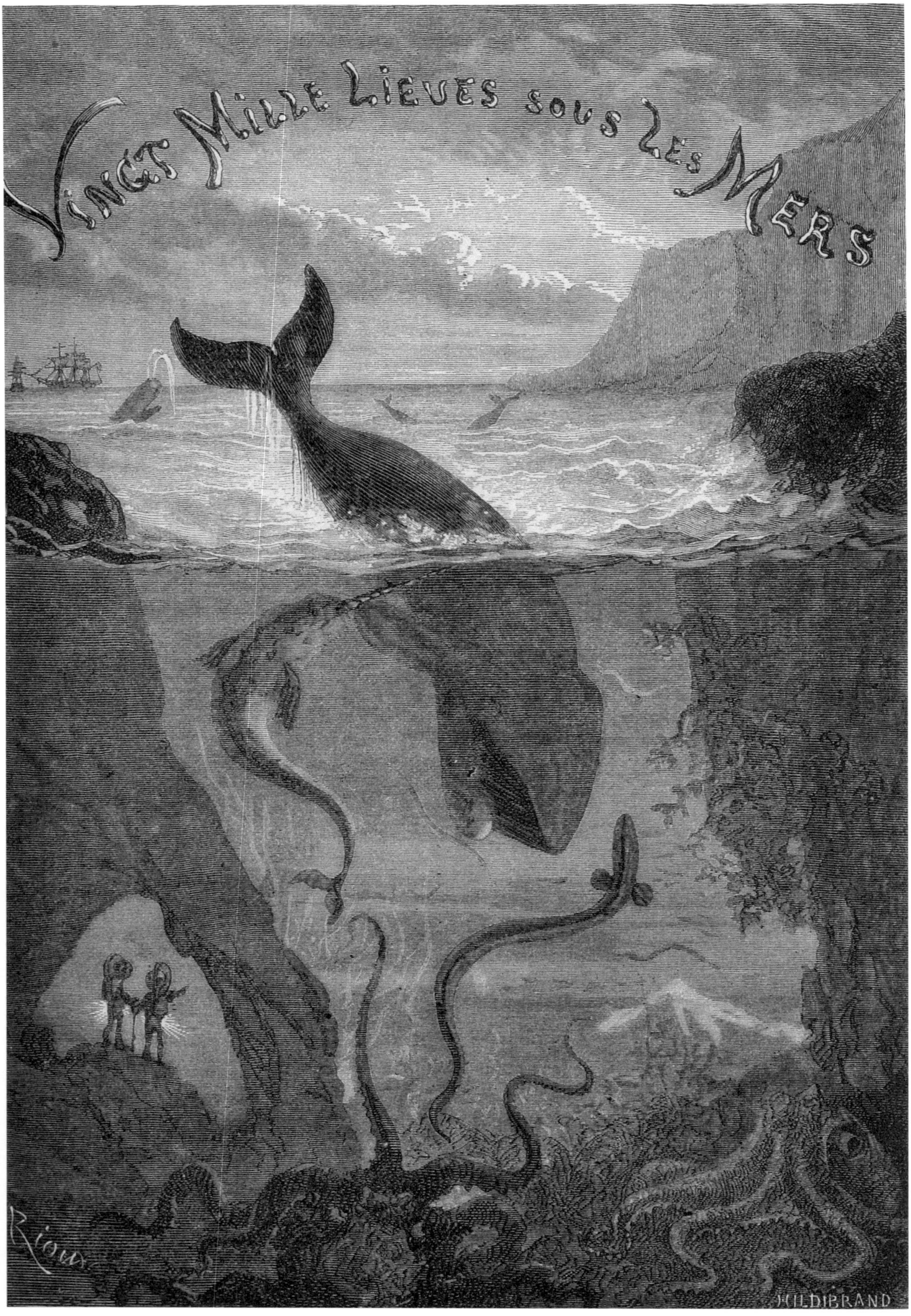

69 A
Jules Verne, Édouard Riou
and Alphonse Marie de Neuville (illustrations),
Twenty Thousand Leagues Under the Sea,
frontispiece, Paris: Hetzel, 1871
Paris, BnF

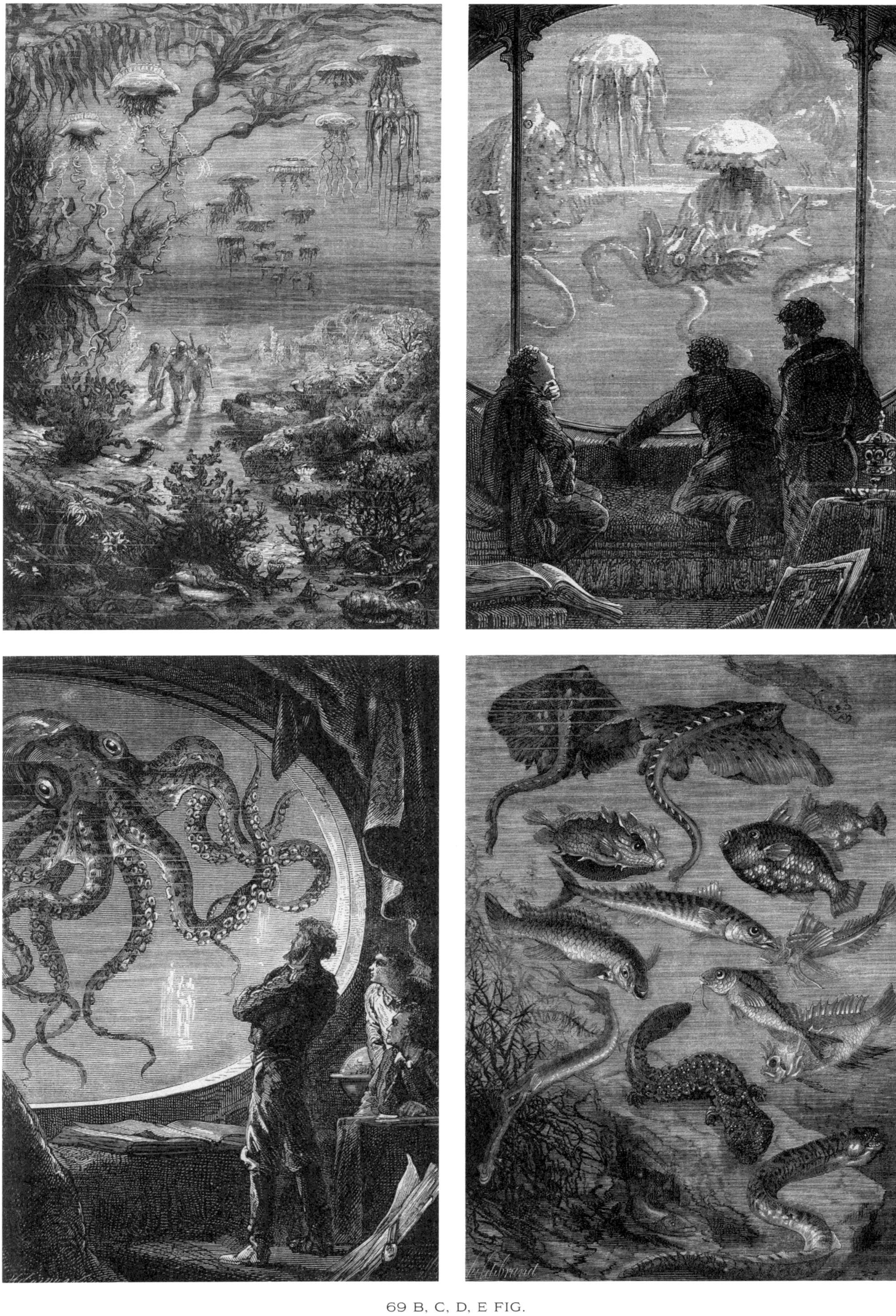

69 B, C, D, E FIG.
Jules Verne, Édouard Riou and Alphonse Marie de Neuville (illustrations),
Twenty Thousand Leagues Under the Sea, Paris: Hetzel, 1871
"Underwater landscape of Crespo Island"
"A window opened into this unexplored abyss"
"It was a squid of colossal dimensions," "Marine animals"
Paris, BnF

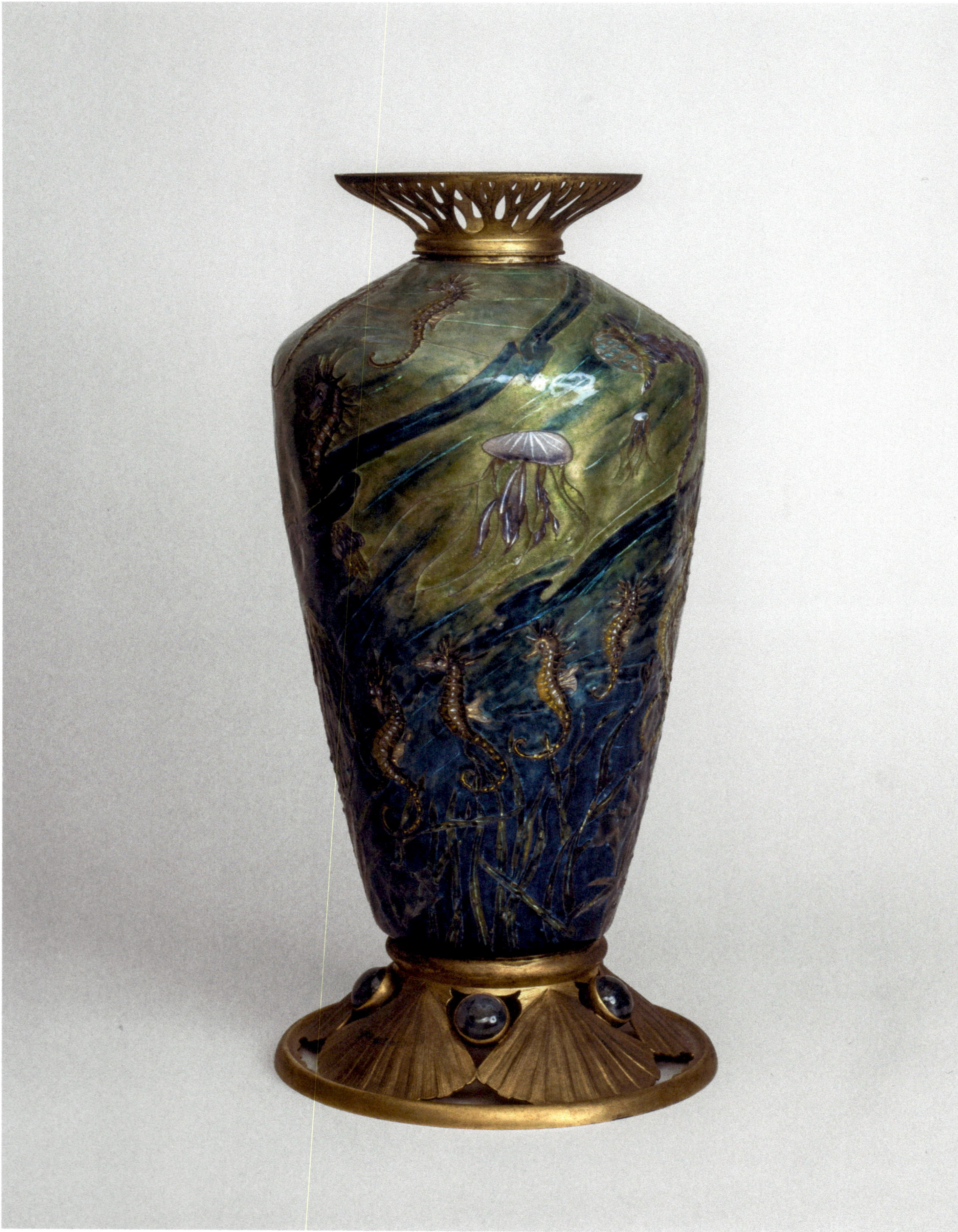

70
Eugène Feuillâtre,
"Sea" Vase, about 1912
Translucent enamel on copper,
37.6 × 17.6 cm
Paris, Petit Palais, musée des
Beaux-Arts de la Ville de Paris

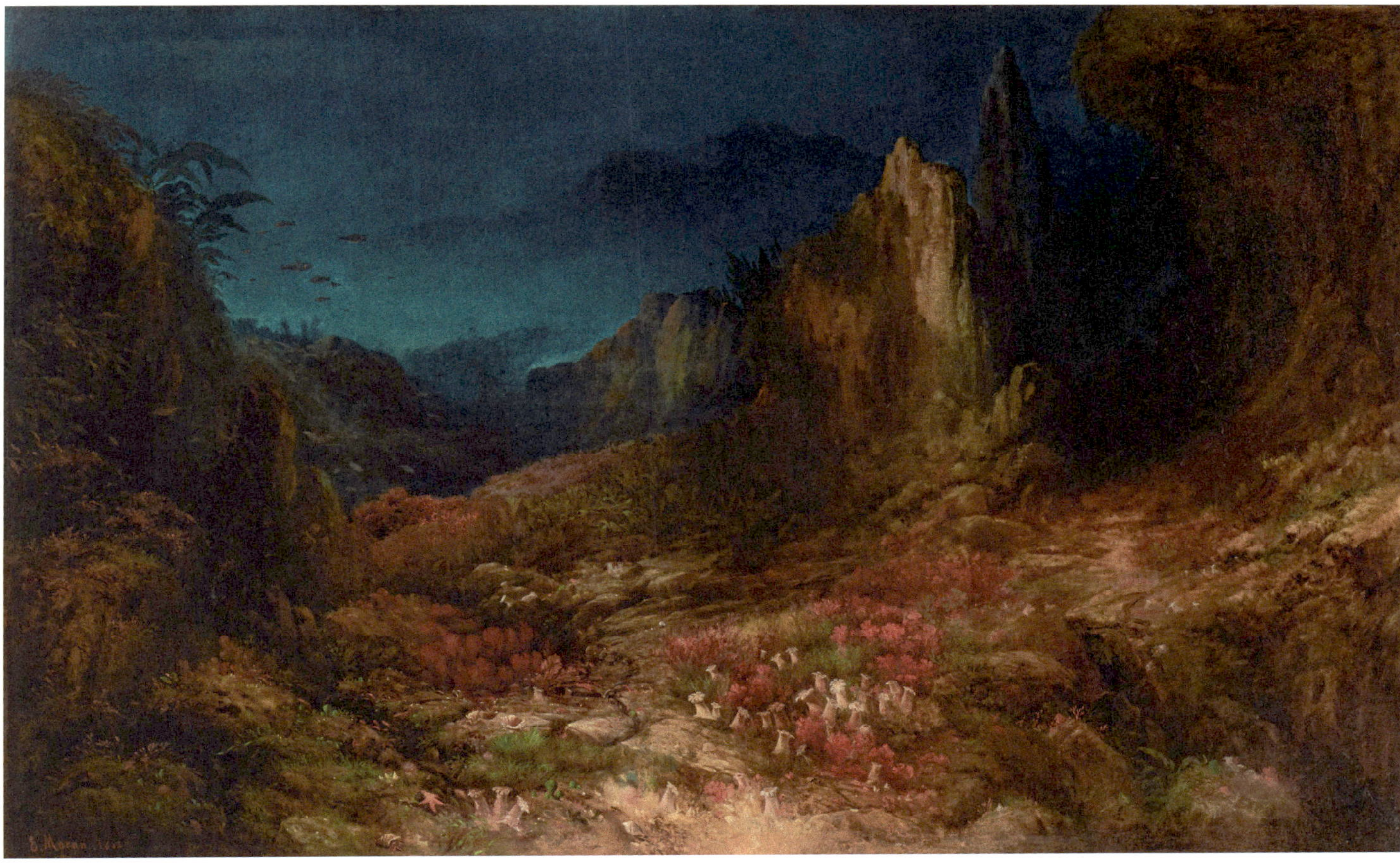

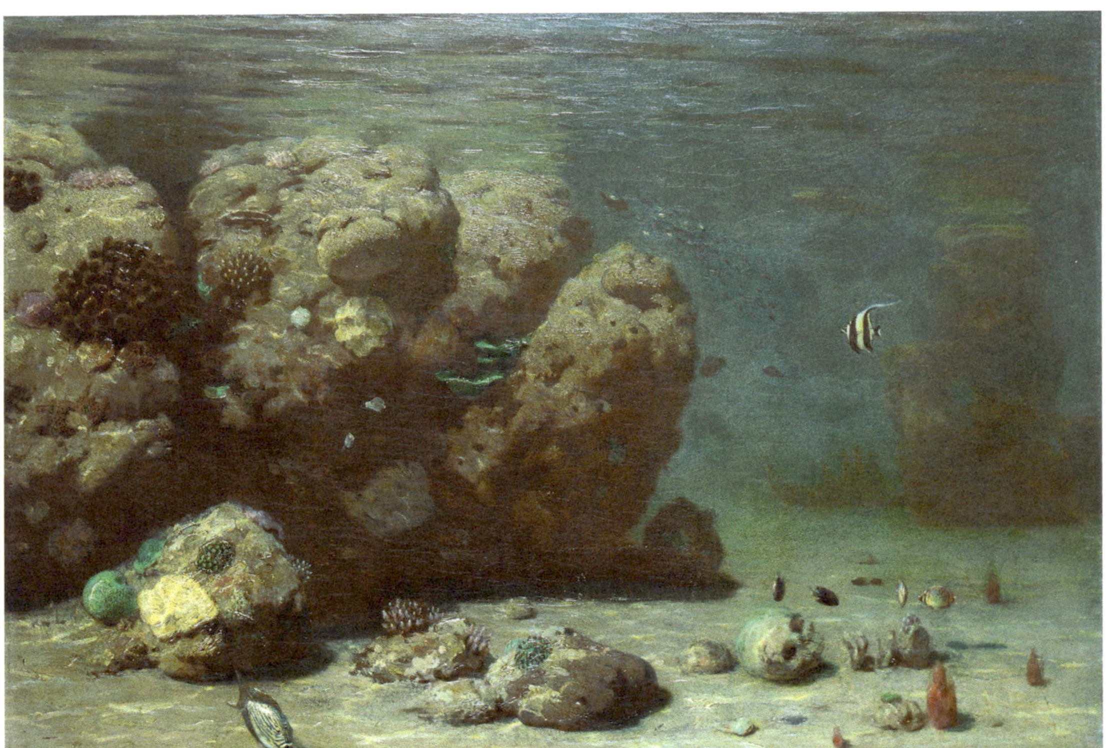

71
Edward Moran,
The Valley in the Sea, 1862
Oil on canvas, 102.9 × 162.6 cm
Indianapolis, Indianapolis
Museum of Art at Newfields

72
Eugen Ransonnet-Villez,
Underwater Landscape, 1864
Oil on canvas, 50 × 70 cm
Vienna, Naturhistorisches Museum

P. 106–107

73 A, B
Philip Henry Gosse, *Actinologia Britannica: A History of the British Sea-Anemones and Corals*, pls. 2 and 6, London: Van Voorst, 1860
Paris, MNHN

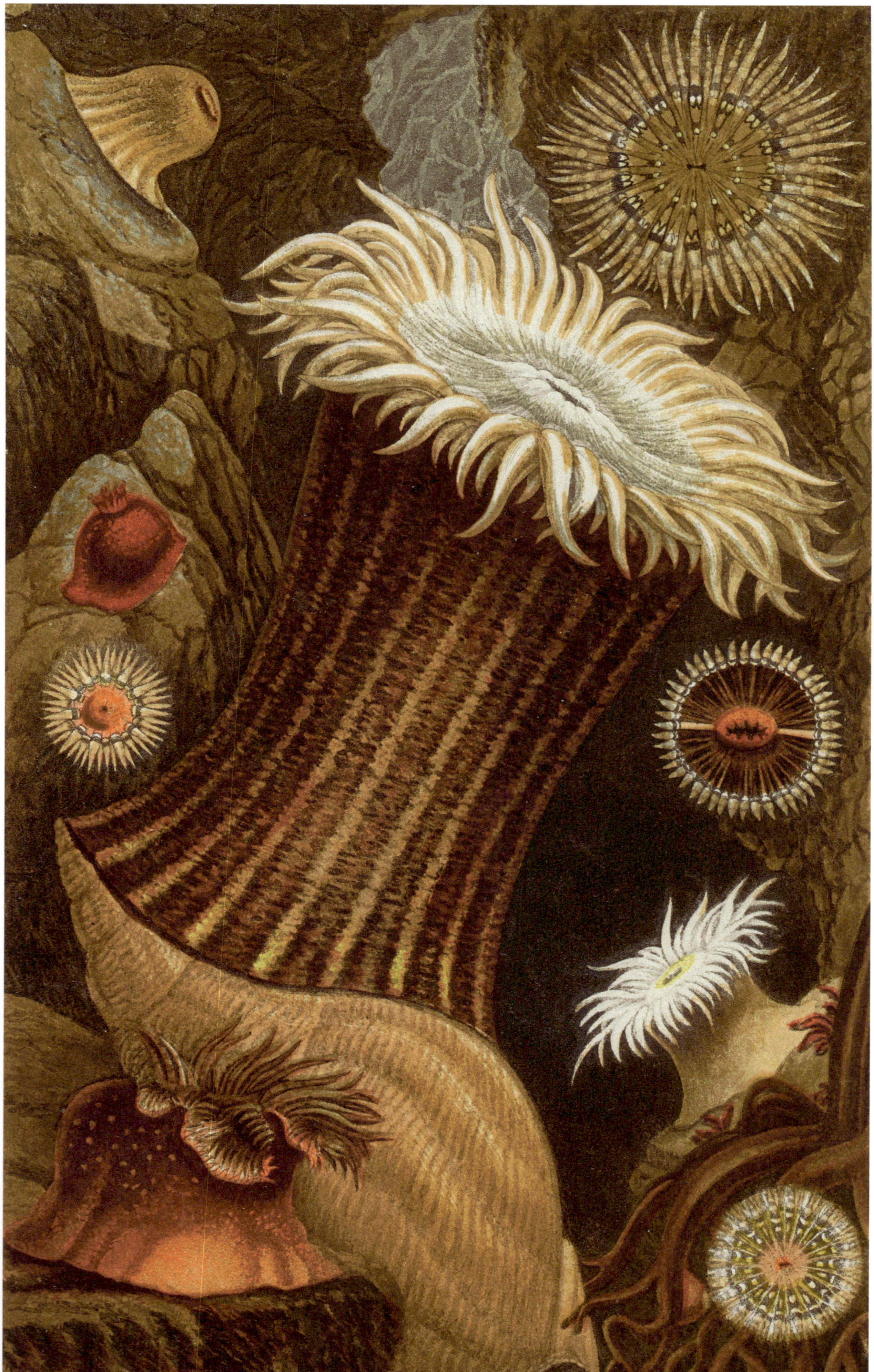

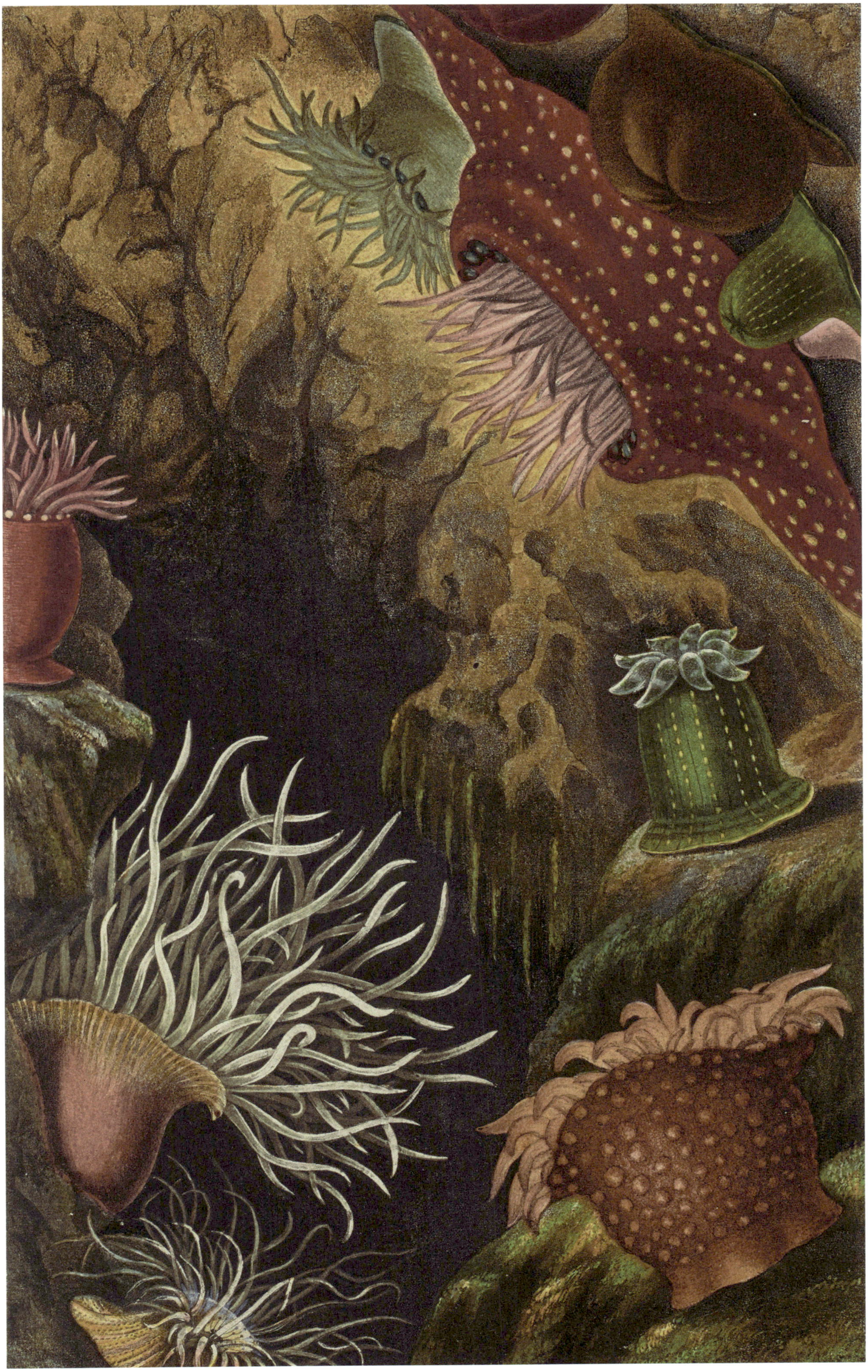

UNLOCKING NATURE'S SECRETS:

TWO ARTISTIC AND SCIENTIFIC COLLECTIONS FEATURING FRUITS FROM THE ÎLE DE FRANCE IN THE EARLY NINETEENTH CENTURY

JOËLLE GARCIA AND ALICE LEMAIRE

With the burgeoning of voyages of exploration and trade in the nineteenth century, Europeans began discovering the flora of faraway lands whose exoticism appealed to their romantic taste for nature. Born of a passion that was both scientific and artistic, the *Carporama* created by Louis-Marc-Antoine Robillard d'Argentelle is a unique collection of 112 life-size wax models of tropical plants and fruits, created between 1802 and 1826 on the Île de France (present-day Mauritius). The series of paintings of fruit[1] in the collections of the Muséum national d'histoire naturelle, which was also executed in Mauritius between 1801 and 1810 by Michel Garnier, reflects the same desire to record the diversity of life. Taking plants in their natural environment as a model, the two travelling artists and enlightened amateur botanists sought to bring to light the secrets of a unique natural world.

THE *CARPORAMA* BY ROBILLARD D'ARGENTELLE

Captain Louis-Marc-Antoine Robillard d'Argentelle, a member of General Claude Théodore Decaen's expedition to India, settled in 1802 on the Île de France, where he marvelled at the spectacular native plants as well as those that were cultivated in the Pamplemousses Botanical Gardens.[2] Having discovered the art of wax modelling during the Italian campaigns, he reproduced plants unknown in Europe that were either useful for food and industry or simply striking in their natural state.

During his excursions into the forest, in the preparation of materials and the creation of waxes, Robillard d'Argentelle was assisted by "a mulatto named Mercury."[3] The botanist-artist represented the elements that characterize each species in life-size cut-outs. Using a secret recipe, he mixed resins, wood and fibres with coloured wax. Naturalist René-Primevère Lesson praised the accuracy of the models: "The artist was able to put together a composition that combined the lustre, freshness and realistic appearance of wax, with the solidity of a metal that would ensure the long-lasting preservation of these fruits. The unique talent of this artist enabled him to render the ephemeral organs of the flowers, the texture and facies of the leaves, the fruits in their various stages of growth, the branches and their patterns, the hues, roughness and nuances of bark, etc. In a word, one could say Mr. Dargentel had unlocked nature's secrets"[4] (74, 75, 77, 78).

When the island became a British Crown Colony in 1810, Robillard d'Argentelle continued his work, intending to show it to the public. However, he returned to France in 1826 and died before achieving his goal. It was his nephew Louis Adolphe Humbert de Molard who took charge of the exhibition, which opened in Paris in December 1829. It displayed 112 plant and fruit models under the title *Carporama*, a word derived from the Greek meaning fruit display. A brochure presented the history, characteristics and properties of each plant.[5] Also on display were objects that reflected the daily life of the Indigenous populations.

Visitors discovered evocative types of fruit previously known only through botanical descriptions, illustrations, travel stories or processed products. The press praised the exhibition as being of interest to botanists, artists in search of realism, and the merely curious alike.

THE PAINTINGS OF MICHEL GARNIER

Michel Garnier had not intended to spend time on the Île de France. In October 1800, he embarked on a scientific expedition to the Austral Lands led by Commander Nicolas Baudin, but left the undertaking in August 1801. This regular visitor to the Salons[6] was not the only artist member to withdraw, as François Péron points out in the expedition diary.[7]

From 1801 to 1810, genre painter Michel Garnier lived the life of an Indian Ocean naturalist and devoted himself to depicting the fruits of the Île de France and the Île Bourbon (present-day Réunion island). Transforming his sense of wonder into a concrete contribution to representing the plant world, he painted more than 200 fruits, evidently wishing to produce a body of scientific work (76, 79).

The paintings were executed on large sheets of paper that were easy to carry into nature for sketching subjects on the spot. They were then (no doubt on their return to Paris and with a view to selling) mounted on canvas by adding strips of paper to fit the size of the sheet to that of the stretcher, as recent restorations have shown.

WORKS OF ART IN THE SERVICE OF SCIENCE

Despite their identical approaches at the same period in time, we can only conjecture that Michel Garnier and Robillard d'Argentelle did actually meet.[8] However, a recent public sale of Garnier's *Study of a Grapefruit*—similar in its composition to the wax conserved at the Muséum and most probably having belonged to Robillard d'Argentelle[9]—provides a significant clue. A comparison of the paints and waxes strongly suggests that the two artists worked together using the same specimens. In representations of both grapefruits and avocados (75, 76), the wax models and the paintings show the fruit attached to the branch on very similar twigs, as well as a cross-sectional detail of the pit.

Beyond being united in their conception, the two groupings are also united in their destinies, having similar histories before entering the Muséum's collections several decades after their creation.

Upon returning to Paris, after a journey during which he lost all his possessions except for his treasured paintings, which he managed to save,[10] Garnier exhibited his work at the Salon of 1814 under the title *Partie d'une collection de 140 tableaux, représentant des fruits de l'Asie, de l'Afrique et de l'Amérique, acclimatés aux îles de France et de Bourbon* [Part of a collection of 140 paintings, representing fruits from Asia, Africa and America, acclimated to the Île de France and the Île

1 The Muséum has 127 paintings out of an estimated 140.
2 The first gardens for testing and acclimatization, created in 1770 by intendant Pierre Poivre.
3 "Collection de plantes et de fruits exotiques," *Le Courrier des tribunaux*, October 29, 1829, p. 1440.
4 René-Primevère Lesson, "Note sur une collection précieuse de fruits équatoriaux, modelés, avec une composition tenue secrète, par M. Dargentel, de l'île de France," *Bulletin des sciences naturelles et de géologie*, vol. 6 (1825), p. 72–75.
5 *Catalogue des fruits et des plantes modelés composant le Carporama*, [Paris], n.d.
6 Elvire de Maintenant, "Michel Garnier, peintre de genre sous la Révolution," *L'Estampille–L'Objet d'art*, no. 370 (2002), p. 76–82.
7 François Péron, *Voyage de découvertes aux Terres australes exécuté par ordre de sa majesté l'empereur et roi, sur les corvettes « Le Géographe », « Le Naturaliste » et la goélette « Le Casuarina », pendant les années 1800, 1801, 1802, 1803 et 1804* (Paris: Imprimerie impériale, 1807), p. 61, and see the essay in this catalogue, p. 86.
8 Gérard Aymonin, "Les tableaux de végétaux tropicaux de Michel Garnier, 1753–1819," in "Mahé de La Bourdonnais. La Compagnie des Indes dans l'océan Indien," special issue, *Cahiers de la Compagnie des Indes*, no. 4 (1999), p. 103–108.
9 Christie's sale, November 12, 2015.
10 Maintenant, "Michel Garnier, peintre de genre sous la Révolution," p. 76–82.

Bourbon].[11] He had tried unsuccessfully in 1812 and again in 1819 to sell the collection to the Muséum, which declined his offer due to lack of means and despite the scientific and artistic interest expressed in particular by the botanists Antoine-Laurent de Jussieu and René Louiche Desfontaines, as well as by Gerard van Spaendonck, the painter of the Vellum Collection. Garnier's collection would be added to the Muséum much later in two stages. In 1851, Adolphe-Théodore Brongniart bought ninety paintings, seen as worthy as the vellums, from a second-hand dealer: "The Professor of Botany requests that the Assembly acquire for the Botany collections and for the sum of 300 francs, to be taken from the vellum fund, a series of paintings representing fruits of the Tropics."[12] Twenty-five years later, Ernest-Théodore Hamy acquired another thirty-seven paintings at auction. It was only in 1898 that he identified their creator when he discovered a list, identical to that of the works kept in the Muséum, of paintings "executed by Mr. Garnier, painter and draughtsman, appointed by order of the French Government in the year 1800 to the long-term expedition commanded by Captain Baudin."[13]

The Académie des sciences recognized the scientific value of the *Carporama* as early as 1826. During the exhibition, the press emphasized the usefulness of the models, "from which students and manufacturers could benefit greatly if they were displayed in the Muséum d'histoire naturelle in Paris, where there is no *carporama* and where only dried fruits that have lost their shapes and colours can be observed."[14] After the July Revolution of 1830, the Robillard d'Argentelle family proposed that the State acquire the collection. The teachers at the Muséum successively declined a number of offers, because of the price being asked and the space needed to preserve it. The family finally donated the 112 wax models to the Muséum in 1887. Since then, this singular collection, which presents the exact morphology of fruits that are impossible to preserve in herbaria or in alcohol, continues to arouse the interest of scientists and art historians.[15]

The paintings and models created by these naturalist artists based on living organisms gave vivid expression to the diversity of tropical flora, and they opened new horizons for the artists' contemporaries. The natural forms, as exuberant as they are suggestive, provide a fertile ground for reflection on the diversity of nature's evolution. At the Muséum, these two series of artistic works have served science and contributed to the dissemination of knowledge since the end of the nineteenth century. Thus, unique perspectives intersect in the vast undertaking that unites scholars and artists over the centuries in recording the diversity of living things.

11 *Explication des ouvrages . . . des artistes vivans, exposés au Musée royal des arts* (Paris: Dubray, 1814).

12 Minutes of the Assemblée des professeurs [Professors running the museum], Archives du Muséum, AM 4. On the Muséum's vellums, see the essay in this catalogue, p. 60.

13 Ernest-Théodore Hamy, "Les peintures de Michel Garnier au Muséum d'histoire naturelle," *Bulletin du Muséum d'histoire naturelle*, vol. 4, no. 8 (1898), p. 336–344.

14 "Carporama," *Le Corsaire*, April 13, 1830, p. 2.

15 Monique Keraudren-Aymonin, "Le *Carporama* de L.M.A. Robillard d'Argentelle," *Bulletin du Muséum national d'histoire naturelle*, series no. 4, no. 1 (1979), p. 117–149; Michel Lemire, *Artistes et mortels* (Paris: Chabaud, 1990).

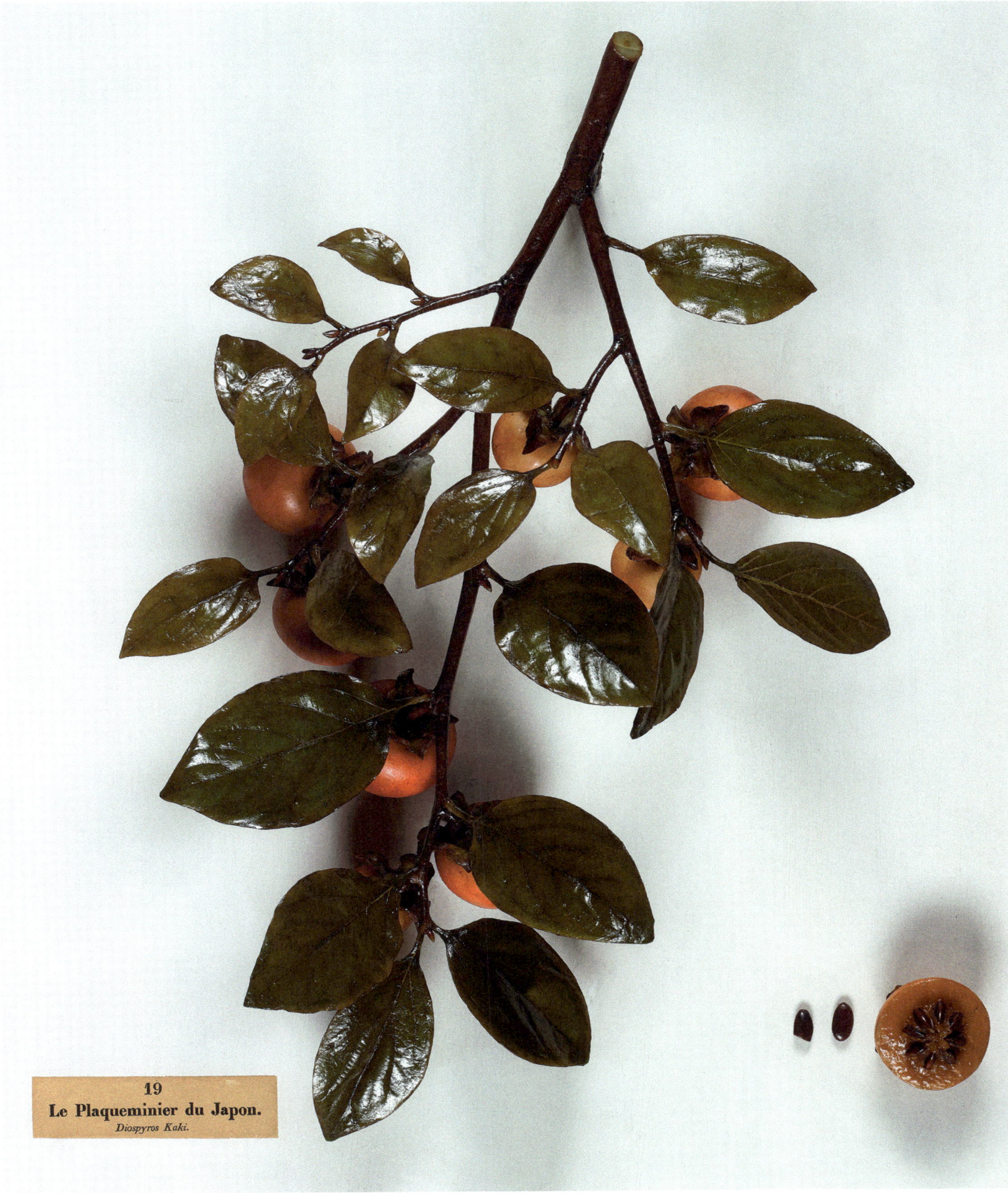

74
Louis-Marc-Antoine
Robillard d'Argentelle,
Japanese Persimmon,
between 1802 and 1826
Wax, wood, metal, 52 × 45 × 16 cm
Paris, MNHN

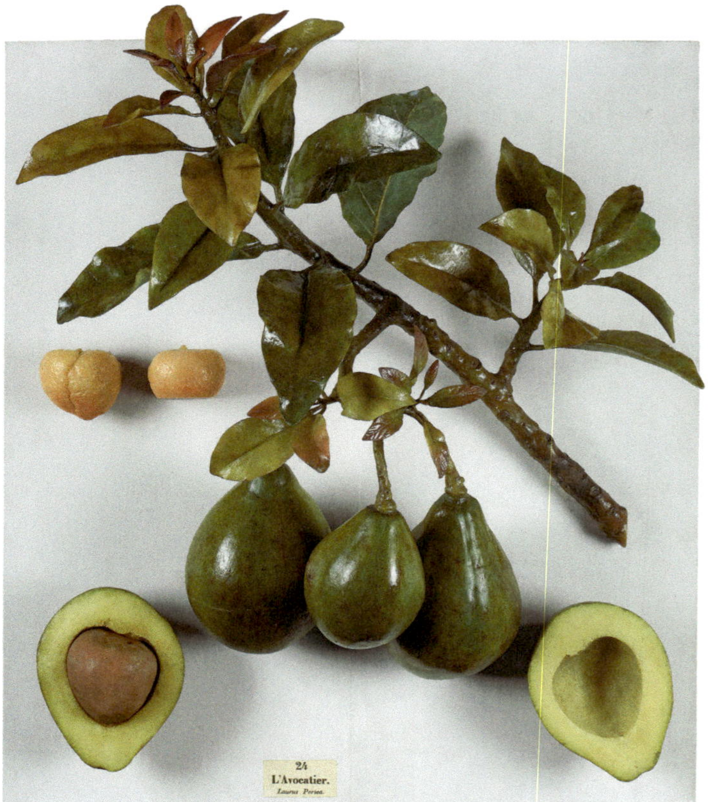

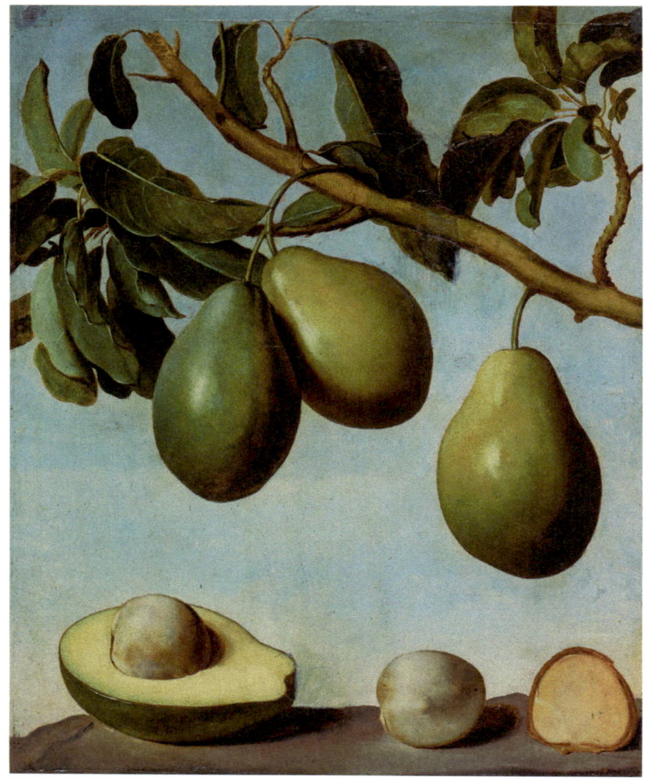

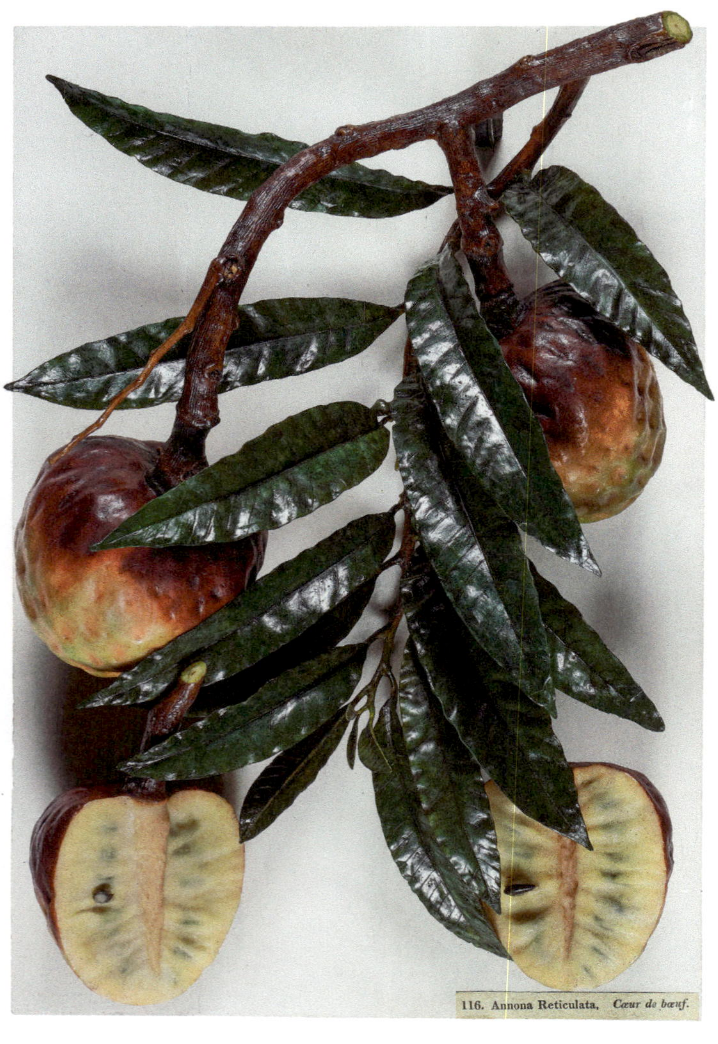

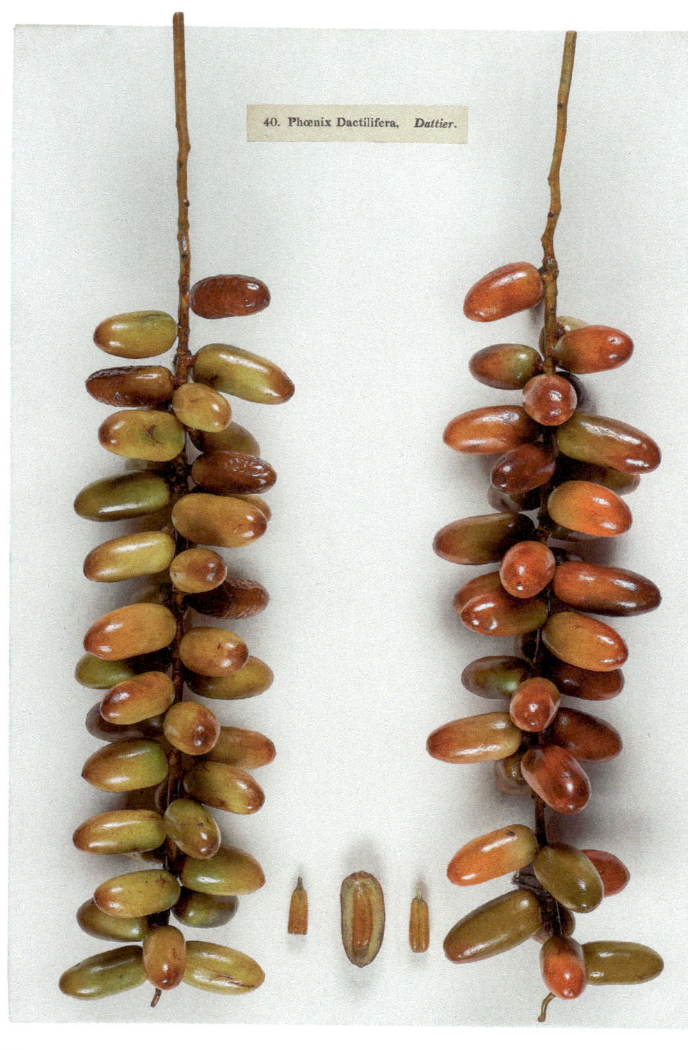

75 FIG., 77, 78
Louis-Marc-Antoine Robillard d'Argentelle,
between 1802 and 1826, Wax, wood, metal
Avocado Tree, 24 × 45.5 × 52 cm,
Bullock's Heart, 45 × 30 × 14 cm,
Two Clusters of Dates, 45 × 30 × 10 cm
Paris, MNHN

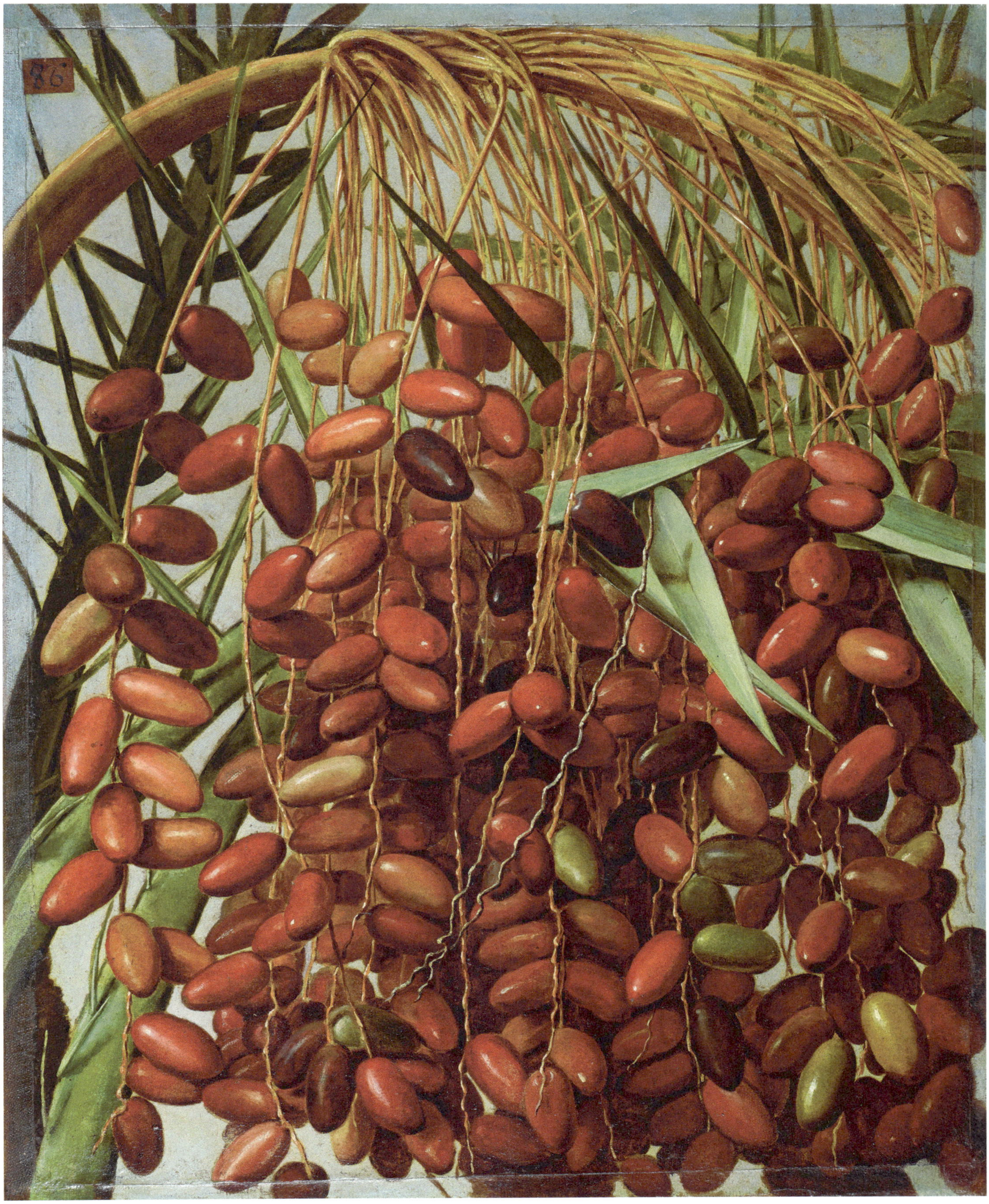

76 FIG. 79
Michel Garnier, between 1801 and 1810
Oil on paper mounted on canvas
Avocado Tree, 58 × 47 cm,
Date Palm, 58.5 × 46.5 cm
Paris, MNHN

VERNE REINVENTED BY MÉLIÈS

MARCOS UZAL

Although cinematography was invented a decade before Jules Verne's death, there is no evidence to suggest that the author ever went to the cinema or that he was even interested in it. And yet, the Lumière brothers' invention, part science and part magic, was to perpetuate his imaginative works in two seemingly contradictory ways: on the one hand, through the exploration of uncharted territories pioneered by the brothers and their cameramen, and on the other hand, through the creation of marvellous worlds in films produced by the wizard Georges Méliès.

Beginning with his stage plays, the filmmaker frequently took inspiration directly from Verne's books but deliberately deviated from the original. Méliès's famous *A Trip to the Moon* of 1902 goes back to the beginning of Verne's *From the Earth to the Moon* of 1865—the Astronomers' Club and the building of the cannon that will fire the rocket vehicle—but completely changes the second part of the book. Whereas in Verne's work the space travellers are content to look at the planet without visiting it, in Méliès's film they land on it and discover a wonderful world inhabited by a hostile people, the Selenites. "I then imagined, in using the process of Jules Verne, (gun and shell) to attain the moon, in order to be able to compose a number of original and amusing fairy pictures outside and inside the moon, and to show some monsters, inhabitants of the moon in adding one or two artistical effects (women representing stars, comets, etc) (snow effect, bottom of the sea, etc)."[1] It should be noted that the second part of the film was probably inspired by another novel—itself possibly conceived as a sequel to Verne's work—*The First Men in the Moon* (1901) by H. G. Wells, in which we find a population of Selenites. In any event, Méliès's statements sum up the cavalier use of fantasy that allowed him to detach Verne's oeuvre from its scientific basis, forgoing any of the verisimilitude required in true science fiction, in favour of the French *féerie* tradition and popular entertainment. The same is true of his *Under the Seas* of 1907, in which the ocean floor resembles the lunar landscape in his *A Trip to the Moon*: a luxuriant baroque landscape populated mainly by fish women in which one might move and dance as if on a stage (81). Likewise, in adapting even more freely Verne's *The Adventures of Captain Hatteras* of 1866 in his film of 1912, *The Conquest of the Pole*, he transformed the author's fully documented and realistic work into a comic spectacle: where the book ends in madness and cannibalism, Méliès gives us a snow giant who is more comical than frightening (80).

Apart from the fact that he never took Verne's novels seriously, Méliès is radically different from the writer in his total lack of interest in anything that is not human. The world that Verne's characters generally have to deal with is, on the contrary, a natural world that has evolved without man, with shapes and proportions all the more extraordinary because they are not remotely adapted to our species. Those not content just to contemplate them but venture into them, dive into a world that is literally inhuman. But in the fantastic de-dramatized

1 Georges Méliès, "Reply to a Questionary," 1930 (in English); quoted in André Gaudreault, "Theatricality, Narrativity, and Trickality: Reevaluating the Cinema of Georges Méliès," in *The Fantastic Voyages of the Cinematic Imagination: George Méliès's Trip to the Moon*, ed. Matthew Solomon (Albany: State University of New York Press, 2011), p. 37. See also http://collections.cinematheque.qc.ca/articles/reply-to-a-questionnary/.

80
Georges Méliès,
The Conquest of the Pole
France, 1912
Short silent film in black and white

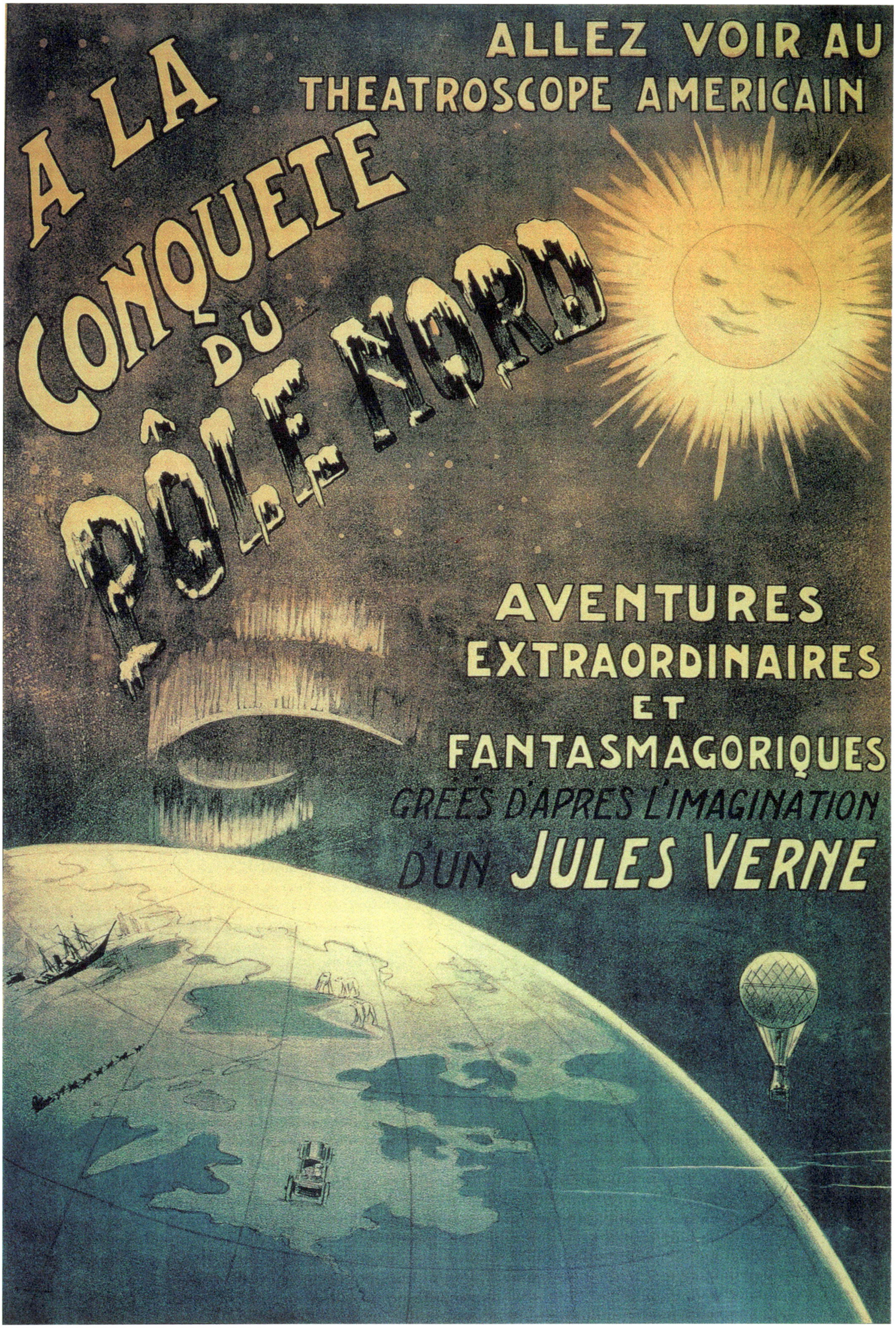
ALLEZ VOIR AU
THEATROSCOPE AMERICAIN
A LA
CONQUETE
DU
PÔLE NORD
AVENTURES
EXTRAORDINAIRES
ET
FANTASMAGORIQUES
CREES D'APRES L'IMAGINATION
D'UN JULES VERNE

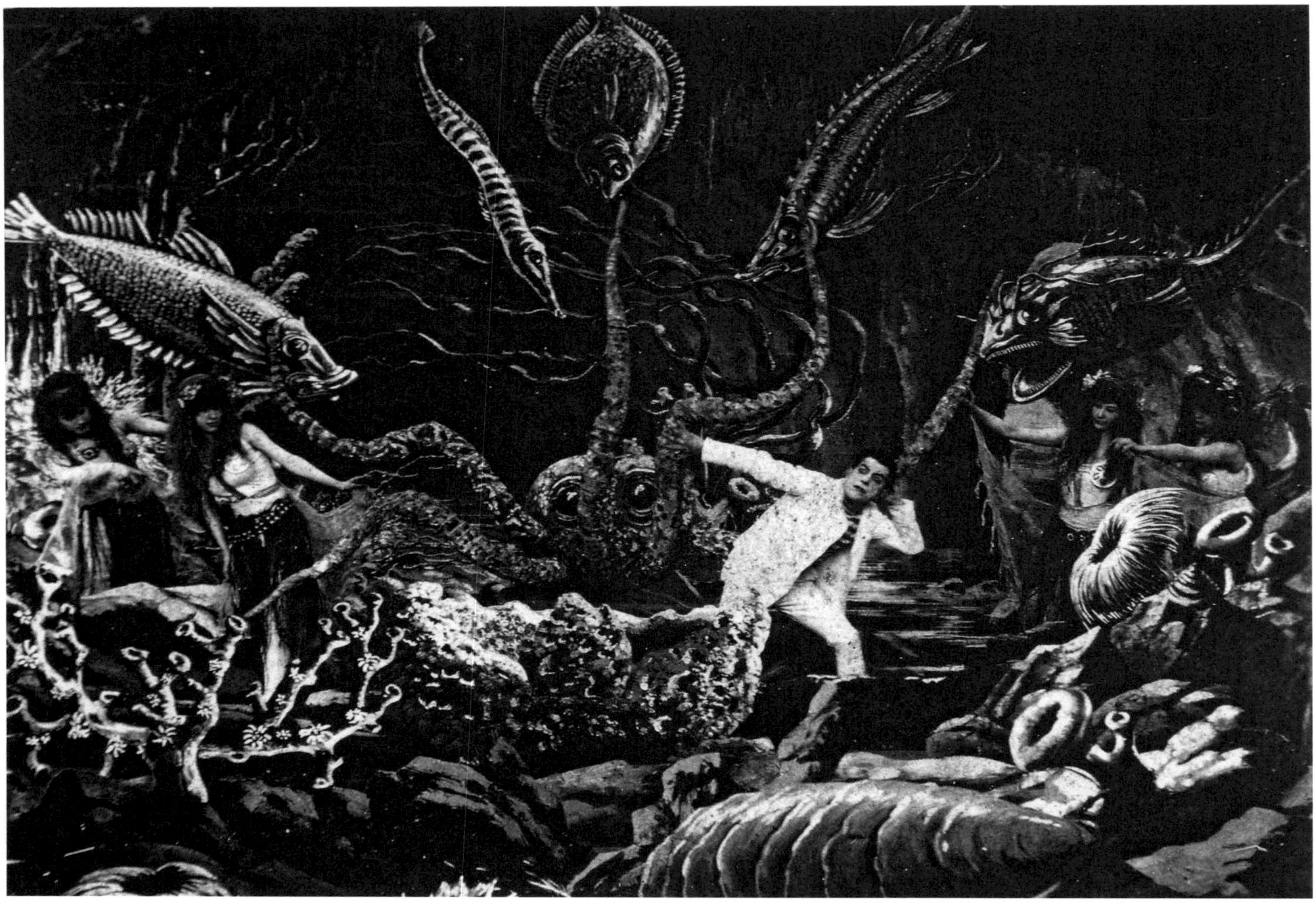

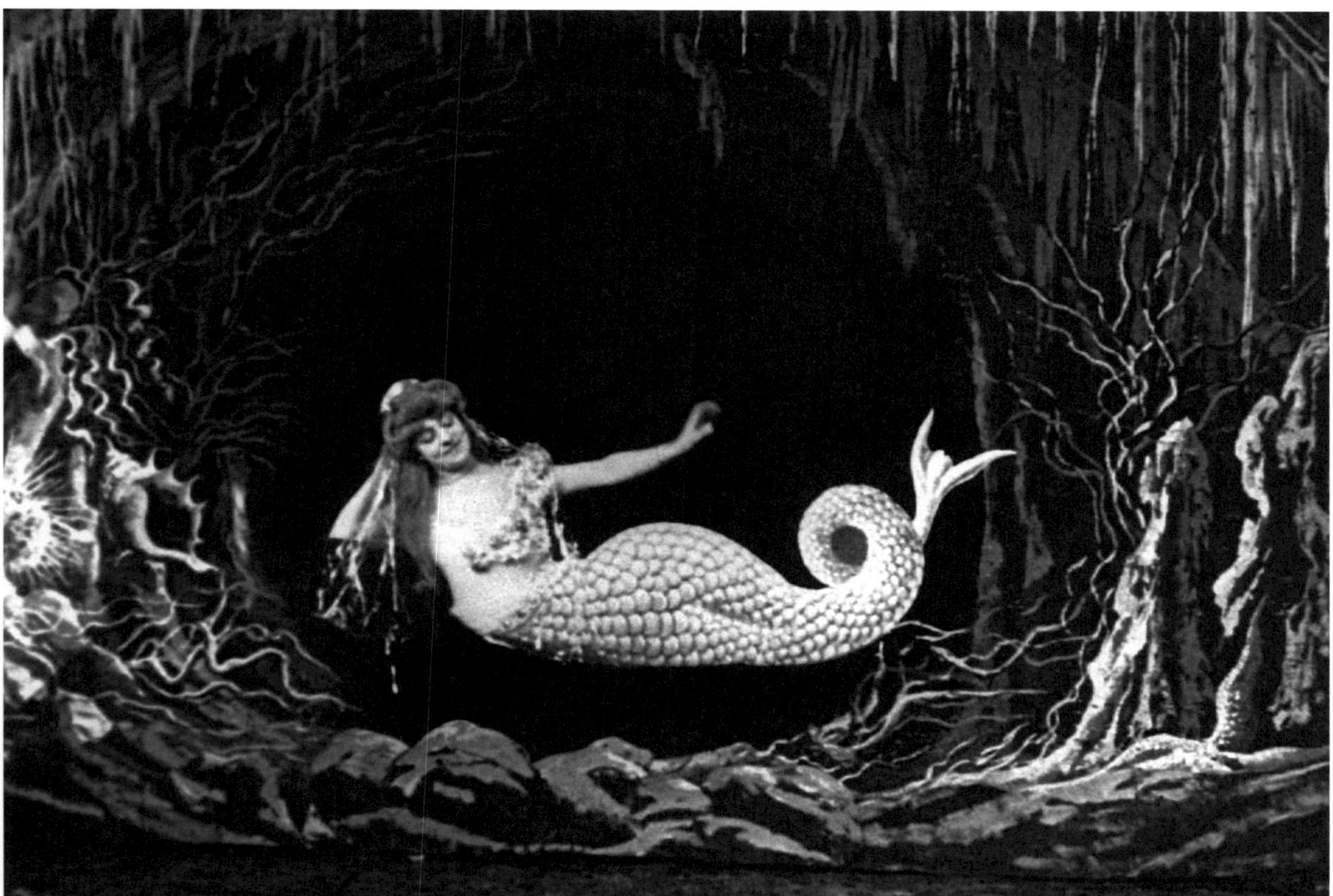

world of Méliès, everything comes back to man. And even though fishes and a giant octopus finally appear in his *Under the Seas*, they turn out to be merely figments of a nightmare. His cosmos and his nature are entirely anthropomorphic: the stars are women, the Moon is a face (that gets a rocket in its eye), the Selenites are masked dancers in tights, the bleakness of the North Pole is personified by a bearded giant, the sea is inhabited by mermaids (81, 82), and so on. In this sense, it might be said that the world of Méliès is more ancient than that of Verne; it is an allegorical world that makes fun of science and invents joyous gods.

The techniques of cinematography would have to evolve further to achieve a balance between science and imagination sufficient to convey something of the spirit of Verne's novels. The adaptations of the 1950s and 1960s—Richard Fleischer's *Twenty Thousand Leagues under the Sea* (1954), Henry Levin's *Journey to the Center of the Earth* (1958), Cy Endfield's *Mysterious Island* (1961)—succeeded in part by combining natural backgrounds with studio sets and creating gigantic monsters simply by changing the proportions of real animals. But the films closest to the world of Jules Verne were perhaps the scientific documentaries that showed us the natural world as we had never seen it before. Essentially, *The Silent World* (1956) by Captain Cousteau and Louis Malle, with its experimental machines and its unprecedented undersea panoramas, might well be a sort of adaptation of Verne's *Twenty Thousand Leagues Under the Sea*. Jean-Luc Godard, excited by Haroun Tazieff's documentary *The Devil's Blast* (1959), saw in these staggering scientific images a supreme form of poetry and in the filmmaker an heir to Phileas Fogg, one of those artist adventurers who "don't know what it is to go too far." Here, the world explored is fascinating because it is indomitable: a place where Méliès went beyond Verne in seeing the cosmos, the oceans and all the creatures of the universe through the eyes of a child.

81
Georges Méliès,
Under the Seas
France, 1907
Short silent film in black and white

82
Georges Méliès,
The Mermaid
France, 1904
Short silent film in black and white

III THE ANTIQUITY OF THE WORLD: THE DISCOVERY OF NATURE'S TIME

THIERRY HOQUET

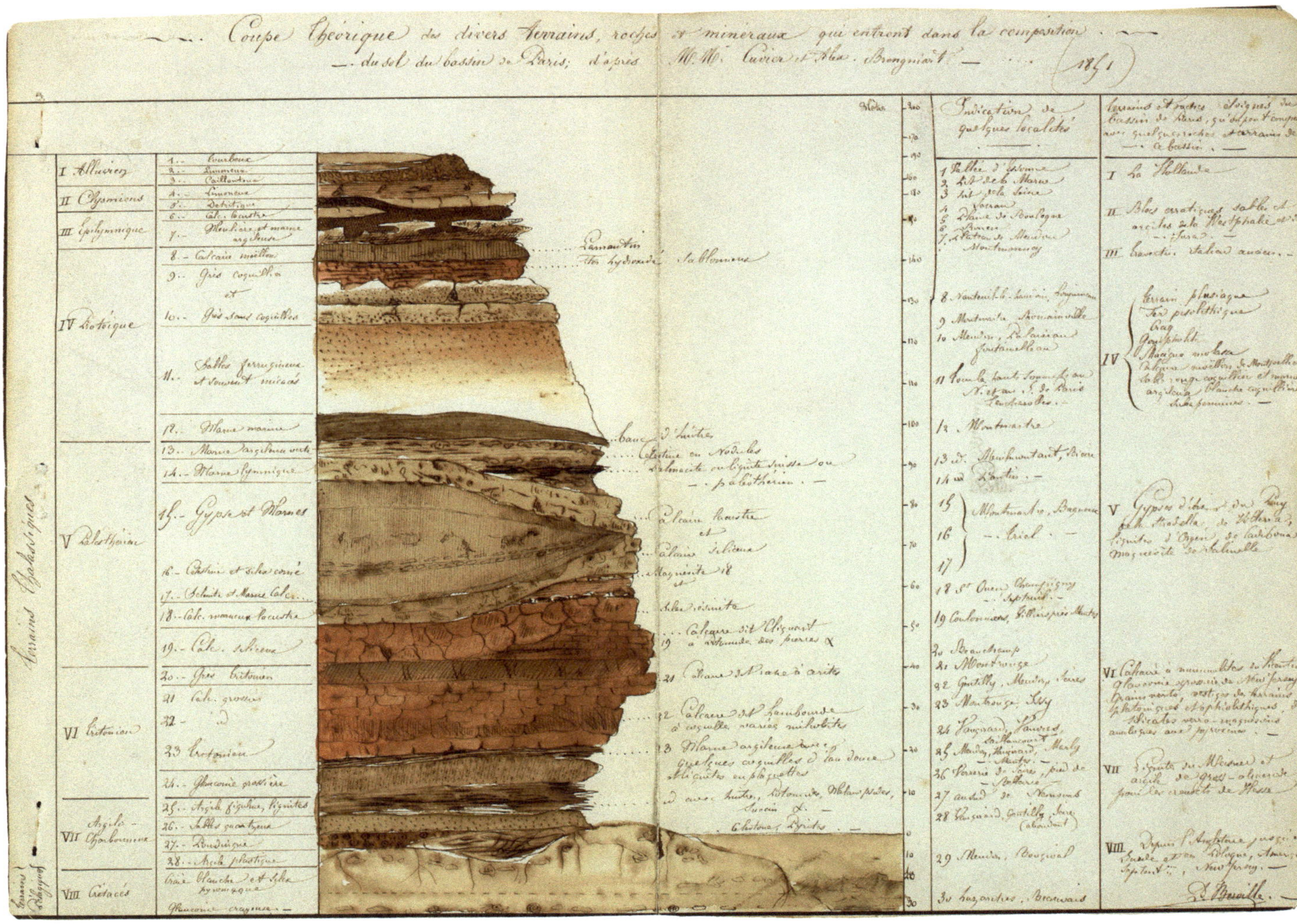

83
Georges Cuvier and Alexandre Brongniart, *Coupe théorique des divers terrains, roches et minéraux qui entrent dans la composition du sol du bassin de Paris*, Paris: Clerget, 1851
Off-print in watercolour, 24 × 32.5 cm
Paris, MNHN

Our vision of an ancient world that saw major changes over the course of its history is the result of the discovery of nature's time—which we will trace here.[1] How did the "time" dimension come to be added to our vision of the physical and biological world? How were scientists able to dispel the clouds of uncertainty that once surrounded any speculation about nature's past and seemed to make it impossible?

Humans began discovering time through the minds of children, who live in the present and can conceive only short spans of time, almost visible time: "A rose, she lived as long as roses live, the span of a morning," says Malherbe in his famous poem. However, such temporality is far from that of the billions of years that separate us from the Big Bang in contemporary cosmology. And that road is strewn with obstacles of several types.

MULTIPLE OBSTACLES

That there is time in nature at first appears evident: the succession of days and the return of the seasons forcefully proclaim it. But, at our level, the world is composed of regularities: its time is cyclical, it is ageless. Its revolutions are understood as rotations that do not make history. It is a time without catastrophes, without discontinuity, without depth. By contrast, the time that concerns us here is a long, uneven time, in which countless small- and large-scale events occur, in which revolutions are ruptures.

Whereas natural time, over the span of a few dozen human generations, is cyclical and thereby trapped in the circle of repetition, linear time appears to be a distinguishing feature of human societies: only they are deemed to have *a history*. Accordingly, the first obstacle to a concept of nature's time stemmed from the opposition between *cyclical time* and *linear time*, *nature's time* and *human time*. In the classic view of time, humans had their reigns, their battles and their generations, but nature remained a static setting. For readers of the Holy Scriptures, only two events had occurred: the Creation and the Flood (5, P. 25; 84). Everything known about the world had to be attributed to one or the other of those events.

But the fixed, orderly conception of nature that prevailed around 1700 was also the product of the mathematical physics of Descartes, Galileo and Newton and its invariably reversible movements. At the end of his *Opticks*, Newton establishes the classic image of nature: solid, "massy" particles whose motion is governed by the laws occasioned by the force of inertia, and by certain active principles such as gravity, magnetism, electricity and fermentation.[2] These principles operate according to general regularities including the inverse square law of gravitation. The stability or uniformity of the physical world is expressed, on the one hand, in the regularity of planetary orbits and, on the other, in the regularity of animal forms through the process of generation.

Finally, it was not enough to accept the idea that "the world," understood as the context in which human history unfolds, had taken shape over a long period of time. There was still a psychic obstacle having to do with the length of time elapsed between the birth of the world and the present: our psychological inability to imagine this time had to be overcome. In 1834, the British geologist Henry Thomas De La Beche affirmed that our relationship to time is distorted by our mental habits: "Measuring time as man does in the minute manner suited to his wants and conveniences, a few thousand revolutions of the earth in its orbit appear to him to comprise a period so considerable, that he feels it difficult to conceive that great lapse of time which geology teaches us has been necessary to produce the present condition of the earth's surface."[3]

In a section devoted to the lapse of time,[4] Darwin invites his readers to imagine "how incomprehensibly vast have been the past periods of time"—a crucial condition of his theory. Take a drop of water falling on a rock: at first glance it has no effect of any kind. But give it enough time and you will see that the continuous fall of drops ultimately digs into the stone. Similarly, waves break against a cliff face only a few hours a day, and yet, in the end, they abrade and erode the rock. Conversely, initially imperceptible deposits eventually form remarkably thick layers.

Thus, geology gave us our first lesson in time: "No one but a practised geologist can really comprehend how old the world is, as the measurements refer not to

1 Stephen Toulmin and June Goodfield, *The Discovery of Time* (London: Hutchinson, 1965).

2 See Isaac Newton, *Opticks or, a Treatise of the Reflections, Refractions, Inflections, and Colours of Light* [1730, 4th ed.] (New York: Dover Publications, 2012), Query 31, p. 376.

3 Henry Thomas De La Beche, *Researches in Theoretical Geology* (London: Knight, 1834), p. 397.

4 Charles Darwin, "On the Vast Lapse of Time, As Inferred from the Rate of Deposition and of Denudation," in chapter 9 of *On the Origin of Species by Means of Natural Selection* (London: John Murray, 1859), p. 282.

5 Charles Darwin, Notebook E, p. 125e, transcribed in *Charles Darwin's Notebooks, 1836–1844*, ed. Paul H. Barrett et al. (London: British Museum; Ithaca, NY: Cornell University Press, 1987), p. 432.

6 Addition to the 5th edition [1869], in Charles Darwin, *The Origin of Species: A Variorum Text*, ed. Morse Peckham (Philadelphia: University of Pennsylvania Press, 1959), p. 485.

7 See, among others, Johannes Marsham, *Canon Chronicus Ægyptiacus, Ebraicus, Graecus* [1672], 2nd ed. (Franeker: Leonard Strick, 1696).

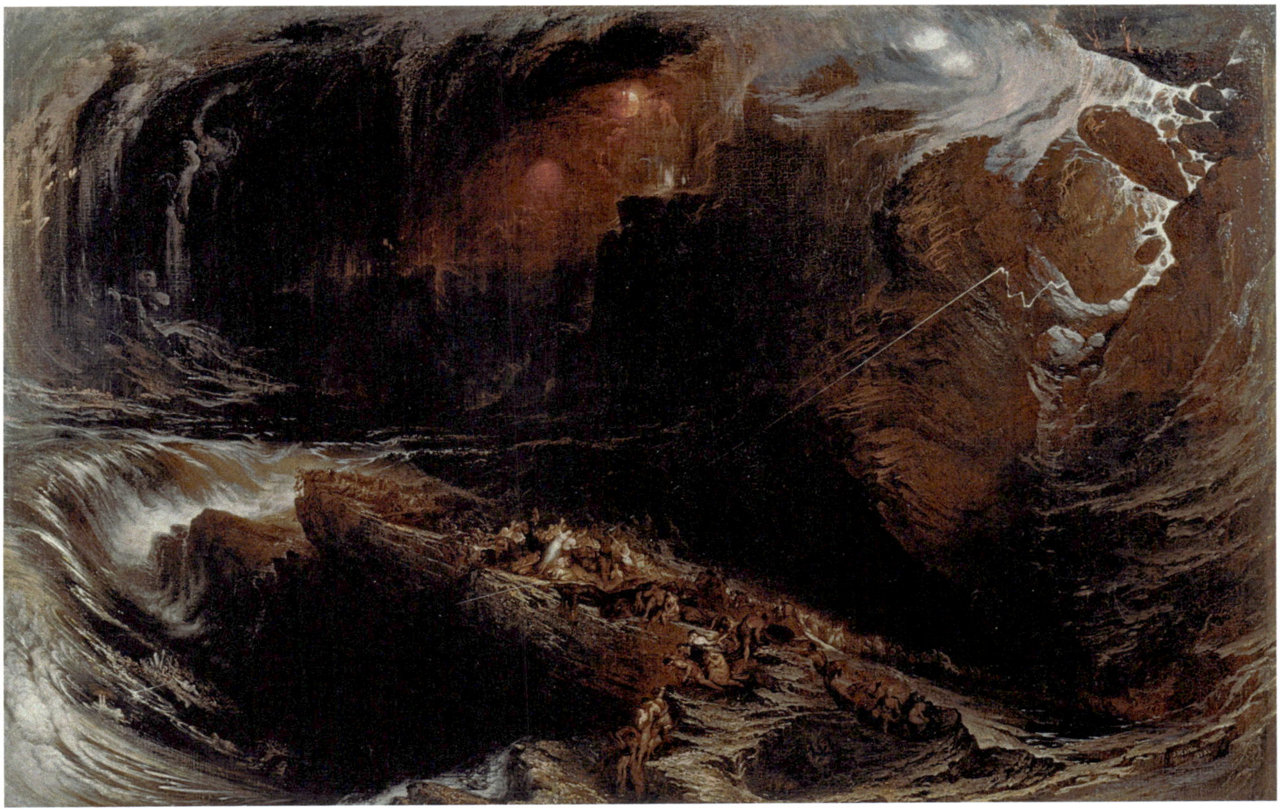

revolutions of the sun & our lives, but to [a] period necessary to form heaps of pebbles &c. &c.," wrote Darwin in one of his notebooks. The time of life forms came second: "The succession of organisms tells nothing about length of time, only order of succession."[5]

For Darwin, the challenge was, of course, to transpose the lessons of geology to biology, where it meant introducing an unthinkable, monstrous idea—the evolution of species—into untutored minds to whom it was obvious that cats do not produce dogs but always cats, and dogs always dogs. It was this idea that had to be overturned, by the concept of the course of descent, of lineage over time: it is through reproduction that species change and new forms eventually appear.

In an 1869 addition to the fifth edition of *The Origin of Species*, Darwin continued to debate these questions and to think up pedagogical devices to make "the lapse of time" discernible to his refractory readers: "Few of us, however, know what a million really means: Mr. Croll gives the following illustration: take a narrow strip of paper, 83 feet 4 inches in length, and stretch it along the wall of a large hall; then mark off at one end the tenth of an inch. This tenth of an inch will represent one hundred years, and the entire strip a million years."[6]

How, with what evidence, did humans manage to overcome the obstacles to nature's long time, to acknowledge and admit the mutability of the physical world?

CHRONOLOGY AND FOSSILS: GIVING EARTH A HISTORY

Early attempts to compile global chronologies were based on data given in the books of the Pentateuch, combined with the dating of major events in the Greek, Egyptian, Hebrew and Sumerian civilizations.[7] Numerous scholars deployed the most advanced methods of comparative philology to establish the age of Earth and calculate the age of the world since its creation—which they agreed to set at about 6,000 years, as in the chronology of James Ussher, Archbishop of Armagh.[8] Following the interpretation of Jacques d'Auzoles, known as "the Prince of

84
John Martin,
The Deluge, 1834
Oil on canvas,
168.3 × 258.4 cm
New Haven, Yale Center
for British Art

Chronologists," Marin Mersenne pegged the age of the world at precisely 5,954 years. According to Dr. Lightfoot, Vice-Chancellor of Cambridge University, it could be said with precision that Man was created at 9 o'clock in the morning on the 23rd of October of the first year of the world.

In the late seventeenth century, the correlation of natural science data and sacred texts became a separate genre known as "theories of the Earth." Basing his argument on astronomical observations, Isaac Newton proposed to shorten the chronology of historical times, incurring the ire of historians.[9] These "theories of the Earth" intended to deduce the history of the globe's formation from the observation of natural phenomena, to identify the changes that had occurred there, and to predict those still to come.

Take, for example, the mountains and oceans so characteristic of certain regions of Earth today. By custom, it was assumed that this landscape was created at the time of Genesis. However, in 1681, Thomas Burnet argued that it was a consequence of the Flood.[10] Adhering as closely as possible to the Genesis story, Burnet posited an initial chaos, a fluid matter in which the materials were mingled in a confused mass. Those elements then precipitated according to their respective weights: the heaviest dropped towards the centre to form the earth's core while the rest, divided by the principle of gravity, separated into ocean and atmosphere, with the earth's first crust formed from this decantation. By this reasoning, the initial Earth issued from the hands of God was structured, by successive deposits, as follows: in the centre, a ball of fire, surrounded by a layer of water and, above that, a superficial layer of earth, perfectly smooth and featureless. It was this layer which, in drying out, eventually cracked and collapsed, provoking the Flood and, at the same time, giving rise to the mountains and seas that we know.

Little by little, objections to the immutability of Earth accumulated. However, as Martin Rudwick shows,[11] the "facts" were not yet established in science as they are for us today. This meant that their "testimony" was still questionable.

In particular, the discovery of "fossils" or "figured stones" had begun to excite curiosity, although, initially, all stones of peculiar shape found in the ground were called "fossils" (from Latin *fossilis*: "dug up"). Both the nature and the origin of these stones were problematic. For some, they had been lodged in place by God during Genesis. For others, they were simply *lusus naturae*, sports or whims of nature. Some saw them as mineral species spontaneously created from germs.[12] Still others conceded that those which clearly resembled shells were most likely of marine origin. In the seventeenth century, Niels Steensen (Nicolas Steno) and Robert Hooke, like Leonardo da Vinci and Bernard Palissy before them, recognized the crucial importance of fossils. But it remained to explain how shells found far from any sea had arrived there. At the time, many people believed they probably had been carried there by a miraculous force: they became "relics of the Flood."[13] Others, though, believed in a "translocation of the lands and the seas," that the seas, or at least salt water lakes, had once been situated in places where we now find solid ground.

Such a view implied that the face of the world had changed, that natural causes, such as the action of the tides, had played a part in the formation of Earth's features. These traces of nature's past came to be designated as "monuments" or "medals of nature."

FROM NATURAL HISTORY TO THE HISTORY OF NATURE

More and more facts were emerging to disturb the diluvian-creationist paradigm and its chronology of nature: the fossils, of course, but also the extinct volcanoes of the Massif Central, recognized as such. Further to the work of Jean-Étienne Guettard, the

8 Martin J. S. Rudwick, *The Meaning of Fossils: Episodes in the History of Palaeontology* [1972] (Chicago: University of Chicago Press, 1985), p. 70. In 1650, Ussher published his *Annals for the Old Covenant from the First Origin of the World.*

9 See the entry "Chronologie" in *L'Encyclopédie*, and Jean-Pierre de Bougainville's preface in Nicolas Fréret, *Défense de la chronologie fondée sur les monuments de l'histoire ancienne, contre le système chronologique de M. Newton* (Paris: Durand, 1758).

10 Thomas Burnet, *Telluris theoria sacra*, 2 vols. (London: Kettilby, 1681).

11 Martin J. S. Rudwick, *Earth's Deep History: How It Was Discovered, and Why It Matters* (Chicago: University of Chicago Press, 2014).

12 Robert Plot, *The Natural History of Oxfordshire, Being an Essay towards the Natural History of England*, 2nd ed. (London: Lichfield, 1705).

13 See the works of Johann Jakob Scheuchzer, *Piscium querelae et vindiciae* (Tiguri: sumt. Authoris, 1708); *Herbarium Diluvianum* (Tiguri: Literis Davidis Gesneri, 1709); *Museum diluvianum* (Tiguri: sumtibus possessoris, 1716); *Physica sacra* (Amsterdam, 1732); *Kupfer-Bibel... oder beheiligte Natur-Wissenschaft... erklärt und bewährt* (Augsburg und Ulm: Wagner, 1731–1735); *Physique sacrée ou Histoire naturelle de la Bible*, trans. Jacques de Varenne (Amsterdam: Schenk, 1732–1737).

14 *Histoire de l'Académie des sciences*, 1752, p. 5, concerning the memoir of J.-É. Guettard, "Mémoire sur quelques montagnes de la France qui ont été des volcans," in *Mémoires de l'Académie royale des sciences*, 1752, p. 27–59.

15 Jacques Roger, "Introduction," in Georges-Louis Leclerc, comte de Buffon, *Les Époques de la Nature*, critical edition with manuscript facsimile, annotated by Jacques Roger [1962] (Paris: Éditions du Muséum, 1988).

16 Jacques Roger, "Buffon et l'introduction de l'histoire dans l'*Histoire naturelle*," in *Buffon 88*, proceedings of the international symposium for the bicentennial of Buffon's death (Paris, Montbard, Dijon, June 14–22, 1988), ed. Jean Gayon (Paris: Vrin; Lyon: Institut interdisciplinaire d'études épistémologiques, 1992), p. 193–205.

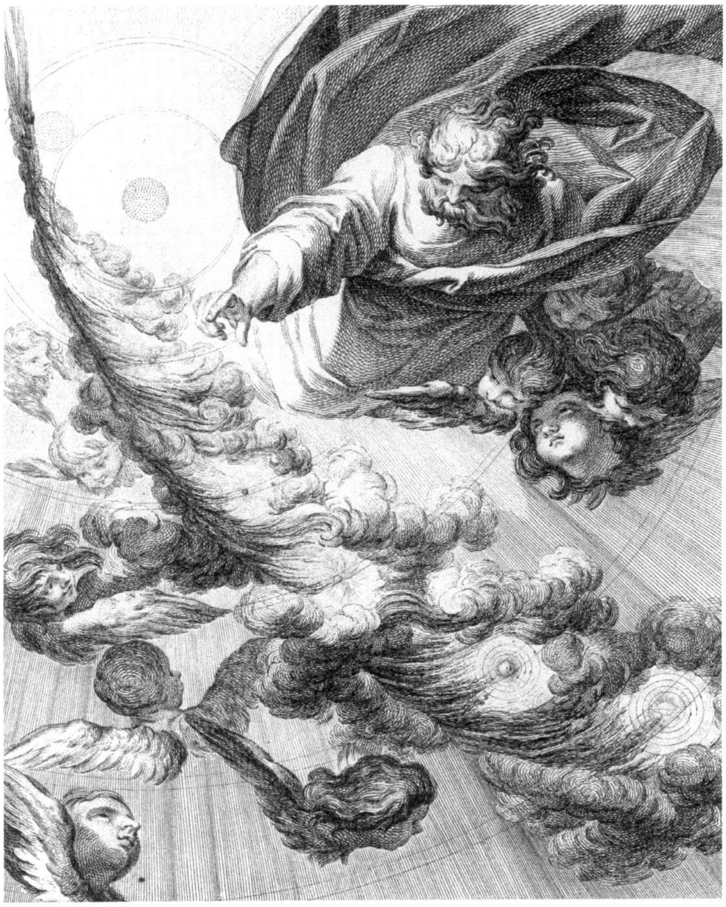

Histoire de l'Académie royale des sciences proposed that mountains be searched for *vestiges* of their former condition as volcanoes: "The Mont d'Or does not offer quite as many volcanic vestiges as the Puy-de-Dôme and the mountain of Volvic."[14]

In 1778, Georges-Louis Leclerc, Count de Buffon, drew a parallel between natural history and civil history in *The Epochs of Nature*. Taking up the torch of the theories of the Earth with a critical eye, he variously speculated about how, by physical and non-miraculous means, Earth could have been formed and acquired its present appearance.

His 1749 hypotheses on the origin of the solar system, connected to the Christian tradition by a drawing by Nicholas Blakey engraved by Fessard (85), became, as further volumes of *Natural History* appeared, an investigation of the history of life that envisioned even the transformation of species—repeatedly raising concern at the Sorbonne's Faculty of Theology.

Buffon no doubt delayed the publication of his reflections, but in 1774, in *Natural History of Minerals*, he described his experiments on the cooling of globes conducted in his forge at Montbard. Then, in 1778, his *Epochs of Nature* clearly posed the problem of natural history, combining different types of facts: the results of experiments along with such "monuments" as fossilized bones and the frozen mammoths of Siberia.

While his work still referred to the world of Genesis, the purpose was to mitigate the opposition of the theologians. In reality, Buffon utterly shattered the biblical framework.[15] He maintained that Genesis had been written not for scholars but for the ignorant populace, meaning that it was a text that had to be interpreted. Each of the six days of Creation did not last twenty-four hours; rather, each was actually an "epoch" that lasted between 3,000 and 35,000 years. Buffon estimated that the world was then 75,000 years old and that it would take 93,000 more years before the earth became completely cold and all life was extinguished. In this way, his *Natural History* made a transition from natural history to the history of nature.[16]

Compared to the results known in his day, it can be said that none of Buffon's theses taken separately was new. Descartes and Leibniz had clearly seen that the planets were incandescent stars. Newton had calculated how much time it would take an iron globe the size of Earth to grow cold. Newton, echoed by Maupertuis, had contended that the flattening of the Earth at the poles was due to the exertion of centrifugal force on a globe in fusion. Burnet had interpreted the days of Creation as epochs. And Benoît de Maillet had imagined humans emerging from the waters.

Based on our present-day knowledge, it also can be said that Buffon's chronology was far too short and that his efforts to win over the theologians were in vain. Although, ultimately, his pioneering work—attempting to calculate the age of the Earth by physical principles—ended in failure, his importance lies elsewhere. His *Natural History* allowed nature's time to free itself from the orthodox framework of biblical chronology. Step by step, the project to explain nature by nature's resources alone was gaining ground, consigning attempts to promote a *physica sacra*, or theology of nature, to the past.

85 FIG.
Georges-Louis Leclerc, comte de Buffon, Nicholas Blakey (draughtsman), "La formation des planètes," in *Histoire naturelle, générale et particulière . . .*, vol. 1, Paris: Imprimerie royale, 1749
Paris, MNHN

His efforts in that direction were not isolated. In Germany, Abraham Gottlob Werner analyzed the superposition of layers and established a chronological scale of rocks: from granite and gneiss, where fossils are never found, to the varying layers of sand, marl and volcanic rock. Some scholars favoured a form of Neptunism, emphasizing the role of waters, which had the advantage of reconciling scientific theory with the Flood narrative. Others supported Plutonist theories, emphasizing the role of volcanoes—in particular James Hutton, whose work, like Buffon's, was an important milestone in the revolution that broke "the historical time-barrier."[17]

The libertine novel reflected these difficulties, in its own way. In Diderot's fable *The Indiscreet Toys*, the story is set in a period that makes no sense within the framework of the Christian chronology: "Thus it was that, in the year of the world 15,000,000,032,000,021 [sic], of the empire of *Congo* 390,000,070,003, began the reign of *Mangogul*, the 1,234,500 of his race in a direct line."[18]

What years were those? If the Congos measured time as we did, the impiety was obvious. If it was a different chronology, the reader was given no yardstick by which to convert. So, it was impossible to establish a concordance between the histories. Furthermore, how long did those people live? Was there room for more than a million generations between Adam and one of the Congos? The biblical framework was cracking on all sides.

A certain number of scientists held out hope that the new discoveries would confirm the Scriptures and foster a new Pentateuch, "written in the rocks, chapter by chapter and verse by verse."[19] In 1785, William Cowper ironically described the hyperactivity of these new-style geologists, who, as they explored the earth, were competing to learn Earth's exact age:

"And thus they spend
The little wick of life's poor shallow lamp
In playing tricks with nature, giving laws
To distant worlds, and trifling in their own."[20]

But, around 1820, geology was nearly constituted as a science and researchers were more concerned with the internal consistency of their results than by consistency with biblical text. Geological science had become autonomous, building itself on the evidence of facts and no longer adhering to the letter of Mosaic text. By the 1870s, Flaubert was able to add this to his list of received, ridiculous and outdated yet persistent ideas: "Fossil: Proof of the Flood. Pleasantry in good taste, when talking about a member of the Academy."[21] Fossils had acquired new meaning.

THE INTRODUCTION OF TIME IN THE LIVING WORLD

The collapse of the time-barrier was not confined to the geological sciences. Through paleontology, a research field in which mineralogy was coupled with zoology, progress in the life sciences was closely linked to progress in the sciences of inert nature.

The question of the origin of species over time took on meaning, and with it the possibility of an evolutionary genealogy arose.

The fixity or immutability of species was the essential dogma of naturalist classifications. Set during the divine creation, the boundaries of the botanical species were clearly defined. Linnaeus is rather confused on the matter. In his 1738 *Classificatio plantarum*, he states: "There are as many species as the Infinite Being produced diverse forms in the beginning [*Species tot sunt quot diversas formas ab initio produxit infinitum Ens*]." Likewise, in the 1736 *Fundamenta botanica* and the 1751 *Philosophia botanica*: "We reckon as many species as there were diverse forms created in the beginning. [*Species tot numeramus quot diversae*

17 Toulmin and Goodfield, *The Discovery of Time*, p. 159.

18 Denis Diderot, *Les Bijoux indiscrets, or, The Indiscreet Toys* [1748]. Ebook translation produced by Project Gutenberg, 2017. https://www.gutenberg.org/files/54672/54672-h/54672-h.htm.

19 Toulmin and Goodfield, *The Discovery of Time*, p. 163.

20 William Cowper, *The Task* [1785], Book 3: "The Garden," in *Poems*, 5th ed. (Dublin: J. Jones, 1790), vol. 2, p. 75.

21 Gustave Flaubert, *Dictionary of Received Ideas* [1911–1913], trans. Gregory Norminton (Richmond: Alma Classics, 2016), p. 38.

22 *Fundamenta botanica*, 1736, § 157. See also *Philosophia botanica*, 1751, § 157. Quoted in John Ramsbottom, "Linnaeus and the Species Concept," *Proceedings of the Linnean Society of London*, vol. 150, no. 4 (December 1938), p. 196.

23 Daniel Rudberg, "Dissertatio botanica de Peloria" [1744], in Carl Linnaeus, *Amoenitates academicae: seu Dissertationes variae, physicae, medicae, botanicae, antehac seorsim editae nunc collectae et auctae cum tabulis aeneis* (Leiden: Cornelium Haak, 1749), vol. 1, p. 280–297.

24 Georges Cuvier, *Discourse on the Revolutionary Upheavals on the Surface of the Earth* [1826], trans. Ian Johnston (Arlington: Richer Resources, 2009), p. 48.

25 Auguste Comte, *Cours de philosophie positive*, vol. 3: *La Philosophie chimique et la philosophie biologique* (Paris: Bachelier, 1838), 42nd lesson, p. 558.

26 Thomas Henry Huxley, "On the Method of Zadig: Retrospective Prophecy as a Function of Science" [1880], in *Collected Essays*, vol. 4 (London: Forgotten Books, 2012).

27 Carlo Ginzburg, "Morelli, Freud, and Sherlock Holmes: Clues and Scientific Method," in *The Sign of Three, Dupin, Holmes, Peirce*, ed. Umberto Eco and Thomas A. Sebeok (Bloomington: Indiana University Press, 1983), p. 81–118.

28 Claudine Cohen, "Stratégies et rhétorique de la preuve dans les *Recherches sur les ossements fossiles de quadrupèdes* de Cuvier," in *Le Muséum au premier siècle de son histoire*, ed. Claude Blanckaert, Claudine Cohen, Pietro Corsi and Jean-Louis Fischert (Paris: Muséum national d'histoire naturelle, 1997), p. 523–539 (specifically, p. 532). The article analyzes in particular the case of the Montmartre opossum.

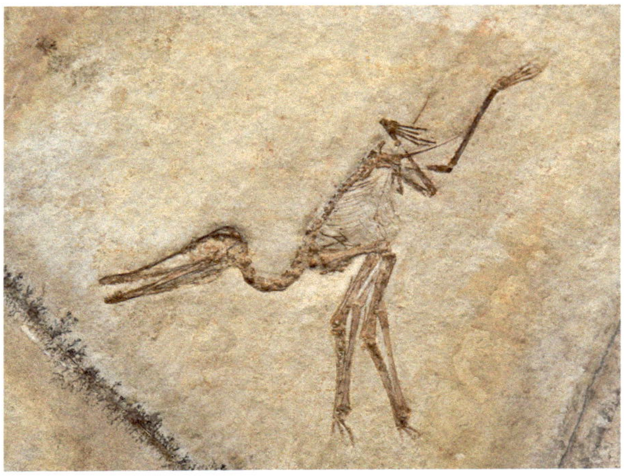

formae in principio sunt creatae.]"[22] Laconic pronouncements in which the species is not defined but simply quantified.

But the diversity in the number of species in each genus is such that it raises doubt about what was created *in principio* or *ab initio* and which Linnaeus does not really define: Is it species? Is it genus? Often held up as the paragon of fixism, Linnaeus nonetheless allows in his text the possibility that some species may have developed their individual characteristics from generic forms prefiguring a common original ancestor (the generic form). He was fascinated by variations in the common toadflax but named the plant *Peloria*—from the Greek for "monster."[23]

Buffon became concerned about the fixity of form when he compared the horse and the ass and concluded that the ass was not a degenerate horse, thus rejecting the hypothesis of their common ancestry. In 1766, however, his essay "Of the Degeneration of Animals" accepted the hypothesis that the 200 known species of mammals had descended, through degeneration, from a lesser number (38) of original forms.

While species have a history and have transformed themselves over time, that history stretches far back in time. Early in the nineteenth century, Georges Cuvier scientifically proved the existence of animals belonging to extinct species. The extinction of species was now a recorded scientific fact (86, 87).

For Cuvier, the reconstruction of extinct animals had to be performed according to two principles that served to connect the scattered fragments to an organized, finalized whole. The first was "the correlation of parts," by which the vertebrate skeleton was interpreted as a series of corresponding parts. According to this principle, "The entirety of an organic being forms a coordinated whole, a unique and closed system, in which the parts mutually correspond and work together in the same specific action through a reciprocal relationship. None of these parts can change without the others changing as well. Consequently, each of them, taken separately, points to and reveals all the others."[24] For example, a herbivorous diet implies a certain sort of teeth, a mode of locomotion and thus a particular skeletal structure. The second principle, called "the subordination of parts," established a hierarchy of physiological functions, the first being eating habits—a determination criticized by Auguste Comte.[25]

Reconstructing an entire skeleton from parts entailed a series of steps: the anatomical comparison that brought to light analogies and differences between living species and fossils. First and foremost, the animal had to be classified in its family, genus and species. It was this first fundamental information that defines the rules applicable to any animal: it was only after that, in a second step, that the animal could be reconstructed in terms of its anatomy and its faunal, floral and geological environment.

At the other end of the nineteenth century, Thomas Huxley was greatly impressed by what he called, in reference to a story by Voltaire, the "method of Zadig": based on a single tooth, it should be possible to identify a particular class of animal.[26] As the historian Carlo Ginzburg jokingly noted, Cuvier's method resembled that of Sherlock Holmes who, working from the slightest of clues (a cigarette butt), could reconstruct an entire world.[27] Cuvier said that he had founded paleontology as an inductive and predictive science. In putting those principles into practice, he employed what Claudine Cohen calls a "rhetoric of proof": "visual and discursive presentation of the 'facts' and arguments, sequencing and style of demonstration."[28] The plates that accompany the articles do not simply show: they demonstrate, enabling reconstruction of the forms.

In addition, the study of animal fossils was included in a geological framework. With Alexandre Brongniart, the comparison of levels—the stratigraphy of the Paris Basin (83)—served to establish a

86
Pterodactyl
Fossil, 26 × 25 × 2.5 cm
Paris, MNHN

P. 126-127

87
Charles Léopold Laurillard (draughtsman), Couët (engraver), "Megatherium," in Georges Cuvier, *Recherches sur les ossemens fossiles de quadrupèdes . . .*, vol. 1, Paris: Déterville, 1812
Paris, MNHN

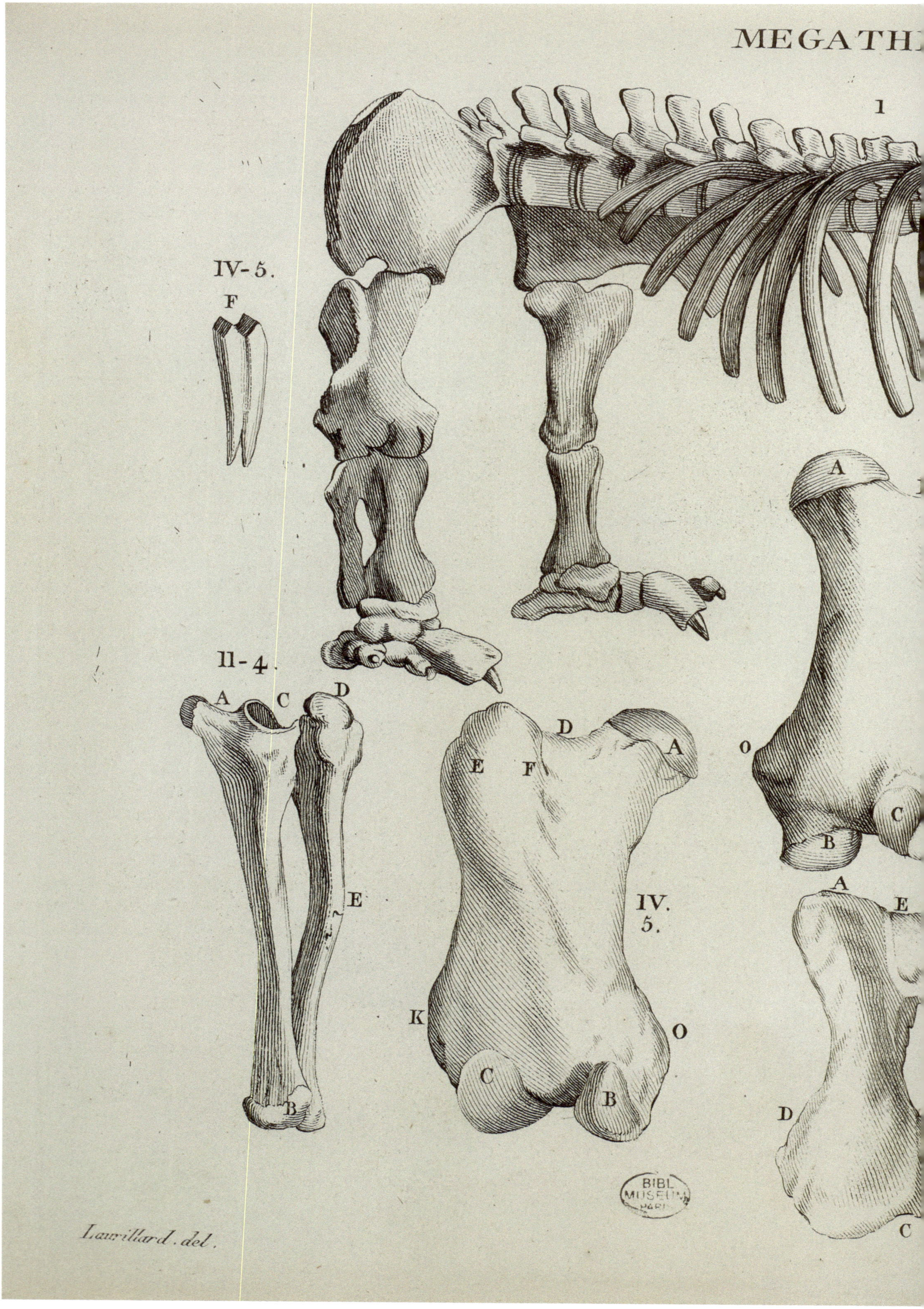
MEGATH
1
IV-5.
F
II-4.
A
C
D
E
B
D
A
E
F
IV.
5.
K
O
C
B
A
O
C
B
A
E
D
C
Laurillard. del.
BIBL MUSEUM PARIS

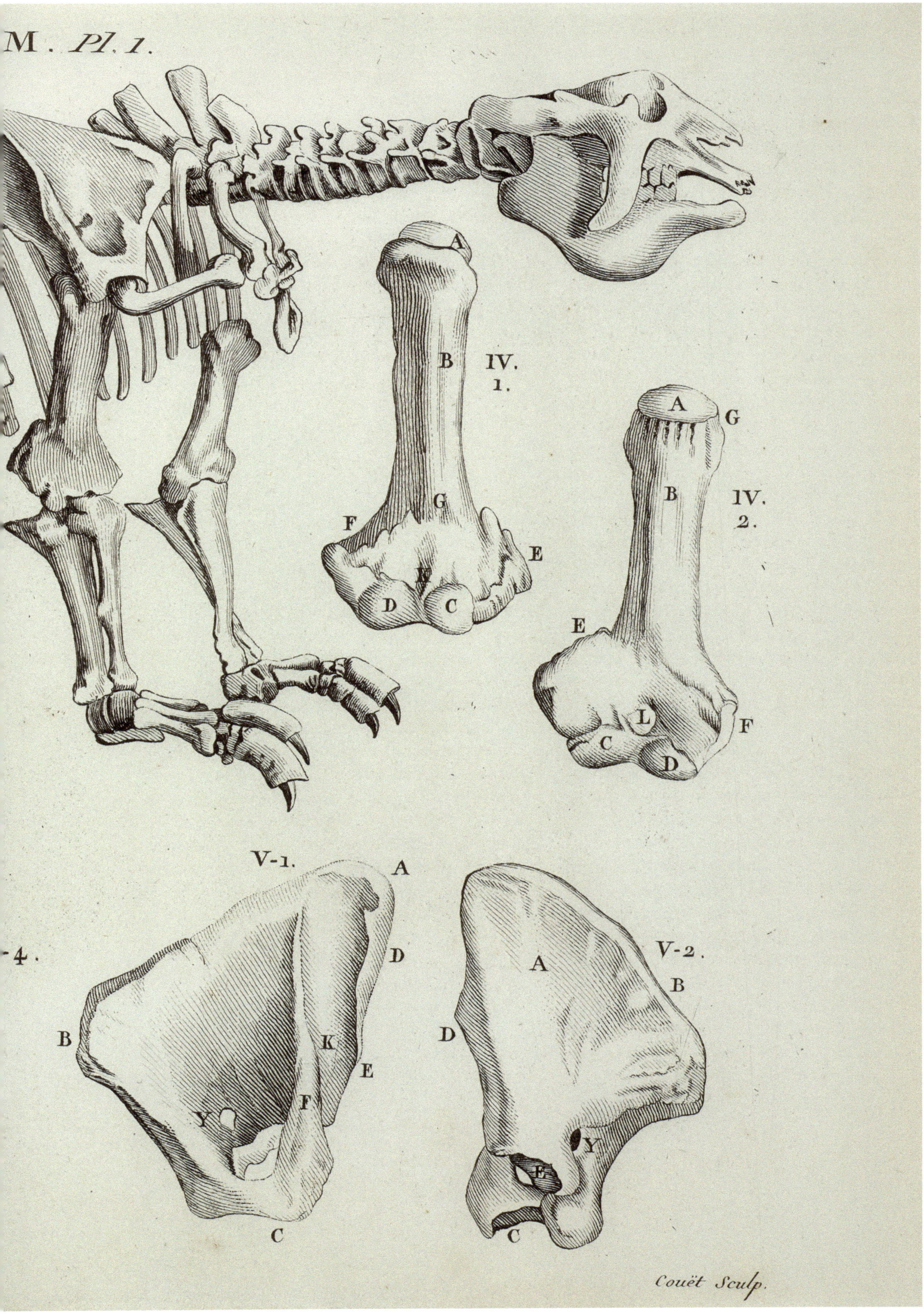
M . Pl. 1.
A
B
IV.
1.
G
F
E
E
D
C
A
G
B
IV.
2.
E
L
C
F
D
V-1.
A
-4.
D
B
K
E
F
Y
C
V-2.
A
B
D
Y
E
C
Couët Sculp.

systematic relationship between the succession of the sedimentary layers and the nature of the fossils they contained.[29] When major discontinuities were revealed, they were interpreted as occurrences of large-scale disasters that had destroyed the successive fauna. This again assumed multiple acts of creation. The succession of fossils became a determining factor in the knowledge of Earth's history and, inversely, the disposition of the fossils within the stratigraphic layers became crucial proof of the discontinuity of the history of life.

Cuvier's achievements were destined to live on, well beyond the strict field of natural sciences. This is seen most strikingly in Honoré de Balzac's repeated references to his work. First, in the introduction to *The Human Comedy*, where he finds in the academic debates between Cuvier and Geoffroy Saint-Hilaire the idea that "the Creator works on a single model for every organized being. 'The Animal' is elementary, and takes its external form, or, to be accurate, the differences in its form, from the environment in which it is obliged to develop."[30] Later, in a passage of *The Magic Skin*: "Have you never launched into the immensity of time and space as you read the geological writings of Cuvier? Carried by his fancy, have you hung as if suspended by a magician's wand over the illimitable abyss of the past? When the fossil bones of animals belonging to civilizations before the Flood are turned up in bed after bed and layer upon layer of the quarries of Montmartre or among the schists of the Ural range, the soul receives with dismay a glimpse of millions of peoples forgotten by feeble human memory and unrecognized by permanent divine tradition, peoples whose ashes cover our globe with two feet of earth that yields bread to us and flowers. Is not Cuvier the great poet of our era?"[31]

Balzac is not concerned with the naturalist's rigorous reconstruction method here; he is applauding the greatness of his vision. And in conclusion, the novelist has the audacity to compare two antithetical figures of the natural sciences—Cuvier and Buffon. Michel Foucault, too, perceived the revolutionary dimension of Cuvier's discourse: while others tended to relegate Cuvier to the conservative rearguard for his fierce belief in the stability of species, Foucault hailed him as a truly innovative thinker for having brought to the natural sciences an authentically historical way of thinking.[32] However, Cuvier was very prudent as to how time should be used.

In his *Discourse on the Revolutionary Upheavals on the Surface of the Earth* (1826), he describes himself as "a new sort of antiquarian" and evokes mummified animals from Egypt. He observes that "there are thus characteristics in the animals which resist all influences, whether natural or human, and nothing indicates that the passage of time has, so far as they are concerned, more effect than the climate and domestication."[33] In other words, Cuvier ascribed long time to the history of living things, but not all power and all licence. Much to the contrary, he severely condemned the casualness of his contemporaries: "I know that some naturalists rely a great deal on the thousands of centuries which they add up with the stroke of a pen. But in such matters we can hardly judge what a lengthy time would produce, except by multiplying mentally what a lesser time produces."[34]

THE POWER OF SMALL CHANGES AND THE LENGTH OF TIME

The rationales that introduced time into nature were equally divided between those who assumed abrupt changes (catastrophism: Cuvier in France, Buckland in England) and those who proceeded by the accumulation of small changes (uniformitarianism: James Hutton and Charles Lyell). The effect of sea level variations on the columns of the Temple of Serapis at Pozzuoli, in Campania, was depicted on the frontispiece of Lyell's *Principles of Geology* and serves as the canonical illustration of this view of time, explaining the geological events of the past by *current* causes, which continue to operate with the same force or intensity still today (88). Darwin embraced this view and unhesitatingly declared the consonance between his theory on the transformation of species and the geology of his colleague and friend: "He who can read Sir Charles Lyell's grand work on the Principles of

29 See Georges Cuvier and Alexandre Brongniart, "Essai sur la géographie minéralogique des environs de Paris," *Annales du Muséum d'histoire naturelle*, vol. 11 [1808), p. 293–326.

30 Honoré de Balzac, "Introduction," in *The Human Comedy* [1829–1847]. Ebook produced by Project Gutenberg, 2010. http://www.gutenberg.org/files/1968/1968-h/1968-h.htm.

31 Honoré de Balzac, *The Magic Skin* [1831], trans. Ellen Marriage, in *The Human Comedy*. Ebook produced by Project Gutenberg, 2010. https://www.gutenberg.org/files/1307/1307-h/1307-h.htm.

32 Michel Foucault, *The Order of Things: An Archaeology of the Human Sciences* [1966] (London: Tavistock; New York: Pantheon Books, 1970).

33 Cuvier, *Discourse*, p. 63.

34 Ibid., p. 73.

35 Darwin, *Origin of Species* [1859], chap. 9, p. 282. See Charles Lyell, *Principles of Geology, Being an Attempt to Explain the Former Changes of the Earth's Surface, by Reference to Causes Now in Operation* [1830–1833], reproduced with an introduction by Martin J. S. Rudwick (Chicago: University of Chicago Press, 1990–1991).

36 Darwin, ibid., p. 287.

37 *Origin of Species* [1872, 6th ed.], p. 409, in *The Origin of Species: A Variorum Text*, ed. Morse Peckham, p. 728.

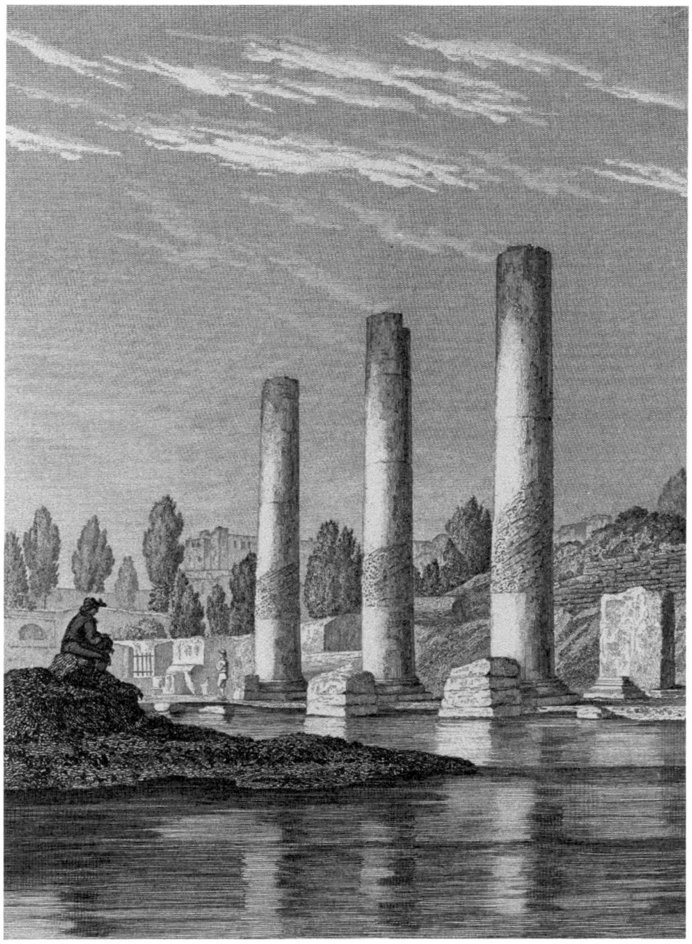

88
Charles Lyell,
Principles of Geology,
London: J. Murray, 1830–1833
Paris, MNHN

Geology, which the future historian will recognise as having produced a revolution in natural science, yet does not admit how incomprehensibly vast have been the past periods of time, may at once close this volume."[35]

However, Darwin and Lyell diverged significantly. In particular, in that the geologist above all opposed the idea of any direction or progression in terrestrial history. Darwin was more ambiguous about the possibility of development over time and continually struggled with the idea of geological time. In the first edition of the *Origin of Species*, he laid out his calculation on the denudation of the Weald, in the southeast of England, which he estimated had taken more than 300 million years.[36] His figures met with immediate reproach and he was forced to revise them. Later, it was Sir William Thomson, the future Lord Kelvin, who took exception to his concept of time—objections that greatly troubled Darwin when the engineer Henry Charles Fleeming Jenkin cited them in a critique of the *Origins* published in 1867 in the *North British Review*. Physicists questioned whether nature's time was sufficient to have allowed species to be produced by the mechanism of natural selection. Darwin, always prepared to acknowledge the qualities of his opponents, saw that as "probably one of the gravest [objections] yet advanced"[37] against his theory. But the physicists went further and judged that time had nothing to do with the matter. Jenkin, for example, challenged the uniformitarian premise that anything is possible, given enough time. To this end, he developed an arsenal of metaphors. For instance: fire a cannonball as hard as you can and give it unlimited time to travel its path—it will fall to earth no less quickly and continue moving for no more than a certain time. Cuvier's objection thus persisted in the physicists' discourse: time may be long but it is not all-powerful. Darwin still had to demonstrate that the time of living things was *efficient* and *creative*: quite a challenge at a time when his thermodynamics-enthusiast contemporaries were increasingly drawn by the concepts of depletion, limitation, reversion and regression.

LANDSCAPE AND GEOLOGICAL IMAGINATION

PIERRE WAT

Of all the pictorial genres, landscape bears the strongest imprint of the development of geology as a science in the late modern era. Since geology is the science of the composition, structure, history and evolution of Earth's inner and outer layers, it is not surprising that the genre undertaking to depict the world and, indeed, humankind's experience and conception of it, was profoundly affected by such a mutation of knowledge. However, the connections between landscape and geology cannot be seen simply as a dissymmetrical relationship, with an agent, geology, *influencing* a genre, landscape painting, responding passively to the advent of new knowledge. Of course, new knowledge engenders new representations. Thus, the development of stratigraphy and volcanology, two branches of geology, affected the way artists imagined evocations of the world's internal structure as depictions of volcanic phenomena. But what is striking in viewing the works is to see how painters, as in other art-science relations, practised a form of appropriation of scientific knowledge for artistic purposes: an appropriation that exceeds, and sometimes transgresses, a concern for accuracy in order to nourish the development of what could be called geological imagination.

STRATIGRAPHIC GAZE

The construction of this imagination, made fecund by new knowledge of the Earth, engaged the relationship between seeing and knowing. Many landscapes produced around the turn of the eighteenth to the nineteenth century exhibit the emergence of a stratigraphic gaze.[1] By stratigraphic gaze I mean a way of grasping the landscape that, instead of focusing on the expanse and horizontality, takes into account the depth, considered not only spatially but also temporally. When Frederic Edwin Church painted *Cotopaxi* (91), an active volcano and one of the highest peaks in Ecuador, the manner in which he associated in this one work the verticality of the smoking crater, the horizon and the chasm into which the dizzying waterfall plunges, owed much to the pioneering work of Niels Steensen (Nicolas Steno) who, in the seventeenth century, laid the foundations of stratigraphy, in particular with the principle of *superposition*, which states that "in any undisturbed sequence of rocks deposited in layers, the youngest layer is on top and the oldest on bottom, each layer being younger than the one beneath it and older than the one above it." Steno's law enables a vertical reading of landscape. Once one accepts the superposition and original horizontality of geological strata and their deformation by a certain number of earthly forces (tectonic, volcanic, etc.), and that a given layer is of the same age throughout its extent, one understands both the rising of mountains and volcanoes and the carving out of valleys. Lastly, the idea that the spatial distribution of ground layers corresponds to a temporal accumulation transforms landscape into a book that recounts Earth's history.

After the development of geology, landscape could no longer be defined solely by its spatial dimension, which had to be perceived in terms of time and duration as well, as did the slow birth of the world, which, henceforth, could not be thought of as having been completed in seven days. Modern archaeology, which inherited stratigraphic science from geology, further strengthened this gaze aware of the belowground, which, from Caspar David Friedrich to John Ruskin (249, P. 384), led to a fascination with the subterranean world, with its hidden structures and the things found there, fossils lying like a sedimented concentrate of the history of the world.

SUBLIME CATASTROPHE

Because it involved a connection between earthly and painterly forms, geology's relationship to landscape art was not spared in the aesthetic debates of the times. The case of volcanology is exemplary in this regard. While a scientific understanding of the formation of volcanoes was gradually emerging, after centuries of pre-scientific theories attempting to find a link between volcanic phenomena and Catholic Church dogma, volcano painting was spreading across Europe. Pierre-Jacques Volaire was one of the artists who offered representations of Vesuvius erupting (90).[2] First-hand observation of volcanic eruptions, as well as a growing understanding, in the nineteenth century, that volcanoes are formed by the successive build-up of materials and not by a sudden swelling of the ground, heightened the verisimilitude of these works, but they still embodied the disaster landscape genre, midway between science and the search for the sublime as defined by Edmund Burke in *A Philosophical Enquiry into the Origin of Our Ideas of the Sublime and Beautiful* (1757). Eruptions and avalanches (J.M.W. Turner), floods (John Martin) (84, 97) and storms (Turner again) (96) are among the natural spectacles that provoke the effect that Burke termed "delightful horror": the pleasure of "self-preservation" one feels when viewing a potentially deadly spectacle, provided one is viewing it without actually being in danger. The terrifying spectacle of nature, even informed by modern geology, satisfied a longing for the sublime astonishment which, in art, offered an attractive alternative to the Beautiful and the Good that had held sway for so long. Furthermore, despite progress in the scientific understanding of volcanic phenomena, the eruption as a motif echoed *catastrophism*, the theory of the origin and evolution of the world that, compatible with the biblical narrative, foregrounds the supposed role of catastrophic events, such as the Flood, in shaping the features of Earth's surface. While *uniformitarianism*—the idea of continuity between the distant past and the present, of Earth changing only under the influence of slow, gradual action—was gradually gaining ground as the basic principle of modern geology through the work of Charles Lyell, who would later support and inspire Darwin, paintings of terrifying geological phenomena restored a metaphysical dimension to a field devoted a priori to rational knowledge.

1 See Pierre Wat, *Pérégrinations. Paysages entre nature et histoire* (Paris: Hazan, 2017).

2 See Alexis Drahos, *Orages et tempêtes, volcans et glaciers. Les peintres et les sciences de la Terre aux XVIIIe et XIXe siècles* (Paris: Hazan, 2014).

GEOGNOSIS AND QUEST FOR ORIGIN

In upending our concept of how the world began, geology, far from assuaging our obsession with origin, gave it new forms and new means. This is manifest in German Romanticism, one of whose founders, August Wilhelm von Schlegel, declared in his *Lectures on Dramatic Art and Literature* that it was "the expression of a secret attraction to a chaos which lies concealed in the very bosom of the ordered universe, and is perpetually striving after new and marvellous births; the life-giving spirit of primal love broods here anew on the face of the waters."[3] An artist is someone who, through natural sciences, moves in a sort of incessant back-and-forth. The return movement towards chaos is also the movement of chaos—always there, latent—towards the ordered surface of creation.

In this quest for origin, painting's assigned task was to give form, or forms, to chaos. This determined the tireless efforts of Turner, whose predilection for disaster landscapes was an attempt to thematize the "secret attraction to a chaos . . . perpetually striving" that Schlegel describes. The turmoil of the elements is a return, through art, to the primitive chaos of nature. The physical material of the works, the coloured impasto that throws nature's order into confusion, is also an attempt to make the original turmoil rise back to the "surface."

In working on the visible part of the world as revealed by the landscape, the painter's task is also to understand and suggest its hidden part. The landscapes of Carl Gustav Carus (89) combine precise presentation of the visible world (the basaltic reliefs painted consistent with observed reality) with a search for its invisible foundation. Carus, who was a doctor (obstetrician-gynecologist) before becoming a painter, wrote, among other things, "Notes toward a Physiognomy of Mountain Ranges"[4] for inclusion in his *Nine Letters on Landscape Painting* and might well have borrowed these words from his contemporary and friend Caspar David Friedrich: "It is not the faithful representation of air, water, rocks and trees which is the task of the artist, but the reflection of his soul and emotion in these objects."[5] After Johann Kaspar Lavater, but before Alexander von Humboldt, Carus applied a physiognomic method to that which in nature pertains to geology, seeking, by observing the visible, to trace things back to their invisible origin.

The role of the artist is to capture the particularity of the living world through the accurate observation and sensitive rendering of its visible form: "As we gain no idea of an animal's inherent character from a lifeless tracing of its outlines, but only through the lively apprehension of an artist's eye, it seems that the true type and individuality of a mountain range can be conveyed only by a genuinely artistic representation: a true *geognostic landscape*."[6]

Landscape painting is that which makes the world visible to us again: comprehensible in its divine organization. In other words, although geological science is necessary to obtain such works, painting

3 Quoted in Martin Meisel, *Chaos Imagined: Literature, Art and Science* (New York: Columbia University Press, 2016), p. 322.

4 Carl Gustav Carus, "Letter XI, Enclosure I," in *Nine Letters on Landscape Painting* [1831], trans. David Britt (Los Angeles: Getty Publications, 2002), p. 136–141.

5 Caspar David Friedrich, quoted in James J. Sheehan, *German History 1770–1866* (Oxford: Clarendon Press, 1993), p. 333.

6 Carus, *Nine Letters on Landscape Painting*, p. 138.

7 Ibid., p. 107.

remains the art of the soul. Nourished by scientific knowledge, the painter accedes to still greater knowledge, for the entire phenomenon of nature is the "true bodily revelation or—in human terms—language of God."[7]

Logically, Carus deemed the word "landscape" [*Landschaft*] inadequate to render such spiritual painting of the world informed by geological expertise. He therefore proposed another word: *Erdlebenbildkunst*, the art of depicting Earth life. Nothing could better describe how geology had amplified the knowledge, the role but also the symbolic significance of landscape painting.

P. 134-135

89
Carl Gustav Carus,
Geognostic Landscape, 1820
Oil on canvas, 91.5 × 133 cm
Stuttgart, Staatsgalerie
Stuttgart

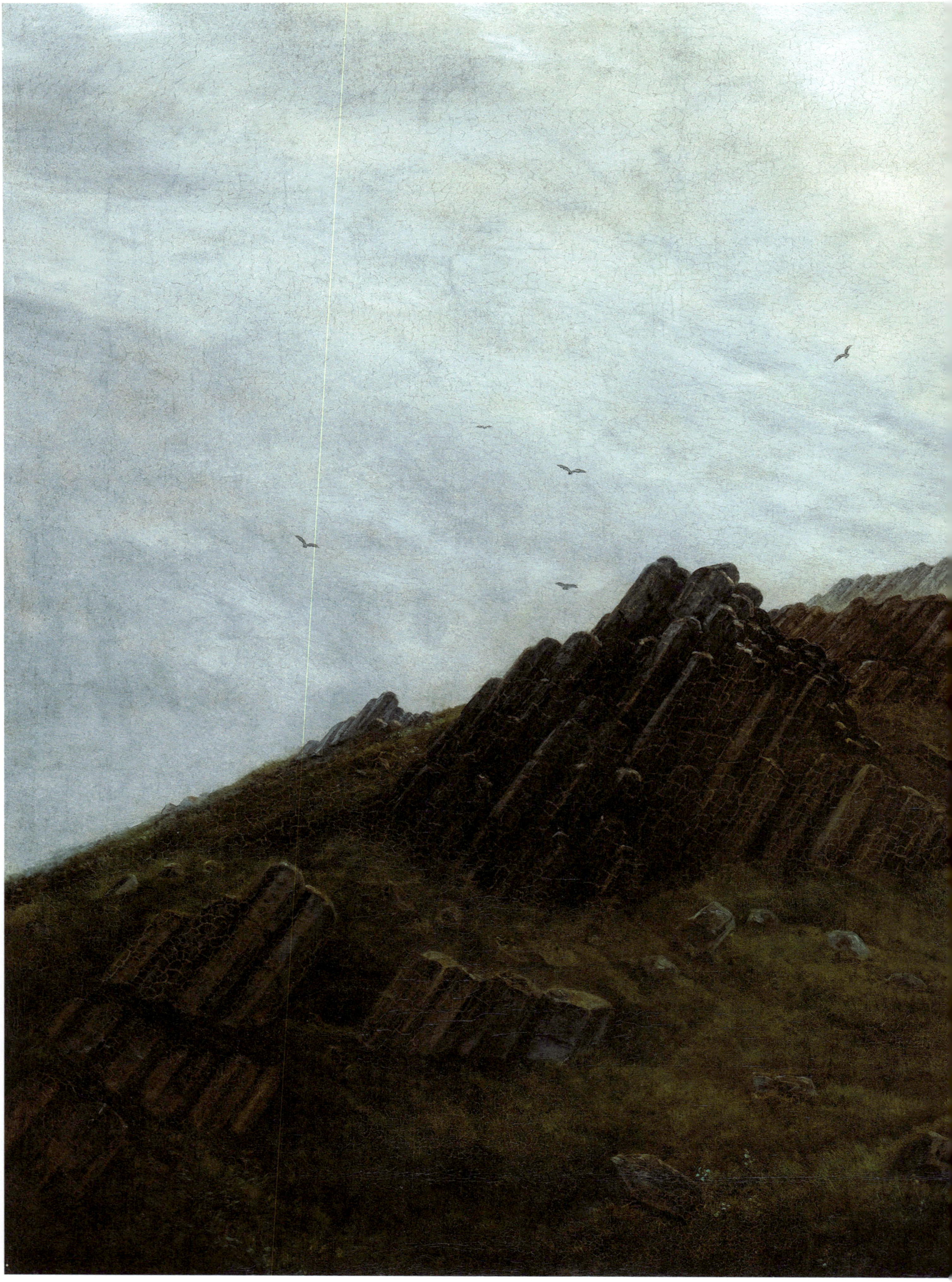

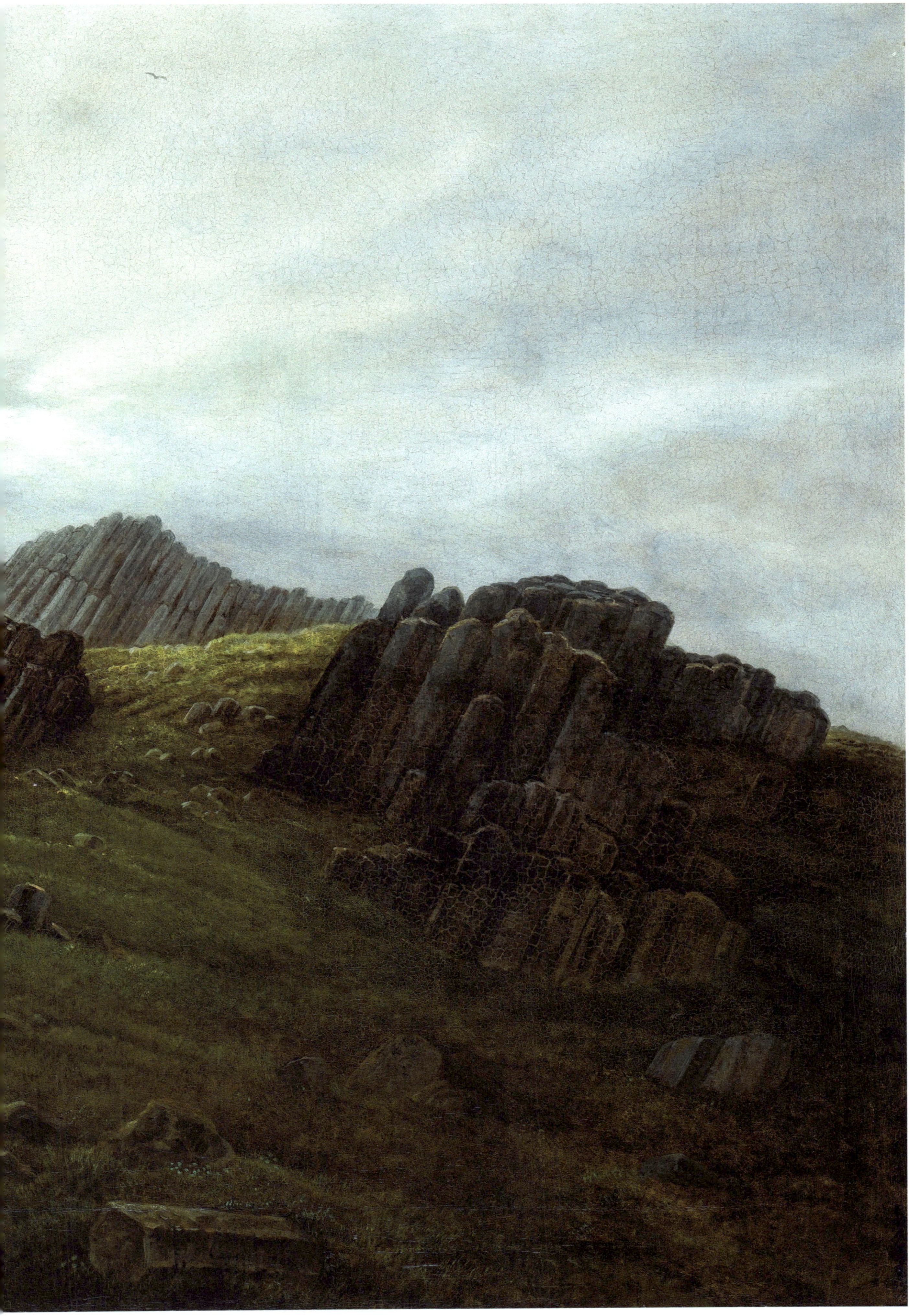

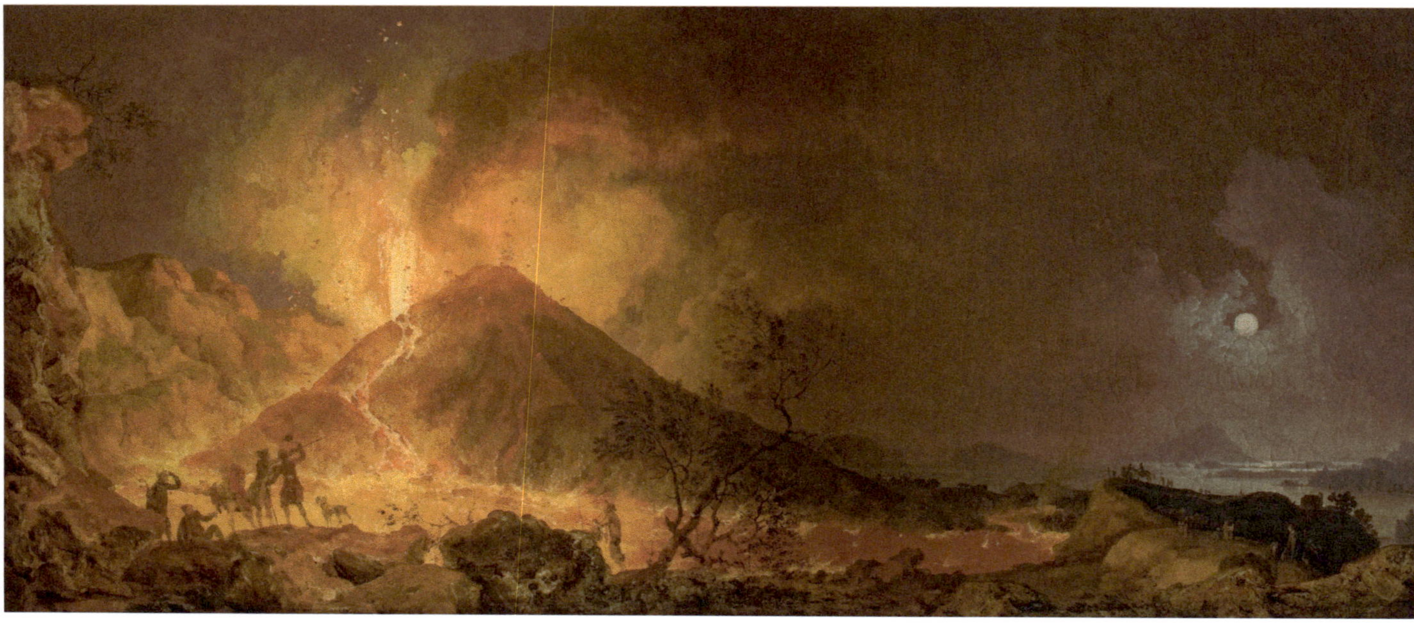

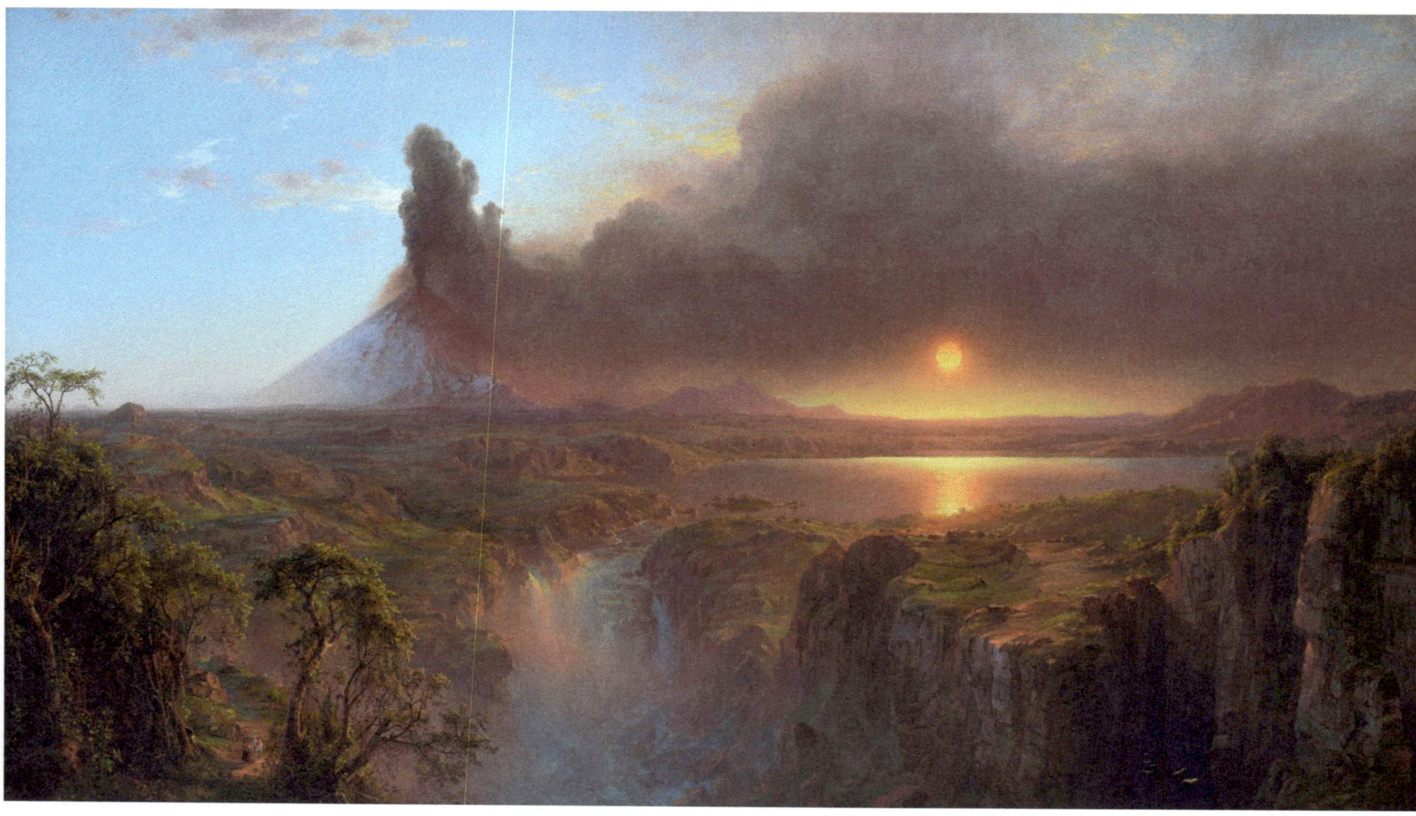

90
Pierre-Jacques Volaire,
The Eruption of Vesuvius, 1771
Oil on canvas, 73.3 × 162 cm
Le Havre, Musée d'Art
moderne André-Malraux

91
Frederic Edwin Church,
DeWitt Clinton Boutelle,
Cotopaxi, Ecuador, 1862
Oil on canvas, 87.6 × 142.2 cm
Reading, Reading Public
Museum

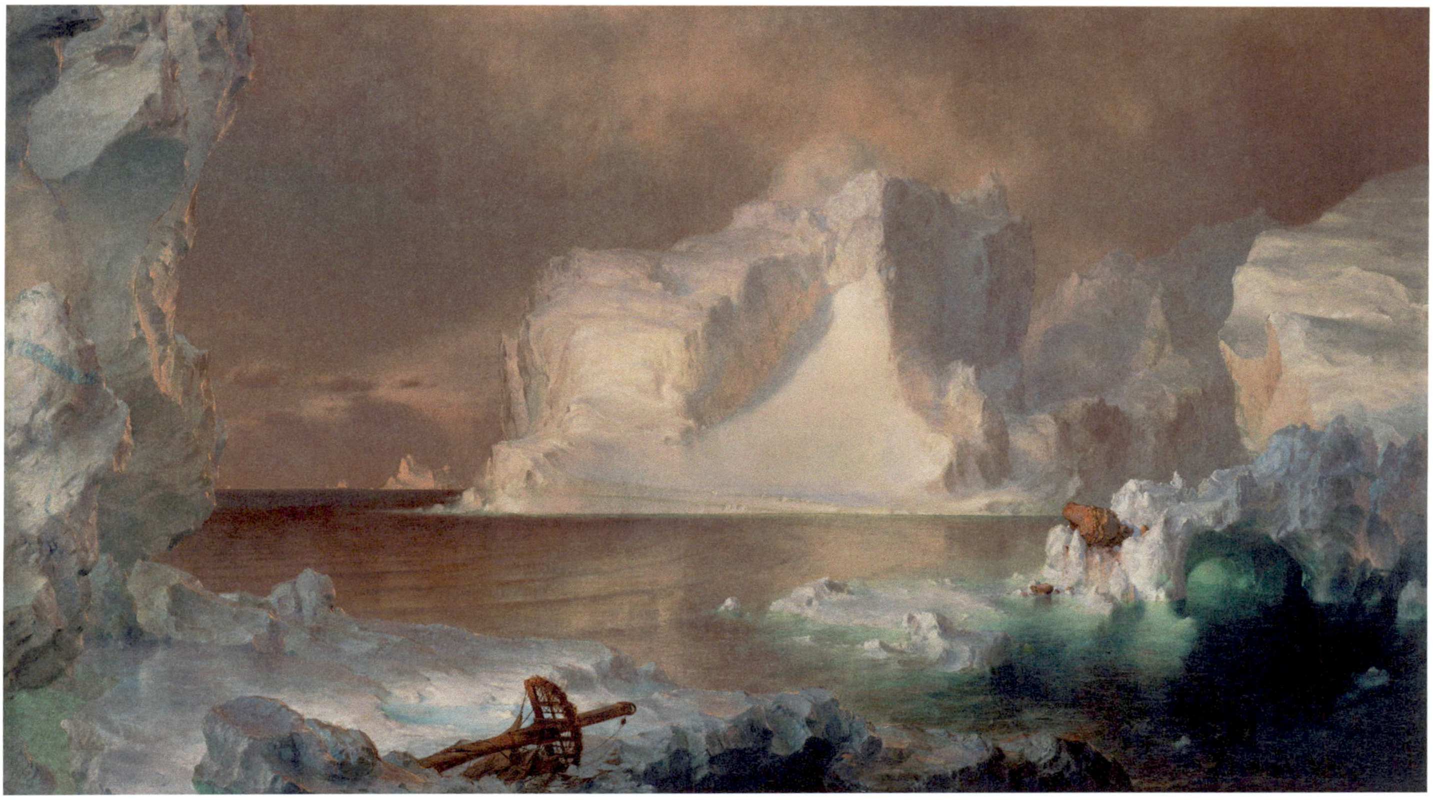

92
Frederic Edwin Church,
The Icebergs, 1861
Oil on canvas,
163.8 × 285.7 cm
Dallas, Dallas Museum of Art

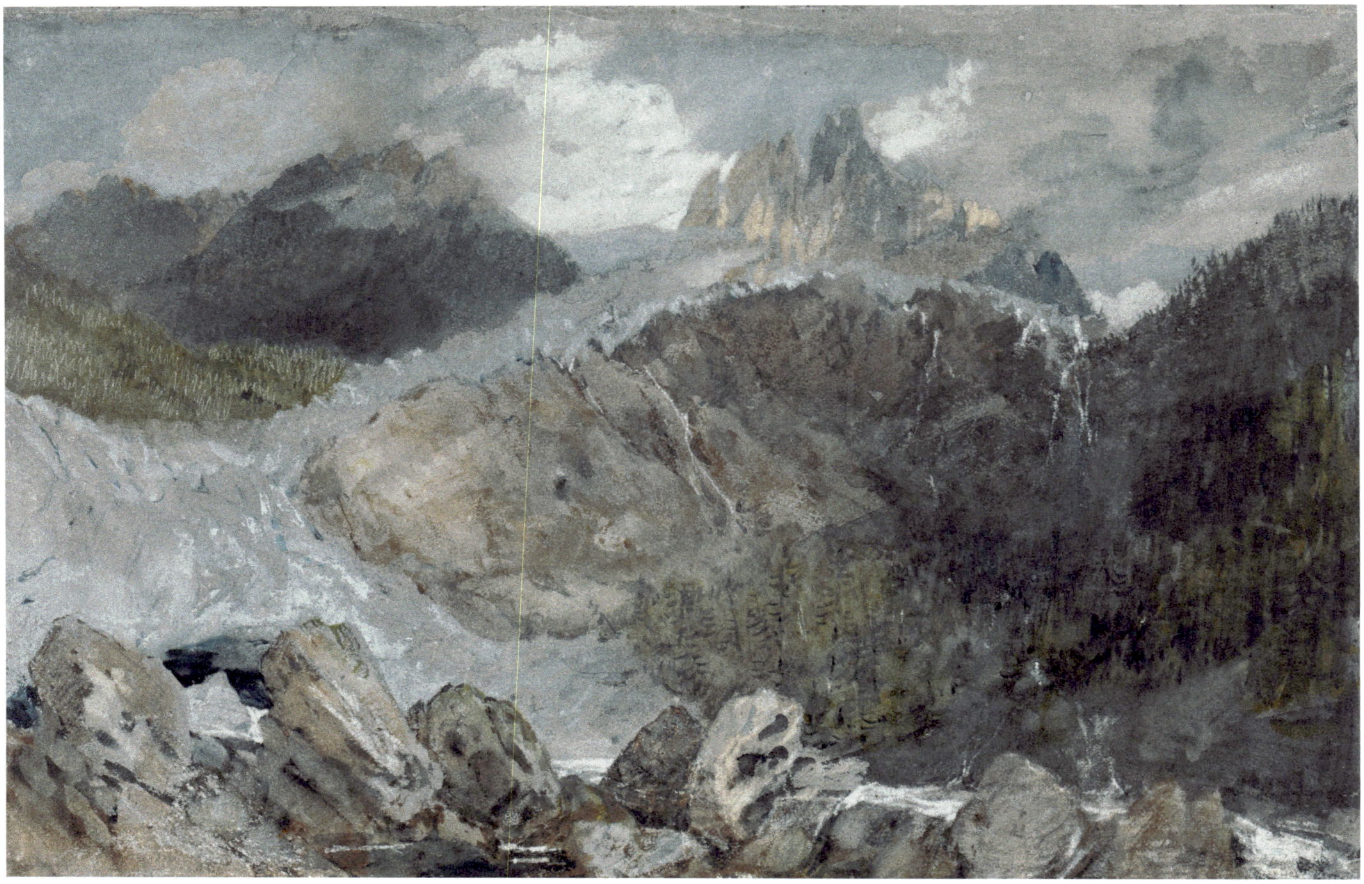

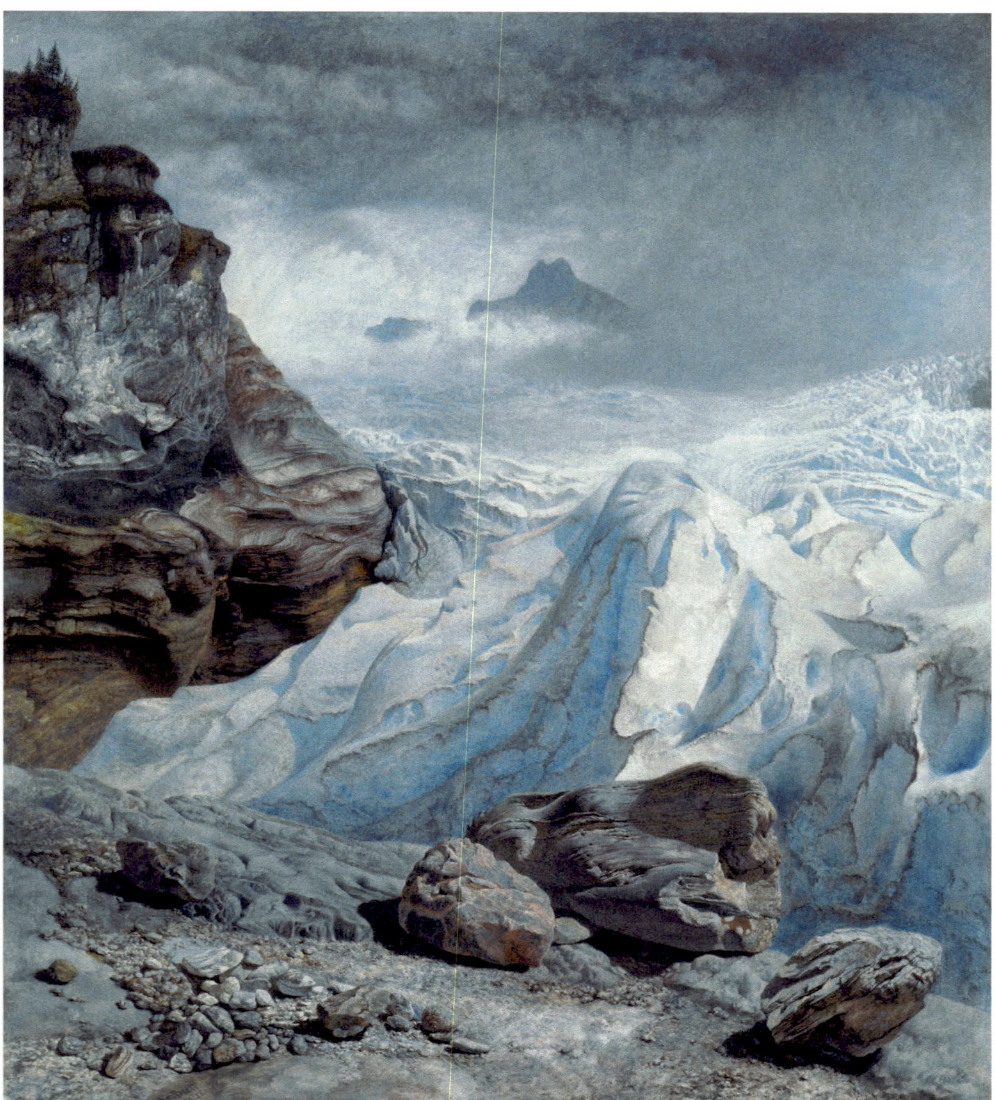

93
Joseph Mallord William Turner,
The Source of the Arveyron below the Glacier du Bois and Mer de Glace, 1802
Graphite, watercolour and gouache, 31.3 × 46.8 cm
London, Tate

94
John Brett,
Glacier of Rosenlaui, 1856
Oil on canvas, 44.5 × 41.9 cm
London, Tate

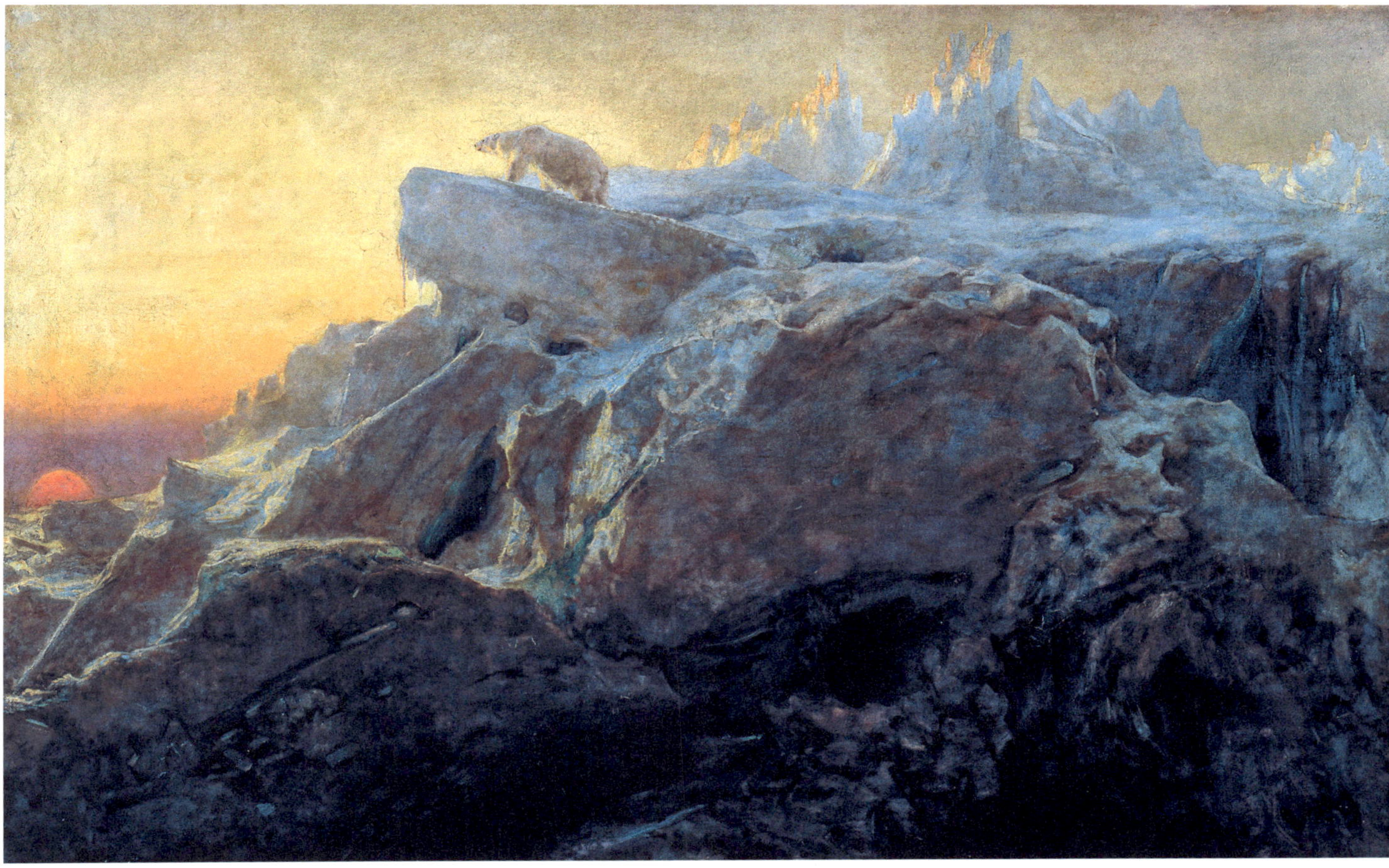

95
Briton Riviere,
Beyond Man's Footsteps,
about 1894
Oil on canvas, 119 × 184.5 cm
London, Tate

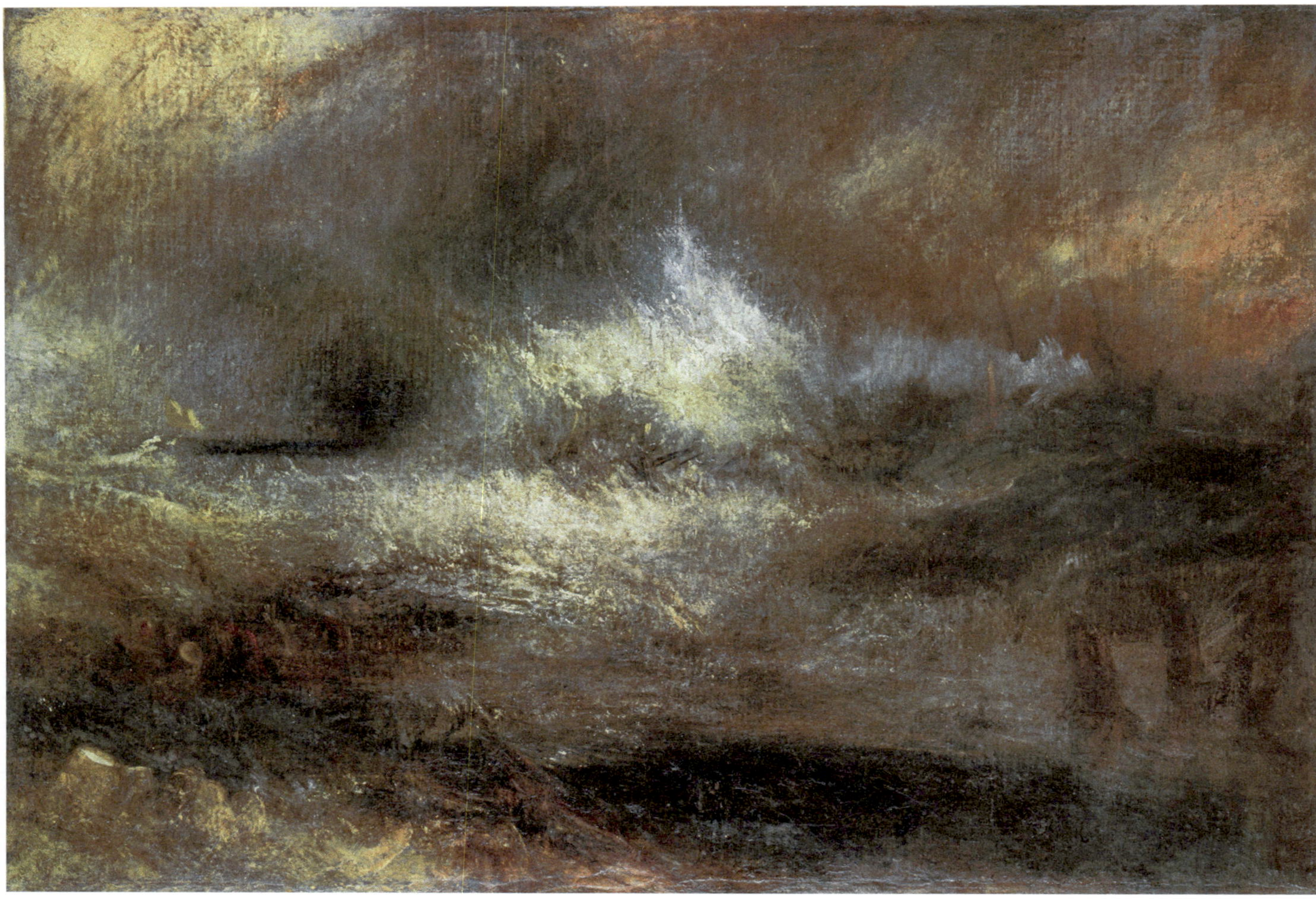

96
Joseph Mallord William Turner,
Stormy Sea with Blazing Wreck, about 1840
Oil on canvas, 99.4 × 141.6 cm
London, Tate

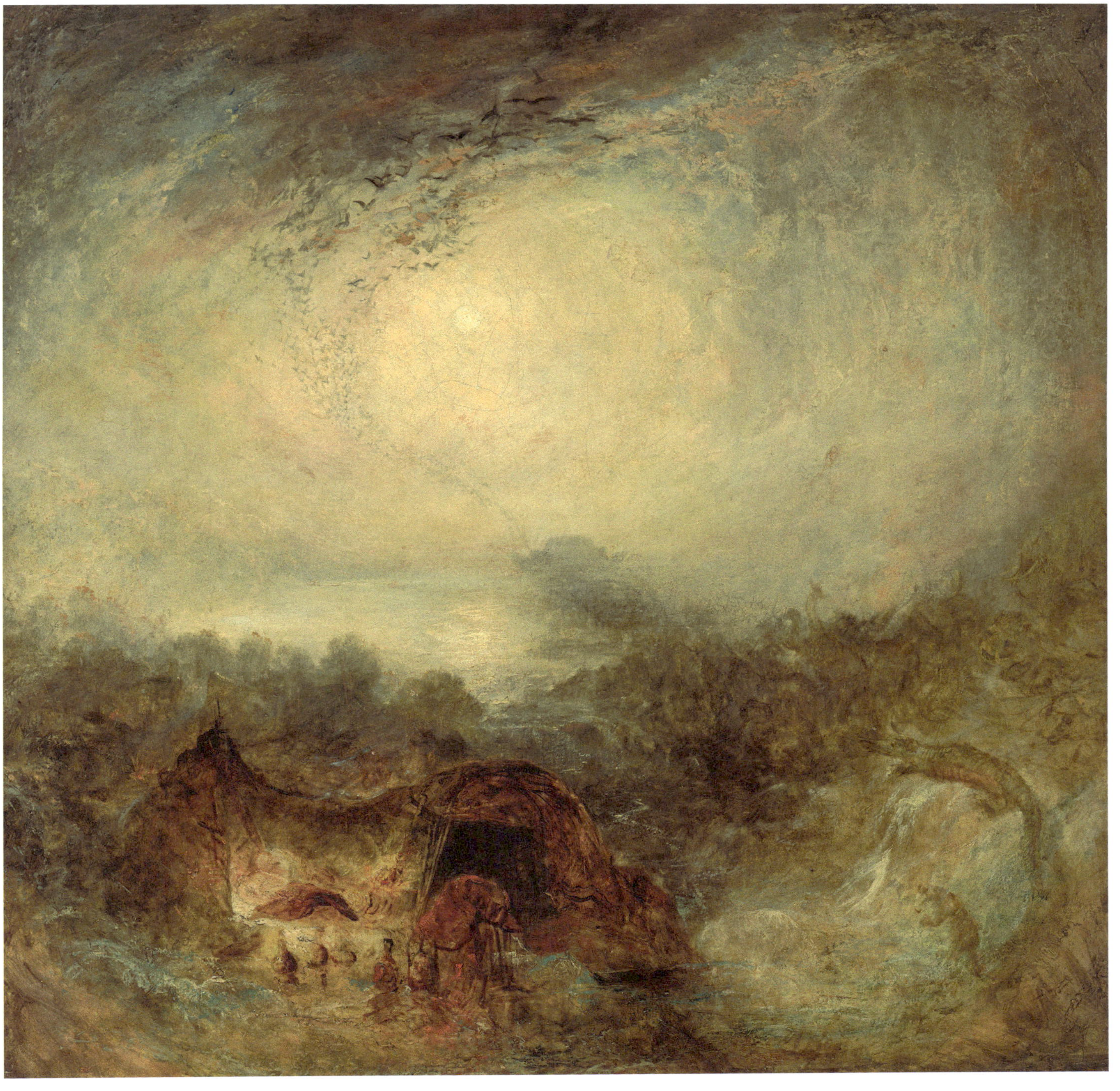

97
Joseph Mallord William Turner,
The Evening of the Deluge,
about 1843
Oil on canvas, 76 × 76 cm
Washington, National
Gallery of Art

THE INVENTION AND REINVENTION OF DINOSAURS:

THE NINETEENTH-CENTURY DISCOVERY

ELLINOR MICHEL
AND MARK P. WITTON

THE REVELATION OF EXTINCTION

Today, extinction as a natural process is part of our world view. Extinctions in the past intrigue us, as we marvel at the implausible lost forms of fossil animals and the forces that extinguished their evolutionary stories. Current extinctions supercharge our political discourse, and the fate of our own species, with its increasingly obvious connection to the changing nature of our planet, colours our views of the future. We learn about extinction as perhaps one of our earliest abstract concepts—ask a five-year-old what has happened to (non-avian) dinosaurs, and they will likely give you a scientifically acceptable answer.

It was not always thus. The eyes of one scientist, the great French anatomist Baron Georges Cuvier, were opened to the possibility that there were animals that had gone extinct when he studied fossil vertebrates of peculiar and unrecognizable forms, such as pterosaurs (86, P. 125). Our divorce from these creatures by eons of geological time was unacceptable to us at some level, however, and we yearned—as we still do—to imagine these species as flesh-and-blood animals rather than as fossil remains. The first paleoart, or scientifically based life reconstruction of an extinct animal, was a sketch of the *Pterodactylus* pterosaur fossil provided to Cuvier by his colleague Professor Jean Hermann from Strasbourg in 1800. Cuvier's studies of extinct mammal fossils clinched the concept of extinction, which he revealed to his scientific colleagues (87, P. 126–127). Extinction created an existential shift in our view of history, as it required recognition that there had been a series of vibrant worlds dominated by big, impressive animals with no record of, or role for, humans.

THE DISCOVERY OF DINOSAURS

Science progresses as an interplay between new data and improved interpretation. The newly born science of paleontology grew from dialogue between scientists in many countries. After its first founding with Cuvier's work, a flourishing of new fossil discoveries, ideas and discussions erupted, particularly in Britain. What is now called the "Jurassic Coast" in Dorset revealed fossils of many strange marine reptiles, such as ichthyosaurs and plesiosaurs, as well as invertebrates, such as ammonites, unlike anything living today. Scientists studying these extinct remains challenged each other's ideas in person and in print. They were constructing puzzles with very few pieces, as fossils are often in terrible shape: mashed, smashed, misplaced partial remains cast in stone. Piecing these elements together revealed whole new categories of animals and plants unlike anything encountered before.

The first reconstruction of an extinct habitat, a life scene, was *Duria Antiquior, a More Ancient Dorset* by Henry De La Beche (98). While the sketch itself is charming and historically critical as it shows ecology, behaviour and interactions in the past, there is an additional human story to reprinted versions. De La Beche sold copies of the sketch to raise funds for Mary Anning, the highly skilled fossil collector who discovered and prepared most of the Dorset Jurassic marine reptiles. Her gender and humble background meant that, despite her impressive contributions to paleontology, Anning received little recognition and struggled with poverty. The reprints were distributed across the scholarly world, and this act of personal kindness to counterbalance an unfair social system had an additional benefit, giving the fossil reconstructions a visual life and immediate recognition from the wide-ranging publicity. "Cleaned up" renditions were drafted, such as the large painting by Robert Farren, which eliminated many busy details, including the amusing and biologically realistic poops (99)!

At the same time, dinosaurs were heaving into view from fossil bones of huge extinct reptiles found in outcrops in southern England. *Megalosaurus*, the first dinosaur to have a scientific description, was named in 1824 by William Buckland, but with input from Cuvier (who visited Buckland in Oxford and commented on the fossils), Richard Owen, Gideon Mantell and other paleontologists of the time. Two additional huge reptiles, *Iguanodon* and *Hylaeosaurus*, were discovered in England shortly thereafter. In 1842, Owen gave the formal scientific name "Dinosauria," meaning "terrible lizard," to the group anchored by this triumvirate of first discoveries.

Naming these animals was the easy part, however, and debate ensued about their size (from 9 to 25 metres long) and stance (two- or four-legged). Owen said they must have stood like a large mammal, such as a rhinoceros, with all four legs straight and firmly underneath.

Although fossil bones could be viewed by paying visitors to museums at the time, the wonders of extinct past worlds did not touch the imaginations of the wider public even with paleoart, which remained a primarily academic fixture. This was about to change with an audacious, large installation featuring the work of a visionary artist, Benjamin Waterhouse Hawkins.

BRINGING THE PAST TO LIFE

As all of us can relate, imagination is fuelled by visuals, and understanding is grounded in a sense of personal experience. In 1854, Waterhouse Hawkins designed and built a grand artistic and scientific tableau for the new sciences of paleontology and geology on the grounds of the Crystal Palace in south London (100). In a walk-through experience of deep (geologic) time, he arrayed thirty-three beautiful and convincingly alive sculptures of extinct animals and a few plants in their appropriate time-place.

The installation, now known as the "Crystal Palace Dinosaurs," was a global first in many respects: it was the first major public outreach on science through "visual education" or "edutainment"; the first major experiential public outreach on natural history; the first reconstruction of extinct organisms in life form; the first depiction of deep time (geologic time) and changes in fauna and flora; and the first staged explanation of Britain's mineral wealth and geological processes.

However, the importance of the site extends far beyond its specific subjects as it was a turning point in democratization of big ideas.

Although there was input into the design and content of the site by many luminaries, Waterhouse Hawkins deserves the major credit. In his reconstructions he reflected the best paleontological knowledge and some of the controversies at the time, albeit mostly leaning towards opinions of his scientific advisor, Richard Owen, in cases of debate. Before the Crystal Palace contract, Waterhouse Hawkins was recognized as especially skilled in giving a sense of life to his illustrations of the living world, a skill apparent when he turned his efforts to extinct animals (102, 103). It should be noted that, despite the installation's name, there are only four sculptures of dinosaurs on the site, representing three species of dinosaurs, as they were the only ones recognized at the time. The rest of the sculptures include marine and flying reptiles and mammals, but an unrealized fifteen-sculpture expansion was to include a moa and a dodo, which as birds, we now recognize as part of the dinosaur clade, or evolutionary group.

Waterhouse Hawkins worked out the likely natural form for the extinct animals in small maquettes, such as the 1:12 scale *Megalosaurus* and *Iguanodon*. He then built the sculptures on site in a workshop, using a wide and seemingly improvised array of mixed materials, including cast clay panels, mortar, bricks, pipes, tiles, iron, lead and even real fossil parts (104). Although these outdoor sculptures have required regular conservation, it is a testament to Waterhouse Hawkins's practical skills that they are still intact and giving edutainment to the public 170 years later.

In a stroke of inspired promotion, Waterhouse Hawkins and the Crystal Palace Company staged a very quirky event that brought lasting fame to the dinosaurs. On New Year's Eve 1853, they hosted a dinner for twenty-one scientists, patrons and publicists inside (or potentially around) the mould of the standing *Iguanodon* (101). The exuberant spirit of the event, evident in the invitation and menu illustrated by Waterhouse Hawkins (1, P. 4), was widely reported in enthusiastic press coverage and is reminisced upon even today.

DINOSAURS IN THE AGE OF DARWIN

Waterhouse Hawkins's "putting flesh on the bones" for his sculptures fed into one of the more interesting shifts in philosophical views on life's changes through time. The "Owenian" view of diversity, to which Waterhouse Hawkins was committed, was heavily Platonic and Romantic, invoking ideal forms and divine processes underlying diversity, and implying that taxa were static. However, the reality of anatomical data, and issues revealed in sculpting convincing-looking animals, gave strength to an argument by Thomas Huxley that pivoted our views on the relationships between dinosaurs and birds. With that, he shifted the majority scientific view on evolution.

Mixing personalities and philosophies, Huxley was a "Darwinian and neo-Lyellian" and "anti-Owenian," who was called "Darwin's Bulldog" for his fierce public defence of natural selection. At first Huxley saw no direct link between birds and dinosaurs, but throughout the 1850s and 1860s, he became increasingly aware of their anatomical similarities and came to realize the possibility that dinosaurs could be reconstructed as strutting forms, like huge ostriches. Subsequently, after reading Ernst Haeckel's *Morphologie*, Huxley wrote to his German colleague about his epiphany, suggesting that there could indeed be evolutionary transitions between higher taxa, as "The road from Reptiles to Birds is by way of Dinosauria to the Ratitaez—the Bird 'Phylum' was Struthious, and wings grew out of rudimentary fore limbs. You see that . . . I have been reading [your] Morphologie."[1]

Huxley gave us the potential for the upright, fast-moving dinosaurs we have today, and moreover, set the foundations for seeing evolution as possible across all taxonomic levels, from species to higher groups.

Somewhat curiously, although Darwin was interested in fossils and collected significant numbers of them in his travels, he seemed to give the whole dinosaur saga a fairly wide berth. He didn't even mention the Crystal Palace site, despite living only sixteen kilometres away and records showing that he both visited and had a year's pass to the park. Was his silence simply due to his profound antipathy towards Richard Owen, which was building through the 1850s to reach a peak with the publication of *On the Origin of Species* in 1859? He had great respect for Waterhouse Hawkins's skills and had hired him to illustrate parts of his records from the *Beagle* expedition decades before, and provided letters of recommendation in the decade following. He did, however, pitch in on the bird-dinosaur debate: "In the sixth and final edition of the *Origin*, Darwin fairly chortled over 'the most unexpected manner' in which *Archaeopteryx*, ostriches and *Compsognathus* [a small dinosaur] had helped bridge an embarrassing gap [in fossil links]."[2]

THE LEGACY OF THE CRYSTAL PALACE DINOSAURS

Away from academia, Waterhouse Hawkins's reconstructions of the first dinosaurs— *Iguanodon*, *Megalosaurus* and *Hylaeosaurus*—continued to influence illustrators and popularizers of science for decades. Artists such as Édouard Riou and Camille Flammarion (105, 106, 107) closely approximated Hawkins's dinosaur work until at least 1884, ensuring that "dinosaur" remained synonymous with visions of large, powerful and undeniably reptilian animals in the public eye. This occurred despite rapid advancements in dinosaur science, which included major revisions in our understanding of their anatomy, a growing appreciation for the true outline of dinosaur diversity, and

1 Written in 1868, as quoted in Brian Switek, "Thomas Henry Huxley and the Reptile to Bird Transition," in *Dinosaurs and Other Extinct Saurians: A Historical Perspective*, ed. R.T.J. Moody, E. Buffetaut, D. Naish and D. M. Martill, Geological Society Special Publications, vol. 343 (London: Geological Society, 2010), p. 251–263, https://doi.org/10.1144/SP343.15.

2 Adrian J. Desmond, *Archetypes and Ancestors: Palaeontology in Victorian London, 1850–1875* (London: Blond & Briggs, 1982), p. 131.

the proposal of an evolutionary link with birds. Waterhouse Hawkins's rhinoceros-like quadrupedal dinosaurs were scientifically superseded within just a few years thanks to the discovery of the bipedal and more bird-like dinosaurs *Hadrosaurus*, *Compsognathus* and *Archaeopteryx*. Although Waterhouse Hawkins illustrated and mounted restored skeletons of *Hadrosaurus* in an appropriately updated fashion, his new work lacked the impact of the Crystal Palace sculptures, and illustrations showing dinosaurs as bird-like animals remained rare prior to the twentieth century.

The lack of traction of bird-like dinosaurs and the evolutionary theory they represented in Victorian paleoart may reflect, in part, their misalignment with cultural values placed on dinosaurs in Europe and the US in the mid-late nineteenth century. American tycoons of this time associated the discovery, display and gifting of spectacular dinosaur fossils with high cultural standing, and it was a major cultural offensive by American investors that saw the final "popular extinction" of Waterhouse Hawkins's dinosaur depictions in the 1880s. Novel American dinosaurs, many represented by far superior fossils to their European counterparts, were installed as museum centrepieces around world and accompanied by life reconstructions by American paleoart masters, such as Charles Knight. These new restorations overshadowed Waterhouse Hawkins's Victorian creations both in science and spectacle and they became major cultural currency for museums and their rich sponsors. This revision is captured perfectly in Flammarion's frontispiece to *Le Monde avant la création de l'homme*, where the spiked and armoured American *Stegosaurus* meets a now antiquated Crystal Palace-style *Iguanodon*. Naturally, the biggest, most unusual and most ferocious dinosaurs carried the most clout, so while dinosaur restorations underwent a scientific overhaul compared to Waterhouse Hawkins's creations, popular attention remained on species which were also the least bird-like in outward appearance. Though transformed from mammal-like quadrupedal reptiles to more life-accurate forms, dinosaurs retained their greatest value as huge, powerful extinct animals entirely unlike birds, and would be presented as such in new forms of media. It would not be until the late twentieth century that birds would finally be incontrovertibly recognized as a group of living dinosaurs, although popular culture and the public at large still struggle to represent and recognize the true origins of the feathered dinosaurs living around us today.

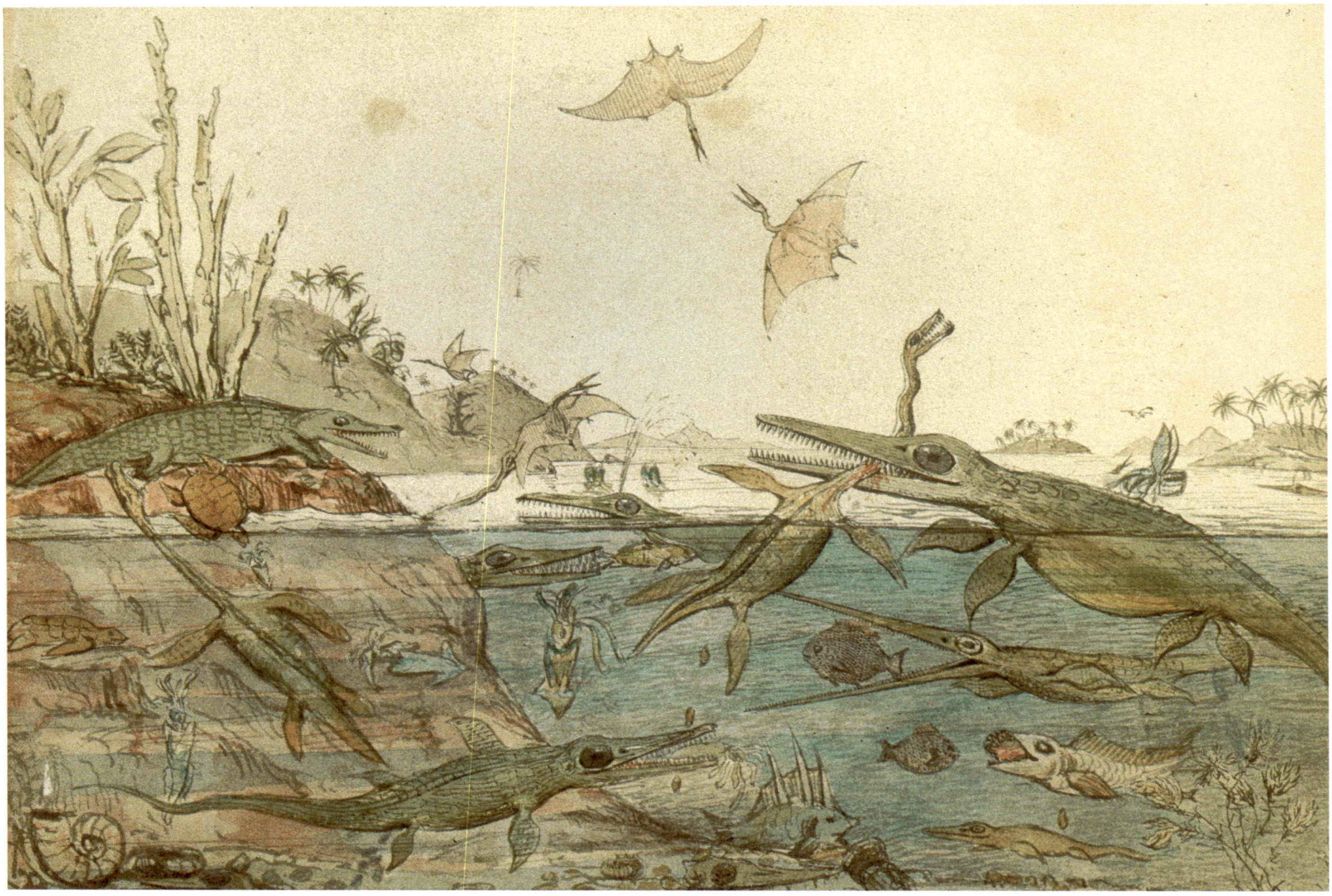

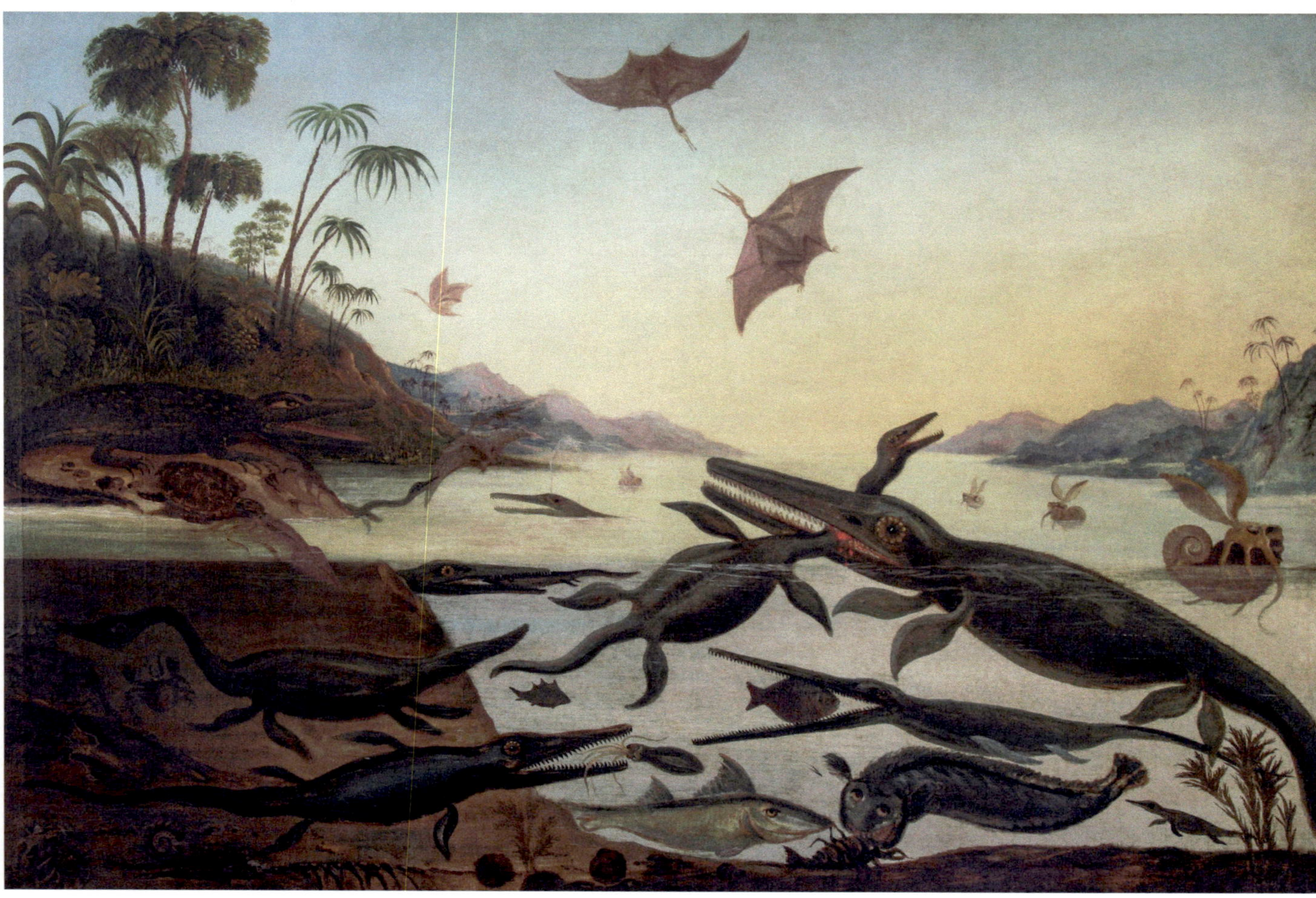

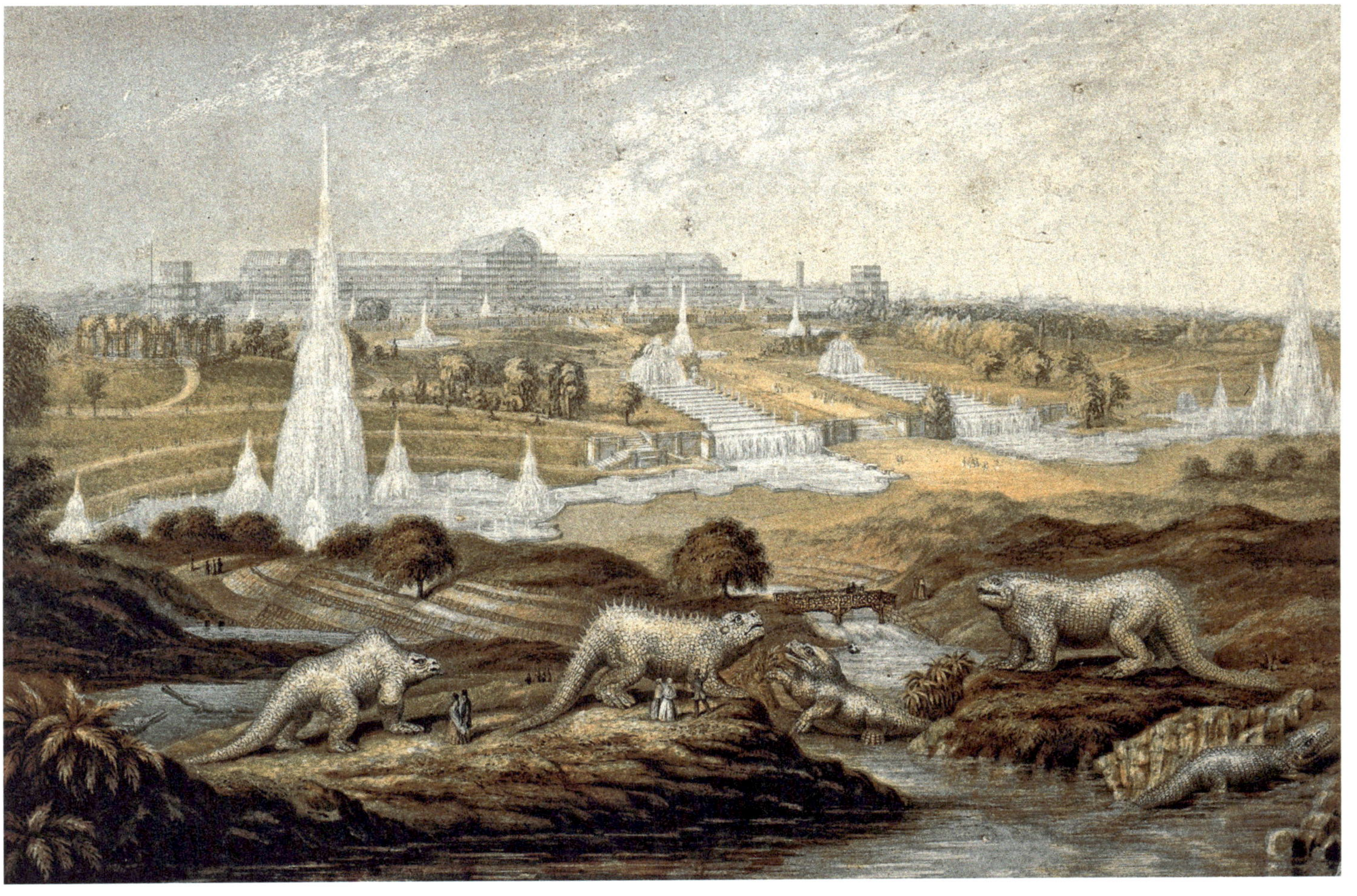

98
Henry De La Beche,
Duria Antiquior, a More Ancient Dorset, 1832
Watercolour, 23 × 32 cm
Cardiff, Amgueddfa Cymru – National Museum Wales

99
Robert Farren,
Duria Antiquior, a More Ancient Dorset, about 1850
Oil on canvas, 166 × 244 cm
Cambridge, Sedgwick Museum of Earth Sciences

100
George Baxter
(after Benjamin Waterhouse Hawkins), *The Crystal Palace and Park in Sydenham*, Day & Son, 1854
Lithograph, 11 × 16 cm
London, The British Library

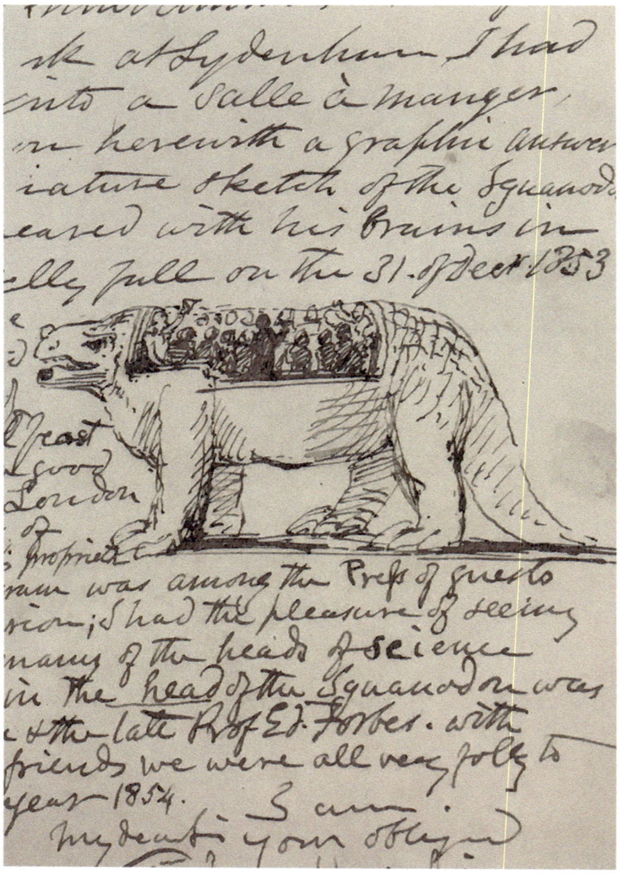
ik at Sydenham I had
into a Salle à manger
n herewith a graphic autocr
iature sketch of the Iguanod
eared with his brains in
lly full on the 31. of Decr 1853
l feast
good
London
of
proprietr
was among the Press of guests
ion; I had the pleasure of seeing
many of the heads of science
in the head of the Iguanodon was
& the late Prof Ed. Forbes. with
friends we were all very jolly to
year 1854. I am
my dear Sir your obliged

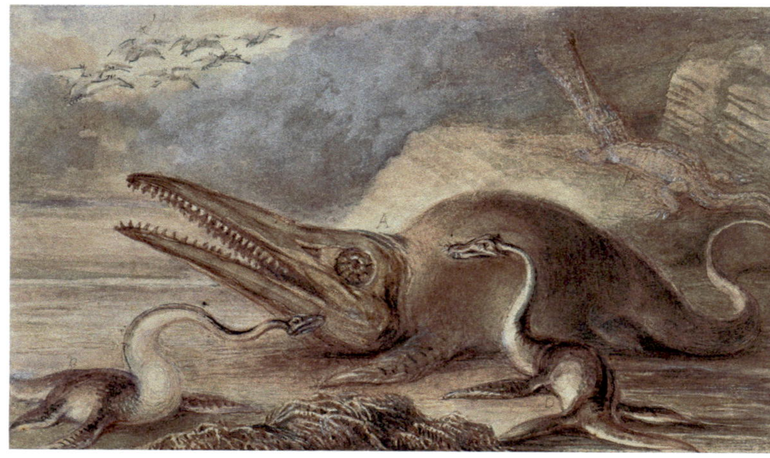

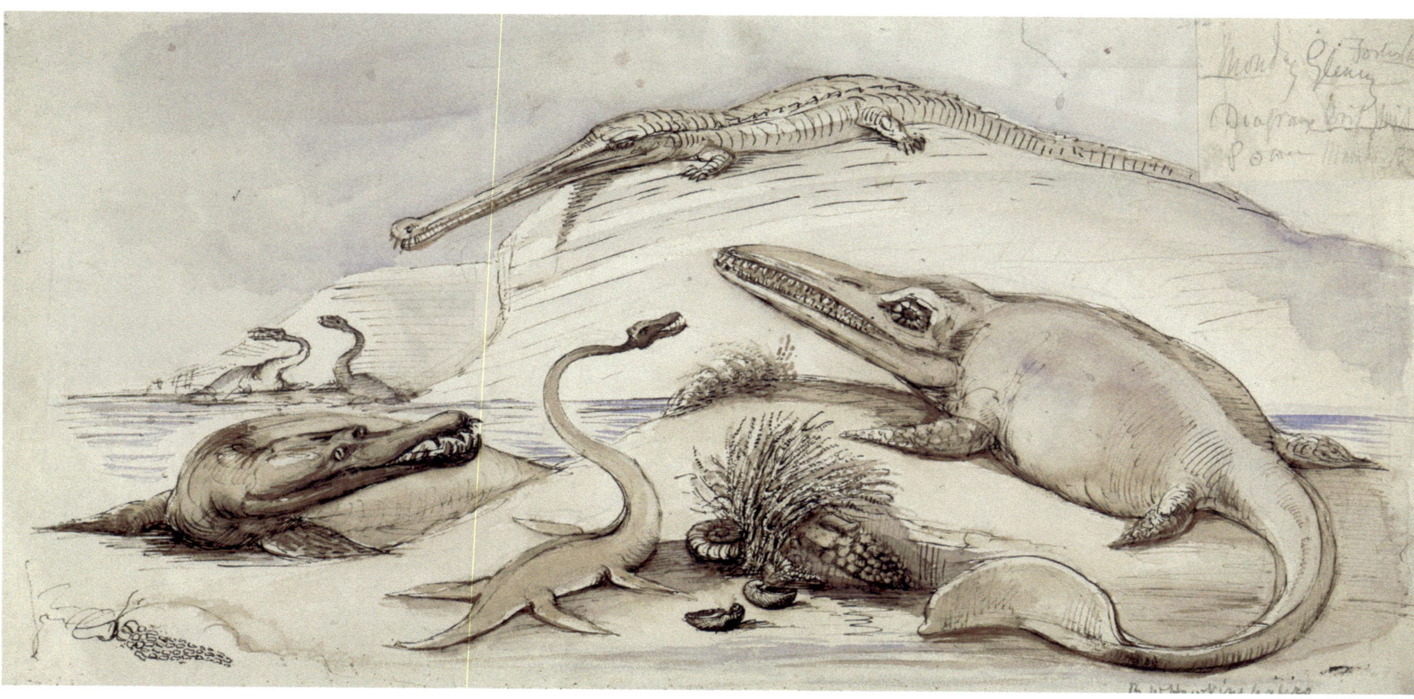

101
Benjamin Waterhouse Hawkins,
Detail of the Dinner Party Held Inside the Iguanodon, after 1854
Handwritten letter, 20.5 × 13.5 cm
London, The British Library

102
Benjamin Waterhouse Hawkins,
Extinct Creatures Including Ichthyosaur, 1858
Watercolour and ink
London, The Natural History Museum

103
Benjamin Waterhouse Hawkins,
Ichthyosaur and Other Extinct Creatures, about 1850
Watercolour and ink,
17.5 × 35.8 cm
London, The Natural History Museum

104
Philip Henry Delamotte,
"Benjamin Waterhouse Hawkins at Work on Dinosaur Models,"
in *The Illustrated London News*,
December 31, 1853
London, The British Library

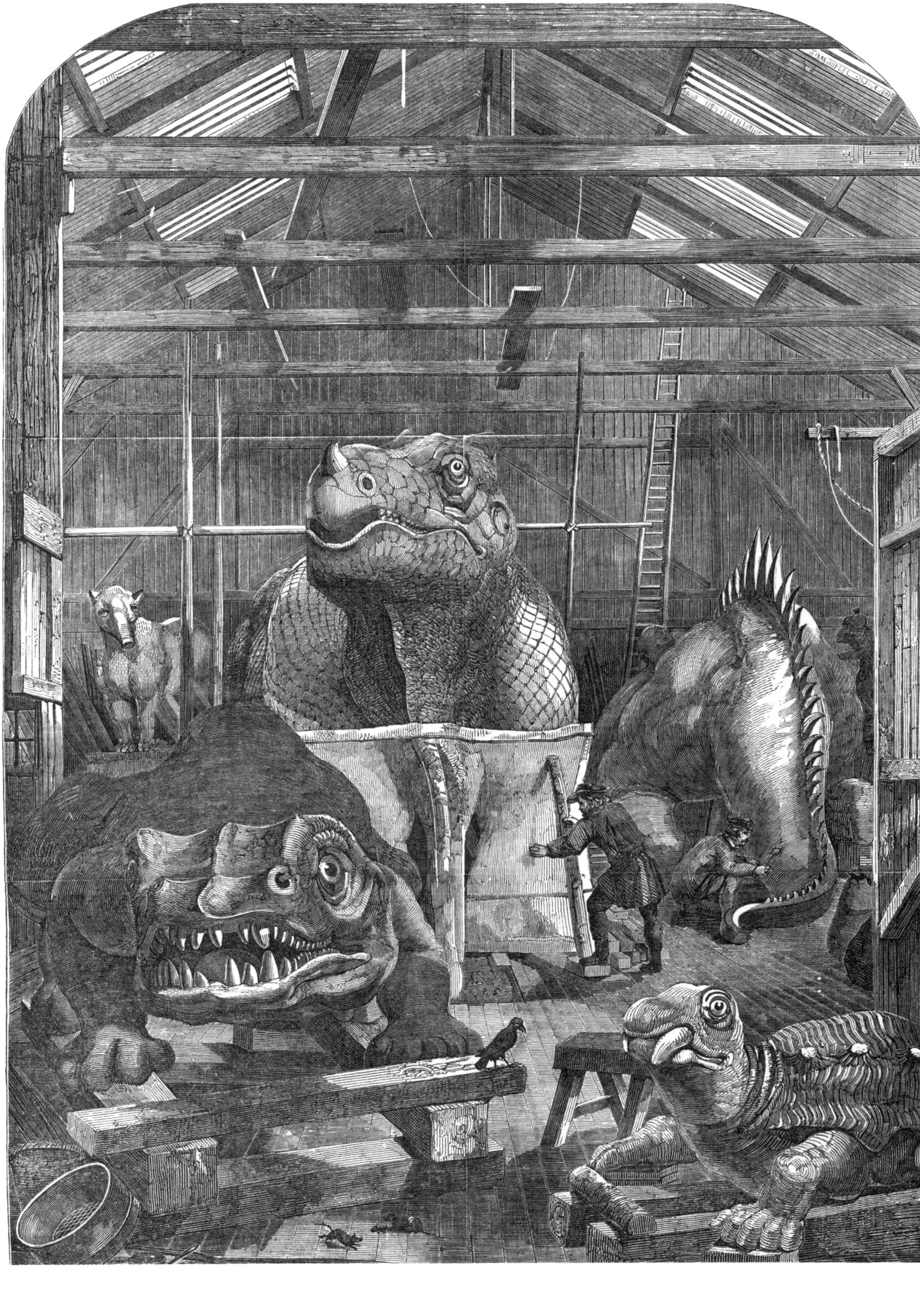

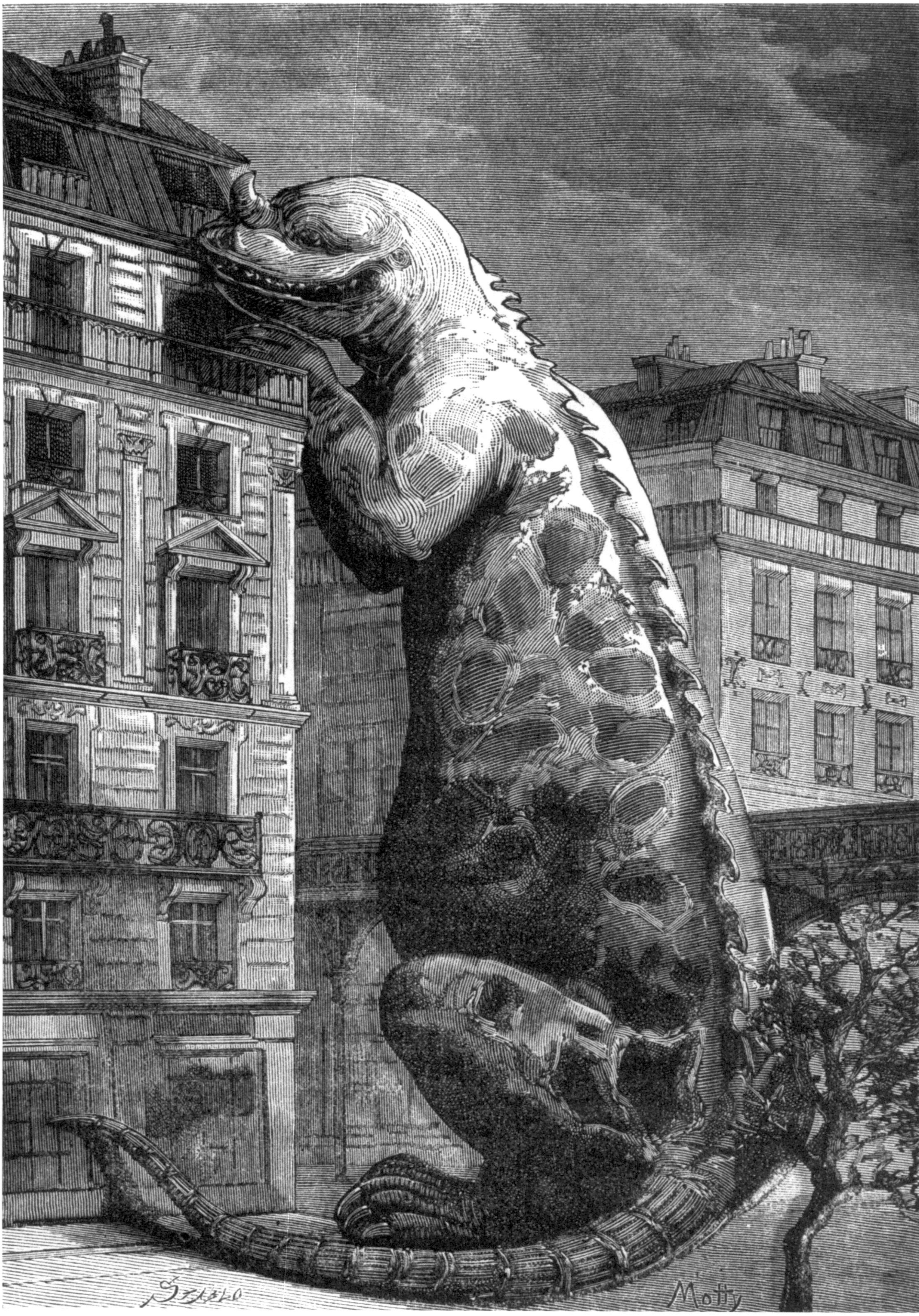

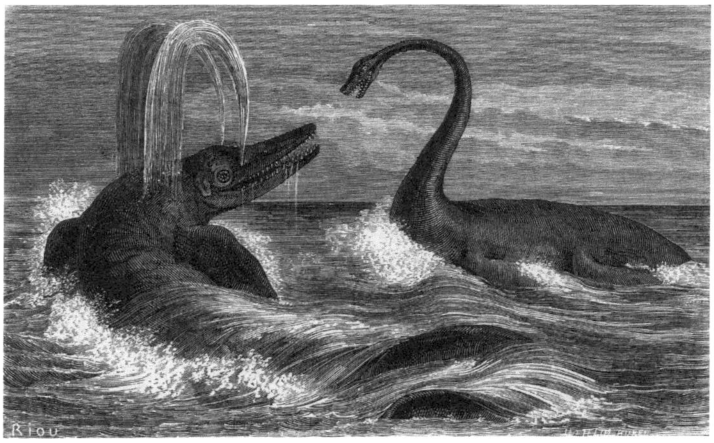

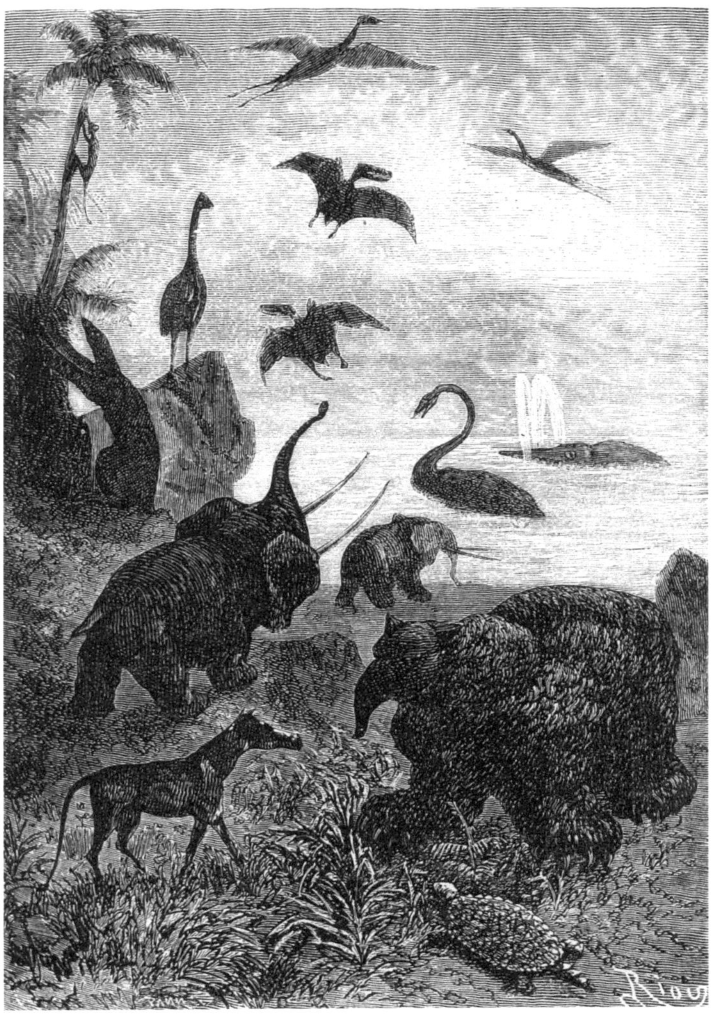

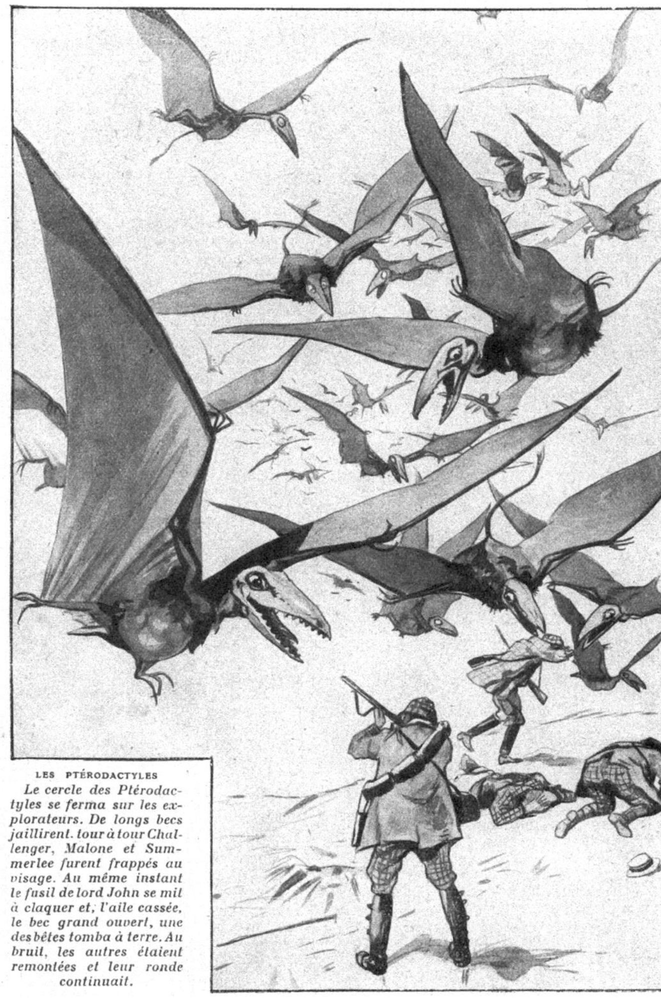

105 FIG.
Anonymous,
"By resting their paws on one of our tallest buildings, they could have eaten on a fifth-storey balcony," in Camille Flammarion, *Le Monde avant la création de l'homme*, Paris: Marpon et Flammarion, 1886
Private collection

106
Louis Figuier, Édouard Riou (illustrations), "Ideal Scene of the Lias with Ichthyosaurus and Plesiosaurus," in *La Terre avant le Déluge*, 2nd edition, Paris: Hachette, 1863
Paris, MNHN

107 FIG.
Jules Verne, Édouard Riou (illustrations), *Journey to the Centre of the Earth*, p. 153, Paris: Hetzel, 1867
Paris, BnF

108 FIG.
Arthur Conan Doyle, *The Lost World*, in *Je sais tout*, January 15, 1914
Paris, BnF

PREHISTORIC HUMANS IN ARCHAEOLOGY AND THE IMAGINATION

NATHALIE RICHARD

The idea of prehistory was not an invention of the nineteenth century, and the belief that human beings existed before the first written sources is an ancient one. But a fundamental breakthrough occurred around 1850: as the deep time of nature was being more comprehensively documented through stratigraphy and fossils, the idea began to gain a foothold in the study of human history.

European antiquarians have been interested in prehistoric remains since the Renaissance.[1] These humanists with a passion for history and archaeology wrote descriptions of monumental sites, such as Stonehenge in England, and discovered graves and objects buried in the ground. They interpreted them in light of ancient sources as the traces of peoples the Greeks and Romans encountered as they advanced through Europe. In the eighteenth century, the uncertainty that still surrounded the flints known as "lightning stones" was lifted. Some doubted that they had been made by humans and attributed the crystallization of these strange stones, believed to be endowed with magical properties, to the impact of lightning hitting the ground. Comparison with the artifacts of Native American peoples helped to settle the question. During the Age of Enlightenment, objects and monuments ceased to be attributed to the Gauls or Germans that Caesar had encountered shortly before the beginning of our era, and a belief emerged that they harked back to an older and more indeterminate time. Their creators were called "Celts," and Celtomania became a veritable vogue in Romantic Europe. "Celtic" or "Druid" monuments were interpreted as the traces of a cult that used dolmens as sacrificial altars.

In the first half of the nineteenth century, the development of archaeology and the popularity of geology, particularly through the efforts of learned societies, slowly changed this mindset. More careful excavations of Neolithic sites, such as those carried out by the Société polymathique du Morbihan (Vannes) in Carnac, France, showed that human sacrifices were a myth and that dolmens were in fact burial monuments. At the same time, naturalists in search of fossils were making enigmatic discoveries. Human remains were unearthed in geological layers that also contained bones of extinct animals, such as mammoths. Were these collections the result of upheavals that mixed together traces of different eras in the same geological strata? Or was it time to acknowledge that humanity was older than we thought, and that there had been a "fossil man" whose history needed to be understood in terms of geological time? This debate began in the 1820s and was not resolved until the 1860s.

Jacques Boucher de Perthes[2] played a key role in this controversy. He was director of the customhouse at Abbeville (Somme), a member of the local learned society and a man of letters. He accumulated a significant collection of prehistoric tools, and in 1847, began publishing arguments for the geological antiquity of humans. In his *Celtic and Antediluvian Antiquities*, he distinguished between a recent Celtic period, identified by polished stone tools and the remains of present-day animals, and an ancient "antediluvian" period, characterized by carved flint tools (112, 113) and the bones of extinct animals (86, P.125). The latter, he believed, was separated from our time by a geological catastrophe originally compared to the biblical Flood. These conclusions, presented by a provincial amateur, did not convince the scientific community. But they were confirmed in 1858 by excavations carried out in England in the Brixham Cave (Devon), following which the famous geologist Charles Lyell, a friend of Darwin's, spoke up in favour of the antiquity of man. The intertwining of new discoveries, Franco-British competitiveness and a quarrel over priorities resulted in 1859 in a general recognition of the geological antiquity of humans.

The modern idea of prehistory emerged from this connection of the period of human history with those of geology and paleontology, and a new scientific discipline was institutionalized in learned societies, museums and, later, universities.[3] Prehistory soon had its own specialized journals and treatises, such as John Lubbock's *Prehistoric Times* (1865) in the United Kingdom and Gabriel de Mortillet's *Le Préhistorique* (1883) in France. International congresses on prehistoric anthropology and archaeology were held regularly from 1866, and as with the dinosaurs, prehistory began to be featured in universal expositions starting in 1867. Arising at a pivotal moment in the evolution of textual and visual mass culture, and contemporaneous with the advent of the modern broadsheet press and the beginnings of the popularization of science, these new theories rapidly made their way into popular culture, literature and the arts.

The concept of a deep human past appeared at the same time as the publication of *The Origin of Species* and as the anthropological science of races was taking shape in Europe. Like biology and anthropology, prehistory raised questions that echoed the social concerns of the nineteenth century. In the absence of definitive dating techniques (carbon-14 was not used until 1951 at Lascaux), the first challenge was to establish the chronology of the archaeological sites. A general framework, developed in Northern Europe in the first half of the nineteenth century, was provided by the "three-age system," which distinguishes between the Stone, Bronze and Iron Ages. The English scientist John Lubbock coined the terms "Paleolithic" and "Neolithic." The French paleontologist Édouard Lartet proposed dating remains based on the fossils of the dominant animal in the archaeological layers: the term "Reindeer Age" is still used today. The system that prevailed, however, is the one developed by Gabriel de Mortillet around 1870: periods were identified by the tools that typified them and were named after the sites where they were found. Paleolithic is thus divided into four main periods: Acheulian (site of Saint-Acheul, Somme), Mousterian (Le Moustier, Dordogne), Solutrean (Solutré, Saône-et-Loire) and Magdalenian (La Madeleine, Dordogne). This chronology was widely accepted in the nineteenth century and was considered universal. It stems from a linear vision of technological progress in accordance with

1 Alain Schnapp, *The Discovery of the Past: The Origins of Archaeology*, trans. Ian Kinnes and Gillian Varndell (London: British Museum Press, 1996); Bruce Trigger, *A History of Archaeological Thought*, 2nd ed. (Cambridge: Cambridge University Press, 2006).

2 Claudine Cohen and Jean-Jacques Hublin, *Boucher de Perthes. Les origines romantiques de la préhistoire* (Paris: Belin, 1989; new revised edition, "Alpha" series, 2017).

3 Marc Groenen, *Pour une histoire de la préhistoire. Le Paléolithique* (Grenoble: Jérôme Millon, 1994); Arnaud Hurel, *La France préhistorienne de 1789 à 1941* (Paris: CNRS Éditions, 2007); Nathalie Richard, *Inventer la préhistoire. Les débuts de l'archéologie préhistorique en France* (Paris: Vuibert, 2008).

the ideology of the time, and raises the question of parallelism between the increasing sophistication of tools, the development of mental and social abilities and the biological evolution of humans.

Could prehistory provide evidence of evolution and eventually document the simian ancestry of humans? The first prehistoric human bones that were determined to be authentic—the remains unearthed by Louis Lartet in the Cro-Magnon cave (Dordogne) in 1868—were those of modern humans. A skullcap discovered by naturalist Carl Fuhlrott in 1856 in the Neander Valley (Germany) and studied by anatomist Hermann Schaaffhausen, launched a debate about the possible existence of an older race that was different from present-day humans. But this bone fragment was initially regarded as distorted by disease. The debate over the presence of a "missing link" between apes and humans then centred on a hypothesis that there was a being who made tools in a geological period preceding the Ice Age, known as the "Tertiary Era." Notched bones and flint objects found in Europe and the United States fuelled these discussions. Some denied that the objects were tools, and maintained that they had been shattered into shards by a huge flood, or by ice or fire. Those who believed that the flints had been carved were divided about who had made them. The Abbé Louis Bourgeois, who discovered the first Tertiary flint tools at Thenay in 1867 (Loir-et-Cher), spoke of a "Tertiary man" indistinguishable from modern humans. Admitting that "days" as used in the Genesis narrative of creation must be understood metaphorically, he insisted that humankind was much older than the nearly 6,000 years that seventeenth-century exegetes had calculated based on the Old Testament. But this Catholic clergyman remained a creationist and refuted the concept of evolution. Others, however, followed the prehistorian Gabriel de Mortillet, one of the champions of evolutionism in France, who spoke of a "precursor of man" or an "anthropopithecus." The physical appearance of this intermediate being was the subject of conjecture, in the absence of any bone fragments to describe it. Representations of it were based upon older iconographic traditions, such as the "men of the woods" that Linnaeus assumed existed in the eighteenth century, or images of the racial physiognomy of black men.[4] Some drew purely on their own imaginations and invented a chimera who was half-man, half-ape (114).

In the late nineteenth century, the debate over the ancestors of humans took a different direction when more complete Neanderthal skeletons were discovered at Spy (Belgium, 1886), Krapina (Croatia, 1899), La Chapelle-aux-Saints and La Ferrassie (France, 1908, 1909). These remains brought to light anatomical features that were distinct from those of modern humans. The term *Homo neandertalensis*, devised in 1864 by the Irishman William King, circulated and marked the beginning of a gradual acceptance of an interpretation of evolution in which the Neanderthal was considered distinct from modern humans. But this thesis itself led to a variety of interpretations. Its most radical iteration was expressed by Frenchman Marcellin Boule: Neanderthals

4 Raymond Corbey and Bert Theunissen, eds., *Ape, Man, Apeman: Changing Views Since 1600. Evaluative* (Leiden: Leiden University Press, 1995); Albert Ducros and Jacqueline Ducros, eds., *L'Homme préhistorique. Images et imaginaire* (Paris: L'Harmattan, 2000).

were a divergent branch that had been killed off by *Homo sapiens*. This idea, fuelled by imagining that Neanderthals shared a bestiality and primitivism (imperfect bipedalism, cannibalism) with the "Tertiary precursor," resulted in an insistence on the "inferiority" of certain anatomical and cultural traits. The question of the origin of modern humans remained unanswered since, in this view, Neanderthals could not be their ancestor. The discovery on the island of Java of a fossil hominid named *Pithecanthropus erectus* (1891–1893) challenged Europe's claim to be the cradle of humanity. This gave rise to the hypothesis of a European *pre-sapiens* from which both Neanderthals and *Homo sapiens* would have originated. The oldest European fossils known at that time (Mauer's jaw, Germany, 1908) were interpreted according to this hypothesis, which also led to the perpetration of a famous fraud. In 1912, a skull fragment found in Piltdown (Sussex) was presented to the Natural History Museum in London. Modern in shape, it was attached to an incomplete jaw that appeared to be simian. Many paleontologists believed this to be the "missing link," and in the context of the rise of nationalism prior to the First World War, some British people saw this as proof that England was in fact the cradle of humanity. Fluorine testing exposed the hoax in 1953: an unknown forger had attached a monkey's jaw to a modern human skull. Arthur Conan Doyle, who was then writing his prehistoric novel *The Lost World* (1912), was among the names put forward as possible perpetrators (108).

A final major debate concerned the use of symbolism among prehistoric humans. The discoveries of Cro-Magnon remains near the village of Les Eyzies (France) in 1868, and in the Balzi Rossi caves near Ventimiglia (Italy) by Émile Rivière in 1872, launched discussions about prehistoric funerary practices. The debate was particularly animated in France, against the backdrop of the policy of secularism instituted by the Third Republic. Some believed that the burials confirmed the universality of religious beliefs across time and space, supporting their own Christian convictions. One of these was the geologist Adrien Arcelin, who thought he had discovered tombs on the excavation site of Solutré (France). Arcelin, a Catholic who made no secret of his faith, played an active role in the International Scientific Congresses of Catholics held between 1888 and 1900. Other freethinkers did not believe that primitive humans could have held such complex beliefs. For them, the burial sites were not prehistoric, but pits that had been recently dug on ancient lands, a position Gabriel de Mortillet, one of the most anticlerical prehistorians of the time and a political radical, held until his death in 1898. Prehistoric art generated similar debates. Bones and engraved tablets were discovered as early as the 1860s and were immediately labelled "furniture" art, interpreted primarily as an art of leisure and imitation, without any symbolic meaning. Cave art, on the other hand, was almost unanimously rejected. The first decorated cave, Altamira, was discovered in 1879 in northwestern Spain by Marcelino Sanz de Sautuola. The beauty and complexity of the polychrome frescoes, their state of conservation, and the need there would have been for lighting in the cave to create them were all arguments against acceptance of their authenticity. It seemed inconceivable that the humans of the Paleolithic period, then perceived as primitive, could have created these works. The vast majority of prehistorians thus refused to accept the concept of cave art until further discoveries, particularly in southwestern France, revived the issue in a new intellectual context. Cultural anthropology brought to light the complexity of the social structures and beliefs of so-called "primitive" peoples, as can be seen in James George Frazer's *The Golden Bough* (1890). The scientific community changed its position on the subject around 1900, and cave art became an essential element in the twentieth-century view of prehistory.

In the nineteenth century, therefore, one was inevitably taking a philosophical and political stance by choosing whether to portray prehistoric humans with modern or ape-like features, whether to represent their religious practices and social structures, and whether to accentuate their artistic activities. It is thus not only aesthetic choices that distinguish the sculptures of Louis Mascré and Aimé Rutot from those of Paul Richer (206, P. 297).

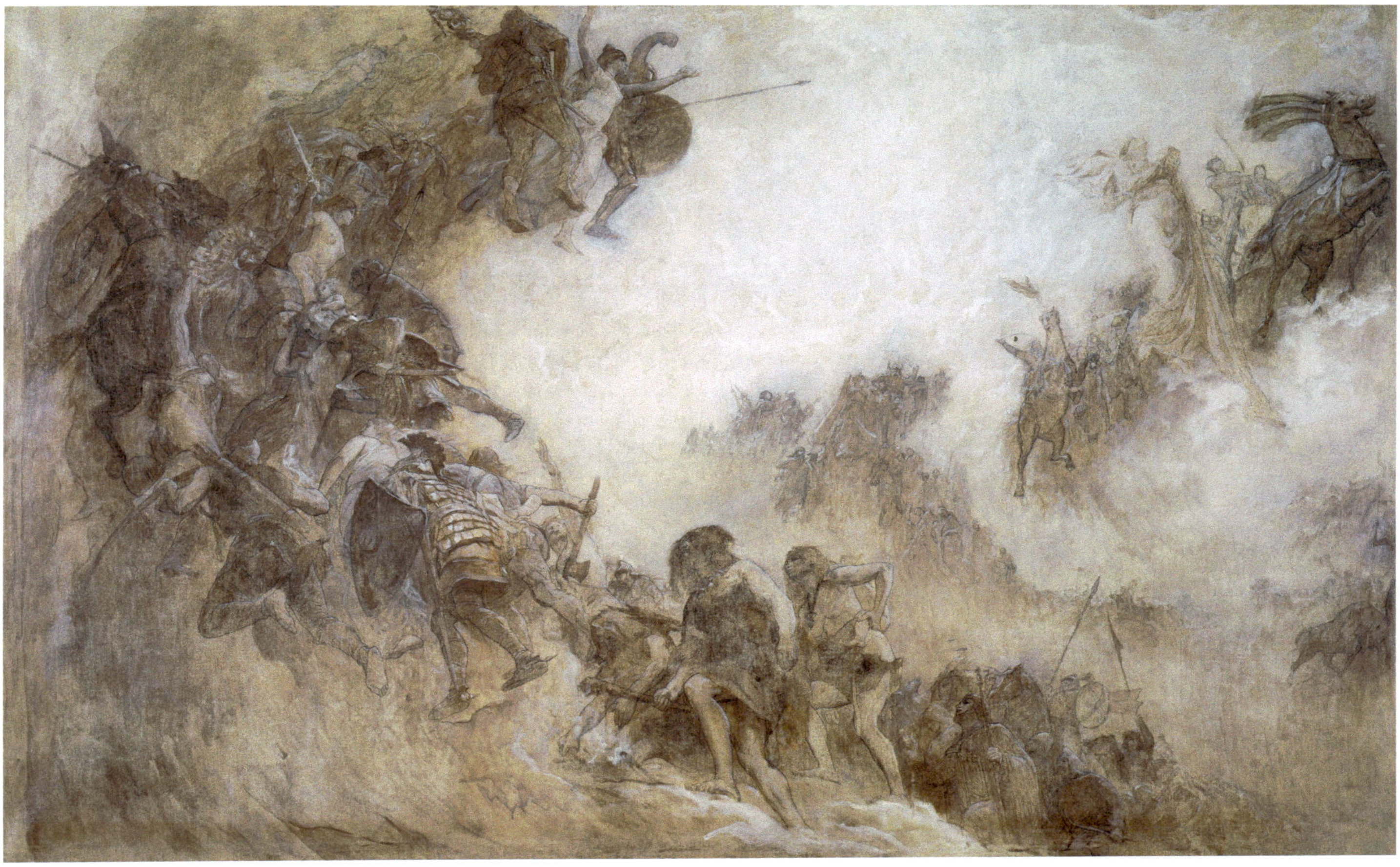

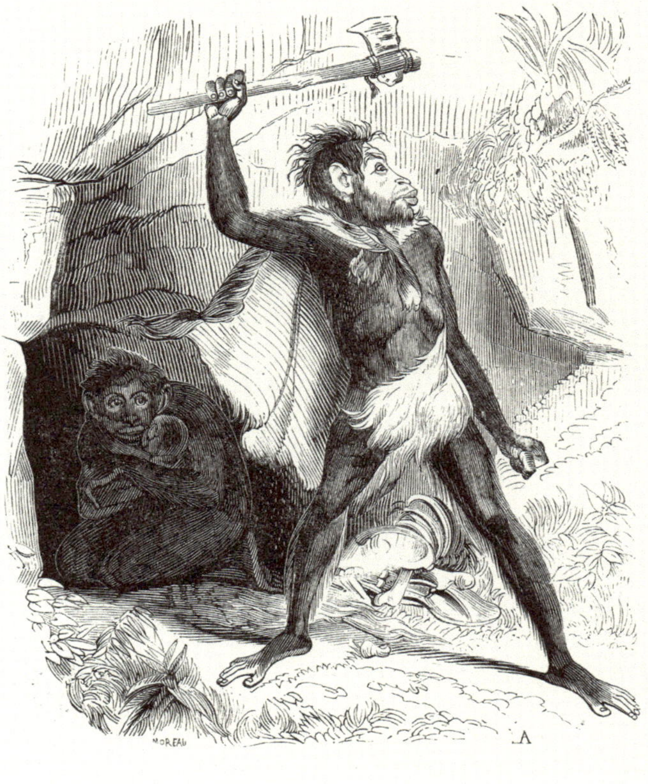

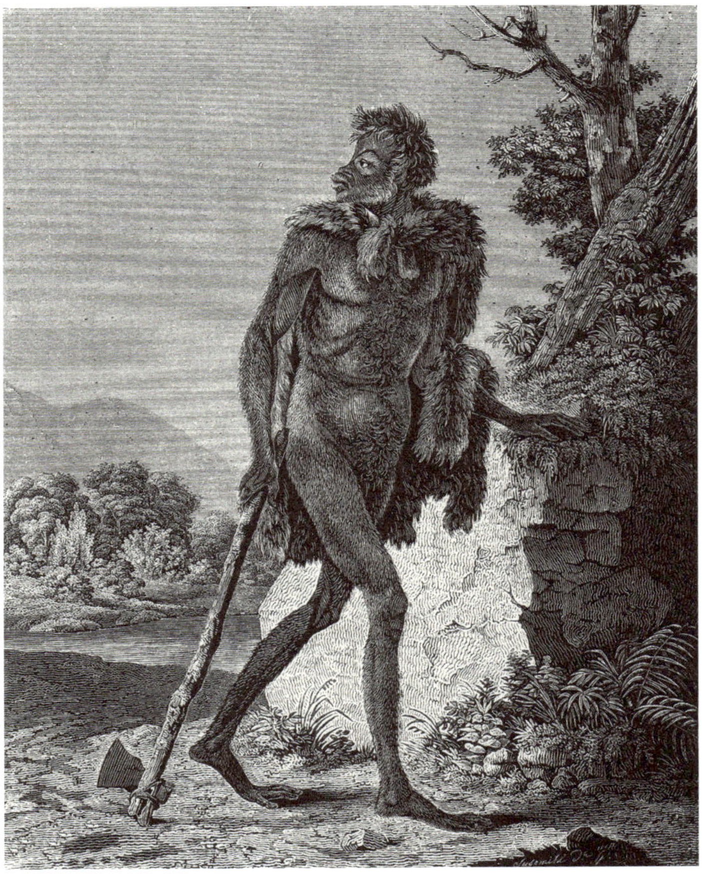

109
Fernand Cormon,
The Human Races, sketch for the ceiling of the amphitheatre of paleontology in the Muséum d'histoire naturelle de Paris, 1897
Oil on canvas, 76 × 118.5 cm
Paris, Petit Palais, musée des Beaux-Arts de la Ville de Paris

110
Pierre Boitard,
"Fossil Man," in
Études antédiluviennes: Paris avant les hommes,
Paris: Passard, 1861
Paris, MNHN

111
Pierre Boitard, Johann Theodor Susemihl (draughtsman),
"Fossil Man," in *Le Magasin universel*, vol. 5, April 1838
Paris, BnF

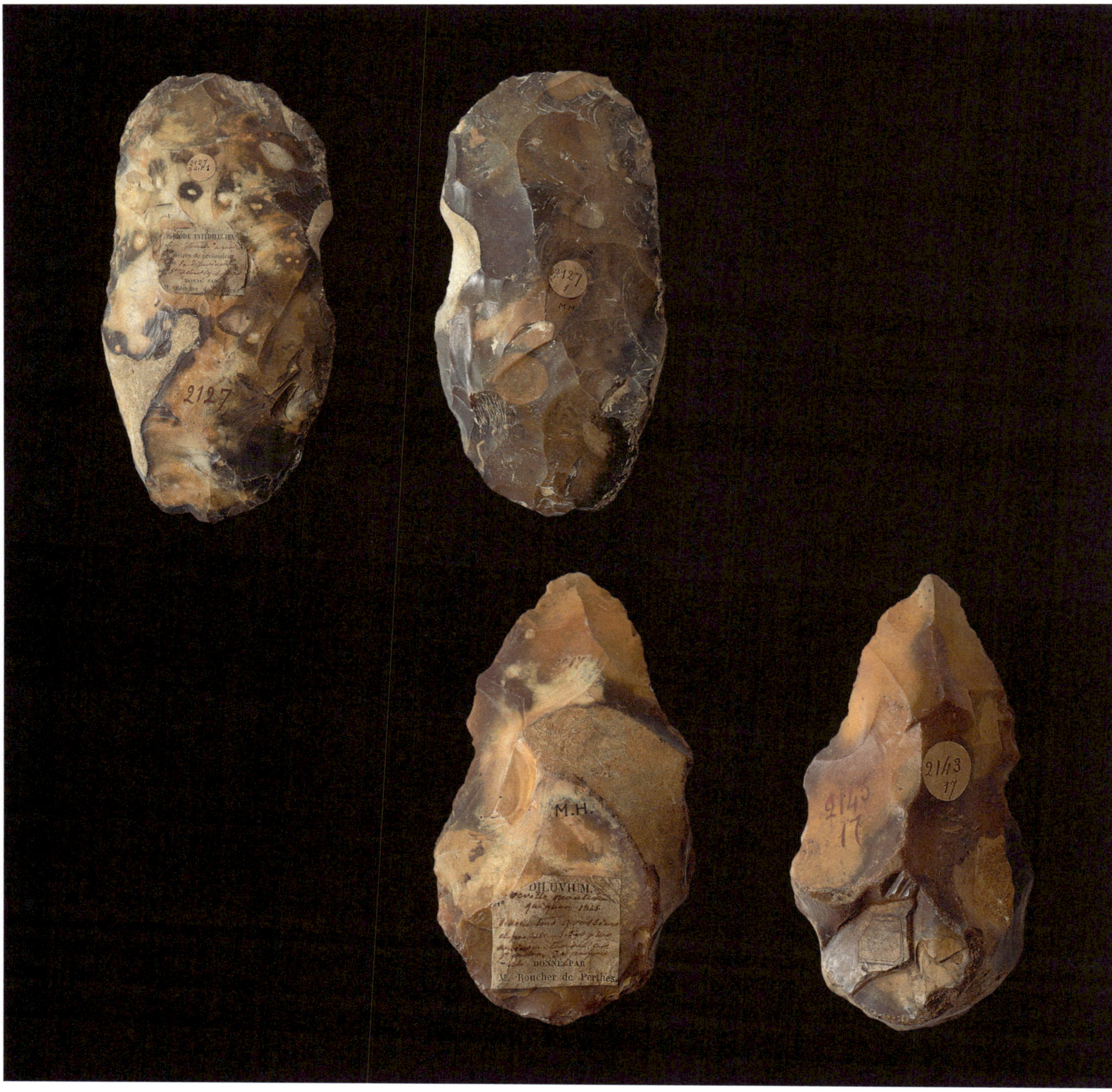

112
Biface, Middle or Late Paleolithic
Cut stone,
22.5 × 10.5 × 4.3 cm
Paris, MNHN

113
Biface, Middle or Late Paleolithic
Cut stone,
14.4 × 7.5 × 5.2 cm
Paris, MNHN

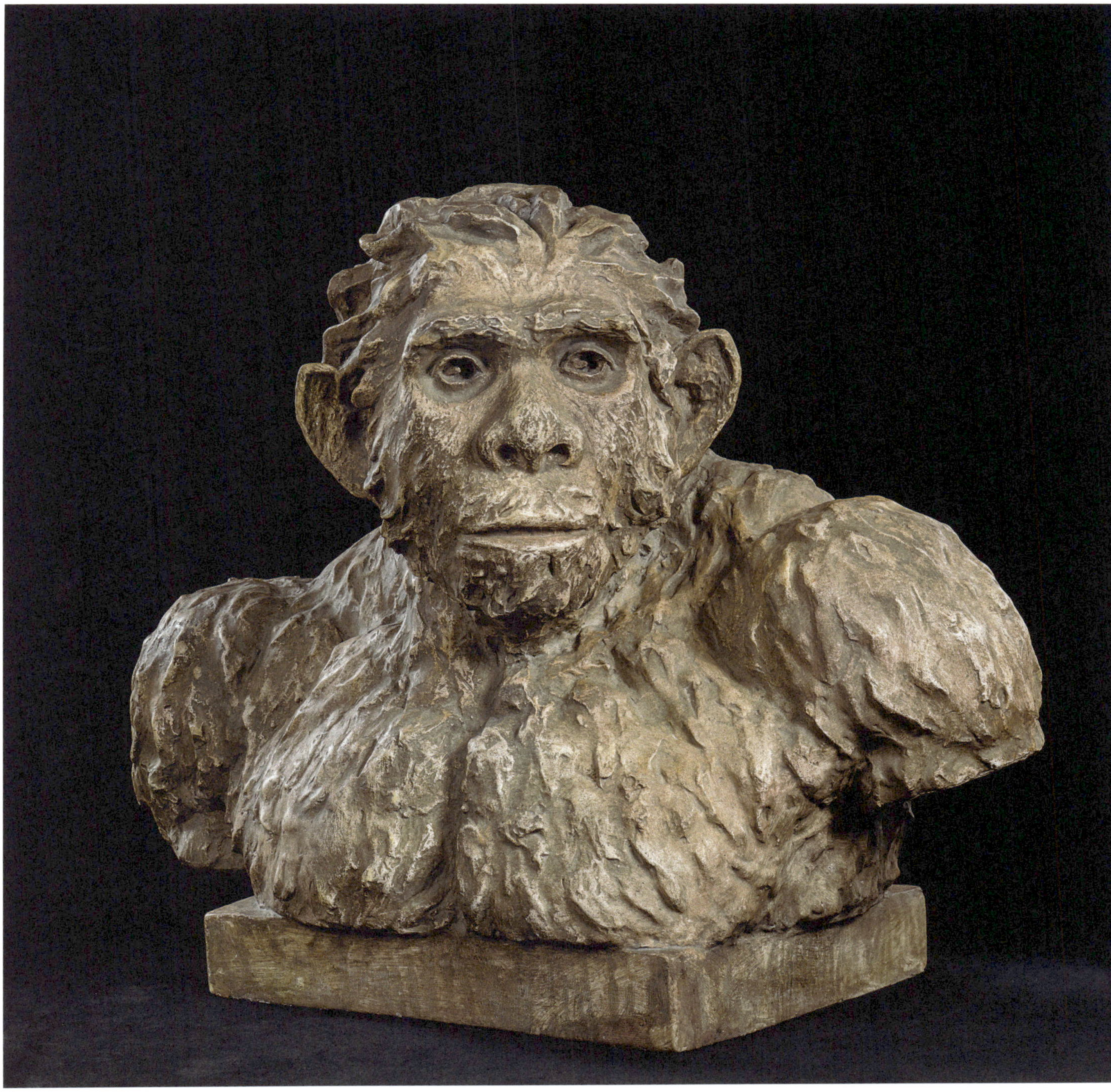

114
The Neanderthal Man from Le Moustier, Reconstruction by Mr. Faure and Miss Parvillée, 1923
Plaster, 50 × 60 × 30 cm
Les Eyzies, Musée national de Préhistoire

ROSNY AÎNÉ'S NOVEL OF ORIGINS

DANIELLE CHAPERON

J.-H. Rosny aîné is the pseudonym of Joseph Henri Honoré Boex. Until 1908 he formed, with his younger brother Séraphin Justin François, a writing partnership whose works appeared under the same pseudonym of J.-H. Rosny. In 1935, the two brothers signed an agreement which settled the division of their titles between the "elder" (l'aîné) and the "younger" (le jeune).[1] J.-H. Rosny aîné sketched the outline for his first "*Cavernes* novel" in 1885, a time when his burgeoning career rested on a series of social novels. Two years later, he broke with Zola and distanced himself from the naturalist model.[2] In 1891, when *Vamireh* was published, as part of Jules Huret's *Enquête sur l'évolution littéraire*, he laid claim, from that point on, to "a more complex, nobler literature . . . through a more profound, more analytical and more just comprehension of the *entire* universe and of the humblest of individuals, acquired through science and the philosophy of modern times."[3] Science was always "a *poetic* passion"[4] for Rosny aîné, but he may have acquired an interest in prehistory from Alphonse Daudet, his new mentor.[5] He believed that this science still left "a great deal to conjecture,"[6] while providing a valuable documentary foundation. For his six novels about primitive times (*Vamireh* [1891], *Eyrimah* [1895], *Nomai* [1897], *The Quest for Fire* [1911], *The Giant Feline* [1916], *Helgvor of the Blue River* [1930]),[7] he referred principally to the works of Gabriel de Mortillet,[8] whose classification of periods[9] he adopted, meticulously following the descriptions regarding fauna and flora. There was no risk (or chance) of encountering, as he turned a page, the fascinating ichthyosaurus or plesiosaurus of Jules Verne's *Journey to the Centre of the Earth*, or the pterodactyl and iguanodon of Arthur Conan Doyle's *The Lost World* (108, P. 153).[10] Only the human ancestors and, consequently, the animals with which they actually coexisted (for instance mammoths) were of interest to him (116, 117).

With regard to hominids, Rosny aîné's imagination was, by contrast, utterly unbridled. The Men-without-Shoulders (the Wah), the Red-Skinned Little Men, the Blue-Skinned Men, the Lemur Men, the Worm Eaters, the Kzamm, the Dhole Men, innumerable and improbable precursors allowed him to fictionalize Darwin's discoveries. Thus the Wah in *The Quest for Fire* and *The Giant Feline*, refined, mutually supportive and hospitable, though frail and slow, exemplify a principle of *The Descent of Man*: "Natural selection acts only in a tentative manner. Individuals and races may have acquired certain indisputable advantages, and yet have perished from failing in other characters."[11] As for Gammla in *The Quest for Fire*, she demonstrates the importance of "the exertion of some choice on the part of the female [that] seems a law almost as general as the eagerness of the male."[12] The female characters prefer hairless individuals (Naoh rather than the hirsute Aghoo), but also those who are slower to anger and gentler with weaker creatures. Rosny aîné's heroes always have qualities that distinguish them among their clan members: curiosity, an artistic sensibility, a taste for exotic loves and odd friendships. This, in contrast to the laws formulated by the followers of Herbert Spencer's social Darwinism, which recommends eliminating the weakest and the prohibition of miscegenation.

From his extensive reading of contemporary psychologists,[13] Rosny ainé would draw material to flesh out his characters' inner natures. In *The Giant Feline*, for example, Zouhr has an intelligence that is full of dreams; he "loved those hazy hours during which memories and images unfurled in his head."[14] Quite to the contrary, his friend Aoun has a mind that is "vivid, bold and succinct"; it is drawn only to the "immediate difficulties and movements of living creatures."[15] Aoun is the son of the hero of *The Quest for Fire* (an Oulhamr) and Zouhr is the last "man-without-shoulders" (a Wah), whose fellow humans were victims of the "continuous slowing down of their thinking and their movements."[16] "Their actions seemed to be full of uncertainty; they often acted like creatures who are fighting off sleep or emerging from a dream; yet they almost never made mistakes."[17] Indeed, Zouhr manages to find time for reflection in between perception and taking action: "We must wait" and "The Urus's son should not fight yet!" he repeats to his friend. The effectiveness of his counsel is evidence that adaptation depends not only on heredity, but also on exchanges that we would describe as cultural.

The points of view of primitive beings—animals or humans—are central to the writer's research. Contrary to *Journey to the Centre of the Earth* or *The Lost World*, his prehistoric novels provide no modern character with which the reader may identify. No intrepid adventurer (laden with guns), no irascible sage (armed with binoculars) whom we can follow, from London or Paris, into the depths of an Icelandic volcano or an Amazonian forest. Rather than being assigned the role of target or specimen, primitive beings become the entire focus of attention (116). "What! Make people speak who resemble the natives of Tierra del Fuego, whose entire language is made up of seventy-five words! Yes, in a *hypothetical historical study*, fine! But a novel!" exclaimed Goncourt.[18] In fact—in keeping with naturalist technique—the narrator's voice takes charge of what the character perceives and senses without their being able to formulate it themselves. But the narrative tone of this voice can be surprising: "Two large clouds drew together, smoky, hemmed in phosphorescence. An anguishing breath, gyrating, ascending, enveloped the plants; thunder rumbled majestically above the forest. Vamireh delighted in this storm, his entire being breathed in the power and the movement, emotions that echoed his state of mind."[19] Goncourt would easily have recognized the mark of an appalling "decadent" influence. The choice of artistic, or impressionist, writing is, however, not so far-fetched if one recalls that the aim of this style is raw sensation, even if one must deconstruct syntax and supersede the word. Nominal sentences, substantive adjectives, neologisms, assonance and alliteration, all the obsessions denounced by Goncourt,[20] are, for Rosny aîné, suitable for expressing the acute and complex perceptions of our ancestors.[21] From *Vamireh* to *The Quest for Fire*, the style would become

1 This contribution benefited from two previous articles: "Préhistoire 1887. L'exotisme antérieur de J.-H. Rosny aîné," in *Crise fin-de-siècle et tentation de l'exotisme*, proceedings of the conference (Université Charles-de-Gaulle – Lille 3, March 16–17, 2001), ed. Guy Ducrey and Jean-Marc Moura (Villeneuve-d'Ascq: Université Charles-de-Gaulle – Lille 3, 2002), p. 143–155; and "La préhistoire expérimentale de J.-H. Rosny aîné," in *Les Écrivains face au savoir*, ed. Véronique Dufief-Sanchez (Dijon: Éd. universitaires de Dijon, 2002), p. 61–74. See also the documentation assembled by Jean-Michel Pottier in *Les Cahiers naturalistes*, no. 70, 1996.

2 He was one of the signatories—and the editor—of the "Manifeste des Cinq," published in *Le Figaro*, August 28, 1887.

3 Jules Huret, *Enquête sur l'évolution littéraire* [1891] (Paris: José Corti, 1999), p. 241.

4 J.-H. Rosny aîné, *Torches et lumignons. Souvenirs de la vie littéraire* (Paris: La Force française, 1921), p. 11.

5 He began frequenting Daudet in 1887. See ibid., p. 148.

6 J.-H. Rosny aîné, *Les Origines* (Paris: Borel, 1895), p. 167.

7 *Elem d'Asie* (1896) is an abridged version of *Vamireh*. Some titles have a generic indication: "Roman des temps primitifs" [A novel of prehistoric times] (*Vamireh*) or "Roman des âges farouches" [A novel of a savage age] (*Helgvor of the Blue River*). See J.-H. Rosny aîné, *Romans préhistoriques* (Paris: Robert Laffont, "Bouquins" series, 1985).

8 Gabriel de Mortillet, *Le Préhistorique. Antiquité de l'homme* (Paris: Reinwald, 1883).

9 *The Quest for Fire* and *The Giant Feline* were set in the Mousterian Period, *Vamireh* and *Elem d'Asie* take place at the end of the Magdalenian, *Helgvor of the Blue River* during the "hiatus" between the Paleolithic and Neolithic, and *Eyrimah* and *Nomai* during the Neolithic.

10 Jules Verne, *Journey to the Centre of the Earth*, trans. William Butcher (Oxford: Oxford University Press, 1992), first published as *Voyage au centre de la Terre* (Paris: Hetzel, 1864); Arthur Conan Doyle, *The Lost World*, first published in *The Strand Magazine* in 1912.

more subdued, but the writer continued to plunge the reader into a primeval world that bears a greater resemblance to the visions of Monet or Redon than to the reconstitutions of Jamin (115, 117) or Rousseau-Decelle (248, P. 383). Let us consider this excerpt:

"He inhaled the night with his entire being. He was an extraordinary creature, penetrated by the subtle things of the Universe: with his sight, he captured phosphorescence, pale shapes, movements of the shadows and he rose up among the stars; with his hearing, he distinguished the voices of the breeze, the creaking of plants, the flight of insects and birds of prey, the tread and crawling of beasts; he discerned the jackal's distant yelping, the hyena's laugh, the howling of wolves, the cry of the osprey, the trilling of locusts; the flowers' perfume, the sweet scent of grass, the stench of wild beasts, the reptiles' stale or musky odour all penetrated his nostrils. His skin quivered with each minute variation of cold and hot, of dampness and dryness. Thus he experienced that which filled Space and Time."[22]

11 Charles Darwin, *The Descent of Man, and Selection in Relation to Sex* (London: John Murray, 1871), vol. 1, p. 178.
12 Ibid., p. 273.
13 In particular, Théodore Ribot's *Psychologie de l'attention* (Paris: Alcan, 1894), but he also quotes William James, Alfred Binet, Maurice Guyau, Gustave Le Bon, Alfred Maury, Bergson and Taine. See J.-H. Rosny aîné, *Les Sciences et le pluralisme* (Paris: Alcan, 1922).
14 Rosny aîné, *Le Félin géant*, in *Romans préhistoriques*, p. 366.
15 Ibid.
16 Ibid.
17 Ibid.
18 Edmond de Goncourt, [November 26, 1891], in Edmond and Jules de Goncourt, *Journal. Mémoires de la vie littéraire*, vol. 3: *1887–1896* (Paris: Laffont, "Bouquins" series, 1989), p. 640.
19 Rosny aîné, *Vamireh*, in *Romans préhistoriques*, p. 39.
20 Goncourt, [October 9, 1888], in *Journal*, vol. 3, p. 163.
21 "When I look back upon my childhood, I see myself, as I draw closer to the unconscious, increasingly incapable of simultaneous perceptions, confusing, increasingly, actions, desires, changes of all kinds as things appear." Rosny aîné, *Les Sciences et le pluralisme*, p. 105).
22 Rosny aîné, *La Guerre du feu*, in *Romans préhistoriques*, p. 314.

115
Paul Jamin, *Stone-age Abduction*, 1888
Oil on canvas, 282 × 200.2 cm
Reims, Musée des Beaux-Arts

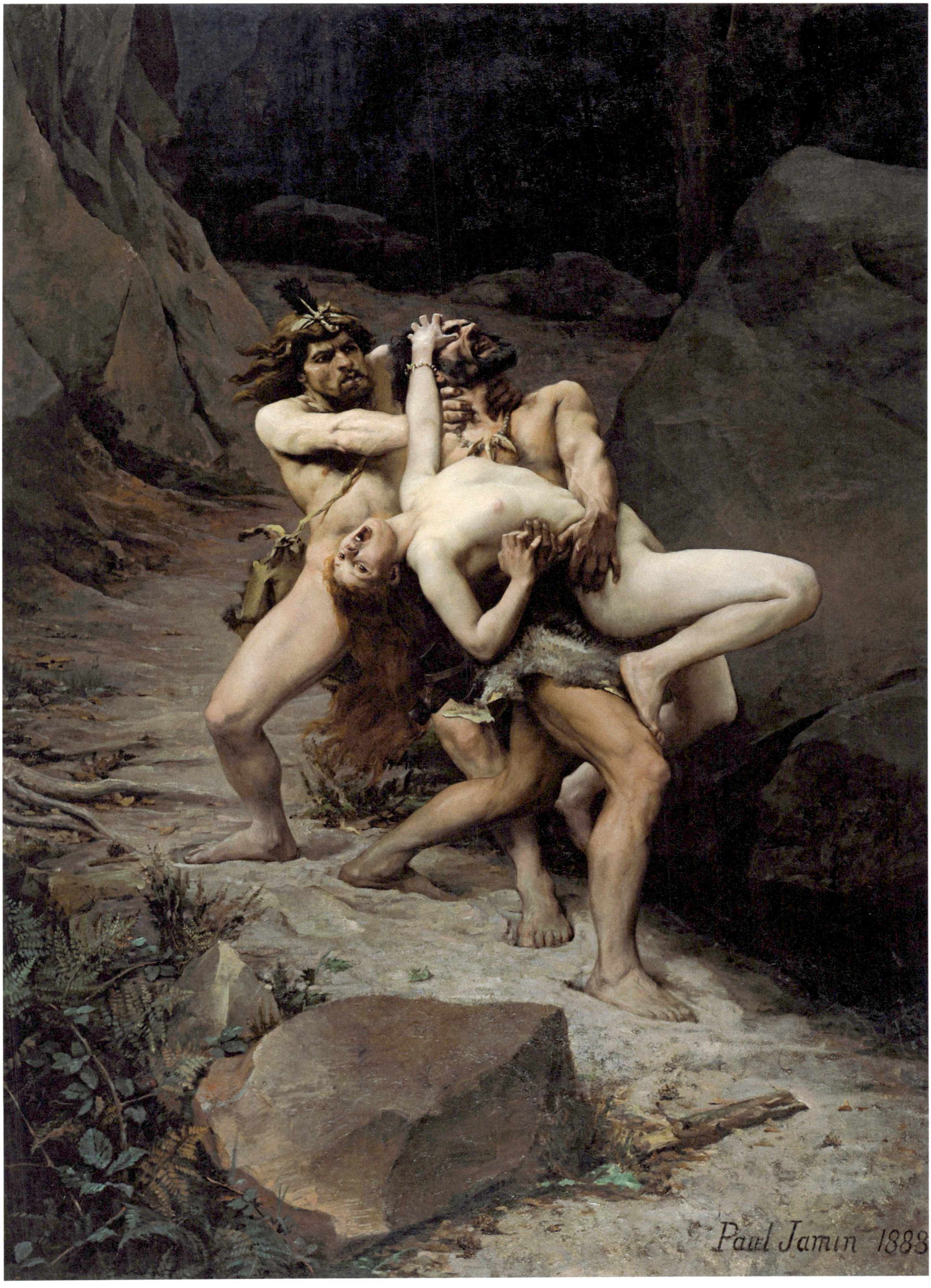
Paul Jamin 1888

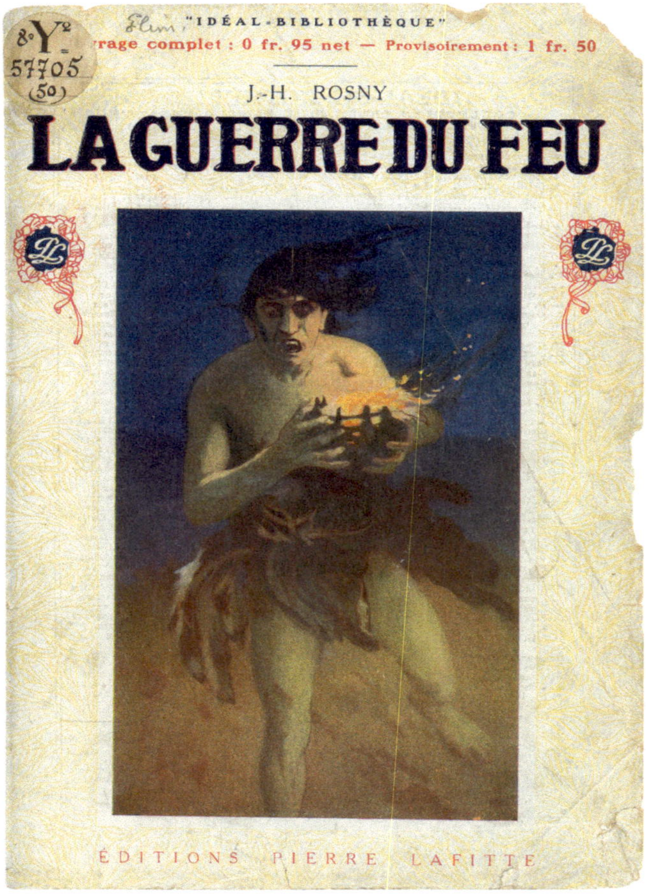

A

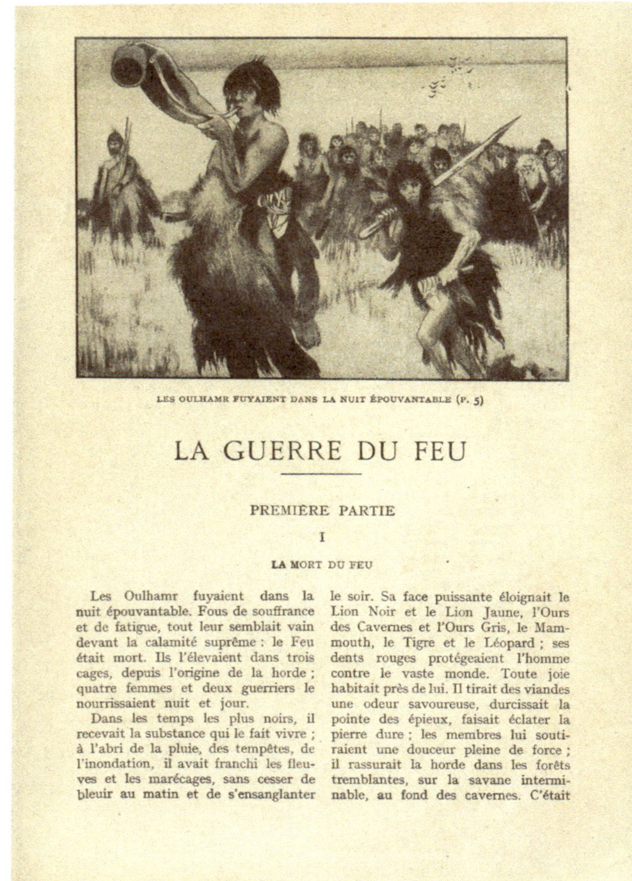

LES OULHAMR FUYAIENT DANS LA NUIT ÉPOUVANTABLE (P. 5)

LA GUERRE DU FEU

PREMIÈRE PARTIE

I

LA MORT DU FEU

Les Oulhamr fuyaient dans la nuit épouvantable. Fous de souffrance et de fatigue, tout leur semblait vain devant la calamité suprême : le Feu était mort. Ils l'élevaient dans trois cages, depuis l'origine de la horde ; quatre femmes et deux guerriers le nourrissaient nuit et jour.

Dans les temps les plus noirs, il recevait la substance qui le fait vivre ; à l'abri de la pluie, des tempêtes, de l'inondation, il avait franchi les fleuves et les marécages, sans cesser de bleuir au matin et de s'ensanglanter le soir. Sa face puissante éloignait le Lion Noir et le Lion Jaune, l'Ours des Cavernes et l'Ours Gris, le Mammouth, le Tigre et le Léopard ; ses dents rouges protégeaient l'homme contre le vaste monde. Toute joie habitait près de lui. Il tirait des viandes une odeur savoureuse, durcissait la pointe des épieux, faisait éclater la pierre dure ; les membres lui soutiraient une douceur pleine de force ; il rassurait la horde dans les forêts tremblantes, sur la savane interminable, au fond des cavernes. C'était

B

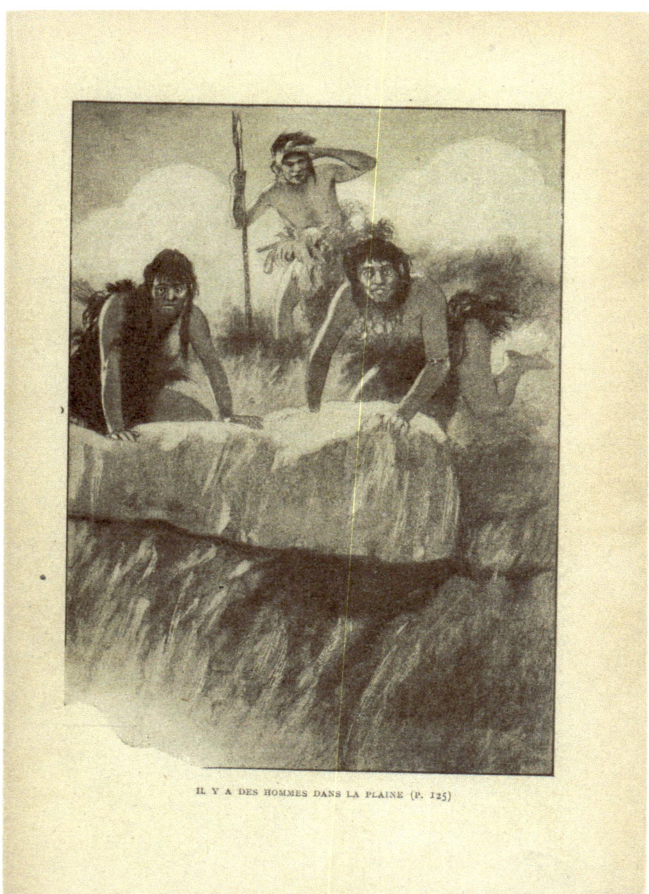

IL Y A DES HOMMES DANS LA PLAINE (P. 125)

C

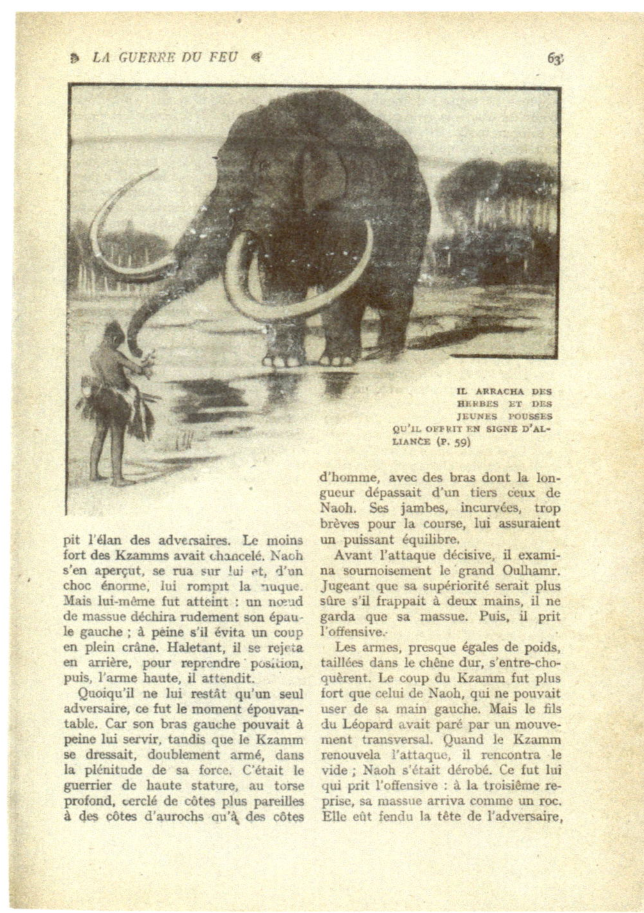

LA GUERRE DU FEU 63

IL ARRACHA DES HERBES ET DES JEUNES POUSSES QU'IL OFFRIT EN SIGNE D'ALLIANCE (P. 59)

pit l'élan des adversaires. Le moins fort des Kzamms avait chancelé. Naoh s'en aperçut, se rua sur lui et, d'un choc énorme, lui rompit la nuque. Mais lui-même fut atteint : un nœud de massue déchira rudement son épaule gauche ; à peine s'il évita un coup en plein crâne. Haletant, il se rejeta en arrière, pour reprendre position, puis, l'arme haute, il attendit.

Quoiqu'il ne lui restât qu'un seul adversaire, ce fut le moment épouvantable. Car son bras gauche pouvait à peine lui servir, tandis que le Kzamm se dressait, doublement armé, dans la plénitude de sa force. C'était le guerrier de haute stature, au torse profond, cerclé de côtes plus pareilles à des côtes d'aurochs qu'à des côtes d'homme, avec des bras dont la longueur dépassait d'un tiers ceux de Naoh. Ses jambes, incurvées, trop brèves pour la course, lui assuraient un puissant équilibre.

Avant l'attaque décisive, il examina sournoisement le grand Oulhamr. Jugeant que sa supériorité serait plus sûre s'il frappait à deux mains, il ne garda que sa massue. Puis, il prit l'offensive.

Les armes, presque égales de poids, taillées dans le chêne dur, s'entre-choquèrent. Le coup du Kzamm fut plus fort que celui de Naoh, qui ne pouvait user de sa main gauche. Mais le fils du Léopard avait paré par un mouvement transversal. Quand le Kzamm renouvela l'attaque, il rencontra le vide ; Naoh s'était dérobé. Ce fut lui qui prit l'offensive : à la troisième reprise, sa massue arriva comme un roc. Elle eût fendu la tête de l'adversaire,

D

116 A, B, C, D FIG.
Manuel Orazi, *The Quest for Fire*, Paris: Éditions Pierre Lafitte, 1913
Cover-frontispiece
"The Oulhamr fled in the terrible night." "There are men on the plain."
"He pulled up grass and young shoots that he offered as a token of alliance."
Paris, BnF

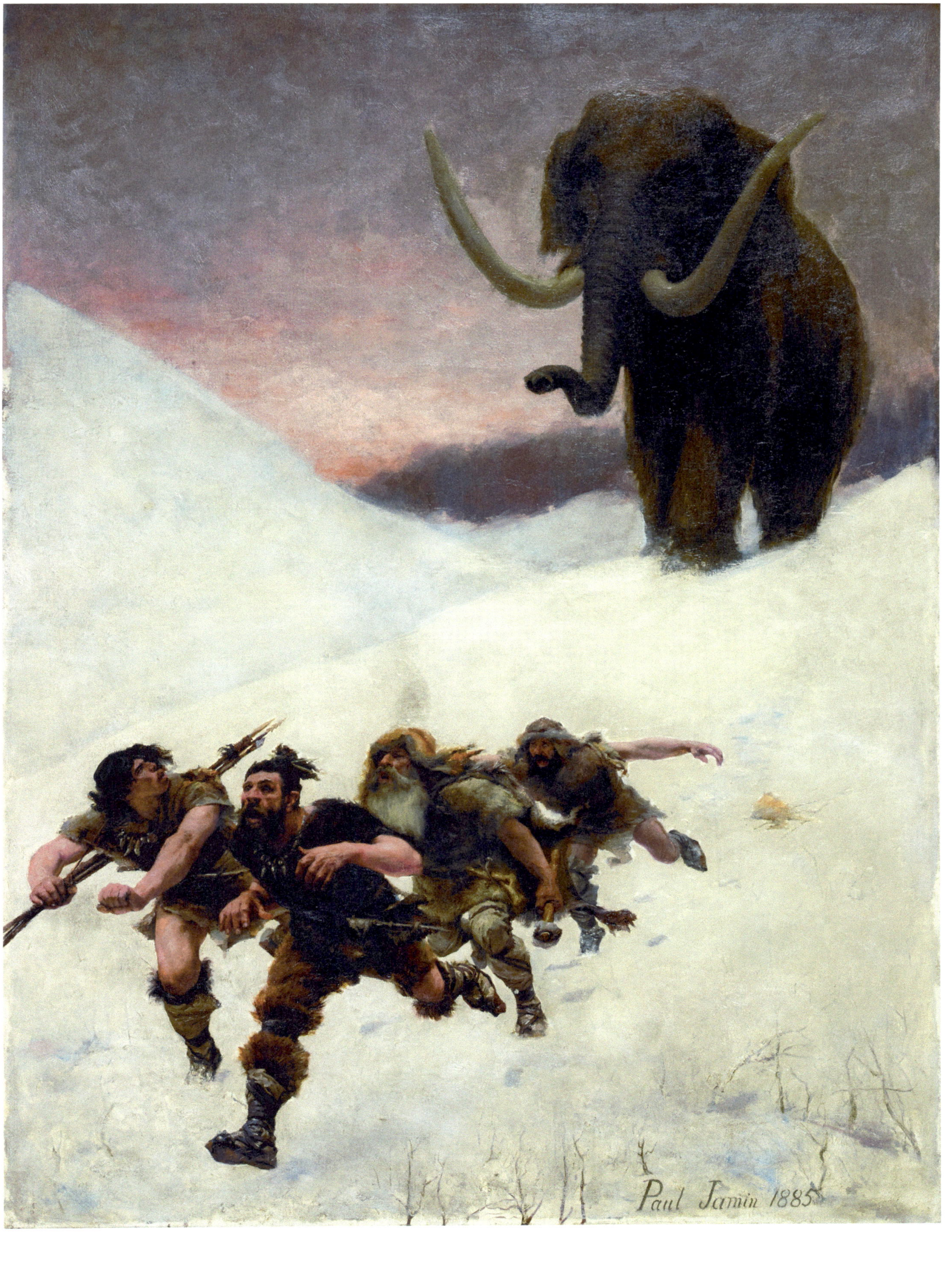

117
Paul Jamin, *Escaping from the Mammoth*, 1885
Oil on canvas, 125 × 94.5 cm
Paris, MNHN

EVOLUTIONS: AUTHORS, THEORIES, READERS

PIETRO CORSI

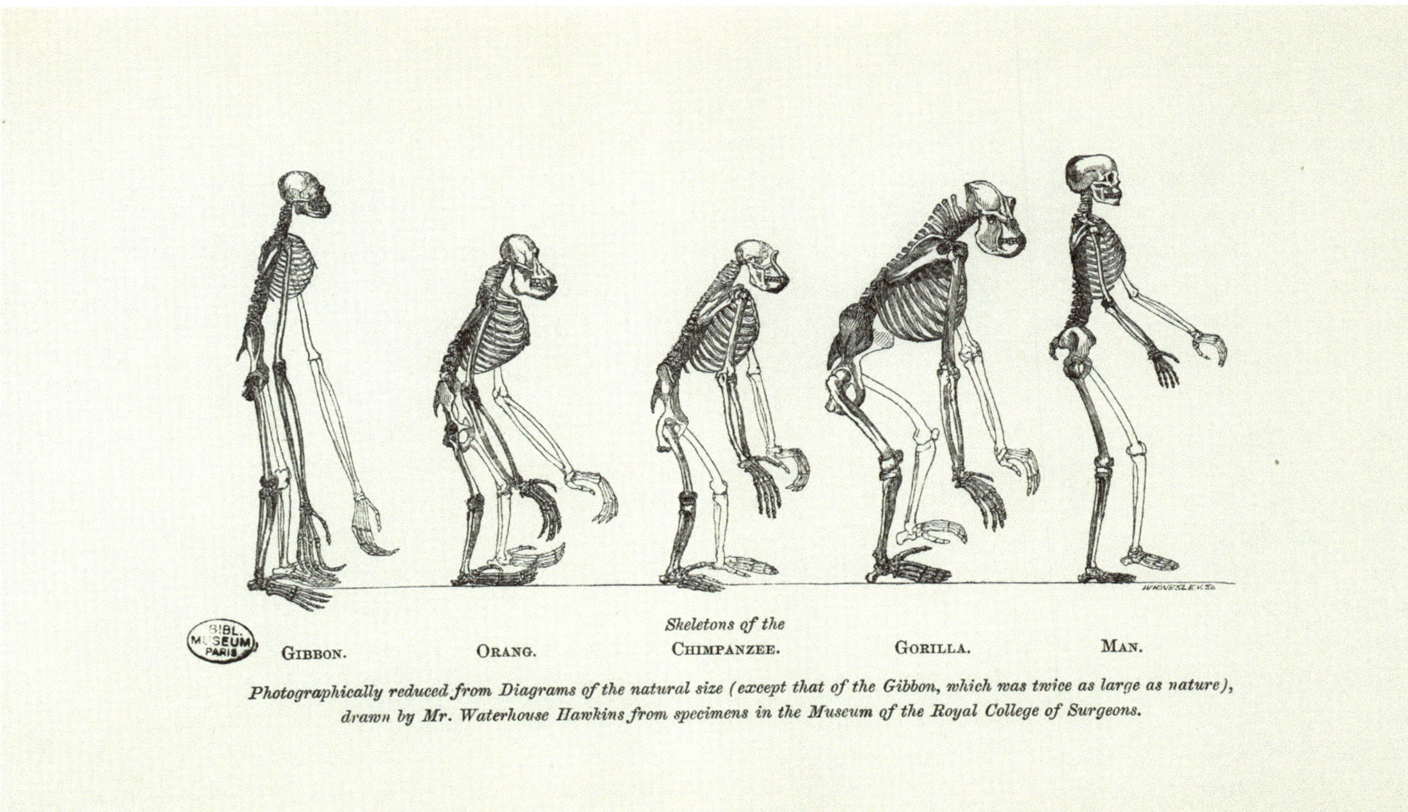

118
Benjamin Waterhouse Hawkins,
"Skeletons of the Gibbon,
Orang, Chimpanzee, Gorilla,
Man," in Thomas Henry Huxley,
*Evidence as to Man's Place
in Nature*, London:
Williams and Norgate, 1863
Diagram, 23 × 14.5 cm
Paris, MNHN

Classic histories of evolution follow a narrative pattern that shows little variation in its main tenets. Jean-Baptiste Lamarck, an isolated and much-derided dreamer (it is claimed), put forward a first evolutionary synthesis in various works published between 1801 and 1820. His theories failed to attract attention, and Lamarck died alone, blind and embittered in late December 1829. Up to the publication of Charles Darwin's *On the Origin of Species by Means of Natural Selection*, in November 1859, only a handful of equally isolated authors insisted that living nature was subjected to processes of change—however interpreted. A notable exception to the isolation narrative was Robert Chambers, whose *Vestiges of the Natural History of Creation*, published anonymously in October 1844, became a veritable "Victorian sensation," as the best available account of his popular success aptly narrates. Yet, it is said, Chambers was strongly opposed by the scientific community in the British Isles and throughout Europe and left little trace after the establishment of the Darwinian synthesis.[1]

Conversely, the long-established tradition of looking for precursors of Darwin in the one or two centuries before 1859 has been mostly abandoned, since commentators have become aware that utterances out of context could make anybody a precursor of Darwin, from the τὰ πάντα ῥεῖ ("all things are in flux") of Heraclitus of Ephesus to a long list of eighteenth-century philosophers and naturalists: Denis Diderot and Georges-Louis Leclerc, Count de Buffon, to name the most prominent ones, have in turn been called upon to anticipate the one or the other tenet Darwin expounded in the *Origin*.[2]

The explosion of evolutionary narratives after 1859, a scientific and cultural phenomenon traditionally and sensibly attributed to the publication of *On the Origin of Species*, does however create a few problems. Indeed, as it has been often pointed out, it is difficult to single out "true" followers of Darwin, and this for two simple yet very strong reasons: firstly, Darwin's many publications reflected the constant theoretical revisions and additions, changes of direction, even of domains of research the British naturalist engaged in. Secondly, even the closest allies of Darwin, Thomas Henry Huxley, "Darwin's bulldog," or Ernst Haeckel, one of the most influential proponents of evolution in the second half of the nineteenth century, felt free to speak their mind to criticize crucial features of their idol's doctrines, or, more diplomatically, they simply ignored them. Thus, to be brief, Huxley never abandoned his belief that mutations giving rise to new species could be sudden and marked. Neither did he hide his deep scepticism concerning the theory of pangenesis, Darwin's attempt to explain the transmission of acquired characteristics to the next generation. For his part, Haeckel expressed his conviction that Goethe and Lamarck contributed in fundamental ways to pave the way to the theory of descent (a comment his British mentor might have found hard to swallow), and engaged in drawing speculative phylogenetic histories of well-marked and hierarchically distributed species (12) and races (36) of man, thus contravening Darwin's belief that all humans belong to the same species. Not to mention the fact that Darwin was suspicious of phylogenetic trees of the living world, and only sketched in words one possible chain of descent in his work on man (127). Even Herbert Spencer, the extremely successful philosopher of the evolution of mind and societies, as of everything else, who coined the problematic expression "the survival of the fittest," wrote pages and chapters of unbridled speculative exuberance that gave Darwin a headache. Spencer, ever a very original if not idiosyncratic mind, owed more to Lamarck and to Chambers, or to German recapitulation theories, than to Darwin himself.[3]

1 James Secord, *Victorian Sensation: The Extraordinary Publication, Reception, and Secret Authorship of* Vestiges of the Natural History of Creation (Chicago: University of Chicago Press, 2003).

2 Bentley Glass et al., eds., *Forerunners of Darwin, 1745–1859* (Baltimore: Johns Hopkins University Press, 1968); for a pioneering critique of the hunt for precursors, see the classic Camille Limoges, *La sélection naturelle. Étude sur la première constitution d'un concept, 1837–1859* (Paris: Presses universitaires de France, 1970).

3 Charles Darwin to Charles Lyell, March 25 [1865], "I have read most of H. Spencer's Biology & agree with you. Some of his remarks are very clever & suggestive, but somehow I seldom feel any wiser after reading him, but often feel mystified. His style is detestable in my opinion; & no wonder as he dictates & never alters"; and to J. D. Hooker, June 30 [1866], "I have almost finished the last number of H. Spencer & am astonished at its prodigality of original thought. But the reflection constantly recurred to me that each suggestion, to be of real value to science, w[d] require years of work," in www.darwinproject.ac.uk, directed by James Secord.

4 Charles Darwin to Chauncey Wright, June 3 [1872], "I have resolved to waste no more time in reading reviews of my works or on evolution, excepting when I hear that they are good & contain new matter, or are written by men whom I respect."

5 See Pietro Corsi, www.lamarck.cnrs.fr; Richard W. Burkhardt, *The Spirit of System: Lamarck and Evolutionary Biology* (Cambridge, MA: Harvard University Press, 1977; new ed., 1995); Pietro Corsi, *The Age of Lamarck: Evolutionary Theories in France, 1790–1830* (Berkeley: University of California Press, 1988), and the new expanded edition, *Lamarck. Genèse et enjeux du transformisme, 1770–1830* (Paris: CNRS Éditions, 2001).

6 Jean-Baptiste Lamarck, *Hydrogéologie, ou Recherches sur l'influence qu'ont les eaux sur la surface du globe terrestre* (Paris: Agasse, 1802).

7 Wiktor Stoczkowski, "Le portrait de l'ancêtre en singe. L'hominisation sans évolutionnisme dans la pensée naturaliste du XVIIIe siècle," in *Ape, Man, Apeman: Changing Views since 1600*, ed. Raymond Corbey and Bert Theunissen (Leiden: Leiden University, 1995), p. 141–155.

As far as the many doubtful supporters and the even more numerous critics were concerned, Darwin soon gave up the hope not so much to convince the latter, but to make even the former read his works for what they said, not for what admirers said they were arguing for.[4]

In the three sections that follow, I will briefly sketch the context and content of the three major evolutionary syntheses produced during the nineteenth century, by Lamarck, by Darwin and—albeit very briefly—by Haeckel.

JEAN-BAPTISTE LAMARCK ON HIS OWN TERMS

When in the spring of 1800 Lamarck abandoned the fixism of his early works and embraced a theory of constant organic change, he was already past his prime.[5] At the age of fifty-six, he enjoyed a considerable reputation as a field botanist, though several leading botanists disliked him. In 1793, he had been given a chair at the newly established Muséum national d'histoire naturelle, not in botany, as he had hoped, but in the zoology of the numberless cohorts of animals he called invertebrates. In 1795, he had been allotted a seat in the botany section of the Institut national, the restored Académie des sciences which had been abolished in 1793. Lamarck never considered his salaries as sufficient to maintain the lifestyle of a gentleman claiming a doubtful aristocratic ancestry. He thus joined the growing number of men of letters and naturalists writing to supplement their income. He also traded in corals and shells and was known as a thriving collector.

From 1794 to the end of the century, he produced a series of works aimed at cutting a space for himself in the debates on the new chemistry of Lavoisier, the various rival theories of the Earth that captivated the attention of the reading public, and the question of the laws governing animal and vegetable life. In January 1802, he presented himself as a naturalist philosopher working on the ambitious project of a new terrestrial physics, articulated in three subdisciplines: hydrogeology, or the dynamics of the aqueous fluids shaping the surface of the earth; meteorology, or the dynamics of the gaseous fluids responsible for all atmospheric phenomena; and biology, the dynamics of all fluids (aqueous, gaseous, but also heat and electricity, to him very subtle and very active fluids), responsible for all movements inside organized beings.[6]

The high scientific ground he claimed for himself allowed Lamarck to ignore the work of contemporaries—mineralogists and geologists, natural philosophers and contributors to encyclopedic works and periodicals—courting the same reading public he was trying to convince. A variety of competing models were put forward to explain the origin and history of life on Earth, but all lacked the systematic rigour Lamarck was trying to introduce in the debate. Reproducing half-digested readings, even a literary man such as Nicolas-Edme Rétif de La Bretonne had devoted pages of his *La Philosophie de Monsieur Nicolas* (1796) to speculating on the steps living nature had passed through before reaching the organic and intellectual perfection of mankind. Lamarck's colleague Jean-Claude de La Métherie freely drew from the *De Rerum Naturae* by the Latin poet Lucretius and Buffon's *Les Époques de la Nature* (1779) to argue that processes akin to crystallization had formed the primary molecular aggregations giving rise to the main types of living organisms, prototypes that proliferated and differentiated throughout the ages of the Earth, endlessly adapting to the expansion of lands emerged from a primeval Ocean. De La Métherie described in detail—adapting passages inspired by the famous *Telliamed* by Benoît de Maillet—the anatomical and behavioural mutations that turned apes into men.[7]

For the most part, authors pondering their systems of nature between the late 1770s and the early 1800s ignored the great tradition of materialist Enlightenment speculations represented by D'Holbach or Diderot, though they generally declared their debt to Buffon, the most prominent naturalist of the century; their philosophical novels and daring speculations proudly vindicated the right of the naturalist and of the cultivated amateur to enter a literary field once reserved to the great luminaries of the social and intellectual elites. They inaugurated a trend that saw the number of men of letters and amateurs engaging with naturalistic speculations grow decade after decade throughout the nineteenth century, in France as well as throughout the Old and New

Continents. This was the public of authors and of readers that kept broad speculative interests in the formation, history and changes of life forms very much alive, in spite of the fierce opposition of a number (albeit not all) of prominent representatives of the scientific establishment. One could say that a significant portion of the career of Georges Cuvier was devoted to fighting and deriding the varied population of system builders, arguing for the exclusive right of the naturalists working in major institutions to deal with topics amateurs should refrain from approaching. He failed to convince the vast majority of literary commentators and readers alike.[8]

Lamarck was in a difficult position during the early decades of the nineteenth century. He was a member of elite institutions (the Muséum and the Institut) yet he opposed the anti-speculative ethos Cuvier argued for. At the same time, he did his best to distance himself from his immediate competitors, thus giving the impression of being isolated and a prophetic pioneer. He adopted a cautious tone when dealing with religion and downplayed his commitment to eternalism and materialism, but contemporaries were not fooled. His many works suffered from reticence, repetitions, unsolved tensions and key assumptions that were left undisclosed. The fear of losing his job in the increasingly conservative climate of the Consulate and the Empire made Lamarck dilute his main theoretical claims: thus, for instance, his hydrogeology was based on the assumption of endless cycles of displacement of the ocean basin around the surface of the Earth, indicating an indefinite age for our planet. In the *Philosophie zoologique* (1809) (126), he denied he was an eternalist, like the followers of epicurean atomists and of Lucretius—well-known enemies of revealed religion.[9] More importantly, he repeatedly denied working on his "Biology" project in vain: no one had missed his intention of providing a thoroughly materialistic interpretation of all organic phenomena, including the mental and moral prerogatives of man.[10] In his last work, *Système analytique des connaissances positives de l'homme* (1820, in fact, a collection of articles published in the *Nouveau Dictionnaire d'histoire naturelle*, 1816–1819), Lamarck denied the existence of a human soul, though he paid lip service to the Creator of Nature. Yet, he specified, nothing could be known and said about Him, since He was completely outside our reach. Humans could only know physical phenomena, being themselves physical entities, and their consciousness was merely the result of a complex set of processes involving the circulation of the nervous fluids.[11]

In the hands of his critics, Lamarck's complex theory of evolution was reduced, and still is today, to the examples of adaptive changes he offered, mainly concerning vertebrates. Though on a corpus of several thousand pages the example of the giraffe stretching its neck to reach higher leaves was only mentioned on

8 Pietro Corsi, "Systèmes de la nature and Theories of Life: Bridging the Eighteenth and Nineteenth Centuries," *Republics of Letters*, vol. 6, no. 1 (2018), https://arcade.stanford.edu/rofl/systèmes-de-la-nature-and-theories-life-bridging-eighteenth-and-nineteenth-centuries.

9 In *Hydrogéologie*, p. 151, Lamarck argued that thanks to his work "la marche éternelle de la Nature peut actuellement se dévoiler sans obstacle à nos yeux"; in *Philosophie zoologique* (Paris: Dentu, 1809), vol. 1, p. 362, he was more cautious: "Regarder la nature comme éternelle, et conséquemment comme ayant existé de tout temps, c'est pour moi une idée abstraite, sans base, sans limite, sans vraisemblance, et dont ma raison ne sauroit se contenter"; but in vol. 2, p. 465, he reiterated his true belief: "La *Nature*, cet ensemble immense d'êtres et de corps divers, dans toutes les parties duquel subsiste un cercle éternel de mouvemens et de changemens que des lois régissent."

10 Lamarck announced his project "Biology" in January 1802 in *Hydrogéologie*, p. 8 and 188; he quickly denied he was working on it in July 1802 in *Recherches sur l'organisation des corps vivans*, p. vi, and in *Philosophie zoologique*, p. xviii; he announced the project again, as if for the first time, in *Histoire naturelle des animaux sans vertèbres*, vol. 1 (Paris: Verdière, 1815), p. 49. See Pietro Corsi, "Biologie," in *Lamarck, Philosophe de la nature*, ed. Pietro Corsi et al. (Paris: Presses universitaires de France, 2006), p. 37–64, and www.lamarck.cnrs.fr.

11 Jean-Baptiste de Lamarck, *Système analytique des connaissances positives de l'homme* (Paris: Chez l'auteur and Belin, 1820), p. 11: "Qu'est-ce qu'un être spirituel ? C'est ce qu'à l'aide de l'imagination l'on voudra supposer. En effet, ce n'est que par le moyen d'une opposition à ce qui est matériel que nous nous sommes formé l'idée d'un esprit; mais comme cet être supposé n'est nullement dans la catégorie des objets qu'il nous soit possible d'observer, nous ne saurions rien connaître à son égard. L'idée que nous en avons est donc absolument sans base."

12 Charles Darwin, "Changes not result of will of animal, but law of adaptation as much as acid and alkali," Notebook B, p. 21 [Transmutation of species (1837–1838)], CUL DAR 121, in http://darwin-online.org.uk/, and "With respect to how species are Lamarck's 'willing' doctrine absurd (as equally are arguments against it—namely how did otter live before being modern otter—why to be sure there were a thousand intermediate forms," p. 216–217, in John van Whye, ed., http://darwin-online.org.uk/; Charles Darwin to Joseph Dalton Hooker [January 11, 1844], "Heaven forfend me from Lamarck nonsense of a 'tendency to progression' 'adaptations from the slow willing of animals' &c,—but the conclusions I am led to are not widely different from his."

13 Richard W. Burkhardt, "Lamarck, Evolution, and the Inheritance of Acquired Characters," *Genetics*, vol. 194, no. 4 (2013), p. 793–805, doi: 10.1534/genetics.113.151852.

14 Erasmus Darwin, *Zoonomia; Or, the Laws of Organic Life*, 2 vols. (London: Johnson, 1794–1796); Giuseppe Gautieri, *Slancio sulla genealogia della terra e sulla costruzione dinamica della organizzazione* (Jena: 1805); Johann Christian Rödig, *Lebende Natur* (Leipzig: Breitkopf and Haertel, 1801). Rödig generously flooded European naturalists, Cuvier, Goethe and Sir Joseph Banks included, with copies of his tract. Pietro Corsi, "Before Darwin: Transformist Concepts in European Natural History," *Journal of the History of Biology*, vol. 38, no. 1 (2005), p. 67–83.

15 "C'est surtout dans les corps vivans, et principalement dans les animaux, qu'on a cru apercevoir un but aux opérations de la nature. Ce but cependant n'y est là, comme ailleur, qu'une simple apparence et non une réalité. En effet, dans chaque organisation particulière de ces corps, un ordre de choses, préparé par les causes qui l'ont graduellement établi, n'a fait qu'amener par des développemens progressifs de parties, régis par les circonstances, ce qui nous paraît être un but, et ce qui n'est réellement qu'une nécessité," in Lamarck, *Histoire naturelles des animaux sans vertèbres*, vol. 1 (1815), p. 323–324.

three occasions, and in passing, this was—and is—what people remembered of his theory. Even Darwin sarcastically commented on Lamarck's believing that organic change was brought about by organisms "willing" to adapt, in spite of the fact that the French naturalist insisted that "will" was a physico-chemical process present only in animals endowed with a developed nervous system—an infinitesimal fraction of the living world.[12]

In plants and most animals deprived of a nervous system, or endowed only with elementary nerve ganglions, organic fluids circulating through tissues were influenced by changes in the environment, at times fractionally modifying their circulation patterns. In very simple organisms, fluids such as electricity and heat made life possible by traversing and shaping the gelatinous bodies. If the environmental action persisted, the change in circulation patterns would be stabilized, and structures would be imperceptibly created or modified as a consequence. Lamarck never used the expression "inheritance of acquired characteristics," simply because what was inherited was the almost imperceptible change of organic fluid distribution patterns, eventually leading to the appearance of a modified or new character.[13] It is appropriate to mention that giraffes stretching their necks, elephants elongating their trunks, flying squirrels on their way to becoming bats, or hippopotamuses learning to become whales were peppering the pages of the Englishman Erasmus Darwin, the Italian Giuseppe Gautieri, or the German Johann Christian Rödig, who had never read Lamarck. Processes of change due to increased use or lack of use of parts were commonplace in natural history and medical literature of the decades bridging the nineteenth century, and were still there in part of Spencer's speculations.[14]

Lamarck did not believe in fossils, since to him very few organisms have perished throughout the endless ages of the Earth. The trees of life (at least two separate ones for animals and one for plants) were to him the only testimony we have of the progressive complication of organic structures. Like Lyell and Darwin after him, albeit in a very different epistemological and scientific context, Lamarck stressed that knowledge was possible only of processes we can witness in today's world. He was almost alone in arguing that histories of the Earth and of life were fantasies, since they started from the unknown—a very distant past no one had ever witnessed—to explain the present. The opposite path had to be followed, from observed change in today's world, to the only evidence we have of changes in an indefinite past, namely, the trees of life.

Lamarck also opposed those who believed that complex organisms could be generated through spontaneous generation. Only extremely simple forms of life could be produced directly by material elements, today as at the beginning of things. Lamarck never discussed one major tension of his theory, namely how to conciliate the assertion that life had an origin and a history, as the taxonomic order suggested, with the epistemological principle he emphatically insisted upon, that a science of origins is precluded to humans. Moreover, his cyclical view of nature assumed the constant presence of a rich biodiversity essential to provide mineral compounds and masses of remains (from corals to coal): spontaneous generations alone would never guarantee the survival of the planet. The Earth would not be what it is, he argued, if only spontaneous generations existed.

Contrary to common interpretations of his thought, then as well as now, Lamarck denied that change was constant, pushed by a quasi-vitalist drive. Several organic structural plans did not go anywhere: infusoria or radiarians did not reach any higher; they were and are still very simple organisms. Nothing pushed them forward. Relatively complex structures such as intestinal worms did ascend the ladder, once the organisms managed to live in the outside world, but nothing pushed them, only an extraordinary set of environmental circumstances. Birds were to him a branch on their own, again not going anywhere. He spoke at times of the "power of life," in the sense that by looking at the trees of life we can admire what the endless repetition of simple physical processes achieved through endless changes of circumstances. To him, there was no "tendency" to complexity in organic phenomena: the only "tendency" was within the human mind, that was used to recognize a finality and a goal where there was only the cumulative action of numberless acts of fluid circulation.[15]

Opposed at the institutional level, Lamarck's doctrines were widely summarized and discussed—

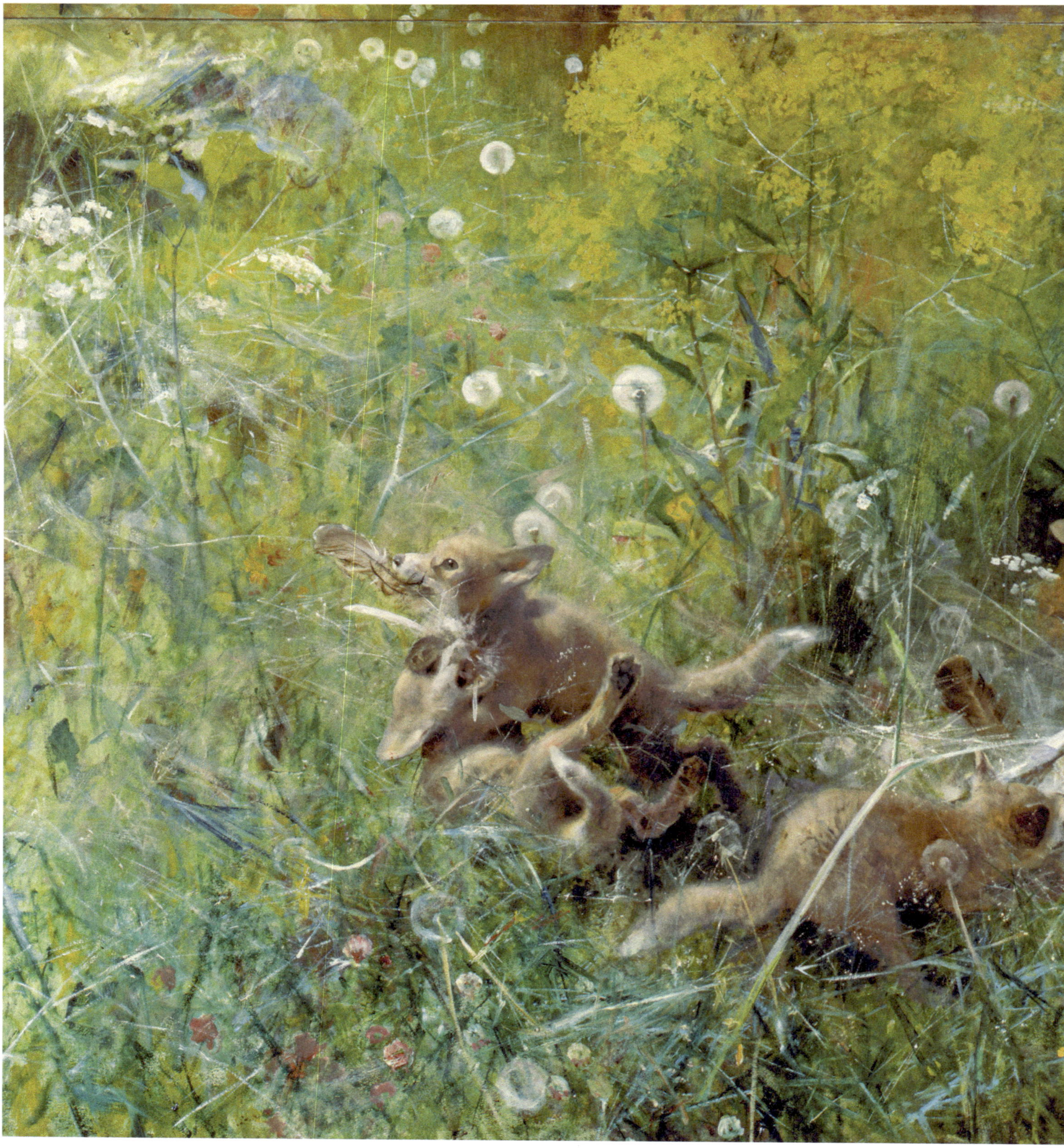

119
Bruno Andreas Liljefors,
A Fox Family, 1886
Oil on canvas, 112 × 218 cm
Stockholm, Nationalmuseum

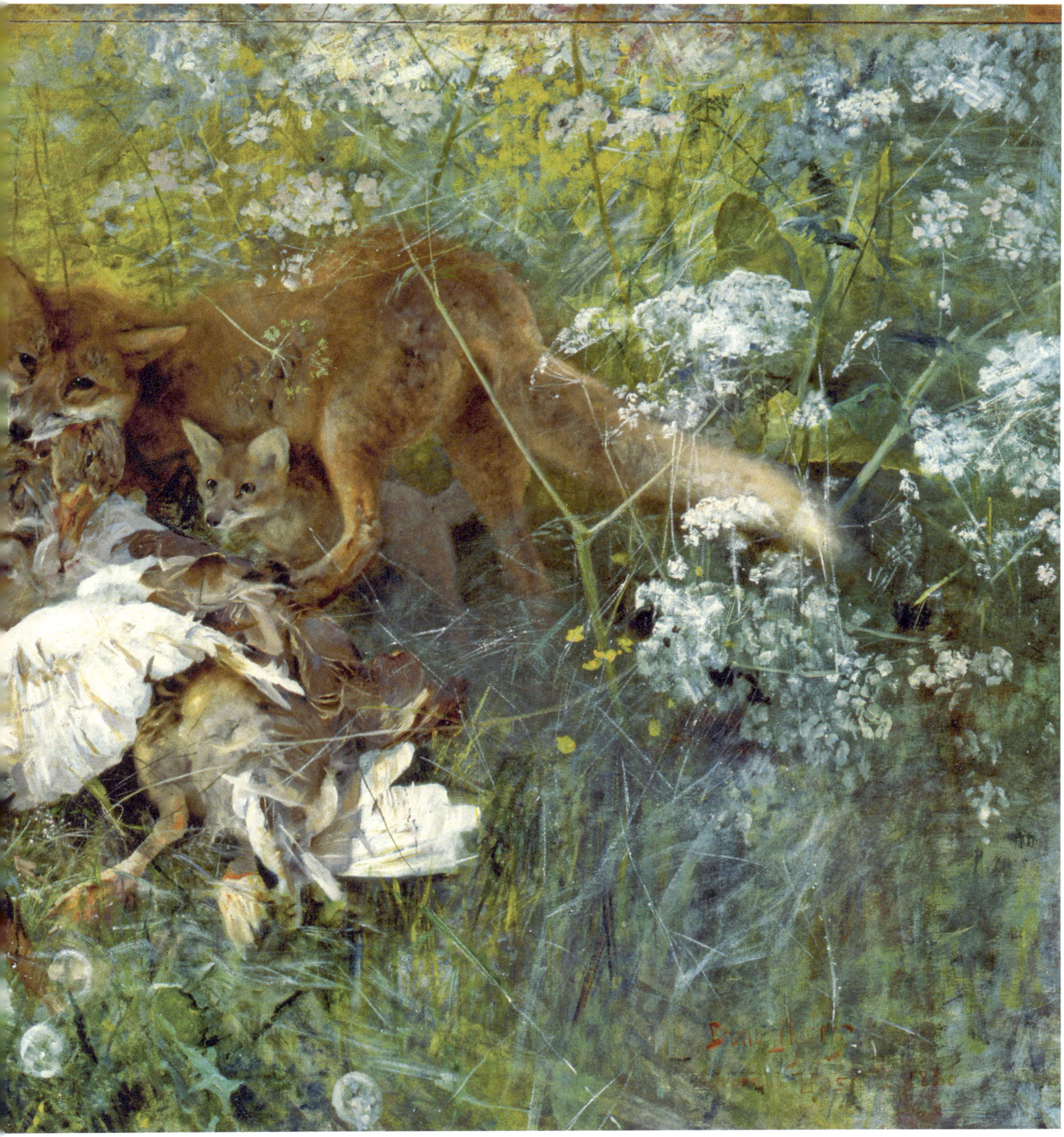

however interpreted—throughout Europe in popular dictionaries and in journals directed to the cultivated public. In France, doctors and amateur naturalists, scientific journalists, social reformers and literary figures, but also a few paleontologists and naturalists were not shocked by Lamarck's attempt to provide an explanation for the endless forms life was exhibiting throughout the surface of the world and its history. To quote one example only, the *Dictionnaire classique d'histoire naturelle* (17 vols., 1822–1831), edited by Jean-Baptiste Bory de Saint-Vincent, published warm eulogies of Lamarck, whose doctrines it praised, summed up and updated—but did not share. Bory's dictionary travelled with Darwin in the cabin of the *Beagle*.[16]

CROSSING THE CHANNEL—CHARLES DARWIN

During the decades bridging the eighteenth and the nineteenth centuries, a majority of British naturalists, often members of the Anglican Church, fiercely opposed the materialistic tendencies they saw as rampant in France. They greeted with enthusiasm Cuvier's onslaught on systems of nature, histories of the Earth and of life, and on Lamarck's doctrine of organic change. Yet, British naturalists active in the early decades of the nineteenth century had become weary of attempts at using science to prove the truth of Scriptures, as many scriptural geologists were doing, and strived to gain their independence from theological supervision. Properly pursued, they argued, independent scientific endeavours increased our understanding of God's benevolence and providential care for His creatures. Natural theology became a sort of trademark of British science throughout the first half of the nineteenth century.

Not everyone agreed, needless to say. In part, also because, undeterred by official sarcasm, theories of organic change appeared to proliferate in France and in Europe, in new and philosophically more acceptable terms. The work of Étienne Geoffroy Saint-Hilaire, for instance, attracted attention in the British Isles. His insistence on a unity of type pervading the animal and vegetable kingdoms could be accommodated to philosophical interpretations of the role of creative ideas in the "evolution" of Nature: as the human embryo unfolded through stages of development finally reaching the perfection of mankind, the organic world as a whole developed through stages of increased complexity. In Germany, Carl Friedrich Kielmeyer and Johann Friedrich Meckel argued that the history of life on Earth was summed up in the "evolution" of the human embryo from a gelatinous mass to the complete individual, taking up the form of worms, fishes, reptilians and mammalians on the way up.[17] In the late 1820s and the early 1830s, Geoffroy declared that extinct animals were the ancestors of living ones, and his views were debated in Edinburgh and London, as well as elsewhere in Europe. In spite of current interpretations, however, Geoffroy was not a follower of Lamarck, whom he praised at times, and at times denounced using Cuvier's insulting tone.[18]

Youth and crucial intellectual influences preserved Charles Darwin from being impacted by the European debates on life and its transformation of

16 Pietro Corsi, "The Revolutions of Evolution: Geoffroy and Lamarck, 1825–1840," *Bulletin du Musée d'Anthropologie préhistorique de Monaco*, no. 51 (2011), p. 97–122. For an innovative and exemplary analysis of the diffusion of broadly "evolutionary" views in French literary and intellectual circles, see Nicolas Wanlin, "Évolutions. Littérature et sciences au XIXe siècle" (habilitation thesis, Université Paris-Diderot, thesis director Mme le pr. Paule Petitier, 2018), and the essay in this catalogue, p. 310.

17 Robert J. Richards, *The Meaning of Evolution: The Morphological Construction and Ideological Reconstruction of Darwin's Theory* (Chicago: University of Chicago Press, 1992); Sander Gliboff, *H.G. Bronn, Ernst Haeckel, and the Origins of German Darwinism: A Study in Translation and Transformation* (Cambridge, MA: MIT Press, 2008); Stéphane Schmitt, "From Eggs to Fossils: Epigenesis and Transformation of Species in Pander's Biology," *International Journal of Developmental Biology*, vol. 49, no. 1 (2005), p. 1–8.

18 Étienne Geoffroy Saint-Hilaire, "Mémoire où l'on se propose de rechercher dans quels rapports de structure organique et de parenté sont entre eux les animaux des âges historiques, et vivant actuellement, et les espèces antédiluviennes et perdues," *Mémoires du Muséum d'histoire naturelle*, vol. 17 (1828), p. 209–229; Étienne Geoffroy Saint-Hilaire, "Of the Continuity of the Animal Kingdom by Means of Generation, from the First Ages of the World, to the Present Times: On the Relations of Organic Structure and Parentage That May Exist Between the Animals of the Historic Ages and Those at Present Living, and the Antediluvian and Extinct Species," *The Edinburgh New Philosophical Journal* (April–June 1829), p. 152–155; and Étienne Geoffroy Saint-Hilaire, "Palaeontographie. Considérations sur les ossemens fossiles la plupart inconnus, trouvés et observés dans les bassins de l'Auvergne," *Revue encyclopédique*, vol. 59 (July–September 1833), p. 76–95, followed Cuvier in equating Lamarck with Rödig and Telliamed.

19 See the outstanding biography of Darwin by Janet Browne, *Charles Darwin: A Biography*, vol. 1: *Voyaging* (London: Jonathan Cape, 1995; Princeton: Princeton University Press, 1996), and vol. 2: *The Power of Place* (London: Jonathan Cape, 2002; Princeton: Princeton University Press, 2003). Jonathan Hodge and Gregory Radick, eds., *The Cambridge Companion to Darwin* (Cambridge: Cambridge University Press, 2009). Patrick Tort, ed., *Dictionnaire du darwinisme et de l'évolution*, 3 vols. (Paris: Presses universitaires de France, 1999).

20 Adrian Desmond, *The Politics of Evolution: Morphology, Medicine, and Reform in Radical London* (Chicago: University of Chicago Press, 1989).

21 Charles Lyell, *Principles of Geology*, 3 vols. (London: John Murray, 1830–1833); Jean-Baptiste-Julien D'Omalius d'Halloy, *Eléments de géologie* (Paris: Levrault, 1831); Aimé Boué, "Résumé des progrès des Sciences Géologiques pendant l'année 1833," *Bulletin de la Société géologique de France*, vol. 5 (1834), p. 1–506.

22 Pietro Corsi, "The Importance of French Transformist Ideas for the Second Volume of Lyell's *Principles of Geology*," *The British Journal for the History of Science*, vol. 11, no. 3 (1978), p. 221–244.

the 1820s. A student at Edinburgh and at Cambridge, on December 27, 1831, he sailed from Plymouth for his journey around the world that lasted until October 2, 1836. His scientific mentors at Cambridge, all members of the Anglican Church, had taught him how to botanize and keep a collection, and instilled a suspicion for the French love of fanciful theorizing. The reading of Alexander von Humboldt's *Personal Narrative of Travels to the Equinoctial Regions of the New Continent* (7 vols., 1818–1829) had fired his love of tropical nature and Charles Lyell's *Principles of Geology* (3 vols., 1830–1833) had taught him how to reason in science.[19]

Like Lamarck, Lyell disliked philosophical cosmogonies and theories of the Earth. Only the knowledge of phenomena observable in today's world can provide a foundation for reliable science: the past can be deduced from the present, not vice versa. And the past was not very different from the present, in a world in which cycles of uniform geo-climatic events repeated each other, with the same intensity—there could be no science of beginnings. Lyell had devoted eleven chapters of the second volume of *Principles* to a fair and dispassionate critique of Lamarck. He, too, like many religious naturalists in the British Isles, was afraid that the increasing success of Lamarck's views—however interpreted—could put scientific practices once again in jeopardy, submerged by the accusation of leading to materialism.[20] In the early 1830s, leading European geologists, such as the Belgian Jean-Baptiste-Julien d'Omalius d'Halloy, or the President of the French geological society for the years 1830–1835, Ami Boué, did not hide their support for theories trying to explain the succession of species in natural terms.[21]

Lyell interpreted Lamarck in his own way—for instance, he stated that the French colleague believed in a constant progress of life from monad to man, which was not exactly Lamarck's view of the matter. Contrary to Lamarck, Lyell argued that species extinction was a natural phenomenon, as was the introduction of new species to replenish the scene of life. Organisms competed with each other for survival, and the close relationship between prey and predator, for instance, required that an extinct predator be replaced by a similar form, playing the same role in the equilibrium of nature. Yet, he had no idea how new species were introduced, and hinted that a creative act might be required to explain the phenomenon. Lyell taught Darwin to consider species as populations occupying a given territory, constantly striving to survive the competition to access limited resources. Any geo-climatic imbalance would lead to extinction.[22]

Darwin travelled around the world as a pupil of Lyell and a firm supporter of his uniformitarian geology. Inspired by his mentor, he solved the puzzle of coralline islands, even though the solution went against Lyell's hypothesis that corals formed on the slopes of volcanoes eroded by the waves. Darwin applied Lyell's geology at the global level, and concluded that vast portions of the Earth were emerging from the sea, while other portions were slowly sinking beneath the waves. Corals formed at the edge of areas of subsidence. The process was so slow that corals could always access light and live, though their debris kept slowly sinking thousands of metres deep together with the earth mass they were attached to. Whereas the Argentinian pampas had emerged in geologically recent times, vast areas of the Pacific were sinking, as demonstrated by the countless coralline islands and the great Australian coralline barrier. Darwin rightly hoped to make a career in science as a geologist.

On his return back to England in October 1836, he met with Lyell and a lifelong friendship started. His mentor's theoretical geology was then under attack because the *Principles* had not solved the problem of species. As William Whewell, a Cambridge scientific star, pointed out, Lyell had been compelled to admit an act of creation to explain the introduction of new species: so, why deny that God could also intervene in the physical history of the Earth? Geology could never become a science independent from theological presuppositions, and indeed required them. Darwin set to work on the "mystery of mysteries," the production of new species through natural means at work in today's world, to fill a weak spot in Lyell's and his own geology.

All the phenomena he had observed during the journey around the world, and in particular the peculiar fauna of the Galapagos Islands, where different species of the same genus of birds or tortoises inhabited islands in sight of each other, or the fossils he had extracted in South America, now took on a new theoretical meaning. The notebooks he started filling in

July 1837 with hasty and at times illegible scribbles allow reconstructing a complex thought process, culminating in the Malthusian insight of September 28, 1838. Darwin reached the conclusion (albeit temporary) that the extreme severity of competition for life allowed very few individuals of each generation to survive, those who were born with a however small advantage over their siblings. The road was still long before Darwin was forced by an unexpected issue of priority to complete a summary of his theory, which he published in November 1859 as *On the Origin of Species by Means of Natural Selection, or The Preservation of Favoured Races in the Struggle for Life* (127).

By 1859, the intellectual and scientific climate of England and of Europe had changed dramatically, with respect to 1830. A scientific explanation for the production of species was now seen by many as a philosophical and even theological necessity, as the Reverend Baden Powell, Professor of Geometry at Oxford, put it—and Darwin quoted him with approval. In 1844, the *Vestiges of the Natural History of Creation* anonymously published by Robert Chambers, however crude, had shown how much support there was among the reading public for a narrative describing the progressive march of everything—from nebulae to man—towards the perfection represented by the white British men, those at least endowed with the superior intellectual organs phrenology had started to investigate. Top naturalists and scientists working in official institutions—including all of Darwin's teachers at Cambridge—continued to blast the various forms of explanations for species change expressed in terms of natural laws, but it was their turn to be increasingly isolated.[23]

During the 1850s, several prominent naturalists throughout Europe expressed strong reservations against the traditional creationist views and started proposing alternative explanations, though to many the role of a Divine architect was not in question. Several naturalists agreed that species within a given genus could be formed through adaptation to local environments and argued for an expanded definition of species to include the potential for the adaptive formation of slightly modified forms. Others advocated a mixture of Lamarckian, embryological or geographical speciation and paleontological factors to sketch more ambitious views of an ever-changing organic life. And, of course, Darwin was not alone in thinking that selective pressures at work on variations (geographical or within a given population) were responsible for the formation of new species. As is well known, Darwin dropped the writing of his long book on species and started drafting *On the Origin* because of the arrival, on June 18, 1858, of a manuscript paper written by one of his correspondents, Alfred Russel Wallace, announcing a theory Darwin thought was very similar to his own. Lyell and Darwin's botanist friend Joseph Dalton Hooker arranged the reading of texts by Darwin and Wallace at a meeting of the Linnaean Society (July 1, 1858), and one year later Darwin published his masterpiece.

Historians have debated whether Darwin had been totally honest with Wallace, and every ten to twenty years a new study is published arguing that Darwin stole his main ideas from the poor Wallace. Apart from the fact that the two theories differed on substantial points, and that Wallace never doubted that Darwin had preceded him, the interesting point to emphasize is that Wallace had become an evolutionist on his own terms, quite independently from Darwin. He had read Lyell's critique of Lamarck and felt sympathy for the French naturalist; he also found *Vestiges* fascinating. Moreover, selective models were not that common, but quite present in French works of the 1830s; the relationship of descent between living and fossil species was widely accepted, though interpreted in different ways, even in terms of the declination through time and space of creative ideas conceived by a superior intelligence. The Wallace-Darwin episode is telling us something important about the state of debates on species in England and Europe: who was really first is not the right question to ask and it simply misses the wider meaning of the encounter.

The reading of *On the Origin of Species*, and of the Darwinian corpus as a whole, highlights the difference between the English naturalist and the scores of contemporary and pre-1859 authors. For the vast majority, endorsement of the one or the other model for organic change was not followed by systematic discussion of actual instances potentially occurring in the natural world. In Darwin's works, a stringent reasoning style was accompanied by endless examples taken from the main disciplines dealing with organic life at the time, from the domestication of

23 Pietro Corsi, *Science and Religion: Baden Powell and the Anglican Debate, 1800–1860* (Cambridge: Cambridge University Press, 1988).

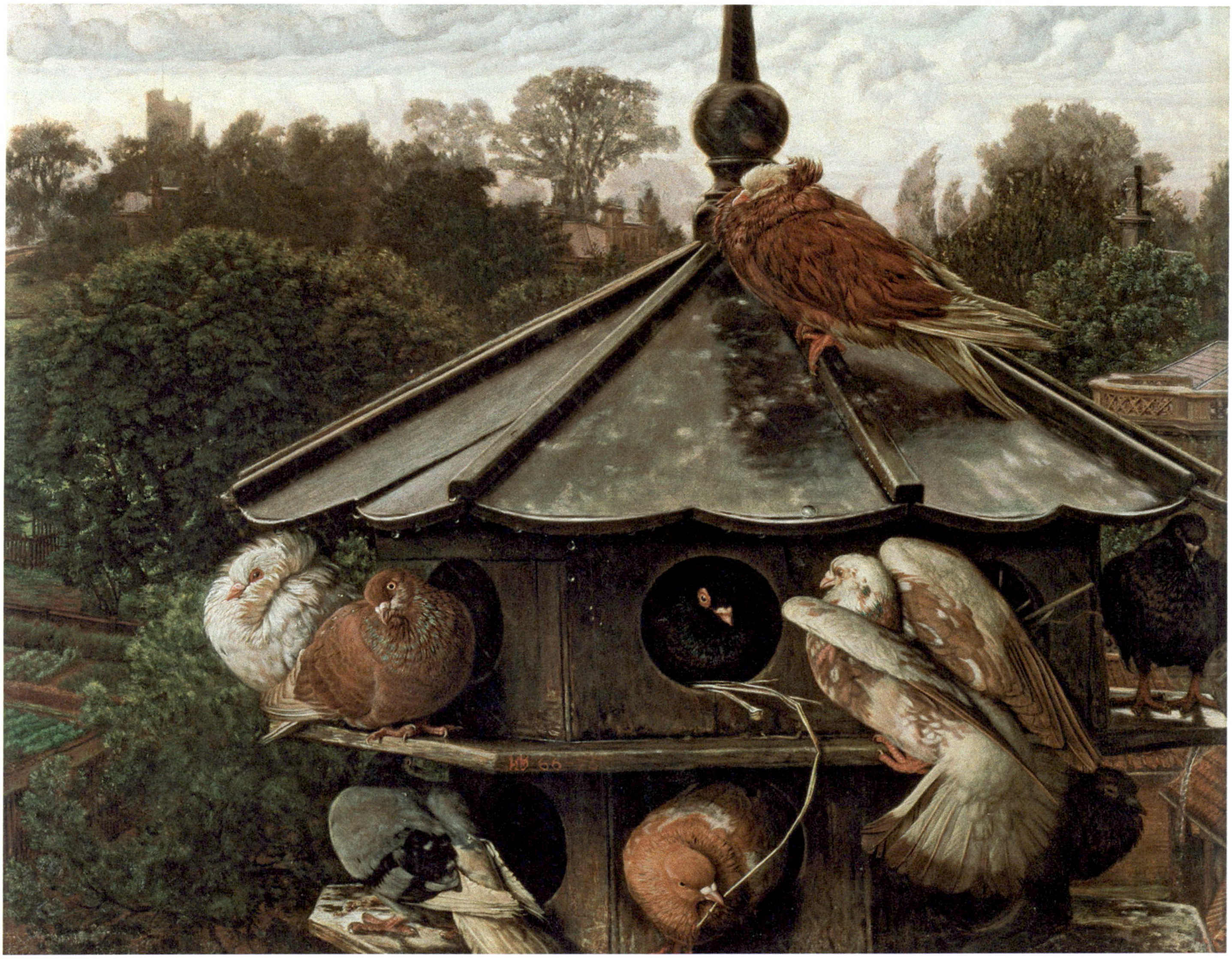

120
William Holman Hunt,
The Festival of St. Swithin
(The Dovecot), 1856–1866
Oil on canvas, 73 × 91 cm
Oxford, Ashmolean Museum,
University of Oxford

animals and plants to geographical distribution, embryology, comparative anatomy, paleontology, and the study of instincts and of generation. Darwin's method of work paid dividends. Since the late 1830s, he had collected scraps of information, reading notes, even pieces of letters cut with scissors, in several folders, each devoted to one of the main disciplinary fields and personal interests he was working on. Thus, in writing each chapter of his works, he could marshal decades of data gathering, each data finalized to highlight a particular theoretical point. The reading of Darwin's works is still today a challenge. With the years, in *The Variation of Animals and Plants* and in some of his botanical works, the listing of examples makes the reading difficult for those who do not master the technical points he hammered in, the one after the other.

It is impossible to account in this context for the many facets of the explanatory strategies Darwin deployed in *On the Origin of Species*, and even more so to detail the slight changes of accent, or the major revisions of his theory he embarked upon in the subsequent works.[24] *On the Origin* went through six editions, the last of which was published in February 1872. Darwin also undertook to publish "Natural Selection," the comprehensive evolutionary treatise he was working on from 1854 and abandoned when the 1858 paper by Wallace reached his doorstep. The manuscript grew in length, and a first part was published in 1868 under the title *The Variation of Animals and Plants under Domestication*, a book for specialists which was not very successful: Darwin abandoned the project altogether. And "Natural Selection" had to wait a long time before seeing the light.[25] *The Descent of Man, and Selection in Relation to Sex* (1871) and the follow-up volume, *The Expression of Emotions in Man and Animals* (1872), broke Darwin's reserve on the subject of mankind. In spite of the fact that in *On the Origin* he had eluded the issue, and referred to a future when "light will be thrown on the origin of man and his history," one of the earliest, famous cartoons making fun of his theory represented him as an ape (153, P. 224). Scores of reviewers and commentators expanded upon Darwin's hint and applied the doctrine of natural selection to introduce narratives on man's different species developing from apes, frequently with overtly racist overtones. Darwin felt he had to state his position and take his distances from extreme interpretations of his thought.

Mankind had been a long-standing interest for Darwin. His manuscripts, especially the notebooks known as "M" and "N" (drafted in 1838), attest the variety of issues the study of humans raised since the early days of his scientific career—their geographical differences (Darwin remained a convinced monogenist), their moral and intellectual propensities, their sexuality and gender differences. The second volume of the *Descent of Man* was the first systematic survey of sexual habits and choices throughout the animal kingdom ever published. Darwin introduced the principle of sexual selection, which to him was essential to account for behaviours and ornaments that would in fact render the organism more exposed to destruction and the action of natural selection, like the famous case of the feathers of the peacock's tail (121).[26] Sexual selection determined differential rates of reproduction among males competing for females, thus shaping their behaviour and appearance.[27] Ultimately, sexual selection was probably responsible for the formation of human groups we call races.

From the early 1860s, and increasingly so during the 1870s, Darwin devoted boundless energy and theoretical work to several fundamental issues in botany. His articles and monographs on the fertilization of

24 Thierry Hoquet, *Darwin contre Darwin. Comment lire* L'Origine des espèces (Paris: Seuil, "L'Ordre philosophique" series, 2009).

25 R. C. Stauffer, ed., *Charles Darwin's Natural Selection* (Cambridge: Cambridge University Press, 1975).

26 See the essay in this catalogue, p. 192.

27 Adrian Desmond and James Moore, *Darwin's Sacred Cause: Race, Slavery and the Quest for Human Origins* (Chicago: University of Chicago Press, 2009); Evelleen Richards, *Darwin and the Making of Sexual Selection* (Chicago: Chicago University Press, 2017).

28 Charles Darwin, *On the Various Contrivances by Which British and Foreign Orchids Are Fertilised by Insects* (London: John Murray, 1862); Charles Darwin, "On the Movements and Habits of Climbing Plants," *Journal of the Linnean Society of London (Botany)*, vol. 9, nos. 33–34 (1865), p. 1–118; Charles Darwin, *The Effects of Cross and Self Fertilisation in the Vegetable Kingdom* (London: John Murray, 1875); and Charles Darwin, *The Power of Movement in Plants* (London: John Murray, 1880).

29 Richard Bellon, "'The Great Question in Agitation': George Bentham and the Origin of Species," *Archives of Natural History*, vol. 30, no. 2 (2003), p. 282–297; Richard Bellon, "Inspiration in the Harness of Daily Labor: Darwin, Botany and the Triumph of Evolution, 1859–1868," *Isis*, vol. 102, no. 3 (2011), p. 392–420; Craig W. Whippo and Roger P. Hangarter, "The 'Sensational' Power of Movement in Plants: A Darwinian System for Studying the Evolution of Behavior," *American Journal of Botany*, vol. 96, no. 12 (2009), p. 2115–2127; Charles Darwin, *The Effects of Cross and Self Fertilisation in the Vegetable Kingdom*, 2nd ed. (London: John Murray, 1878): "From the facts given in this volume we may infer that with mankind the marriages of nearly related persons, some of whose parents and ancestors had lived under very different conditions, would be much less injurious than that of persons who had always lived in the same place and followed the same habits of life. Nor can I see reason to doubt that the widely different habits of life of men and women in civilised nations, especially amongst the upper classes, would tend to counter-balance any evil from marriages between healthy and somewhat closely related persons," p. 465–466.

30 Charles Darwin, *The Formation of Vegetable Mould, Through the Action of Worms* (London: John Murray, 1881).

31 Mark Francis, *Herbert Spencer and the Invention of Modern Life* (Ithaca: Cornell University Press, 2007).

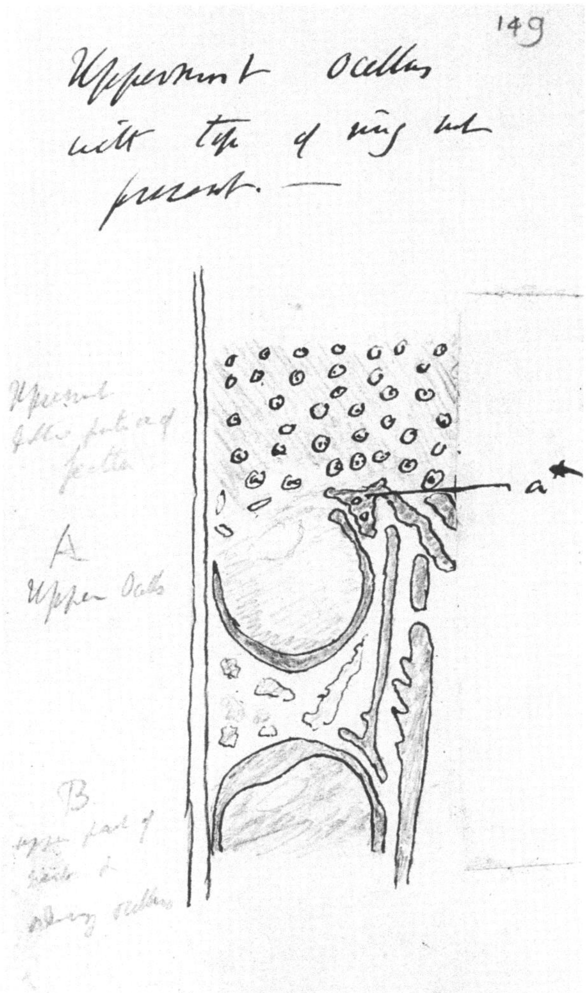

121
Charles Darwin,
Argus Pheasant Feathers, 1868
Ink and graphite, 16 × 8 cm
Cambridge, Cambridge
University Library

orchids by insects, on climbing plants, on the fertilization and the movement of plants established him as a leading European botanist, capable of devising sophisticated yet homemade (so to speak) experiments, carried out in the greenhouses he built in his country house.[28] His theoretical daring showed no sign of relenting. Darwin assumed and tried to prove that plants enjoyed a degree of sensibility far superior to the one that characterized animal forms endowed with nervous fibres and organs. He proved to his own satisfaction that cross-fertilization was not always harmful to descendants—his children were not doomed for being born from two first cousins, an idea that persecuted him for three decades.[29] The final work, on the action of earthworms in rendering a vast area of the Earth's surface fertile, showed his endless capacity for turning the seemingly simplest subject into a major, ambitious field of research.[30] He died at his home in Downe, Kent, on April 19, 1882.

Darwin had mixed feelings concerning the immense popularity of his work. Throughout Europe, colleagues and the growing army of commentators writing for the cultivated generalist public selectively adopted and adapted his sophisticated research questions into narratives that often contradicted the very foundations and epistemological guidelines of all his undertakings. This was the case with his views of the origin of mankind, or of the sophisticated nuances he introduced to explain the action of natural selection, and indeed the modest success the principle of sexual selection was met with during his lifetime. Darwin insisted that natural selection was not to be interpreted as a bloody war of all against all, even though some of his passages could be construed as endorsing so extreme a view.

One could argue that nineteenth-century-world Darwinism had only something to do with Darwin: the major Darwinists of the century showed a remarkable capacity to make Darwin say what he never said or dared to say. Two brief illustrations will suffice: Herbert Spencer and Ernst Haeckel. Their books largely outsold all Darwin's works put together, and both enjoyed an immense reputation worldwide, which contrasted with the oblivion they fell into during the twentieth century. Spencer pleased the general public by boldly providing answers for wide social, psychological and political questions Darwin had never dared to ask in print. Moreover, Spencer, like Wallace, had independently embraced a form of evolutionism during the late 1840s and the early 1850s that freely drew from Lamarck and Chambers.[31]

ERNST HAECKEL, THE FATHER OF "DARWINISMUS"

Ernst Haeckel was undoubtedly the most interesting case of problematic Darwinian, though certainly a major figure of nineteenth-century evolutionism. His scientific output was extensive and highly diversified: from his inaugural, innovative taxonomical work, *Die Radiolarien* (1862, enriched by beautiful illustrations drawn by the author himself), and his ambitious *Generelle Morphologie der Organismen* (1866) (128) that sold rather poorly, to *Systematische Phylogenie* (1894) and the extremely successful philosophical synthesis *Die Welträtsel* (1899, *The Riddle of the Universe*). He also combined his morphological work and his artistic talent in producing the remarkable *Kunstformen der Natur* (1904) (142, 144, 145, 150). Disappointed by the lack of success of his *Generelle Morphologie*, the first monograph of many to come, where he expanded upon his own version of Darwin's contribution to evolutionary theory, Haeckel decided to address the reading public with a series of popular lectures, which turned into the extremely successful *Natürliche Schöpfungsgeschichte* (1868, 2 vols., *Natural History of Creation*) (122).[32]

Haeckel, as many evolutionary naturalists before and after him, well understood that the support of the reading public, that is, of the cultivated elites, was essential to gather momentum and convince reluctant colleagues. In his *Natural History of Creation*, he offered a partial albeit revealing synopsis of the authors that had interested him, and probably many of his readers: Goethe and Lamarck in particular. In the late 1780s, Goethe had announced the poetical and philosophical intuition of the *Urpflanze*, the prototype for all vegetable life, and had put forward the equally fundamental view that the vertebra was the fundamental unit of the spine of animals we appropriately call vertebrates. The skull of animals endowed with a central nervous system was in fact made up of vertebrae that had adapted their shape to protect the brain. Haeckel lavished praise on Lamarck, though he did not refrain from criticizing his French colleague. He also called attention to German authors (Gottfried Reinhold Treviranus, Franz von Unger, Hermann Schaaffhausen) who had doubted the constancy of species and helped to pave the way for the acceptance of Darwin's work, to him the major achievement ever within natural sciences.[33]

His own task was to demonstrate the fruitfulness of Darwin's insights and to make explicit all the consequences to be drawn from his idol's works—taxonomy, morphology, paleontology, embryology, the study of man and of languages, philosophy and even religion were going to be dramatically revolutionized. Haeckel had none of Darwin's prudence and caution, in science as well as in religion: caution is "almost the soul of science," Darwin had written to Anton Dohrn on January 4, 1870. Over the years, Darwin had become a moderate atheist, officially an agnostic, but kept his ideas as much as possible to himself—he did not wish to antagonize readers and get distracted by unnecessary controversy, nor was he prepared to increase his wife's worries about his eternal destiny, or lack of it. Haeckel considered the defeat of religious superstition the ultimate achievement of his many scientific battles, propelled by the triumph of his monist philosophy. Monism was predicated on the assumption of the unity of matter and spirit, "the oneness of the cosmos, of the indissoluble connection between energy and matter, between mind and embodiment—or, as we may also say, between God and the world." Haeckel saw himself as the last representative of a tradition that spanned Western thought, from Empedocles and Lucretius to Lamarck and David Strauss, though the laws of evolution now made monism an inescapable conclusion of the history of man's search for ultimate truths.[34]

If Darwin, following Lyell's advice and his own gentle nature, did his best to avoid controversy, and often enrolled pupils and friends to fight battles on his behalf, Haeckel was at his best in the firing line, incessantly blasting opponents and lukewarm commentators. Darwin constantly tried to show that

32 For a perceptive summary and discussion of *Generelle Morphologie*, and of Haeckel's intellectual trajectory, see Mario di Gregorio, *From Here to Eternity: Ernst Haeckel and Scientific Faith* (Göttingen: Vandenhoeck & Ruprecht, 2005).

33 Henning Hufnagel, Frank Jäger and Nicolas Wanlin, "Une approche multidisciplinaire d'Ernst Haeckel," in "Ernst Haeckel entre science et esthétique," special issue, *Arts et Savoirs*, no. 9 (2018), https://journals.openedition.org/aes/1086; Robert Richards, *The Tragic Sense of Life: Ernst Haeckel and the Struggle Over Evolutionary Thought* (Chicago: University of Chicago Press, 2009); Joan Steigerwald, *Experimenting at the Boundaries of Life: Organic Vitality in Germany around 1800* (Pittsburgh: University of Pittsburgh Press, 2019); Sander Gliboff, "Evolution, Revolution and Reform in Vienna: Franz Unger's Ideas on Descent and Their Post-1848 Reception," *Journal of the History of Biology*, no. 31 (1998), p. 179–209; Hermann Schaaffhausen, "Über Beständigkeit und Umwandlung der Arten," *Verhandlungen des Naturhistorischen Vereins der preussischen Rheinlande und Westphalens*, no. 10 (1853), p. 420–451.

34 Ernst Haeckel, *Der Monismus als Band zwischen Religion und Wissenschaft. Glaubensbekenntniss eines Naturforschers* (Bonn: Verlag von Emil Strauss, 1892); in English, *Monism as Connecting Religion and Science: The Confession of Faith of a Man of Science* (London: Adam and Charles Black, 1895).

35 Charles Darwin to Asa Gray, September 10 [1860], "Embryology is to me by far the strongest single class of facts in favour of change of form, and no one of my reviewers has alluded to this."

36 See Richards, *The Tragic Sense of Life*, chap. 8, for a sympathetic assessment of the polemics over Haeckel's plates representing embryos.

37 Daniel Gasman, *The Scientific Origins of National Socialism: Social Darwinism in Ernst Haeckel and the German Monist League* (New York: American Elsevier, 1971; new ed., London: Routledge, 2017). Richard Weikart, *From Darwin to Hitler: Evolutionary Ethics, Eugenics and Racism in Germany* (Basingstoke: Palgrave MacMillan, 2004); Robert Richards, *Was Hitler a Darwinian? Disputed Questions in the History of Evolutionary Theory* (Chicago: University of Chicago Press, 2013).

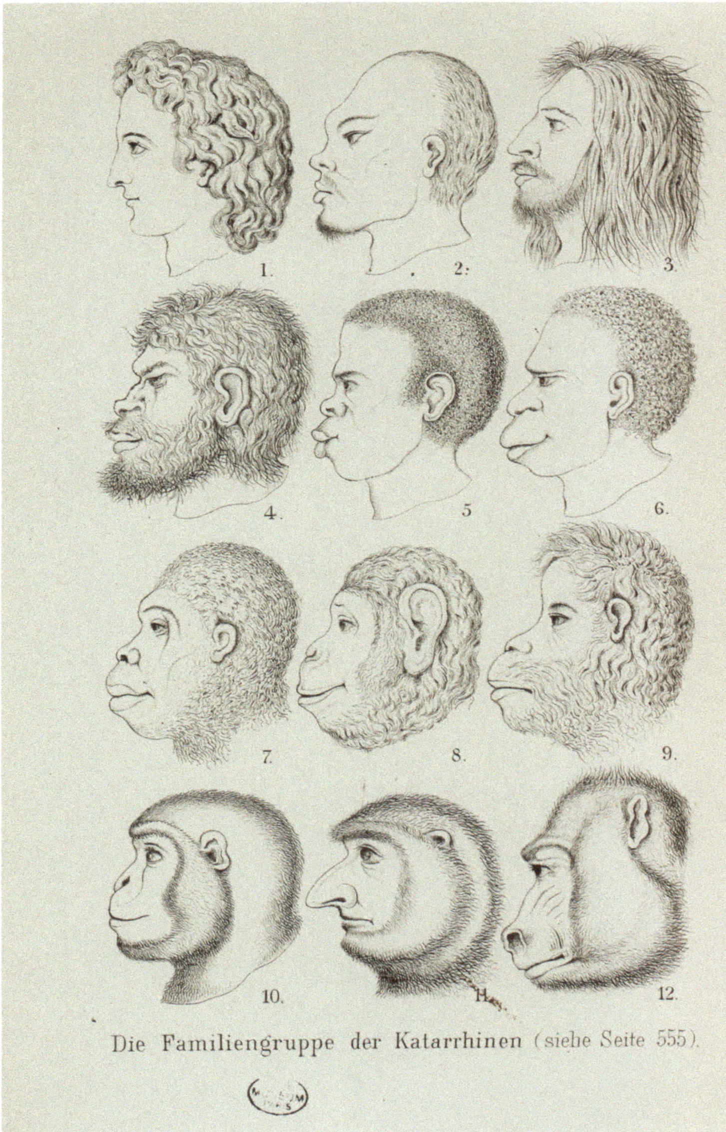

theories at first sight opposed to evolution could in fact be turned upside down and be enlisted in support. Thus, he was keen to prove that embryological recapitulation could still be of service, even though a leading embryologist, Ernst von Baer, had produced ponderous experimental evidence showing that very early in development embryos of individuals showed the traits of the class they belonged to. He was very unhappy with reviewers for missing the importance of the chapter he had devoted to the issue.[35] Haeckel boldly advanced the biogenetic law of organic recapitulation, and produced endless heuristic tables of embryos climbing the ladder towards higher forms, though at times his zeal led to painful accusations of scientific fraud from opponents and former pupils.[36]

To Darwin's chagrin, Haeckel made free use of anatomical, craniological and linguistic arguments to propose various hierarchical classifications of the many independent species and races of mankind living at the surface of the Earth, though positions on the ladder to perfection, increasingly represented by the Teutonic populations, were allotted to different groups at different times. Phylogenetic trees (however tentative and hypothetical, as he often stressed) became his passion; they also became a graphically effective way to communicate his faith in development and progress, from the simple spontaneously generated organisms to the highest forms of life. In spite of Haeckel's caution, many saw the elegantly drawn trees (Haeckel was an accomplished artist) as the objective representation of natural phenomena (128).

Historians have disputed whether Haeckel had indeed been one of the most accomplished continuators of Darwin, or someone who used Darwin's name to forward his own—at times idiosyncratic—agendas. Some have even suggested that, together with Darwin, he helped to establish deadly ideas of merciless conflict for survival and paved the way to Nazism and the twentieth-century colonial policies of extermination.[37] As we have suggested, it is more fruitful and historically more convincing to consider Haeckel as one of the many naturalists and public intellectuals who sincerely acknowledged the role Darwin had played in powerfully arguing for organic and human evolution, and felt it was their duty to further his work. To reduce the staggering complexity of evolutionism after Darwin and well into our own times to the question of measuring the degree of faithfulness to the teaching (one should say "teachings") of the English naturalist risks missing the actual articulations of the theories put forward.

As often repeated in this contribution, an unprejudiced reading of works published before and after Darwin reveals the many ways the heritage of eighteenth-century materialism, of Lamarck or Geoffroy Saint-Hilaire, of Chambers and later on of Darwin, Huxley, Haeckel or Spencer, was adopted and adapted by a rich population of naturalists, philosophers,

122
Ernst Haeckel, "Die Familiengruppe der Katarrhinen," in *Natürliche Schöpfungsgeschichte*, frontispiece, Berlin: G. Reimer, 1868
Paris, MNHN

public intellectuals and journalists, and indeed even theologians. Many of them undoubtedly referred to higher scientific authorities to forward their own views of nature, of racial superiority, of the natural foundations of social and gender inequalities. To impute to Darwin or to Haeckel the exclusive paternity of the extreme political doctrines that have dramatically marked the twentieth century is to ignore the growing popularity of racism throughout the nineteenth century. Darwin was surely not a racist, though, as a Victorian man, he did not like the Irish very much and did not think intelligence was necessary to women.[38] He did, however, concede that ladies of the upper classes, educated over a long sequence of generations, could certainly improve their intellectual performance.

Haeckel was undoubtedly convinced that the white man was superior to all other races, and during the First World War, now an old and rather isolated man, he joined the army of pamphleteers extolling the superiority of the German *Volk* now threatened by the unholy alliance of the envious British with inferior people like the French and the Italians. Yet, the literature on racial doctrines vastly exceeds the pages Haeckel devoted to the issue. Extermination policies actively pursued throughout the nineteenth century against Indigenous populations in North America, Africa or Asia did not need "scientific" endorsement or teaching. It is within the history of a dreadful century that lie the premises of the horrors to come, when the same policies, economic motivations and extreme violence were turned against the very European populations convinced of representing the pinnacle of humanity and of civilization.

---

38 Charles Darwin, *The Descent of Man, and Selection in Relation to Sex* (London: John Murray, 1871), vol. 2, p. 329: "In order that woman should reach the same standard as man, she ought, when nearly adult, to be trained to energy and perseverance, and to have her reason and imagination exercised to the highest point; and then she would probably transmit these qualities chiefly to her adult daughters."

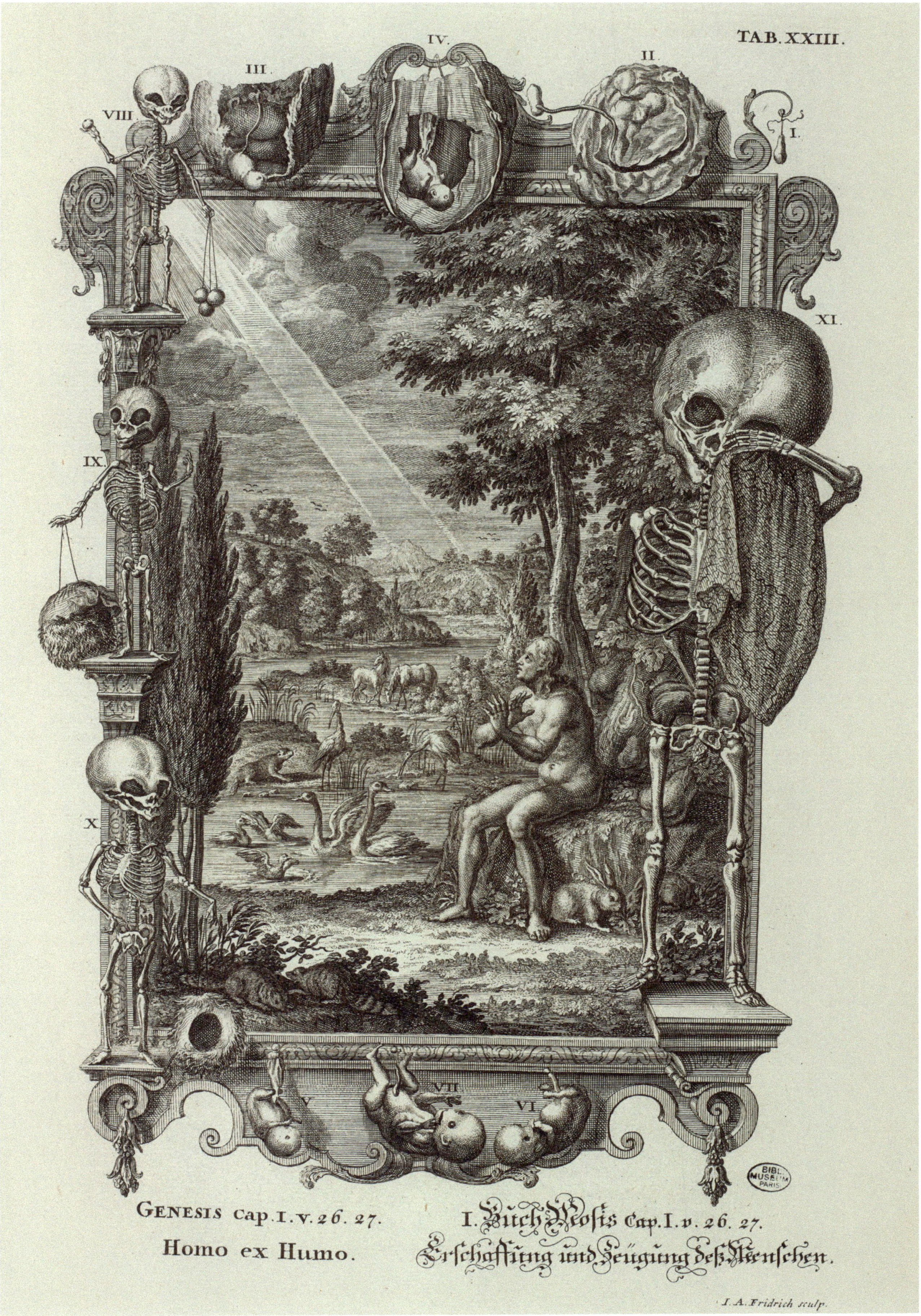

123
Johannes Scheuchzer,
Physica sacra, pl. 23,
Augsburg and Ulm, 1731–1733
Paris, MNHN

FROM THE GREAT CHAIN OF BEING TO THE TREE OF LIFE

GUILLAUME LECOINTRE

The desire to explore the real world rationally is inconceivable unless one has previously established categories. In addition to scientific categories, which have attained greater or lesser degrees of coherence depending on the epoch in question, there are also, in the background, "images of Nature,"[1] the organizing schemas of these categories, representations of the theoretical framework used to account for the apparent natural order.

Note that we are speaking here only of those sciences that strive to encompass all living things by managing the relationships between words, beings and categories through the use of systematics and taxonomy. Since Aristotle, the dominant theoretical paradigm has been the figure of continuity called the "chain of being," or the "scale of being," which, as far as we know, goes back to Democritus and Plato.[2] This chain of being comes to us from the Middle Ages by way of authors like Ramon Llull and Charles de Bovelles, who were mainly concerned with theological matters. This figure of continuity would assume its scientific forms in the work of Gottfried Wilhelm Leibniz and, especially, in that of Charles Bonnet, even if it was a reflection on values that largely accounted for the fact that human beings were ultimately placed at the top of all these scales of being (124). These epochs were characterized by relative dissociation between the theoretical schema that described the organization of nature, on the one hand, and the creation of categories, on the other. In the second half of the eighteenth century, the map replaced the scale as the organizing paradigm of nature in the work of Vitaliano Donati and Buffon (as of 1755), and it remained the predominant model in the 1770s and 1780s. The tree made its appearance in 1766 (163, P. 237).[3] We must not forget, however, the distributive power of tables and how certain trees functioned as tables (126).[4]

In the late Renaissance, scientists began developing the concept of natural classification, which would have things to say about nature quite different from simple retranscriptions of humans' material needs. The goal of classifiers was, therefore, to reconstruct an order that was inherent in nature and should lead to unity in its classification. The late eighteenth century witnessed the beginnings of what would become an increasingly widespread tendency to draw correspondences between classificatory logic and what Bertrand Daugeron called *raison graphique*:[5] the requirement of natural classification logically involved matching the taxa that were created with the organizing theoretical framework, even if some scattered inconsistencies would later continue to be found. Within this framework, the fixist schema of the scale served as the graphic expression of the divine plan. Towards the end of the eighteenth century, science, having become experimental as the result of a movement begun a century earlier, freed itself from theology and considerations of value. It was at this point that the tree metaphor began to predominate in life sciences. The tree image appeared in 1766 in the work of the agronomist Antoine-Nicolas Duchesne, who used it to represent a genealogy; in 1801, it reappeared in the work of Augustin Augier, who used it to underscore the discontinuities of nature, not to suggest transformation (125). The power of the hierarchical organization of attributes was clearly what made possible the "tree-branch" order used primarily by Augier. The successive branches formed a nested hierarchy going from the most general (at the tree roots) to the most particular (on the terminal branches). Between pure ordering and pure genealogy, there have been many trees bearing very different meanings.[6] In 1798, the tree clearly functioned as a genealogical metaphor for Erasmus Darwin (even if he did not draw one). The inscription of the tree figure within a temporal and "phylogenetic" dimension occurred in the work of Jean-Baptiste Lamarck in 1809, in that of Charles Darwin in 1837, and in that of Alfred Russel Wallace in 1856. The tree that took hold, especially in the final third of the nineteenth century, was called "phylogenetic," after the German term coined by Haeckel in 1866.[7] The tree-shaped figure that Darwin drew in 1859[8] was purely theoretical, dealt with abstract taxonomic entities and functioned as a veritable research plan (not completed in his lifetime) within a nominalist framework (127); Haeckel's tree, on the other hand, reflected an interest in the genealogy of concrete organisms whose successive phases, along the branching structure, constituted ideal forms that stemmed from the German morphological tradition and thus bore certain echoes of Goethe (128). Generally speaking, the idealistic character of the Haeckelian program and the unfinished character of Darwin's theoretical program made the relationship between the taxa and phyletic lineages complex in practice during the century between 1859 and the spread of phylogenetic analysis during the 1970s.

It was Darwin who, in 1859, would finally establish—explicitly and lastingly—the specifications for classification in natural history. The result of evolutionary mechanisms is that living beings, over the course of their genealogy, undergo transformations while leaving their hereditary attributes, new and old, to their descendants. Consequently, an attribute found to be in a similar state in several contemporary species was probably inherited from a hypothetical common ancestor exclusive to those species. And this ancestor was not that of other contemporary species in whom the attribute occurred in a slightly different form. Thus, the order of attributes would serve to establish degrees of relative kinship among living beings. But it did not stop there, because the degree of kinship came to serve as the basis for a classification program. The apparent hierarchy in the distribution of attributes was no longer seen as evidence of a divine plan: for Darwin, classification had to reflect genealogical development, i.e. that of biological evolution, as faithfully as possible. Yet Darwin did not have a working method capable of reconstructing what Haeckel had come to call "phylogeny," in other words, the graphic representation of degrees of relatedness. Willi Hennig was to provide this method a century later. Meanwhile, imprecise post-Darwinian classification systems, including the eclectic systematics of the years from 1940 to 1970, made it possible to dress

1 Bertrand Daugeron, *Collections naturalistes, entre science et empires, 1763–1804* (Paris: Muséum national d'histoire naturelle, 2009), p. 199.

2 Arthur O. Lovejoy, *The Great Chain of Being: A Study of the History of an Idea* (Cambridge, MA: Harvard University Press, 1976); Pascal Duris and Gabriel Gohau, *Histoire des sciences de la vie* (Paris: Belin, 2011).

3 Giulio Barsanti, *La scala, la mappa, l'albero: immagini e classificazioni della natura fra Sei e Ottocento* (Florence: Sansoni, 1992); Daugeron, *Collections naturalistes, entre science et empires, 1763–1804*, p. 199.

4 Daugeron, ibid.

5 Ibid., p. 225.

6 See Pascal Tassy, *L'Arbre à remonter le temps* (Paris: Bourgois, 1991); Marie Fisler and Guillaume Lecointre, "Categorizing Ideas about Trees: A Tree of Trees," *PLoS ONE*, vol. 8, no. 8 (August 7, 2013), https://doi.org/10.1371/journal.pone.0068814; Marie Fisler, Cédric Crémière, Pierre Darlu and Guillaume Lecointre, "The Treeness of the Tree of Historical Trees of Life," *PLoS ONE*, vol. 15, no. 1 (January 15, 2020), https://doi.org/10.1371/journal.pone.0226567.

7 Ernst Haeckel, *Generelle Morphologie der Organismen* (Berlin: Reimer, 1866).

8 On Haeckel's trees, see Benoît Dayrat, "The Roots of Phylogeny: How Did Haeckel Build His Trees?" *Systematic Biology*, vol. 52, no. 4 (2003), p. 515–527; on Darwin's tree, see Jean Gayon, "Mort ou persistance du darwinisme ? Regard d'un épistémologue," *Comptes rendus Palevol*, no. 8 (2009), p. 321–340. See also Guillaume Lecointre, "'Arbori-culture': une typologie des 'arbres' dans la culture phylogénétique," *Biosystema*, no. 30 (2015), an issue devoted to "L'arbre du vivant, trente ans de systématique," ed. Patrick Martin, Sophie Nadot and Christophe Daugeron, p. 113–132.

up a persistent if not rampant anthropocentricism in evolutionary garb. This would endure in the world of research up to the translation of Hennig's work in 1966, which sparked a scientific revolution in the field of systematics. This revolution stemmed from many sources, but one of the most important was the transparency of classification procedures. In fact, the relevancy of results was measured by way of the tree's consistency, which became mathematically measurable in the early 1970s.[9] Since 1966, when the tree of life became accessible to us, *the tree became synonymous with classification*. Whether one looks at anatomical or molecular characters, the latter being available since 1967, they remain "characters"—which changes nothing about the graphics used in the associated classification program.

The proponents of molecular systematics, which proved particularly fecund between 1985 and 2005, wagered that, by obtaining DNA sequences of one or more genes from different species, they could make them into markers of the species' histories. This paid off for most eukaryotes, as long as the scales of comparison remained at a distance from the perispecific level. But for a large part of the bacterial world, horizontal gene transfer (i.e. cross-species transfer) was frequent even in the case of very old divergences between lineages. Indeed it was so common that, when these exchanges attained a certain level of intensity, the genes told only their own individual histories. One no longer knew which entities one ought to use to establish a phylogeny—unless one went with genes, and with genes that were not, moreover, too composite. As tools of investigation and representation, the phylogenies of archaebacteria and eubacteria undoubtedly should represent themselves with the help of numerous trees or graphic structure systems.[10] Thus, genealogical theory would no longer be divergent, as in Darwin's schema (1859), but possibly reticulate—in other words, divergent and then convergent. Nonetheless, the tree as a theoretical framework did not fade away; it remains still valid for most multicellular organisms.[11]

9 James S. Farris, "The Information Content of the Phylogenetic System," *Systematic Zoology*, vol. 28, no. 4 (1979), p. 483–519.

10 Daniel H. Huson and David Bryant, "Application of Phylogenetic Networks in Evolutionary Studies," *Molecular Biology and Evolution*, vol. 23, no. 2 (2006), p. 254–267.

11 Guillaume Lecointre and Hervé Le Guyader, *Classification phylogénétique du vivant*, 4th ed., 2 vols. (Paris: Belin, 2016–2017).

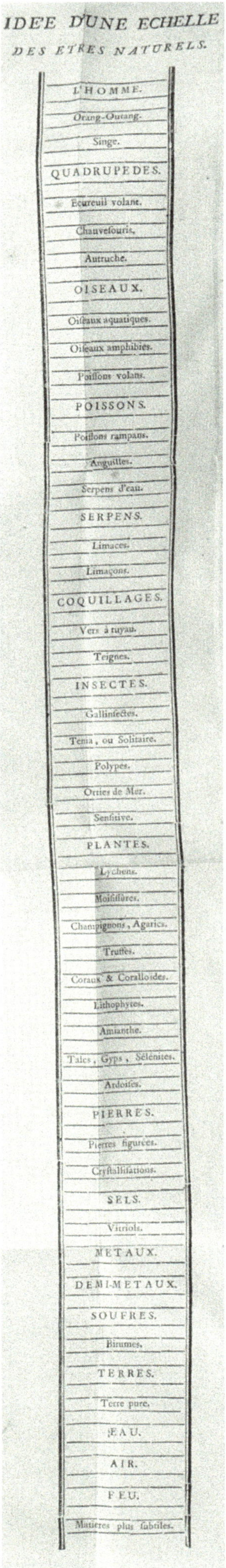

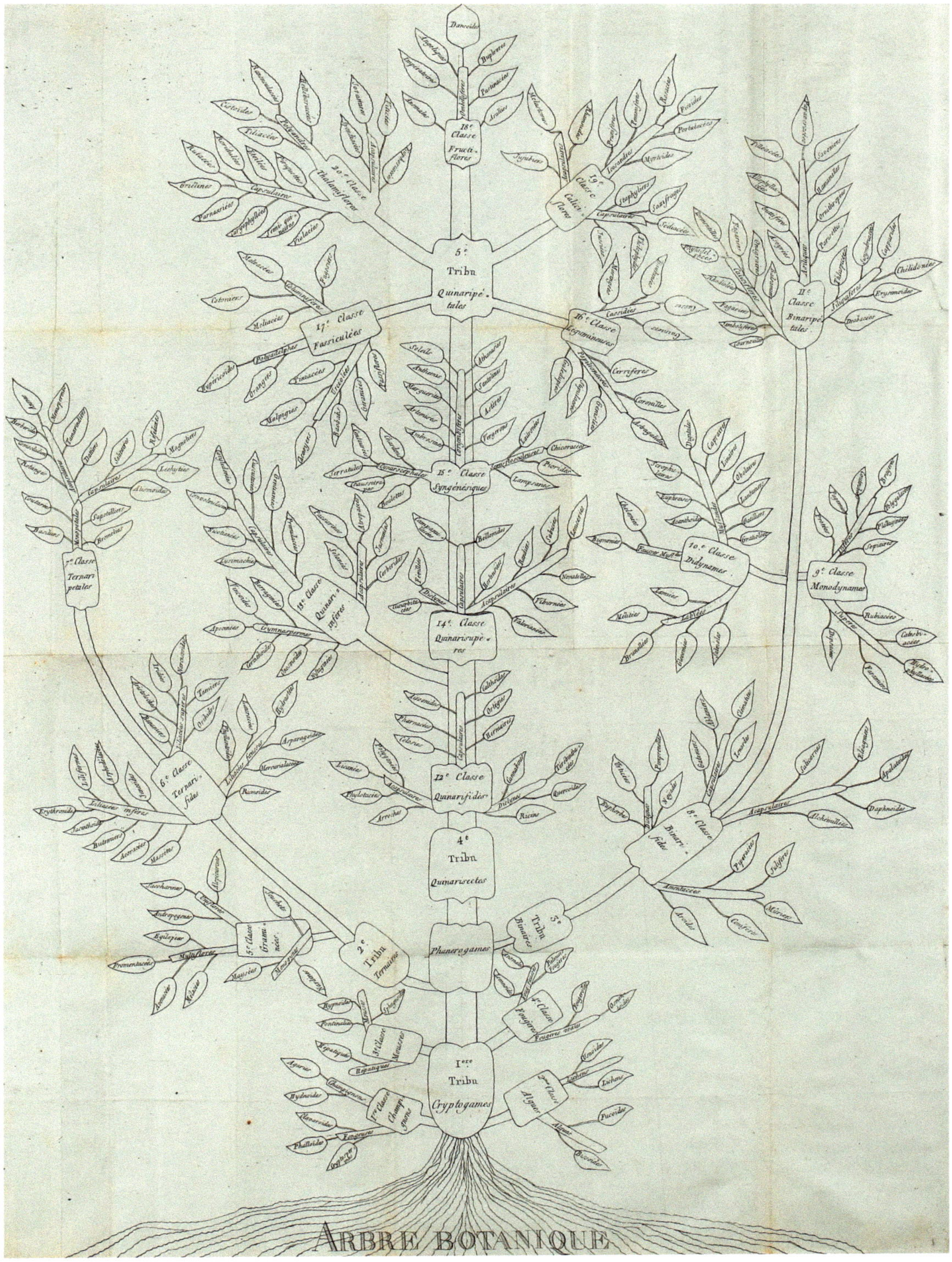

124
Charles Bonnet,
"Idea for a Scale
of Natural Beings,"
in *Traité d'insectologie
ou Observations sur
les pucerons*, Paris:
Durand, 1745
Paris, MNHN

125 FIG.
Augustin Augier,
"Botanical Tree,"
in *Essai d'une nouvelle
classification des végétaux*,
Lyon: Bruyset Ainé,
an IX – 1801
Cambridge, MA, Harvard
University Botany Libraries

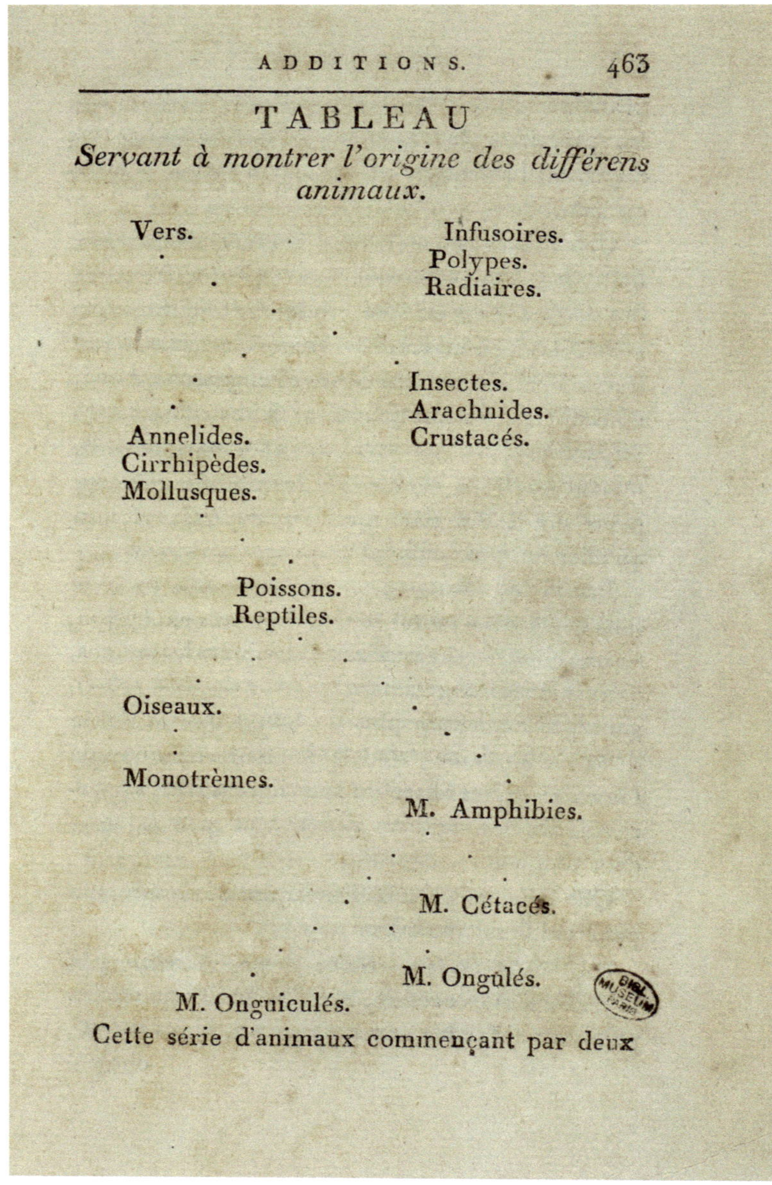

ADDITIONS. 463

TABLEAU

Servant à montrer l'origine des différens animaux.

Vers.
Infusoires.
Polypes.
Radiaires.

Insectes.
Arachnides.
Crustacés.

Annelides.
Cirrhipèdes.
Mollusques.

Poissons.
Reptiles.

Oiseaux.

Monotrèmes.

M. Amphibies.

M. Cétacés.

M. Ongulés.

M. Onguiculés.

Cette série d'animaux commençant par deux

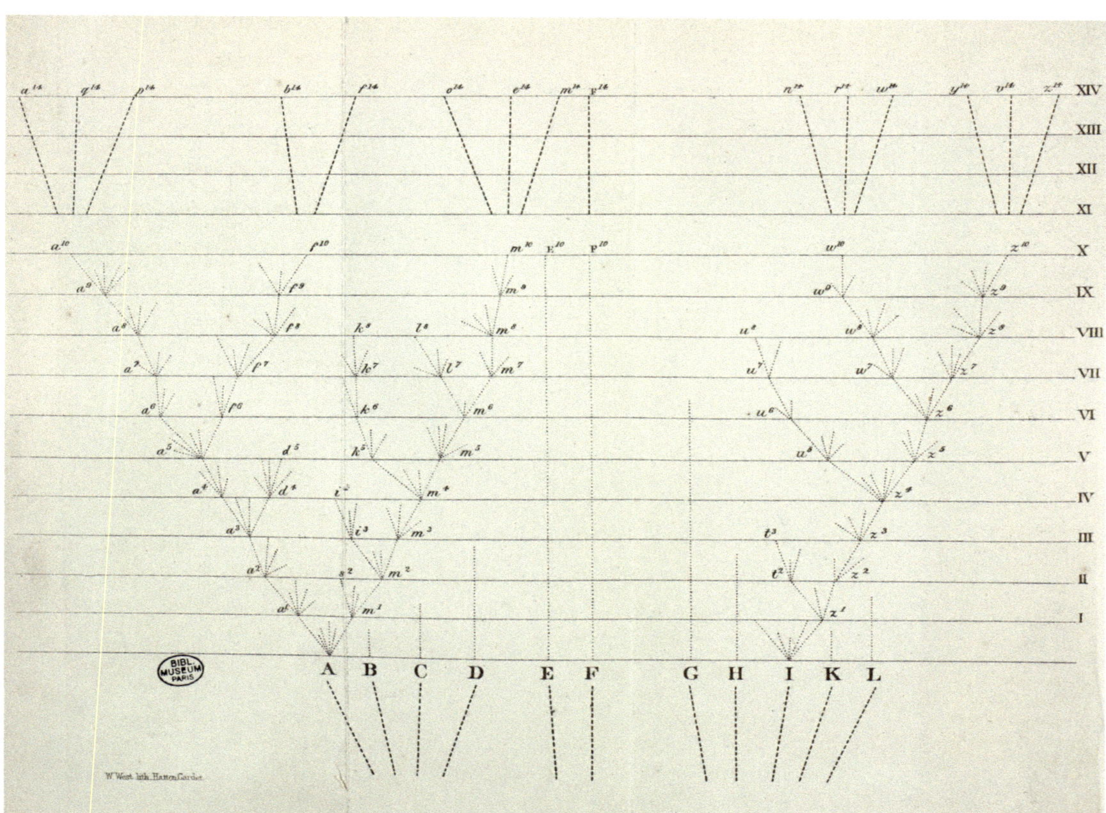

126
Jean-Baptiste Lamarck,
"Table Showing the Origin
of the Various Animals,"
in *Philosophie zoologique
ou exposition des
considérations relatives
à l'histoire naturelle
des animaux*,
Paris: Dentu, 1809
Paris, MNHN

127
Charles Darwin,
"The Probable Effects
of the Action of Natural
Selection through Divergence
of Character and Extinction,
on the Descendants
of a Common Ancestor,"
in *On the Origin of Species
by Means of Natural
Selection . . .*, London:
J. Murray, 1860
Paris, MNHN

128
Ernst Haeckel,
*Generelle Morphologie
der Organismen*, pl. 1, vol. 2,
Berlin: G. Reimer, 1866
Paris, MNHN

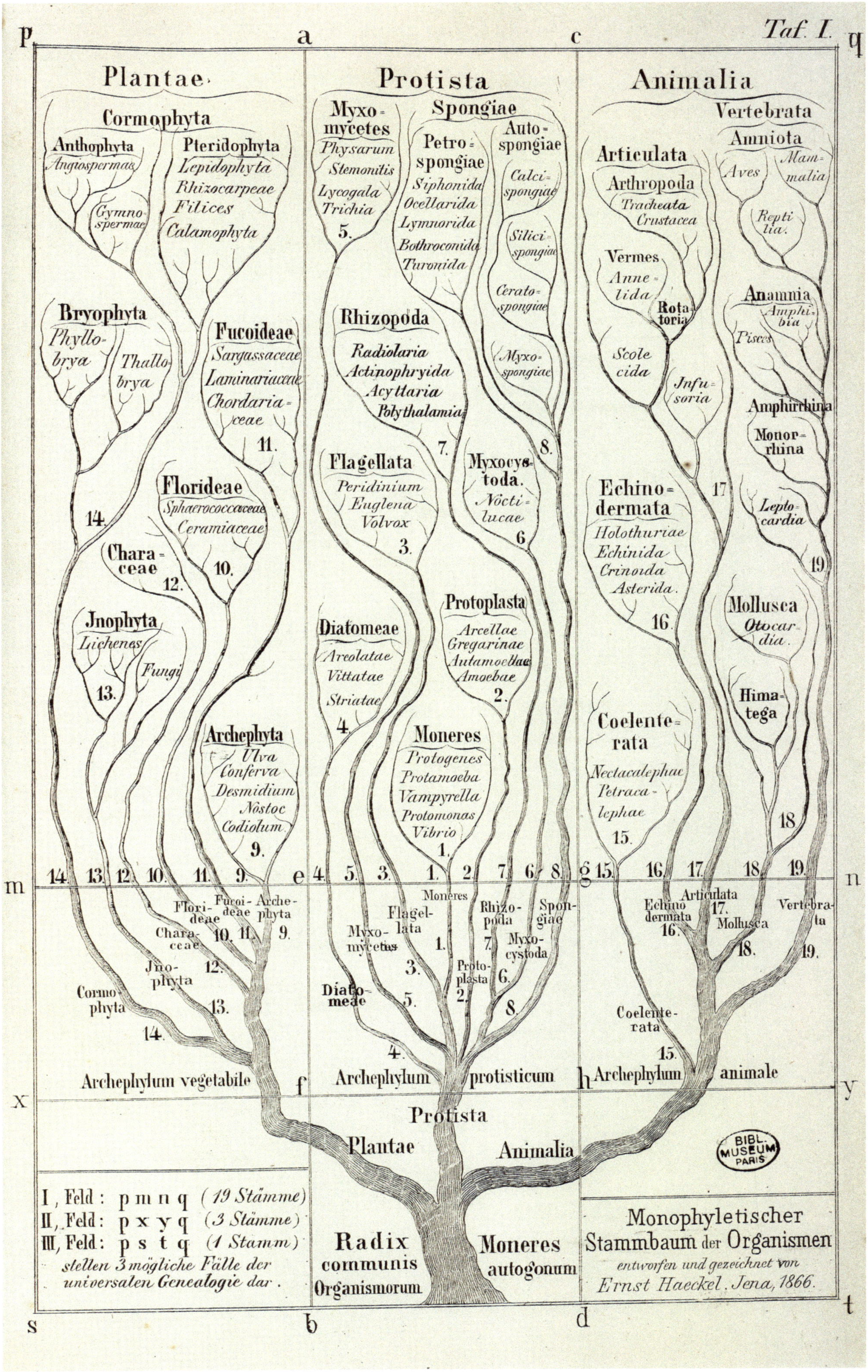
Taf. I.
p
a
c
q
Plantae
Protista
Animalia
Cormophyta
Anthophyta
Angiospermae
Gymnospermae
Pteridophyta
Lepidophyta
Rhizocarpeae
Filices
Calamophyta
Bryophyta
Phyllobrya
Thallobrya
Fucoideae
Sargassaceae
Laminariaceae
Chordariaceae
11.
Florideae
Sphaerococcaceae
Ceramiaceae
10.
14.
Characeae
12.
Jnophyta
Lichenes
Fungi
13.
Archephyta
Ulva
Conferva
Desmidium
Nostoc
Codiolum
9.
Myxomycetes
Physarum
Stemonitis
Lycogala
Trichia
5.
Spongiae
Petrospongiae
Siphonida
Ocellarida
Lymnorida
Bothroconida
Turonida
Autospongiae
Calcispongiae
Silicispongiae
Ceratospongiae
Myxospongiae
8.
Rhizopoda
Radiolaria
Actinophryida
Acyttaria
Polythalamia
7.
Flagellata
Peridinium
Euglena
Volvox
3.
Myxocystoda.
Noctilucae
6.
Protoplasta
Arcellae
Gregarinae
Autamoebae
Amoebae
2.
Diatomeae
Areolatae
Vittatae
Striatae
4.
Moneres
Protogenes
Protamoeba
Vampyrella
Protomonas
Vibrio
1.
Vertebrata
Amniota
Aves
Mammalia
Reptilia.
Articulata
Arthropoda
Tracheata
Crustacea
Vermes
Annelida
Rotatoria
Scolecida
Infusoria
Anamnia
Amphibia
Pisces
Amphirrhina
Monorrhina
Leptocardia
19.
17
Echinodermata
Holothuriae
Echinida
Crinoida
Asterida.
16.
Mollusca
Otocardia.
Himatega
18.
Coelenterata
Nectacalephae
Petracalephae
15.
m
14. 13. 12. 10. 11. 9.
e
4. 5. 3. 1. 2. 7. 6. 8.
g
15. 16. 17. 18. 19.
n
Florideae
Fucoideae
Archephyta
10. 11. 9.
Characeae
Jnophyta
12.
Cormophyta
13.
14.
Moneres
Flagellata
Rhizopoda
Spongiae
Myxomycetes
1.
7.
Myxocystoda
3.
Protoplasta
6.
Diatomeae
5.
2.
8.
4.
Articulata
Echinodermata
17.
16.
Mollusca
Vertebrata
18.
19.
Coelenterata
15.
Archephylum vegetabile
f
Archephylum protisticum
h
Archephylum animale
x
y
Protista
Plantae
Animalia
BIBL. MUSEUM PARIS
I, Feld: p m n q (19 Stämme)
II, Feld: p x y q (3 Stämme)
III, Feld: p s t q (1 Stamm)
stellen 3 mögliche Fälle der universalen Genealogie dar.
Radix communis Organismorum
Moneres autogonum
Monophyletischer Stammbaum der Organismen
entworfen und gezeichnet von Ernst Haeckel. Jena, 1866.
s
b
d
t

WHY IS THE PEACOCK BEAUTIFUL?

LAURA BOSSI

"The sight of a feather in a peacock's tail, whenever I gaze at it, makes me sick!"

Charles Darwin, 1860[1]

The peacock is the most beautiful of birds. Its image decorated homes in ancient Rome and appeared in the depictions of the Garden of Eden by medieval miniaturists and Baroque painters (8, P. 29; 10, P. 30). Its feathers served as a model for the wings of angels, its train was compared to the heavens, and in its iridescent plumage alchemists saw the colours of the spectrum.[2]

In India, its native land, it is associated with Krishna; in Greece, it was the bird cherished by Hera, the wife of Zeus, who placed the one hundred eyes of the giant Argus, her faithful watchman, on its tail to create the ocelli, or eyespots;[3] and, among the Romans and early Christians, it was a symbol of immortality and resurrection.

Jean de La Fontaine compared its plumage to the "... wearing / Of rainbow silks a hundred different dyes."[4] Robert de Montesquiou entitled his affectedly refined collection of poems *Les Paons* [*Peacocks*] in the hope of "clothing it in the winged orbs of the bird whose plumage is an ambulant setting, and which along its train bears opalescent eyes where all sorts of cabochon gems seem to glisten."[5]

However, in bestiaries of the Middle Ages, the peacock was associated with the sin of pride, or that of vanity, most likely due to its propensity to "strut about" during the mating season. In the time of the Renaissance, it often embodied pride in illustrations of the seven deadly sins.[6] The French caricaturist J. J. Grandville would choose the bird to portray "the illustrious Marmus," a charlatan in a Balzac story who is covered in honours as a result of painting his donkey with the stripes of a zebra.[7]

BUT WHY IS THE PEACOCK BEAUTIFUL?

When a peacock (an inveterate polygamist) "fans its tail," raising and spreading out the long feathers of its train to form an arc up to two metres wide, and then shaking them to produce an astonishing rattling noise, it invariably makes a deep impression on us. While the white peacocks that can be seen in the gardens of Italy's Borromean Islands are remarkable, blue peacocks are particularly fascinating, undoubtedly because of their ocelli (129, 130),[8] which seem to watch us ... eye to eye.

Although peahens seem to take no notice of male displays, continuing to calmly peck about, they do watch them out of the corner of their eyes, and scholars have long suspected that they are not insensitive to such ostentatious beauty.

Buffon devoted a lengthy chapter of his *Natural History of Birds* to the peacock and its beauty, detailing its "nobleness of demeanour," its "light waving tuft, painted with the richest colours," its "matchless plumage" that dazzles the eye "in the sparkling lustre of the gems" and "astonishes it in the grand display of the rainbow." He was probably not mistaken in attributing its display to "the fires of love, joined to the secret influence of the season" inspiring a "new ardour and new desires."[9]

For Darwin, on the other hand, the peacock and its plumage were not a subject of observation and wonder, but a problem, even an obsession. As he wrote in April 1860, "The sight of a feather in a peacock's tail, whenever I gaze at it, makes me sick!" Its heavy, showy train did not seem to him compatible with his theory of natural selection—according to which variations likely to improve adaptation to the environment, and therefore survival, are passed on to succeeding generations—as it could attract the attention of predators, hinder movement and impede flight,[10] in short, to be counter to natural selection. What purpose, then, does the peacock's train have, if not to please the peahen? In addition, how could the formation of a structure as complex as the ocellus of the feathers of the peacock or its cousin, the Argus pheasant (*Argusianus argus*), be explained by small successive variations over generations (121)?

"[These] ocelli . . . are shaded in so wonderful a manner as to resemble balls lying within sockets, and consequently differ from ordinary ocelli. No one, I presume, will attribute the shading, which has excited the admiration of many experienced artists, to chance—to the fortuitous concourse of atoms of colouring matter. That these ornaments should have been formed through the selection of many successive variations, not one of which was originally intended to produce the ball-and-socket effect, seems as incredible, as that one of Raphael's Madonnas should have been formed by the selection of chance daubs of paint made by a long succession of young artists, not one of whom intended at first to draw the human figure. In order to discover how the ocelli have been developed, we cannot look to a long line of progenitors, nor to various closely allied forms, for such do not now exist. But fortunately the several feathers on the wing suffice to give us a clue to the problem, and they prove to demonstration that a gradation is at least possible from a mere spot to a finished ball-and-socket ocellus."[11]

Such thoughts on the beauty of the peacock would lead Darwin to develop his theory of sexual selection, which he based on two forces of selection: "the power to charm the female" and "the power to conquer other males in battle."[12] Those powers are expressed in "ornaments" that attract the opposite sex, such as the peacock's feathers or the wings of butterflies, and "armaments," such as the horns, antlers or tusks of many mammals (132).

The theory was already outlined—although only in connection with the battle of males—in the 1848 essay that would be read, along with one by Alfred Russel Wallace,[13] in 1858 at the Linnean Society; it would be worked out in detail in *The Descent of Man*, where Darwin clearly distinguished natural selection, which is "adaptive," from sexual selection, which applies only to characteristics that give one male an advantage over another.

In respect to the "ornaments," Darwin declared that they served solely to charm females—therefore having an aesthetic significance—and that beautiful males have evolved towards beauty, that is, in literal terms, "have been rendered beautiful for beauty's sake."[14]

1 Letter to Asa Gray, April 3, 1860; https://www.darwinproject.ac.uk/letter/DCP-LETT-2743.xml.

2 Ernst Thomas Reimbold, *Der Pfau: Mythologie und Symbolik* (Munich: Callwey, 1983).

3 Ovid, *Metamorphoses*, Book I, ll. 622–723. Hera transferring Argus's eyes onto the peacock's feathers is depicted in the painting by Peter Paul Rubens (1611) in the collection of Cologne's Wallraf-Richartz-Museum.

4 Jean de La Fontaine, "The Peacock Complaining to Juno," in *The Fables of La Fontaine*, trans. Walter Thornbury (New York: Cassell & Company, 1873), p. 90.

5 Robert de Montesquiou, *Les Paons* (Paris: G. Richard, 1908), p. VI.

6 See, for example, the "Superbia" (Pride) drawing from Pieter Brueghel the Elder's series "The Seven Deadly Sins," 1557.

7 J. J. Grandville (Jean Ignace Isidore Gérard), "Mon maître, obstinément appelé l'illustre Marmus" [My master, persistently called the illustrious Marmus], illustration for the satirical story by Honoré de Balzac "Guide-âne à l'usage des animaux qui veulent parvenir aux honneurs" [The Donkey's Handbook for Animals Wishing to Achieve Honours], *Scènes de la vie privée et publique des animaux* (Paris: J. Hetzel et Paulin, 1842), vol. 1, p. 206.

8 Blue peacocks have from 165 to 170 ocelli in their trains.

9 Georges-Louis Leclerc, Count de Buffon, "The Peacock," in *The Natural History of Birds*, trans. William Smellie (London: A. Strahan, T. Cadell and J. Murray, 1793), vol. 2, p. 253–254.

10 See the section "Display by Male Birds of their Plumage" in chap. 13 of Charles Darwin, *The Descent of Man, and Selection in Relation to Sex* (London: John Murray, 1871), vol. 2, p. 97–98.

11 Darwin, *The Descent of Man*, vol. 2, p. 141–142.

12 Darwin, *The Descent of Man*, vol. 1, p. 279.

13 See http://darwin-online.org.uk/content/frameset?itemID=F350&viewtype=text&pageseq=1.

14 Charles Darwin, *The Origin of Species By Means of Natural Selection*, 6th ed. (London: John Murray, 1872), p. 161.

That is not only a scientific theory, but also a theory of art. Beauty would have a basis in nature, related to sexual attraction, and does not require great powers of reasoning. To a certain extent, the peahen's choice indicates that the capacity to perceive beauty creates it.

Darwin anticipated the objections to such a theory, particularly in regard to the aesthetic sense of peahens that, like chickens, were generally deemed lacking in developed intellectual faculties. He maintained that "low powers of reasoning . . . are compatible, as we see with mankind, with strong affections, acute perception, and a taste for the beautiful."[15]

Darwin's theory of sexual selection failed to achieve any success at the time of its publication. Even Alfred Russel Wallace, who considered himself "more Darwinian than Darwin," after initially giving it a favourable welcome, abandoned it. He found it unnecessary, considering natural selection sufficient explanation for the peacock's plumage and, more generally, the "ornaments" of males, the indicators of their "vigour," activity and vitality. By putting sexual selection, understood as competition between males, back into the framework of natural selection, the latter would be relieved of an "abnormal excrescence."[16] As for female choice, he did not believe in it, thinking that peahens were incapable of appreciating the variety and infinite beauty of the delicate contrasts and subtle harmonies of peacock feathers' colours. Ernst Haeckel, the main advocate of Darwinism in the German-speaking world, gave the theory a polite mention in his books, but devoted no discussion to it.

The most virulent criticism arguably came from Saint George Jackson Mivart, a former protege of Thomas Henry Huxley, who had become an influential opponent of Darwin's theories. He accepted the principle of male combat, but denied the power of choice of females, including human ones.[17]

Above all, it was the fact that Darwin attributed the power of choice and an artistic sense to peahens that curbed adherence to his theory, and he could be accused of anthropocentrism when he stated that it must be assumed that peahens admire the peacock's train as much as we do. However, if Darwin tried to see things from the peahen's viewpoint, it was because he thought that we, too, are animals, and our capacities differ from those of so-called lowly creatures only in degree. *The Expression of the Emotions in Man and Animals*[18] would deal specifically with that.

The peacock was also the focus of heated contention with the painter and art critic John Ruskin, who considered Darwin's theories stamped by a disguised atheism and chose the question of natural beauty as his line of attack.[19] Ruskin had studied natural history at Oxford and was well acquainted with Darwin, whom he held in high esteem. Nonetheless, he rejected the idea that the elaborate forms and the beauty of peacock and Argus pheasant feathers could be explained by natural selection and sexual selection alone. He would deal with that point in two publications. In *The Eagle's Nest*, he related his disappointment with Darwin's

15 Darwin, *The Descent of Man*, vol. 2, p. 108; see also p. 111 and p. 123–124.

16 Alfred Russel Wallace, "The Colours of Animals and Plants: I and II," *Macmillan's Magazine*, vol. 36 (September and October 1877), p. 384–408 and p. 464–471; quoted in Evelleen Richards, *Darwin and the Making of Sexual Selection* (Chicago: University of Chicago Press, 2017), p. 466.

17 St. George Jackson Mivart, "[Review of] Darwin's *Descent of Man*," *Quarterly Review*, vol. 131 (July 1871), p. 47–90. See Richards, *Darwin and the Making of Sexual Selection*, p. 468 and following.

18 Charles Darwin, *The Expression of the Emotions in Man and Animals* (London: John Murray, 1872).

19 Phillip Prodger, "Ugly Disagreements: Darwin and Ruskin Discuss Sex and Beauty," in *The Art of Evolution: Darwin, Darwinisms, and Visual Culture*, ed. Barbara Larson and Fae Brauer (Hanover, NH: Dartmouth College Press, 2009), p. 40–58.

explanations concerning such feathers' intricate patterns and magnificent colours; he concluded by suggesting that Darwin draw a peacock feather himself: "When you have drawn some of the actual patterns of plume and scale with attention, I believe you will see reason to think that spectra of organic species may be at least as distinct as those of metals or gases; but learn at all events what they are now, and never mind what they have been."[20]

In *Proserpina*, he described the "materialist" explanations of chromatic evolution as the work of "human bats . . . incapable of . . . seeing, much less thinking about, colour." He put Darwin's "vespertilian treatise" on the peacock's feathers in that category, but hoped to make him change his mind, if he could have "him here in Oxford for a week, and . . . force him . . . to copy a [peacock] feather."[21]

Ruskin himself drew a number of peacock feathers (130), as well as other natural objects such as stones, shells and plants (249, P. 384). Those splendid works show the importance of drawing from life in grasping the complexity and beauty, in addition to the uniqueness, of every being and every feather. Although he was a poor draughtsman, Darwin also took the trouble to copy, most likely using a camera lucida, the feather of a peacock from Java as well as that of an Argus pheasant, which would be reproduced in *The Descent of Man*. However, he did so in an entirely different spirit: his aim was to show how even very complex ornaments could be acquired gradually, through successive small changes.

The Darwinian idea of sexual selection inspired generations of artists in the late nineteenth century.[22] Many painters from various schools, like František Kupka (174, P. 253), and the Symbolists Max Klinger, Franz von Stuck and Alfred Kubin, depicted battling males. For its part, the peacock feather would be associated with the trope of the femme fatale, becoming an emblem of seduction, especially among the Pre-Raphaelites (133). It would become ubiquitous in the decorative arts (134), from the works of Tiffany to those of the Arts and Crafts movement, not to mention the magnificent peacocks that filled Ludwig II of Bavaria's castles. Archibald Thorburn would provide a more understated version (135), yet one that was perfectly informed by the scientific issues: his peacock fans its tail, and at its feet alights a butterfly proudly showing off its own ocelli.

The theory of sexual selection has been well-nigh forgotten for a century. Ernst Mayr, for example, did not mention it in his identification of Darwin's five theories: evolution as such, or descent with change, common descent, gradualism, multiplication of species, and natural selection.[23] However, it is now once again the subject of studies and, at times, stormy debates,[24] as scientists have turned their attention to the old questions raised by the beauty of the peacock. Is it one of the criteria females use in choosing their partner? Are those criteria purely utilitarian, as Wallace thought, or is beauty indeed chosen for its own sake, as Darwin saw it? Is there a natural basis for beauty? We can hardly ask the peahen for an answer.

20 John Ruskin, *The Eagle's Nest: Ten Lectures on the Relation of Natural Science to Art* (New York: Maynard, Merrill & Co., 1893), p. 200; quoted in Prodger, "Ugly Disagreements: Darwin and Ruskin Discuss Sex and Beauty," p. 43, whose argument I am following here.

21 John Ruskin, *Proserpina: Studies of Wayside Flowers, While the Air Was Yet Pure Among the Alps, and in the Scotland and England Which My Father Knew*, 3rd ed. (Orpington and London: George Allen, 1897), vol. 1, p. 94.

22 See, for example, Barbara Larson, "Darwin's Sexual Selection and the Jealous Male in Fin-de-Siècle Art," in *The Art of Evolution*, p. 173–193.

23 Ernst Mayr, "Darwin's Five Theories of Evolution," in *The Darwinian Heritage*, ed. David Kohn (Princeton: Princeton University Press, 1985), p. 755–772.

24 See Thierry Hoquet, ed., *Current Perspectives on Sexual Selection: What's Left after Darwin?* (Dordrecht, the Netherlands: Springer, 2015); Richard O. Prum, *The Evolution of Beauty: How Darwin's Forgotten Theory of Mate Choice Shapes the Animal World—and Us* (New York: Doubleday, 2017); Michael J. Ryan, *A Taste for the Beautiful: The Evolution of Attraction* (Princeton: Princeton University Press, 2018).

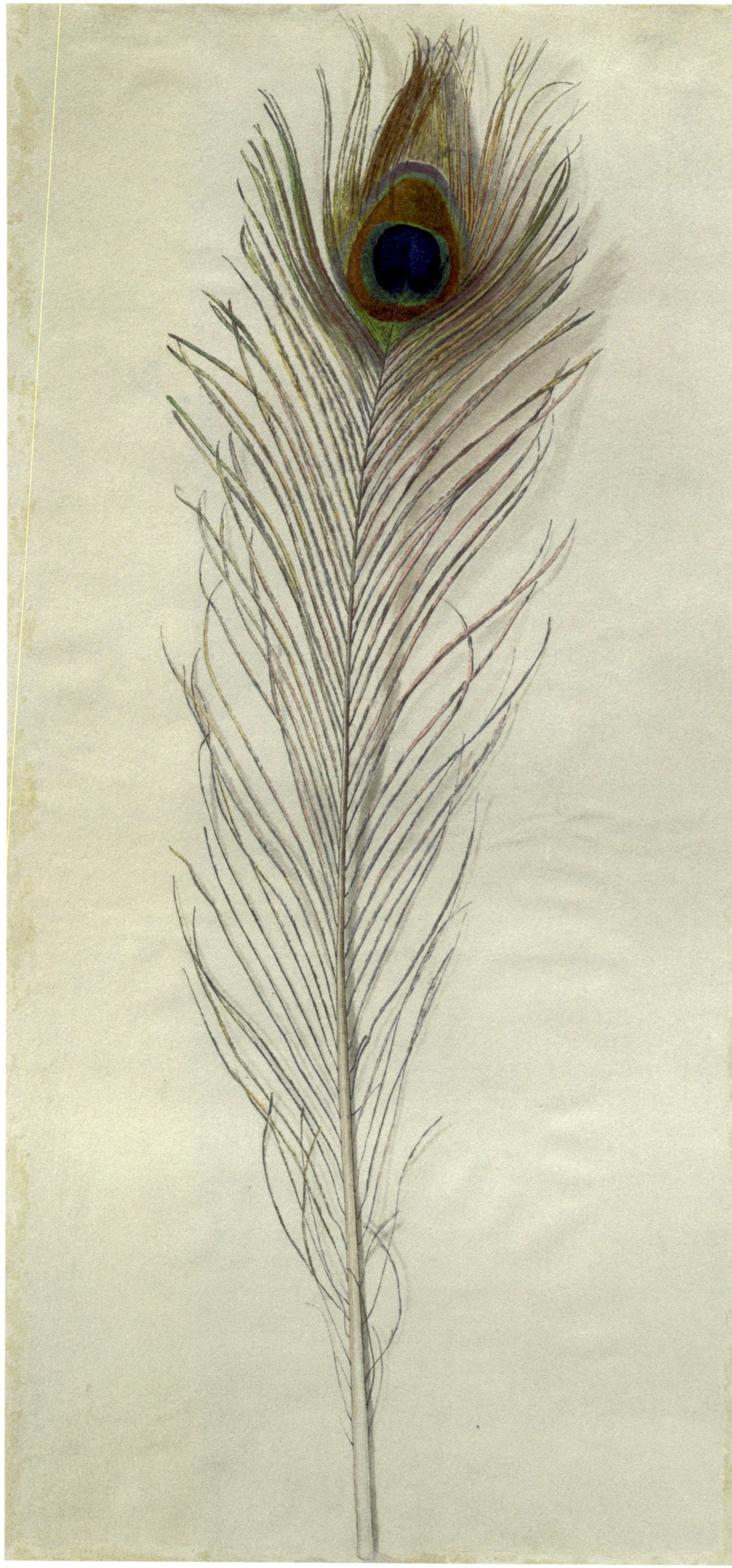

129
Anonymous,
Study of a Peacock's Feather, n.d.
Watercolour over graphite on wove paper, 64.3 × 30.7 cm
Oxford, Ashmolean Museum, University of Oxford

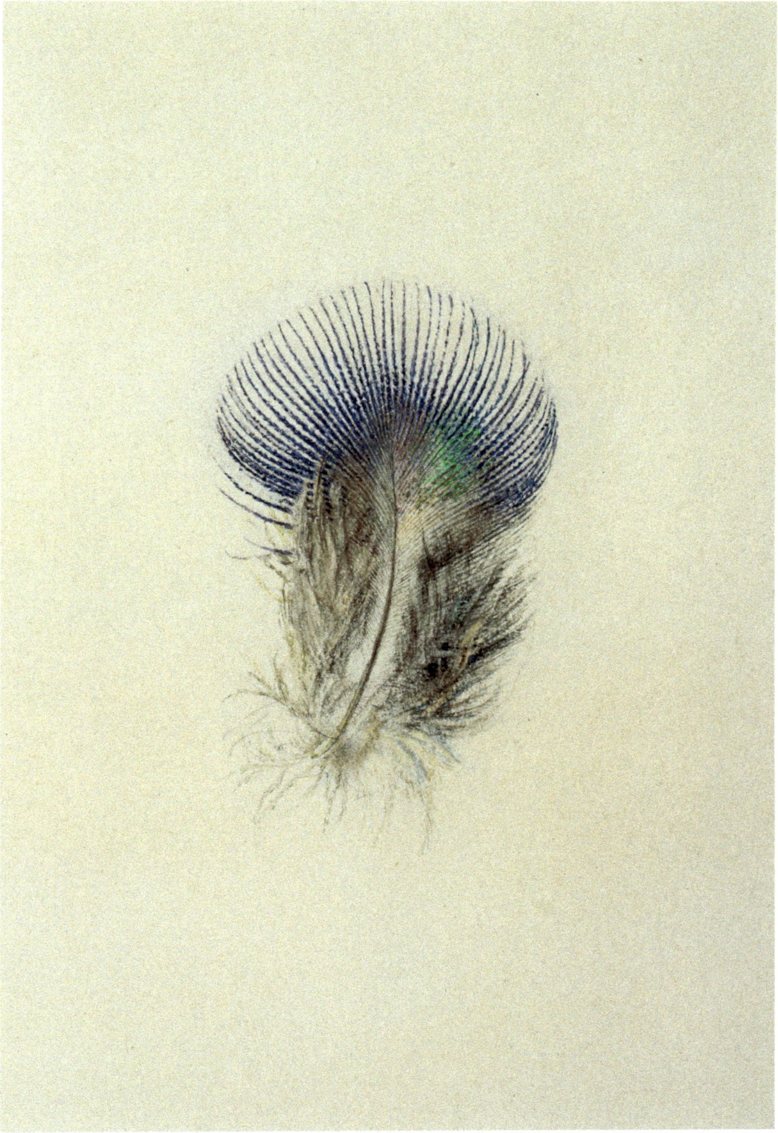

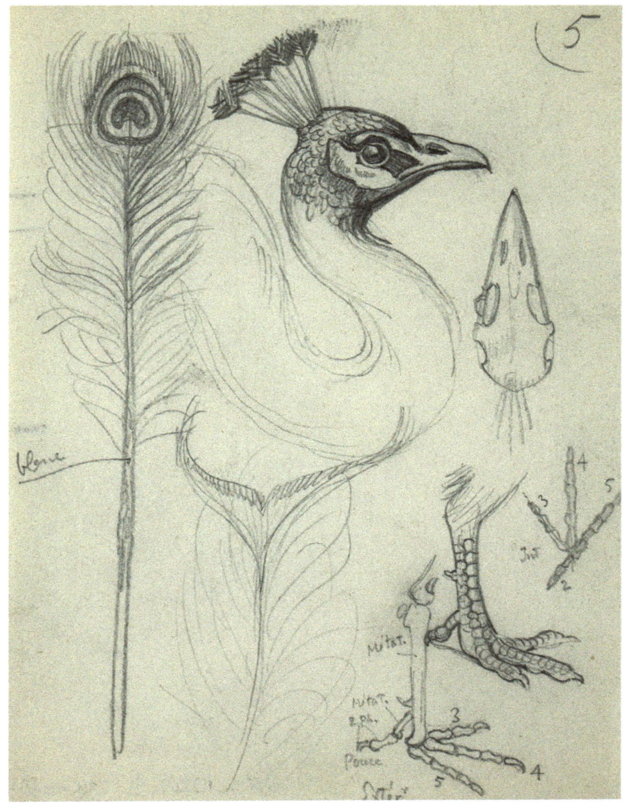

130 FIG.
John Ruskin,
Study of a Peacock's Breast Feather, 1873
Watercolour, 28.9 × 19.8 cm
Sheffield, Sheffield Galleries and Museums Trust

131
Eugène Grasset,
Peacock Feet, Feathers and Heads, in Profile,
between 1890 and 1903
Pencil, 18.7 × 13.9 cm
Paris, Musée d'Orsay

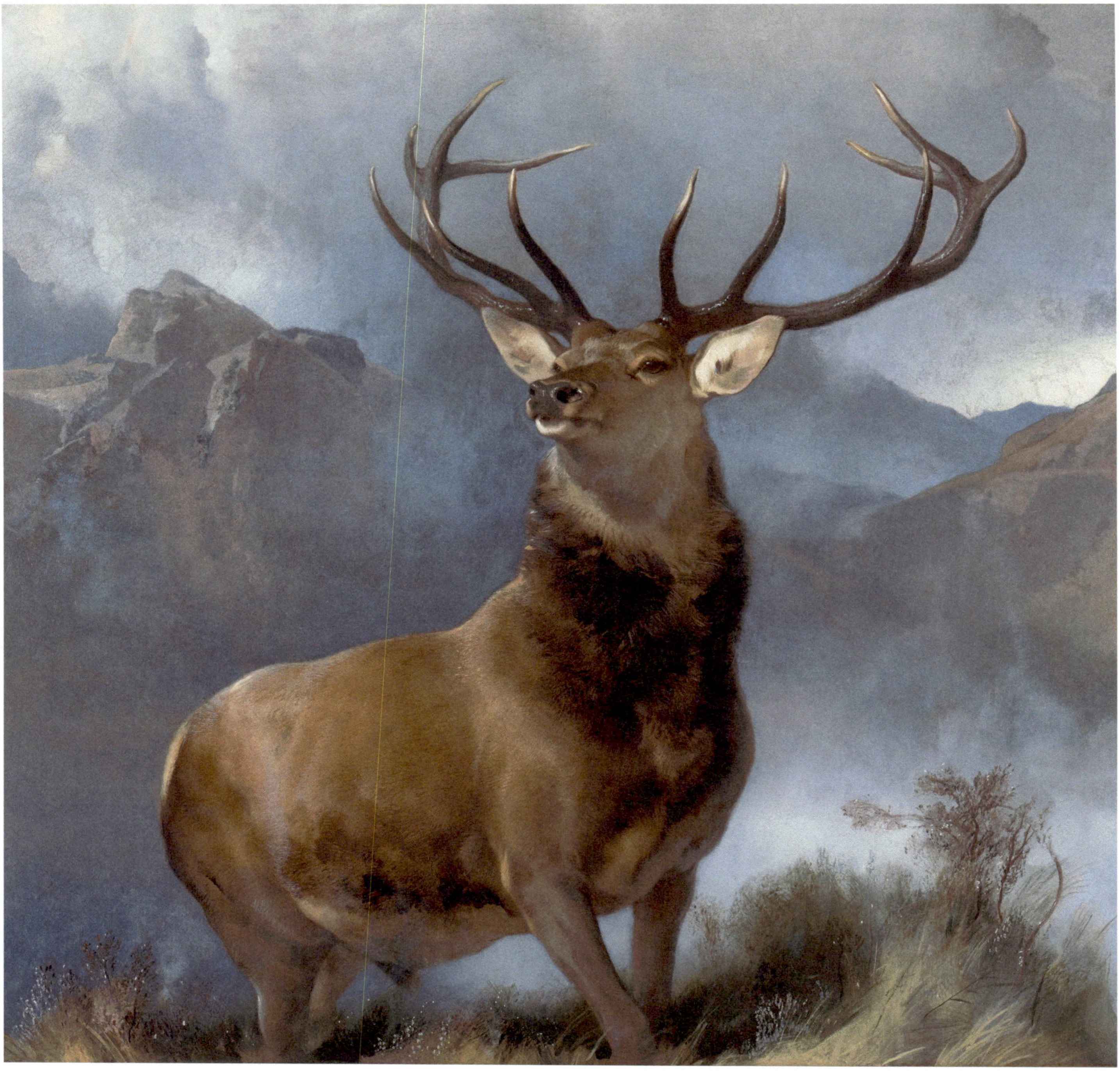

132 FIG.
Sir Edwin Henry Landseer,
The Monarch of the Glen,
about 1851
Oil on canvas,
163.8 × 168.9 cm
Edinburgh, Scottish
National Gallery

133
Frederic Leighton,
Pavonia, 1858-1859
Oil on canvas, 53 × 41.5 cm
Private collection

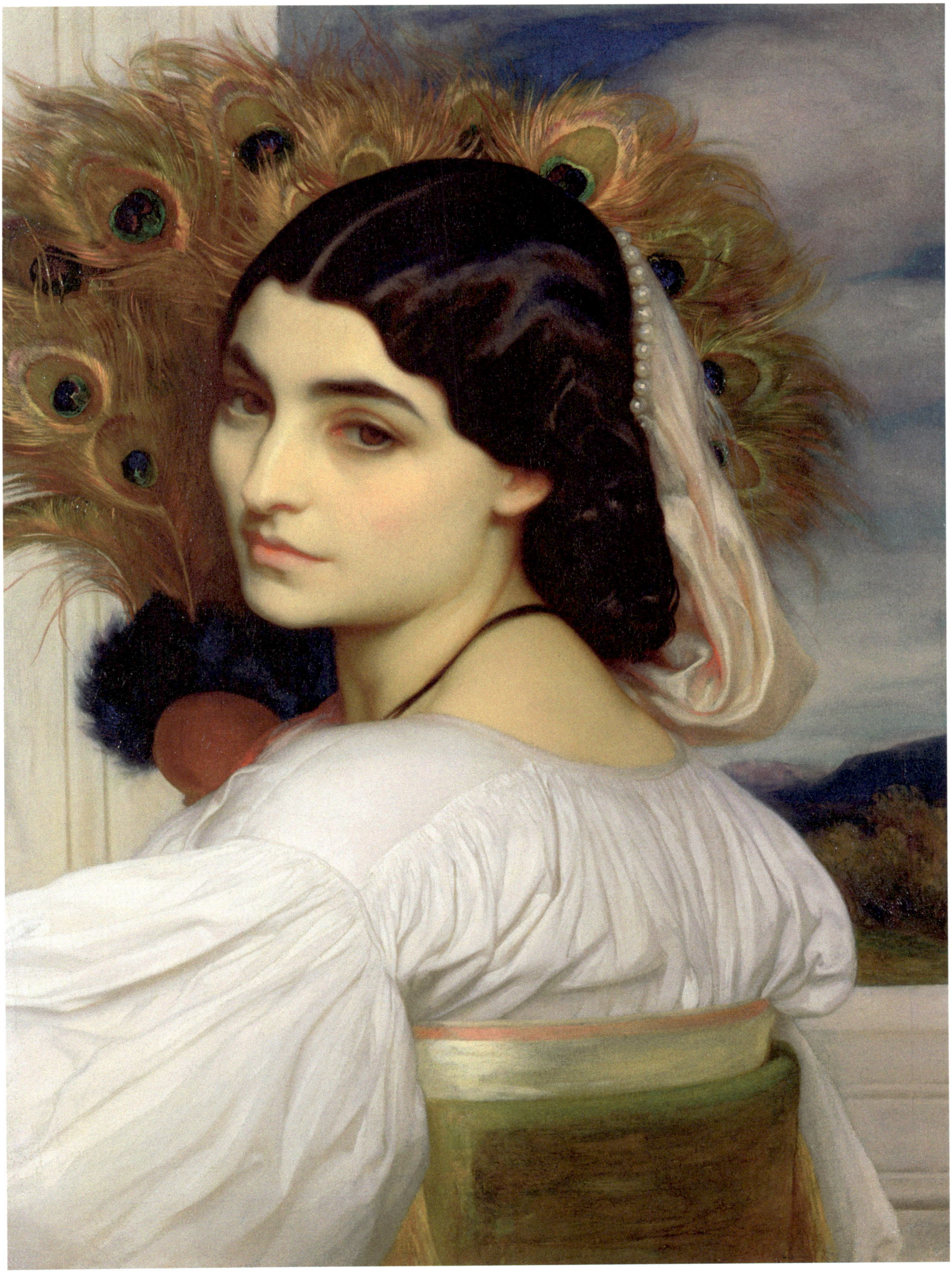

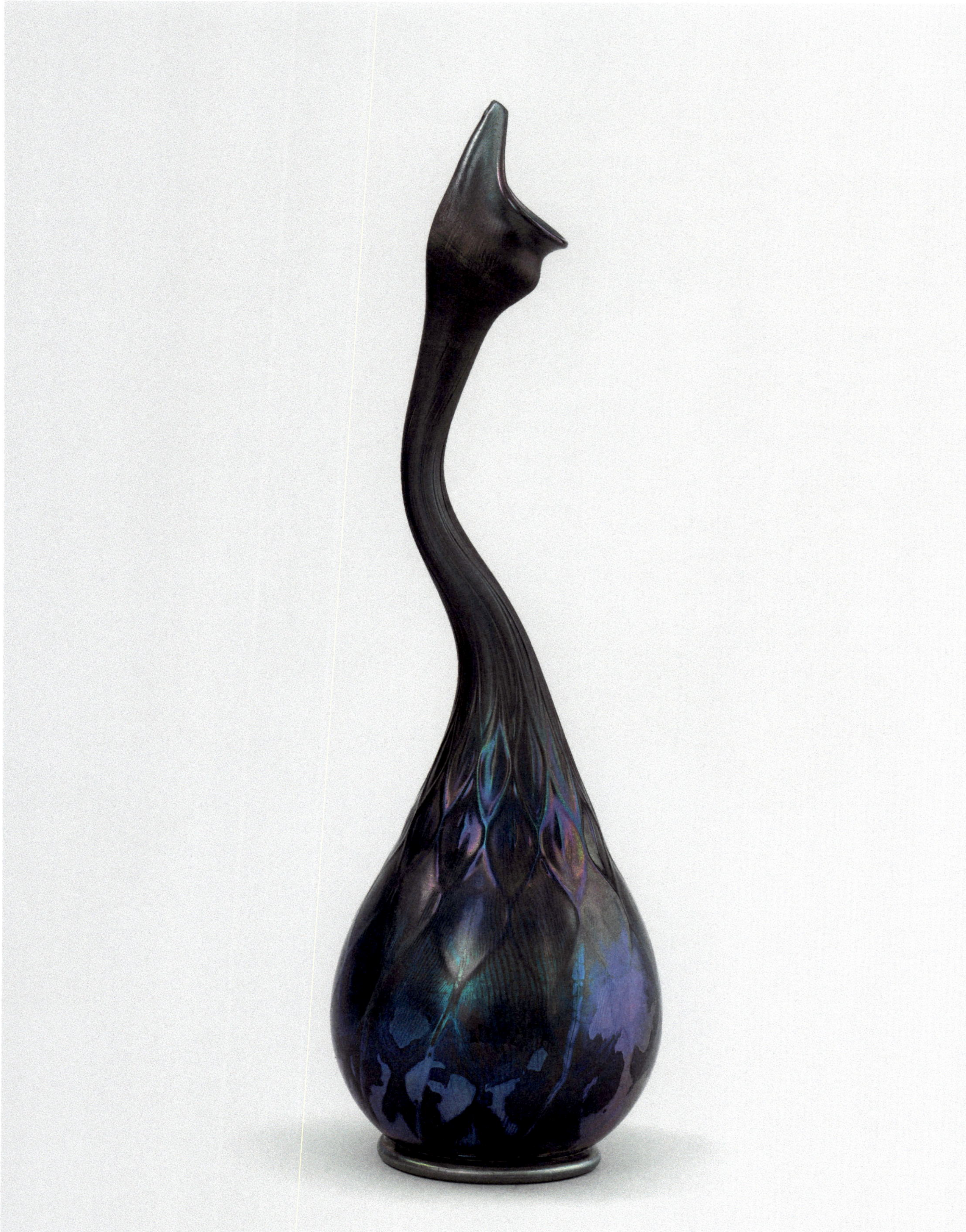

134
Louis Comfort Tiffany,
"Peacock" Bottle, about 1897
Glass, 41.2 × 12.5 cm
Paris, Petit Palais,
musée des Beaux-Arts
de la Ville de Paris

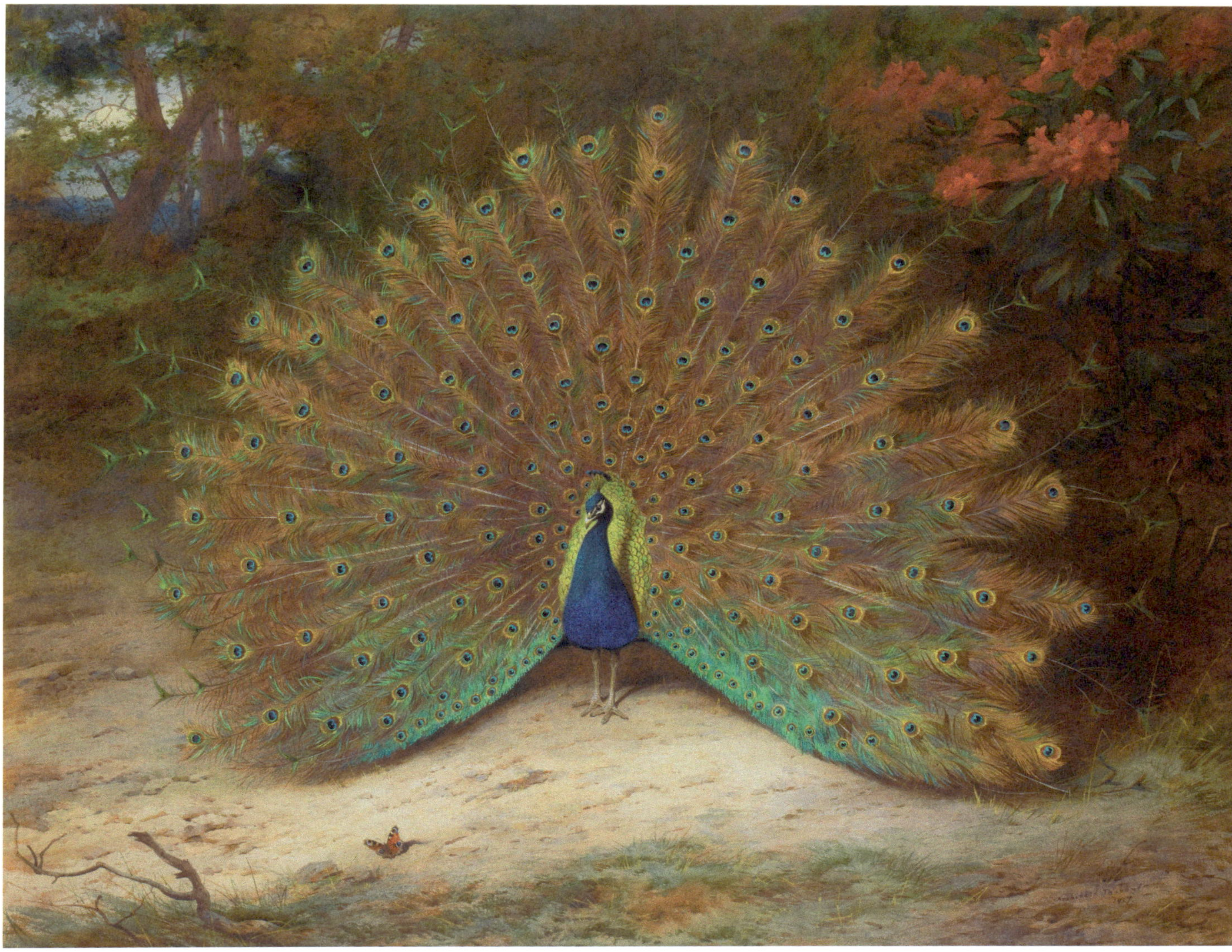

135
Archibald Thorburn,
Peacock and Peacock Butterfly, 1917
Oil on canvas, 87.5 × 111.5 cm
Private collection

LAUGHTER AND DREAMS

JULIA VOSS

In 1871, a few months after Charles Darwin published his voluminous *The Descent of Man, and Selection in Relation to Sex*, the New York magazine *Harper's Weekly* featured a caricature showing the British scientist alongside an anthropoid ape (152, P. 224). This was the work of Thomas Nast, one of New York's most famous cartoonists, who had made his fortune with the refined humour of his drawings. But he also inspired some measure of fear because his art enabled him to reach the illiterate strata of the population. To depict Darwin, Nast had performed a delicious role reversal: instead of showing a human scandalized by his kinship with the animal, he depicted a gorilla breaking into tears at the sight of such an unbecoming relative. "That Man wants to claim my Pedigree. He says he is one of my Descendents," says the gorilla, turning to Henry Berg, the President of the American Society for the Prevention of Cruelty to Animals. Bergh relays the reproach to Darwin, saying: "Now, Mr. Darwin, how could you insult him so?"[1]

Convinced, like Nast, that humour could serve as a tool for raising awareness, Darwin had contacted the illustrators of the British satirical magazine *Punch;* he saw no harm in using humour to alert the general public to his arguments—quite the opposite. He was persuaded that laughter would be useful in disseminating his theory of evolution.

Moreover, Darwin had a somewhat unorthodox view of laughter. He did not see it as the prerogative of man and, in 1872, he even gave voice to his ideas on the subject in *The Expression of the Emotions in Man and Animals,* where he did not mention the gorilla portrayed by Nast; nor did he mention it in *The Descent of Man,* where he refused to speculate on the degree of kinship with the larger hominoid, apart from in this single passage: "We should, however, bear in mind that an animal possessing great size, strength, and ferocity, and which, like the gorilla, could defend itself from all enemies, would not perhaps have become social: and this would most effectually have checked the acquirement of the higher mental qualities, such as sympathy and the love of his fellows. Hence it might have been an immense advantage to man to have sprung from some comparatively weak creature."[2]

Darwin distanced himself from those who affirmed that the theory of evolution made man into a cruel animal, like the gorilla. The image of the gorilla baring its fangs, which Nast also seized upon, appeared for the first time in 1858 in a science magazine published by the Muséum national d'histoire naturelle in Paris, where the world's first naturalized specimen of this animal was conserved in a vat filled with ethyl alcohol. The illustrator Firmin Bocourt had portrayed the gorilla as a kind of frightening Adam, as in Albrecht Dürer's depiction of Paradise. All subsequent illustrations imitated this model, up to the twentieth century's King Kong (169, P. 245). Darwin never had an opportunity to see a live gorilla because he passed away before the first specimen arrived at the London Zoo in 1887.

But in the spring of 1871, a zoo attendant at Regent's Park Zoological Gardens made a curious observation about a much smaller animal, a monkey, the *Cynopithecus niger*, or jet-black crested macaque, which had been brought to London from Celebes Island shortly before. The macaque was laughing, or more precisely, sniggering. It cocked back its diminutive ears and emitted a low cry like a kind of jabbering, drew back the corners of its lips and bared its front teeth. The zoo administrators immediately notified Darwin, a regular visitor to the garden whose interest in animal behaviour was well known. The *Cynopithecus niger*, it was explained to him, laughed from pleasure when petted.[3] Darwin wanted to have a portrait of the animal done without delay and the name of Joseph Wolf, an animal painter, was recommended to him. "The eye of Mr. Wolf is like photographic paper," said the zoo superintendent Abraham Bartlett, "it will seize on anything" (138).

In *The Expression of the Emotions in Man and Animals* (1872), the *Cynopithecus niger* is shown laughing, and three young girls and an old man all wear the same expression (136). The portrait format, a pictorial convention, underscores the closeness of man and animal. In successive depictions, the little monkey tilts back its crest and ears, raises its brows slightly, opens its mouth, draws the corners of its lips backwards and upwards, and utters a low cry—the sound of its laughter. The threatening large black visage of the gorilla had been replaced by a small laughing face. It was not, therefore, a ferocious wild animal that created the link between man and animal, but a laughing monkey.

Apart from the monkey, Darwin adduced a second animal, the dog, as the essential proof of the kinship between men and animals. "A dog, whilst dreaming," Darwin had written in his notebook almost thirty years earlier, twitches its paws, "which they do only when considerably excited, [which] shows their power of imagination—for it will not be allowed that they can dream, and not have day-dreams—think well over this; it shows similarity in mind"[4] (140).

The British painter Briton Riviere had already noticed this. In 1865, he had made a painting of an English Greyhound sleeping, eyes closed and head resting on its front paws, one of which is laid over the other. The canine motif had long been used in art, where it served essentially as an allegory of fidelity and loyalty. This was the function of the dogs shown at the foot of their dead master's bed in many paintings from the sixteenth century onwards.

In the summer of 1871, when Darwin contacted Riviere, their correspondence began with the question of the dog's "grimace" or "laughter."[5] They finally agreed that, for *The Expression of the Emotions in Man and Animals*, Riviere would draw one dog whose body language expressed hostility, and another demonstrating joyful excitement. Before the artist and scientist were satisfied, however, Riviere's drawings had made the rounds four times between London, where Riviere lived, and Downe in Kent, where Darwin had taken up residence. In the end, one of the book's illustrations showed an English Pointer, ears folded back and shoulders drooped, its head extended towards its master and its gaze fixed on his eyes (137). Describing the muzzle of an

1 The American Society for the Prevention of Cruelty to Animals was founded in 1866 by Henry Bergh. On this caricature and Charles Darwin, see Janet Browne, "Darwin in Caricature: A Study in the Popularisation and Dissemination of Evolution," *Proceedings of the American Philosophical Society*, vol. 145, no. 4 (December 2001), p. 496–509. See also the essay in this catalogue, p. 220.

2 See Charles Darwin, *The Descent of Man, and Selection in Relation to Sex* [1871] (New York: D. Appleton and Company, 1875), p. 64.

3 This event is related in a letter that Darwin wrote on March 3, 1871, to Joseph Wolf, asking him for a sketch of the macaque. The letter is reproduced in Alfred H. Palmer, *The Life of Joseph Wolf, Animal Painter* (London: Longmans, 1895), p. 193 and ff.

4 See Charles Darwin, "Are Monkeys Right-Handed?," Notebooks M and N; and Charles Darwin, "A Biographical Sketch of an Infant," *Mind: A Quarterly Review of Psychology and Philosophy*, vol. 2, no. 7 (July 1877), p. 285–294.

5 Unpublished correspondence. See https://www.darwinproject.ac.uk/ search/ ?keyword=briton &tab=, DAR 176: 174 to 182.

animal excited with joy, Darwin notes: "The upper lip during the act of grinning is retracted, as in snarling, so that the canines are exposed." As to the frequency with which dogs grin, Darwin also fell back on Riviere's observation and experience: "Mr. Riviere, who has particularly attended to this expression, informed me that it is rarely displayed in a perfect manner, but is quite common in a lesser degree."[6]

Darwin's argument for the theory of evolution was equally surprising and convincing: we human beings resemble the animals around us. Dog owners know this well.

6 On the canine grin, see Charles Darwin, *The Expression of the Emotions in Man and Animals* [1872] (New York: D. Appleton and Company, 1897), p. 120; with regard to the mention of Riviere, see ibid., p. 25.

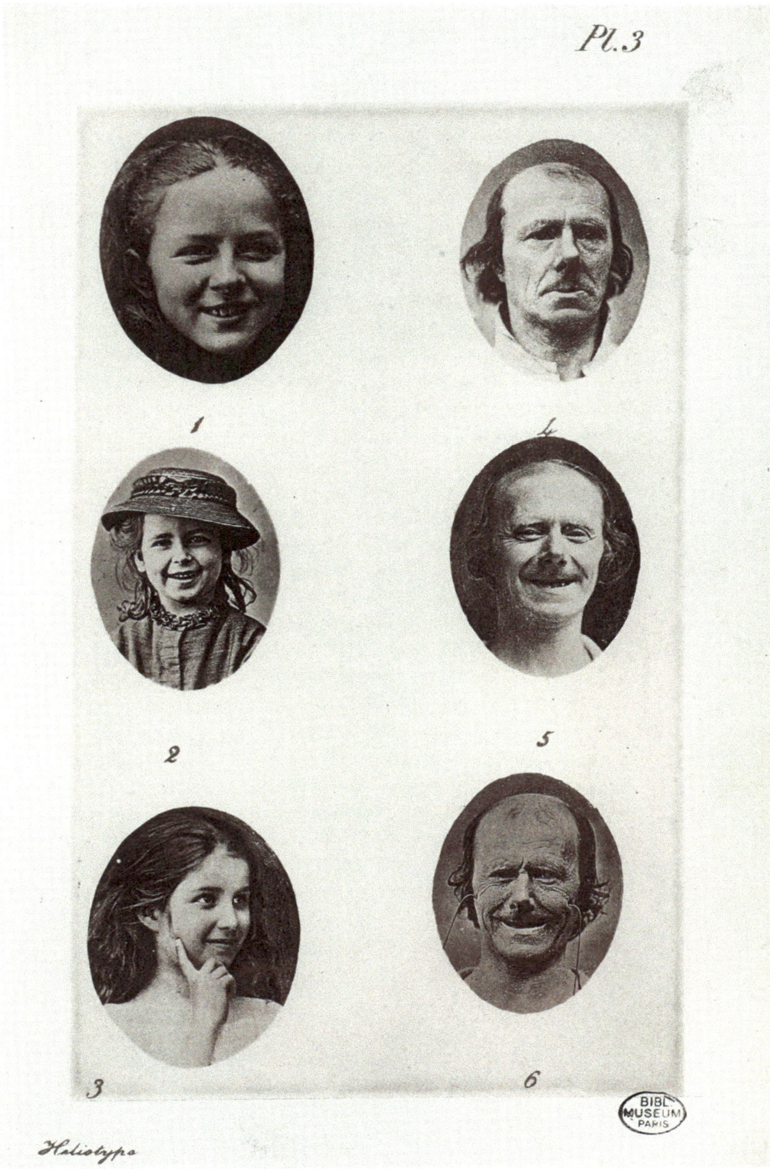

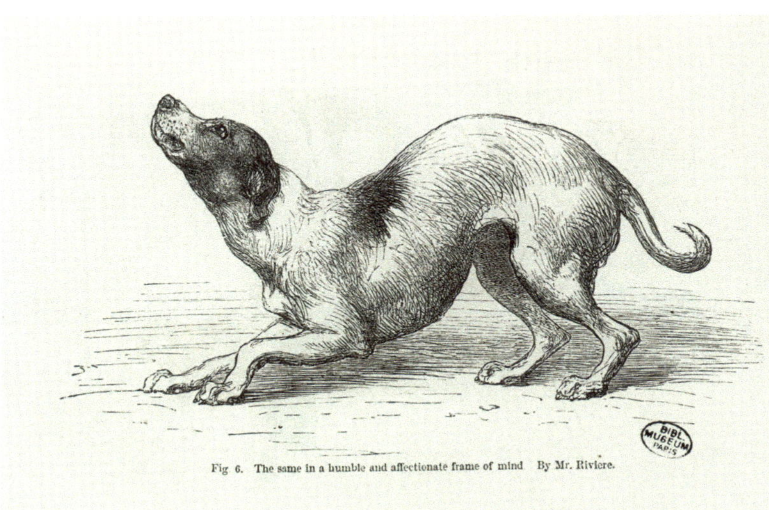

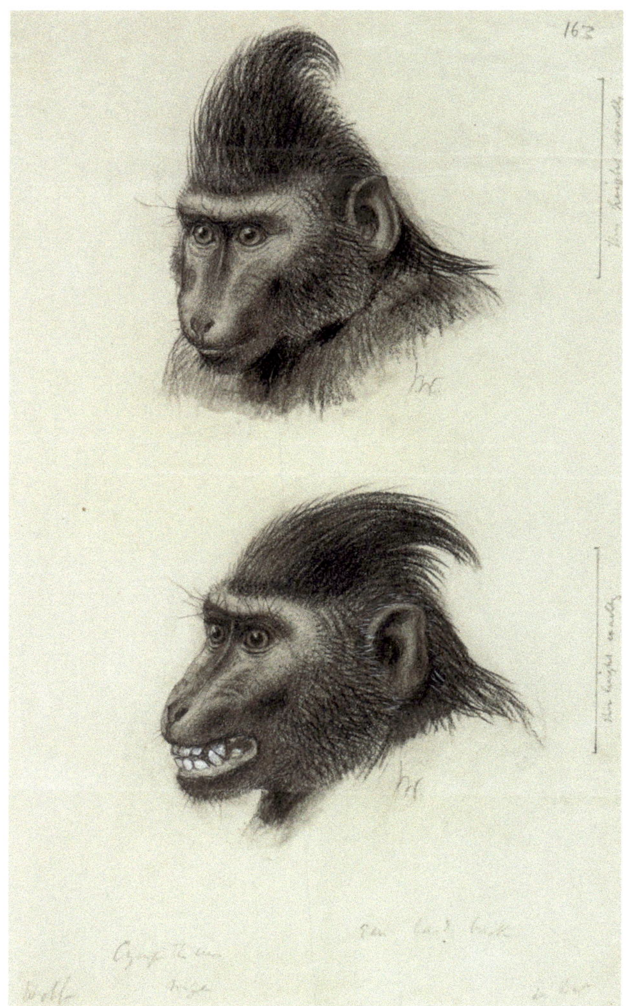

136 FIG.
Charles Darwin,
The Expression of the Emotions in Man and Animals,
pl. 3, London: J. Murray, 1872
Paris, MNHN

137 FIG.
Briton Riviere, *Dog "in a humble and affectionate frame of mind,"* in Charles Darwin, *The Expression of the Emotions of Man and Animals*,
London: J. Murray, 1872
Paris, MNHN

138
Joseph Wolf,
Cynopithecus niger, in a Placid Condition; The Same, When Pleased by Being Caressed,
about 1871–1872
Black chalk, white highlights,
32.5 × 19.5 cm
Cambridge, Cambridge University Library

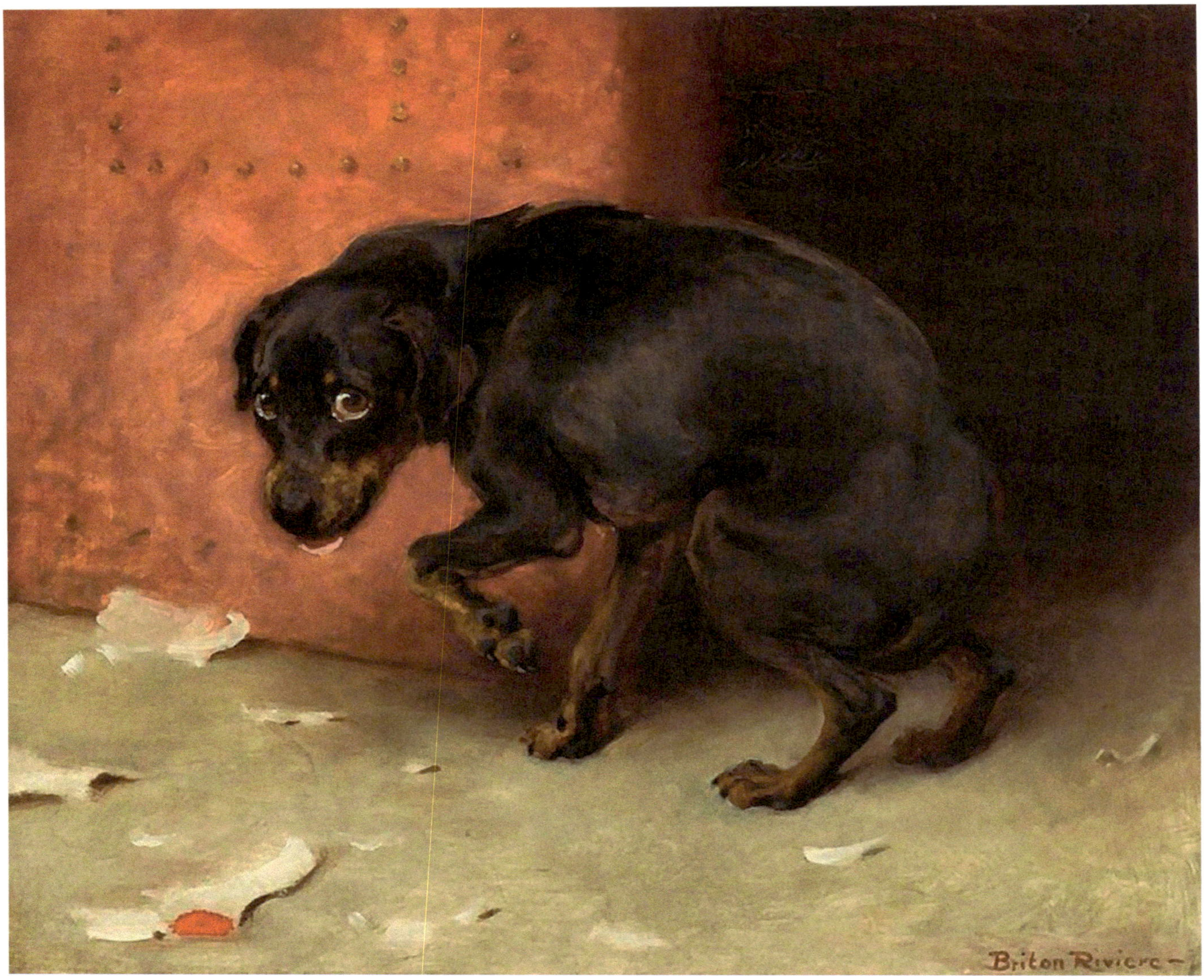

139
Briton Riviere,
Conscience Makes Cowards of Us All, about 1871–1872
Oil on canvas, 29.7 × 39.8 cm
Manchester, The Whitworth, The University of Manchester

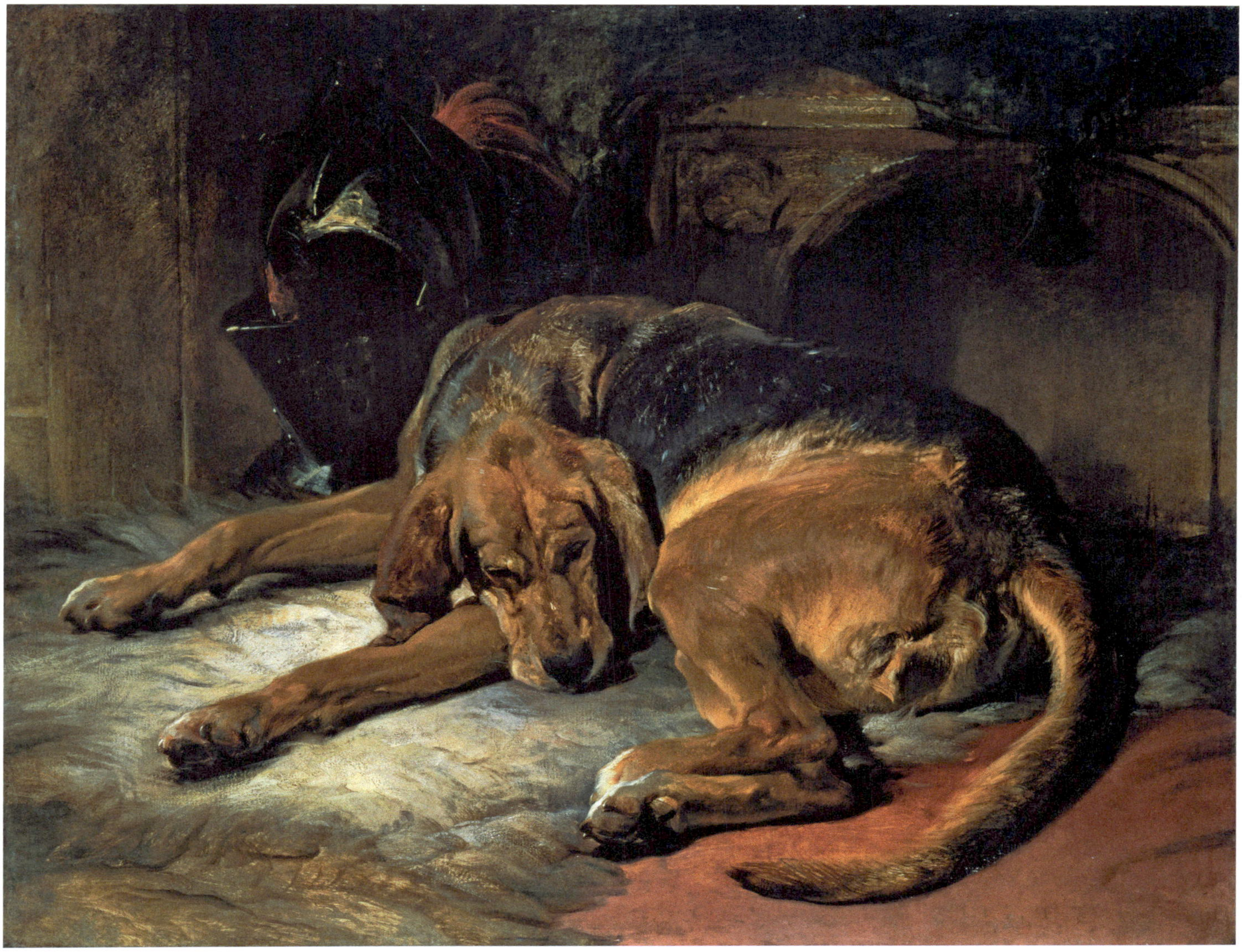

140
Sir Edwin Henry Landseer,
Sleeping Bloodhound, 1835
Oil on canvas, 101.6 × 127 cm
London, Tate

ERNST HAECKEL AND BEAUTY:

NATURE'S ART FORMS

LAURA BOSSI

Ernst Haeckel was the most important conveyor of evolutionism in Europe and, indeed, throughout the world.[1] A specialist in marine invertebrates and an immediate convert to Darwinism upon his reading *The Origin of Species* in 1860, he undertook to establish a new classification of living things on the basis of genealogy and proposed the first evolutionary synthesis grounded in morphology, comparative anatomy and embryology.[2] However, his evolutionism was consistent with the cultural context in the Germany of his time, with its faith in progress and the pantheistic tendencies that were the inheritance of *Naturphilosophie,* a philosophically oriented approach to natural history. In his work he combined four key ideas that had been put forth by four major figures: Goethe, the harbinger of a vision of the natural world as a continual creator of forms through its metamorphoses; Lamarck, the true originator of transformism; Darwin, who, through natural selection, had established the mechanism of evolution; and Haeckel himself, who, by reason of the "great biogenetic law" according to which "ontogeny recapitulates phylogeny," stressed the "threefold parallel between the embryological, systematic and paleontological development of organisms."[3]

A remarkable and multi-faceted personality, Haeckel played a significant role in both the history of ideas and the history of art. His evolutionism represented a *Weltanschauung,* a "world view" that compelled a radical philosophical and religious change. The word *Entwicklung* (meaning "development," but also "evolution") was, for him, "the magic word that would enable every riddle to be solved," with everything that lives—or has lived—conjoined in a single great genealogical tree rooted in a distant past. Through constant transformations, the gradual improvement of organisms would be brought about by the natural world; such progress would be "a law that neither priests nor tyrants can quell." Through his "monism," the "connection between religion and science" based on "the True, the Beautiful, and the Good"[4] and driven by an unshakable faith in progress, he would offer an antidote to the dark side of evolution (struggle for survival, extinctions, degeneration). In the late nineteenth century, that substitute religion, combining science with the worship of the natural world, would appeal to many free thinkers, progressives, materialists and socialists in the *Kulturkampf* that pitted Bismarck against Catholics in Germany, as well as in Europe as a whole. Haeckel's monism also involved a natural foundation for beauty that became a theory of art: every cell is a basic organism endowed with a "cell-soul," and all of animate nature, beginning with the simplest creatures like single-cell organisms, possesses a *Kunsttrieb,* a drive for beauty that leads it to create an astonishing variety of forms.

Haeckel became interested in the art forms of nature early on. An avid botanist in his youth, his bedside reading included Alexander von Humboldt's *Ansichten der Natur* [*Aspects of Nature*], Darwin's journal of his voyage on board the *Beagle,* and the botanist Matthias Jakob Schleiden's *Die Pflanze und ihr Leben* [*The Plant, a Biography*]. He drew and painted his entire life and, during his major 1859–1860 trip to Italy, he wavered between a career in science and one in art. His own inclinations led him to particularly favour landscape painters whose work was in the spirit of Humboldt: Friedrich Preller the Elder, Ernest Carl Eugen Koerner, Hermann Eschke, who would give him lessons in painting, and Hermann von Königsbrunn.

His discovery of the radiolarians—single-cell organisms of the marine plankton that have siliceous skeletons of incredibly varied shapes—in the waters off Messina in Sicily would lead him to a consideration of the geometry of living beings that became a lifelong obsession. His first monograph dealing with those microscopic creatures, which he deemed "the greatest artists amongst the protists," was written when he was twenty-eight for the habilitation that would qualify him for a university professorship. Its thirty-five plates, stylized reproductions of his own drawings, would enchant Darwin.[5] He devoted 211 pages in his magnum opus, the 1866 *Generelle Morphologie*, to an attempt to reduce the variety of life forms to a limited number of geometric shapes[6] that could be expressed in mathematical formulas; with forty basic forms classified according to the relation of parts of organisms to an axis of symmetry, in it he developed a veritable organic stereometry. His work on radiolarians, jellyfish, siphonophores, sponges and corals would result in numerous publications—among them his contributions to the reports on the HMS *Challenger* oceanographic expedition[7]—illustrated with superb lithographic plates.

Haeckel was able to share his wonderment at the beauty of such creatures with a wide audience and numerous artists thanks to a forty-year-long collaboration with the talented lithographer Adolf Giltsch[8] that culminated in the collection *Kunstformen der Natur* (1899–1904). This compendium of one hundred large plates, which enabled their viewers "to see the marvellous treasures of beauty carpeting the depths of the oceans or visible only through a microscope," would exert a lasting fascination, one that has ensured it is still published today.[9] In selecting the creatures depicted, Haeckel gave priority to "the hidden beauties of nature," the microscopic animals and marine invertebrates he specialized in (144, 145, 150 A), but he also included plants and a few vertebrates. In the afterword, Haeckel declared that he had refrained from any idealization, stressed the importance of geometry in natural forms, and proposed their classification into four categories: *centrostigma* (sphere), whose geometric centre is a point; centraxonia, whose geometric axis is a straight line; *centroplana* (or *bilateralia*), whose axis is a sagittal plane; and centraporia or irregular (like amoebae). His work would inspire D'Arcy Wentworth Thompson, who used a number of Haeckel's plates in his book *On Growth and Form*.[10] However, while he was equally concerned with applying a mathematical approach to morphology, D'Arcy Thompson did not take into consideration the life and ancestry of geometric forms but solely the forms' timelessness.

In *The Wonders of Life*,[11] Haeckel would return to those considerations and attempt to define a "science of the Beautiful," with a gradation

1 See Mario Di Gregorio, *From Here to Eternity: Ernst Haeckel and Scientific Faith* (Göttingen: Vandenhoeck & Ruprecht, 2005); Robert J. Richards, *The Tragic Sense of Life: Ernst Haeckel and the Struggle over Evolutionary Thought* (Chicago: University of Chicago Press, 2008).

2 Ernst Haeckel, *Generelle Morphologie der Organismen*, 2 vols. (Berlin: G. Reimer, 1866).

3 The theory of recapitulation postulates that, during embryonic development (ontogeny), every individual organism repeats, in an abbreviated fashion, the evolution of species (phylogeny); see Laura Bossi, "Après Darwin, Haeckel," in *Darwin au Collège de France*, ed. Antoine Compagnon and Céline Surprenant (Paris: Collège de France, "Passage des disciplines" series, 2020), https://books.openedition.org/cdf/7356.

4 Ernst Haeckel, *Der Monismus als Band zwischen Religion und Wissenschaft: Glaubensbekenntnis eines Naturforschers*, 2nd ed. (Bonn: Emil Strauss, 1892), p. 36; published in English as *Monism as Connecting Religion and Science: The Confession of Faith of a Man of Science*, trans. J. Gilchrist (London: Adam and Charles Black, 1894).

5 Ernst Haeckel, *Die Radiolarien (Rhizopoda radiaria): Eine Monographie*, 2 vols. (Berlin: G. Reimer, 1862).

6 See the chapter "Promorphologie" in Haeckel, *Generelle Morphologie der Organismen*.

7 *Report of the Scientific Results of the Voyage of H.M.S. Challenger During the Years 1873–76*. The report comprised 50 volumes issued from 1880 to 1895 by an international team of scientists that based their work on the samples collected during the *Challenger* expedition, the first major oceanographic research endeavour.

8 See the obituary Haeckel wrote upon the lithographer's death: "Adolf Giltsch. Ein Nachruf," *Jenaer Zeitung*, June 28, 1911, https://www.mediamatic.net/en/page/369299/obituary-of-adolf-giltsch-by-ernst-haeckel.

9 Ernst Haeckel, *Kunstformen der Natur* (Leipzig and Vienna: Bibliographisches Institut, 1899–1904), 100 pls. in 1 vol.

10 D'Arcy Wentworth Thompson, *On Growth and Form* (Cambridge: Cambridge University Press, 1917).

11 Ernst Haeckel, "Lebensformen," in *Die Lebenswunder: gemeinverständliche Studien über biologische Philosophie* [1904] (Leipzig: Alfred Kröner, 1923), p. 145–148; published in English as "Forms of Life," in *The Wonders of Life: A Popular Study of Biological Philosophy*, trans. Joseph McCabe (New York and London: Harper and Brothers, 1905), p. 170, 184–187.

of the "beauty of natural forms" according to an increasing perfection, moving from a "direct" beauty, the subject of a sensual aesthetic in which pleasure is registered through the "aesthetic" or sensory neurons, to an "indirect" beauty, the subject of an associational and symbolic aesthetic related to the activation of the neurons enabling imagination and thought. He thus identified eight ascending levels of beauty: simple, rhythmic, "actinal" (radial) and symmetrical, and then biological, anthropistic, sexual and, at the highest level, landscape beauty, which requires variety and irregularity, and which is similar to the "physiognomic" beauty of landscape described by Humboldt.

In his little book of 1913, *Die Natur als Künstlerin* [*Nature as artist*],[12] Haeckel would respond to those who had criticized him for exaggerating the symmetry of his natural forms—some even spoke of a "troublesome symmetry"—by referring to the rigorous symmetry of ancient Greek temples and Gothic domes. He would come back to the *Kunsttrieb*, the unconscious memory, the artworks of "animate protoplasm" and his vision of art, which was not a supernatural gift, but the product of the human brain and the work of its nerve cells, and which had developed, as had the other operations of the soul, from the *Kunsttrieb* of our ancestors.

Haeckel's recently restored home in Jena, the Villa Medusa, as well as the Phyletic Museum he founded in 1907, indicate his love of jellyfish and the elegant radiolarians. The villa's ceilings are decorated with jellyfish, and in his office, which has been kept as it was during his lifetime, can be seen a bookcase featuring radiolarian and jellyfish motifs created by Ella von Crompton, a planter with radiolarians, and a wooden chest with mother-of-pearl inlays forming a jellyfish, a gift to Haeckel for his seventieth birthday.

A number of the scientist's friends and disciples, such as his biographer Wilhelm Bölsche, the biologist and writer Ernst Krause (whose pen name was Carus Sterne) and the publisher Wilhelm Breitenbach, contributed to the dissemination of Haeckel's illustrations of life forms.

Haeckel also acted as an advisor to the Dresden "scientific artists" Leopold Blaschka and his son Rudolf, who are most known for their glass flowers in the Harvard Museum of Natural History's Ware Collection of Blaschka Glass Models of Plants. Unrivalled masters of glassmaking, they also created a wide variety of models of marine invertebrates, jellyfish, octopi and even radiolarians that can be found in museums around the world. Through their exactitude and beauty, those works fall within the realms of both scientific education and art (148, 149).

Many were the artists and architects who studied Haeckel's plates and drew inspiration from them for their own work (142, 144, 145, 150).[13]

Inaugurated in 1910 with Haeckel present, Monaco's Oceanographic Museum, founded by the principality's Prince Albert I, an explorer and scholar, is a veritable palace of the sea decorated entirely with Haeckelian motifs. Immediately upon entering can be seen a huge chandelier in the form of a jellyfish, which was designed by Constant

12 Ernst Haeckel, *Die Natur als Künstlerin*, with Dr. Wilhelm Breitenbach, *Formenschatz der Schöpfung* (Berlin: Vita, 1913).

13 See Christoph Kockerbeck, *Ernst Haeckels* Kunstformen der Natur *und ihr Einfluß auf die deutsche bildende Kunst der Jahrhundertwende* (Frankfurt am Main: Peter Lang, 1986).

Roux based on the lovely *Rhopilema frida* that Haeckel named in memory of his mistress, Frida von Uslar-Gleichen (147). Radiolarian-shaped chandeliers (146) in the museum's ceremonial hall celebrate the beauty of those micro-organisms of plankton. In the same vein, Haeckel's plates directly inspired the designs of windows and stucco walls in the Institut océanographique in Paris, initiated by Albert I in 1908 and completed in 1911.[14]

Haeckelian motifs can also be found in numerous pieces of furniture designed by the Antwerp painter, designer and architect Henry van de Velde.[15] This major exponent of Art Nouveau in Belgium, as well as in Germany (he founded the Kunstgewerbeschule in Weimar, the school of applied arts that would become the Bauhaus), the Netherlands and France, knew Haeckel well (he had married a distant cousin of the scientist, Maria Sèthe) and was a monist sympathizer.

Before devoting himself to sculpture, ceramics and tapestry, the artist Hermann Obrist had been a student of botany. He met Haeckel and made a study of his *Kunstformen*; the radiolarians, jellyfish and spiral-shaped shells of its plates had a particular influence on his drawings, vases, fountains and maquettes (237, P. 339).

The presence of Haeckelian patterns can also be traced in the work of August Endell, the architect of the Hackesche Höfe complex in Berlin; Joseph Maria Olbrich, who was responsible for the Secession Building in Vienna and many architectural works in Darmstadt; Hermann Finsterlin, the creator of visionary, yet never realized, Expressionist architecture projects (*Formspiele*, or form games); and Rudolf Steiner, who built the Goetheanum in Dornach, near Basel in Switzerland, as the headquarters of the Anthroposophical Society.[16]

In Amsterdam, the interior decoration of the Het Scheepvaarthuis building (now the Grand Hotel Amrâth), conceived by the architect Joan Melchior van der Meij and built in 1913 as the head office for six shipping companies, was very largely inspired by *Kunstformen*. Theodorus Willem Nieuwenhuis was responsible for its wallpaper, Michel de Klerk designed chandeliers with sea creatures, and Willem Bogtman adorned the windows with jellyfish, starfish and sea horses.[17] The architect Hendrik Petrus Berlage, best known for the Amsterdam Commodities Exchange that opened in 1903, designed wallpaper patterns that were directly inspired by *Kunstformen*, four of which were produced by the Amsterdam company Rath & Doodeheefver in 1927.

In Austria, the publication of *Kunstformen* over the years 1899 to 1904 came just after the Secession, a time when the teaching of art and architecture was being revitalized. Otto Wagner had been head of architecture at the Vienna Academy of Fine Arts since 1894, and Felician von Myrbach-Rheinfeld was the director of the School of Applied Arts, where the teaching staff included people like Josef Hoffmann and Koloman Moser. In a period that saw the enervation of historicism and its reproduction of the decorative forms and styles of the past, inspiration was sought in the very oldest of sources—the natural world.

14 See Christelle Inizan, "L'Institut océanographique de Paris," *In Situ*, no. 17 (2011), http://journals.openedition.org/insitu/865.

15 See *Henry van de Velde: Passion, Function, Beauty, 1863–1957*, exh. cat., ed. Thomas Föhl, Sabine Walter and Werner Adriaenssens (Tielt, Belgium: Lannoo, 2013).

16 See Wolfgang Pehnt, *Expressionist Architecture* (London: Thames & Hudson, 1973); Siegfried Wichmann, *Jugendstil floral funktional in Deutschland und Österreich und den Einflussgebieten* (Zürich: Atlantis, 1984); published in English as *Jugendstil Art Nouveau: Floral and Functional Forms* (Boston: Little, Brown, 1984).

17 In 2013–2014, the Netherlands' National Glass Museum held an exhibition, *A Sea of Glass*, that examined Haeckel's influence on the Amsterdam School's decorative arts.

Franz Sales Meyer's 1888 compendium, *Illustriertes Handbuch der Ornamente*, which presents three hundred plates depicting plants, animals, humans and man-made objects (emblems, trophies), with natural forms shown next to stylized versions of them, still adhered to the historicist viewpoint in making designs available for copying.[18] In contrast, in his preface to *Kunstformen*, Haeckel explicitly rejected stylization, proposing a synthesis of scientific stringency and artistic model: "In these true 'Art Forms in Nature' modern fine art and the modern, greatly flourishing arts and crafts will find a wealth of new and attractive motifs. In compiling them I have confined myself to a true-to-life depiction of products of nature that actually exist, and have desisted from a stylistic modeling or decorative exploitation; I leave this to the artists themselves."[19]

Other publications used photography to offer artists natural forms they could use as models. The goal of Moritz Meurer's twenty-volume work[20] was analogous to that of Haeckel's *Kunstformen*: to teach young people to "look at nature by thinking."[21] The work of his best-known student and assistant, Karl Blossfeldt, considered a major Neue Sachlichkeit [New Objectivity] figure, would be in the same spirit: "A plant is formed according to the same principles that any builder has to consider . . . It forms and grows according to logic and function, and, with primeval power, forces everything to the most sublime artistic form."[22] His book *Urformen der Kunst* (1928),[23] a compilation of photographs of plants that included shots of their details enlarged up to forty times actual size, would have a similar function.

In Vienna, the art historian Alois Riegl incorporated a Haeckelian Darwinism in his methodology and, according to Marsha Morton,[24] even in the conceptualization and language of his theories. Evolutionary themes are present in his major works, from *Stilfragen* (1893) right up to his magnum opus *Historische Grammatik der bildenden Künste* (1897–1899). Like Haeckel, Riegl regarded evolution as involving constant progress. His *Kunstwollen* (artistic volition, or formative will of art), which calls to mind Haeckel's *Kunsttrieb*, was exerted through the artist; again like Haeckel, Riegl gave precedence to landscape painting, which visualizes the connections among all organisms, whether plants or animals, and shows the harmony of the world in which everything is subject to continual flux. The alliance in Riegl of the admiration of scientific progress and the pantheistic view of nature is close to Haeckel's monism, in which spirit is immanent in matter and the human soul is subsumed in the world soul.[25]

The Austrian painter and printmaker Alfred Kubin was familiar with *Kunstformen der Natur*, as well as with Max Klinger's and Odilon Redon's plates; however, his repertoire of crystals, microscopic organisms, marine animals, embryos, apes and primitive humans had none of the geometric beauty of Haeckel's depictions. While he reprised a Haeckelian black background in his 1906 work in tempera *From the Deep Sea*,[26] his ocean depths teemed with strange, indeterminate creatures that certainly took their inspiration from the radiolarians, octopi,

18 See *Welträtsel und Lebenswunder: Ernst Haeckel—Werk, Wirkung und Folgen*, exh. cat., ed. Erna Aescht et al. (Linz, Austria: Oberösterreichisches Landesmuseum, 1998).

19 Translated in Rudolf Finsterwalder, ed., *Form Follows Nature: A History of Nature as Model for Design in Engineering, Architecture and Art*, trans. Jeremy Gain (Vienna: SpringerWienNewYork, 2015), p. 441.

20 Moritz Meurer, *Meurer's Pflanzenbilder: Ornamental verwerthbare Naturstudien für Architekten, Kunsthandwerker, Musterzeichner*, 20 vols. (Dresden: G. Kühtmann, 1896–1903); *Form Follows Flower: Moritz Meurer, Karl Blossfeldt & Co.*, exh. cat., ed. Angela Nikolai and Sabine Thümmler (Berlin: Deutscher Kunstverlag, 2017).

21 That calls to mind the *denkende Betrachtung*, or "thoughtful observation," advocated by Aristotle, as well as Hegel, Cassirer and Haeckel's mentor, Johannes Müller.

22 Karl Blossfeldt, "Vorwort," in *Wundergarten der Natur* (Berlin: Ernst Wasmuth, 1932), p. 5; translated in Rolf Sachsse, "Karl Blossfeldt," in *Encyclopedia of Twentieth-Century Photography*, ed. Lynne Warren (New York: Routledge, 2006), vol. 1, p. 138.

23 Karl Blossfeldt, *Urformen der Kunst: Photographische Pflanzenbilder* (Berlin: Ernst Wasmuth, 1928); published in English as *Art Forms in the Plant World: 120 Full-Page Photographs*, with an introduction by Karl Nierendorf (New York: Dover Publications, 1985).

24 Marsha Morton, "Art's 'Contest with Nature': Darwin, Haeckel, and the Scientific Art History of Alois Riegl," in *Darwin and Theories of Aesthetics and Cultural History*, ed. Barbara Larson and Sabine Flach (Burlington, VT: Ashgate, 2013), p. 53–68.

25 Haeckel, *Monism as Connecting Religion and Science*, p. 16.

jellyfish, insects and plants of the natural world, but did not correspond to any specific organic forms (210, P. 302; 213, P. 303).

Gustav Klimt was acquainted with Haeckel's version of Darwinism, likely, among other reasons, through his attendance at Berta Zuckerkandl's salon, which brought together artists and scientists. In the Vienna Secession building's *Beethoven Frieze* (which presents the composer's *Ninth Symphony* as a way of transcending animality's dark instincts in order to attain a pure love—an "embrace for all the world"), he portrayed the giant Typhon as a gorilla (209). The mosaics Klimt created to decorate the Stoclet House near Brussels represent a *Tree of Life* (1909). And the tentacular tresses strewn with tiny stars and flowers of the women in his paintings *Goldfish* (1901–1902) and *Water Serpents* II (1904–1907) cannot fail to bring to mind Haeckel's sea creatures.[27]

In France, a number of Art Nouveau artists and master glassmakers such as Émile Gallé[28] (201, P. 292) and René Lalique[29] also read and studied *Kunstformen*. Yet, it was above all on the work of architect René Binet that Haeckel's plates had a direct influence.[30] Binet corresponded with Haeckel,[31] who in turn sent the architect his monograph on the radiolarians, *Kunstformen* and *Arabische Korallen*, as well as his 1898 book on the origins of mankind. In his letters, Binet expresses his admiration of Haeckel and how important the latter's depictions of radiolarians were to the creation of the monumental entrance gate for the Paris Universal Exposition of 1900 in the Place de la Concorde (143): "At present, I am realizing the monumental entranceway for the 1900 exposition, in which everything, from the general composition to the smallest details, is inspired by your studies.[32] The main model for the monumental gate was the radiolarian *Clathrocanium reginae* (142); other motifs were taken from Plate 46, *Porodiscida*. Binet's compendium of decorative designs, *Esquisses décoratives*,[33] was also directly inspired by the work of Haeckel: "From your discoveries, I wished to draw out the wonderful principles of order and balance that appear in the infinite variety of microscopic animals that you have brought to light."[34] While Binet no doubt had other sources, such as the illustrations of fossils in Alcide d'Orbigny's multi-volume *Paléontologie française* (1842–1874), or Orientalism, several of his plates bear striking similarities to those of Haeckel. For example, his designs for electric lights are modelled on the radiolarians depicted in Plate 71 of Haeckel's *Kunstformen* (150 B).

Finally, the dome topping of what was the central hall (and is now the main restaurant) of Le Printemps department store in Paris, a late work by Binet that was built between 1907 and 1911,[35] reprises the geometric form of a radiolarian, the *Litharachnium eupilium*, described by Haeckel under the name *Sethophormis eupilium* (Plate 56).

That is not the end of the story. Haeckel's natural art forms have found a new life in the world of design, fashion and even pop music. They decorate mugs, mouse pads, cushions and T-shirts; fashion designer Vivienne Westwood has had them printed on jeans, and the Icelandic singer Björk used them in the stage designs of her world tours in the 2000s.

---

26 Marsha Morton, "From *Monera* to Man: Ernst Haeckel, Darwinismus, and Nineteenth-Century German Art," in *The Art of Evolution: Darwin, Darwinisms, and Visual Culture*, ed. Barbara Larson and Fae Brauer (Hanover, NH: Dartmouth College Press, 2009), p. 59–91.

27 Emily Braun, "Ornament as Evolution: Gustav Klimt and Berta Zuckerkandl," in *Gustav Klimt: The Ronald S. Lauder and Serge Sabarsky Collections*, exh. cat., ed. Renée Price (New York: Neue Galerie; Munich: Prestel, 2007).

28 See the essay in this catalogue, p. 288.

29 Wichmann, *Jugendstil Art Nouveau: Floral and Functional Forms*.

30 Olaf Breidbach and Robert Proctor, *René Binet: Natur und Kunst* (Munich: Prestel, 2007); published in English as *René Binet: From Nature to Form* (Munich and London: Prestel, 2007); Erika Krauße, "L'influence d'Ernst Haeckel sur l'Art nouveau," in *L'Âme au corps. Arts et sciences, 1793–1993*, exh. cat., ed. Jean Clair (Paris: Réunion des musées nationaux/ Gallimard/Electa, 1993), p. 343–351.

31 Their correspondence (letters nos. 6933 to 6945, dated 1899 to 1904) is kept in the Ernst-Haeckel-Haus archives in Jena. Concerning Haeckel and Binet's relationship, see also Henri de Pesquidoux, "René Binet," *Revue illustrée* (December 20, 1906), p. 127–132.

32 Letter from Binet to Haeckel, March 21, 1899.

33 René Binet, *Esquisses décoratives*, preface by Gustave Geffroy (Paris: Librairie centrale des beaux-arts, 1902).

34 Letter from Binet to Haeckel, July 10, 1903.

35 Bernard Marrey, *Les Grands Magasins, des origines à 1939* (Paris: Picard, 1979), p. 173–183.

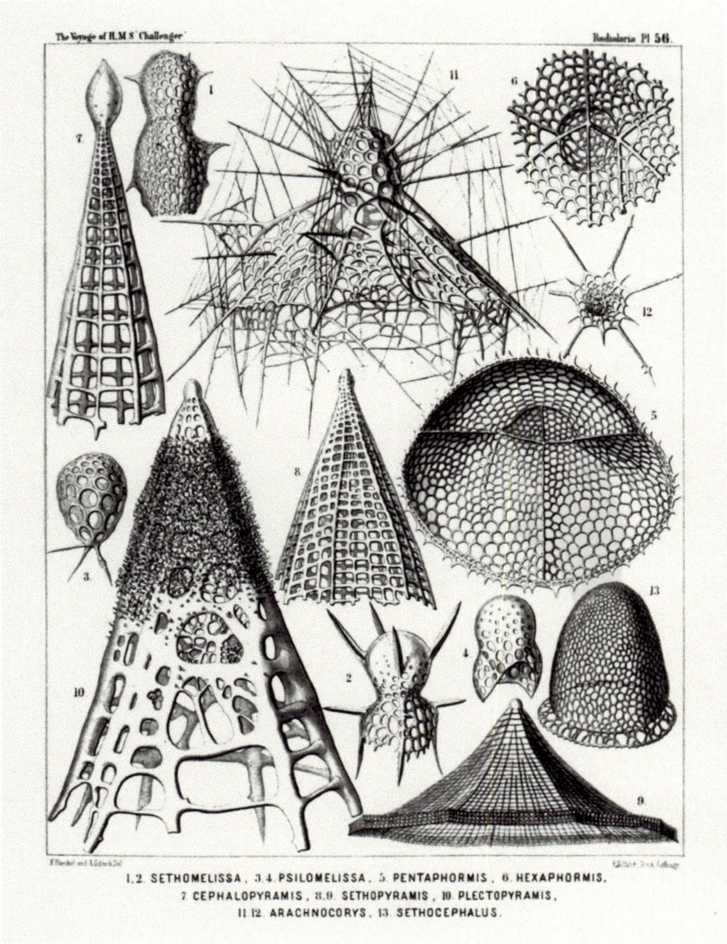

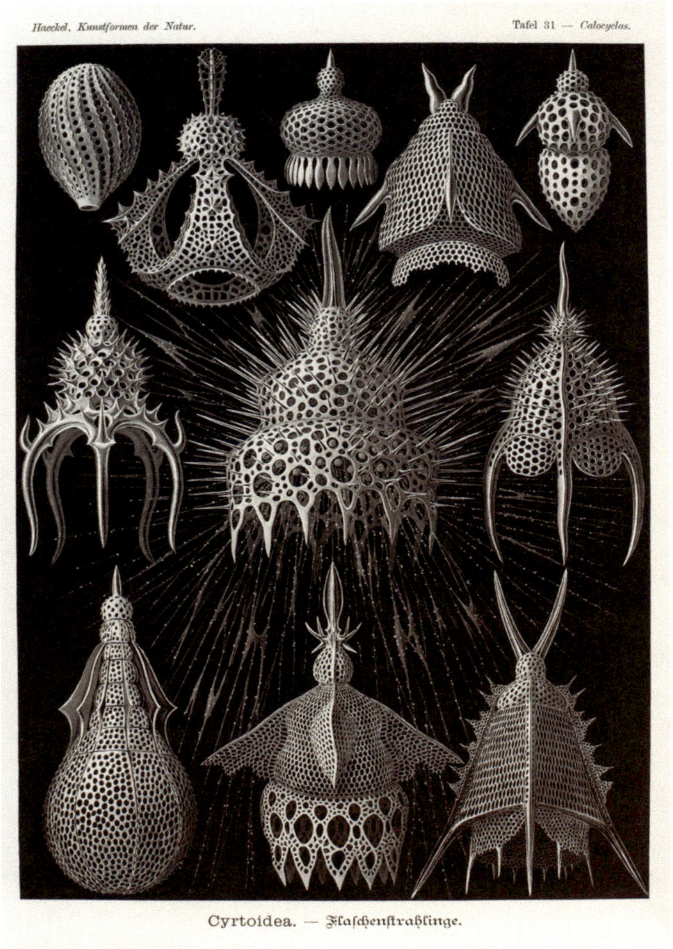

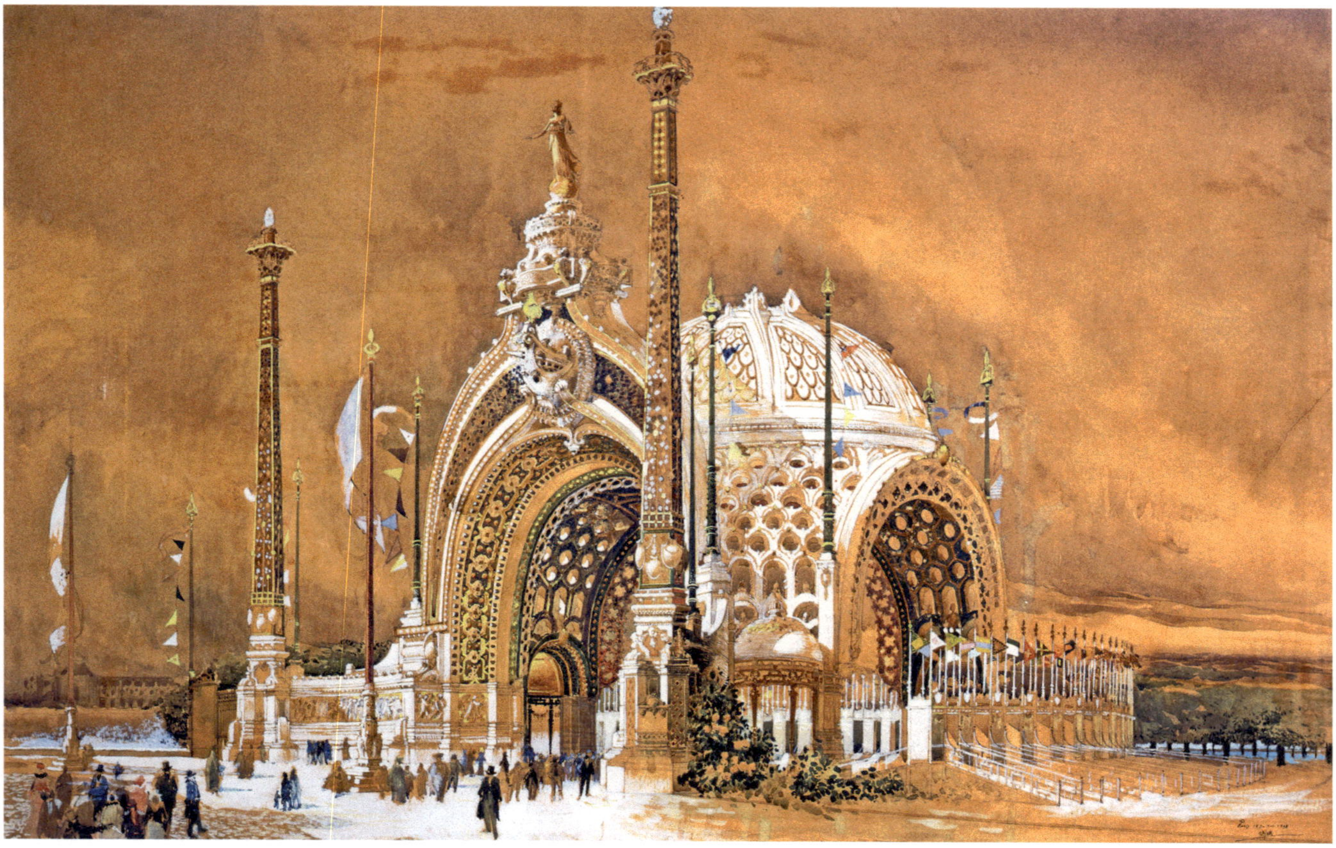

141 FIG.
Ernst Haeckel,
"Radiolara," in *Report on the Scientific Results of the Voyage of H.M.S. Challenger*, pl. 56,
Edinburgh: Neill, 1887
University of California Library

142
Ernst Haeckel,
Kunstformen der Natur, pl. 31,
Leipzig, Vienna: Bibliographisches Institut, 1899–1904
Private collection

143
René Binet,
Design for the Monumental Entrance Gate of the 1900 Universal Exposition, 1898
Watercolour, 62 × 95 cm
Sens, Musées de Sens

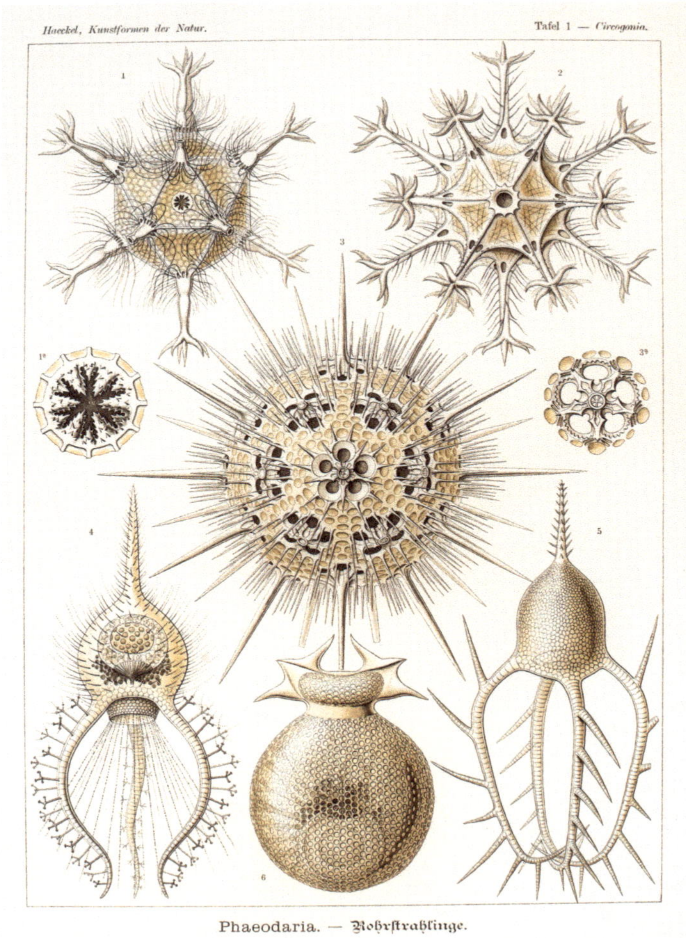

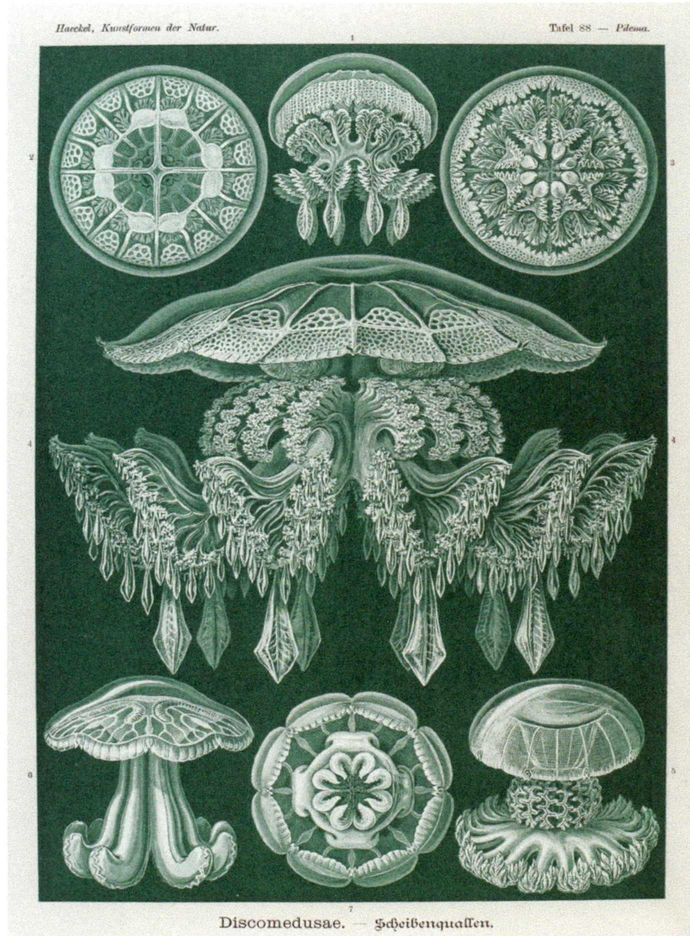

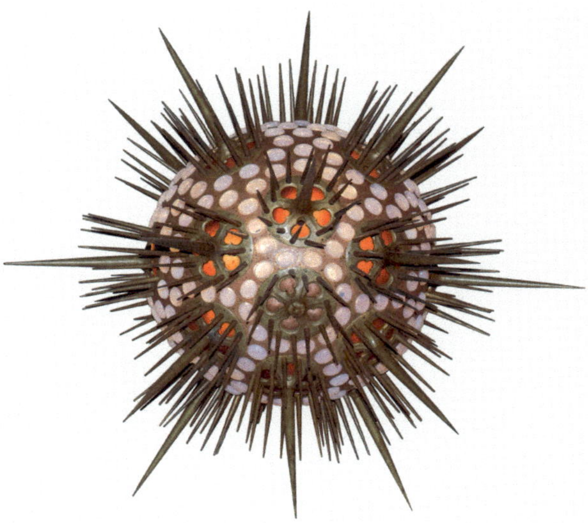

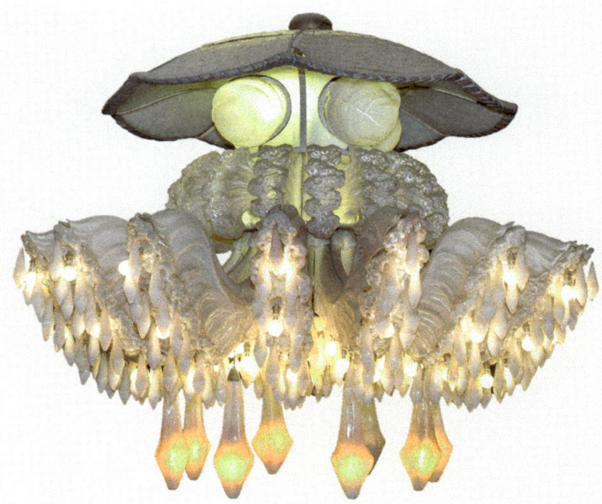

144
Ernst Haeckel,
Kunstformen der Natur, pl. 1,
Leipzig, Vienna: Bibliographisches
Institut, 1899–1904
Private collection

145
Ernst Haeckel,
Kunstformen der Natur, pl. 88,
Leipzig, Vienna: Bibliographisches
Institut, 1899–1904
Private collection

146
Constant Roux (drawing),
for the Maison Baguès,
Radiolarian Chandelier, 1910
Hammered copper with sheets
of opalescent and red glass,
140 × 80 cm
Monaco, Institut océanographique

147 FIG.
Constant Roux (drawing),
for Maison Baguès,
"Jellyfish" Chandelier, 1908
Glass paste, bronze, 200 × 200 cm
Monaco, Institut océanographique

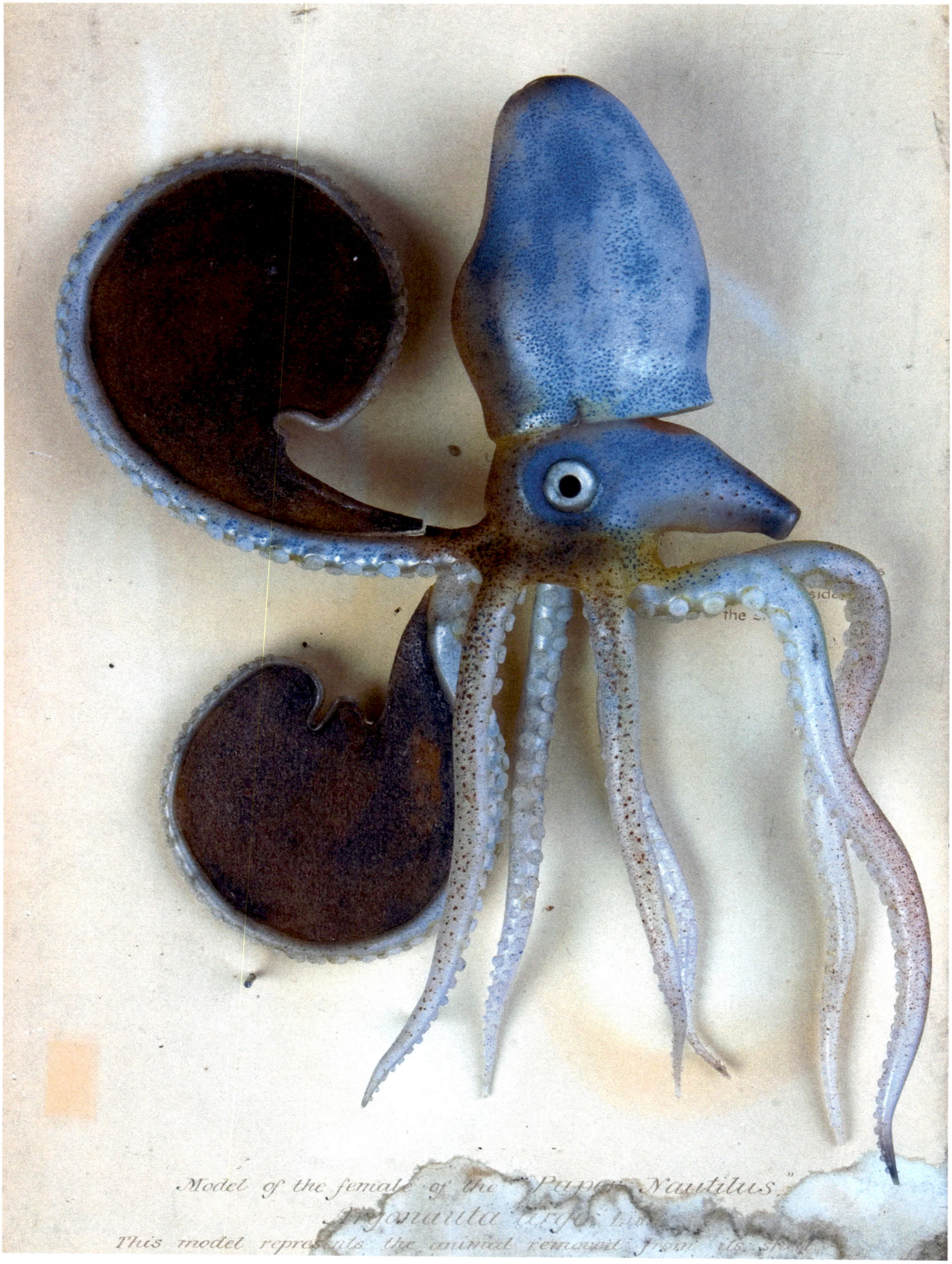

148
Leopold and Rudolf Blaschka,
Argonauta argo, 1866
Glass
London, The Natural History Museum

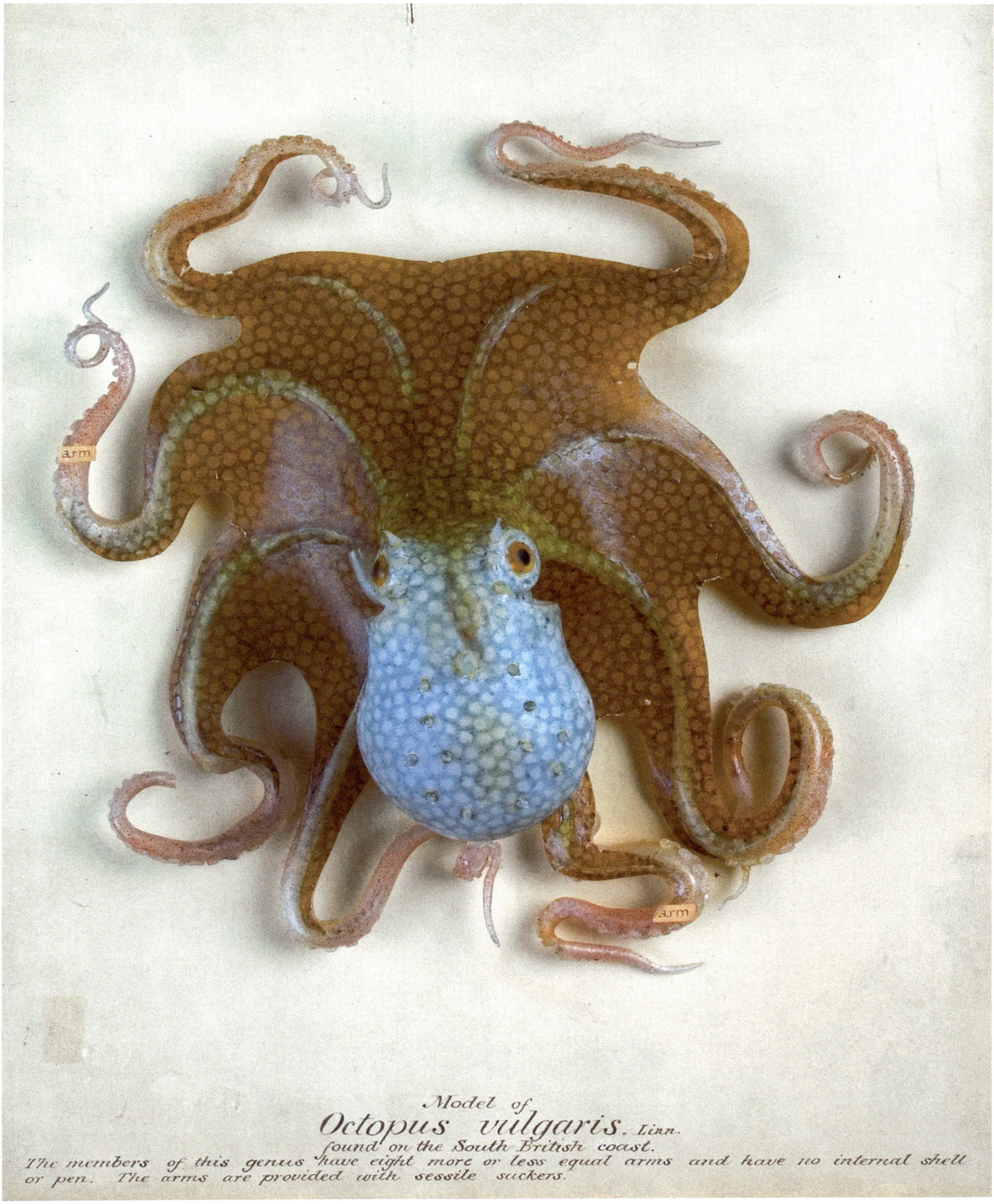

149
Leopold and Rudolf Blaschka,
Octopus vulgaris, 1866
Glass
London, The Natural History
Museum

P. 218-219

150 A, B
Ernst Haeckel, *Kunstformen der Natur*, pl. 8 and 71, Leipzig,
Vienna: Bibliographisches
Institut, 1899-1904
Private collection

Haeckel, Kunstformen der Natur. Tafel 8 — *Desmonema.*

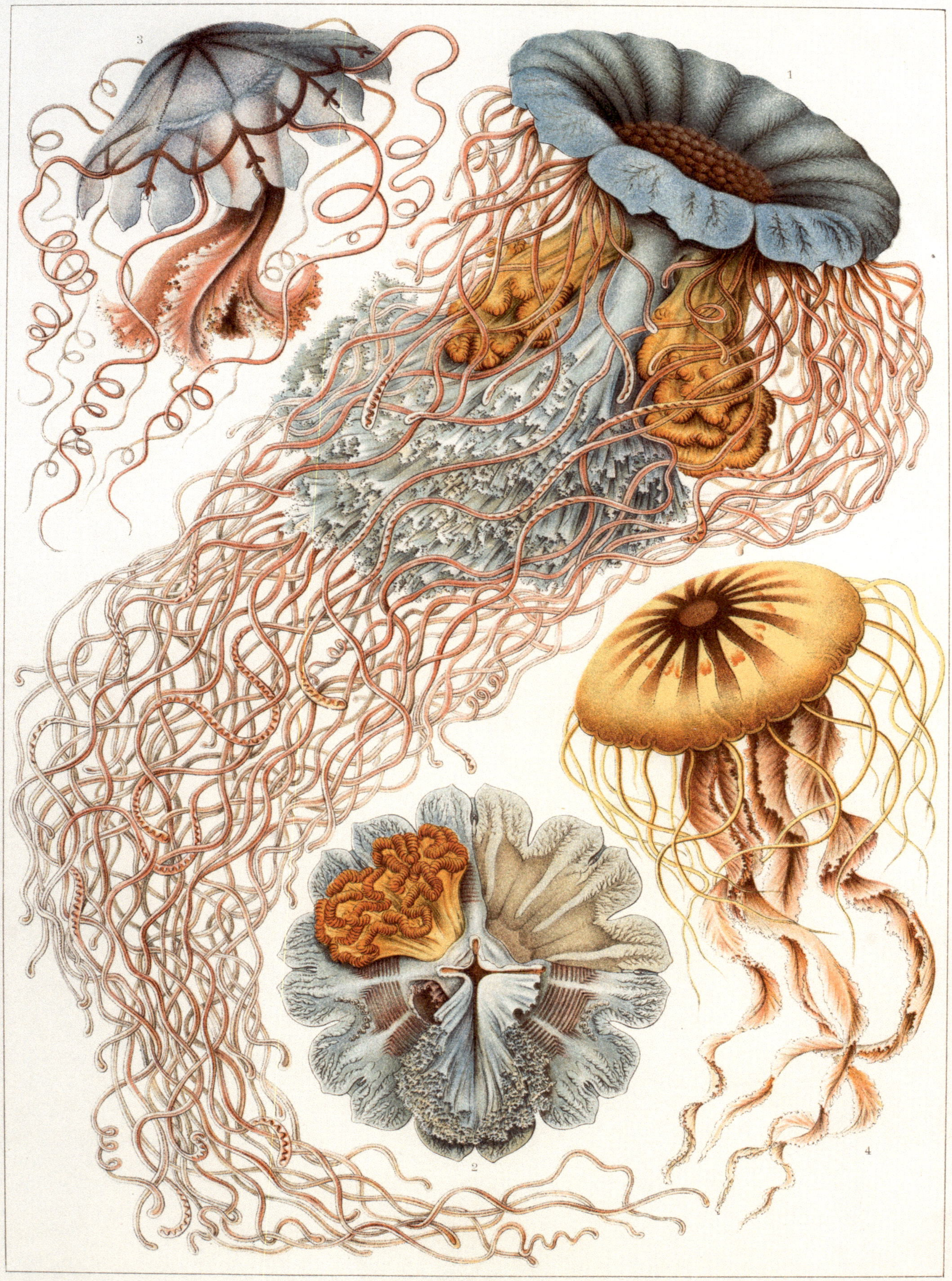

Discomedusae. — Scheibenquallen.

1

2 8

9 4

5

12 6

7

3

11 13

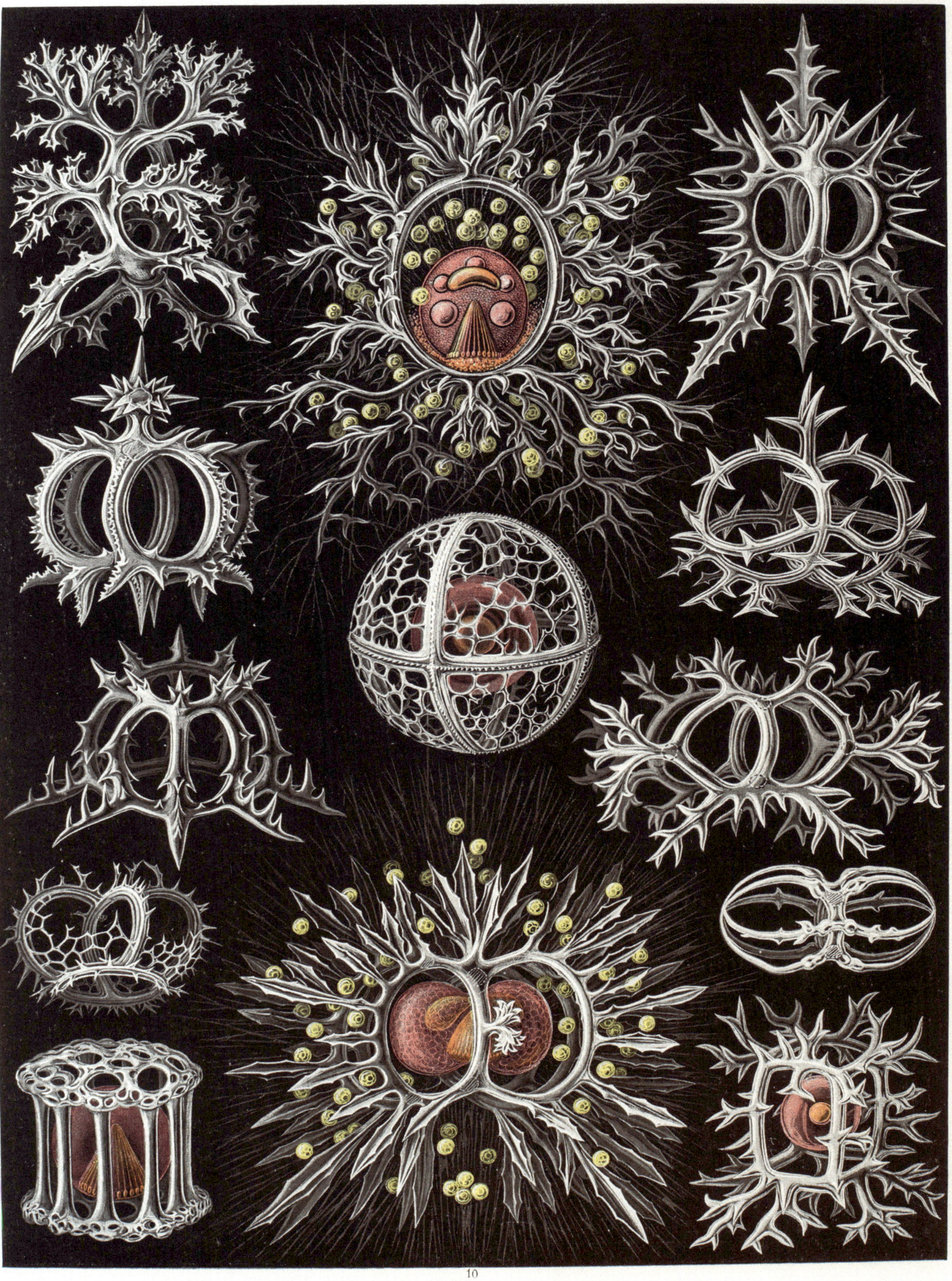

10

Stephoidea. — Ringelstrahlinge.

DARWIN, EVOLUTION AND CARICATURE

BERTRAND TILLIER

It was at the age of fifty, with the publication in 1859 of *On the Origin of Species* and its cohort of editions and translations, that Charles Darwin entered the realm of caricature, at the same time as his theory of evolution by means of natural selection spread within nineteenth-century popular culture thanks to the print media that—often through misinterpretations and exaggerations—disseminated and debated it.[1] Depicted in print, the subject of lampoons and humorous songs, Darwin became the target of caricaturists, whose intrinsic idiom helped publicize and ridicule his theories, while simultaneously shaping his celebrity.[2] The successive publication of *The Variation of Animals and Plants under Domestication* (1868), *The Descent of Man, and Selection in Relation to Sex* (1871) and *The Expression of the Emotions in Man and Animals* (1872) established him, well beyond the confines of scientific circles, as one of the most renowned naturalists of his time. In Victorian Britain, and more widely in Europe as in the United States, caricaturists would continue to link his name and person, through the lens of ideas selected for their startling and simplified nature, to the origins and descent of humankind by transforming him, through condensing and turning his theory around, into its embodiment and consequence. In doing so, cartoonists seemed to respond to the invitation of the scholar himself who, in the final pages of *The Descent of Man*, confided: "For my own part I would as soon be descended from that heroic little monkey, who braved his dreaded enemy in order to save the life of his keeper; or from that old baboon, who, descending from the mountains, carried away in triumph his young comrade from a crowd of astonished dogs—as from a savage who delights to torture his enemies, offers up bloody sacrifices, practises infanticide without remorse, treats his wives like slaves, knows no decency, and is haunted by the grossest superstitions."[3]

Artists since antiquity have continuously explored, with various aims, the relationship between animals and human beings.[4] On the grounds that, to some extent, an animal always lies dormant deep within every human being, metamorphosis and its endless interplay of physical and psychological combinations haunted the imaginative universes that nourished the bestiaries of fables, proverbs and beliefs,[5] and would engender the political and social imagery of innumerable caricaturists in the nineteenth century. Well before the disclosure of Darwin's theories, in one lithograph representative of many others, Honoré Daumier had placed the simian, hunchbacked character Monsieur Mayeux face to face with an ape, as if they were looking at themselves in a mirror, human and animal cancelling each other out to merge into a being "midway between man and ape" (151). That topos was clearly reprised in 1874 by a *London Sketch Book* caricature in which Darwin, depicted with an ape's body, offers a mirror to the actual primate beside him in order to gaze at its reflection and compare themselves (156).

"What place does Mayeux occupy on the scale of being? Is he on the same level as man, or the orang-outang?"[6] When, in 1865, the historian of caricature Champfleury (the pen name of Jules François Félix

Husson) wondered about the graphic character created some thirty years earlier, he was no doubt aware of the Darwinian theories that had arisen in the meantime. Such questions could also have guided the caricatures that, throughout the 1860s and 1870s, would consistently associate Darwin with the great apes, illustrating, at times interpreting them from a humorous perspective that did not refrain from literalness, the naturalist's assertions: "[M]an is descended from a hairy quadruped, furnished with a tail and pointed ears, probably arboreal in its habits, and an inhabitant of the Old World. This creature, if its whole structure had been examined by a naturalist, would have been classed amongst the Quadrumana . . . "[7] Prompted by the publication of *The Descent of Man,* that is what is behind the scene drawn by the American Thomas Nast in which a gorilla is reduced to tears by Darwin—who maintains he is one of his descendants—asking about his pedigree. A third party, taking him to task, interjects: "Now, Mr. Darwin, how could you insult him so?" (152). In the majority of the caricatures alluding to Darwin's theories published in Britain, France, Germany and the United States, evolution was addressed through the figure of the ape, which was in keeping with a long-standing comic and satirical tradition, as well as tinged with sensationalism through the revelations of the explorer Paul Du Chaillu.[8] With the publication of *The Descent of Man,* followed by *The Expression of the Emotions in Man and Animals,* it was Darwin himself who became a primate in the caricatures of him. Cartoonists took mischievous delight in fulfilling the wishes of their model by using his conclusions disparagingly against him: "We must acknowledge . . . that . . . man still bears in his bodily frame the indelible stamp of his lowly origin."[9] To that end, they created a hybrid being that had Darwin's face—with his long beard, bald head and deep-set eyes—as it was shown in the contemporary photographs that followed Julia Margaret Cameron's 1868 portrait of him, and the hairy anatomy of an ape. The archetype for that was provided by the caricature "A Venerable Orang-outang: A Contribution to Unnatural History" that appeared in *The Hornet* in 1871, which mocked intellectuality through animality and which Darwin seems to have been aware of (153).

The effectiveness of that image of Darwin as an ape is not only measured by its frequency, but also by its suggestive power, as demonstrated, for example, in a caricature of the scholar by James Tissot entitled "Natural Selection," published in *Vanity Fair* under the pseudonym Coïdé.[10] It shows a very thin Darwin sitting in an armchair that is too low and too small for him, his arms and legs tightly crossed. The figure's pose, the wily expression in his eyes and the smile revealing his teeth, as well as the indistinct appearance of his right hand that, obscured by shadow, looks like a paw, is reminiscent of the naturalistic sketches of monkeys found during the same era in the work of an artist such as Gustave Moreau.

A caricature in the British satirical weekly *Fun*[11] again depicted Darwin as an ape conversing with an elegant young woman—whose

1 Yvette Conry, *L'Introduction du darwinisme en France au* XIX*e siècle* (Paris: Vrin, 1974).

2 Janet Browne, "Darwin in Caricature: A Study in the Popularisation and Dissemination of Evolution," *Proceedings of the American Philosophical Society*, vol. 145, no. 4 (December 2001), p. 496–509; Janet Browne, "Charles Darwin as a Celebrity," *Science in Context*, vol. 16, nos. 1–2 (March 2003), p. 175–194.

3 Charles Darwin, *The Descent of Man, and Selection in Relation to Sex* (London: John Murray, 1871), vol. 2, p. 404–405.

4 See Laurent Baridon and Martial Guédron, eds. *Homme-animal. Histoires d'un face à face*, exh. cat. (Strasbourg: Musées de Strasbourg; Paris: Adam Biro, 2004).

5 See the works of Michel Pastoureau.

6 Champfleury, *Histoire de la caricature moderne* (Paris: Dentu, 1865), p. 194.

7 Darwin, *The Descent of Man*, vol. 2, p. 389.

8 Stuart McCook, "'It May Be Truth, But It Is Not Evidence': Paul du Chaillu and the Legitimation of Evidence in the Field Sciences," *Osiris*, no. 11 (1996), p. 177–197; on Paul Du Chaillu's revelations, see the essay in this catalogue, p. 227.

9 Darwin, *The Descent of Man*, vol. 2, p. 405.

10 Coïdé, "Natural Selection," no. 33 in the series "Men of the Day," *Vanity Fair* [London], September 30, 1871.

11 "That Troubles Our Monkey Again," *Fun*, November 16, 1872.

hand he holds as if reading her palm—who is referred to as the female descendant of a marine ascidian, a small, headless animal with a short intestine that is at the threshold of the vertebrate group. "Really, Mr. Darwin, say what you like about man . . . ," she entreats him. That image, which reshuffles the cards by playing with human/animal and male/female oppositions, reveals the cultural and sexual tensions that permeated Victorian society, and within which, by creating a kind of uneasiness and its share of farce fuelled by the conflict between science and religion, the reception of the theory of evolution was caught.[12]

That type of strange combination also guided the caricature of Darwin the Frenchman André Gill published on the front page of *La Petite Lune* in August 1878 (157). The image was created against the background of the Barré-Lebiez case that rattled Paris from April to September of that year, in which Darwinism played a part.[13] It involved the murder of a milkwoman, the Widow Gillet, whose carefully dismembered body had been found in Rue Poliveau, not far from the Muséum d'histoire naturelle. A week after committing the crime, one of its two perpetrators, Paul-Louis Lebiez, a medical student and teaching assistant in natural sciences, gave a public lecture entitled "Darwinism and the Church," in which he explained that life belonged to the young and strong, rather than the old and weak. Following the two being sentenced to death, the Bishop of Angers appealed to France's president, Patrice de Mac Mahon, to grant them a reprieve, on the grounds that the young men had been led astray by Darwin's theories. But it was to no avail; Lebiez, whom the press described as the "first scientific killer," as well as his accomplice Aimé Barré, went to the guillotine. "Many people don't know anything about Darwin. If it hadn't been for the slasher Lebiez, who thought he was a Darwinian, it's possible you would not have known his name. Well, here's what he's about. He's a learned Englishman, a naturalist—a right swotter. He has said that man descended from the apes, that man is an ape that has been transformed, perfected, *selected*, as people who are educated say. You understand that annoys the big talkers who consider themselves of a higher order! . . . So, this is what they've come up with: that, while Lebiez had murdered the Widow Gillet, it was Darwin's fault!" said *La Petite Lune*[14] on the overleaf of the caricature in which Gill depicted Darwin as a great ape hanging from a branch of the tree of science. Two weeks earlier, in the same satirical magazine, Gill had published the caricature of a Kanak from New Caledonia holding up a poster showing the faces of Lebiez and Barré, indignantly saying in the caption: "And you know? . . . I'm the one they call a savage!"[15] Issues concerning humanity, bestiality and savagery innervate such images and their connections, in which Darwin and his theories, by revealing their fissures and contradictions, enabled the shaking up of philosophical and moral certainties beyond biology.

In the corpus of Darwinian caricatures alluding to evolution, the return of the missing link amounted to an obsession.[16] Therefore,

12 John Hedley Brooke, "Darwin and Religion: Correcting the Caricatures," *Science & Education*, vol. 19, no. 4–5 (May 2010), p. 391–405.

13 I would like to thank Daniel Raichvarg, emeritus professor at the Université de Bourgogne, for his insights on the Barré-Lebiez case and its connection with the caricature of Darwin by André Gill.

14 La Mère Plutarque, "Les Hommes illustres, Darwin," *La Petite Lune*, no. 10 (August 1878), p. 2.

15 "Et puis, vous savez ? . . . C'est moi qu'on appelle sauvage ! . . ." André Gill, "Le mot du Canaque," *La Petite Lune*, no. 8 (July 1878).

16 Constance Areson Clark, "'You Are Here': Missing Links, Chains of Being, and the Language of Cartoons," *Isis*, vol. 100, no. 3 (September 2009), p. 571–589.

cartoons of Darwin set the old scholar's head, with its resemblance to the traditional portrayals of Plato or Socrates, upon a primate's body. The use of trees as the characteristic location of apes and as the literal representation of mankind's descent was also a part of that lexicon. However, it is to the long tail of the quadrumane that particular attention must be paid. In most of the satirical images, it coils and ripples, filling the space; in a highly sexualized vein, it creates connections with the humans that it catches by the arm or waist. In the structuring of such caricatures, the serpentine tail—which the great apes lack in actual fact—has the same function as the ellipse arranging the gradual transformation of animals into human beings in a series of illustrations published in 1863 by Charles Bennett that speculated about the transitional, evolutionary and mutational nature of biological life. The same dynamic of circularity was called upon by Edward Linley Sambourne in a caricature from 1881[17]—the year that Darwin published *The Formation of Vegetable Mould, Through the Action of Worms*—that depicted an earthworm as man's ancestor, supporting a vision of evolution from Chaos following a spiral path plotted by the earthworm's rings to the old naturalist, sitting enthroned in the midst of the process. By concealing the missing link through the artifice of continuity, that framework enabled biological evolution to be shown in a transformist way, from ostensibly "lowly" to what were called "superior" forms of life.

Just like Victor Hugo, Émile Zola and Louis Pasteur,[18] men whose work was of seminal importance for their time, Darwin became practically the sole personification of evolution in caricatures. Nonetheless, another figure—in that instance in France, in the person of Émile Littré (154)—was subject to similar graphic critiques. Among some of the cartoons of the positivist thinker as a primate, a caricature by André Gill showed two simian characters in the ring of a hippodrome circus (155): Littré, the lexicographer—who proclaimed the evolutionary nature of the French language and was often accused of aping Darwin's hypotheses—has his mentor jump through paper-covered hoops, thereby puncturing "credulity" and "ignorance," "superstitions" and "errors." By means of this contemporary caricature relating to the Barré-Lebiez case, Gill showed his adherence to Darwin's theories and which side he was on. While not all caricaturists did so as clearly, many of them demonstrated their amused fascination with evolutionism and its imagery by publishing humorous pictures whose power for promotion Darwin was not impervious to,[19] as he himself was concerned with constructing his media celebrity. As he inquired of a friend, "Ah, has *Punch* taken me up? . . . I keep all those things. Have you seen me in *The Hornet*?"[20]

17 Edward Linley Sambourne, "Man Is But a Worm," *Punch's Almanack for 1882*, December 6, 1881.

18 Bertrand Tillier, "Pastorales satiriques," in *Caricaturer Pasteur*, exh. cat. (Besançon: Éditions du Sekoya; Dole: Musée Pasteur, 2014), p. 24–29.

19 Janet Browne, "Looking at Darwin: Portraits and the Making of an Icon," *Isis*, vol. 100, no. 3 (September 2009), p. 542–570.

20 James Hague, "A Reminiscence of Mr. Darwin," *Harper's New Monthly Magazine*, vol. 69, no. 413 (October 1884), p. 760.

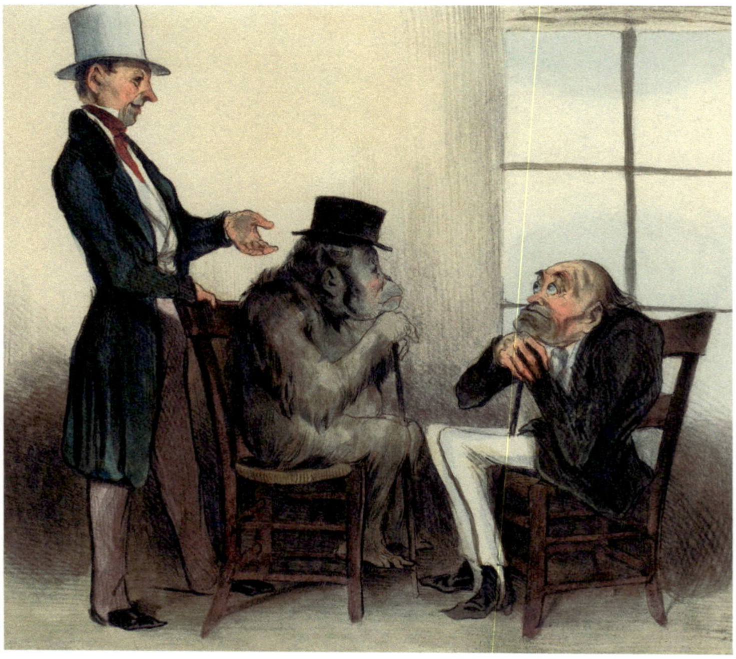

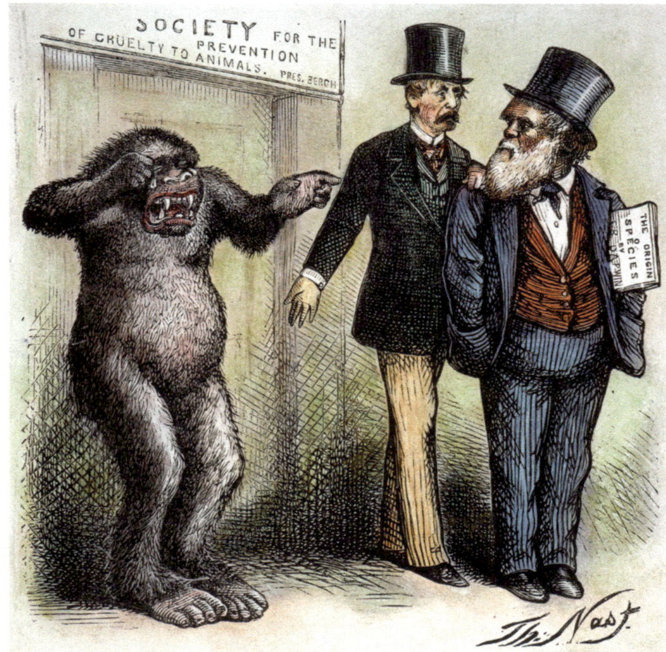

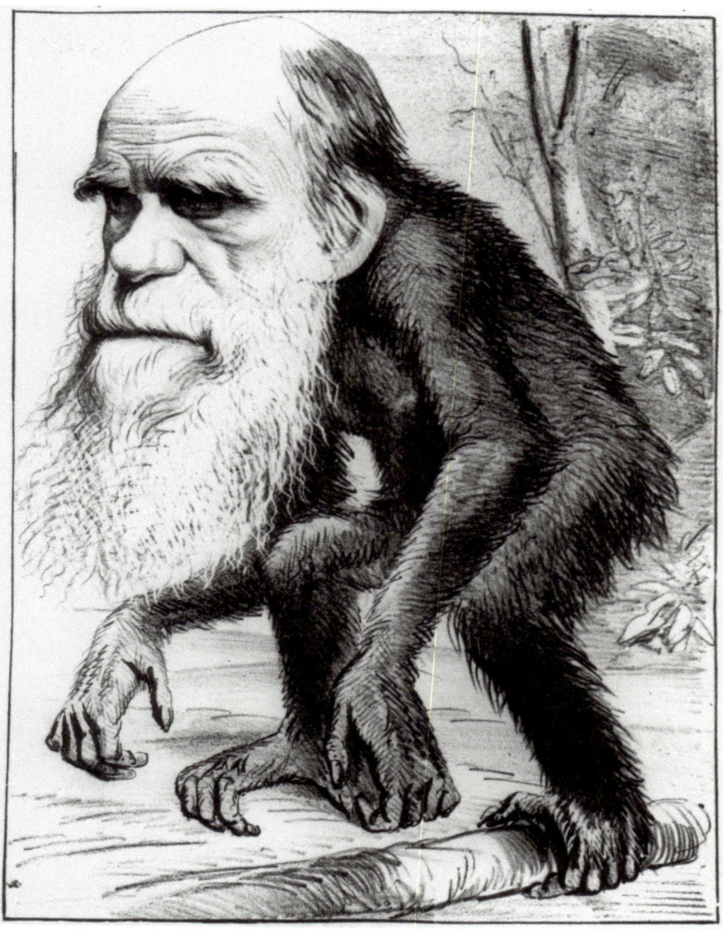

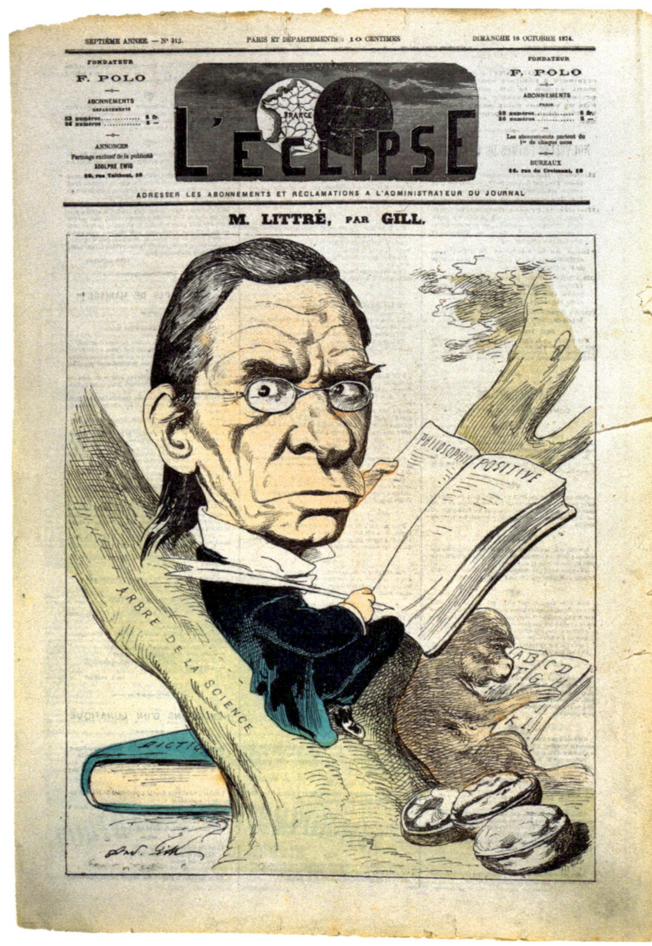

151 FIG.
Honoré Daumier,
"Look Mr. Mayeux, this animal is halfway between man and monkey. – Good God! He can boast of being d . . . ned ugly!,"
in *Charivari*, October 6, 1836

152 FIG.
Thomas Nast,
"Mr Bergh to the Rescue,"
in *Harper's Weekly*, August 19, 1871
London, The British Library

153 FIG.
Anonymous,
"A Venerable Orang-outang: A Contribution to Unnatural History," in *The Hornet*,
March 22, 1871
Private collection

154 FIG.
André Gill,
"M. Littré," in *L'Éclipse*,
October 18, 1874

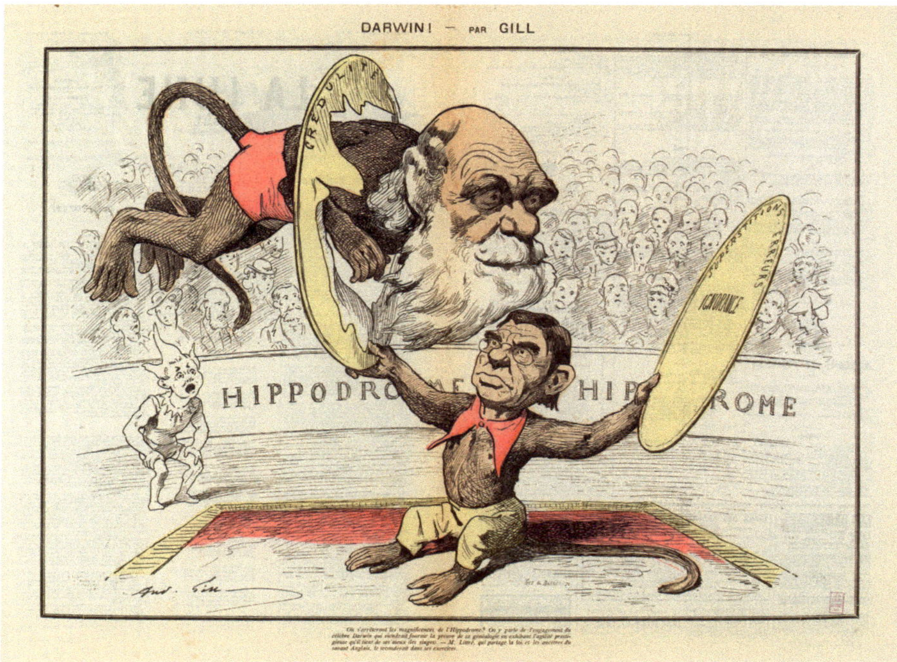

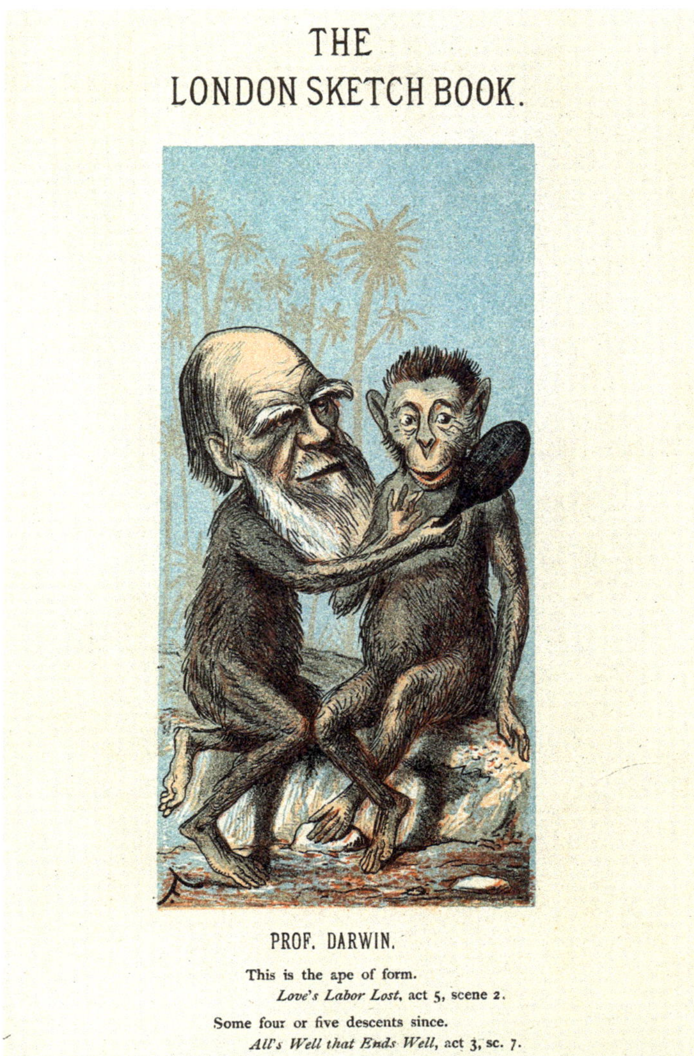

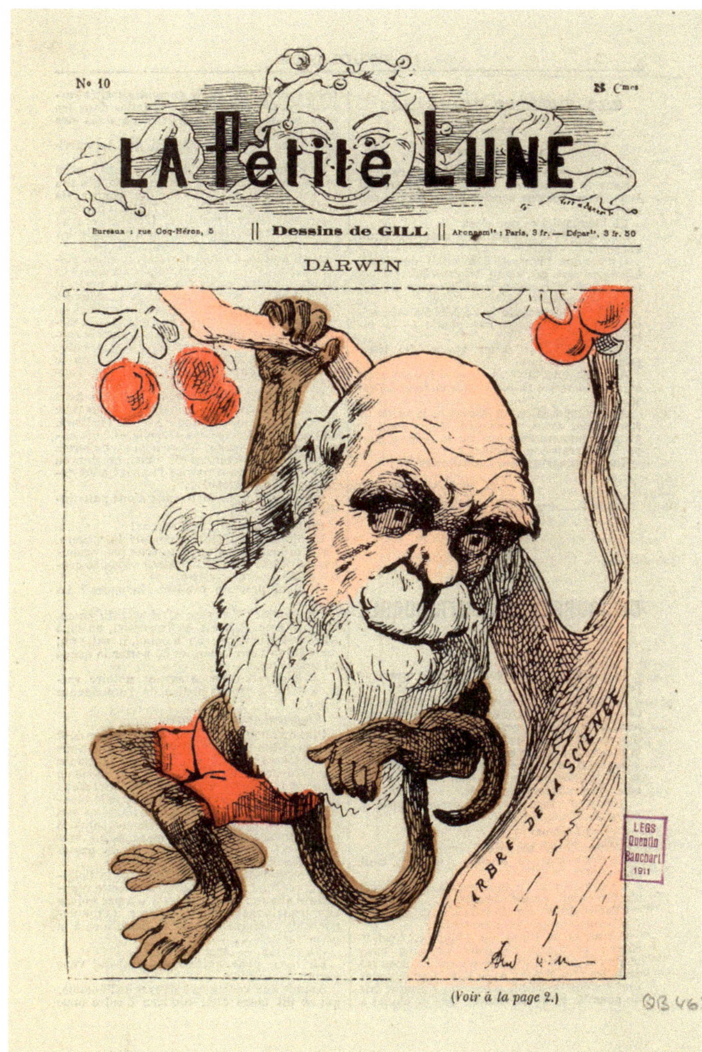

155 FIG.
André Gill,
"Darwin !" in *La Lune rousse*,
August 18, 1878
Paris, Musée Carnavalet

156 FIG.
Faustin Betbeder,
"Prof. Darwin," in *Figaro's London Sketch Book of Celebrities*,
February 18, 1874
Wellcome Collection

157 FIG.
André Gill,
"Darwin," in *La Petite Lune*,
no. 10, 1878–1879
Paris, BnF

V THE DOUBLE: VARIATIONS ON THE APE

CLAUDE BLANCKAERT

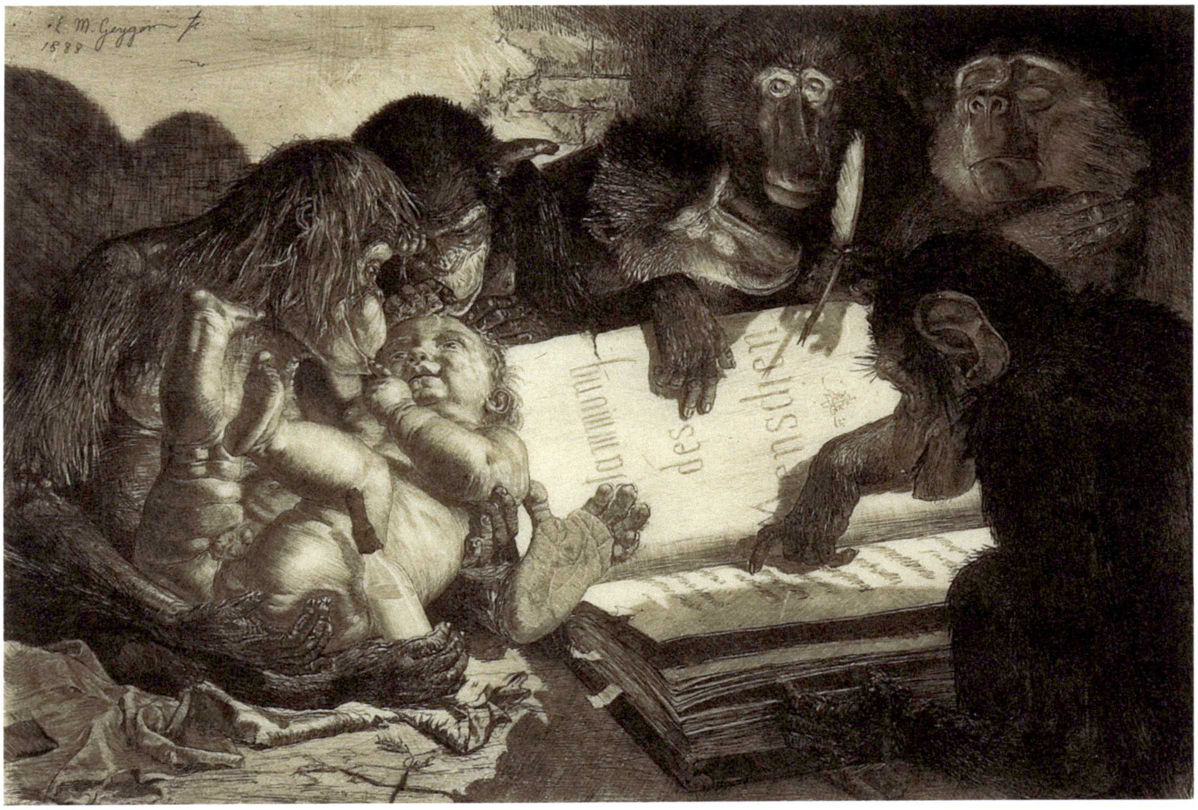

158
Ernst Moritz Geyger,
Darwinian Disputation, 1888
Etching, 24 × 33.5 cm
Frankfurt, Städel Museum

In the West, apes and artists have been associated ever since the Minoan "Blue Monkey" frescoes in the palaces at Knossos and Akrotiri. However, in Ancient Greece, only the Barbary ape, the baboon and some species of Old World monkeys in nearby Africa were recorded in writing. For over a millennium, naturalists would return to this first list without adding to it. But scholars were not unaware of the potential significance of a surprising class of animals that Aristotle had already recorded as being "intermediate in shape between man and quadrupeds."[1] The physician Galen, whose theories were influential until the seventeenth century, declared that a resemblance to humanity does not make a man. Far from resembling us "in everything," apes are the objects of mockery. Trying to copy the actions of men but not succeeding, they are "ludicrous toys for children." For the rest, their "distorted" and clumsy gait reveals the inadequacy of their minds. The inner self corresponds to the caricature. "Having absurd spirits," apes therefore have "completely ridiculous bodies."[2] According to the principle of final causes, Galen reaffirmed the divine privilege of man, the only thinking two-legged being, and simultaneously opposed the dubious etymology of the term *Simia*, thought to have derived from *similitudo*. The analogy was unsustainable. By its very duplicity, the animal that apes us reveals its real nature and its abjectness. It is certainly an unkind portrait, unrelated to the actual, totally unknown behaviours of the species in question. Indeed, usually denoting exoticism or intractable savagery, the tamed ape, the only image familiar in Europe, looks at first sight like the counterpart of a fairground entertainer, the stooge of a travelling barber or a sign of high rank on a great lady's muff.

According to the classical authorities, the ape's bipolar identity, sometimes wonderful, sometimes repulsive, inspired a prolific output of images that gave the illusion of naturalness without reproducing its character. The animal was still rare, an object of curiosity, and seemed spontaneously predisposed to mimic human failings with its grimaces. It cheated, pilfered, got drunk at times and gave free rein to its lusts. While remaining mysterious, the ape depicted ran the gamut of deviant behaviours, disapproved of and therefore desirable. Totally vulgarized, its bad reputation was also its glory. Without any doubt, its escapades denoted intelligence and artists could celebrate, under its mask, the improbable union of contraries. From medieval illuminations parodying chivalry to the variations of the heterogeneous couple of Beauty and the Beast in the work of Pierre-Yves Trémois, the ape plays the role of an equal. Its borrowed gestures allow for many symbolic permutations and break down the barriers between species. From imaginary images to allegories, shifting from identification to revulsion, the world of the arts displays the whole range of anthropomorphisms and, in this unusual case, offers one of the most durable examples of the prolific persistence of representations.

THE DEVIL IN THE FLESH

In biblical illustrations, the "ape," here a generic term, has existed since the Creation and is seen leaping about in the trees of the Garden of Eden. However, the pictures of Eve being tempted by the serpent, the eating of the fruit from the tree of knowledge and God's punishment show the reverse: the ape, fallen back to earth and apparently stunned, walks hunched over behind Adam and Eve as they are banished from the Earthly Paradise. It seems stricken in its turn by the weight of a fatal error, which it may have caused.[3] In reality, it revels in the contrast and disguises its treachery. Like the he-goat, the cat or the snake, the ape is part of a diabolic arsenal and shakes all our certainties. Far from consoling the first humans in their exile, it brands them as infamous. It is an evil creature, and like the fallen angel who loathes the human race, it seeks to delude souls, to lead them astray. Shamelessly copying an ancient Alexandrian encyclopedia, the *Physiologus* [the Naturalist], the moralizing bestiaries of the twelfth and thirteenth centuries made similar use of the analogy. "*Physiologus* says that the ape symbolizes the Devil." It is, like him, "totally ugly and repulsive both in front and from behind."[4] The exegesis here does not rely on the subconscious but on an endorsed suggestion. With its hairiness, its fidgetiness, its screeching and its ungainly figure, the simian shows all the outward signs of an inner ugliness. A tail indicates its brutality, an absence of tail its lack of symmetry, its deformity and consequently its degradation.[5]

1 Aristotle, "Parts of Animals," in *The Complete Works of Aristotle: The Revised Oxford Translation*, ed. Jonathan Barnes, vol. 1 (Princeton: Princeton University Press, 2014), p. 1076.

2 Claude Galien, *De l'utilité des parties du corps humain*, in Claude Galien, *Œuvres anatomiques, physiologiques et médicales*, trans. Charles Daremberg (Paris: Baillière, 1854–1856), vol. 1, p. 162–163; vol. 2, p. 79–80.

3 See Laura Bossi, *Histoire naturelle de l'âme* (Paris: Presses universitaires de France, 2003), p. 241–246.

4 Pierre de Beauvais, *Bestiaire*, in *Bestiaires du Moyen Âge*, Gabriel Bianciotto, ed. (Paris: Stock, 1980), p. 44.

5 Jacques Voisenet, *Bêtes et hommes dans le monde médiéval. Le bestiaire des clercs du Ve au XIIe siècle* (Turnhout: Brepols, 2000), p. 65–67, 317, 361–362.

6 Isidore Geoffroy Saint-Hilaire, *Histoire naturelle générale des règnes organiques, principalement étudiée chez l'homme et les animaux*, vol. 2 (Paris: Masson, 1859), p. 251.

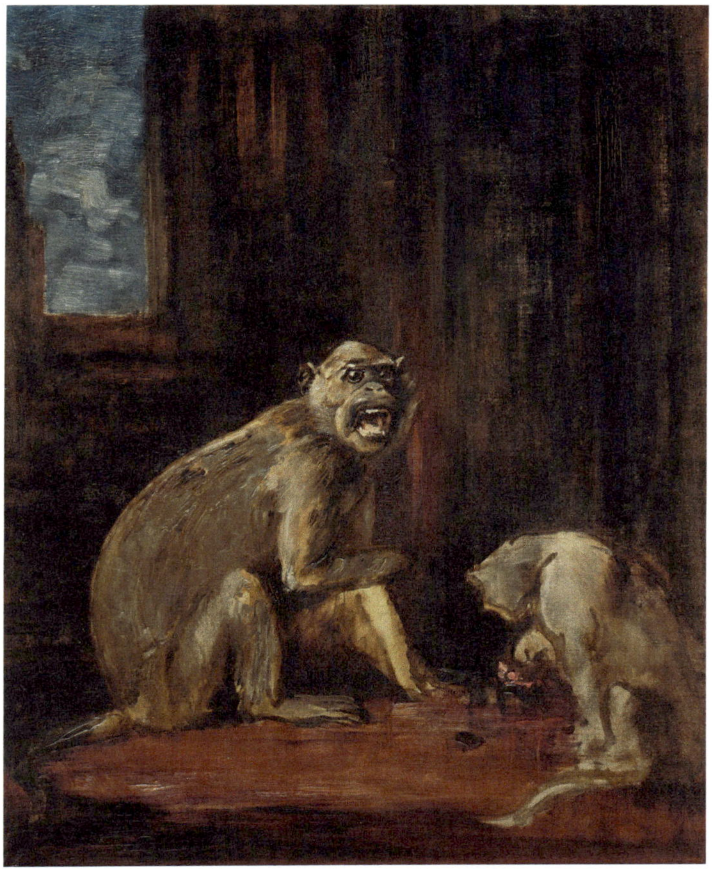

Alongside the hybrid monsters with human features—men without heads or with gigantic ears (panotti), pygmies and sirens—which can be found in the cosmographies or the edifying books of the modern era, the ape bears witness to a world disrupted by the assaults of the Devil, and conversely, by the anger or the power of the Almighty. In medieval folklore and psalters, the humanized ape embodies this "inverted" world—or double creation—which Satan imposes on innocent nature. It stands in for man in numerous comical engravings in which it poses as a barber shaving a cat, as a ludicrous schoolteacher, a tonsured monk or a carter flogging horses. More "realistic" in their style, painters were not to be outdone. At first sight, scenes of popular festivals assign a modest role to apes. But in the corner of a painting an ape will appear nonetheless, tricking a merchant in order to gorge on apples or using a strolling player as its sidekick. Associated with despicable acts and crude amusements, the ape is notable for repulsive traits (hairiness, blood, teeth) when there appears on the platform a surgeon flanked by his macaque. At the other end of the social spectrum, the marmoset of courts and salons looks refined. Peeping out from the folds of a gown or perched on a shoulder, it exemplifies the vanity of wealth, the pretentious and superficial character of anyone who turns away from the glory of God to worship the pleasures of this earth. That way lies danger. No one is spared. Laughter, jollity, jitteriness, malice: the ape's ambiguous diabolic stigmata give a structure to the eternal conflict between vice and virtue. Even concealed under its comical side, which accentuates the imitation, the image contains a clear moral message: man must overcome his demons.

The allegory of the ape chained or tethered by a leash in a state of subjection symbolizes the defeat of Satan. His plans are foiled. In other words, cosmic order is restored with his capitulation. The antitheses depicted continually bring good men back to their birthright. Ultimately their positions are indicated by the ape's chain and the individual who holds it, who may or may not be accompanied by an owl, the symbol of wisdom. Mind governs matter. The captive ape, led about like a pet dog or riveted to a ball and chain, was to be a constant theme from the tapestries of the fifteenth century to Georges Seurat. The starting point for any different reading of this analogy has been lost over the years. However, although it is difficult to estimate the level of comprehension or even of mere intelligence of the onlookers, the recurrence of this witty motif demonstrates the persistence of the aesthetic codes.

In another way, the ape questions our (in)humanity. Even in the mid-nineteenth century, serious naturalists still agreed that, however humiliating it was for man, "the more we study his organs, the more we realize that the being 'made in the image of God' has the same physical characteristics as this hideous animal."[6] "Hideous," perhaps. Others might say "unclean." Despite these negative traits, its exuberance is proportionate to the robustness of its appetites. It is generally agreed that apes are highly sensual creatures. Their gluttony and the stratagems they employ in order to satisfy themselves have struck many commentators. Often depicted by painters and engravers is La Fontaine's fable "The Monkey and the Cat" (i.e. Bertrand and Raton), two arrant rascals who love

159
Gustave Moreau,
Bertrand and Raton,
1879–1885
Oil on canvas, 82 × 65 cm
Paris, Musée national
Gustave-Moreau

chestnuts. In this story, their malfeasance includes the selfish gluttony of Bertrand the monkey, which seizes their booty for itself (159). It is the cat that has "pulled the chestnuts out of the fire." The ape's affect, whetted by its intelligence, seems to be always in search of immediate gratification. A painting by Johann Rudolf Bys shows a golden lion tamarin with silky fur perched on a heap of fruits and vegetables. Cantankerous green monkeys play the same greedy role in many paintings by Frans Snyders.

Epitomizing all the seven deadly sins, incapable of repentance, irreverent, the ape figure plays a prominent role, fettered or not, in numerous allegories of the senses or the four elements. In the latter case, it is assigned to the earth, not to the air, the kingdom of the birds. In short, the material element predominates. In terms of the senses, the ape mainly symbolizes taste, alluding discreetly to the attraction of forbidden fruit. But, almost equally, both its habits and its instinct predispose it to touch things. The hand was often thought to be "the philosophical sense," and quadrumanes, or non-human primates, dexterous with feet as well as hands and curious about everything though fickle, are frustrated by the isolating chain. This is immediately apparent in one of the tapestries of *The Lady and the Unicorn—Touch*—in which the metal bridle and the collar prevent the animal from reaching the nearby tree with its tempting fruit. The tether here is a reminder that the body is the prison of the soul.

Likewise, in a number of images we see the unexpected alliance of the ape and the parrot, both of which are expert at mimicry, one copying human movements and the other the human voice. But even there a thought must be found to inspire the word or the action. A puppet babbling nonsense is a travesty of a human, the only being worthy of "the name of an animal that reasons."[7] The naturalist Buffon, entirely Cartesian on this point of doctrine, explained in 1766 that an ape cannot even imitate a man, for imitation presupposes an intention to imitate, a voluntary act of which an animal is incapable. This saving distance, it might be said, has the advantage of appeasing our consciences. It nevertheless betrays the real vexation felt by the observer on seeing his sinister reflection. Hence, the representations are intertwined with the mentalities. They persist and have scarcely evolved even in the modern world. The mythology of the capricious and miscreant ape was not at all affected by the development of knowledge about primates. Indeed, it became a pretext for updating the myths. While Ancient Egypt worshipped the dog-faced baboon divinity Thoth and India included Hanuman in the Hindu pantheon, Charles Darwin suggested the converse, that our simian ancestry was the source of our worst passions: "The Devil under form of Baboon is our grandfather!"[8] Thus, whether we see it as a deceptive imitation, a metaphor for original sin, or, more insidiously, a cruel truth that overturns the principles of a human identity, the ape will long display the ambiguities of nature, the accursed side of creation.

AN INTERMEDIATE LINK

Like actors in a bad cosmic drama, apes were credited with metamorphoses and dual personalities as colonial conquests increased in the East Indies and Africa from the seventeenth century on. In the earlier period, the discovery of the little quadrumanes of tropical America (marmosets, squirrel monkeys, etc.) had already appeared in the cabinets of naturalists and sparked the interest of painters in these new models. Their palettes were enriched with many more bright hues, since these apes could compete with the vivid colours of parrots. Imagination and admiration soon outweighed moral disapproval. Naturalists thus distinguished between the Devil, the father of lies and destroyer of the soul, and his so-called "incarnation." The Jesuit José de Acosta, who lived in the New World from 1571 to 1588 and saw in Peru both wild and tame tamarins, was impressed by their dexterity and good behaviour. The cleverness shown by apes once tamed, he remarked, seemed "not to come from brute beasts, but from a mankind understanding." This is no small admission. He found their attitude so amiable that he went so far as to thank the Lord for having created so comical a creature "only for the recreation and delight of man."[9] This is only one example of the initial rehabilitation, an echo of which was to continue into the next century with the highly popular fashion for caricatures of the human comedy. Freed from his pact

7 François Hédelin, abbé d'Aubignac, *Des satyres brutes, monstres et démons* [1627], ed. Gilles Banderier (Grenoble: Jérôme Millon, 2003), p. 79.

8 Charles Darwin, *Metaphysics, Materialism, and the Evolution of Mind: Early Writings of Charles Darwin*, ed. Paul H. Barrett (Chicago: University of Chicago Press, 1980), Notebook M, p. 29.

9 Joseph de Acosta, *The Natural and Moral History of the Indies* [1589], trans. Edward Grimeston (1604; repr., London: Hakluyt Society, 1880), book 4, chap. 39, p. 285–286.

10 Hédelin, *Des satyres brutes, monstres et démons*, p. 96.

11 Ibid., p. 191ff.

12 See Ernest-Théodore Hamy, "Documents inédits sur l'*Homo sylvestris* rapporté d'Angola en 1630," *Bulletin du Muséum d'histoire naturelle*, no. 7 (1897), p. 277–282.

13 See Bernard Heuvelmans, *Les Bêtes humaines d'Afrique* (Paris: Plon, 1980), p. 248–249.

14 Ibid., p. 160.

15 See Lucienne Strivay, *Enfants sauvages. Approches anthropologiques* (Paris: Gallimard, 2006), p. 102.

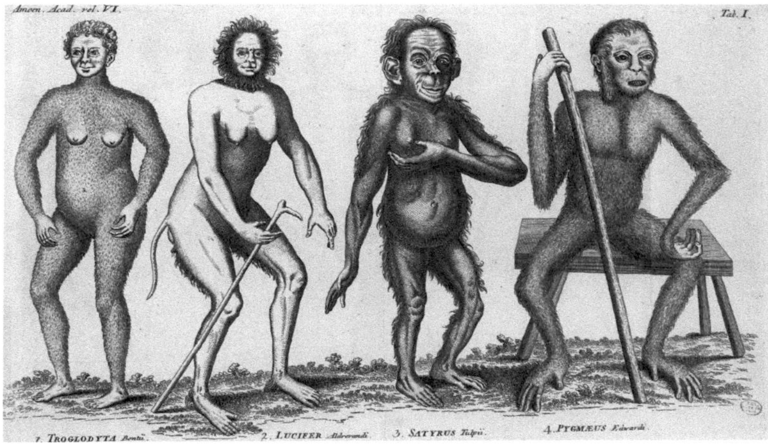

with the Devil, the miscreant ape took on a more pleasing role as a joker and libertine.

This era was also less inclined to believe in the old fables that peopled the antipodes with monsters and prodigies. Despite their unaltered transmission from classical culture onwards, these mythologies were scrutinized in the light of philological reasoning. Scholarship therefore cut down on the burgeoning of mythical half-and-half creatures: centaurs, satyrs with claw-like feet, fauns with horns and men with tails. In 1627, the Abbé d'Aubignac, grandson of Ambroise Paré, distinguished three categories of acknowledged "satyrs": the "monsters" in the strict sense, products of disruptions in nature or mixtures of species; the demons and deceits of the Devil, propagators of idolatry; and lastly the "hairy creatures," irrational animals such as apes. His work aimed at clarification, seeking mainly to demonstrate that a scholarly exegesis showed that satyrs "could not be men" and that there was only one immortal species, "the descendants of Adam." He nevertheless conceded that the "hairy creatures," a term taken from Albert Le Grand, had the lower bodies of goats and upper bodies almost human in appearance except that their ruddy faces are surmounted by two frontal horns![10] The smaller apes, known as "homunculi," may have been the source of pagan cults and even of stories of demons. Brute satyrs and demons have similar attributes, the hairiness and the horns, and analogous, invariably outrageous, behaviours: furious lustfulness, grotesque dancing and "frenzied passion." Not surprisingly, satyrs suffer from satyriasis.[11]

Far from fading away with the increase in knowledge, this timeless mythical figure reappears in all its poetic depiction in the earliest images of "sylvan satyrs" from Africa and the Dutch East Indies. The latter are known to us today as *anthropoids*, the tailless great apes, our modern pongids. But this word, dating only from the nineteenth century, is inadequate. It would betray the astonishment and fascination felt by all those privileged to see these unclassifiable creatures. In June 1630, the Dutch erudite Willem de Groot told his correspondents in Paris that the private menagerie of Frederick Henry, Prince of Orange-Nassau, had just acquired "a monster, man or beast, I cannot tell which," that had come from Guinea.[12] The examination carried out shortly afterwards by Nicolaas Tulp, the founder of the Amsterdam College of Medicine, did not dispel the doubts. With some generalizations about its regular proportions and human form, the rounded shape of the breast and the daintiness of its manners, in 1641 Tulp published his conviction that the young female confirmed the existence of Pliny's satyrs. Indeed, he called it an "Indian Satyr" without any reference to its geographical origin, adding to the mystery that the East Indians called it *orang-outang*, a term in the Malay vernacular (wrongly spelled) meaning "man of the woods." This name became known all over scholarly Europe in its Latin translation, *Homo sylvestris*. The Dutch anatomist found it expedient to assert, on the strength of local rumours, that in Borneo the males of this species seemed "libidinous and lecherous" and that they lusted after women and girls.[13]

This short description, accompanied by an engraving in a realistic style showing the specimen seated, with the eyes either closed or modestly looking downwards, was to fix the scholarly archetype of *Homo sylvestris* for generations to come. Sometimes feminized by an alteration to the silhouette or long hair, or at other times depicted in movement, Tulp's small female was continually reproduced. Under the name of *ouranus outanus*, it appears in 1696 in a plate by Johann Zahn alongside "different" men without mouths or acephalous (headless men), the hairy "savage man" of medieval illuminations and the man with a tail.[14] In the *Physica sacra* of the Swiss naturalist Johann Jakob Scheuchzer, dating from the 1730s, an engraving shows it beside Esau, the elder son of Isaac, of whom Genesis (25:25) reports that he was born "red and hairy all over" (123, P. 185).[15] It is found again

160
Carl Linnaeus,
"Troglodita Bontii, Lucifer Aldrovandi, Satyrus Tulpii, Pygmaeus Edwardi,"
in *Amoenitates academicae, seu Dissertationes variæ physicae, medicae, botanicae . . .* , vol. 6,
Leiden: Cornelius Haak, 1749
Paris, MNHN

in 1765 with the depiction of a *kimpezé*, a chimpanzee, in a work by the surgeon Claude-Nicolas Le Cat, who suggests, in line with an accepted explanation, that the "Ourang-Outang" represents "a species midway between brute and man."[16]

A living paradox, the Indian satyr—sometimes renamed pongo, pygmy, sylvan or troglodyte—was long a mystery to scholars. Later accounts by men who had travelled to Africa or by eyewitnesses from the Dutch East Indies confirmed the too quickly repeated stereotypes. These were probably designed to please European readers eager for novelty. The ambiguity remained. Most of the stories dwelt on the one hand on the creature's delicacy of feeling and astonishing ability to learn, and on the other, its undesirable qualities, primarily its rebelliousness and the immoderate lust of the males for the so-called weaker sex. The importation of live, always young, specimens followed by their dissection did not silence the tall stories repeated by compilers. The naturalists themselves were stymied. The subject became a metaphysical one, questioning what is specific to humans. Could the orangutan, much more frequently compared to man than to apes with tails, be the result of interbreeding? "If these animals did not belong to a different race that perpetuated itself," one of Buffon's correspondents living in Batavia (Jakarta) remarked in 1770, "we might call them *monsters of human nature*."[17] In any case, this ape in the process of becoming human fits between the ordinary ape and man to the highest degree of animal perfection in the scale of being (124, P. 189).

The year 1698 saw the death in London of another juvenile individual from Angola that looked very much like a man. The great English anatomist Edward Tyson, who meticulously compared their anatomies and noted their similarity, described the orangutan, or "pygmy," to use his word, as an *intermediate link*.[18] The formula became fashionable. Tyson also concluded after a painstaking examination that nature had meant it to be a biped, and that its quadruped posture, visible when it was alive, was only due to its very young age or to the weakness of a disease.[19] The standing portrait executed by William Cowper that accompanied Tyson's monograph, in which the upright pygmy is leaning on a walking stick, harks back to the approach of earlier artists who sought to use some distinctive prop to characterize the silhouette of a "hairy wild man." Of course the pygmy's walking stick is just an artifice. However, in the Christianized iconography of the later Middle Ages, the stick or club carried by the hairy men has many other conventional associations: aggressiveness, madness, uncouthness, force. The use of a weapon also helped to dissociate the "wild man," no matter how hairy, from the brute beast (111, P. 159).[20] The canes or walking sticks adopted by Tyson became essential and were not abandoned for some time. By transposition, the attribute of the "pygmy" became almost indissociable from the description of the red or brown "orang-outangs." In naturalist illustrations, the stick emphasized the straightness of the creature's body in an unrealistic manner. The anthropomorphism of the portraits ended in total confusion. This tended to increase during the Enlightenment with the intertwining of colonial hearsay and the homogeneous and repetitive image library.[21]

The splitting and reintegration of man and his simian counterpart commenced with an iconoclastic statement by the principal reformer of taxonomy, Carl Linnaeus. In 1735, the Swedish scholar published the first in-folio edition of his classification, *Systema naturae*, in which the table ranks the genera Man, Ape and Bradypus (sloth) in the order of *Anthropomorpha*, the first class of Quadrupeds. If it does exist, he admitted, the human-shaped hairy "satyr," prone to lively

16 Heuvelmans, *Les Bêtes humaines d'Afrique*, p. 266.

17 Georges-Louis Leclerc de Buffon, "Les orangs-outangs, ou le Pongo et le Jocko," in *Histoire naturelle générale et particulière*, ed. Charles-Nicolas-Sigisbert Sonnini (Paris: Dufart, an IX), vol. 35, p. 127–128.

18 Edward Tyson, *Orang-outang, sive Homo Sylvestris: or, The Anatomy of a Pygmie Compared with That of a Monkey, an Ape, and a Man*... (London: Thomas Bennet & Daniel Brown, 1699), p. 5 and 91.

19 Ibid., p. 49–50, 82.

20 See Florent Pouvreau, *Du poil et de la bête. Iconographie du corps sauvage en Occident à la fin du Moyen Âge, XIIIe–XVIe siècle* (Paris: Comité des travaux historiques et scientifiques, 2014), p. 129–131; Strivay, *Enfants sauvages*, p. 198–199.

21 See Heuvelmans, *Les Bêtes humaines d'Afrique*; Giulio Barsanti, "Storia naturale delle scimmie, 1600–1800," *Nuncius. Annali di storia della scienza*, vol. 5, no. 2 (1990), p. 99–165; Londa Schiebinger, *Nature's Body: Gender in the Making of Modern Science* (Boston: Beacon Press, 1993), chap. 3; Franck Tinland, *L'Homme sauvage. Homo ferus et Homo sylvestris* [1968] (Paris: L'Harmattan, 2003); Strivay, *Enfants sauvages*; and Giacomo Giacobini and Riccarda Giraudi, "Y el hombre encontró al simio," *Elementos: Ciencia y cultura*, vol. 14, no. 68 (2007), p. 3–23.

22 See Thomas Bendyshe, "The History of Anthropology," in *Memoirs Read before the Anthropological Society of London, 1863–1864*, vol. 1 (London: Trübner, 1865), in particular, part 2, p. 421–458.

23 See Luca Zucchi, "Linneo e Parkinson: il botanico e le scimmie nel giardino dell'Eden," *Nuncius*, vol. 16, no. 1 (2001), p. 85–152.

24 See Anton Leikola, "The Development of the Species Concept in the Thinking of Linnaeus," in *Histoire du concept d'espèce dans les sciences de la vie*, proceedings of the international colloquium (Paris, Fondation Singer-Polignac, May 1985) (Paris: Fondation Singer-Polignac, 1987), p. 54; and Gunnar Broberg, "*Homo sapiens*: Linnaeus's Classification of Man," in *Linnaeus: The Man and His Work*, ed. Tore Frängsmyr (Berkeley: University of California Press, 1983), p. 179ff.

25 Claude Blanckaert, "Frontières de l'humanité. Le 'satyre' des Lumières entre science et fiction critique," in "L'homme et la nature," special issue, *Histoire et anthropologie*, no. 25 (2002), p. 13–32; and Claude Blanckaert, "'Premier des singes, dernier des hommes'? Les métamorphoses de l'homme-singe aux XVIIe–XVIIIe siècles," *Alliage*, nos. 7–8 (1991), p. 113–129.

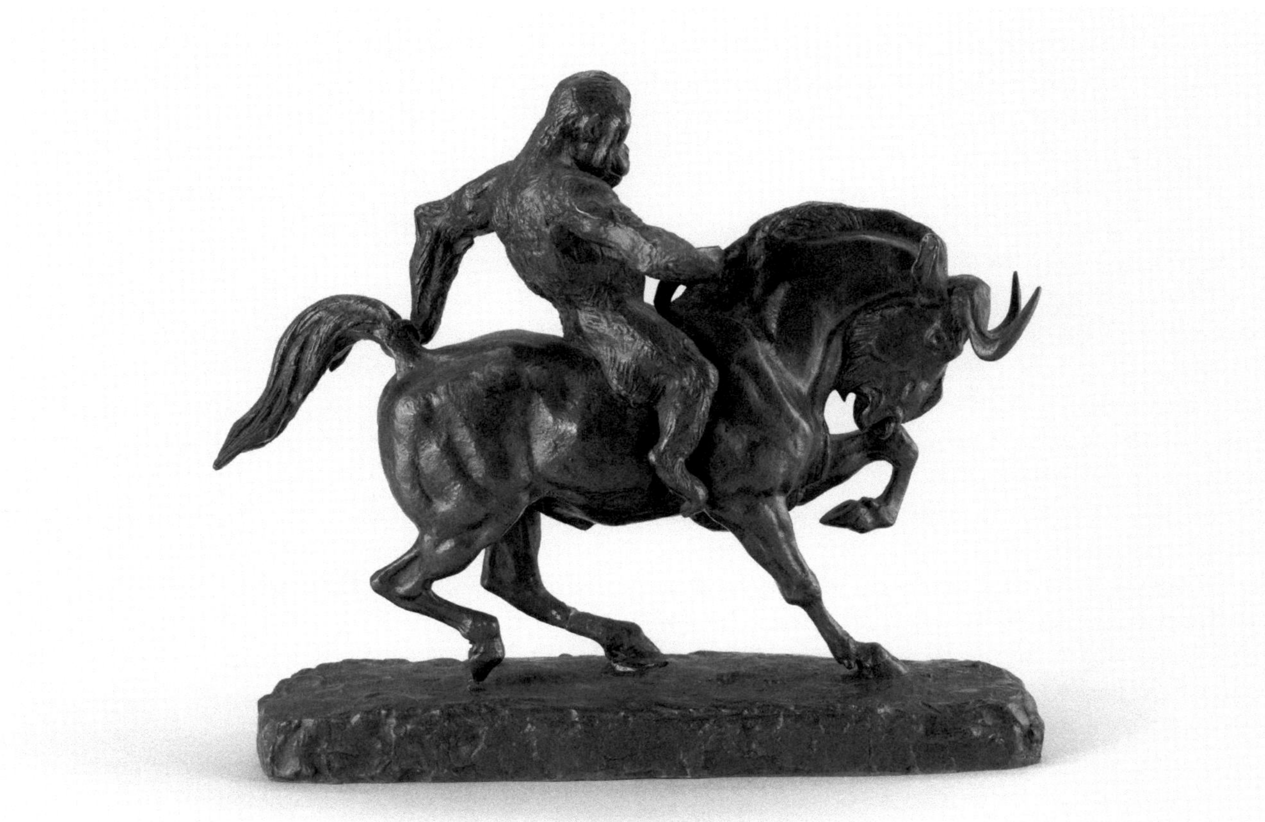

gestures and extremely lascivious, should be included in the genus Ape (*Simia*) with the "men with tails." This pairing, already considered scandalous in itself, was to appear even more concerned with the sovereignty of humankind after the publication of the tenth, totally revised edition of *Systema naturae* in 1758. Among the "Primates," as they were now called, Linnaeus made a clear distinction between the genera *Homo* and *Simia*. Only the taxon *Homo* was divided in its turn into two species, the "day man" (*Homo sapiens*, who now makes his appearance under the nickname *diurnus*) and the "night man" (*Homo nocturnus* or *Homo troglodytes*), a combination of the sources of classic erudition and a recollection of the sylvan orangutan and allusions to travellers' tales of albino peoples with red eyes. Anything could be expected at the edges of the civilized world. Hidden in caves until nightfall, with white bodies and half the size of the average man, "troglodytes" were, according to Linnaeus, able to think and spoke by means of whistling, and imagined, if travellers' tales are to be believed, that the Earth was created for them and that one day they would be the masters![22]

The day man, the night man . . . the frontispiece of the *Systema naturae* (German edition of 1760) presents Adam and, as Tulp points out, the Indian satyr as symmetrical images of each other, framing the many-breasted goddess Artemis, symbol of the diversity and fertility of creation, who seems to invite mankind to know itself.[23] This is the great task of the century. Nature never makes leaps. The notion of a combination of the two species, *sapiens* and *nocturnus*, although of different origins, naturally rekindled concern. In his Swedish manuscripts, Linnaeus refers to the possibility that the Hottentot peoples of southern Africa sprang from the hybridization of these two "human" species and called the "anthropomorphs" "cousins of man."[24] Nature, one might say, sought to mortify mankind. These brief notations seem to suggest that the night man challenged the hegemony of his daytime pendant. This perception reverses two apparently contradictory rationales, absolute otherness and reciprocally the most worrying of proximities. As might be expected, the libertines of the eighteenth century looked favourably on this "philosopher" of a new kind,[25] while the

161
Antoine-Louis Barye,
Ape Riding a Gnu, about 1840
Patinated bronze,
23.2 × 25.8 × 10 cm
Paris, Musée du Louvre

naturalists cold-shouldered Linnaeus as a storyteller. "I very much doubt the truth of the description of this *nocturnal man*," warned Buffon;[26] "I don't even believe he exists." To begin with, Linnaeus's "night man" was scarcely ever depicted by the illustrators of the time.[27] But it should perhaps be noted that, just as it was appreciated by contemporaries, this "laughable chimera" was to be seen in a plate, standing up with a commander's baton in his hand, even in the works of the contradicters. Buffon would also approve of images of the terrestrial biped ape, confident and cocky, whose members, according to Tulp, are "as like those of humans as two eggs are alike." Thereafter, many other well-known publications starting with Diderot and D'Alembert's *Encyclopédie*, would repeat this "commonplace." This model only disappeared at the turn of the nineteenth century, coinciding—not by chance—with the recognition that the red orangutans of Malaysia were primarily tree-dwellers. Then the famous "great apes," the other dark side of our humanity, entered the lists.

BETWEEN THE SINGERIE AND THE MENAGERIE

The artistic depictions of simians and actual knowledge of them rarely coincided. The former reconciles the studio tradition with the sensibilities fashionable at the time, whereas the latter adapts to all the requirements of observation and comparison. In contrast to the amusing, always original works by the animal painters of the eighteenth century, in which the features are less than objectively portrayed, zoological vignettes often have the look of tailor's dummies, static and stuffed. Surprisingly, sketches from life seem to be rare. The vignettes primarily typify the species in order to distinguish them by shape or colouring, as may be seen in the instalments of the *Histoire naturelle des singes et des makis* by Jean-Baptiste Audebert, published between 1797 and 1800. Even in their obvious discrepancies, the standing portraits of the so-called "anthropomorphs" did not go beyond the very real limitations of the academic genre. These were inanimate bodies. On occasion, a background landscape, whether rural or rocky, was meant to compensate for the unlikely poses and situations. It was not easy to reconcile aesthetic values with the expressiveness proper to naturalist illustration. Few artists could excel in both these styles. The name of George Stubbs is scarcely remembered in this context, and yet despite his marked preference for portraying horses and dogs, for the Muséum of the anatomist John Hunter he depicted several portraits and dissections of exotic animals—including some striking dog-faced baboons and lemurs.[28]

The divergence in the vocational training of artists and, of course, in the expectations of their patrons explains why these depictions were no longer common. During the Enlightenment, ornamental "singeries," popular in salons, featured libertine and burlesque scenes. Combining chinoiseries and dressed-up female monkeys, arabesques and apes busy with human occupations both practical and contemplative, these images displayed great freedom in both motif and execution. The *Singeries* in the Château de Chantilly and in the Hôtel de Rohan-Strasbourg in Paris, outstanding in every respect for the delicacy of the little faces, their knowing expressions and the overall rendering, present pastiches of aristocratic behaviours and allegories of the senses, the seasons, the arts and the sciences.[29] Their decorative function makes movement more important than authenticity. In this context of luxury, these attractive and much-admired animals could be found on innumerable surfaces, from wall paintings and ceilings to screen, furniture and crockery. Thus tamed, they became part of home decoration. Celebrated painters specialized in the genre: Claude Gillot, Claude III Audran, Antoine Watteau and Christophe Huet—the last-named probably executed the singeries at Chantilly.

The animal's gift for mimicry also lends itself to visual paraphrase, criticism or an excellent illustration of the Renaissance maxim, "Art apes nature." The metaphor of the ape, the born imitator, allowed for many subversions of a servile, uninspired art devoted solely to the transcription of reality.[30] From David Teniers II to Watteau, Chardin and Alexandre-Gabriel Decamps, the deliberately humanized ape musician, sculptor or painter was the counterpart of the histrionic artist.

26 Buffon, "Les orangs-outangs, ou le Pongo et le Jocko," p. 84.

27 A couple appears, however, under this name, on print no. 3 of *La Découverte australe par un homme-volant* by Rétif de La Bretonne (Leipzig, 1781, vol. 1, p. 183). A sketch by the painter Jean-Baptiste Huet might also be considered (see Barsanti, "Storia naturale delle scimmie, 1600–1800," p. 134 and fig. 25).

28 Basil Taylor, *Stubbs* (London: Phaidon, 1971), especially p. 30–31 and figs. 73, 74 and 75.

29 Nicole Garnier-Pelle, Anne Forray-Carlier and Marie-Christine Anselm, *Singeries et exotisme chez Christophe Huet* (Saint-Rémy-en-l'Eau: Éditions Monelle Hayot, 2010); and Philippe Béchu and Christian Taillard, *Les Hôtels de Soubise et de Rohan-Strasbourg. Marchés de construction et de décor* (Paris: Somogy/Centre historique des Archives nationales, 2004), p. 432–433, ills. p. 423, 436–439.

30 See Bertrand Marret, *Portraits de l'artiste en singe. Les singeries dans la peinture* (Paris: Somogy, 2001).

31 See Claude Blanckaert, "'Produire l'être singe'. Langage du corps et harmonies spirituelles," *Annales historiques de la Révolution française*, no. 377 (2014), p. 9–35.

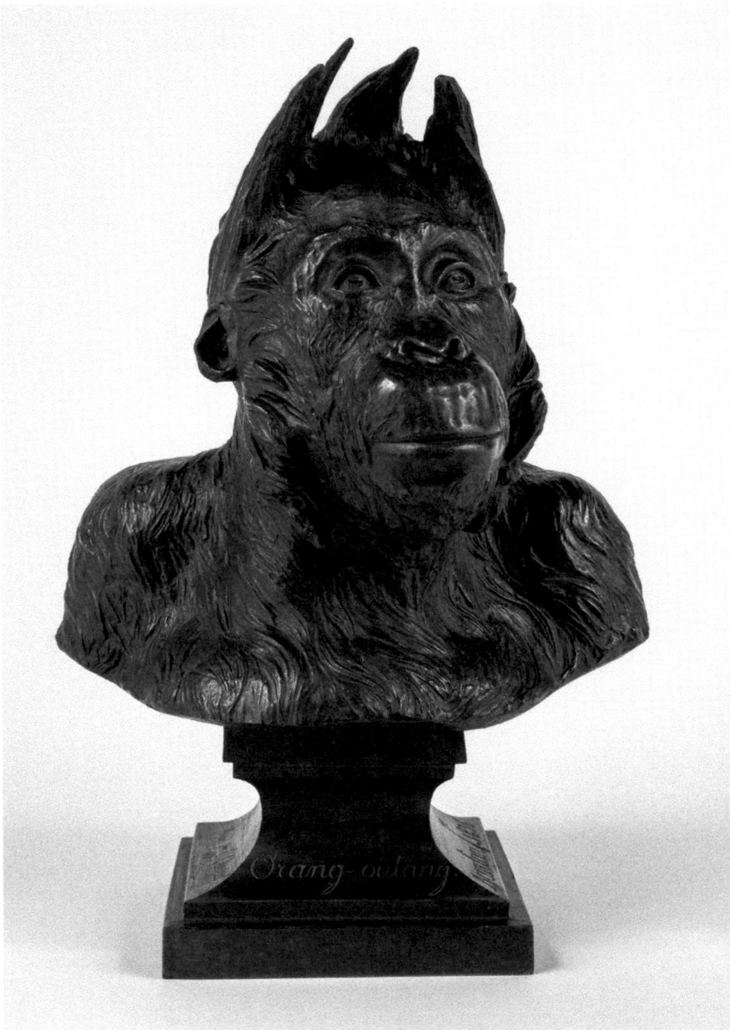

162
Jean-Pierre Dantan,
known as Dantan the Younger,
Portrait of the Orangutan Called "Jack," 1836
Bronze, 42 × 28 × 23 cm
Paris, Musée Carnavalet

Self-mockery? Not necessarily. It all depended on the intention. With this subterfuge, the artist taught viewers to look beyond appearances, put an end to a misunderstanding and claimed his share of truth. The joke always had a philosophical side to it, requiring some imagination. For caricaturists, this role-playing became a mockery or a savage satire of fashionable schools of art. The critical ape came into its own again. It is noteworthy that the orangutan, Rétif de La Bretonne's "ape-man," man's spitting image, is nowhere to be found in the singeries, although it is included in some naturalists' depictions. The two iconographic series thus multiplied and diverged during the first quarter of the nineteenth century.

Also in this period, zoology revised its initial judgements on the advantageous parallel between the ape designated as "extraordinary" and man.[31] Deep Africa had not yet yielded up all its secrets, but the Southeast Asian orangutan, now better understood in its true morphology if not in its development, was subjected to a scrutiny of its anatomy. It turned out that it was certainly not a biped designed to live at ground level. From its conformation and its forest habitat, it was suited to an evolution in the trees, where its four "hands" could hold on to branches. Its outsized arms, according to Professor Georges Cuvier, gave it "the hideous look of a spider," while its feet, when clenched into fists, recalled the claws of a lobster. The grotesque ape, unable to stand up straight, stooping over his cane, cannot put up a good front. It is a high-flying acrobat, an entirely physical being, as different from the two-handed human as a shapeless sketch is from the finished type. On the same template, man is transfigured. Tasked with putting them together, Linnaeus's order of Primates was soon replaced by two distinct orders, Bimanes and Quadrumanes, which would only be re-evaluated two generations later by Darwin's adherents. After 1800 or 1810, this new view of simians made a nonsense of earlier descriptions and of the hasty opinions of half-educated or credulous pedants.

Furthermore, it was necessary to establish an equal balance between the very young orangutans, the only living ones brought to Europe, which resembled human infants and were also gentle creatures, and their adult forms, known from a few skeletons sent to museums in London and Paris. In the Paris Muséum, Cuvier and Étienne Geoffroy Saint-Hilaire taught the same lesson. Appearances can be deceiving. In 1836, there came from Nantes to the Royal Ménagerie in the Jardin des Plantes a nine-month-old orangutan from Sumatra nicknamed Jack, which "attracted the attention of the Parisians." Curious citizens flocked to watch it and its antics became the talk of the city. The draughtsman Jean-Charles Werner painted the creature in real life, holding a branch with a hand and a foot (44, P. 66). The naturalist Geoffroy Saint-Hilaire praised his work and the "exquisite grace" of the rendering as opposed to the detestable figures like those by Buffon that had been tolerated hitherto (163), none of which had captured the distinctive traits of the original. However, Geoffroy insisted that the species was truly indeterminate. The unanimous verdict of the visitors was that the animal from Sumatra was "neither a man nor an ape." For Geoffroy, it was "obviously an intermediate being," a special combination, in short the equivalent of "Linnaeus's *homo-troglodytes*." On that basis, the bronze bust of Jack of which Jean-Pierre Dantan made a study in the Jardin des Plantes in 1836 still conforms to the pattern of the old, placid and magisterial ape (162). Its gaze questions the passerby, its youth is less apparent; Jack seems to have retained, from a distant mythological ancestry, the horns and pointed ears of the old-stock satyr. An anachronism or a harbinger? In reality, nothing was written about it. Shortly after this panegyric, Geoffroy Saint-Hilaire warned that the creature's engaging behaviour in childhood would change radically over time. The resemblance to a man would end with the spectacular, indeed monstrous, metamorphosis of its head as it grew. After studying a series of orangutans of different ages, naturalists had remarked that the face lengthened into a muzzle, the skull became thicker with "defacing protrusions" and the brow arch ossified into bony bulges, crushing the forehead. Finally, the emergence of the murderous canine teeth changed the face completely. The bestial mask lost all its affability.[32] The failure of a potential humanization could still be read in the "lustful expression" on its grimacing face. The creature regressed into its instinctual being and became resistant to change. In maturity the initially promising behaviour of orangutans was subdued and reverted to its biological tendencies.

Honour was saved, cried the more religious scholars. While the divine essence of man permeates his physical being, the animal now personifies the underground domain of uncontrollable passions, emotions and earthly satisfactions. Their destinies both intersect and conflict. Now that the infamous "confraternity" of man and beast had been abolished, the ape-man was reinvented but under a specialized nomenclature. In about 1830, the orangutan of the Sunda Islands still kept the designation of Linnaeus as *Simia satyrus*; the recently differentiated chimpanzee was to be called *Troglodytes* or, even worse, *Anthropopithecus*.[33] Popularized by the illustrated press, novels and soon statuary, the myth of the violent and instinctual anthropoid, prone to raping Indigenous women, did not spring only from the doctrines of Darwin. We can also see here the unexpected result of earlier divisions tending to separate ape from man like flesh from spirit. The visual arts would seize on this contrast, representing in their forecasts feelings of anguish, a taste for the fantastic and a fear of commonality.

THE ROOTS OF THE HUMAN WORLD

"You've made him look terribly like an ape!—That's the way he was!"[34] The first reconstitution of the ancestral ape-man, executed by the painter and engraver Johann Theodor Susemihl, was a full-page illustration for an article by the polygraph naturalist Pierre Boitard published in *Le Magasin universel* in April 1838 (111). The "fossil" man, as it was entitled, is seen in his geological setting. Highly original, the scene shows great expressive power and all the vigour of a new creation. Without disrupting the verisimilitude, the figure covered with hair, with skinny legs and long prehensile feet, is walking towards a shelter under a rock, leaning on a clumsy flint weapon. The half-turned body shows pouting lips and an angular head with heavy jaws that obliterate the nose. Exorcism is no longer required. The skill of the drawing is impressive. Boitard and Susemihl mobilized multiple semiotic sources, linking the medieval "wild man" to copies of the striking portraits of Aboriginal Australians executed by Nicolas-Martin Petit during Captain Baudin's scientific expedition to the Austral Lands (63, 64, P. 90–91).[35] The sharp angle of the head was inspired by an orangutan profile, and the thick-lipped mouth reproduces that of an "Ebo Negro," which a plate by the racist anthropologist Julien-Joseph Virey[36] outrageously compared to an orangutan. The chain of being remained ever-present.

Would hairiness appear improper? Who cares! retorted Boitard: "Read Scripture, sir, and you will see that Esau was hairy like a goat." The borrowings overlapped and formed an astonishing patchwork. This was not Boitard's first attempt. Three years earlier, with the collaboration of Susemihl, he had published an article on the "Chimpanzee," illustrated with a figure "after Tyson," boasting about its accuracy. The composition, mostly taken from poor copies of Buffon and Tulp,

---

32 See in particular Étienne Geoffroy Saint-Hilaire, "Études sur l'orang-outang de la Ménagerie," *Comptes rendus de l'Académie des sciences*, vol. 3 (1836), p. 1–8.

33 The synonymy and instability of these terms are disconcerting. Until the mid-nineteenth century, all of the great apes were generally grouped under the category of "Orang" or "Orang-outang" (later to lose the final "g"). The main terms found in the literature were collected in Isidore Geoffroy Saint-Hilaire's "Description des mammifères nouveaux ou imparfaitement connus de la collection du Muséum d'histoire naturelle. Quatrième mémoire. Famille des singes," *Archives du Muséum d'histoire naturelle*, vol. 10 (1858–1861), p. 78–81.

34 Pierre Boitard, "L'homme fossile. Étude paléontologique," *Le Magasin universel*, vol. 5 (1838), p. 240.

35 See *L'Autre. Les Naturels vus par l'Occident, 1800–1804*, exh. cat. (Le Havre: Éditions du Muséum d'histoire naturelle, 2008), p. 49. See also the essay in this catalogue, p. 86.

36 Julien-Joseph Virey, *Histoire naturelle du genre humain* (Paris: Crochard, 1824), vol. 2, p. 42.

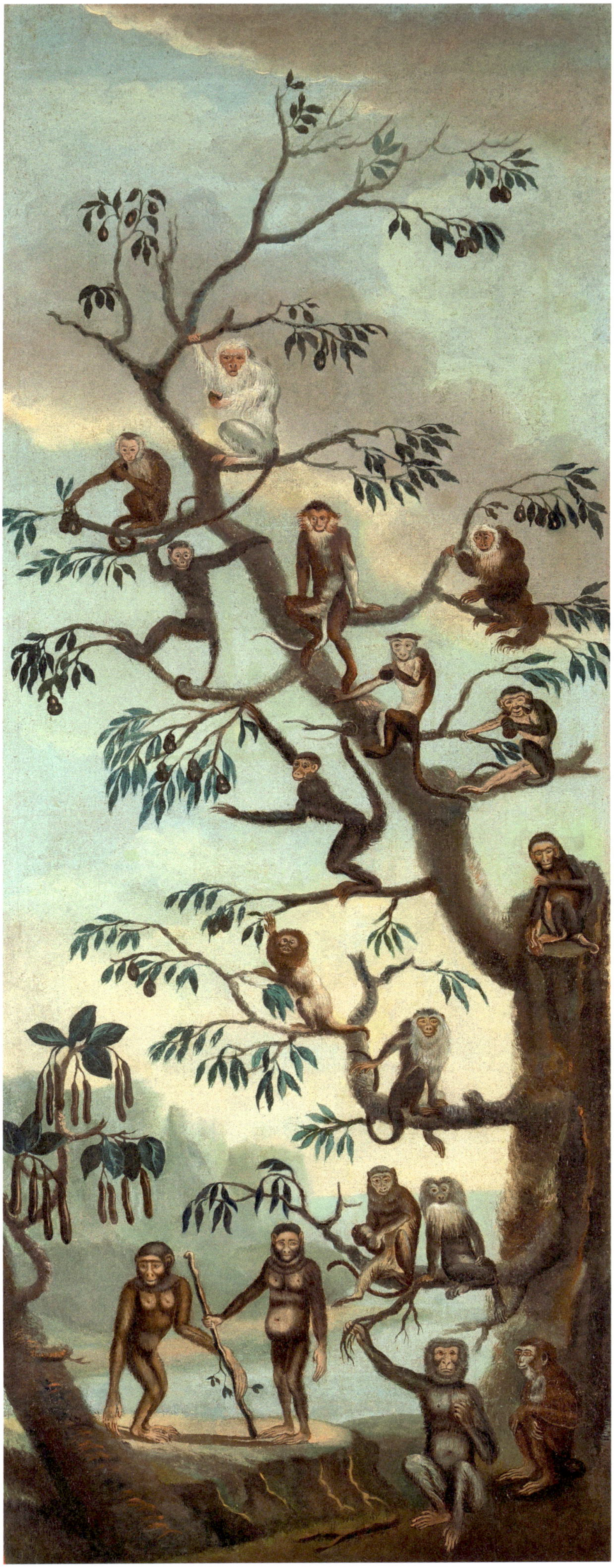

163
French school,
Buffon's Tree of Apes,
after 1766
Oil on canvas, 106 × 44 cm
Montbard, Musée Buffon

represented one of the apes standing, its hands grasping the compulsory cane. Two other apes in the background are running, although the text states that this position was unusual for them. Huts leaning against a sort of palm tree serve them as refuges, but Boitard states that normally they dwelt "in caves in the rocks." Troglodytes could work communally to hunt elephants, if necessary, and reacted violently to aggression by humans. The author concluded his brief description by writing that "it is obvious that the stories our ancestors passed down to us as probable about fauns, pygmies, satyrs pursuing women … took their origin from the history of this animal."[37] Boitard's caveman occupied the same mental space as the "man of the woods." The rhetoric of the images confuses the two, and we are not surprised by the disarming continuities that doom all these prehistoric fictions to be repeated. Furthermore, the iconography tells its own tale and does not keep up with the rapid evolution of knowledge.

Boitard's "fossil man" with his blasphemous reputation had been circulating in the press in England and Spain since the 1840s, but Susemihl's engraving launched its second and bolder career through the plagiarism of popularizers. After 1860, under the impact of official recognition of prehistoric science followed by the bitter controversies sparked by Darwin's theories about the transformation of species, this image revived the traditional idea of the intermediate and now missing link, which both connected and divided present-day man from the lost worlds of paleontology. In the absence of fossils, this was an epiphany. It compelled conviction and filled the gap. However, in a reversal of fortune, the popularizers soon forgot the circumstances in which it was produced. In their accounts, the reconstruction by Boitard and Susemihl became the living approximation of *Pithecanthropus*, the mute "pithecoid man" of the Tertiary age postulated by Ernst Haeckel, the naturalist from Jena, in 1868. It was attributed to him until the 1920s in the publications of Paul Rivet and René Verneau, professors at the Muséum national.

Haeckel would probably not have refuted this affiliation. An eloquent and pugnacious disciple of Darwin, he was also convinced that there was "such an analogy" between the "first of the anthropoid apes" and the "last of mankind" (meaning the contemporary "savages" of Africa and Oceania), "that it does not require a great mental effort to imagine an intermediate type, an approximate and probable portrait of the primitive man or ape-man."[38] In his publications, Haeckel was to demonstrate this in a series of fanciful drawings symbolizing the hierarchy of the species from the mandrill to the "Mediterranean" man—the finished type of beauty and intelligence—in order to relativize human insularity and the nobility of his creation (122, P. 183). He would find an even better interpreter of his evolutionist views in the work of Gabriel von Max, a painter friend and skilled anatomist very interested in primates, and a gifted draughtsman (164, 165, 166). Max executed numerous studies and paintings of his ape "family."[39] He also gave material form to Haeckel's intuition by depicting a reasoning being: the pithecanthropus. Some of his sketches of our proto-human ancestors opted for gracefulness of shape, others for corpulence. The most remarkable of them are the most disconcerting. They showcase the perfect hybrid. The animal fades away as humanity hesitantly appears in its anxious gaze. The pithecanthropus is not only a late avatar of the ape-man of the Enlightenment. From now on he would appear in new dimensions, those of genealogy and interiority. Haeckel saw in "cretins" and "microcephalic" individuals stricken by arrested cerebral development, living analogues of the original man. Von Max, on the other hand, challenged degeneration. He captured visually the node in time between bridge and separation, limit and opening, in which man tore himself away from his brutish straitjacket and developed a conscience. This is the fate of the missing link, wavering between two worlds, scarcely finding a balance, with the result that he must, by his behaviour, oscillate constantly between muscle on one side and reflective thought on the other. For some artists, the monogamous family with its male protector had already been instituted. Gabriel von Max's pithecanthropus, the troglodyte, sets up a household (4, P. 23), as will man's "precursor," the inventor of "Work" by František Kupka. On the other hand, the violent/violator great ape never reaches the introspective stage. Unable to visualize the future, he gives in to his instinct (167, 168). In a later edition, Boitard's fossil man drinks the blood of a hyena …

The gorilla, the "giant of the quadrumanes," was unknown to scientists before 1847. In a long monograph

37 Pierre Boitard, "Le chimpansé," *Musée des familles – Lectures du soir*, vol. 3 (1835–1836), p. 126–128.

38 Ernst Haeckel, *Histoire de la création des êtres organisés d'après les lois naturelles*, 2nd ed., trans. Charles Letourneau (Paris: Reinwald, 1877), p. 615.

39 See *Gabriel von Max: Malerstar, Darwinist, Spiritist*, exh. cat., ed. Karin Althaus (Munich: Hirmer, 2010).

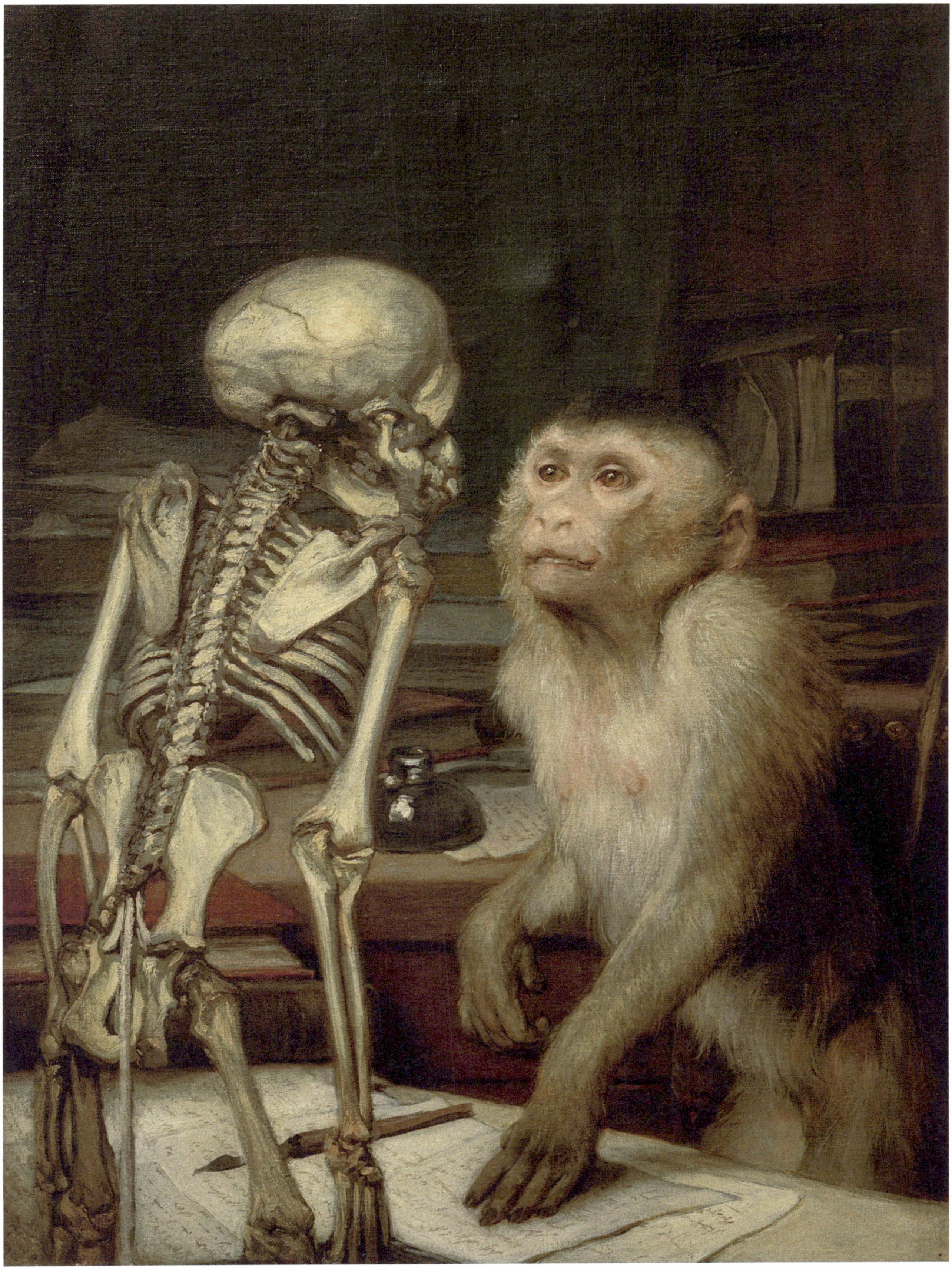

164
Gabriel von Max,
Monkey Before Skeleton,
about 1900
Oil on canvas, 61 × 44.5 cm
Private collection

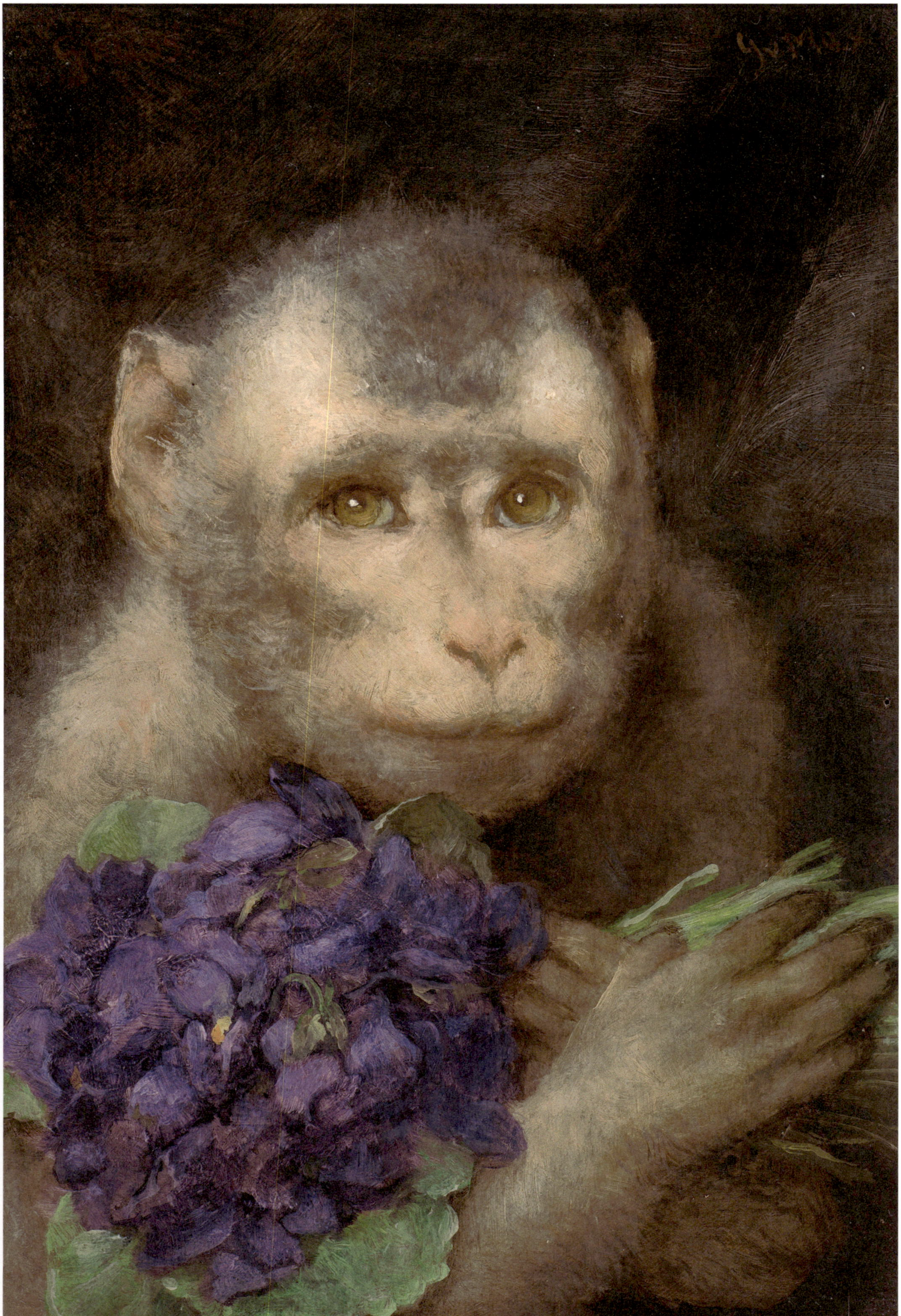

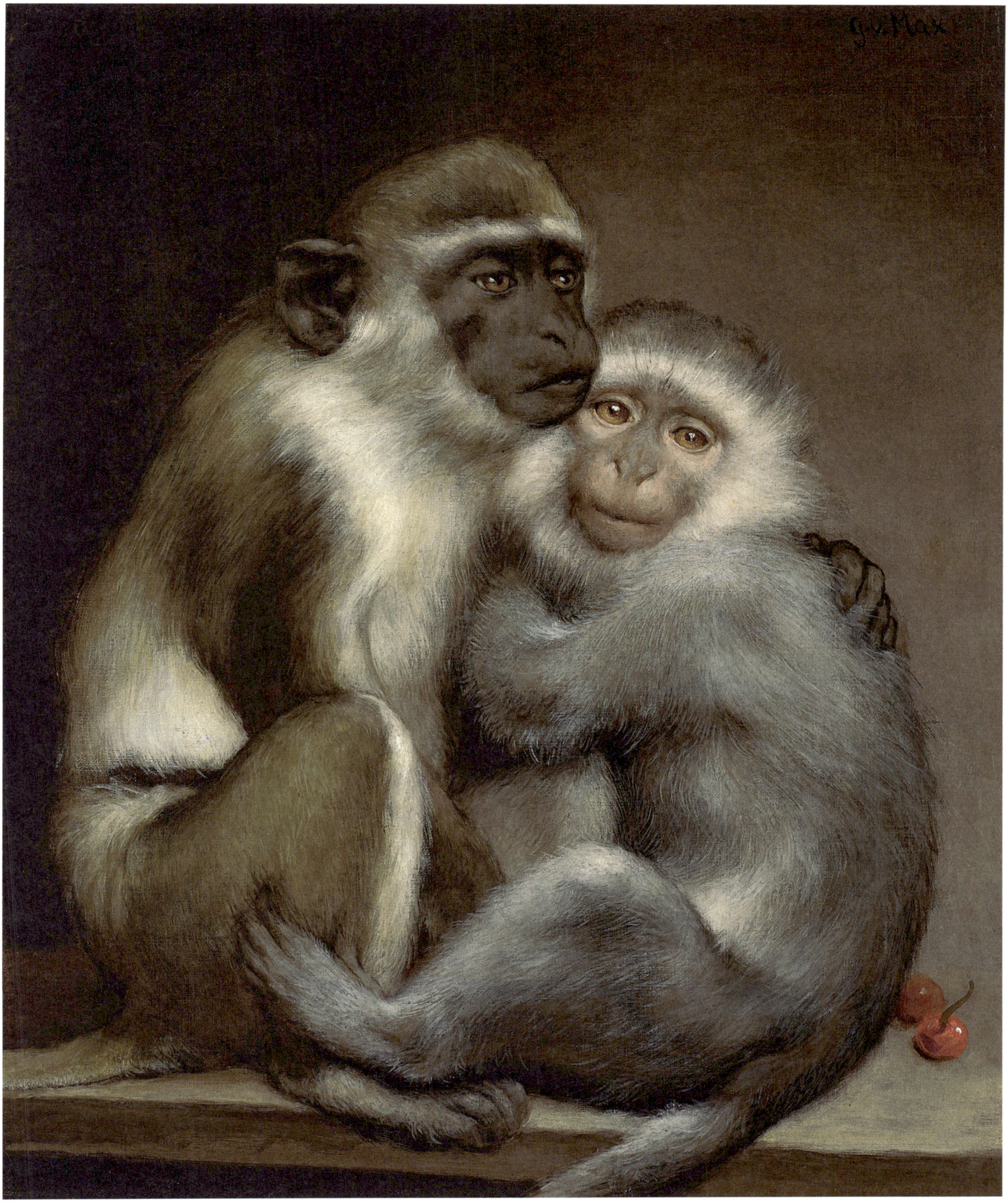

165
Gabriel von Max,
Gruss (Monkey with a Bouquet, a Posy of Pansies), 1901–1915
Oil on wood, 24 × 16 cm
The Jack Daulton Collection

166
Gabriel von Max,
Abelard and Heloise,
after 1900
Oil on canvas, 41 × 36 cm
The Jack Daulton Collection

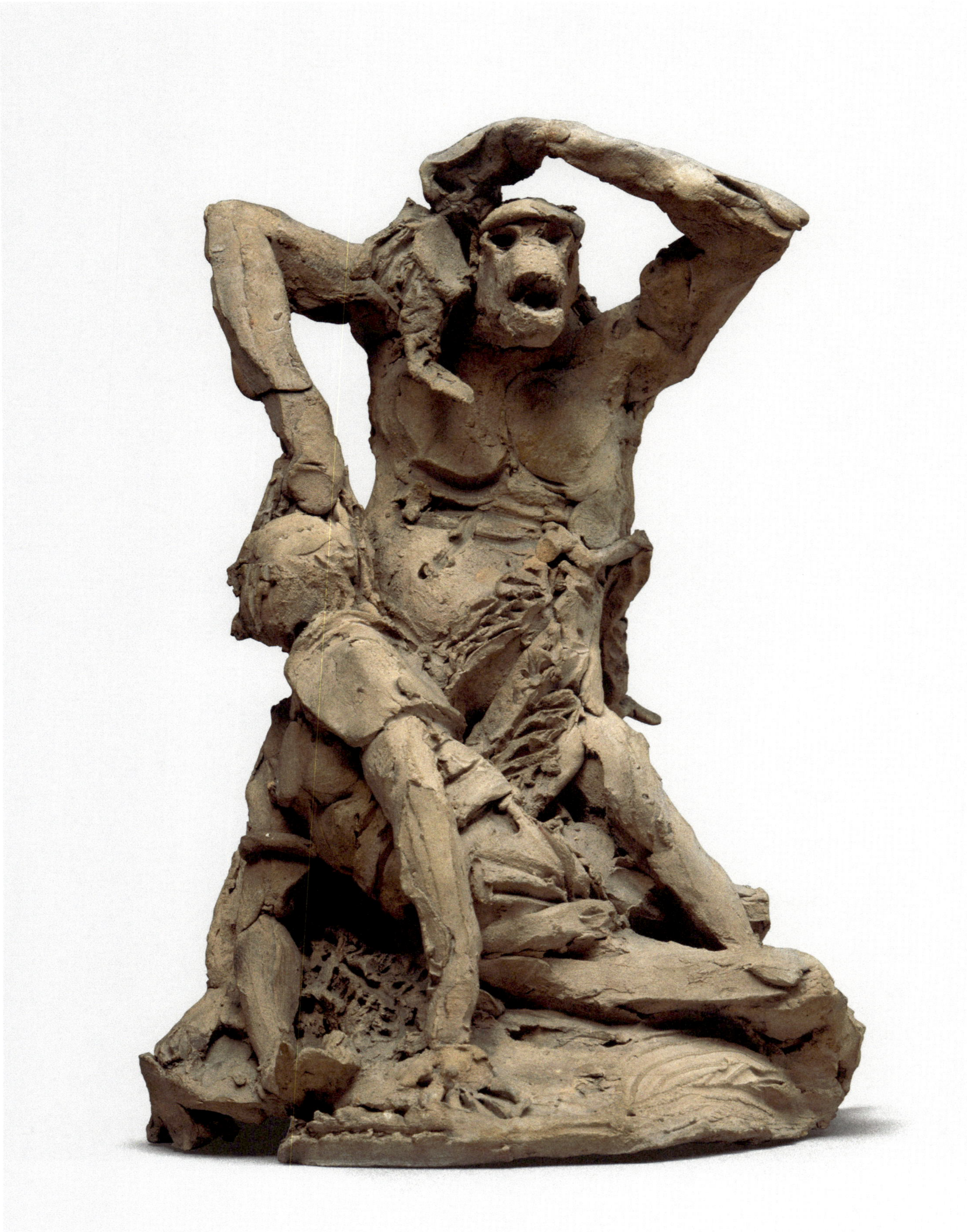

167
Emmanuel Fremiet,
Gorilla Dragging a Warrior by the Hair, about 1859
Terracotta bozzetto,
24 × 17.4 × 15.3 cm
Paris, Musée d'Orsay

168
Emmanuel Fremiet,
Orangutan Strangling a Savage of Borneo, 1893
Polychromed plaster,
58 × 55.5 × 27 cm
Dijon, Musée des Beaux-Arts

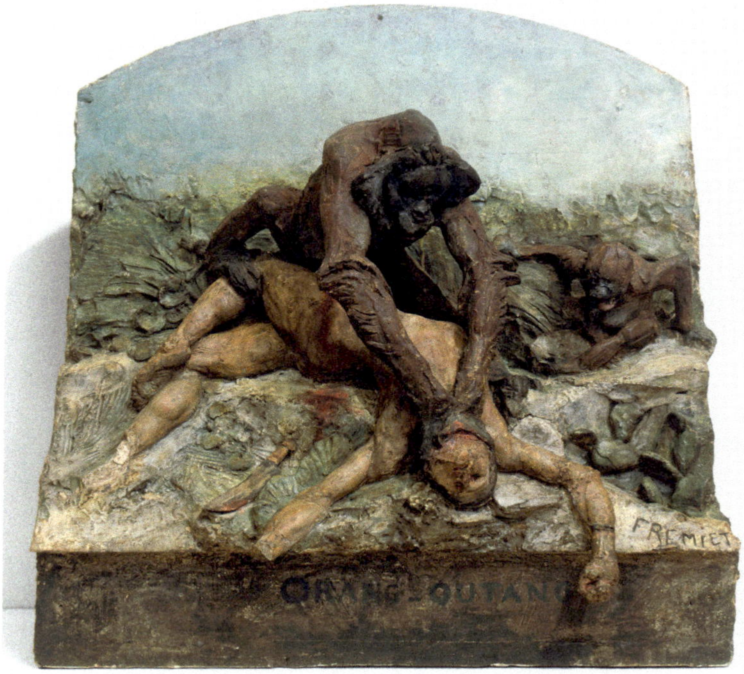

devoted to the creature in 1861, Professor Isidore Geoffroy Saint-Hilaire[40] was still refuting travellers' tales about the murderous fury and lustfulness of the "gigantic ape." And yet illustrations that were circulating, taken from Paul Du Chaillu's accounts of exploration in Gabon published that same year, very soon implied even worse connotations. Du Chaillu compared the gorilla to an enraged "monster." When he goes on the offensive, he says, his grin and his enormous fangs give "his hideous face a diabolical expression": "I can guarantee that no description could give an adequate idea of the horror inspired by its appearance, the ferocity of its attack and the implacable wickedness of its nature!"[41] Paul Du Chaillu was not a Darwinist. Hardly had he been discovered than the gorilla began a slow descent into hell to become, in the words by the Abbé d'Aubignac[42] to describe the he-goat of Satan, "hieroglyphic for all filthiness." The sculptor Emmanuel Fremiet, Professor of Drawing at the Muséum in Paris, confirmed this hyperbole when he submitted the first version of a baroque work of scabrous realism variously entitled *Gorilla Carrying Off a Woman* (or *a Negress*[43]) (170) to the Paris Salon in 1859. "Bad drama," wrote Baudelaire,[44] which seeks to arouse a dubious feeling of terror and "priapic curiosity." The sculpture was rejected but the intention persisted. Fremiet's group was executed, possibly in 1854, modelled after an adult male specimen that had arrived two years before pickled in alcohol at the Jardin des Plantes. As well as the allusion to rape denounced by Baudelaire, tyrannical aggressiveness is conveyed by the tensed body, the menacing face and the heavy carved flint weapon in its free hand. The "orang-outang," as Baudelaire called it, is thus seen as "both more and less than a man." From this indecipherable mixture, the illustrators of adventure novels would dwell on the stereotype of "the most ferocious killer in the world," uncontrollable, so that the rape of young women appears nothing out of the ordinary.[45]

The myth of the ultimate ape is of ancient origin. Despite its polysemy and the obvious impact of the Darwinian concepts, the image of King Kong (169) and his emotional drama are scarcely surprising.[46] King of a prehistoric jungle, faithless and lawless, master of the elements, Kong, yielding to impulse, contracts a doomed alliance and defies what is forbidden. He thus breaks down the sacred barrier that separates the social being from the savage, culture from hostile nature, and finds himself bound hand and foot like a madman and thrust into the heart of the urban theatre. The ape must die so that the human can come. However, through the atavism of heredity, the ape is still with us. He can always resurface in our actions. This idea has been and remains a powerful motor for fiction, widely exploited by humorists and artists. Simply through confrontation, the man and the ape determine each other. We seem to see in one the metaphysical anti-nature of the other. A useless effort. Hence the strange fascination of contemporary artists with the forced cohabitation of two incompatible beings in the unity of a shared nature. The ape and the man facing each other, the two faces of Janus, the god of transitions, endings and beginnings.

40 Geoffroy Saint-Hilaire, "Description des mammifères nouveaux . . . ," p. 56ff.

41 Paul Du Chaillu, *Voyages et aventures dans l'Afrique équatoriale* (Paris: Michel Lévy, 1863), p. 310 and 392.

42 Hédelin, *Des satyres brutes, monstres et démons*, p. 140.

43 See Philippe Dagen, "Images et légendes de la préhistoire," in *Vénus et Caïn. Figures de la préhistoire, 1830–1930*, exh. cat., ed. Hélène Lafont-Couturier, Philippe Dagen, *et al.* (Bordeaux: Musée d'Aquitaine; Paris: Réunion des musées nationaux, 2003), p. 24–25 and 33.

44 Charles Baudelaire, *Salon de 1859*, in Charles Baudelaire, *Œuvres complètes*, ed. Marcel A. Ruff (Paris: Seuil, 1968), p. 421.

45 See Heuvelmans, *Les Bêtes humaines d'Afrique*, ills. 28–29.

46 See René Chateau and Marielle de Lesseps, *King Kong Story* (Paris: Éditions René Chateau, 1976); and David Annan, *King Kong. Les singes au cinéma* (Paris: Éditions Marc Minoustchine, 1976).

APES IN OUR IMAGE

MARCOS UZAL

In our collective psyche, apes stand at the borderline between animals and humans, the wild beast and the civilized being. The first great simian hero in the history of film, the giant gorilla in *King Kong* (169), embodies this ambivalence in every respect. He first appears as a prehistoric monster living on an island still populated by dinosaurs, until his encounter with civilization in the person of a woman awakes in him feelings and desires that will be his downfall. Described by the explorers who discover him as being "neither an animal nor a man," *King Kong* epitomizes the tragedy of the animal overcome by human emotions. Since then, films have rarely depicted the ape as it is but rather in terms of this fascinating ambiguity, either amusing or terrifying, which give it an inconclusive status between beast and human. It is thought of as a cousin, taking us back to our animal origins and presenting a fascinating reflection, whether disturbing or comical, of the human condition (167, 168, 170). We attribute to it our impulses, desires and affections in order to put our own animality into perspective.

In films, therefore, apes are naturally anthropomorphic. There is the child ape, like the one dressed up in a little sailor suit in Buster Keaton's *The Cameraman* (1928), the archetype of which is Tarzan's Cheeta: clever little beasts that amuse us because they behave like kids. At the other end of the fantasy we find the rebellious ape which, becoming aware of its "subhuman" or "enslaved" condition, tries to dominate in its turn—this is the ape that wants to be a man, as the orangutan Louie sings in Walt Disney's version of *The Jungle Book* (1967). This idea is taken to its extreme in the adaptations and spin-offs of Pierre Boulle's *Planet of the Apes*, where humans see their domination threatened by a species of simians capable of waging war and enslaving them. The first adaptation, directed by Franklin Schaffner in 1968, is based on a simple reversal of the dominated and the dominant, whereby humans undergo the suffering they have inflicted on other species, in a society peopled by ape-men wearing clothes and standing upright. The recent film *Rise of the Planet of the Apes* (Rupert Wyatt, 2011) is less concerned with this satirical approach; it is, rather, a reflection on the dangers of science and of advances in genetics, whereby man is overwhelmed by the rapid evolution of a new species. In films ever since *King Kong*, apes have been the animals capable of passionate feelings about human beings. In *Monkey Shines* by George Romero (1988), for example, a gifted capuchin monkey becomes increasingly possessive about her master, a quadriplegic who has trained her to help him, and is dangerously aggressive towards anyone who approaches him. Nagisa Oshima's film *Max, mon amour* (1986) is a satirical take on a romance between a woman (Charlotte Rampling) and a great ape, a sort of comical version of *King Kong*.

The anthropomorphic nature of cinematic apes is apparent not only in the screenplays but also in the way the apes are portrayed. From shaggy costumes to the most complex digital techniques, by way of animation and assembly, they resemble men, even in their gestures

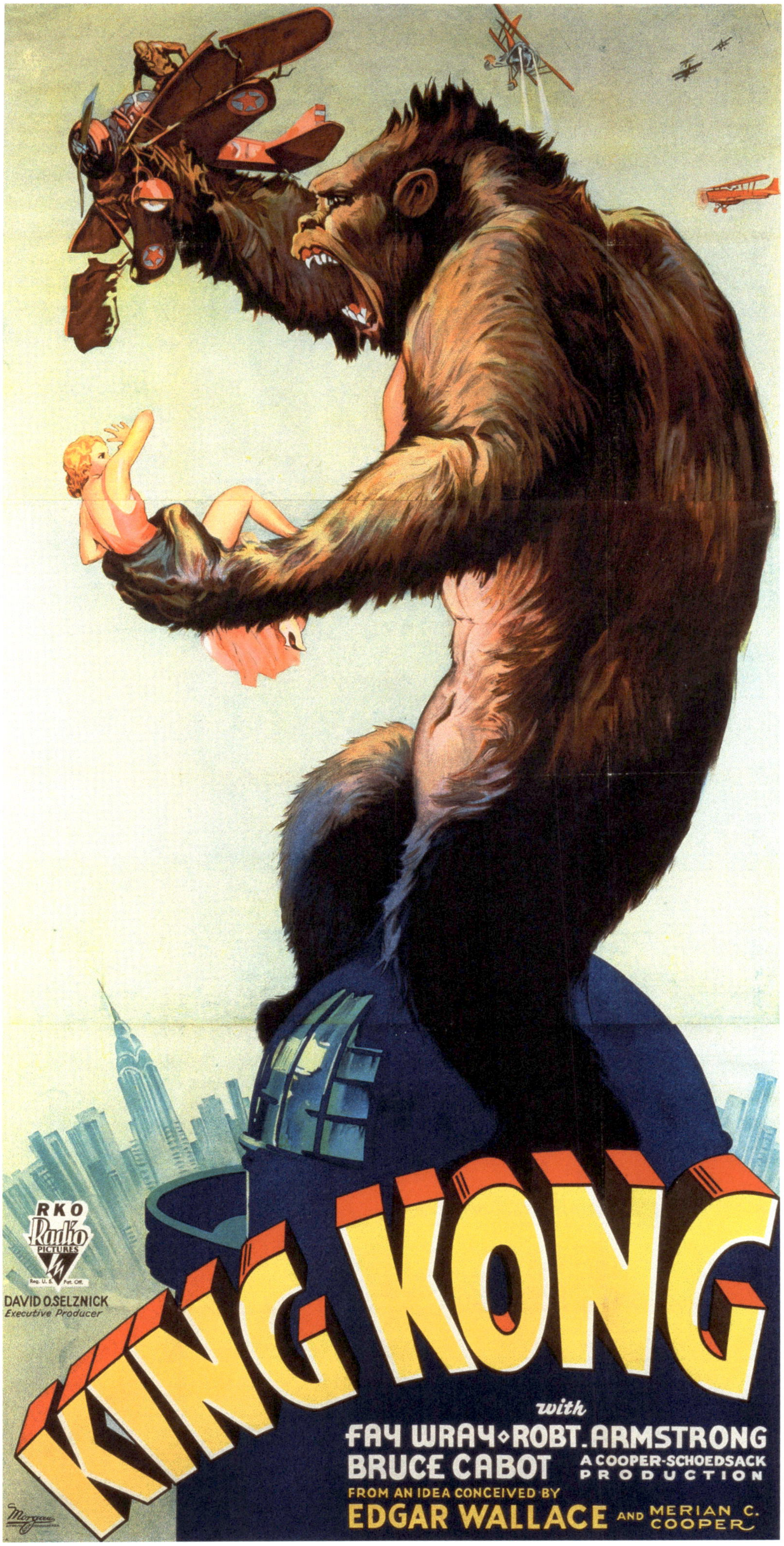

169
Merian C. Cooper,
Ernest B. Schoedsack,
King Kong
United States, 1933
Full-length silent film
in black and white

and postures. In this respect, the transformation of film technology is in itself a small Darwinian phenomenon, as the bodies and movements of these reinvented animals continue to evolve: the simian creatures in films from the 1930s are very different from those of the twenty-first century. But man remains the standard. King Kong's movements, for example, created using figurines of different sizes through the stop-motion technique (the image-by-image animation of three-dimensional objects), were not based on those of a gorilla but came from a mixture of human and animal movements that made King Kong an unclassifiable hybrid monster. Today, actors in monkey suits have been replaced by animatronic technology (robotized creatures) and above all by motion capture, which consists of recording the movements and expressions of a real actor (using numerous markers placed on his or her face and body) and retranscribing them onto a virtual creation. In every era, regardless of technique, the ape's role in film can be basically summed up as follows: man either imitates the ape or requires the ape to imitate him, blurring the logic of regression and evolution.

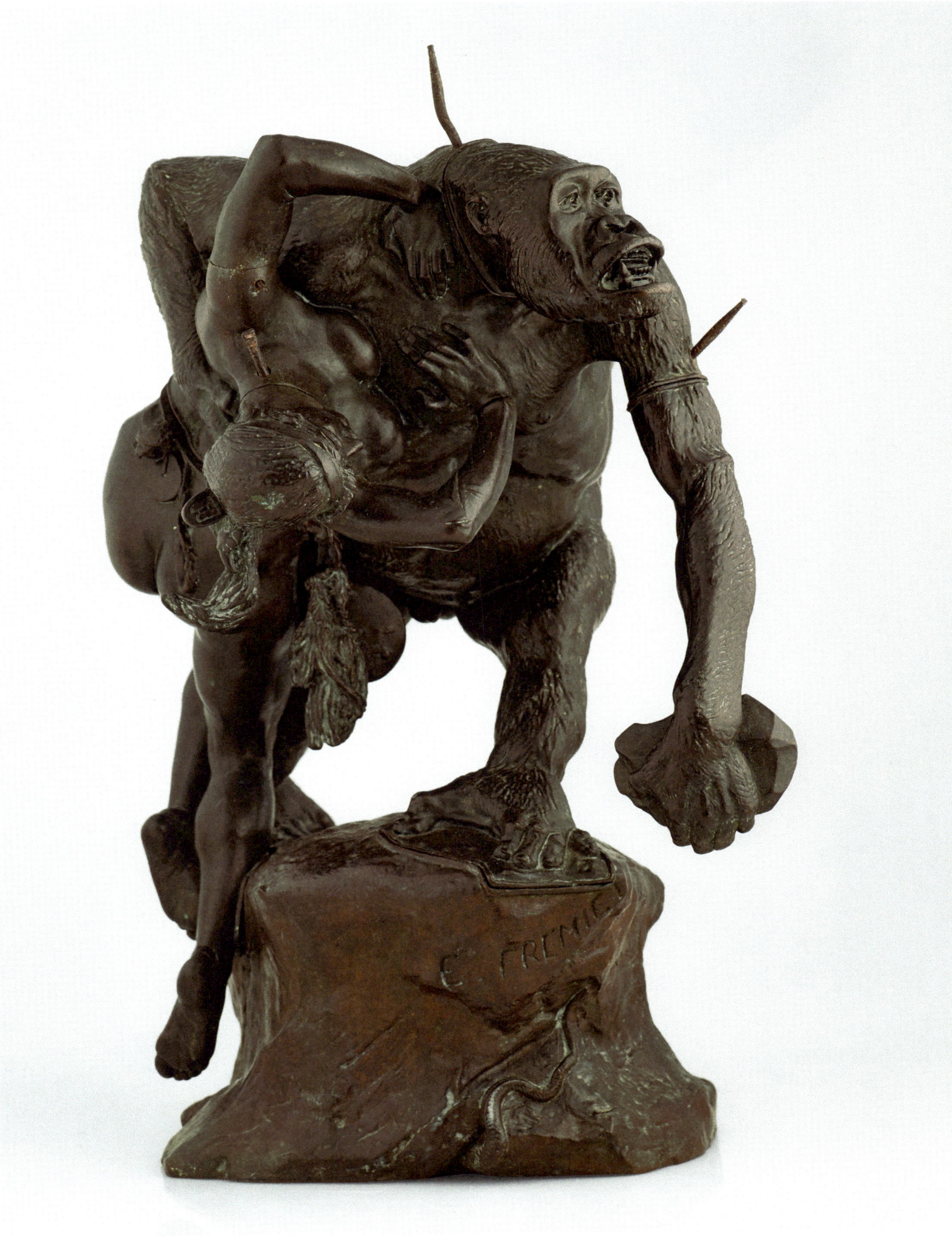

170
Emmanuel Fremiet,
Gorilla Carrying Off a Woman, 1893
Master model, bronze,
45 × 26 × 40 cm
Dijon, Musée des Beaux-Arts

VI

THE BEAST WITHIN: HYBRIDS AND CHIMERAS

BARBARA LARSON

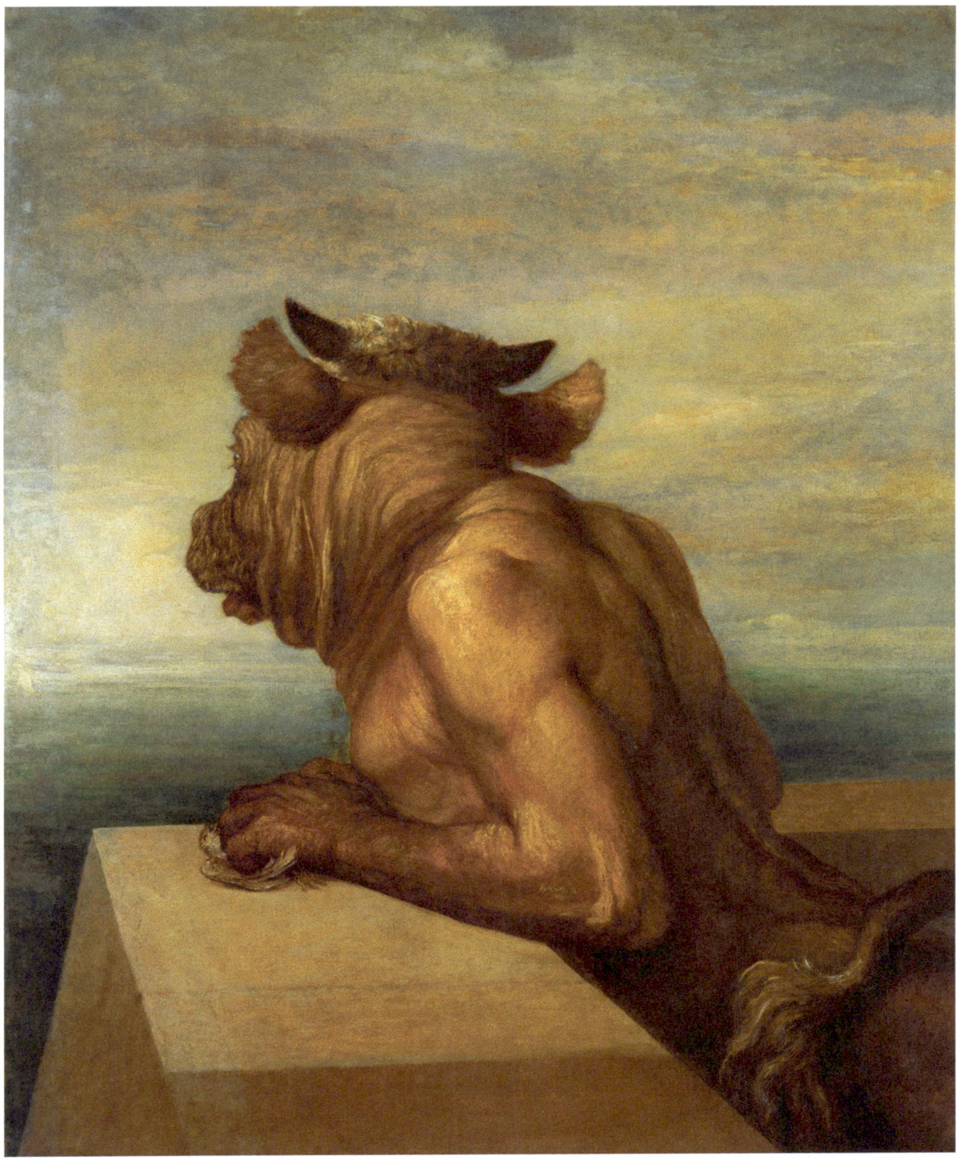

171
George Frederic Watts,
The Minotaur, 1885
Oil on canvas, 118.1 × 94.5 cm
London, Tate

The proliferation of hybrid creatures in the work of artists in the last quarter of the nineteenth century and beyond was an imaginative reaction to controversies circulating over evolutionism. In addition to the unsurprising ape-men, these creatures ranged from the historically familiar—centaurs, mermaids and sphinxes—to the unusual and hypothetical (frog-rabbits and flowers with human heads), to those that evoked recent changes to the fossil record and the search for missing links (winged reptiles). Though they responded to recent developments in science, artists' image sources were mined from Romantic literature, which sketched out early-nineteenth-century considerations of the workings of the unconscious through nightmarish creatures; ancient folk tales with grotesque beings that warned of dangers lurking in unknown corners of nature; classical mythology with its Dionysian side of libidinal fauns and unruly centaurs; bestiaries wherein shrewd monsters were theological symbols of sin; and even the current science fiction of Jules Verne or H. G. Wells (a former student of Thomas Huxley, friend and supporter of Darwin), with their mysterious gigantic and threatening creatures that had somehow managed to escape time itself. (In the 1860s, the recent confirmation of the existence of gorillas by zoologist and explorer Paul Du Chaillu, who claimed to be the first European to see them in the wild, seemed to demonstrate to many that enormous, knowing creatures might still exist in unexplored nature.[1])

The artists who took up composite creatures came from countries represented by a range of evolutionary theorists, not all of whom embraced all aspects of Darwinism.[2] And they were also inspired by influential philosophers such as Eduard von Hartmann, Schopenhauer, Nietzsche and Élisée Reclus, who had shown an interest in ways in which evolutionism supported their view of humanity. But the public identification of Darwin with the idea of descent from apes, the unplanned, even random variations of life forms, and the principle of competition provided imaginative fodder for the artists who created hybrid beings. These artists considered the grotesque within (skeletal and mental similarities with other creatures, most notably the ape); the crucial unknown (missing links or that which lurks in the depths of the ocean); romance or sexuality (now encoded with new meanings); the struggle to survive; and origins (all the way back to the dissolution of kingdoms—the plant world and the animal world). The hybrids point just as often to the dual nature of mankind (the irrational and dangerous as well as the intellectual and empathetic) as they do to transitional forms. Given the lengthy history of evolution itself, the most immediate hybrid reference that implicated mankind in a radical reversal of his formerly understood dominion over nature is the reference to the ape, previously considered an unrelated species, but now overshadowing humans as a close cousin and, as Huxley noted publicly more than a decade before Darwin did, intimately involved in the history of human evolution.[3]

APE-MEN AND ATAVISM

Though Darwin avoided the question of human evolution in *On the Origin of Species by Means of Natural Selection* of 1859, Huxley not only debated it, but included it along with a now famous chart demonstrating the similarities between human and ape skeletons in his 1863 *Evidence as to Man's Place in Nature* (118, P. 169). Humans were most closely related to the gorilla or chimpanzee. The prominent German Darwin follower Ernst Haeckel, who agreed with the general subject of ape ancestry, discussed it in his *Generelle Morphologie der Organismen* of 1866. Darwin only addressed the subject directly in *The Descent of Man* of 1871 in which he acknowledged that man had descended from a "hairy, tailed quadruped, probably arboreal in its habits."[4] Although many understood the subject as implied in *Origin of Species*, it would be nearly two decades before artists took up human evolutionism as a valid subject. Indeed, Paul Meyerheim's *A Darwinian Prehistoric Social Party* of 1865 with its drunken debonair monkeys has more in

1 Paul Du Chaillu, *Explorations and Adventures in Equatorial Africa* (London: John Murray, 1861). See also the essay in this catalogue, p. 227.

2 Huxley, for example, despite his courage in being the public face of Darwinism in England, did not believe that evolution had to be a gradual process as did Darwin and had reservations about Darwin's central mechanism of natural selection. The French of this period were often more interested in reviving the early-nineteenth-century ideas of their native evolutionist, Jean-Baptiste Lamarck. And in Germany, Ernst Haeckel, as much of a Darwinist as Huxley, also diverged from Darwin significantly, in his case through a monist belief that spirit was immanent in matter.

3 In 1860, Huxley engaged in a now famous debate over evolutionism at Oxford. He was supposedly asked by Bishop Samuel Wilberforce whether it was through his grandmother or his grandfather that he claimed descent from a monkey.

4 Charles Darwin, *The Descent of Man, and Selection in Relation to Sex*, 2nd ed. (London: John Murray, 1874), p. 391.

5 That Cormon was a reader of Darwin is noted in Émile Michel, "Les peintures décoratives de M. Cormon au Muséum," *Revue de l'art ancien et moderne*, no. 3 (January 10, 1898), p. 3.

6 The Salon livret for the exhibition included the first three lines of Hugo's poem: "Lorsqu'avec ses enfants couverts de peaux de bêtes / Échevelé, livide au milieu des tempêtes, / Caïn se fut enfui de devant Jéhovah."

7 Darwin, *The Descent of Man*, p. 619 and 125.

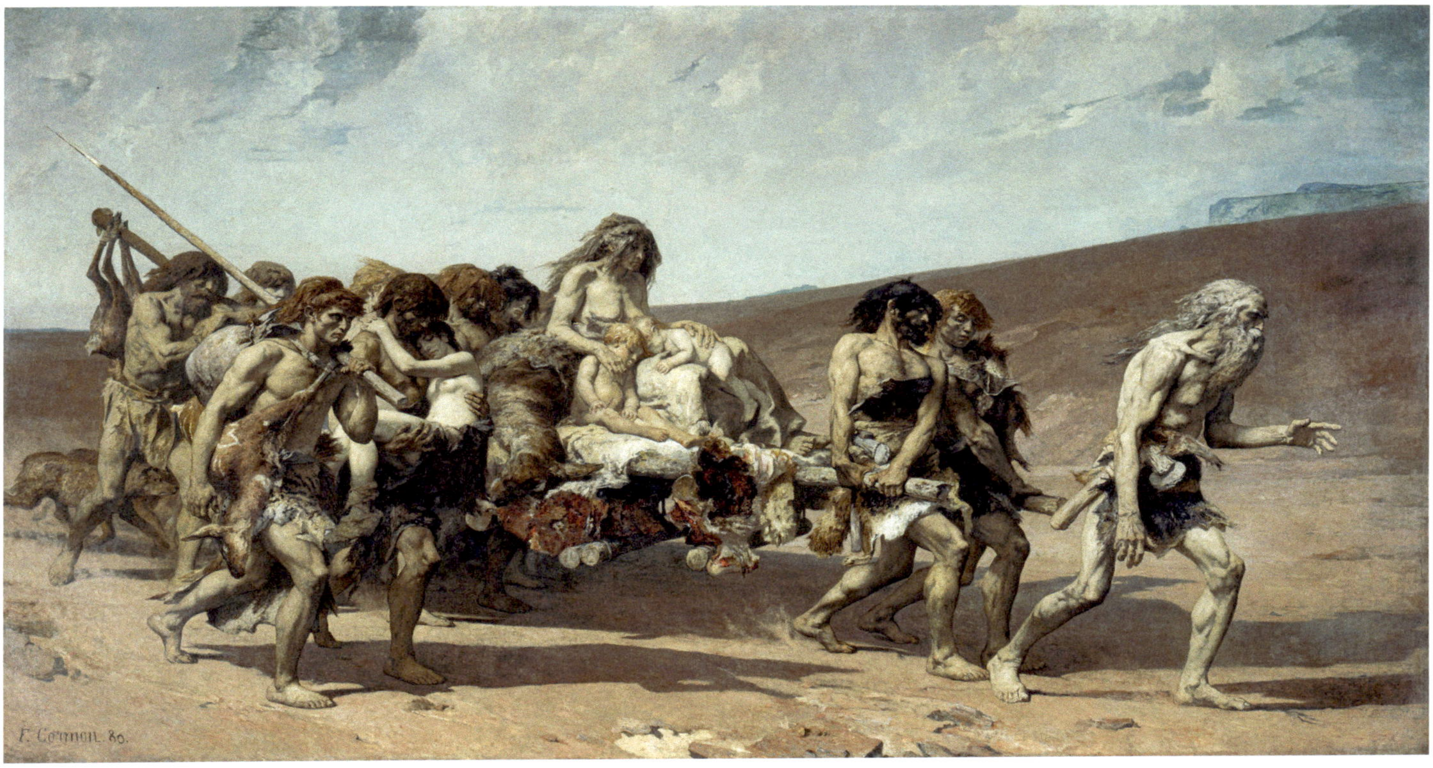

common with eighteenth-century singeries than with evolutionism. The decade of the 1870s was replete with caricatures of evolutionists as monkey-men composites (153, 155, 156, 157), and there was the occasional attempt to represent prehistoric anthropoids as ape-men in scientific illustrations, but these were invariably without any commitment to the real possibility of evolutionism. However, widely attended scientific exhibitions—such as the one in the Pavilion of Anthropological Sciences at the 1878 Universal Exposition in Paris, with its visible fossils, Neanderthal skulls, prehistoric tools, and ape and human skeletons—made the seriousness of the subject unavoidable.

The French artist Fernand Cormon, who read Darwin, brought the human animal into public view in 1880 at the Paris Salon with his huge canvas *Cain*; here, the world's first biblical murderer is depicted as a loping, bent-kneed, long-limbed, ape-like figure running from his crime (172).[5] Inspired by Victor Hugo's poem "Conscience" published in 1859, Cormon's *Cain* attempts to escape the all-seeing eye of God, unaware that it is his own conscience from which he cannot hide.[6] But with no visible reference to biblical history beyond the title, the subject of origins was removed from Judeo-Christian references in favour of the paleontological record, as patriarch Cain leads his pelt-wearing, shaggy-haired descendants with their prehistoric tools through bleak terrain. The "ape within," now topical, brought about discussions not only of skeletal similarities but also of mental attributes. Of the former, Darwin's concluding words of *The Descent of Man* were, "Man with all of his noble qualities . . . still bears in his bodily frame the indelible stamp of his lowly origin," and of the latter he warned, there is "a struggle in man between his social instincts . . . and his lower, though momentarily stronger impulses or desires . . . as man has emerged from a state of barbarism within a comparatively recent period."[7] By 1880, the year *Cain* was first exhibited, pardons had been issued to many Communards who had been engaged in a fratricidal conflict in the streets of Paris in 1871 and subsequently exiled; total amnesty soon followed. The subject of human regression and aggression even against one's own (as was the case with Cain's murder of his brother) was current.

172 FIG.
Fernand Cormon,
Cain, 1880
Oil on canvas, 400 × 700 cm
Paris, Musée d'Orsay

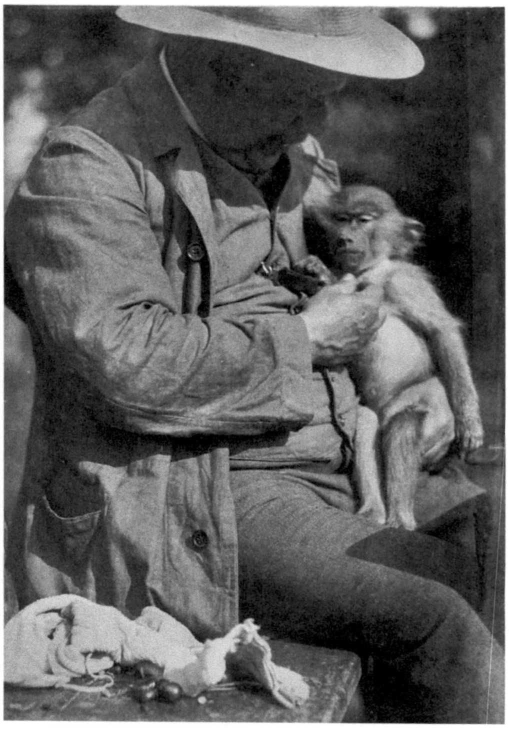

The German artist and evolutionary enthusiast Gabriel von Max, who bred monkeys (173) and had a museum-quality collection of animal skeletons and fossils, often painted amusing works of simian and human reversals that suggest the shared ability to reason. Such is the case with *Apes as Art Critics* (1889) and *Monkey Before Skeleton* (164), in which the creature appears to have just been taking notes. Max also painted what he thought of as a missing link based on newly discovered fossil remains, *Pithecanthropus alalus*, which he then presented as a gift to his friend Haeckel.[8] Here, a "civilizing" family unit, with a female anthropoid sitting cross-legged and a placid male standing nearly upright and content to watch over the female and child, verges into specifically human territory (4, P. 23). Also suggesting the origins of the family unit, the Czech artist František Kupka illustrated such a scene for the anarchist geographer Élisée Reclus's *L'Homme et la Terre* (1905–1908), a multi-volume series on history and geography, which contained many references to the evolutionary origins of social behaviours (231, P. 333).

Living atavistic throwbacks as "missing links" were also part of the search for scientific evidence of evolutionism, and circus impresarios promoted individuals with unusual genetic conditions that seemed to demonstrate this possibility. For example, "Krao," a young woman from Asia with a condition that rendered her hairy throughout her body, was exhibited internationally as "A Living Proof of Darwin's Theory of the Descent of Man." The science popularizer and Darwin follower Carus Sterne included a photo of a boy from India born with a tail in his *Evolving and Vanishing* of 1876. Scientists and anthropologists, including Darwin and Haeckel, tended to think of living tribal populations as remnants of past evolutionary history, not only given their physical aspects, but also in terms of their customs and use of tools and other objects. For example, Haeckel placed an African in a tree with other "catarrhines"—chimp, orangutan and gorilla—in his text *Anthropogenie*, implying close relationship with the apes (122).

Ape and human pairings were sometimes used to represent the battle for survival, as in Fremiet's *Orangutan Strangling a Savage of Borneo* (168), or the battle over a woman, as in his *Gorilla Carrying Off a Woman* (170), but the beast within becomes more seamless in Kupka's *Anthropoids: Fight for a Woman* (174). Here, the direct allusion to human evolutionism is clear. In this case, Kupka explores jealousy and male competition, conditions Darwin wrote about extensively, devoting the second volume of *The Descent* largely to sexual selection (competition for mates). The female in Kupka's work holds a flower, tying her to the long-held Western notion that women belong ultimately to nature and fecundity. Darwin believed human evolution and differences between the sexes owed a great deal to what he referred to as the "law of battle," necessary to drive off rivals. He stated the following: "The greater size, strength, courage, pugnacity, and energy of man, in comparison with woman were acquired during primeval times, and have subsequently been augmented, chiefly through the contests of rival males for the possession of females."[9] Ape imagery was often tied to the sex drive as the ultimate force behind the survival of species (more so than the intellect, as Schopenhauer pointed out), as is referenced in the Austrian Alfred Kubin's libidinal *The Ape* in which an ape overpowers a woman (207).

8 The painting was based on the recent discovery of "Java man" in Southeast Asia by one of Haeckel's followers, Eugène Dubois, who was intrigued by Haeckel's hunch that "the missing link" would be found in Asia. Though the fossils were designated *Pithecanthropus erectus*, Haeckel proclaimed it was the very anthropoid he had discussed as *Pithecanthropus alalus*, an ape-man lacking speech. See Ernst Haeckel, *The Last Link: Our Present Knowledge of the Descent of Man* (London: Adam and Charles Black, 1898), p. 26. Haeckel diverged from Darwin by positing that human evolution took place largely in Southeast Asia, not Africa. Margaret Richter has claimed Max was reading Darwin as early as the 1860s. See Margaret Richter, "Gabriel Max: The Artist, the Darwinist, the Spiritualist" (PhD diss., New York University, 1998), vol. 1, p. 59, vol. 2, p. 306. Haeckel is known to have visited Max's collections beginning in 1892.

9 Darwin, *The Descent of Man*, p. 609.

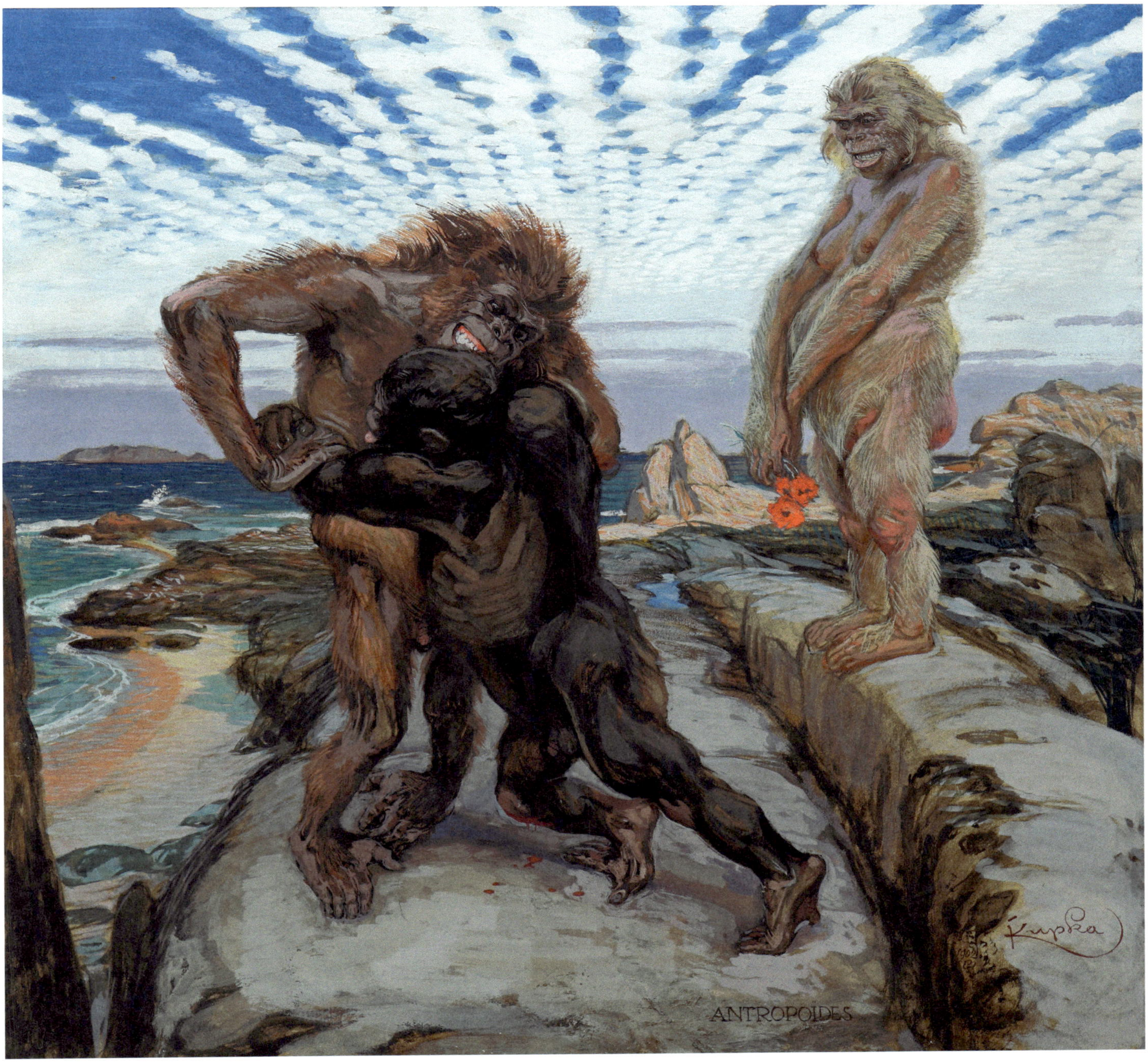

173 FIG.
Anonymous,
Portrait of Gabriel von Max Holding a Monkey, n.d.
Silver print mounted on cardboard, 17.7 × 12.3 cm
The Jack Daulton Collection

174
František Kupka,
Anthropoids: Fight for a Woman, 1902
Watercolour and gouache, 59.5 × 62 cm
Prague, Galerie Zdeněk Sklenář

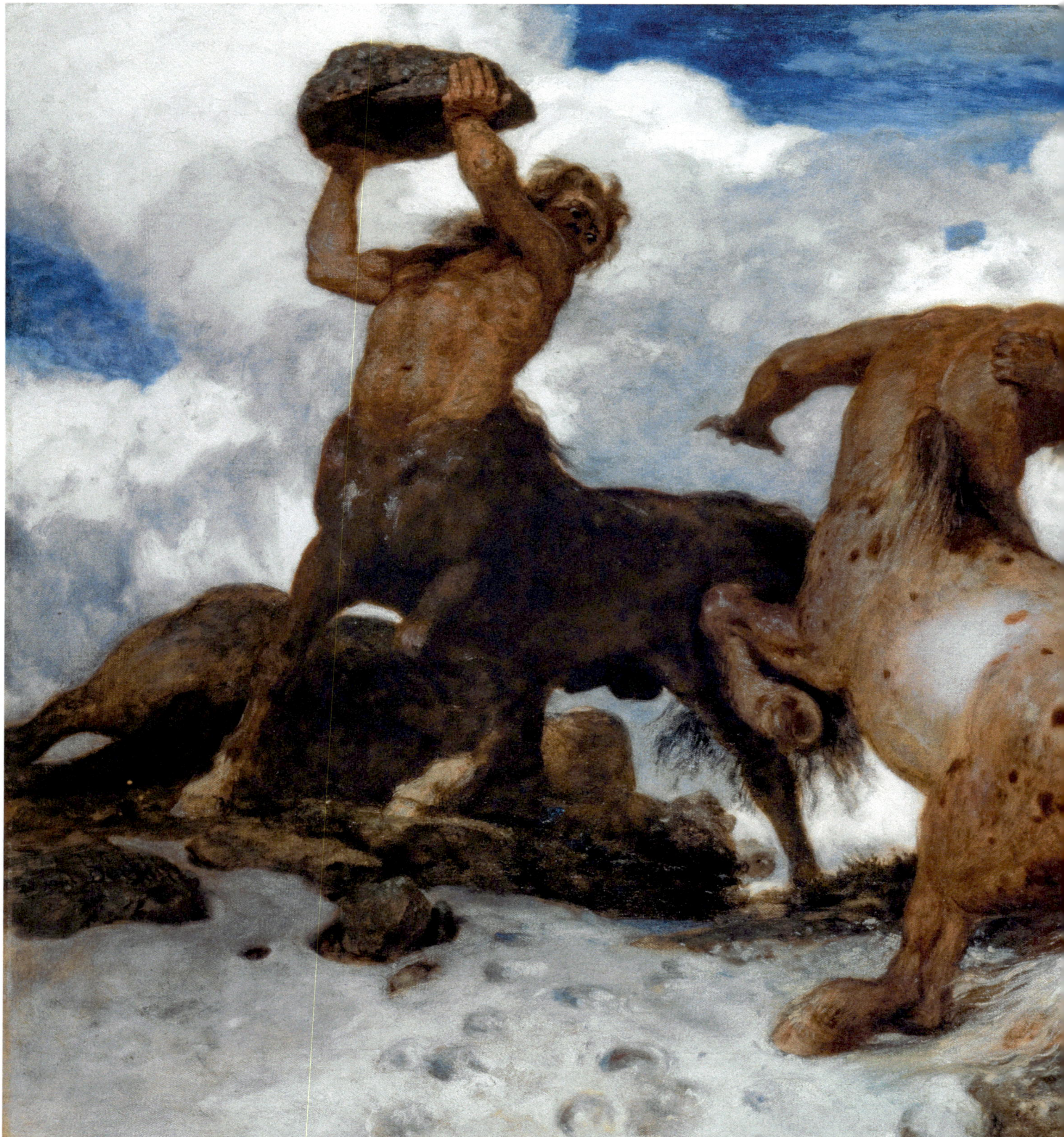

175
Arnold Böcklin,
The Battle of the Centaurs,
1872–1873
Oil on canvas, 104.2 × 194.3 cm
Basel, Kunstmuseum Basel

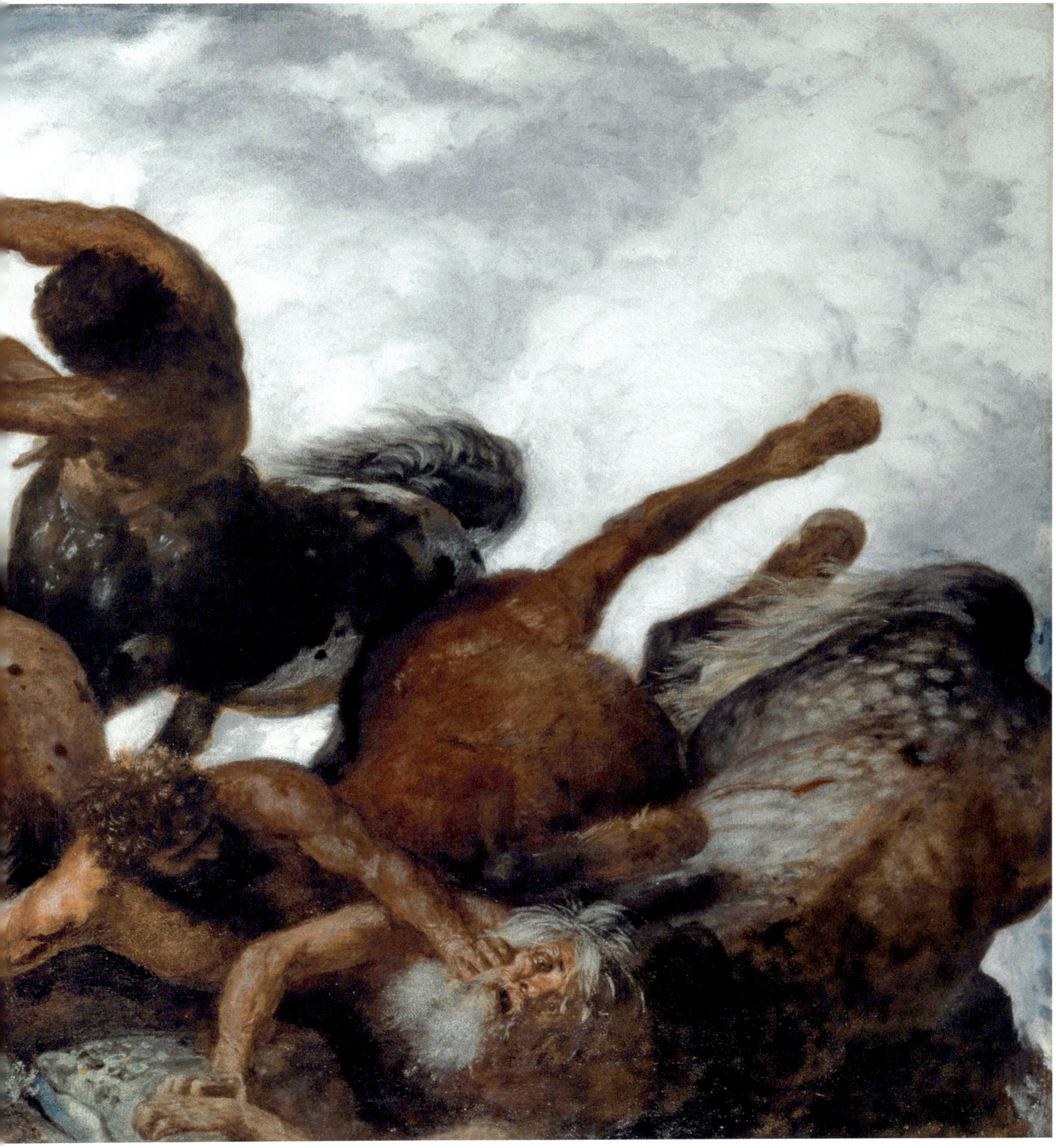

MYTHS, FOLK TALES AND BESTIARIES

Neither scientists nor artists had a perfect language to picture the fossil record of the evolutionary past with its many missing components and its mysterious impressions left in deep time. Scientists often used the analogy of myth in their writings, as was the case with the geologist Lyell in his influential *Principles of Geology* or Huxley who cited Ovid's *Metamorphosis* in the first page of *Man's Place in Nature* wherein "Ovid foreshadowed the discoveries of geologists . . . and though the quaint forms of Centaurs and Satyrs have an existence only in the realms of Art, creatures approaching man more nearly than they in essential structure, and yet as thoroughly brutal as the goat's or horse's half of the mythical compound, are now not only known, but notorious."[10]

While classical subjects remained deeply entrenched in conservative academic art and were referenced by illustrators who opposed evolutionism, Nietzsche's *Birth of Tragedy* of 1872 provided artists interested in evolutionism with a vital avenue to pursue the realm of myth: the inner dichotomy between the rational self and the irrational self (the human and the beast) could be expressed through remembering the different natures of Apollo (logic) and Dionysus (instinct), both sons of Zeus. Life embodies this perpetual dynamic. Centaurs and satyrs, popular during the Romantic period, took on new meaning in the age of Darwin.

The full title of Darwin's 1859 publication was *On the Origin of Species by Means of Natural Selection, or the Preservation of Favoured Races in the Struggle for Life*. The English sociologist Herbert Spencer first coined the phrase "survival of the fittest" out of Darwin's struggle and applied it to his own theories of social and economic pressures (*Principles of Biology*, 1864). It was a catchphrase that seemed to encapsulate many aspects of ongoing human behaviour from Western imperialism and capitalism to warfare in the West. The Swiss artist Arnold Böcklin, who had created calm, classical scenes when living in Italy before 1870, took up the subject of battle to the death in his Ovid-derived *Battle of the Centaurs*, created shortly after the Franco-Prussian War (in which he fought), at a time when evolutionary explanations were circulating regarding the reasons behind ongoing warfare (175). Set on a mountaintop or cliff, clouds hang low causing a confusing backdrop, suggestive of elemental forces. In another version of this scene, a black male fights against the centaurs, alluding to the influence of Haeckel and other scientists on the racial rungs of the evolutionary ladder. Centaurs are often portrayed as male and embody the idea of an incipient aggressive nature in a world of competition and struggle. Auguste Rodin, a former student of the animal artist Antoine-Louis Barye who specialized in scenes of fighting creatures (161), created aggressive centaurs for the first version of *Gates of Hell*.[11] The artist also depicted a solitary *Centauress*, a composite of various studies (176). In this case, Rodin, a long-time admirer of Michelangelo, may have had the Renaissance artist's unfinished *Slaves* in mind, which had come to symbolize the struggle to free spirit from matter. Indeed, an earlier title for this artwork was *Soul and Body*.[12] Paul Gsell in conversation with Rodin commented that *Centauress* represents "a monster . . . composed of this frightful pulling apart of the two natures . . . the soul whose ethereal impulses remain miserably captive in the vile body!"[13] The dual nature of this female is the spiritual self and the material self.

Male satyrs were often portrayed as engaged in battle as well, but often in competition for females, as found in the work of the German artist Franz von Stuck (*Fighting Fauns*, 1889, for example). Satyrs in Greek mythology were semi-human libidinal nature spirits with hooves, often with the haunches of goats. Böcklin frequently positioned them as closely observing a sleeping female (as in *Sleeping Diana*, 1877).

Usually satyrs, centaurs and other male fabulous composite creatures have human-like heads, which in the late nineteenth century symbolized mental processes potentially leading humans beyond their material natures. On occasion, however, artists represented the tragic state of material entrapment through a hybrid with a fully animal head, as the English painter George Frederic Watts did in *The Minotaur* (171). The minotaur is the mythic offspring of a bull and Pasiphae, who was married to the king of Crete—the treacherous liaison is ironic, for the king himself was the offspring of Europa and Zeus, who came to her in

10 Thomas Henry Huxley, *Evidence as to Man's Place in Nature* (London: Williams and Norgate, 1863), p. 1.

11 Rodin studied with Antoine-Louis Barye at the Jardin des Plantes in Paris where he made use of "pieces of animals from the amphitheaters." Rodin stated of Barye, "It was from him I learned most." Quoted from Henri Dujardin-Beaumetz, "Rodin's Reflections on Art," in *Auguste Rodin: Readings on His Life and Work*, ed. Albert Elsen (Englewood Cliffs, NJ: Prentice-Hall, 1965), p. 147.

12 Georges Grappe, *Catalogue du musée Rodin: I. Hôtel Biron. Essai de classement chronologique des œuvres d'Auguste Rodin* (Paris: Musée Rodin, 1944), p. 83.

13 Auguste Rodin, in Auguste Rodin, *Art: Conversations with Paul Gsell* [1911], trans. Jacques de Caso and Patricia B. Sanders (Berkeley: University of California Press, 1984), p. 74.

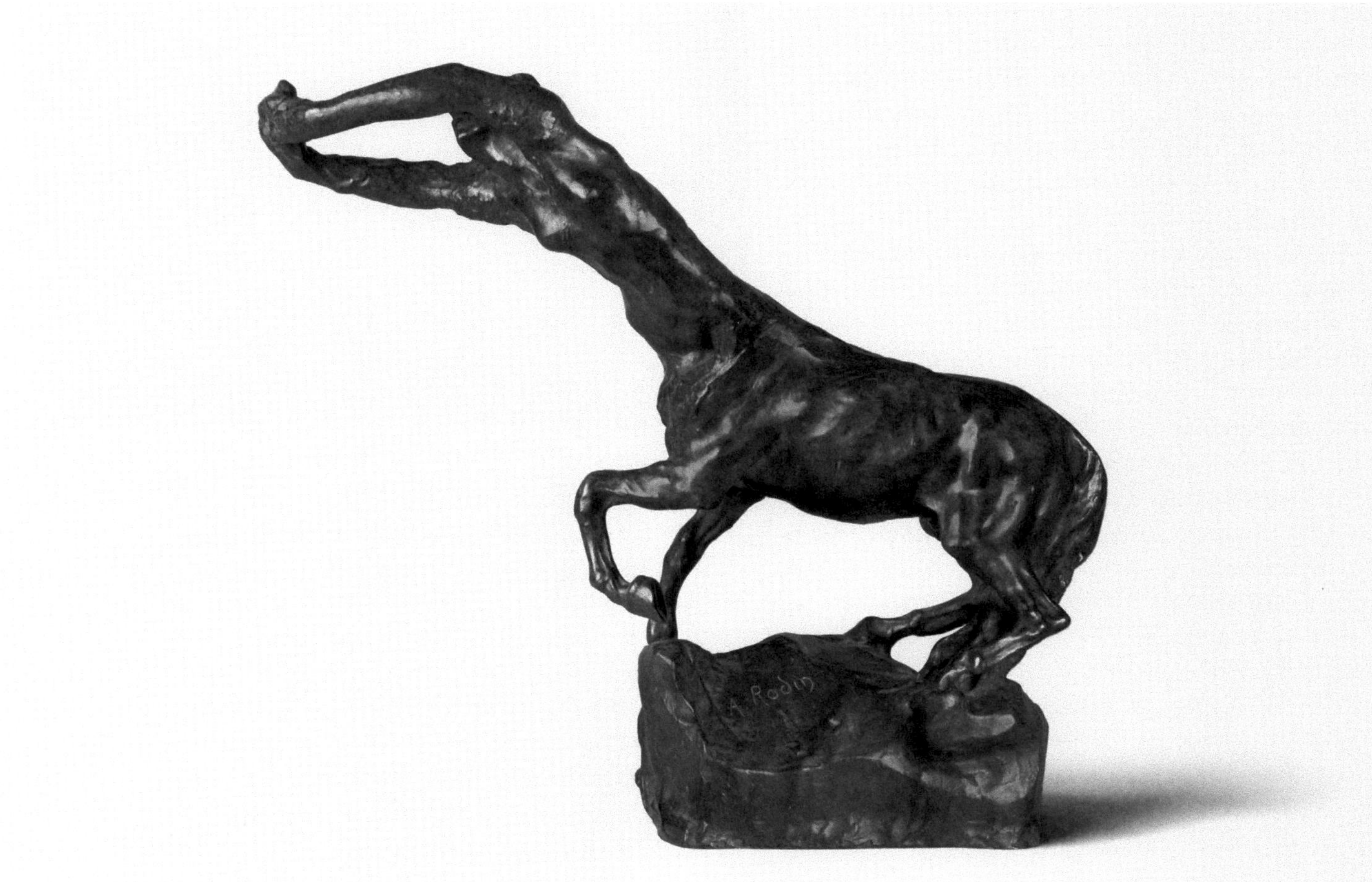

176
Auguste Rodin, Émile Godard (founder), *Centauress*, about 1887, 1969 (cast)
Bronze, lost-wax casting, 45.8 × 49 × 16.7 cm
Paris, Musée Rodin

the guise of a bull; the human-animal sexual act is thus repeated, with repercussions. Watts's minotaur gazes over a parapet of the labyrinth which encloses his dangerous nature in a space of rational confinement—he awaits the delivery of the humans he is to devour, but for the moment crushes a small bird he has instinctively captured.[14]

Sphinxes are composites often associated with women, usually fatal and sexual in nature. The sphinx is a chimera with a classical pedigree, threatening to kill Oedipus, king of Thebes, if he cannot answer the riddle she poses. Her feline body suggests her power and rapaciousness. While Paul Meyerheim's *The Jealous Lioness* of 1880 separates the realm of woman and feline physically if not by shared emotions (178), the sphinx makes them one and the same. Artists including the Belgian Fernand Khnopff and Von Stuck made the sphinx as libidinal a female as the satyr was a male. In Von Stuck's *The Kiss of the Sphinx* (1895), the hybrid brings her human companion to his knees with a devouring and deadly embrace. There is no possibility of Oedipus's rational response saving him from his fate here. Of his erotic mythology, Von Stuck stated, "In choosing my themes, I set out to treat only the purely human, the eternally valid, such as the relationship between man and woman. 'He' and 'she' are present in most of my paintings."[15] The animal nature of the intoxicating female was associated with Darwinian ideas on the overriding drive to procreate and sustain the species, which made any notion of romantic love obsolete.

Late-nineteenth-century artists were not the first to make use of composite monsters or threatening skeletal arrangements. Monsters that invade human domains were explored by the Romantics, most notably in literature. Authors, including Eduard von Hartmann, Charles Nodier, E.T.A. Hoffmann, Edgar Allan Poe, Mary Shelley, Jules Michelet and Heinrich Heine, were interested in the biological destabilization of body and mind or the legacy of nature's experimental forms. Francisco de Goya, Rodolphe Bresdin, Moritz von Schwind and Henry Fuseli mined the Romantic grotesque in art, examples of which might be produced out of fairy tales. In the case of Fuseli, his famous incubus (a lustful villain from folk tales) sitting atop a woman in *Nightmare* of 1781 was interpreted as ape-like by his evolutionist friend Erasmus

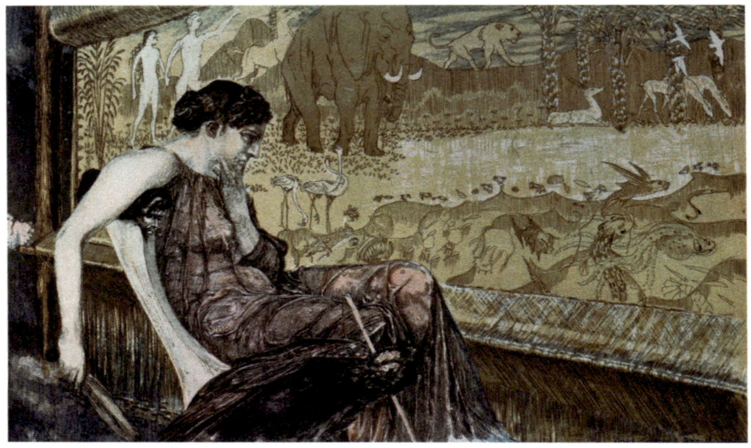

Darwin (grandfather of Charles Darwin), who composed verses along these lines that accompanied the related print.[16]

In the late nineteenth century, folk tales came into focus among evolutionary anthropologists like E. B. Tylor and John MacLennan, who followed a developmental approach of the mind and civilization and saw these superstitious beliefs as part of the "doctrine of survivals"—proof of former belief systems that carry forward into the present. On the other hand, Haeckel saw in the odd creatures of folk tales a past intuition of evolutionism itself. While German fairy tales are full of lustful, threatening male creatures, a taste in the late nineteenth century for female "nixies" led to an interpretation of these aquatic composites as part

14 While Watts stated that the two great ideas of mankind are gravitation and evolution, he did not discuss the type of evolutionism he had in mind (Mary Seton Watts, *George Frederic Watts: The Annals of an Artist's Life*, vol. 3: *His Writings* [London: Macmillan, 1912], p. 296). His large-scale canvas *Chaos* (1873–1882), with its human figures arising from water and mountains, has been interpreted as "Darwinian" by Barbara Bryant in *The Age of Rossetti, Burne-Jones and Watts: Symbolism in Britain, 1860–1910*, exh. cat., ed. Andrew Wilton and Robert Upstone (London: Tate Gallery, 1997), p. 165.

15 Quoted in *Franz von Stuck, 1863–1928: Eros and Pathos*, exh. cat., ed. Edwin Becker (Amsterdam: Van Gogh Museum; Zwolle: Waanders, 1995), p. 12.

16 The print of 1783 was accompanied by these verses by Erasmus Darwin: "On her fair bosom sits the Demon-Ape / Erect, and balances his bloated shape; / Rolls in their marble orbs his Gorgon eyes, / And drinks with leathern ears her tender cries." For more on Fuseli and the Darwins, see Barbara Larson, "Darwin, Burke and the Biological Sublime," in *Darwin and Theories of Aesthetics and Cultural History*, ed. Barbara Larson and Sabine Flach (Burlington, VT: Ashgate, 2013), p. 21–26.

17 Alfred Kubin, *Aus meinem Leben* [1911], ed. Ulrich Riemerschmidt (Munich: Ellerman, 1974), p. 9. Marsha Morton has noted that Kubin owned Franz Otto's *Wunderglaube und Wirklichkeit* (1885), which compares folk creatures and beliefs with evolutionary theory (Marsha Morton, "Nature and Soul: Austrian Responses to Ernst Haeckel's Evolutionary Monism," in *Darwin and the Search for Origins*, ed. Pamela Kort and Max Hollein [Cologne: Wienand, 2009], p. 139).

18 See Amélie Simier, "Jean Carriès, imagier lyonnais," in *Jean Carriès, 1855–1894. La matière de l'étrange*, exh. cat., ed. Amélie Simier (Paris: Paris-Musées/Nicholas Chaudun, 2007), p. 32–33.

19 Klinger read Darwin as early as the mid-1870s. Marsha Morton, "From Monera to Man: Ernst Haeckel, Darwinismus, and Nineteenth-Century German Art," in *The Art of Evolution: Darwin, Darwinisms, and Visual Culture*, ed. Barbara Larson and Fae Brauer (Hanover, NH: Dartmouth College Press, 2009), p. 87, n. 33.

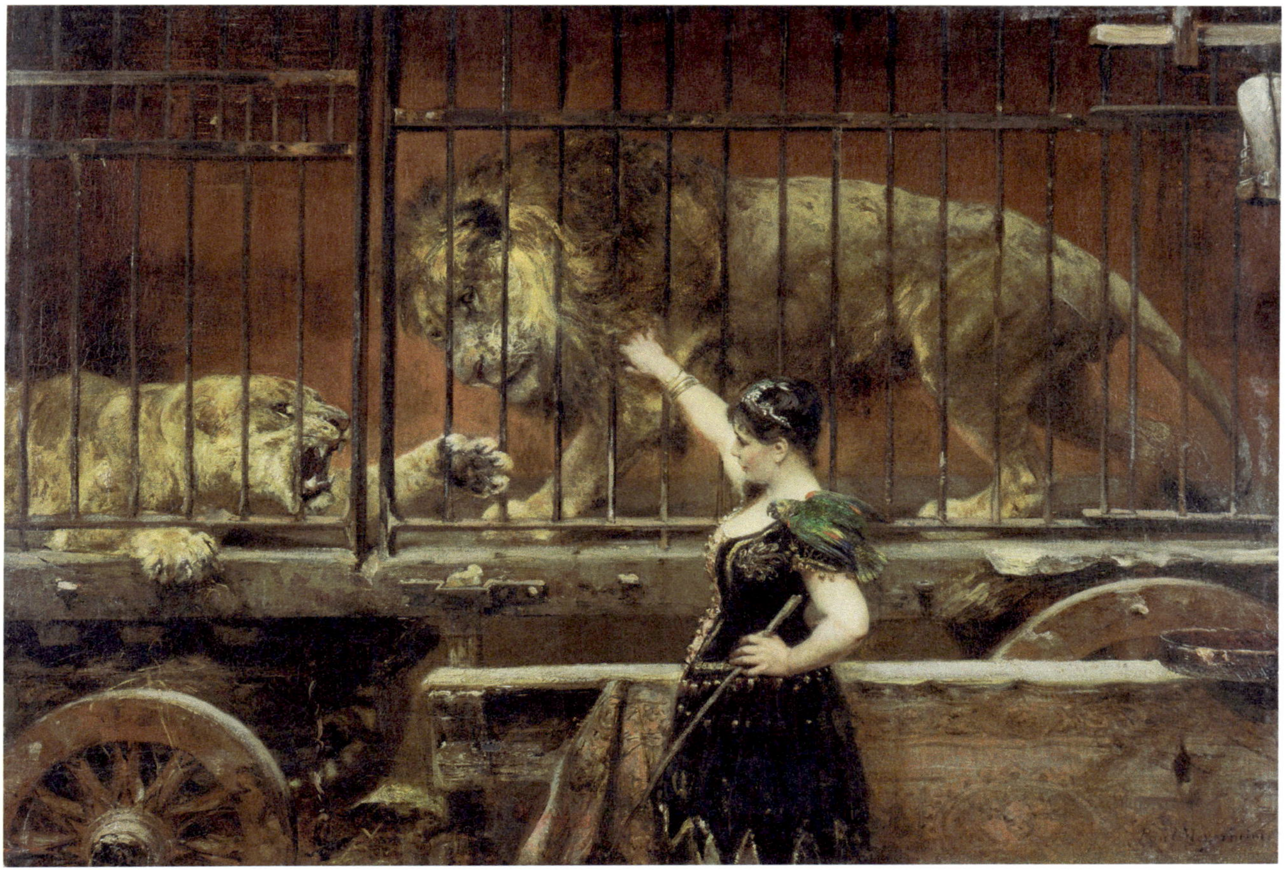

snake and part fish that populated the rivers of this country, setting their sights on hapless fishermen, as in Reinhold Eichler's *Festival of Nature* (1903). Kubin, who claimed both fairy tales and natural science books as an early source of inspiration, put a macabre evolutionary spin on Henry De La Beche's widely circulating antediluvian print *Duria Antiquior* in *Night of Ghosts* (98, P.148); his hybrid creatures crawl out of dried up water beds and cliff faces like so many disturbing fossils being spit out by the earth.[17]

Christian medieval bestiaries included dangerous composite creatures, like dragons, griffins and gargoyles, which belonged at once to different realms (such as the sea and the sky). This source was an inspiration for the French artist Jean Carriès, who in his youth was apprenticed to a sculptor at the Saint-Jean Cathedral in Lyon.[18] The discovery in Bavaria of the sensational and near complete fossils of pterosaurs, a kind of flying lizard, and the archaeopteryx, a small, feathered flying dinosaur, the former in 1859 and the latter in 1861 and 1877, may have encouraged a revival of dragon imagery, notably in the work of Böcklin. The German artist Max Klinger, an engaged reader of Darwin, clearly took these fossils up as the subject of nightmares in his print "Abduction" from the series *A Glove* (1881).[19] Gauguin's winged lizard in *Te Nave Nave Fenua* (*Delightful Land*, 1892), with its recasting of the theme of the temptress, is suggestive of such fossils as well as the role of the miniature dragons that often replaced snakes as embodying sin in medieval art. His "Eve" has two extra toes on her left foot that do not appear to be human, suggesting her atavistic animal origins. Certain of Gauguin's Nabi friends and followers, such as Paul Ranson, were interested in medievalism and transforming composite creatures that suggest an evil presence.

177 FIG.
Max Klinger,
Leuckart Certificate (Penelope), 1895
Coloured lithographic print, etching, aquatint, 18.7 × 30.1 cm
Berlin, Sammlung Archiv für Kunst und Geschichte

178
Paul Meyerheim,
The Jealous Lioness,
1885–1890
Oil on canvas, 49.6 × 69 cm
Frankfurt, Städel Museum

INTO THE DEEP

Many evolutionary works of art call the liminal space of the shore to one's attention and, beyond that, the depths of the sea, the new site of life's origins as opposed to the biblical garden. As Darwin said, "[The origins of mankind] go back to some amphibian-like creature and this again from some fish-like animal."[20] Carriès chose the frog, a small vertebrate that lives between the water and the land, to explore evolutionism. His creatures include a frog-rabbit, a frog-man and a fish-frog (179, 180, 181). In Kubin's *The Swamp*, a woman emerges from a murky body of water, gigantic bizarre fish-like creatures rising behind her (215, P. 304). Mermaids appear in the shallows, notably in the work of Böcklin, but also in that of Edvard Munch and other artists such as John Collier, Huxley's son-in-law. In Böcklin's *The Calm Sea*, a mermaid sits atop a layer of sedimentary rocks indicating the geological time of the earth, while below her is a tailed serpent with lizard-like feet and a simian face (182). In a transitional state and partly submerged, this creature is threatened by the water snakes that slip over its neck. The figure in Rodin's evolutionary *Fish-Woman* gasps for air (183), like her siren sisters in the *Gates of Hell*, or perhaps calls men to their deaths. Sirens or mermaids, like nixies and not unlike sphinxes, were often conceived of as predatory in nature. More sexually violent marine creatures are the bottom-dwelling crustaceans and other aquatic creatures that invade the space of humans in the work of Klinger and Kubin (210, 211, 213, 214, P. 302–303).

The mythical Penelope in Klinger's 1895 eponymous coloured lithograph, intended as a gift to a Darwinian zoologist, weaves a tapestry with what at first appears to be a conventional representation of the natural order (177).[21]

Upon closer inspection, the many shells washed ashore are hints of the churning of the fecund life of the sea, and turtles and crabs crawl onto the land, depending at once on the dual realms of water and earth. However, it is the fantastic creatures of the ocean that seem to have sent Penelope into a meditative state, including a flying fish at right. Submarine visions of hybrids and indeterminate creatures abound in the work of Kubin and the French artist Odilon Redon. Redon's transitional underwater creature in "When Life Was Awakening in the Depths of Obscure Matter" from his lithographic series *Origins* of 1883 partly bears the skeletal configuration of a land-dwelling mammal (187). Redon was influenced by the comparative anatomy cabinets at Paris's Muséum d'histoire naturelle, where Georges Cuvier's formidable work on the subject dominated early paleontology; by Armand Clavaud, a scientific mentor who was an evolutionist;[22] and by literary sources.[23] His many recastings of Oannes, a Mesopotamian god with a fish body who emerges from the deep to bring knowledge to humans (184), were inspired by Flaubert's 1874 *Temptation of St. Anthony*, which itself owed to the author's trips to the Muséum. Flaubert's Oannes confronts religion (St. Anthony) and implied creationism with these words: "I am coeval with all origins. I have inhabited the amorphous world where hermaphroditic beasts lay slumbering . . . where fin, finger and wing are confounded . . . !"[24] At the end of the novel, Flaubert's saint stands on the shore of the primordial sea and sees its monstrous hybrid creatures; he watches animals and plants converge before a full vision of evolutionism unfolds.

References to the fusion or intersection of kingdoms—the plant world and the animal world—implying primal evolutionary history and shared functions (metabolism, protoplasm, respiration) can be found in the work of Scottish Symbolists, Art Nouveau artists such as Émile Gallé and Redon himself, who never forgot the scientific experiments his botanist friend Armand Clavaud conducted on a type of algae that had animal and plant characteristics.[25] Redon often depicted plant-humans—trees along a shore with human faces or swamp-flowers with drooping heads—dissolving two kingdoms into one being. Redon's primitive-looking *Plante grasse* (*Cactus Man*) of 1881 may have been a response to seeing natives of Tierra del Fuego on view at Paris's Jardin d'acclimatation

20 Darwin, *The Descent of Man*, p. 607.

21 On the lithograph and Klinger's Darwinism, see Marsha Morton, *Max Klinger and Wilhelmine Culture: On the Threshold of German Modernism* (Burlington, VT: Ashgate, 2014), p. 93–164.

22 See the essay in this catalogue, p. 267.

23 On Redon and Darwin, see Sven Sandström, "Odilon Redon: A Question of Symbols," *Print Collector's Newsletter*, vol. 6, no. 2 (May–June 1975), p. 31; and André Mellerio, *Odilon Redon* [1913] (New York: DaCapo Press, 1968), p. 16. For more on Redon, evolutionism and the Muséum d'histoire naturelle, see Barbara Larson, *The Dark Side of Nature: Science, Society and the Fantastic in the Work of Odilon Redon* (University Park, PA: Pennsylvania State University Press, 2005), p. 49–84.

24 Gustave Flaubert, *The Temptation of St. Anthony* [1874], trans. Kitty Mrosovsky (Ithaca: Cornell University Press, 1980), p. 171.

25 Redon said of Clavaud, "He worked with the infinitely small. He searched . . . at the edge of the imperceptible world, that life which lies between plant and animal." Odilon Redon, *To Myself: Notes on Life, Art, and Artists* [1922], trans. Mira Jacob and Jeanne L. Wasserman (New York: George Braziller, 1986), p. 14.

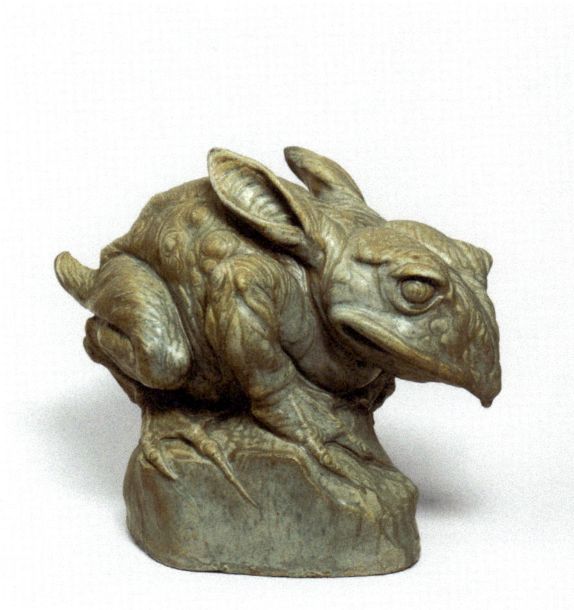

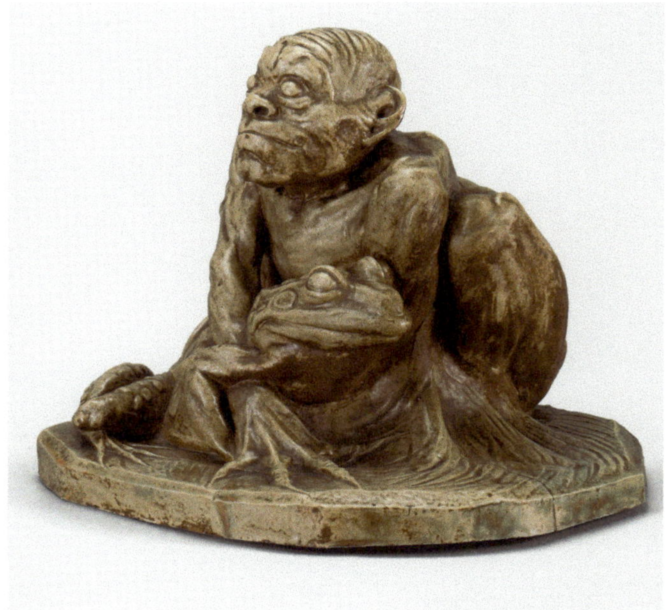

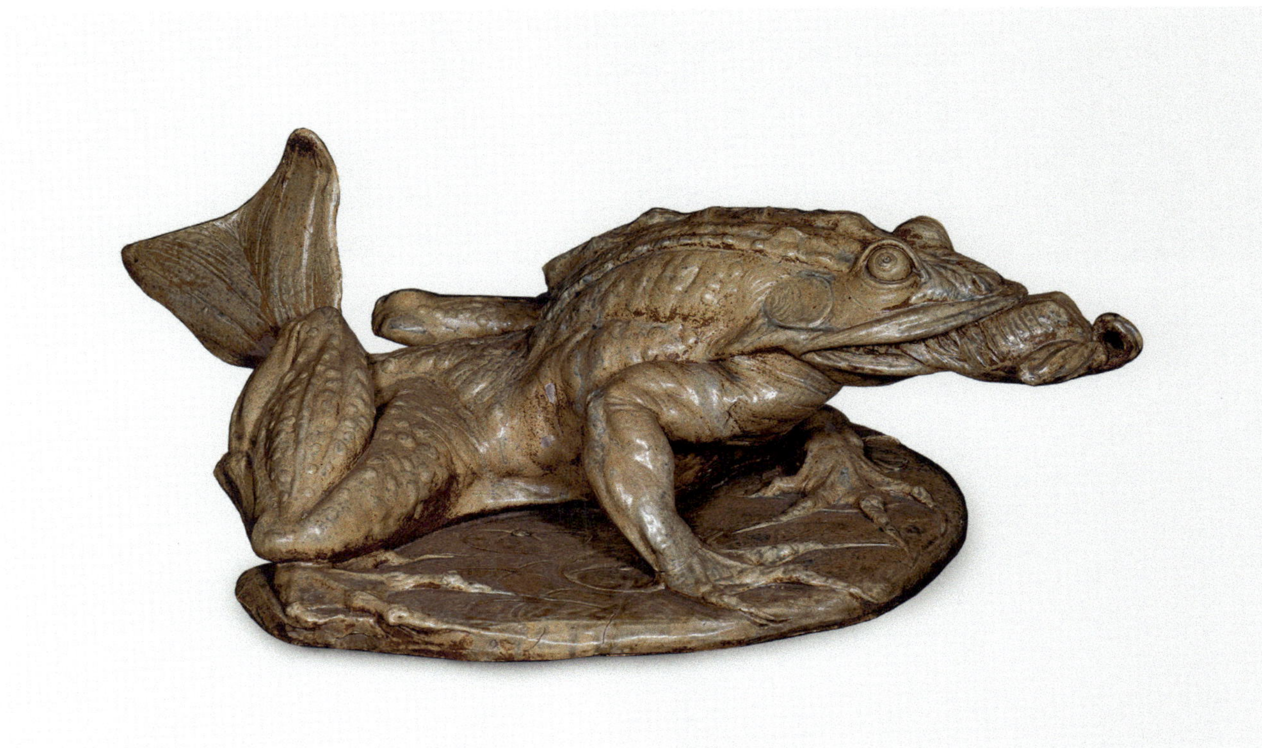

179
Jean Carriès,
Frog with Rabbit Ears, 1891
Glazed stoneware,
34.4 × 26.8 × 41 cm
Paris, Petit Palais, musée des Beaux-Arts de la Ville de Paris

180
Jean Carriès,
Frog-Man, about 1891
Glazed stoneware,
31.5 × 36 × 38 cm
Paris, Musée d'Orsay

181
Jean Carriès,
Fish-tailed Amphibian, 1892
Glazed stoneware,
30 × 84 × 69 cm
Paris, Petit Palais, musée des Beaux-Arts de la Ville de Paris

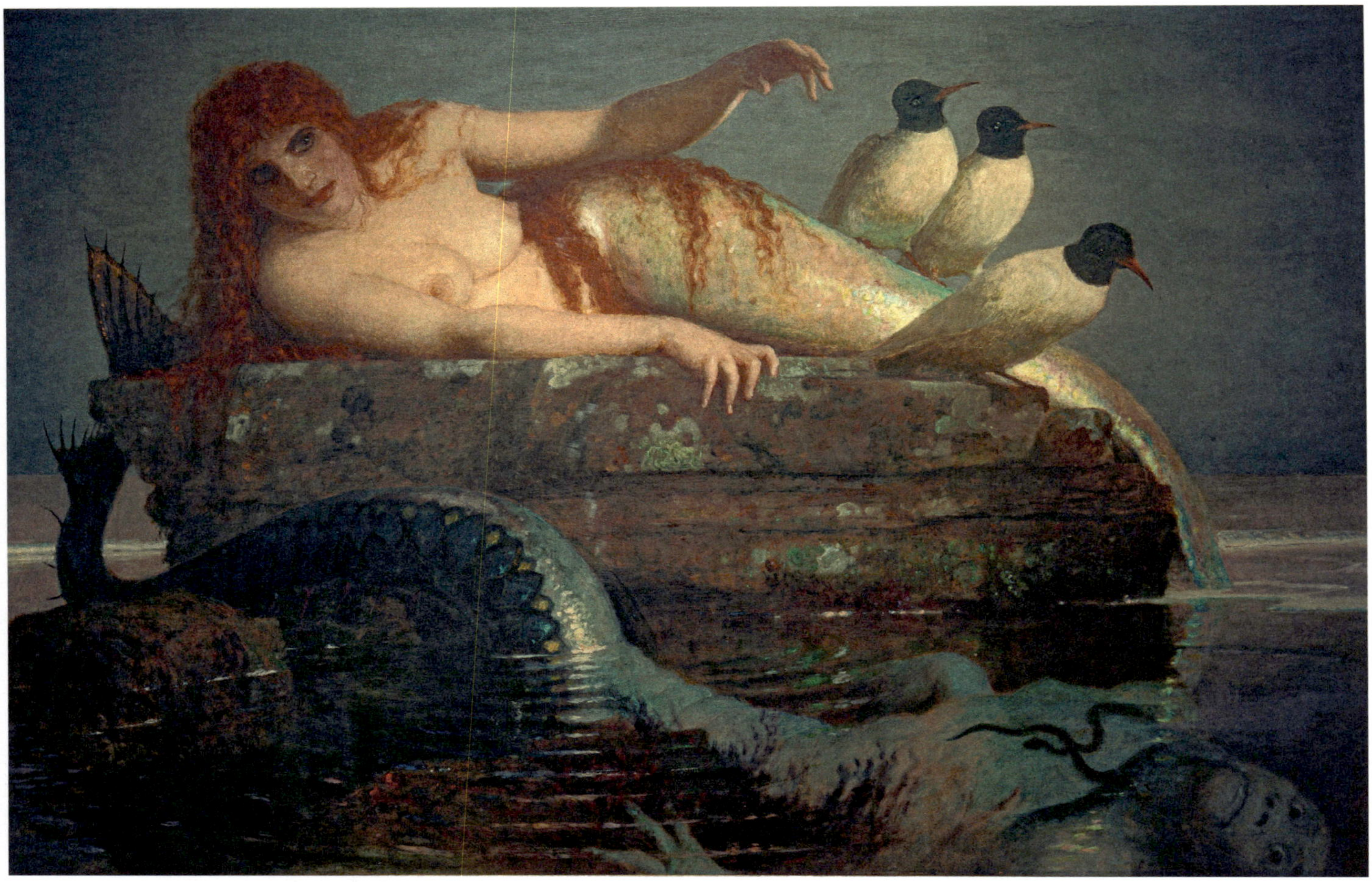

182
Arnold Böcklin,
The Calm Sea, 1887
Oil on wood, 103 × 150 cm
Bern, Kunstmuseum Bern

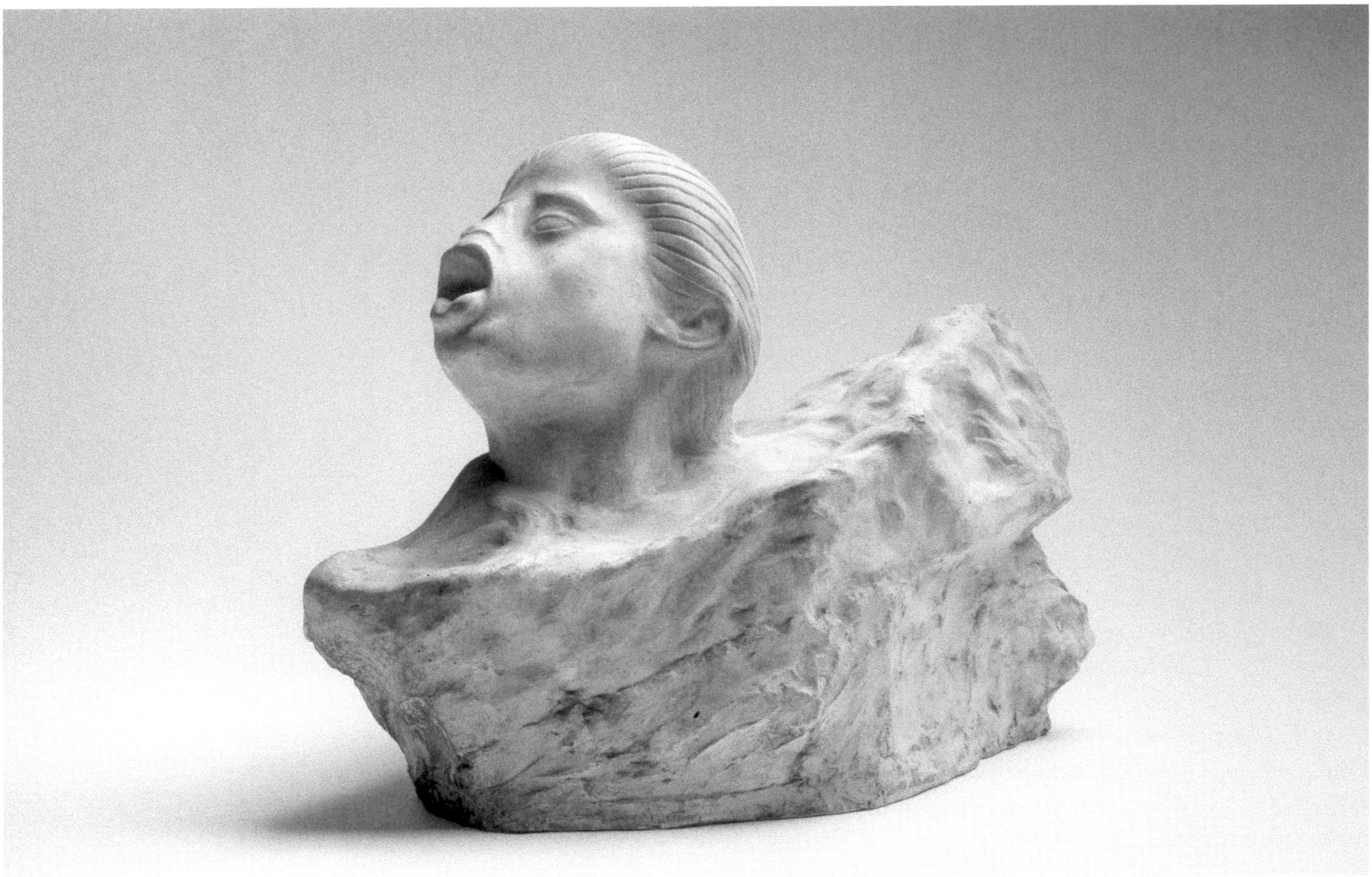

183
Auguste Rodin,
carved by Victor Peter,
Fish-Woman, 1915–1917
Marble, 34 × 32 × 50 cm
Paris, Musée Rodin

that year, one of the many exhibitions of "savages" in Europe: "They gave me a dream of primitive life, a nostalgia for the pure and simple life of our origins. I have never felt with such force the distance our own nature creates between the crawling beast and our highest goal... It is the animal in the complete mightiness of its instinct... Their nudity emerges from the earth like a flower... One might see that rigid flesh... lying on the golden sand of the desert [or] on immaculate shores."[26] After reiterating that humans had evolved from a "lowly organized form," Darwin used the Fuegians, whom he had seen on the *Beagle* voyage (52, 53, P. 74–75), the "lowest of barbarians," as the example of immediate descent in the final chapter of *Descent of Man*: "The astonishment which I felt on first seeing a party of Fuegians on a wild and broken shore will never be forgotten by me, for the reflection at once rushed into my mind—such were our ancestors."[27]

Some artists are known to have read Darwin (Klinger, Cormon), others were close friends with evolutionary scientists (Redon, Max), but evolutionism was also one of the great topics of the day, popularized in journals and scientific exhibitions, and even at the fairground. The human hybrid directly expressed the most heated aspect of the controversies surrounding evolutionism—mankind was embedded within nature and could never fully rise above it; humanity was, in fact, a dispassionate result of its processes.

26 Ibid., p. 70–71.
27 Darwin, *The Descent of Man*, p. 618.

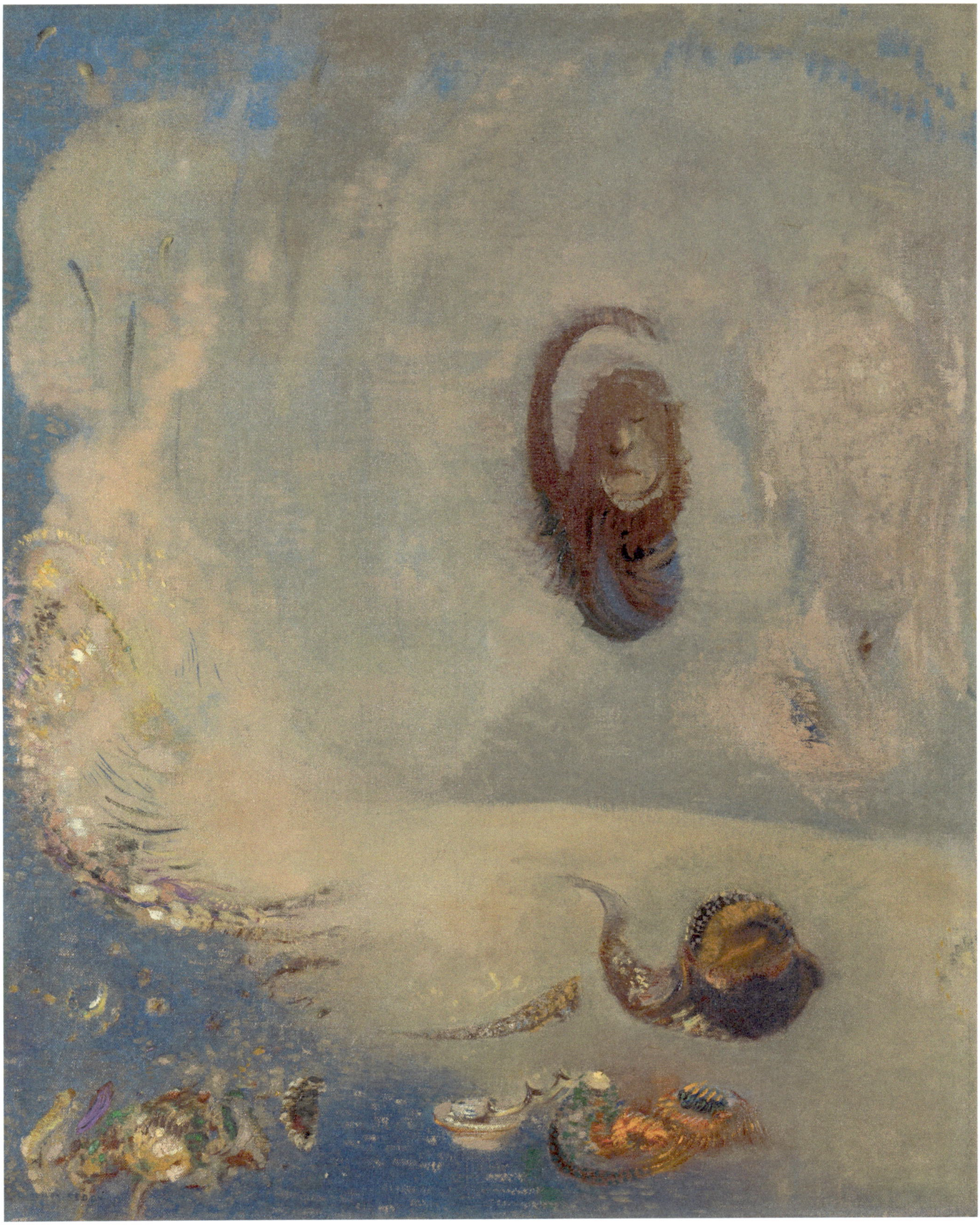

184
Odilon Redon,
Oannes, 1900–1910
Oil on canvas, 66 × 51.5 cm
Otterlo, Stichting
Kröller-Müller Museum

VII

ARS SIVE NATURA: ART, NATURE AND THE "POWER OF IMAGINATION" IN THE WORKS OF ODILON REDON

DARIO GAMBONI

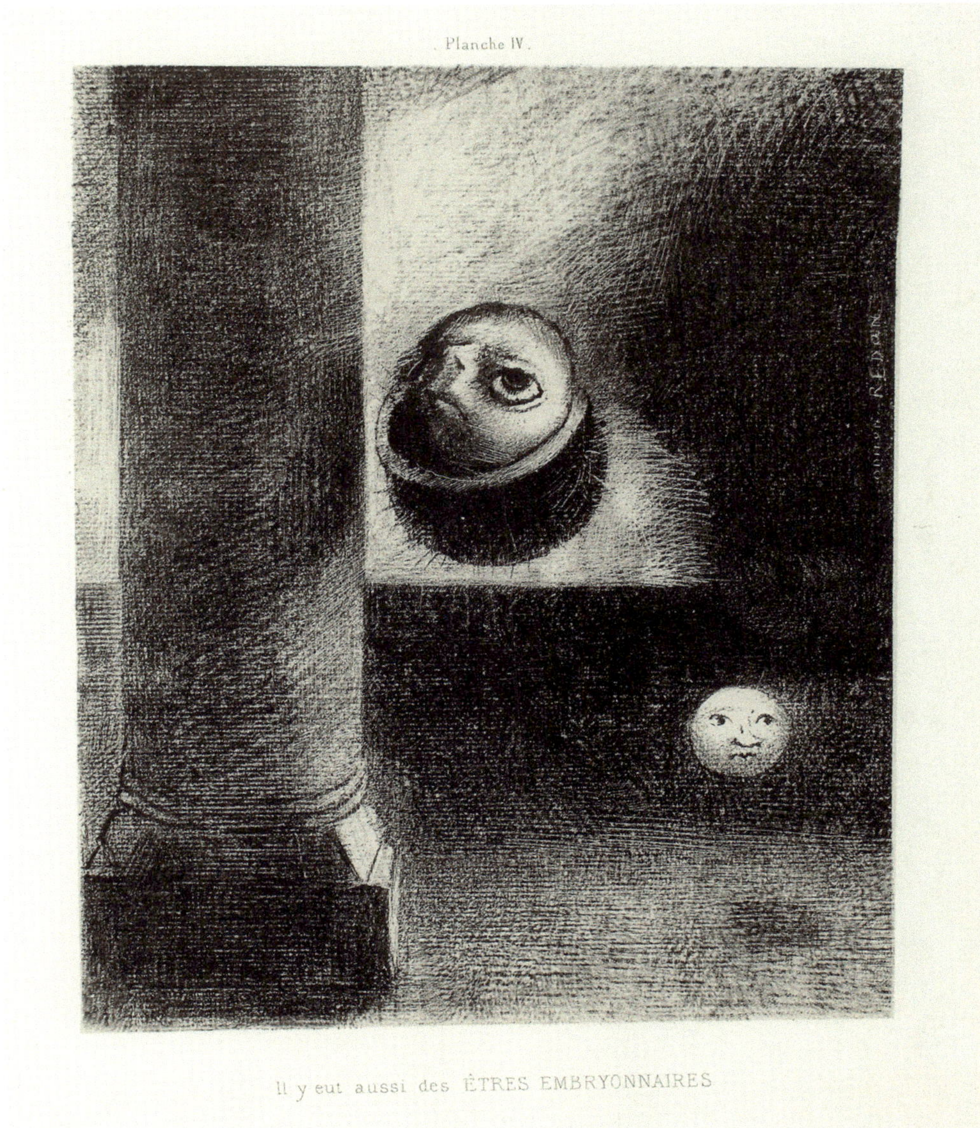

185 FIG.
Odilon Redon,
"There Were Also Embryonic Beings,"
in *Homage to Goya*, pl. 4, Paris:
Lemercier et Cie, 1885
Lithograph on China paper mounted
on wove paper, 23.8 × 19.7 cm (image)
Chicago, The Art Institute of Chicago

The relationship between the arts and sciences is a fascinating yet complex subject. And when the two fields, studied separately, are considered together at a general level, one is tempted to recognize the advances of one in the expressions of the other, and to postulate a causal relationship between the two.[1] However, instead of this rather unidirectional diffusionist model, it might be preferable to examine relationships involving exchange, mediation and interaction that have existed at the empirical level, within the framework of certain well-chosen case studies, thereby circumventing the generalizations. In the late-nineteenth-century period, the figure of Odilon Redon holds particular interest, as this draughtsman, printmaker and painter was known to emphasize the importance of the natural sciences in his work.

The natural sciences held special importance for the arts world, especially in the sphere of painting, which had been referred to since the Renaissance as the "monkey of nature." Darwinism added an ontological dimension to the ambivalence of the monkey figure, while Romanticism promoted the purity of landscapes as a competitor to history painting. Capturing a fragment of nature was seen as expressing a "state of mind," to use Henri-Frédéric Amiel's term.[2] The relationship between artist and nature thus raised not only aesthetic issues, but also fundamental epistemological questions. In his review of the Salon of 1859, Baudelaire compared two types of artists with regard to landscape painting: "There are those who call themselves 'realists'—a word with a double meaning, whose sense has not been properly defined, and so in order the better to characterize their error, I propose to call them 'positivists'; and they say, 'I want to represent things as they are, or rather as they would be, supposing that I did not exist.' In other words, the universe without man. The others however—the 'imaginatives'—say, 'I want to illuminate things with my mind, and to project their reflection upon other minds."[3] Baudelaire wanted to affirm the superiority of Romanticism over Realism, but the awareness of an interdependence between subject and object and an equivalence between art and nature, around the notion of imagination, is not limited to one style or school.

NATURA NATURANS

Two years earlier, Baudelaire had defined imagination as "an almost divine faculty which perceives immediately and without philosophical methods the inner and secret relations of things, the correspondences and analogies."[4] In 1875, Redon attributed the same power to the sense of sight: "Seeing is spontaneously grasping the interrelatedness of things."[5] The role of similarities in the observation, representation and creation of forms helps to explain the affinities between the transformist hypothesis and the visual arts. The entry of nature into history, brought about by transformism, was familiar to artists, especially landscape artists; and metamorphosis had already demonstrated its effectiveness as a model for understanding the genesis of forms, for example in the visual reception of Ovid or in ornamentation and caricature. Rather than a revolution and a "humiliation of human narcissism,"[6] Darwin's evolutionism represented, for some artists, a scientific legitimation of the intuition of *natura naturans* [nature naturing] that Romantic *Naturphilosophie* had already encouraged.

Such intuition, while opposing all fixism, nevertheless left open the question of the agents involved

1 See Christian Joschke, "Images et savoirs au XIXe siècle," *Perspective*, no. 3 (2007), p. 443–458.

2 Henri-Frédéric Amiel, *Journal intime*, vol. 2: *Janvier 1852 – mars 1856*, ed. Bernard Gagnebin and Philippe Monnier (Lausanne: L'Âge d'homme, 1978), p. 295.

3 Quoted in *The Mirror of Art: Critical Studies by Charles Baudelaire*, ed. and trans. Jonathan Mayne (London: Phaidon Press, 1955), p. 239–240.

4 Charles Baudelaire, "Notes nouvelles sur Edgar Poe" [1857], in Edgar Allan Poe, *Œuvres en prose*, trans. Charles Baudelaire, ed. Yves-Gérard Le Dantec (Paris: Gallimard, "Bibliothèque de la Pléiade" series, 1951), p. 1056; quoted in *Baudelaire As a Literary Critic*, trans. Lois Boe Hyslop and Francis E. Hyslop, Jr. (University Park: Pennsylvania State University Press, 1964), p. 127.

5 Odilon Redon, *À soi-même. Journal, 1867–1915. Notes sur la vie, l'art et les artistes* [1922] (Paris: José Corti, 1979), p. 48.

6 Sigmund Freud, "Une difficulté de la psychanalyse" [1917], in *Essais de psychanalyse appliquée*, trans. Marie Bonaparte and E. Marty (Paris: Gallimard, 1933), p. 142–143.

7 Camille Flammarion, *Dieu dans la nature*, vol. 1 (Paris: Ernest Flammarion, 1866), p. XIV.

8 Ibid., p. 88–89.

9 Ibid., p. 93, 191 and 193. See Charles Darwin, *The Origin of Species* [1859], ed. Gillian Beer (Oxford: Oxford University Press, 1996), p. 395.

10 See Sonja Hildebrand, "Gottfried Semper, 1803–1879," in Ulrich Pfisterer, ed., *Klassiker der Kunstgeschichte*, vol. 1: *Von Winckelmann bis Warburg* (Munich: Beck, 2007), p. 62–75.

11 *August Strindbergs brev*, vol. 10: *Februari 1894 – april 1895*, ed. Torsten Eklund (Stockholm: Albert Bonnier, 1968), p. 215 (letter 2897, August 13, 1894). See also August Strindberg, *Du hasard dans la production artistique* (Caen: L'Échoppe, 1990), p. 39.

12 Charles Rosen and Henri Zerner, *Romanticism and Realism: The Mythology of Nineteenth-Century Art* (New York: Viking Press, 1984), p. 229. See Petra Ten-Doesschate Chu, "It Took Millions of Years to Compose That Picture," in *Courbet Reconsidered*, exh. cat., ed. Sarah Faunce and Linda Nochlin (New Haven: Yale University Press, 1988), p. 55–66.

13 Jean Loize, "Un inédit de Gauguin," *Nouvelles littéraires*, May 7, 1953 (text partially reproduced in Paul Gauguin, *Oviri, écrits d'un sauvage*, ed. Daniel Guérin [Paris: Gallimard, 1974], p. 59–61).

14 See Alice Lamy, "*Ars sive natura*. La fortune des Êtres artificiels à l'université de Paris au XIVe siècle," *Vita Latina*, no. 187–188 (2013), specifically p. 320 and 333–336.

15 Redon, *À soi-même*, p. 121.

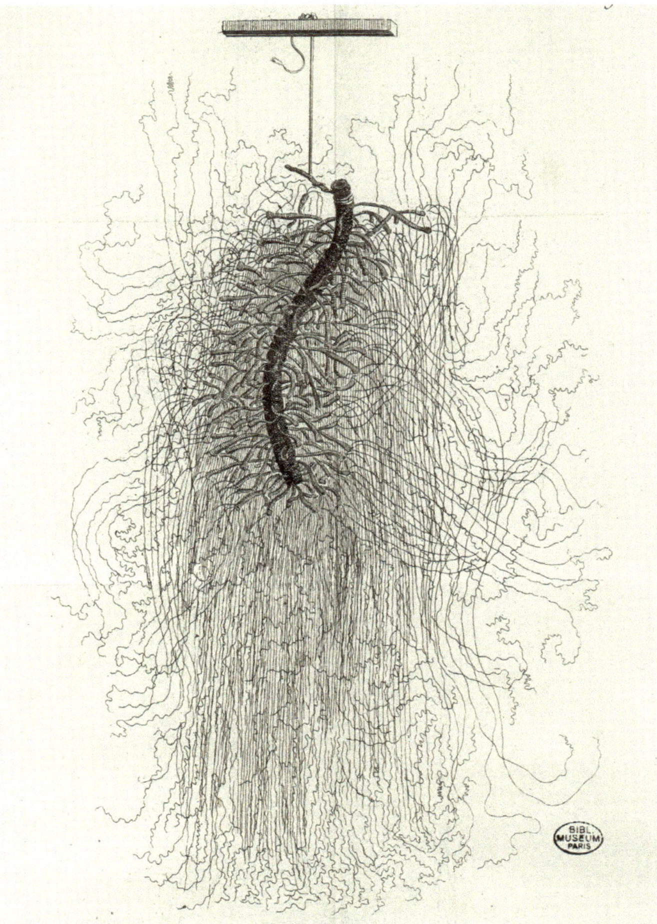

186 FIG.
Abraham Trembley,
"Figure Representing a Twig Covered in Long-Armed Polyps," in *Mémoires, pour servir à l'histoire d'un genre de polype d'eau douce, à bras en forme de cornes*, Leiden: Jean & Herman Verbeek, 1744
Print, etching and copperplate, 16.5 × 10 cm
Paris, MNHN

in the transformation process and of its eventual purpose. In one of his popular science books, *Dieu dans la nature*, the French astronomer Camille Flammarion sought in 1866 to "turn scientific materialism against itself"[7] and to reconcile evolutionism and spiritualism. Science, according to him, revealed nature "as a kind of immense dynamism in which the active forces of physics and chemistry combine and transform themselves," resulting in life.[8] He invoked Alexander von Humboldt's *Cosmos* in support of the "impression of the great unity that presides over the course of things," and quoted with satisfaction *The Origin of Species*, in which Darwin concluded that his explanation was eminently compatible with "what we know of the laws impressed on matter by the Creator."[9]

The idea of a parallelism between the "fundamental laws of nature" and those considered to govern the creation of forms was already current in architecture, where it had been expressed by Gottfried Semper, among others.[10] At the end of the century, the playwright August Strindberg, who had moved to France in the hope of starting a career as a painter, extended this idea to the fine arts and asserted that artists should "work like nature, not according to nature."[11] The way to achieve this was to abandon a preconceived pictorial objective in favour of the "teleology of chance"—a path taken by many professional artists in the twentieth century. By entrusting the direction of the painting to the pictorial material itself, Strindberg seemed to break with any realistic objective, but he was continuing what Henri Zerner and Charles Rosen called "phenomenological realism," as illustrated by Courbet's landscapes and their rocks, which the artist sculpted with a palette knife (217, P. 308).[12]

Paul Gauguin, in defence of Redon, proposed another way to blend artistic creation with nature. Commenting in late 1889 on *Certains*, a collection of art criticism published by Joris-Karl Huysmans, Gauguin contested the characterization of Redon as a creator of monsters: "Nature has infinite mysteries, and imaginative power. It is continuously varying its offerings. The artist himself is one of its means, and for me, Odilon Redon is one of the chosen ones in this continuation of creation. For him, dreams become a reality because of the verisimilitude he gives to them."[13] According to this view, the imagination of the artist, true to its nature, turns his works into products of nature. This idea harks back to the Nominalists' argument in the fourteenth century which defended identifying art with nature, while adapting it to the era of transformism through an allusion to an infinite variety of these creations.[14] Such a concept was in keeping with the views of Redon, for whom the role of the artist was to bring to light "the rare flower of originality."[15]

REDON AND NATURE

Born in Bordeaux and brought up on the family estate of Peyrelebade in the Médoc near the Landes, Redon always emphasized the importance for his art of the intimate relationship with nature he had forged as a child. He returned to Peyrelebade every summer to do his charcoal drawings, until the estate was sold in 1898. The cultural milieu of Bordeaux, where

Romanticism was resisting the rise of Realism, also played an important role for him, in particular when he met the botanist Armand Clavaud and the draughtsman and engraver Rodolphe Bresdin. In his pen lithographs, such as *The Good Samaritan*, Bresdin was evoking, or rather, creating a world that burgeoned with rocks, clouds and vegetation, and where tree branches turned into animals. Literature, which Redon enjoyed as much as music, also offered him models for a dynamic understanding of and union with nature, such as the pantheism expressed by young Maurice de Guérin in his *Journal intime* of the 1830s: "I live with the inner elements of things, I ascend the rays of stars and the courses of rivers to the mysteries of their origins. Nature grants me entry into the most secluded of its divine dwellings, to the starting point of universal life; there I perceive movement and hear the first song of beings in all its freshness."[16] This passage can also be compared to an introspective text written by Redon in 1877: "In the universe too, I contemplate him: the great Being so certain, present and mysterious, whose secrets afflict me. I see him in all of nature, on this first day of Spring, so full, so pure. To him my heart rises, ever higher and further, into the depths of the firmament. My eyes are lost; they are focused on him."[17]

Redon's real intellectual mentor, however, was Armand Clavaud, a self-taught man of modest origin whose scientific skills were recognized in Bordeaux when he became curator of the municipal botanical library and, from 1874, lecturer of the municipal botany course.[18] Reflecting on Clavaud, in 1909, Redon described him as "sympathetic to the revelations of the microscope" and seeking "in the confines of the imperceptible, that intermediate state between animal and plant life, of flower or being, that mysterious element which is animal for a few hours of the day but only under the action of light."[19] In a letter to his friend and collector Andries Bonger, Redon added that the botanist, who had authored in 1882–1884 *Flore de la Gironde*, owed his reputation to an earlier pamphlet dealing with "an imperceptible plant, or animal, of the *Chara* species," which "caused a sensation in Germany in the scholarly world": "It supported Darwin's, thinking, I hear. Clavaud devoted his life to researching the link between the animal and plant worlds."[20] It was characteristic of Redon, for whom ambiguity was a fundamental aesthetic and poetic principle, to focus on the transition between plants and animals, something that was of only marginal interest to Darwin. The affirmation of a vital force transcending the boundaries of taxonomy had already led to the success of Abraham Trembley's research into freshwater hydra in the eighteenth century (186).[21] Also, Horst Bredekamp showed that coral, whose shape served as a model for Darwin to schematically depict the process of evolution, was a favourite item in Renaissance curiosity cabinets precisely because of the perception of such ambiguity.[22] This still had relevance in the nineteenth century, as shown in a passage from *The Sea* in which Jules Michelet, in 1861, placed the algae known as coralline algae at the limits of the "three kingdoms" and at the origin of life: "They tend towards the animal, they tend towards the mineral, and, finally, are assigned to the vegetable. Perchance they form the real point at which Life obscurely and mysteriously rises from the slumber of the stone, without utterly

16 Maurice de Guérin, *Œuvres complètes*, vol. 1, ed. Bernard d'Harcourt (Paris: Les Belles Lettres, 1947), p. 224 (December 1834). Redon had his own copy of the 1868 edition of Guerin's writing that contained the *Journal*.

17 Redon, *À soi-même*, p. 58.

18 See Ariane Durand, "Odilon Redon et Armand Clavaud ou les étroits contacts entre la botanique, l'art et la philosophie," *Revue historique de Bordeaux*, no. 24 (1975), p. 149–182; Sven Sandström, *Le Monde imaginaire d'Odilon Redon. Étude iconologique* (Lund: Berlingska Boktryckeriet, 1955), p. 64–68.

19 Redon, *À soi-même*, p. 18. Quoted in *Odilon Redon, Gustave Moreau, Rodolphe Bresdin*, exh. cat. (New York: Museum of Modern Art, 1961), p. 12.

20 *Lettres d'Odilon Redon, 1878–1916* (Brussels: G. Van Oest, 1923), p. 84 (January 2, 1909; Redon writes "Claveau" and "Cara"). The study he is referring to may be Armand Clavaud, "Sur le *Nitella stelligera* des auteurs," *Actes de la Société linnéenne de Bordeaux*, vol. 25 (1866), p. 1–5, in which the botanist demonstrated that *Nitella stelligera* actually belonged to the genus *Chara*. See Sandström, *Le Monde imaginaire d'Odilon Redon*, p. 67–68.

21 See Elizabeth Athens, "The Vital Ornament: Natural Philosophy in William Hogarth's *The Analysis of Beauty* (1753)," *Oxford Art Journal*, vol. 40, no. 3 (2017), p. 397–418.

22 Horst Bredekamp, *Darwins Korallen. Die frühen Evolutionsdiagramme und die Tradition der Naturgeschichte* (Berlin: Wagenbach, 2005), p. 10, 20 and 62–67.

23 M. J. Michelet, *The Sea*, no trans. (New York: Rudd & Carleton, 1861), p. 141. See also the passage from *La Fée aux miettes* by Charles Nodier quoted in Rodolphe Rapetti, "Redon et l'espace," in *La Nature silencieuse. Paysages d'Odilon Redon*, exh. cat., ed. Sophie Barthélémy et al. (Bordeaux: Musée des Beaux-Arts; Ghent: Snoeck, 2016), p. 14.

24 Jerome Viola, "Redon, Darwin and the Ascent of Man," *Marsyas*, vol. 11 (1962–1964), p. 42–57. See Arthur O. Lovejoy, *The Great Chain of Being: A Study of the History of an Idea* (Cambridge, MA: Harvard University Press, 1978).

25 Sandström, *Le Monde imaginaire d'Odilon Redon*, p. 77–78.

26 See Barbara Larson, "Evolution and Degeneration in the Early Work of Odilon Redon," *Nineteenth-Century Art Worldwide*, vol. 2, no. 2 (spring 2003), http://www.19thc-artworldwide.org/spring03/220-evolution-and-degeneration-in-the-early-work-of-odilon-redon; Barbara Larson, *The Dark Side of Nature: Science, Society and the Fantastic in the Work of Odilon Redon* (University Park: Pennsylvania State University Press, 2005), p. 6 and 49–84.

27 Odilon Redon, *Critiques d'art. Salon de 1868, Rodolphe Bresdin, Paul Gauguin; précédées de Confidences d'artiste*, ed. Robert Coustet (Bordeaux: William Blake & Co., 1987), p. 35 (1894); Michelet, *The Sea*, p. 191.

28 Redon, *Critiques d'art*, p. 36.

29 Sven Sandström distinguishes two "suites" within the album, which he calls "the suite on creation" and "the mythological suite" (Sandström, *Le Monde imaginaire d'Odilon Redon*, p. 78).

30 See Dario Gamboni, *La Plume et le pinceau. Odilon Redon et la littérature* (Paris: Minuit, 1989), p. 99–100.

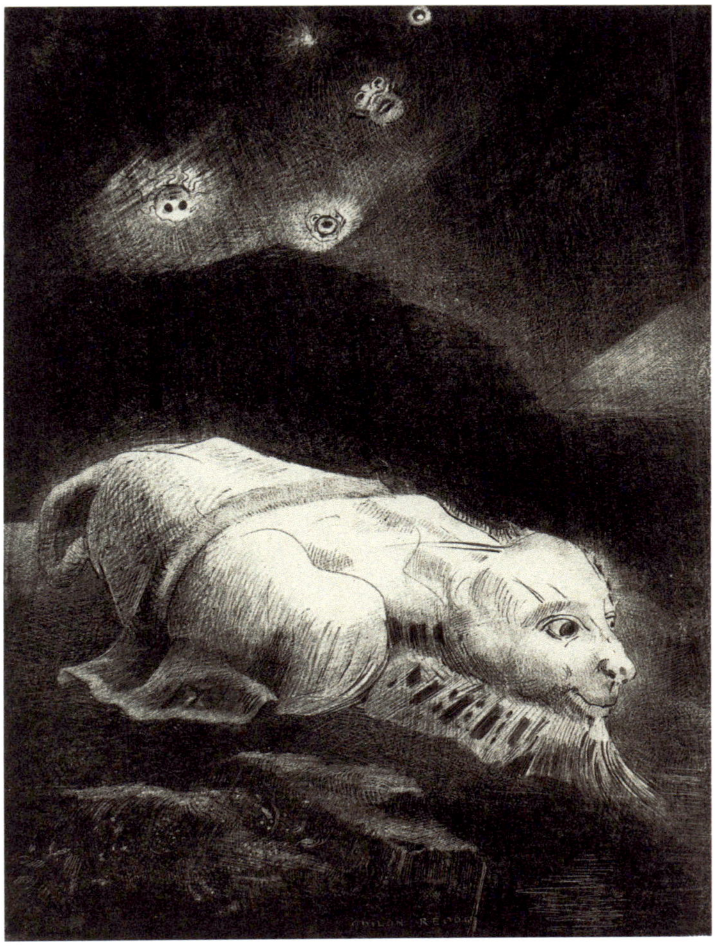

187 FIG.
Odilon Redon,
"When Life Was Awakening in the Depths of Obscure Matter," in *The Origins*, pl. 1, Paris, Lemercier et Cie, 1883
Lithograph on China paper mounted on wove paper,
27.5 × 20.3 cm
Washington, National Gallery of Art

quitting that rude starting-point, as if to remind us, so high placed and so haughty, of the right of even the humble mineral to rise into animation, and of the deep eternal aspiration that lies buried, but busy, in the bosom of Nature."[23]

This "aspiration" evokes Lamarck rather than Darwin, and the same is true of Redon's view, as can be inferred from his works and statements. For him, as for so many artists and writers who did not have to adopt a clear and coherent position on the debates shaking up scientific circles, many diverse notions coexisted and intermingled, which is reflected in the bibliography. Jerome Viola showed what Redon and Clavaud owed to older concepts, including that of the "great chain of being."[24] For Sven Sandström, Redon "started with Darwinism" while seeking to "preserve a spiritualist concept," in the manner of Flammarion.[25] Barbara Larson connected Clavaud's work on algae to the draining of the Gironde marshes and highlighted the artist's interest in the comparative anatomy and paleontology work of Étienne Geoffroy Saint-Hilaire.[26] In his 1894 letter to Edmond Picard, published first by Bonger and then by Picard, Redon said that from 1872 onwards he had conducted "osteological studies in the natural history galleries" in Paris—that is, at the Muséum d'histoire naturelle, which Michelet had described as a "faëry palace in every part of which we see the genius of metamorphoses of Lamarck and Geoffroy."[27] Redon claimed it was there, thanks to "a little comparison," that he discovered "an idea of the relative contexture of all beings," and that he somehow naturally engaged in the transformative process: "It soon occurred to me to create according to my fancy. It was simply a matter of atrophying, reducing or developing [anatomical] parts according to my whim. I would not want to use the word 'monsters' but those of human fantasy on the keyboard of osteology, with an almost Christian underpinning."[28] He was also fascinated by paleontology for its historical dimension, or at least for the light it could shed on the question of the origins of humans and life, which was as intriguing to him as the hypothesis of the porous nature of the boundaries between species and kingdoms.

THE ORIGINS

In 1883, Redon devoted his lithographic album to the theme of biology and, significantly, called it *The Origins*. Several of the imaginary beings depicted in it derive from mythology. These include the cyclops (188 D), the mermaid, the satyr (188 B), the female centaur, the winged horse and the male centaur. But others are inspired by microbiology and embryology.[29] The legends, added later, elucidate a temporal dimension suggested visually by kinetic effects and the top-bottom opposition.[30] The two images that correspond most closely to the album's title are the frontispiece (188 A), with its assembly of embryonic creatures, some reduced to a kind of visual wordplay between single cell and face, almost comically turning back the viewer's gaze, and the first plate, titled "When Life Was Awakening in the Depths of Obscure Matter" (187). Although the face attached to the enigmatic body of the creature in the foreground suggests a lung and not gills, the scene could be underwater,[31] and the ligaments connecting its lower parts to the ground conjure up Michelet's "real point at which Life obscurely and mysteriously rises from the slumber of the stone." In 1885, in reference to Redon's *Homage to Goya* album, and specifically a plate titled "There Were Also Embryonic Beings" (185), Joris-Karl Huysmans spoke of "the waltz of other almost amorphous heads, some like embryos hinting at skulls, some like blurred infusoria, vague flagellates, indefinite monera and bizarre protoplasms like Haeckel's Bathybius, though less gelatinous and less unformed!"[32] The lunar face floating on the right does not justify this description, which must be based on the *Origins* album and the frequent exchanges between Huysmans and Redon at the time. It is therefore possible that the main figure in "When Life Was Awakening in the Depths of Obscure Matter" owes something to *Bathybius haeckelii*, a substance found at the bottom of the sea in which Thomas Henry Huxley thought he recognized the original protoplasm about which Haeckel had theorized.[33]

The most interesting plate in the album from our point of view is the second one, "There Was Perhaps a First Vision Attempted in the Flower" (188 C), whose title clearly alludes to the evolutionary process. By superimposing and combining the features of a flower and an eye, as well as a hint of a face and vague organic elements, the image conjures up a dizzying fusion by incorporating into a primitive plant, like a fern ready to unfurl, an eye, the organ responsible for the sense regarded as most advanced: sight. It is probably not fortuitous that the "extreme perfection and complication" of the eye was examined first by Darwin and then by Flammarion—with opposite conclusions—as a possible objection to the explanation of evolution by natural selection. For Darwin, this objection was disposed of by concluding that the eye could result from a succession of minute variations useful to the animal in the conditions in which it lived. But for Flammarion, the perfection of the eye meant it could not possibly "pass for the work of matter or chance," on the contrary, it presupposed "the existence of an intelligence not only aware of the laws of optics, but also capable of bending all the movements of matter under these laws."[34] If Redon actually read *Dieu dans la nature*, he probably agreed with another of Flammarion's criticisms: "Mr. Darwin is wrong to claim that the origin of the eye is as unimportant as the origin of life itself: we are interested in both."[35] The word "vision" used in his title indicates that he intended to place at the end of his evolutionary shortcut a faculty that was not only sensory, but also cognitive, and probably spiritual. The direction of the gaze, recalling the iconography of the holy martyrs, confirms this interpretation, and suggests that evolution is not finished. The idea of an "attempted" vision also suggests a creative intelligence, proceeding by trial and error, or the Lamarckian concept that organisms spontaneously evolve towards perfection.

In his 1909 autobiography, Redon described Clavaud as "by nature a scholar as much as an artist" and referred in particular to his readings: "He spoke to me of Hindu poems, which he admired and adored above all . . . He professed an almost religious admiration for Spinoza."[36] These diverse interests were consistent, since the equivalence between God and nature (*Deus sive natura*) posited by Spinoza seemed at the time to be in line with traditional Indian thought as well as with the scientific exploration of the natural world. In 1866, in a popular science book titled *Les Merveilles du monde invisible*, the publicist Wilfrid de Fonvielle likened microbiology to Vedantism, according to which "the eternal and infinite Being manifested himself to the sentient world" in the form of an egg: "The microscope seems to have confirmed these teachings of the Brahmanic theocracy, for a deep anatomy has taught us that the little God of this world begins in the same way as the incarnation of the infinite Being Himself."[37] As for Redon, he was in agreement with the principle of Romantic Spinozism, which James Engell sees in the "idea of a transforming and active imagination [at work] in man, God, and nature."[38] Redon had a copy of the first French translation of Spinoza's works, which was published in 1842 by Émile Saisset. In its introduction he had underlined a passage where the translator explains that the philosopher refuted the idea of revelation and believed that the prophets differed from other men only by "an extraordinary power of imagination."[39]

31 See Ursula Harter, "Les jardins océaniques," in *Odilon Redon. Le ciel, la terre, la mer*, exh. cat., ed. Laurence Madeline (Paris: Réunion des musées nationaux, 2007), p. 128–143.

32 Joris-Karl Huysmans, "Le nouvel album d'Odilon Redon," *La Revue indépendante*, February 1885, p. 291–296; see Gamboni, *La Plume et le pinceau*, p. 110–124. Quoted in Joris-Karl Huysmans, *Parisian Sketches*, trans. Brendan King (Porvoo, Finland: WS Bookwell, 2004), n.p.

33 See Philip F. Rehbock, "Huxley, Haeckel, and the Oceanographers: The Case of *Bathybius haeckelii*," *Isis*, vol. 66, no. 4 (1975), p. 504–533.

34 Darwin, *The Origin of Species*, p. 152; Flammarion, *Dieu dans la nature*, p. 419.

35 Ibid., p. 421.

36 Redon, *À soi-même*, p. 18.

37 Wilfrid de Fonvielle, *Les Merveilles du monde invisible* (Paris: Hachette, 1866), p. 188–189.

38 James Engell, *The Creative Imagination: Enlightenment to Romanticism* (Cambridge, MA: Harvard University Press, 1981), p. 116.

39 *Œuvres de Spinoza*, vol. 1, trans. Émile Saisset (Paris: Charpentier, 1842), p. CLXXXIV–CLXXXV.

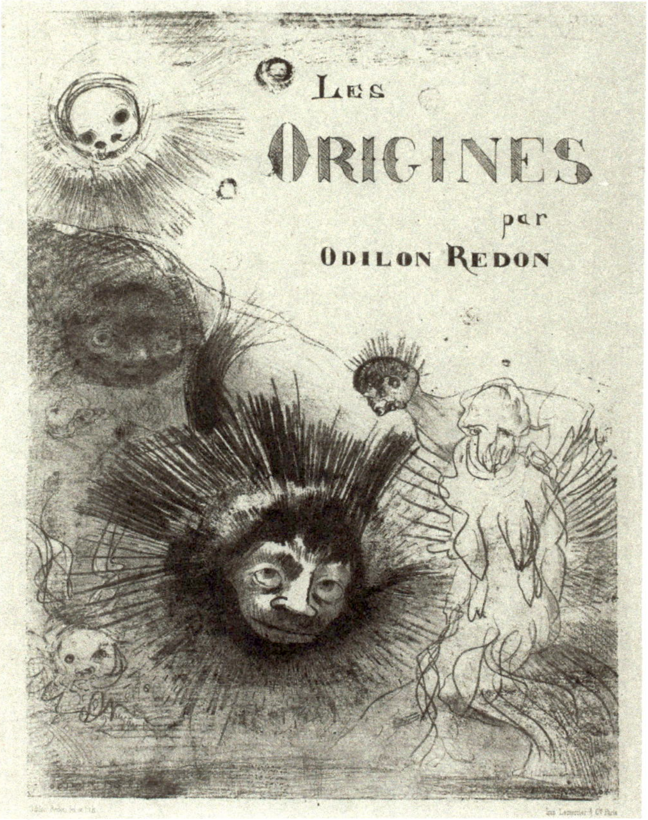

A

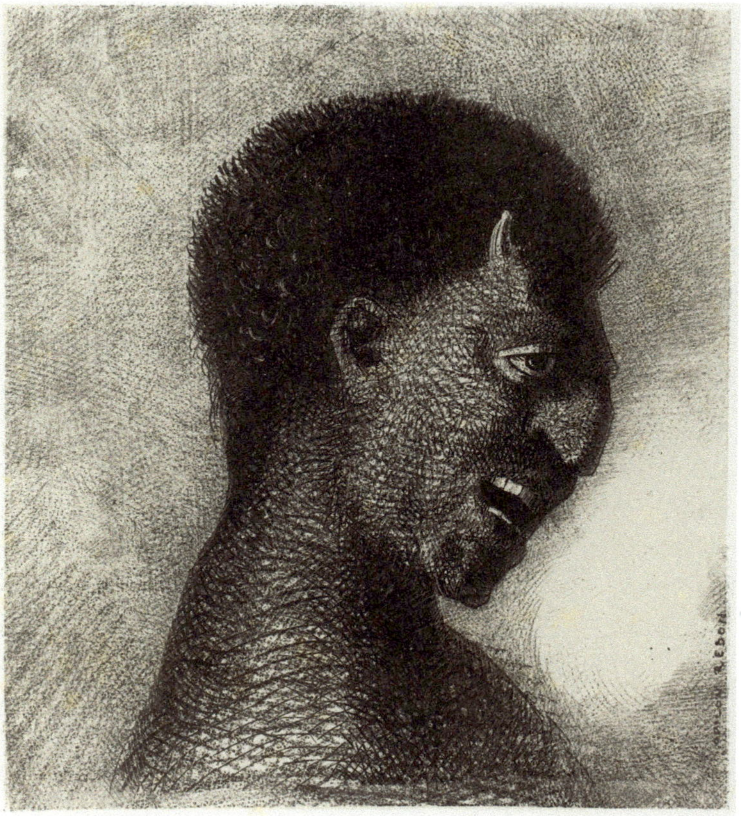

B

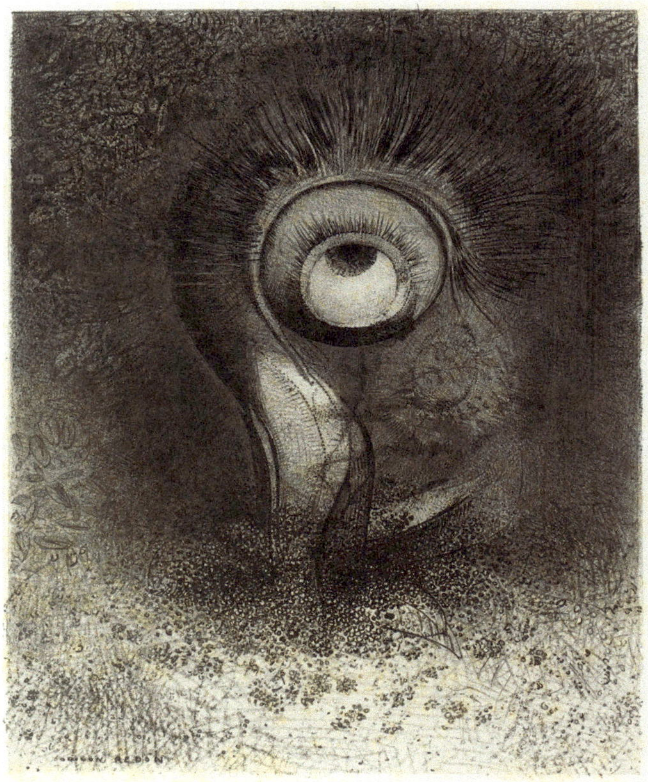

C

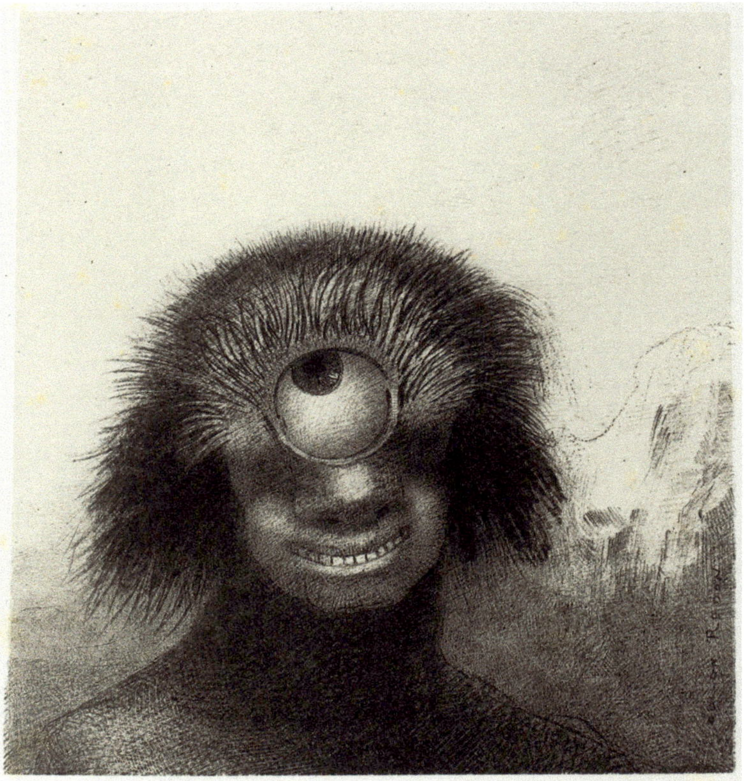

D

188 A, B, C, D
Odilon Redon, *The Origins*, Paris, Lemercier & Cie, 1883
Lithographs on China paper laid down on wove paper

Cover-frontispiece, 30.5 × 22.5 cm (image)
Chicago, The Art Institute of Chicago

"The Satyr with the Cynical Smile," pl. 5, 24 × 20.7 cm (image)
"There Was Perhaps a First Vision Attempted in the Flower," pl. 2, 22.3 × 17.2 cm (image)
"The Misshapen Polyp Floated on the Shores, a Sort of Smiling and Hideous Cyclops," pl. 3, 22.3 × 17.2 cm (image)
Washington, National Gallery of Art

THE DECORATION OF THE CHÂTEAU DE DOMECY

In the early twentieth century, Redon's sense of the world's unity and of the dynamic continuity of creation found unexpected expression in the decorative works he created for certain patrons, at the instigation of the first one, Baron Robert de Denesvre de Domecy. In 1893, Domecy had acquired a charcoal titled *Primitive Man* (189), which evokes the beginnings of humanity through a posture borrowed from pre-Columbian mummies. He later assembled one of the major collections of Redon's work.[40] In 1896, he commissioned the artist to create decorative works for his château in Domecy-sur-le-Vault, in Burgundy, starting with a panel for the chimney breast depicting "a vengeful angel with two figures inciting one to do good or better, the other to do evil."[41] The large panels for the dining room, commissioned in 1900 and completed in 1901, were designed by Redon and Domecy, with a view to their dominant red and yellow colours illuminating the room and matching the images to the tapestry.[42] The principal element was a giant pastel depicting a bouquet of flowers in a blue urn. This now hangs at the Mitsubishi Ichigokan Museum in Tokyo. The fifteen panels that have been donated to the Musée d'Orsay (190–195) also feature the plant world, but two of the vertical panels portray two female figures and landscape features. Working on these panels, Redon wrote to Bonger in early 1901: "I am covering the walls of a dining room with flowers, dream flowers, imaginary fauna."[43]

The "imaginary fauna" is somehow contained in the flora, "on the edge of an imperceptible world" and dwells in "this intermediate life between the animal and the plant" that Redon said he had glimpsed thanks to Clavaud. Branches, leaves and petals detach from one another to become insects, while flowers float like spores, and the corollas of giant sunflowers resemble planets. The infinitely large and the infinitely small, whose simultaneity in Pascal's *Pensées* had influenced Redon, come together within an indeterminate space.[44] Although spherical monads appeared early in Redon's work, their proliferation on his decorative surfaces may owe something to Haeckel's *Kunstformen der Natur* (144, 145, 150 B), the publication of which had begun in 1899. Furthermore, the ability to change scale had just been demonstrated in Paris with René Binet's monumental gate for the 1900 Paris Universal Exposition (143).[45] The abundance of the colour yellow, its powdery, pastel-like appearance and the impression of dispersal produced by its application are reminiscent of pollen and its fertilizing power. This effect can be seen in the panel *Day* in the decor of Fontfroide Abbey, where a golden trail follows the movement of a quadriga and is surrounded by flora with tropical accents.[46] In both cases, a Tahitian painting by Gauguin, *Parahi te marae*, may have served as a model, as suggested by the yellow background on which Redon painted the *Black Profile* in homage to Gauguin after his death.

The two figures in Domecy's vertical panels, with their bowed heads and closed eyes, may hint at the traditional association of women with nature and with

---

40 See *Odilon Redon: Prince of Dreams, 1840–1916*, exh. cat., ed. Douglas Druick et al. (Chicago: The Art Institute of Chicago; New York: Abrams, 1994), p. 82–83; Hiroo Yasui, "Les œuvres de Redon dans l'ancienne collection du baron de Domecy" [in Japanese], *Jissen Women's University Aesthetics and Art History*, vol. 33 (2019), p. 23–53 (in French catalogue, p. 47–53).

41 See Roseline Bacou, "La décoration d'Odilon Redon pour le château de Domecy, 1900–1901," *La Revue du Louvre et des musées de France*, vol. 42, no. 2 (1992), p. 43; Marie-Pierre Salé, "Décor de la salle à manger du château de Domecy," in *Odilon Redon, prince du rêve, 1840–1916*, exh. cat., ed. Rodolphe Rapetti (Paris: Réunion des musées nationaux, 2011), p. 304. A line drawing based on this chimney decoration, still in situ, is reproduced in Yasui, "Les œuvres de Redon dans l'ancienne collection du baron de Domecy," fig. 3. See also Alec Wildenstein, *Odilon Redon. Catalogue raisonné de l'œuvre peint et dessiné*, vol. 4: Études et grandes décorations. Supplément (Paris: Wildenstein Intitute/Bibliothèque des arts, 1998), p. 207–222.

42 See Bacou, "La décoration d'Odilon Redon pour le château de Domecy," p. 43.

43 *Lettres d'Odilon Redon*, p. 43–44 (January 17, 1901).

44 See Gamboni, *La Plume et le pinceau*, p. 136–137; Dario Gamboni, *The Brush and the Pen: Odilon Redon and Literature*, trans. Mary Whittall (Chicago: University of Chicago Press, 2011), p. 173–174.

45 See Robert Proctor, "A World of Things in Emergence and Growth: René Binet's Porte monumentale at the 1900 Paris Exposition," in *Symbolist Objects: Materiality and Subjectivity at the Fin de Siècle*, ed. Claire I. R. O'Mahony (High Wycombe: Rivendale Press, 2009), p. 221–244.

46 See *Odilon Redon, prince du rêve, 1840–1916*, p. 75–85.

47 See Gloria Groom, "The Late Work," in *Odilon Redon: Prince of Dreams*, p. 305–352. Gloria Groom compares these two figures with those in the painting *Mystical Conversation* (about 1896, Museum of Fine Arts, Gifu), acquired by Domecy in 1897, and sees the one on the left as a prelude to Redon's later mythological nudes (p. 310 and p. 420 note 22).

48 L.-F.-Alfred Maury, *Le Sommeil et les rêves. Études psychologiques sur ces phénomènes et les divers états qui s'y rattachent, suivies de Recherches sur le développement de l'instinct et de l'intelligence dans leurs rapports avec le phénomène du sommeil* (Paris: Didier, 1861), p. 326.

49 Salé, "Décor de la salle à manger du château de Domecy," *Odilon Redon, prince du rêve, 1840–1916*, p. 310.

50 See Dario Gamboni, *Paul Gauguin au « centre mystérieux de la pensée »* (Dijon: Les Presses du réel, 2013), p. 176–180 and 374–376.

51 Redon, *À soi-même*, p. 120; Édouard Le Roy, *Une philosophie nouvelle. Henri Bergson*, 3rd ed. (Paris: Alcan, 1913), p. 30. See Roseline Bacou, "La bibliothèque d'Odilon Redon," in *Festschrift to Erik Fischer: European Drawings from Six Centuries* (Copenhagen: Royal Museum of Fine Arts, 1990), p. 29–37.

52 Henri Bergson, *Creative Evolution* [1907], trans. Arthur Mitchell (New York: Henry Holt and Company, 1911), p. 126.

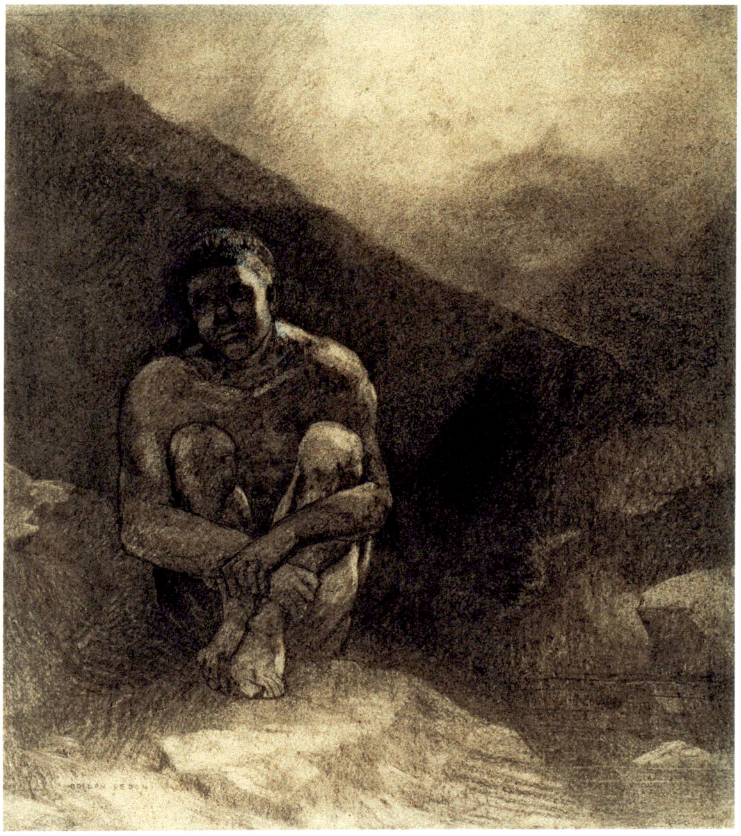

plant life or the unconscious (194, 195).[47] Haeckel's "biogenetic law," according to which ontogeny recapitulates phylogeny, had indeed been extended to the psychic domain by French oneirologists before Sigmund Freud. Alfred Maury had written in *Le Sommeil et les rêves*: "Man, therefore, from the time of his conception to the time of his maturity, ascends all the ladders of animal life, and his soul passes through states analogous to those of the sleeper who is gradually brought from complete sleep to a state of complete wakefulness."[48] Although symmetrical, the two women in Domecy's panels are quite different: the one turning to the right, evoking Salome or Shulamite, is swathed in transparent veils that expose her sensual body. Her counterpart wears an opaque tunic, and the giant plant-like lyre next to her conjures up the Thracian girl carrying the head of Orpheus on her lyre in the painting by Gustave Moreau (*Orpheus*). This is therefore, respectively, a Mad Virgin and a Wise Virgin, which in turn harks back to the Christian iconography of the panel for the chimney breast with its "two figures inciting one to do good or better, the other to do evil." The *Grand Bouquet* itself, whose asymmetrical and centrifugal composition caused Marie-Pierre Salé[49] to speak of an "anti-bouquet," is characterized by a similar contrast of cold and warm colours, ascending and descending movements. At the top, a sunflower contemplates the zenith and is robustly supported by a real stem; further down, at least three flowers evoke the lithograph "There Was Perhaps a First Vision Attempted in the Flower" (188 C). Since 1883, Redon's flower-eye had inspired a number of artists, including Vincent van Gogh and Gauguin.[50] The polarity that informs the decoration of the Château de Domecy signals the enduring presence of an axiomatic and spiritual dimension in Redon's interpretation of transformism.

The upward dynamic, which constitutes the essential tropism of this interpretation, expresses for the artist the "power of imagination" that he shares with nature. It anticipates the notion of "vital impulse," proposed in 1907 in *Creative Evolution* by Henri Bergson, a notion that allowed the philosopher to conceive of evolution as uninterrupted progress. Redon's library contains a copy of Édouard Le Roy's 1913 study of Bergson, in which Redon underlined many passages, including those dealing with the relationship between perception, memory and understanding. The artist, who in 1912 spoke of his art as "flowers at the confluence of two riverbanks, that of representation and that of memory," had to agree with the idea that "perception... is always complicated by memory" and that reality "is becoming and fleeing."[51] If Redon actually read *Creative Evolution*, he probably found confirmation of his quest for indeterminacy in the view that "the role of life is to insert some *indetermination* into matter," in unpredictable forms but in an increasing proportion, until it reaches the nervous system as "a veritable *reservoir of indeterminacy*."[52]

189 FIG.
Odilon Redon,
Primitive Man, 1872
Charcoal, black chalk, stumping, wiping and erasing, highlights of white and ochre gouache, on cream wove paper, 39.3 × 33.8 cm
Chicago, The Art Institute of Chicago

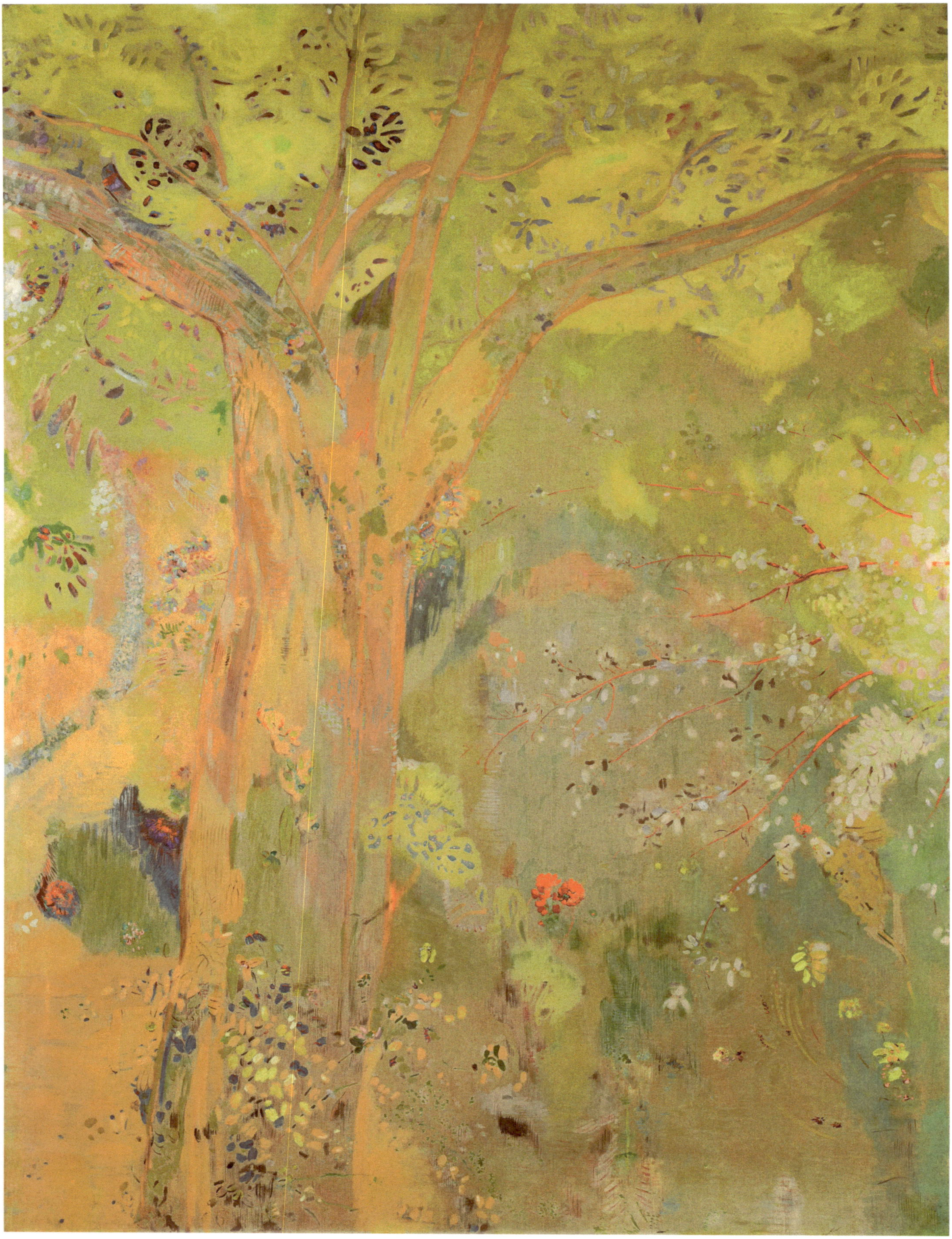

190
Odilon Redon,
Trees on a Yellow Background, 1901
Oil, tempera, charcoal and pastel on canvas, 249.5 × 185.5 cm
Paris, Musée d'Orsay

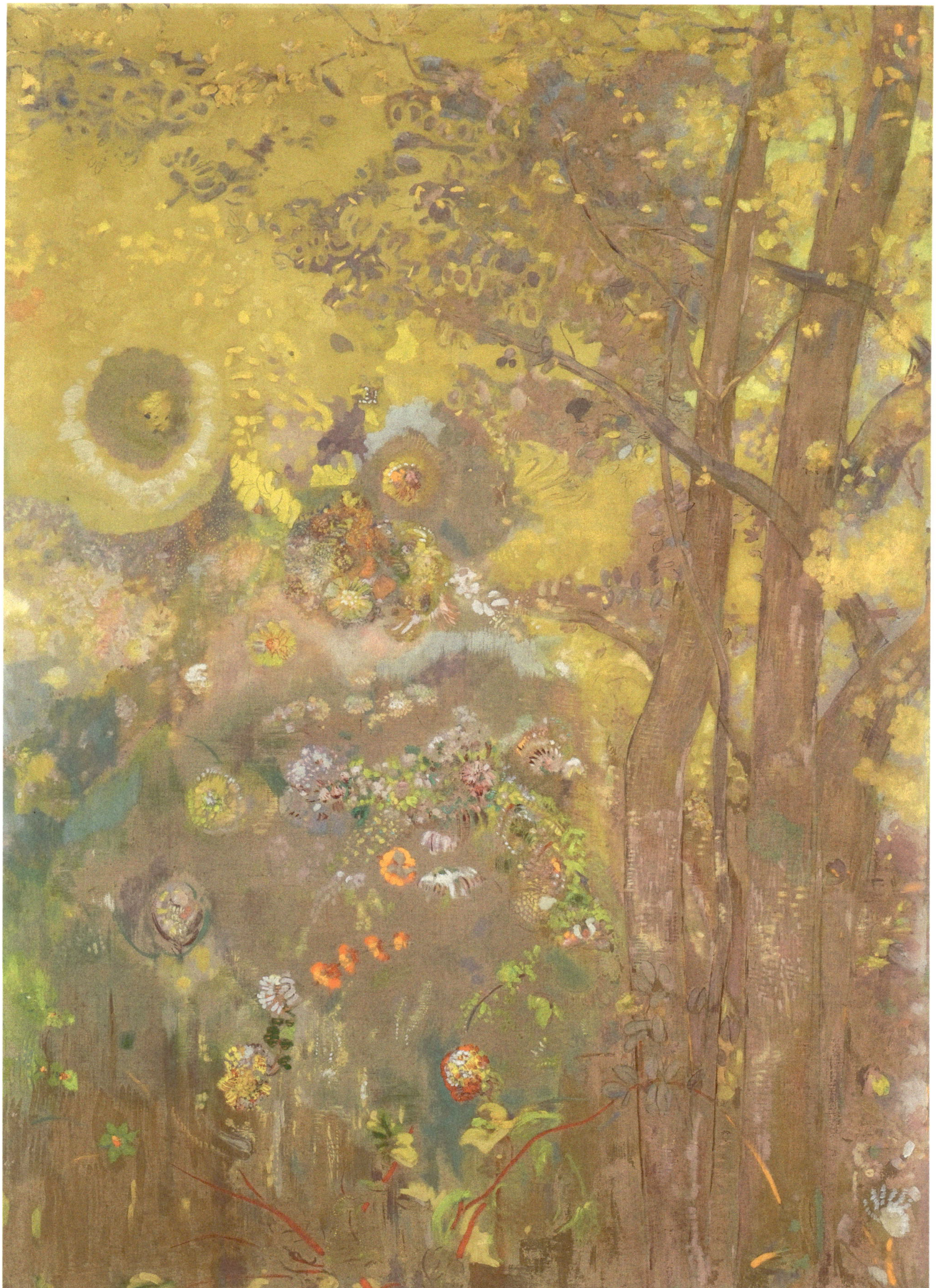

191
Odilon Redon,
Trees on a Yellow Background, 1901
Oil, tempera, charcoal and pastel
on canvas, 247.5 × 173 cm
Paris, Musée d'Orsay

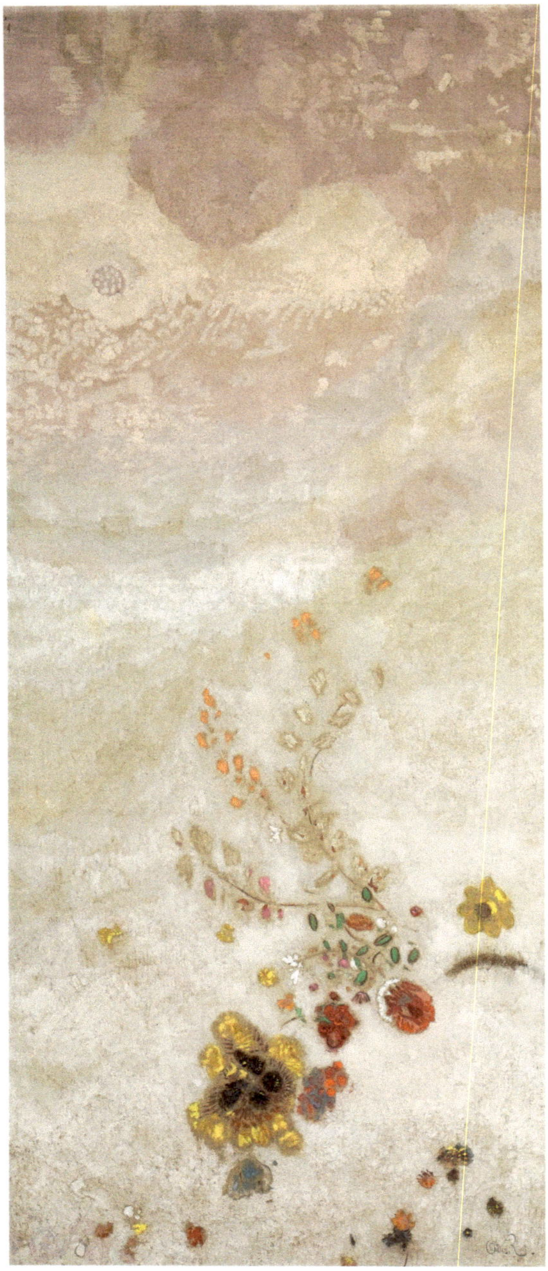

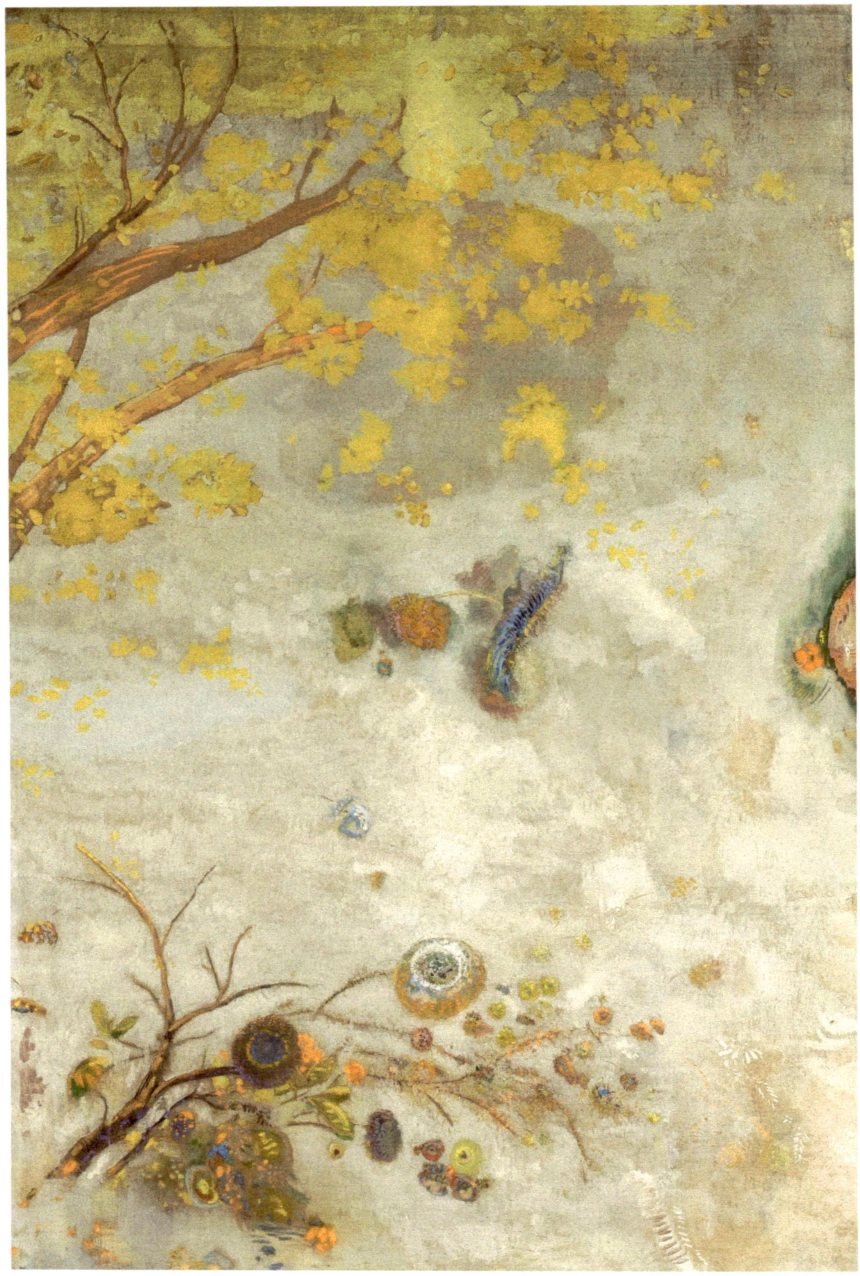

192 FIG., 193 FIG.
Odilon Redon, 1901
Large Decorative Panel with Plant Motifs (Pale Background), 244 × 103.7 cm
Yellow Flowering Branch, 247.5 × 143.5 cm
Oil, tempera, charcoal and pastel on canvas, Paris, Musée d'Orsay

194 FIG., 195 FIG.
Odilon Redon, 1901
Oil, tempera, charcoal and pastel on canvas
Figure, 244.5 × 58.5 cm
Figure (Yellow Flower), 243.8 × 58.3 cm
Paris, Musée d'Orsay

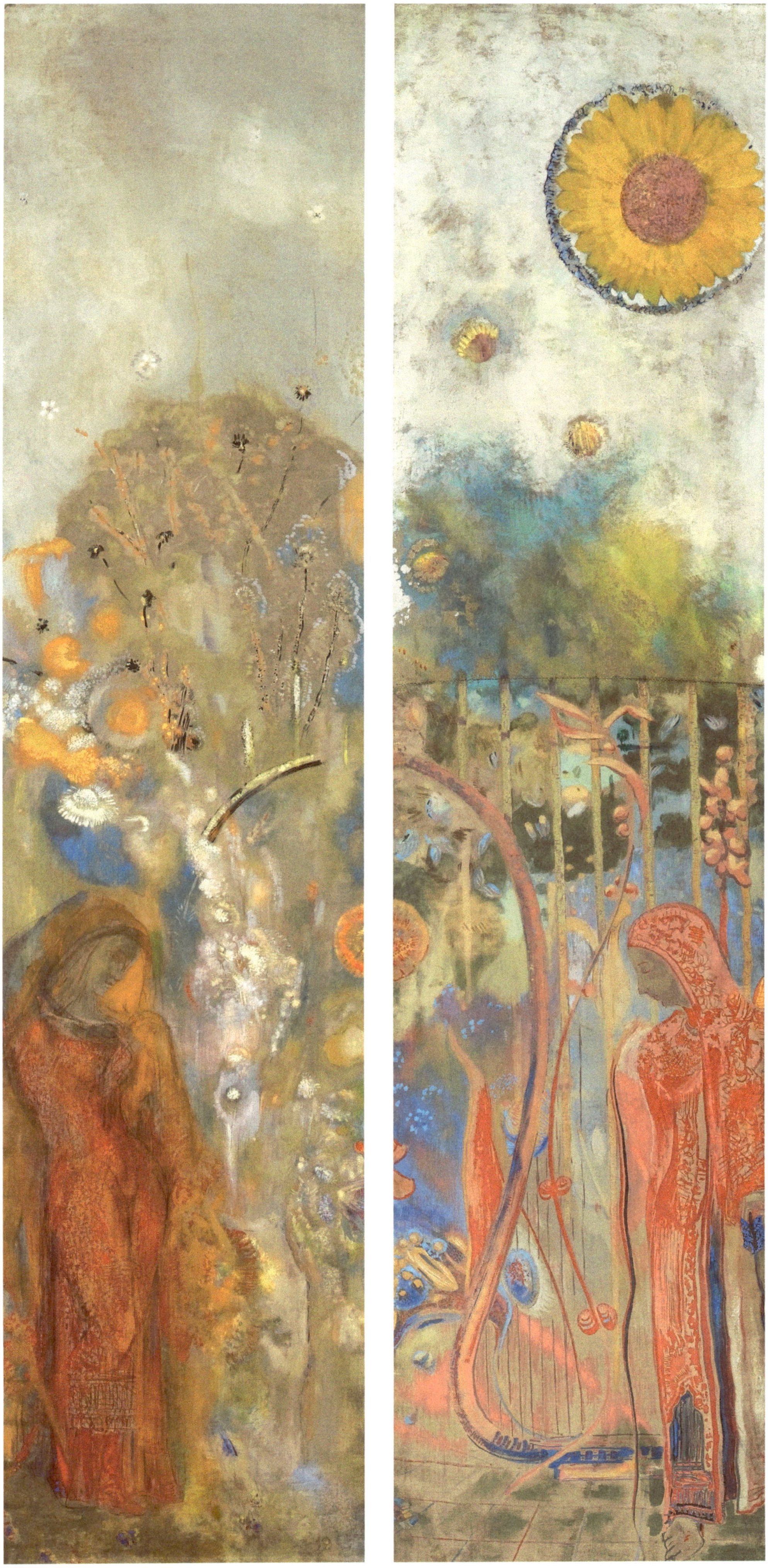

MONET, *WATER LILIES*:

ART, PHILOSOPHY, SCIENCE

NINA ATHANASSOGLOU-KALLMYER

In 1890, Claude Monet purchased the large country house surrounded by a garden which he had been renting since 1883 in Giverny, a village on the Epte, a tributary of the Seine, to the west of Paris. Three years later, he extended his property by purchasing a nearby field in which he created a small lake, fed partly by diverting some of the water of the Epte. From then on, his main occupation, after painting, was gardening. Unlike the garden around the house, which was divided into regular flowerbeds planted with common flowers—roses, tulips and forget-me-nots—the grounds around the lake, conceived as an irregular arrangement of winding paths, were planted with mostly exotic trees and flowers: weeping willows and red daylilies from China, stands of bamboo, reeds and multicoloured irises from Japan and North African lilies. A wooden bridge in the Japanese style arched over the central pool (which Monet enlarged in 1901). Multicoloured water lilies of a new variety, achieved by hybridizing white French water lilies and tropical specimens from the jungles of the Amazon, floated on the calm water. The journalist Maurice Guillemot, visiting Giverny in 1897, was astonished by these "disconcerting, strangely exotic flowers."[1] In fact, Monet had come across these water lilies when they were displayed for the first time at the Universal Exposition of 1889 in an unevenly shaped pool bordered—like the lake at Giverny—with "clumps of reeds," "stinking irises" and "wild sorrel," and above the vegetation, thatched huts like the primitive lakeside dwellings of the Bronze Age.[2] Obviously designed in counterpoint to each other, the two gardens form a whole: the one in the ordered and rational "French" style is counterbalanced by the garden around the lake, outlandish, unexpected, the image of a universal "natura," as exotic as the Orient, as wild as the jungle of the Amazon and as primordial as at the dawn of time.

During the years that followed, the lake was to be Monet's favourite subject. From 1897 until his death in 1926, he went back to it continually. "I took the time to understand my water lilies," he told Marc Elder. "And then, all of a sudden, I saw the light, the wonderland of my pool. I took up my palette. Since that moment I have scarcely had any other model."[3] Some 150 canvases were inspired by it, but only 80 of them have come down to us. The descriptive naturalism of the earliest paintings (1897 and 1900) was gradually replaced, in the following series (1903–1909), by a wider conceptual dimension. Monet transformed the modest lakeside landscape into a vision of nature with cosmic resonances evoked by means of innovative visual stratagems. Striking cut-outs, dizzying perspectives and the fragmentary appearance of the views glimpsed high up on the pictorial surface suggest notions of the infinite, of unity and of universality. The lines of horizon and land, the clusters of flowers, the stands of rushes and bamboo gradually fade away. The surface of the water and the underwater depths combine in a homogeneous whole of colour, pigment and brushstrokes. Boughs of weeping willows and swaying water weeds mingle with the frayed shadows of water-lily roots, like the fluid shapes of dishevelled nymphs rising from the depths of the blue-green water (199). The sky reappears on the water no longer in the form of filmy reflections but rather as mirages of voluminous masses of clouds. Water and sky combine, the depths of an underwater world are transformed into visions of inaccessible aerial spaces. Aquatic, celestial or plant-like, this world is shaped into a single animated material, one substance rendered in a tactile materiality of long, smooth calligraphic brushstrokes and ever brighter colours, expressions of nature, wild, inexhaustible, generating luxuriant life (197, 198, 199).

In his account of the 48 canvases of the *Water Lilies* series exhibited by the gallery owner Durand-Ruel in 1909, the art critic Roger Marx summed up the painter's thinking. Rather than rendering real nature palpable, Monet evokes the insubstantial web of air, light and reflections that transform it and arouse in him "the idea of the infinite . . . as in a microcosm, the existence of the elements and the instability of the universe which at every minute changes under our gaze." His sole desire would be to enter closely into "the grandeur, the power and the immortality" of this immaterial nature.[4] The reading of it presented by two of Monet's closest friends, the writer and art critic Gustave Geffroy and Georges Clemenceau, the politician and art lover, adds a philosophical dimension to the approach. In their view, Monet's lake with its water lilies is the crucible of a universal nature, the image of a microcosm which reflects a world unified by a common substance that pulses with life in "all things": "Then every canvas becomes a world, one of those synoptic instants in which Claude Monet's impressionism evokes the universality of things . . . Everything moves and lives in these admirable works which reveal the 'law of unity' that governs the manifestations of life,"[5] as Geffroy asserts with enthusiasm. For Clemenceau, the views of the lake are timeless and universal depictions of nature as the "condensation of the Universe": "Through their aquatic surfaces we are asked to contemplate the eternal movement of the planets. The sky, the flowers, the rippling surfaces of water mirroring the clouds could be the image 'of a world upside down,' of a cosmic whole undergoing a process of perpetual transformation."[6] A republican atheist with a scientific mindset and a love of classical culture, Clemenceau invoked the vision of a universal Arcadian utopia made up of a single living substance combining the senses and the mind, the domain of a pagan pantheism founded on the harmonious coexistence of man and nature, symbolized by the mythical god Pan, half man and half animal, a hybrid deity.[7]

Indeed, under the Third Republic, for those committed to a socialist ideology, a faith in a materialist, monistic and pantheistic spirituality handed down from the Enlightenment gradually replaced the Cartesian, dualist (mind versus body) and metaphysical thinking dating from the seventeenth century. Nature became the physical space in which the universal spirit resided. Science replaced metaphysics. Body and soul, science and art, spiritual beauty and physical beauty were all one in an integrated universe rethought in hylozoist and pantheist terms in

1 Maurice Guillemot, "Claude Monet," *La Revue illustrée*, no. 13, March 15, 1898.

2 C. L., "Les jardins du Champ-de-Mars," in *L'Exposition de Paris* (1889), published with the collaboration of expert writers (Paris: Librairie illustrée, 1889), vol. 1, p. 87. The garden was one of a series, overhung by aedicules designed by the architect Laforcade to illustrate the history of human habitation throughout the ages from prehistory to modern times. That Monet placed his orders with the inventor of the multicoloured water lilies, the horticulturalist Joseph Bory Latour-Marliac, is confirmed by the invoices addressed to the painter conserved in the archives of the greenhouse Temple-sur-Lot, which is still in operation. See http://latour-marliac.com/en/content/category/4-history.

3 Marc Elder, *À Giverny, chez Claude Monet* (Paris: Bernheim-Jeune, 1924), p. 12.

4 Roger Marx, "Les *Nymphéas* de Claude Monet," *La Gazette des beaux-arts*, vol. 1, no. 4 (June 1909), p. 528–530.

5 Gustave Geffroy, *Claude Monet. Sa vie, son œuvre* (Paris: Crès, 1924), vol. 2, p. 141.

6 Georges Clemenceau, *Claude Monet. Les "Nymphéas"* (Paris: Plon, 1928).

7 Georges Clemenceau, *Le Grand Pan* (Paris: Charpentier & Fasquelle, 1896). Monet owned this book. See Ségolène Le Men, ed., *La Bibliothèque de Monet* (Paris: Citadelles & Mazenod, 2013), p. 242.

accordance with the doctrines of Aristotle, Lucretius and Spinoza.[8] "Matter contains soul," stated the art critic Léon Bazalgette in his book *L'Esprit nouveau* (1898), which he sent to Monet along with an admiring dedication.[9] He announced the advent of "new times," of a modernism steeped in "a new pantheism infinitely broader than the old conception, wholly impregnated with reality and science." According to the critic, Monet's painting achieved this new ideal. It was "the vision of a modern man" who painted life "stripped of all symbols, all artifice, all lies."[10]

In this respect Bazalgette emphasized the influence of the thinking of the German zoologist and philosopher Ernst Haeckel, a professor at the University of Jena, and an evolutionist disciple of Lamarck and Darwin, whose books, translated into French in 1879, became internationally famous in about 1900.[11] Unlike Darwin, whose research was devoted above all to the mechanism of natural selection, Haeckel was primarily interested in the story of life. He believed that everything in nature was alive, from the unicellular "monerons" he discovered in fresh water and in the sea to man himself. "For [Haeckel], an immaterial living spirit was as inconceivable as matter without spirit and without life. In every atom the two are inseparably linked," explains Bazalgette.[12] Between 1899 and 1904—at the very time that Monet was working on his earliest *Water Lilies*—Haeckel published *Kunstformen der Natur* [*Art Forms in Nature*] (142, 144, 145, P. 214–215; 150, P. 218–219). He postulated the diversity of nature as the multiple derivative of primary organisms ("protists") and coined the term *Einheit* [unity] to signify the monist law that governs the universe. An accomplished amateur artist, he illustrated his text with 100 coloured lithograph plates after his drawings. In France, the hybrid world of his protists was to haunt the imaginations of the Art Nouveau artists and decorators, especially the master glassmaker Émile Gallé (203) and Hector Guimard, the architect of the Paris subway entrances inspired by plant motifs. Haeckel also influenced the shape and ornamentation of René Binet's monumental gate to the Universal Exposition of 1900, surmounted by the sculpture of *La Parisienne*, a symbol of modernism (143, P. 214). Gustave Geffroy, a great admirer of Haeckel, wrote the preface, illustrated with vignettes, to Binet's book *Esquisses décoratives* (1900). To fulfill this task to the best of his ability, Geffroy visited the home of the geologist Lucien Cayeux, where he saw through a microscope Haeckel's primary organisms. He came away astonished, having discovered "a universal beauty and harmony," and also having recognized the point "at which science enables us to see one of the stages of the evolution of the species, where it monitors the conjectured unity of matter."[13]

On one of his frequent visits to Giverny, Geffroy may well have shared with the painter his enthusiasm for Haeckel. Did Monet conceive the notion of his lake as a primordial aquatic landscape teeming with life, modelled on Haeckel's monist world? It is tempting to think so when we look at some plates in the biologist's book. The paintings and the lithographs share the same tendency to stylization and symmetry for decorative purposes. Like Haeckel, Monet creates an ambiguous microcosm with primordial, cosmic implications in which his water lilies, hybrid creatures half aquatic and half terrestrial, replace Haeckel's protists. Are we looking at a submarine landscape or a strange scene from overseas? Is this radiant blue a part of the horizon or is it the opaque light in the depths of the sea?

The same questions might be asked in relation to the large panels of the "Great Decorations," the last series of *Water Lilies* on which Monet was working furiously from 1913 until his death. The violence of the First World War and the losses suffered by his family distressed him but did not deter him from focusing on his great project. In his new studio at Giverny he recreated his flowery lake on enormous canvases (200 × 425 cm) and posed surrounded by them as if to underline his belief in the essential unity of man and nature (196). Did he reach his ultimate goal? The large-format canvases of his *Water Lilies*, displayed a year after his death in the specially constructed galleries of the Orangerie, were to be the proof of his faith and its apotheosis.

8 On this subject see *L'Âme au corps. Arts et sciences, 1793–1993*, exh. cat., ed. Jean Clair (Paris: Réunion des musées nationaux/Gallimard/Electa, 1993).

9 Léon Bazalgette, *L'Esprit nouveau dans la vie artistique, sociale et religieuse* (Paris: Société d'éditions littéraires, 1898), p. 38; Le Men, *La Bibliothèque de Monet*, p. 241.

10 Bazalgette, ibid., p. 387–388.

11 Jacques Roger, "Darwin, Haeckel et les Français," in *De Darwin au darwinisme. Science et idéologie*, proceedings of the international congress for the centenary of Darwin's death (Paris and Chantilly, September 13–16, 1982), ed. Yvette Conry (Paris: Vrin, 1983), p. 158–160; Yvette Conry, *L'Introduction du darwinisme en France au XIX*[e] *siècle* (Paris: Vrin, 1974), p. 12–13; Robert Richards, *The Tragic Sense of Life: Ernst Haeckel and the Struggle over Evolutionary Thought* (Chicago: University of Chicago Press, 2008).

12 Bazalgette, *L'Esprit nouveau dans la vie artistique*, p. 39.

13 René Binet, *Esquisses décoratives* (Paris: Librairie centrale des beaux-arts, 1900), p. 13. Binet dedicated his book to Geffroy.

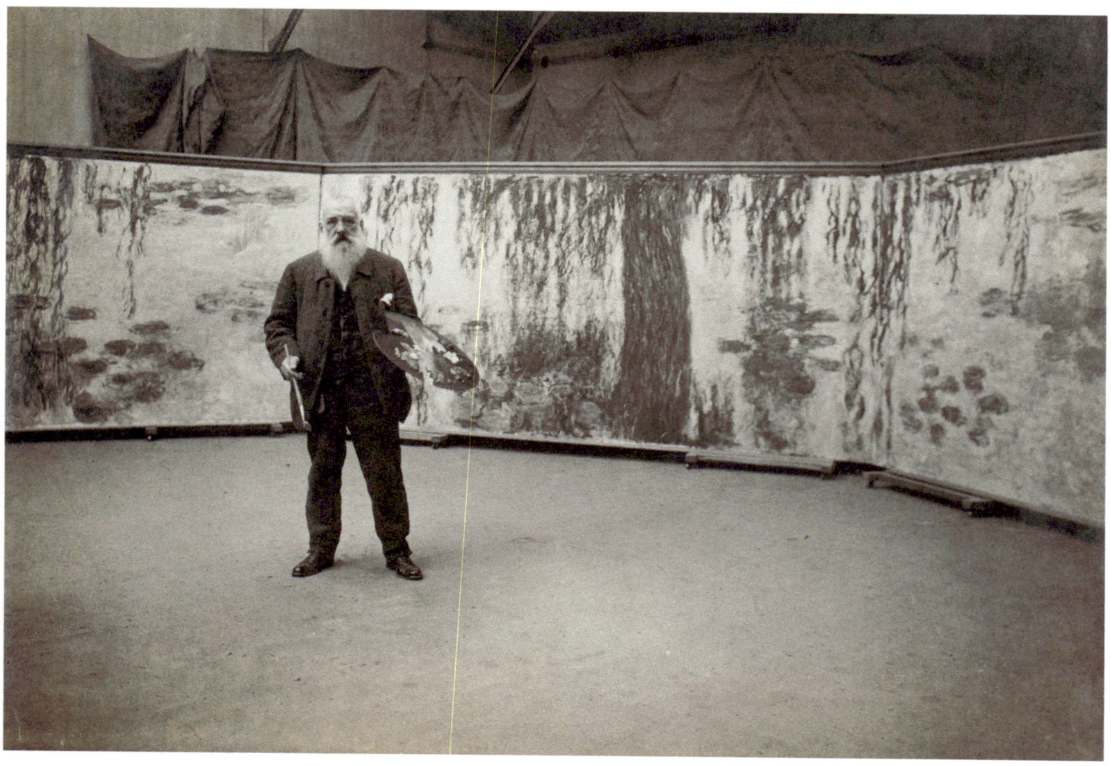

196 FIG.
Attributed to Henry Manuel,
Monet in the "Great Decorations" Studio,
about 1920 or 1922
Paris, Musée d'Orsay

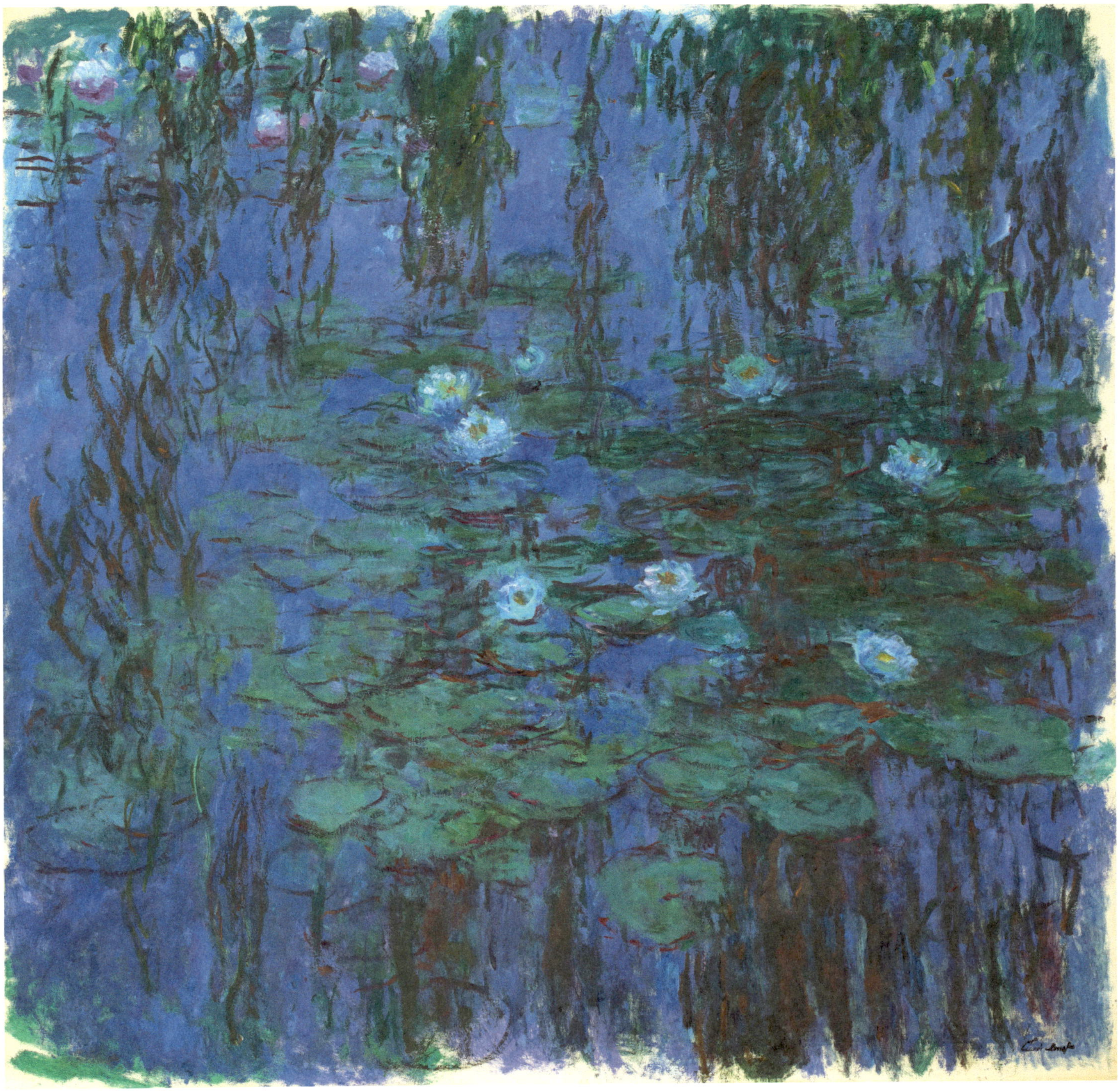

197
Claude Monet,
Blue Water Lilies, 1914-1919
Oil on canvas, 200 × 200 cm
Paris, Musée d'Orsay

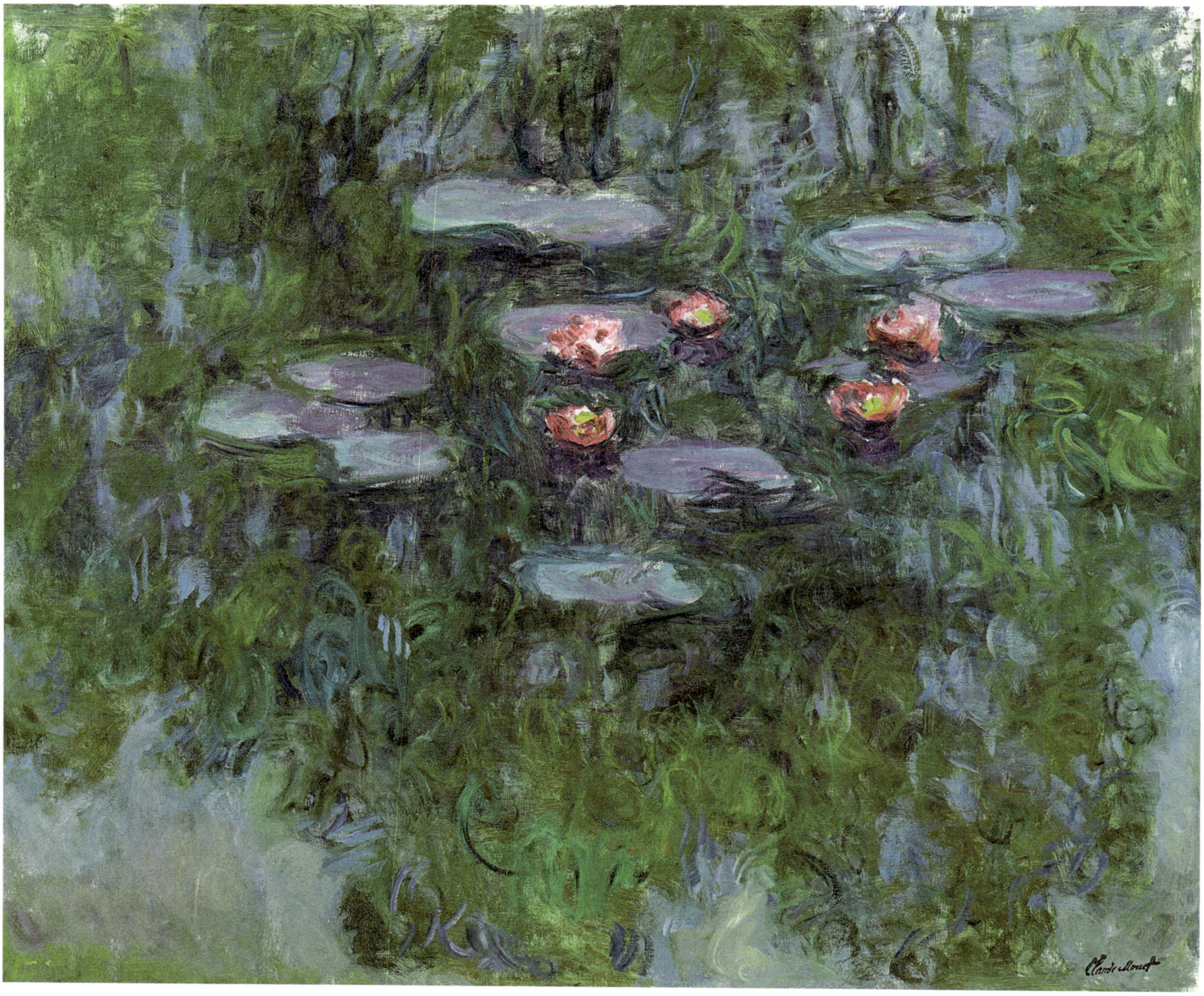

198
Claude Monet,
Water Lilies,
between 1916 and 1919
Oil on canvas, 130 × 152 cm
Paris, Musée
Marmottan Monet

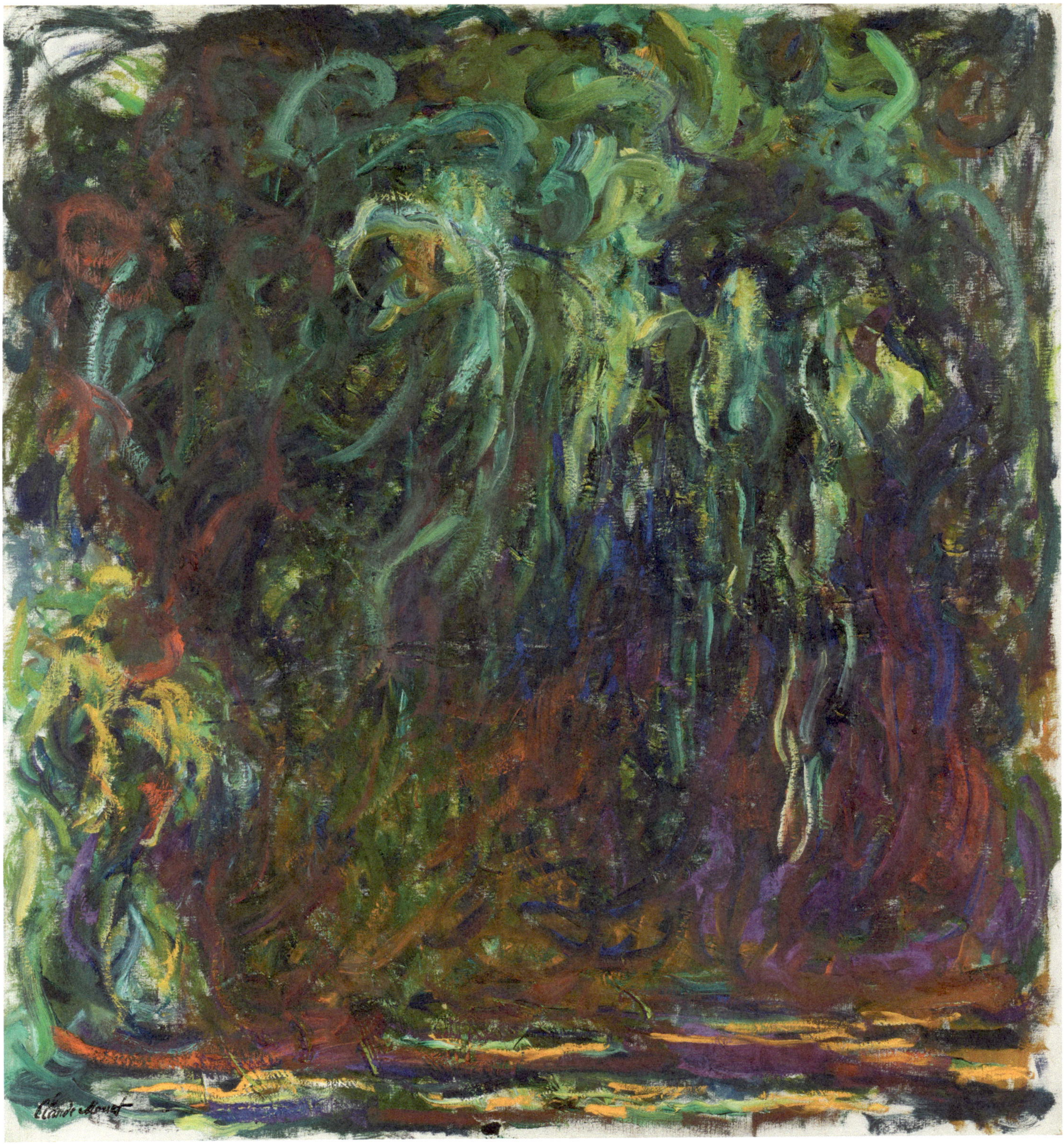

199 FIG.
Claude Monet,
Weeping Willow, 1920–1922
Oil on canvas, 110 × 100 cm
Paris, Musée d'Orsay

EVOLUTION AND ART NOUVEAU:

THE EXAMPLE OF ÉMILE GALLÉ

ÉLISE DUBREUIL

The emergence of Art Nouveau, a short-lived style of the final decade of the nineteenth century, can perhaps be best understood in terms of the evolution in the production of design that occurred during that century and in the development of a critical discourse that placed the question of ornamentation at the heart of the definition of a modern way of life. For the defenders of good taste, the major exhibitions and in particular the first universal expositions[1] were the sign of a decline. The accumulation of "industrial arts" productions, in ever-increasing quantities due to the development of industrial processes, suffered from waning inspiration among those who designed forms, despite often impressive technical virtuosity. This observation occurred simultaneously on both sides of the English Channel, in France and in England, the two European industrial, technical and artistic centres of the time. In 1856, two books appeared that were fundamentally different in spirit from one another, yet similar in their attempt to diagnose this inability of the contemporary arts to reinvent forms and to propose solutions: in France, Count Léon de Laborde wrote *De l'union des arts et de l'industrie*,[2] a work that was informed by his political experience and his role in France's participation in the Great Exhibition of 1851; in England, Owen Jones published his famous *Grammar of Ornament*,[3] with the goal of classifying and organizing a system of decorative vocabulary, inspired by earlier achievements, non-European cultures and natural forms. In both countries, the public authorities, concerned as much with maintaining quality in industrial production, which was a source of wealth, as with encouraging sound artistic practices, which were a source of prestige, implemented measures to promote better training for the draughtsmen and designers of objects, notably by reforming the way in which drawing was taught—industrial drawing in particular[4]—and by establishing museums that offered visitors direct exposure to past, recent and current examples of industrial art and the applied arts.[5] In this context, a theoretical crucible of Art Nouveau, observing the degeneration of art and its inability to invent new forms for modern life led the designers to investigate both models and a theoretical structure with the potential to guide the renewal of forms. Their interest was based on Asian art, Japanese art in particular, on a rereading of the past devoid of the historicist picturesque and on a renewed attention to natural forms. It can be argued that the contemporary development and dissemination of the sciences made their way into the artistic process of the protagonists of Art Nouveau. The very concept of a necessary regeneration appears to be directly related to the dissemination of the theories of evolution. The importance of the reference to the English Arts and Crafts movement, whose development coincides with the dissemination of the ideas of Charles Darwin, underscores the diffuse yet significant part played by these ideas in Art Nouveau. The belief in nature's vital energy and in its ability to produce well-proportioned and beautiful forms served as a model in designing a comprehensive system for the creation of artistic forms. This approach was particularly prevalent among the artists of the School of Nancy, of which Émile Gallé was at the forefront.

Established on February 13, 1901,[6] and spearheaded notably by Émile Gallé, the Alliance provinciale des industries d'art, more commonly referred to as the "School of Nancy," consolidated the city's reputation as a major centre for Art Nouveau. In the particular context of the annexation of Alsace-Moselle, the school also had the more political aspect of promoting Lorraine's French identity and the vitality of its artistic output. Indeed, as an artistic centre, Nancy was particularly active in the exploration of new styles and the revitalization of the decorative arts and architecture. In relying on well-established craft traditions, Nancy's artisans endeavoured to advocate for a revival that was stylistic as well as industrial and economic.[7]

The work of Gallé, a leader of the School of Nancy, is particularly revealing when it comes to the relationship between scientific culture and Art Nouveau. From an early age, he was immersed in the two worlds, artistic and scientific, botany in particular. Before long, his parents, who strove to make a success of their glassware and ceramics business, had made plans for him to pursue a career as a manufacturer and merchant. Upon graduating from high school, and following the example of his father, who trained as a painter on porcelain, he became a pattern designer for the family's faience works. It is said that he also learned to read from Grandville's *Les Fleurs animées*[8] and to collect plants with a number of Nancy botanists, including his friend René Zeiller, the future paleobotanist, and Dominique-Alexandre Godron, a professor of botany in Nancy's Faculty of Science.[9] Through his connection with the local scientific community, Gallé was aware of the latest international scientific developments. It was therefore as a connoisseur, almost as an experimenter, that he told his followers of the School of Nancy: "The modern setting must be in harmony with the dissemination of the biological sciences."[10] Along with Godron, Gallé was interested in teratology and the theory of evolution, producing field-based empirical observations and surveys of endemic orchids.[11] When Godron died, he took over his position on the supervisory committee of the Jardins botaniques de Nancy and pursued his study of orchids.[12] This led him, notably, to give a talk on the orchids of Lorraine during the International Botanical Congress, which took place in Paris, from October 1 to 10, 1900, as part of the Universal Exposition, where, incidentally, he exhibited his most accomplished works. A marked permeability existed between this botanical activity and Gallé's creative work, which, over time, became increasingly personal. He subsequently developed his own symbolism inspired by both literature and science. In the mid-1880s, Gallé increased his glassmaking activities to the detriment of his faience work, for which he had, undoubtedly, exhausted the technical abilities to render his artistic intentions. The plant-inspired ornamentation, which was present from his earliest works, then fully manifested itself in spectacular creations. His major

1 The first in 1851, in London, the second in 1855, in Paris.

2 In 1851, he was named a member of the French commission of London's Great Exhibition.

3 Owen Jones, *The Grammar of Ornament* (London: Day & Son, 1856).

4 See in particular Stéphane Laurent, *L'Art utile. Les écoles d'arts appliqués sous le Second Empire et la Troisième République* (Paris: L'Harmattan, 1998).

5 The Museum of Manufactures, the future Victoria and Albert Museum, was established in London in 1852. In Paris, the Union centrale des beaux-arts appliqués à l'industrie, established in 1864, became the Union centrale des arts décoratifs (UCAD) in 1882, when it merged with the Société du musée des Arts décoratifs, established in 1877.

6 The association had an office made up of a president, Émile Gallé, and three vice-presidents, Louis Majorelle, Antonin Daum and Eugène Vallin.

7 *L'École de Nancy. Art nouveau et industrie d'art*, exh. cat., ed. Nicolas Newmann (Nancy: Musée de l'École de Nancy; Paris: Somogy, 2018).

8 Jean-Jacques Grandville, *Les Fleurs animées*, introduction by Alphonse Karr, text by Taxile Delord (Paris: Gabriel de Gonnet, 1847), 2 vols.

9 Dominique-Alexandre Godron is the author of *De l'espèce et des races dans les êtres organisés et spécialement de l'unité de l'espèce humaine* (Paris: Baillière, 1859). His writing reflects the numerous conferences presented, in particular at the Société des sciences, lettres et arts de Nancy.

10 Statement transcribed by Émile Nicolas, quoted by François Le Tacon in *Emile Gallé, l'Amour de la fleur. Les écrits horticoles et botaniques du maître de l'Art nouveau* (Nancy: Place Stanislas, 2007), p. 324.

11 The article by Godron, titled "Note sur un fait remarquable de tératologie végétale," appeared in the *Bulletin de la Société centrale d'horticulture de Nancy*, in 1879.

12 On Gallé's "Darwinism," see in particular François Le Tacon, "Émile Gallé, botaniste et scientifique," *M/S Médecine Sciences* (online journal), vol. 21, no. 12 (December 2005), p. 1096–1100, https://id.erudit.org/iderudit/012018ar.

works of cabinetmaking and marquetry also pay homage to the flora of Lorraine, revitalizing heraldic and traditional script, which he knew well, having painted it on faience in his early years. The recurring theme of the orchid is particularly important in his works and is a result of the meticulous attention paid to these plants during his sessions in botany. For Gallé, this was also a subtle and symbolic way of stating his convictions: the orchid became a symbol of the region of Lorraine, scarred by annexation, but whose vital force, a guarantee of adaptation and renewal, lives on.

Ultimately, in his later years, Gallé developed, with great artistic power, a theme that until then had been a discreet presence in his work: the sea and the underwater world. Contrary to the undergrowth in the forests of Lorraine, the maritime environment was an unknown territory for the artist, who for many years preferred the dormant waters of ponds and pools, shelters for a plant and animal life that inspired him. In 1889, at the Universal Exposition, he presented *Jardinière Flora marina, Flora exotica,* whose marquetry design was created in collaboration with Victor Prouvé (200). This work marked the beginning of an immersion in the underwater world that provided inspiration for some of his most powerful works (201, 202, 203), including his final masterpiece, *Hand with Seaweed and Shells* (204).[13] During these years, a surprising synthesis existed between the literary and symbolist sensibility expressed by Gallé, who had discovered the power of the marine element through the writings of Charles Baudelaire and Victor Hugo, and his scientific knowledge. Certainly he was interested in oceanography: perhaps, like Julien Thoulet,[14] he followed Nancy's oceanography conferences; in any case, he knew the works of Ernst Haeckel and, at the end of his life, had an opportunity to become acquainted with the illustrated plates of the *Kunstformen der Natur* (142, 144, 145, P. 214–215; 150, P. 218–219).[15] This ability to blend, in very personal works, scientific knowledge and symbolist inspiration contributed to the power of his works. *Hand with Seaweed and Shells* necessarily poses questions due to its strangeness and its very incongruity within the artist's body of work. A final submerging or emerging from the primordial waters, the effect of this work is doubly significant, recalling the ambiguity of life, a force destined to erupt and then die out, inscribed in a nature that is both creator and devourer.

13 On this topic, see Philippe Thiébaut, ed., *Gallé, le testament artistique*, exh. cat. (Paris: Musée d'Orsay/Hazan, 2004).

14 Julien Thoulet, oceanographer, naturalist and hydrographer. Professor at the Université de Nancy, speaker, notably, on Newfoundland (in 1887) and on the Antarctic (in 1894).

15 Ernst Haeckel, *Kunstformen der Natur*, eleven pamphlets published between 1889 and 1904.

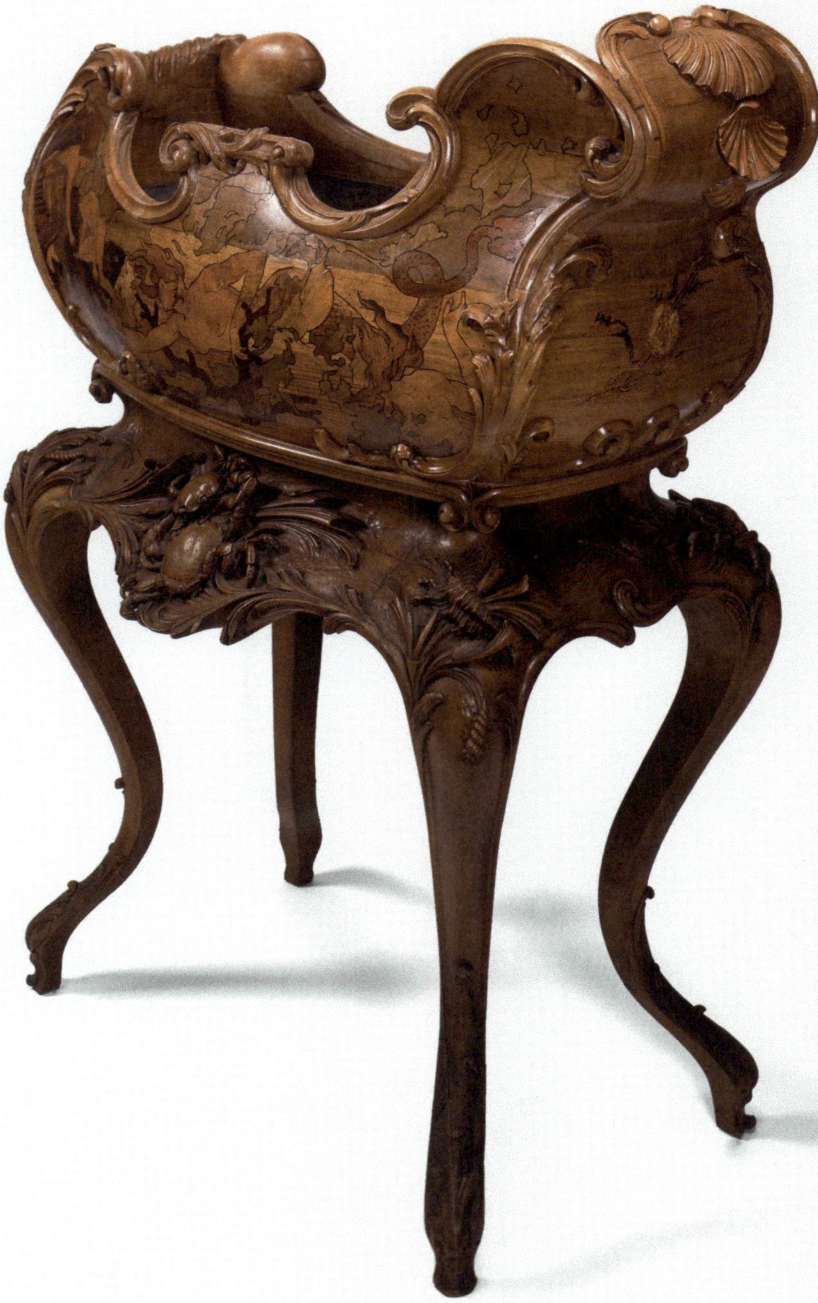

200
Émile Gallé, Victor Prouvé
and Louis Hestaux,
Jardinière Flora marina,
Flora exotica, 1889
Sculpted pear wood,
marquetry in different woods,
117 × 95 × 56 cm
Nancy, Musée de l'École de Nancy

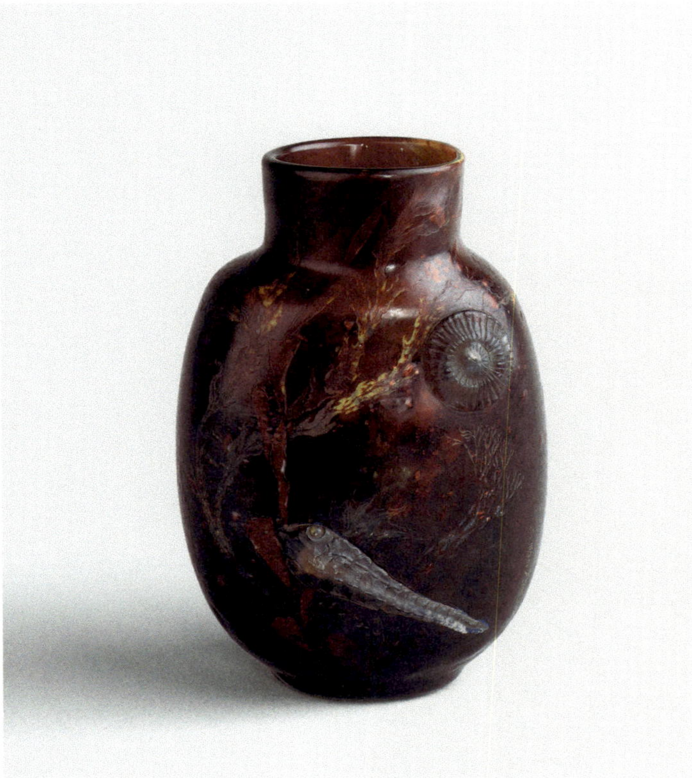

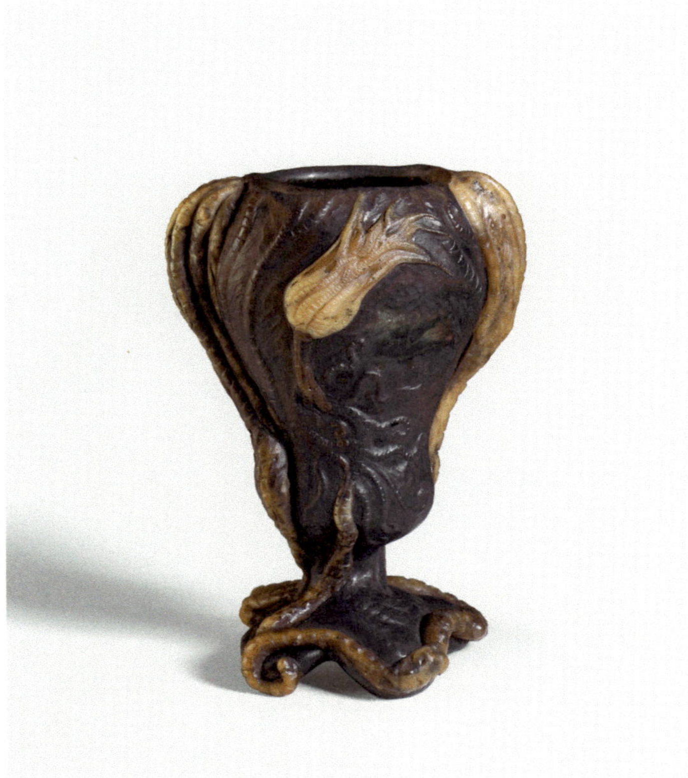

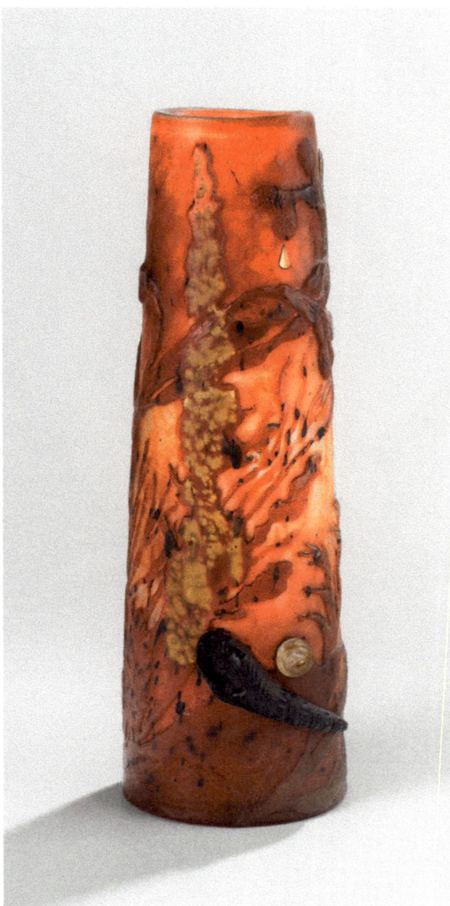

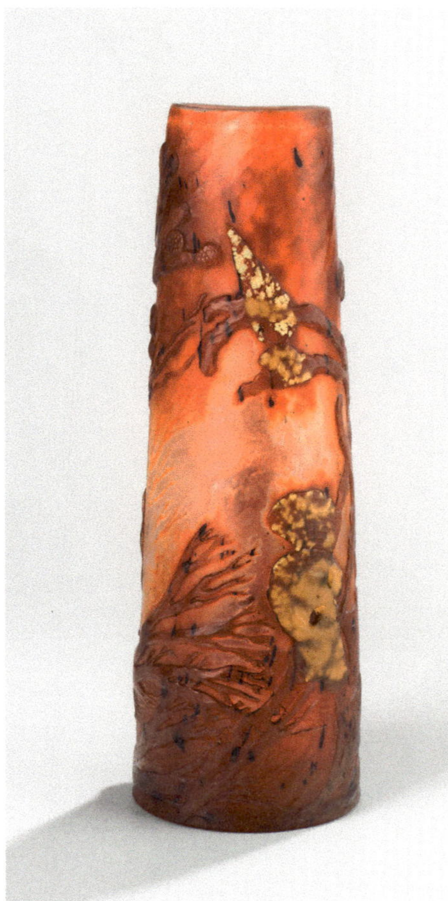

201
Émile Gallé,
The Bottom of the Sea, 1889–1903
Multi-layered glass,
metal inclusions and intercalated bubbles, two wheel-cut applications, 21.8 × 14.3 cm
Nancy, Musée de l'École de Nancy

202
Émile Gallé,
Sea Lily, about 1903
Double-layered glass,
metal inclusions, engraved decoration, 14.5 × 10.5 × 8 cm
Nancy, Musée de l'École de Nancy

203 A, B
Émile Gallé,
Rouleau Vase with Marine Decoration, 1902–1903
Blown glass in several layers, metal inclusions, wheel-cut decoration, applications,
28.2 × 9 cm (base)
Paris, Musée d'Orsay

204
Émile Gallé,
Hand with Seaweed and Shells, 1904
Glass modelled under heat, inclusions, applications,
wheel engraving, 33.4 × 13.4 cm
Paris, Musée d'Orsay

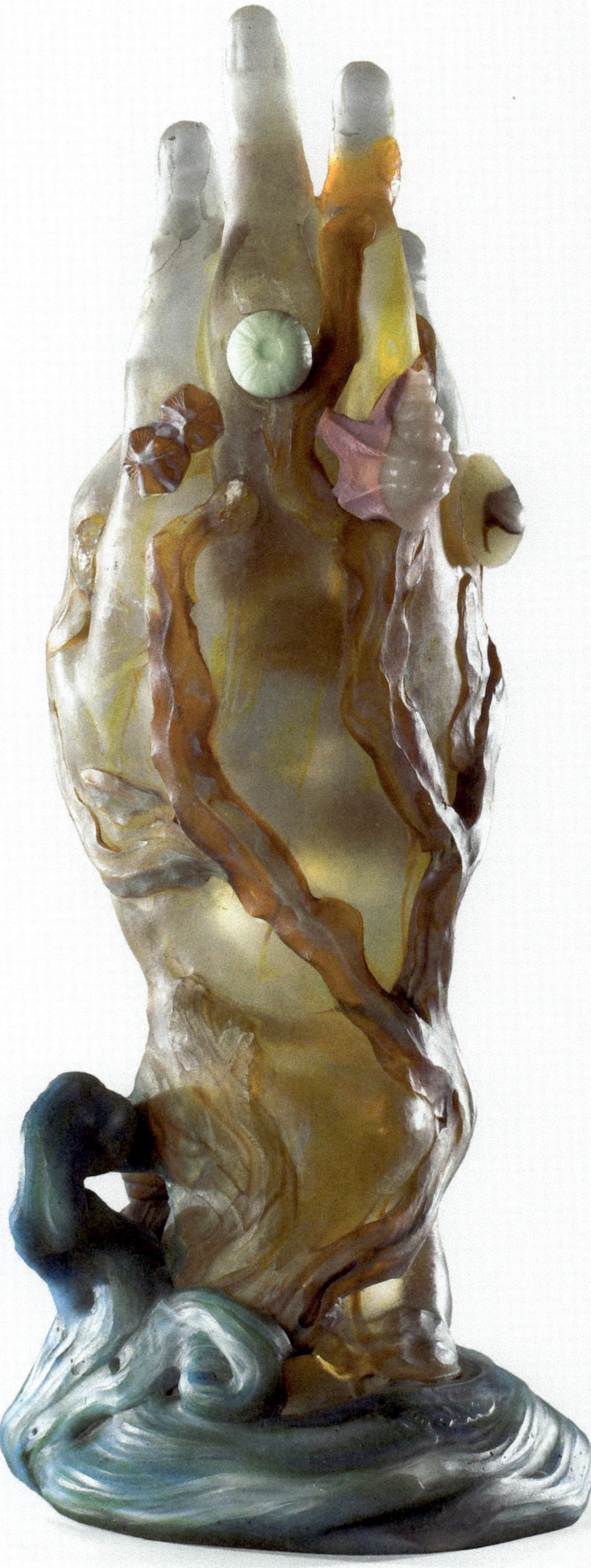

VIII FATAL MIRROR OF THE AGES

PHILIPPE COMAR

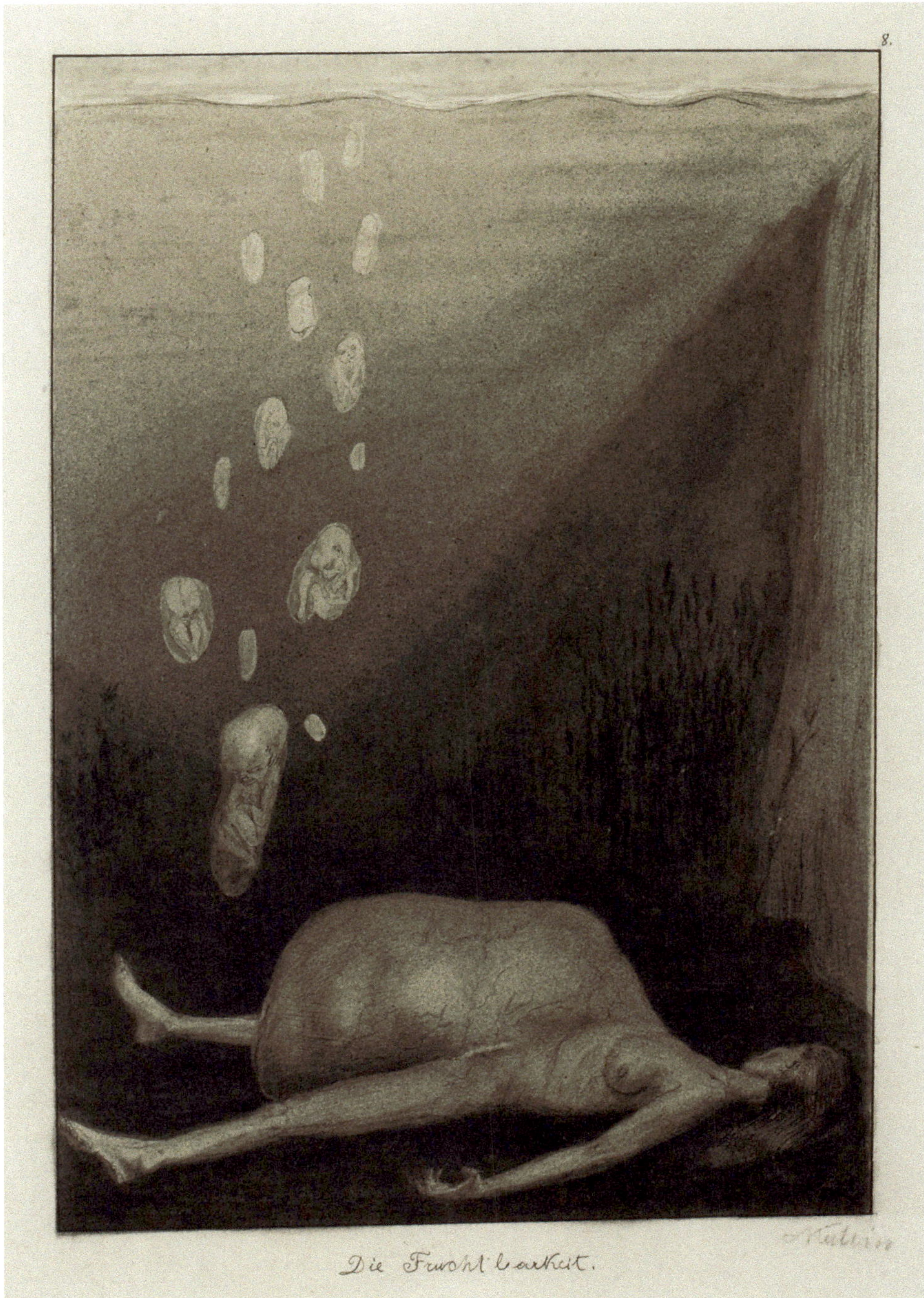

205
Alfred Kubin,
Fertility (Version I), 1901–1902
India ink, pencil, watercolour, spray, on cadastral paper,
37.2 × 27.2 cm
Vienna, Leopold Museum

When in 1879 Marcelino Sanz de Sautuola discovered the wall paintings in the caves of Altamira, scholars and historians could not believe their eyes: the images were too beautiful to be true. How could Stone Age savages have learned to draw so superbly well? It would take a decade to convince them that the cave paintings were not fraudulent. In thirty thousand years, what will happen? Will scholars and historians believe their eyes when they discover that an entire civilization, our own, felt that the quintessence of its culture could be embodied in a black square, splashes on the ground, a split canvas or a urinal? How could an educated, sophisticated civilization recognize itself in such rudimentary productions, which seemed to hark back to the infancy or the prehistory of art?

The triple parallel of the origins of man, presented by Ernst Haeckel in 1863, between *embryological, systematic* and *paleontological* development, corresponds to the triple parallel of the origins of art, that is, between children's drawings, primitive art and prehistoric art.

It was actually at the time when Haeckel was conceiving what would later become his "theory of recapitulation"[1] that attention was first being paid to children's drawings. Aside from Jean-Jacques Rousseau, who in *Émile, ou De l'éducation* considered them worthwhile studying, it was not until the second half of the nineteenth century that scholars and art historians became interested in them and acknowledged their real importance. Ebenezer Cooke, a pupil of John Ruskin, published *Our Art Teaching and Child Nature* in 1885; Corrado Ricci, *L'arte dei bambini*, in 1887; Bernard Perez, *L'Art et la poésie chez l'enfant*, in 1888. At the same time, the development of ethnography saw in so-called "primitive" peoples an image of our ancestors. Of his voyage aboard the *Beagle*, Charles Darwin reported, "The astonishment which I felt on first seeing a party of Fuegians on a wild and broken shore will never be forgotten by me, for the reflection at once rushed into my mind—such were our ancestors"[2] (52, 53, P. 74–75). Since the way of life of these Indigenous peoples was then compared to that of our ancestors, their art was seen as a surviving form of archaic art. Finally, prehistoric art, the existence of which was only discovered with the wall paintings at Altamira, and then at Font-de-Gaume, Pech-Merle and Lascaux—the Sistine Chapel of cave paintings—revealed a body of unexpected images that shattered prevailing notions on the origins of art.

To sum up, *the child, the savage* and *prehistoric man* were not only assimilated into the theory of recapitulation, which viewed them as representing initial stages of development but also were linked to the forms of art they were capable of producing.

Now, modern art was to lay claim to a recurrent kinship, to this triple return to its origins.

First, the return to childhood, a harbinger of which can be seen in Gustave Courbet's *The Artist's Studio*. Standing before the painting depicted is a young boy, suggesting that the artist should look at the world with the eyes of a child. In addition, we see an urchin lying on his stomach scribbling on a sheet of paper. Genius is, according to Baudelaire, "childhood recovered."[3] This return to their early years became a leitmotif for artists from the 1880s on. Paul Gauguin was the leader of this art of ingenuousness. "I retreated far away, farther than the horses of the Parthenon . . . to the *dada*, the gee-gee of my infancy, the wooden horse."[4] We find ourselves thinking about the vicissitudes of the word "dada," which was taken up some years later by an art movement determined to anchor itself in the values of childhood. To return to an early age was to become a precept for the Surrealists: "If he remains clear-headed, [man] can only return to his childhood" (*Manifeste du surréalisme*, 1924). Like his friend Gauguin, Émile Bernard sought to wet-nurse art and make "child-like paintings."[5] André Derain also advocated this ingenuous approach, stating in a letter to Maurice de Vlaminck: "I want to study kids' drawings. That is undoubtedly where truth lies."[6] For his part, Henri Matisse asserted that he had "understood painting" by looking at the drawings of his five-year-old son.[7] The new babyish ambitions of art were not without their critics. August Strindberg compared modern artists to "children dismantling their toys."[8] Paul Valéry made fun of painters whose world boiled

1 Ernst Haeckel, *Generelle Morphologie der Organismen* (Berlin: G. Reimer, 1866), 2 vols.
2 Charles Darwin, *The Descent of Man, and Selection in Relation to Sex* (London: John Murray, 1871), vol. 2, p. 404.
3 Charles Baudelaire, *Le Peintre de la vie moderne*, 1868, chap. 3.
4 Paul Gauguin, *Avant et après* [1903] (Paris: La Table Ronde, 2017), p. 47.
5 Quoted by Vincent van Gogh, *Correspondance générale*, vol. 3: *28 février 1886–29 juillet 1890* (Paris: Gallimard, 1990), p. 269.
6 André Derain, *Lettres à Vlaminck, suivies de la correspondance de guerre* (Paris: Flammarion, 1994), p. 89.
7 Quoted by Hilary Spurling, *The Unknown Matisse*, vol. 1: *1869–1908* (London: Hamish Hamilton, 1998).
8 Quoted by Gauguin, *Avant et après*, p. 55.
9 Paul Valéry, *Degas, danse, dessin* [1936] (Paris: Gallimard, "Folio-Essais" series, 2017), p. 195.
10 William Hogarth, *The Analysis of Beauty* (London: J. Reeves, 1753), p. 124–125.
11 Joan Miró, letter to Christian Zervos, January 20, 1932.
12 Joachim Gasquet, *Cézanne* [1921] (Paris: Encre marine, 2012), p. 87.
13 Marcel Réja, *L'Art chez les fous. Le dessin, la prose, la poésie* (Paris: Mercure de France, 1907), p. 63.
14 Pierre Cabanne, *Le Siècle de Picasso*, vol. 1: *La Naissance du cubisme, 1881–1912* (Paris: Gallimard, "Folio-Essais" series, 1992), p. 286.

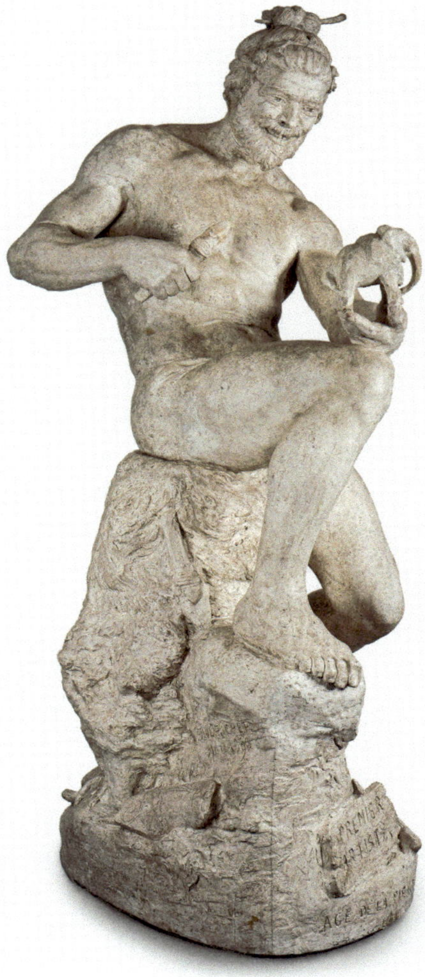

down to a game of cubes and who, like Cézanne, rendered nature as spheres, cones and cylinders and sought to rebuild everything with "playthings of infant geometricians."[9] In earlier times it would have been unthinkable to regard any example of this schoolchildren's art as a masterpiece. William Hogarth, in *The Analysis of Beauty* (1753), did not believe that children had any artistic sensibility, and found no beauty at all in "such plain lines as children make, when of themselves they begin to imitate in drawing a human face."[10] While the classical age saw excellence—the outcome of a long learning process—as the goal of art, the modern era was to make unlearning its ultimate aim. We know Picasso's quip: "As a child, I drew like Raphael, but it has taken me a lifetime to draw like a child," that is to say, to go back to the form of art that precedes learning, which nevertheless, as we know, is mainly composed of stereotypes. The fascination of scholars and men of letters with this ingenuousness reached its apogee when Picasso, Apollinaire and their group gave an uproarious welcome to Le Douanier Rousseau—whose paintings were mocked for their childlike appearance—at the banquet they organized in his honour at the Bateau-Lavoir in 1908, that strange spectacle of high culture celebrating innocence and naivety. From then on the values of childhood would rhyme with those of modernism. This return to the cradle of creation was to find its loveliest expression when Joan Miró, after spending twelve years in Paris, came back to live in Barcelona in 1932: "If you like to learn about odd things, accidents or coincidences, I must tell you that the room that from now on will be my studio is the very bedroom where I was born."[11]

Then came the return to primitive art. There again, the master of the tropics played an important role as a leader: "I shall go to Tahiti and I hope to end my days there. I judge that my art, which you like, is only a seedling and I hope to cultivate it for myself in its primitive and savage state" (letter from Paul Gauguin to Odilon Redon, 1890). Primeval man, accessible by way of an escape to distant lands, became the starting point for a search into oneself. "I am the primitive of my own way of painting," asserted Cézanne.[12] The child's drawing and that of the savage were often perceived as alike. In a chapter entitled "Drawing of Children and Savages," Marcel Réja, a poet and historian of asylum art, saw similarities between children's art and primitive art: "This similarity often makes them indistinguishable, so that one type sheds light on the other."[13] Picasso, for his part, in 1906 visited the Musée d'Ethnographie du Trocadéro, where he found African art, which he began to collect as Matisse, Derain, Vlaminck and Braque were doing. He purchased items most notably from the dealer Émile Heymann, nicknamed "le Père sauvage" and also "le négrier de la rue de Rennes." Like Gauguin, Picasso was turning towards "barbarism."[14] He also assembled some 200 colonial photographs published in nude magazines such as *L'Humanité féminine*, taking with him an issue devoted to "the woman of Africa" when he travelled to Horta de Ebro in 1909. At the same time, Matisse was making use of ethnographic documents for a number of his works, such as

206
Paul Richer,
First Artist, 1890
Plaster, 179 × 67 × 80 cm
Le Puy-en-Velay,
Musée Crozatier

Two Negresses and *Standing Nude*, assimilating the sources of his art from this generic humanity. Constantin Brancusi also frequented the collections in the Trocadéro while repudiating the "beefy sculpture"[15] of Auguste Rodin, in whose studio he had worked on first arriving in Paris in 1904. For the avant-garde, archaic art became the archetype of art.

Finally, we see the return to prehistoric art, initially apparent in the calling into question of the "historic" art exhibited in the museums. Modern art was characterized by iconoclasm and desacralization. Innumerable inflammatory statements were issued by artists condemning museums and inciting the destruction of masterpieces of Western art. As a young man, Cézanne launched the slogan "Burn down the Louvre!" In 1909, the Futurists wanted to "free Italy from her innumerable museums which cover her like countless cemeteries."[16] Marcel Duchamp boasted about never having set foot in the Louvre, but he painted moustaches on the *Mona Lisa*. The *Dictionnaire abrégé du surréalisme* suggested that a canvas by Rembrandt could be used as an ironing board. Joan Miró wanted to "assassinate painting."[17] Even Pierre Bonnard, although not noted for extremism, claimed that the most beautiful thing about museums was what one could see through the windows. Together with these fearless nihilists we find the inevitable cohort of followers. Guy Debord, in 1958, was to defend, in the name of modernism, a young artist who had tried, in the picture gallery of Brera, to break the glass protecting Raphael's *Marriage of the Virgin*. In May 1968, Jean-Jacques Lebel was still calling upon artists to burn down the Louvre, forgetting that Cézanne, late in life, had amended his youthful slogan: "Before, my ideas were confused. Life! Life! I only talked about this one thing. I wanted to burn down the Louvre, poor fool!"[18] Statuettes, carvings on bone, prehistoric drawings and imprints were acclaimed by the artists and theoreticians of the avant-garde, who saw in them a counterpoint to high culture, even though some of them treated this fossil art with a lack of deference that gives food for thought. In 1952, André Breton, who had chosen to live at the foot of the cave of Pech-Merle at Saint-Cirq-Lapopie, was prosecuted for having rubbed out a mammoth's trunk with a finger, in order, he claimed, to check its authenticity. By dint of scratching to discover the origin, he returned to the bare surface.

This search for modernism in prehistory was also apparent in official art. In 1890, Paul Richer, using bones from a Cro-Magnon man, created a model in clay—Adamic earth—of the *First Artist* engaged in carving a mammoth (206), a *mise en abîme* of the sculptor's creative gesture, astounded by his increasing power. A few years later, Paul Jamin depicted *A Painter Decorator in the Stone Age*. Fernand Cormon, Léon Maxime Faivre and others also succumbed to the fashion for prehistory and illustrated the epic of early man (109, p. 159; 172, p. 251). In 1895, the Bal des Quat'z'arts at the École des beaux-arts in Paris was an opportunity to enact prehistoric scenes in which students and models brandishing cudgels and flint axes posed, naturally, in the nude. And the postcards and albums of nude photographs for the use of artists that proliferated in the early twentieth century featured recreations of prehistoric scenes, such as *Cave-dwelling Woman* and *Stone Age Woman*.[19] By 1906, all these photographs formed "a documentary tableau vivant of the process of natural selection through which the female body developed down the centuries, from the origin of the human species to our own day," wrote Amédée Vignola,[20] one of the promoters of prehistoric photography, a new genre.

To this triple artistic parallel linking the art of children, that of savages and that of cave men, was added another, more formal parallel linked to the very genesis of the work of art, which goes through a certain number of stages from the sketch to the finished work. Here again we go back to the newborn stage of creation: the quick sketch and the sketch in colour proved their worth at the expense of the finished work. This return to the beginning was marked by the attention paid to the first draft, without reworking or correcting, by the use of flat areas of pure colour, by the contempt felt for skill and the rejection of ingenious construction in favour of a sort of deliberate naivety.

In formal terms, Cloisonnism, a movement launched by the school of Pont-Aven and subscribed to by Maurice Denis, Van Gogh and Toulouse-Lautrec, consisting of outlining the contours of the forms in a dark shade and filling the flat surfaces with colour, undeniably referred back to Japanese prints but also to the practice of colouring in albums of drawings for children. For their part, the French Fauve painters

15 Georges Duby and Jean-Luc Daval, eds., *La Sculpture, de la Renaissance au xxe siècle* (Paris: Taschen, 2006), p. 954.
16 Filippo Tommaso Marinetti, *Le Figaro*, February 20, 1909.
17 Rémi Labrusse, *Miró, un feu dans les ruines* (Paris: Hazan, 2004), p. 166.
18 Gasquet, *Cézanne*, p. 159.
19 *Vénus et Caïn, figures de la préhistoire, 1830–1930*, exh. cat., ed. Hélène Lafont-Couturier et al. (Bordeaux: Musée d'Aquitaine; Paris: Réunion des musées nationaux, 2003), p. 145.
20 Amédée Vignola, *L'Idéal féminin* (Paris: Librairie documentaire, 1906), p. 109.

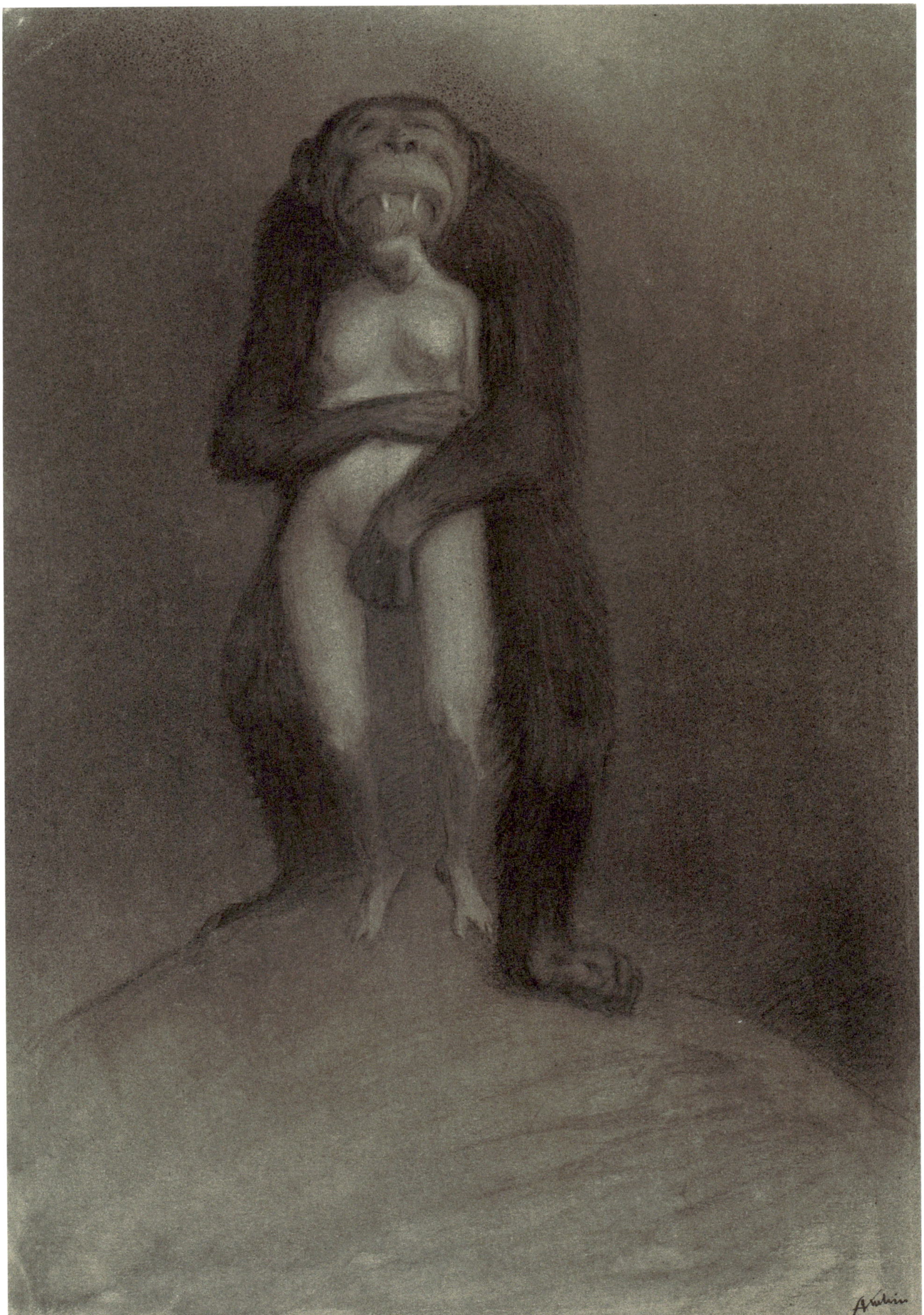

207
Alfred Kubin,
The Ape, about 1903–1904
Chalk and pencil,
35.7 × 23.8 cm
Vienna, Albertina

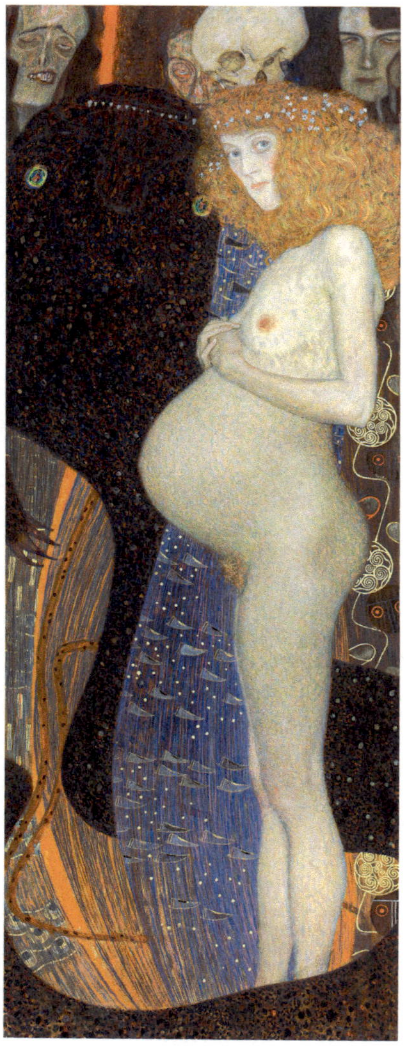

208
Gustave Klimt,
Hope I, 1903
Oil on canvas, 189.2 × 67 cm
Ottawa, National Gallery
of Canada

used colour extravagantly: "How do you see that tree? Is it green? Well then, put down green, the most beautiful green in your palette" (Gauguin).[21] A return to the fundamentals of painting. Twenty years later, in a portrait entitled *The Lipstick* (1908), František Kupka, who painted numerous works featuring anthropoids (174, P. 253), rendered his model's mouth with a "local tone" that came straight out of a tube . . . of lipstick. The literalness of painting. The image and the object merge. It is no longer a representation, only a presentation.

The painters, moving to an exaggerated pictorial simplification, abandoned what until then had pertained to the traditional craft: the glazes, the relief, the stumps, the *non finito* of the contours, the shading and above all the linear and aerial perspectives. The assertion of the flatness of the pictorial surface harked back to the art of cave painting and ushered in the art of the *tache*. Rather than patiently working at making the shapes appear on the canvas, bringing them to light by the application of successive layers of paint, many artists, like Matisse, preferred paper cut-outs and childish collages. Paul Valéry derided the embryonic art of his contemporaries: "Almost everything is done without studies; or rather, almost everything is only a study."[22] An opinion shared by Van Gogh in a letter to Théo: "At the moment I'm really in the shit, studies, studies, studies, and that'll go on for some time yet."[23] A month later he was committed to an asylum.

But this return to the infancy of art is above all apparent in the artists' refusal to finish their works. In 1880, a critic nicknamed Berthe Morisot "the angel of the uncompleted."[24] As for Cézanne, he did not allow himself to "reach the end, which is only admired by imbeciles."[25] The areas of his paintings in which the unpainted canvas looks naked were to be perceived as a formal audacity worthy of admiration. Non-completion became the criterion of great art, especially when it sprang from a creative failure. The painter Claude Lantier, the hero of Zola's *L'Œuvre*, despairing of being able to conclude his masterpiece, destroys the canvas in a rage. The admission of powerlessness that haunts the modern artist becomes the sign of his genius. One day, when Émile Bernard was speaking in the presence of Cézanne about Balzac's *Le Chef-d'œuvre inconnu*, the painter from Aix, pointing his finger to his chest, said: "Frenhofer, that's me!"[26] Return to the original chaos. The subject of the artist's powerlessness is ubiquitous in the literature of the nineteenth century. In addition to Claude Lantier and Frenhofer, Coriolis in the Goncourt brothers' *Manette Salomon*, Pellerin in Flaubert's *L'Éducation sentimentale* and Cyprien in Huysmans's *En ménage*, all of them are confronted by failure. Painting has become impossible. "Painter's block"—which foreshadowed the era of the monochrome—goes back to the birth of the artwork, to the moment preceding it when the artist drew the first line.

The dissolution of the subject entailed another: that of the genres of painting. Maurice Denis's celebrated

21 Quoted by Maurice Denis, *Théories, 1890–1910* (Paris: Bibliothèque de l'Occident, 1913), p. 162.

22 Valéry, *Degas, danse, dessin*, p. 139.

23 Letter to Theo van Gogh, on or about December 1, 1888. See http://www.vangoghletters.org/vg/letters/let723/letter.html.

24 Henri Esmond, "Les expositions : les Artistes indépendants," *L'Europe des artistes*, May 2, 1880, p. 1.

25 Letter from Paul Cézanne to his mother, September 26, 1874. Cézanne, *Correspondance* (Paris: Grasset, 1978), p. 148.

26 Émile Bernard, *Souvenirs sur Paul Cézanne et lettres* (Paris: À la Rénovation esthétique, 1921), p. 44.

27 Maurice Denis [under the pen name Pierre Louis], *Art et critique*, no. 65, August 23, 1890.

28 Eugène Delacroix, *Journal, 1822–1863* (Paris: Plon, 1996), p. 168.

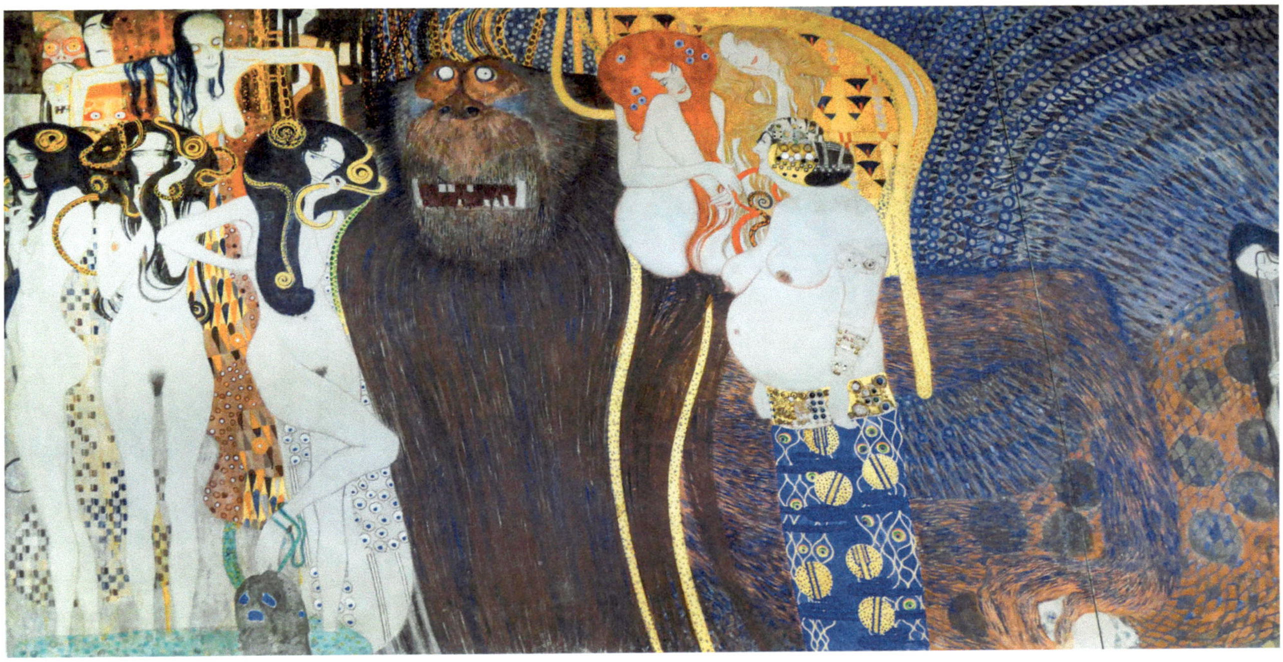

formula has a programmatic value: "Remember that a picture—before being a battle horse, a female nude or some sort of anecdote—is essentially a flat surface covered with colours assembled in a certain order."[27] Since Leon Battista Alberti, the *historia*—what a painting tells—had been its raison d'être. The disappearance of the subject resulted in the emergence of an anthology of titles which in themselves typified modern art: *Soft Construction with Boiled Beans, Metaphysical Interior, Woman 100 Heads, The King and the Queen Surrounded by Nudes, The Stabbed Duration*, not forgetting the commonest of all of them: *Untitled*. The prestige accorded to this floating multitude of images born of daydreams, with their glittering enigmas, raised the monsters engendered by the sleep of reason to become the new muses of art.

For Sigmund Freud, dreams, the symbols of which are almost always sexual, take us back not only to childhood but also to the prehistory of the human race. In art, prehistoric man and woman are for the nineteenth century what Adam and Eve had been for the Renaissance or Apollo and Aphrodite for Ancient Greece. But the defining feature of Darwin's century was to have turned its attention not to the mythological or symbolic foundations of humanity but to its organic origin. It was no longer the origin of culture, the essence of which is transmitted by education, that was emphasized, but rather the biological concept of the species, which involved the recognition of its finiteness. This fascination with beginnings could well be, in fact, the mark of a civilization that, knowing itself to be in decline, has turned back to its past. The iconoclastic, sacrilegious attitude adopted by so many artists after 1860, which made deconstruction, even destruction, the watchwords of modernism, suggests that *The Death of Sardanapalus* by Eugène Delacroix was actually a prediction. Here, man, faced with his imminent end, is bent on destroying the most refined manifestations of culture. The scene, a hellish ferment of human and animal flesh, shrouded in noxious effluvia adding to a powerful aura of sexuality, clearly evokes the end of the world and also its distant origins, what biologists call a "mixed brew of cultures," in which we see in its harshest form the struggle for existence, the death of the weak, the triumph of the strong. Dawn and dusk mirror each other. "The savage," wrote Delacroix, "always returns. The most extreme civilization cannot banish from our cities the atrocious crimes that seem the lot of people blinded by barbarism. In the same way, the human spirit left to itself always falls back into a stupid infancy."[28]

209 FIG.
Gustav Klimt,
The Hostile Forces and the Three Gorgons, 1902
Casein, stucco, graphite, application of glass, mother-of-pearl . . ., application of gold on mortar, 215 × 314 cm
Vienna, Österreichische Galerie Belvedere, on loan to the Secession

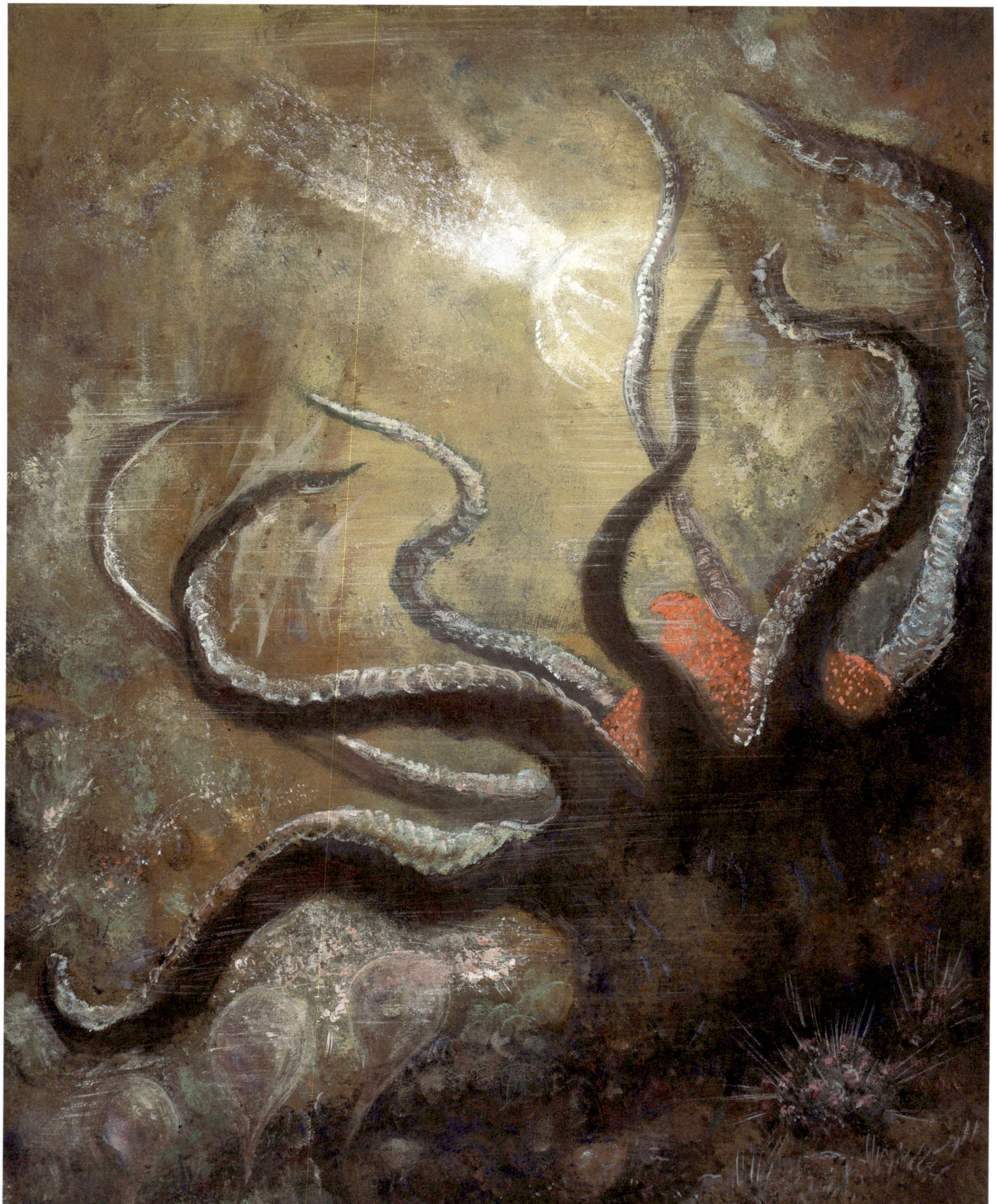

210
Alfred Kubin,
Underwater Scene, 1906
Tempera on cardboard,
39.8 × 31.3 cm
Linz, Oberösterreichische
Landesmuseum

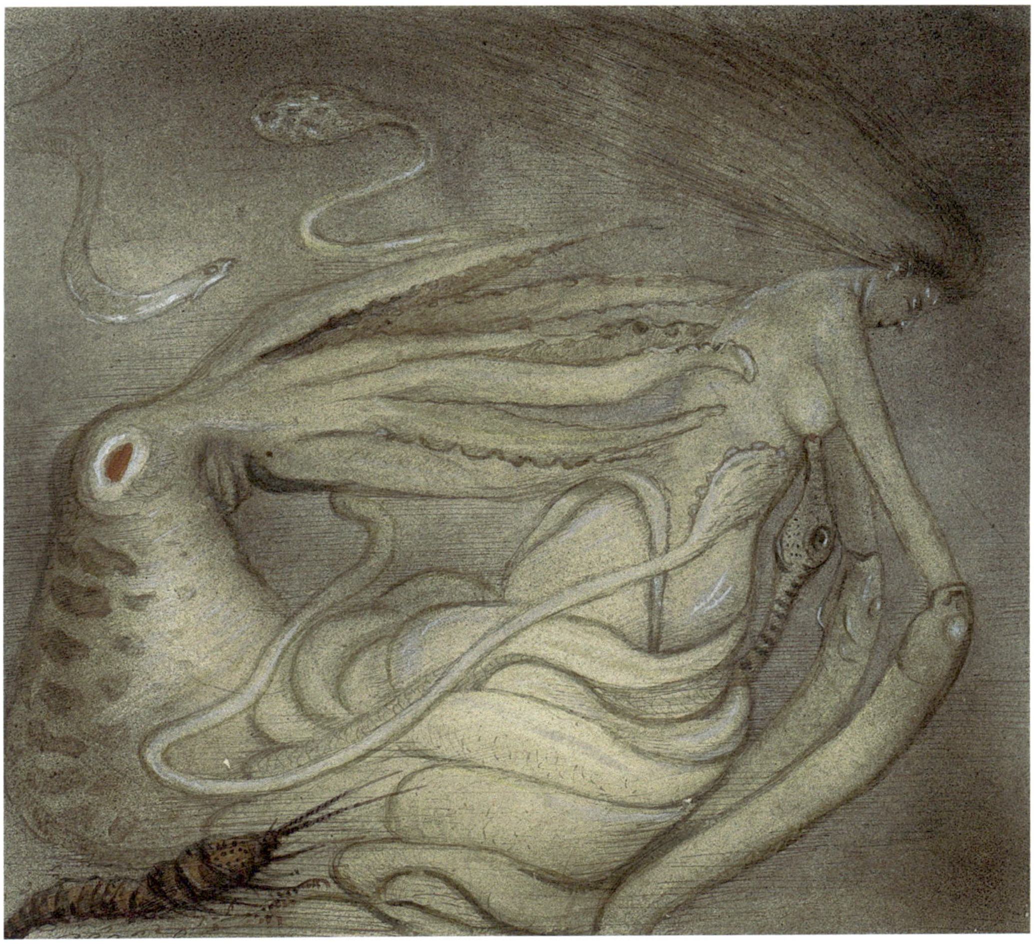

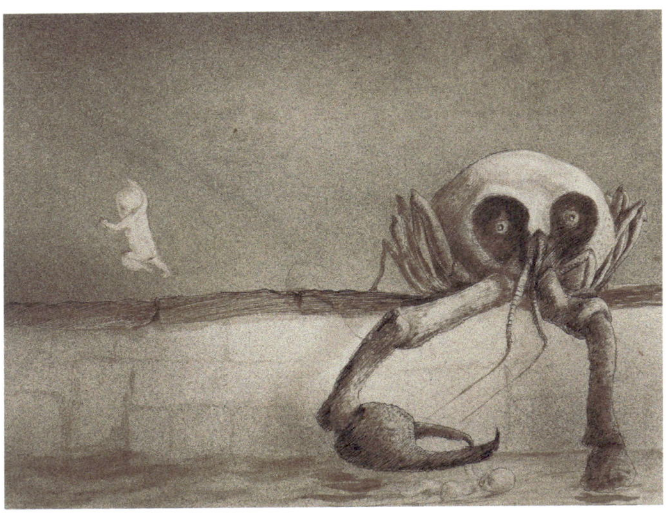

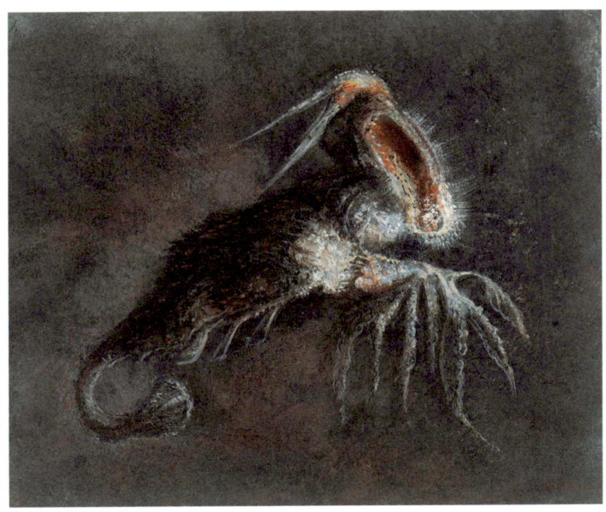

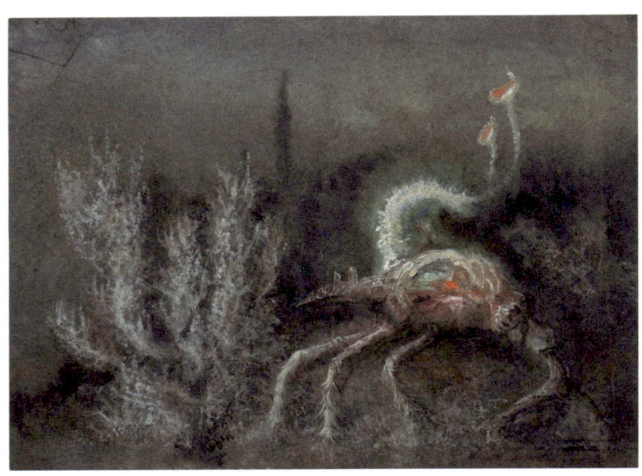

211
Alfred Kubin,
Suckers, 1905
Watercolour, 31.6 × 39.2 cm
Linz, Oberösterreichische Landesmuseum

212
Alfred Kubin,
The Hour of Birth, 1901–1902
India ink, wash, crachis,
25.5 × 30.6 cm
Vienna, Leopold Museum

213
Alfred Kubin,
Untitled, about 1906
Gouache on Bristol board,
25.2 × 29.5 cm
Zug, Kunsthaus Zug

214
Alfred Kubin,
Sea Floor, 1906
Tempera on black cardboard,
23.6 × 51.5 cm
Linz, Oberösterreichische Landesmuseum

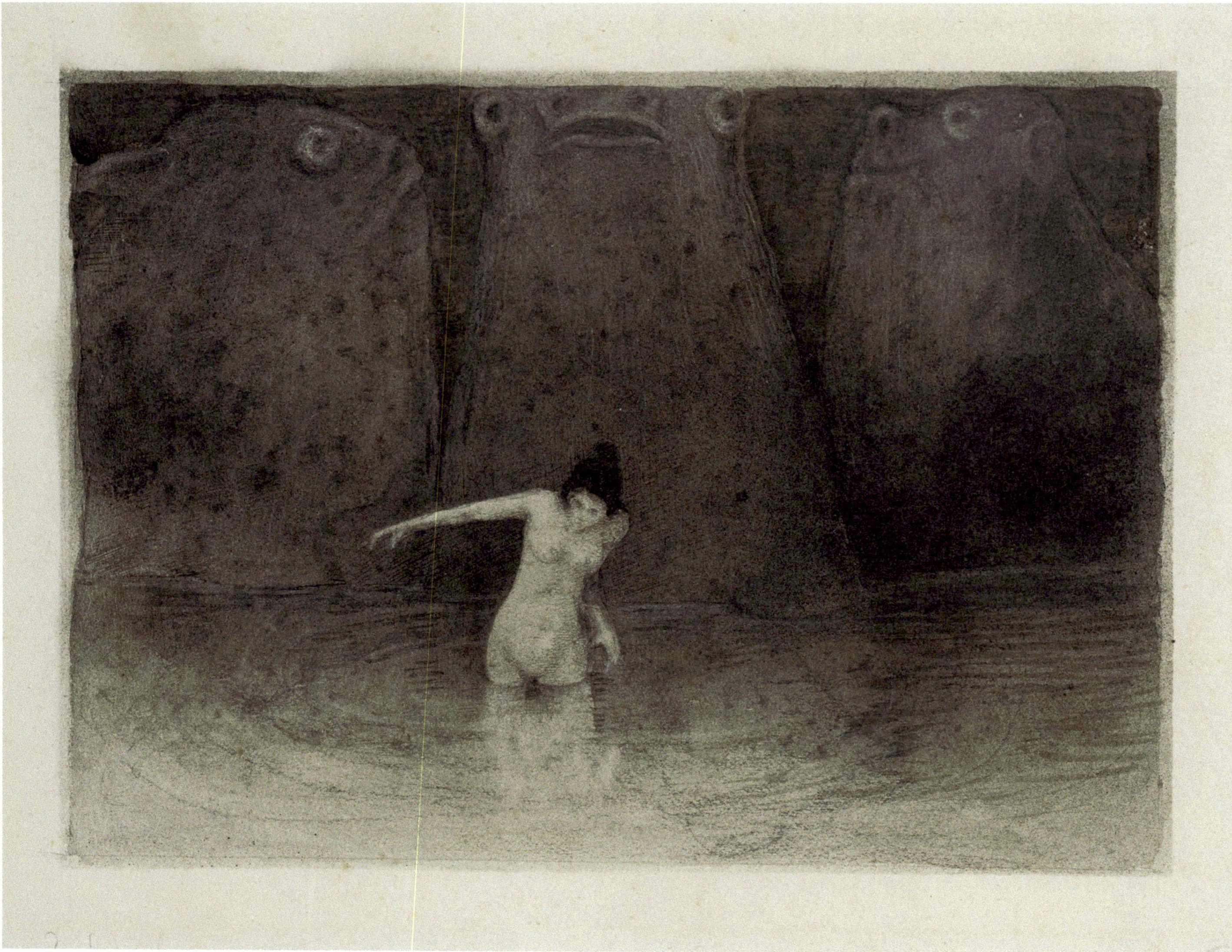

215
Alfred Kubin,
The Swamp, 1903–1905
Pen, ink and crachis,
30.7 × 37.9 cm
Linz, Lentos Kunstmuseum
Linz

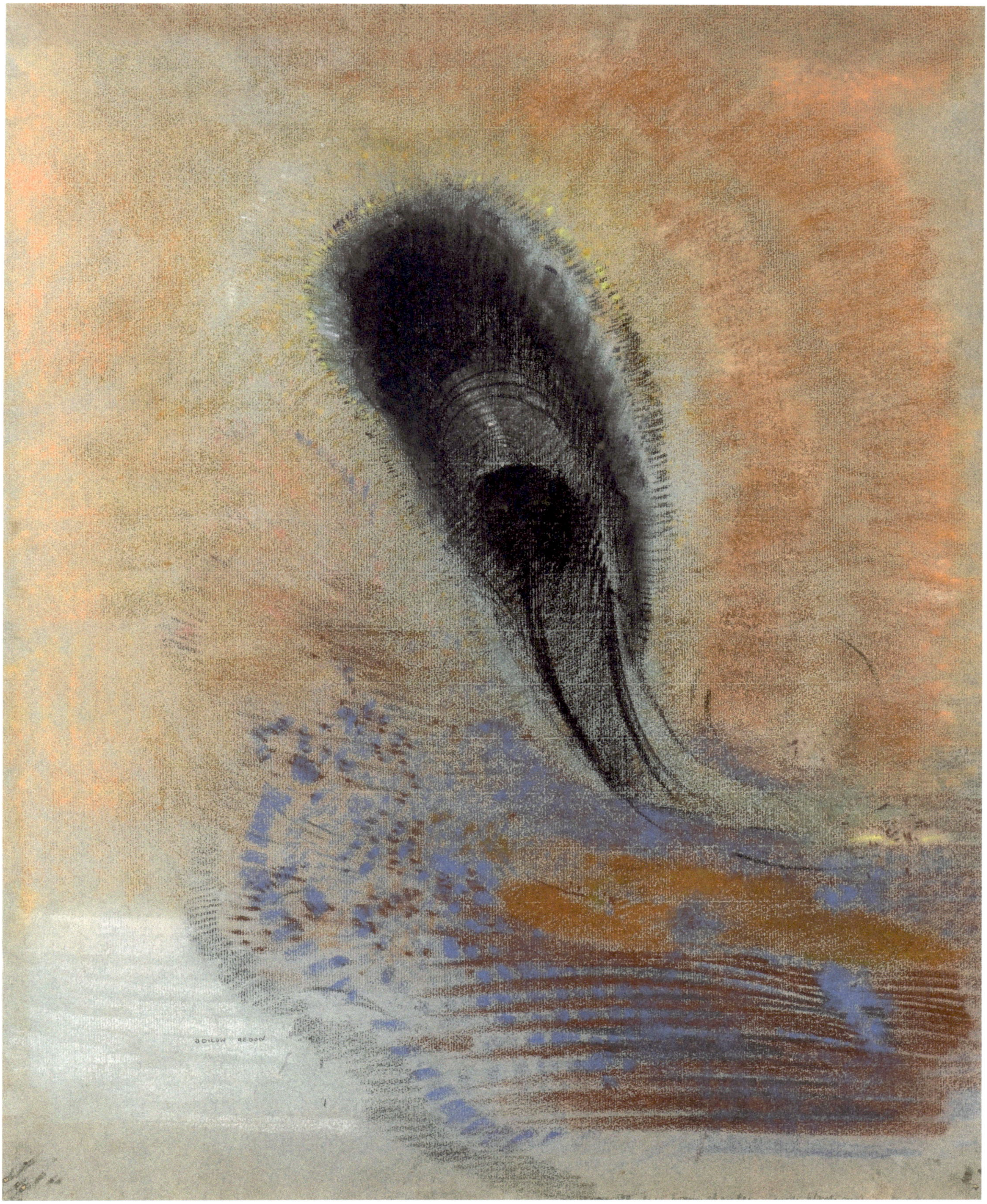

216
Odilon Redon,
Underwater Vision, 1900
Pastel, 66 × 54 cm
Paris, Musée d'Orsay

This return to the dawn of life, which replaces man before his animal ancestors, refers as much to the "primitive scene" of our concept as to the simian origin of our species. Flaubert's novel *Quidquid volueris*, written when he was sixteen, in which a scientist, in order to recreate the missing link between animal and man, mates an orangutan with a black slave girl, is testimony to the fascination exerted by the idea of our genesis. Carnal commonplace or the divine All-powerful? "Evolution or Miracle, we must choose."[29] The curiosity and fear felt about the night of our conception—this irrepressible desire to see what we do not want to see—haunted the imagination of artists in the second half of the nineteenth century: Emmanuel Fremiet and his *Gorilla Carrying Off a Woman* (170, P. 247), Gustave Moreau and his bestial Cyclops in love with the nymph Galatée, Gabriel von Max and his *Pithecanthropus alalus* enjoying the first fruits of married life (4, P. 23), Gustav Klimt and his Typhon, father of all the monsters (209), not forgetting the Beauty and the Beast and King Kong kidnapping a human woman (169, P. 245). Voyeurism and priapic curiosity for some, catharsis aimed at keeping animal impulses at a distance for others. It is difficult to distinguish between the two, since with human beings the lusts of the flesh and the desires of the eyes are interlinked. At the very least, these idylls between primates, in their real or sublimated violence, satisfy the most archaic desires of human beings. "The feeling of happiness derived from the satisfaction of a wild instinctive impulse untamed by the ego," wrote Freud, "is incomparably more intense than that derived from sating an instinct that has been tamed."[30] Édouard Manet's *Olympia* was decried because it crystallized this return to animality. When it was presented at the Salon in 1865, one of the critics had compared the sitter to a "sort of female gorilla."[31]

Beauty left the heights of Parnassus to rejoin the muddy soil of our humanity, the primeval stew of its genesis when protozoa, amoebae, jellyfish and Kafkaesque woodlice struggled and fought for life against a background of crime and rape. For Darwin, and for Freud later on, beauty had its roots in the area of reproductive arousal. It is the word that human beings give to the sexual attraction that stimulates the mechanisms of reproduction. Beauty is what charms, casts a spell, fascinates. In this it recalls the terrifying reign of Medusa—the reign which preceded that of the culture incarnated in the birth of Pegasus, sprung from the blood of the decapitated Gorgon. Subject to this physiological aesthetic, works of art no longer provided delight but sided with the dark sources of the underworld. Beauty—which was for Saint Augustine the other name of God—had become "bitter" for Rimbaud (*Une saison en enfer*), "terrible" for Rilke (*The Duino Elegies*), "convulsive" for Breton (*Nadja*). And the assertion by Dostoyevsky in *The Idiot*, "beauty will save the world," seems like a cry of despair.

Confined by the restraint of a central perspective inherited from the Renaissance, realism and naturalism undoubtedly caused works of art to lose much of their speculative dimension. Like civilizations, art goes through cycles. We move from a simple art to a scholarly one, from a scholarly art to a mannered one, from a mannered art to an art marked by a desire to refresh old formulas, to give it back its omnipotence. The return to archaic forms of representation by educated artists from Gauguin to Dubuffet by way of Picasso can largely be explained by a wish to break out of the impasse in which painting was trapped. But other factors, less inevitable, also played a role. In the second half of the nineteenth century, the demand for high culture to revert to childlike, primitive or prehistoric forms was developing at the same time that scientists were establishing man's animal ancestry and destroying his last illusions about his divine origin. Incredulity and fascination. This disturbing coincidence did not go unnoticed by the theoreticians of the day. In 1869, ten years after the publication of *The Origin of Species*, Charles Rochet in his *Cours d'anthropologie appliquée à l'enseignement des beaux-arts* suggested an audacious parallel between the quest for origins in biology and formal deconstruction in art. For Rochet, Darwinism, which in physical terms reduced man to his animal components and in moral terms to his instincts, could only "demean the arts by reducing them all to the cruder elements of their formation."[32] A few years later, the physician Max Nordau developed the concept of "degeneration" in art, calling the output of his contemporaries a regression. These criticisms—beyond their moralizing tone—posed a question that would be central to modernism: why had art, considered the highest manifestation of

29 Gustave Flaubert, *Lettre à Maxime Du Camp*, November 13, 1879.

30 Sigmund Freud, *Civilization and Its Discontents* [1930], trans. James Strachey (New York: W. W. Norton, 1961), p. 29.

31 Amédée Cantaloube, "Le Salon de 1865," *Le Grand Journal*, May 21, 1865.

32 Charles Rochet, *Cours d'anthropologie appliquée à l'enseignement des beaux-arts. Discours d'ouverture prononcé le 30 mai 1869* (Paris: Renouard, 1869), p. 24.

33 Roger Caillois, *Le Fleuve Alphée* (Paris: Gallimard, 1983), p. 194.

34 Carl Einstein, "André Masson, étude ethnologique," *Documents*, no. 2 (May 1929).

humanity, become indissociable from what was lowest, most suppressed and most primitive in man?

The question is too complex to be easily answered, but one thing is clear: the story of modernism is closely tied to the upheaval sparked by *The Origin of Species* which, having deprived humanity of all transcendence, left it to the existential anguish of its own end and in turn to the obsession with its origins. The album of lithographs by Odilon Redon entitled *The Origins* (188) and Alfred Kubin's series of depictions of molluscs (213, 214) present a troubling primitive bestiary which, far from flourishing in the Garden of Eden, evolves in the sea-green depths of the abysses.

In his series of paintings of the cave where the river Loue rises, near Ornans, his native city, which was a troglodyte habitation in the Neolithic era, Gustave Courbet shows his fascination with prehistory and his fascination with the maternal hearth (217). This depiction of the motherland in which the primordial stream flows is a pendant to *The Origin of the World* (218). This recovery from the rift behind which there remains the nagging question of our existence—*Where do we come from? What are we? Where are we going?* the title of one of Gauguin's most famous paintings—is also the subject of a group of canvases that Cézanne executed in the quarries at Bibémus, where the orangey-pink rock is cut into shady gorges bordered with lips that seem to be made of flesh.

In 1978, a few months before he died, Roger Caillois, who forty years earlier had turned away from the Surrealists, whom he reproached for their childishness, published *The River Alph*. It evoked the Ancient Greek fable about a stream that ran into the sea and having crossed it came again to land and rejoined its symmetrical source, by which it disappeared into a tiny rock cleft. In life as in culture, a return to our initial condition is always possible, and even, ultimately, inevitable. "Innumerable fertile consequences have allowed for life, then intelligence, then coherent and verifiable reasoning to flourish. The road has a similar chance to be travelled in the opposite direction."[33]

The history of the avant-gardes is paradoxically that of a temporality running backwards. Art was used to flow back again to its hypothetical source. Its relationship with history developed negatively, until all heritage was abjured. An accident along the way? The end of art? A delusional parenthesis? In 1929, the writer Carl Einstein, the shrewd art critic of the periodical *Documents*, postulated that modern art—which had seen the reversal of values, the collapse of *mimesis*, the repression of consciousness and the return to childish forms—was comparable, historically speaking, to a "hallucinatory interlude."[34] A mirage of the origins. Seeds are fascinating, for they contain *in posse* all future developments, but only *in posse*. If every block of raw marble contains within it a potential sculpture by a Michelangelo, without an artist to liberate it the block remains just a simple shapeless rock. It took millions of years for life to appear, and tens of thousands more for art to develop; a few moments would suffice to wipe it all out. If today, thanks to Jean-Baptiste Lamarck, Charles Darwin, Ernst Haeckel and some others, man can face up to his genesis, accept his animal origins and admit that he is the result of a felicitous combination of certain amino acids, which are themselves an accidental arrangement of a handful of atoms, he has nothing to gain by taking comfort in his origins or seeking an impossible reply to the reason for his existence. From the primeval Nothing he is interrogating, he cannot hear anything save the echo of his own imminent disappearance. The fascination with origins is deadly. In the Greek fable, Medusa turns to stone those who allow themselves to be stupefied. She is part and parcel of the death-wish that haunts human beings, the desire to go back to the equilibrium that preceded the world of the living, when everything was still amorphous, impassive and immortal, in the bowels of the earth.

The myth of Orpheus who, in order to get out of those bowels from which we come, recommends that we go forward without looking back, is full of wisdom—truly alive.

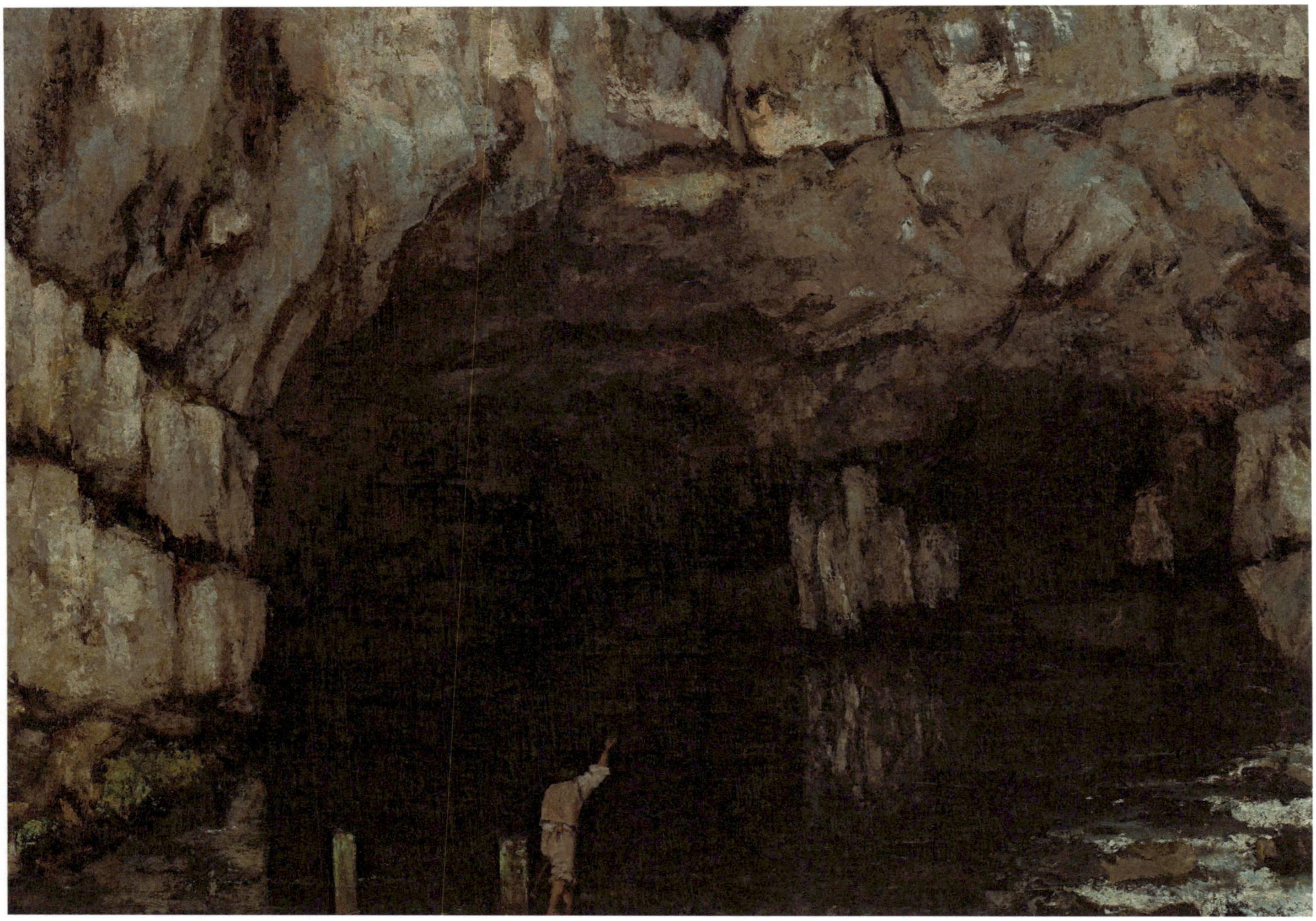

217
Gustave Courbet,
The Grotto of the Loue, 1864
Oil on canvas, 98.4 × 130.4 cm
Washington, National Gallery
of Art

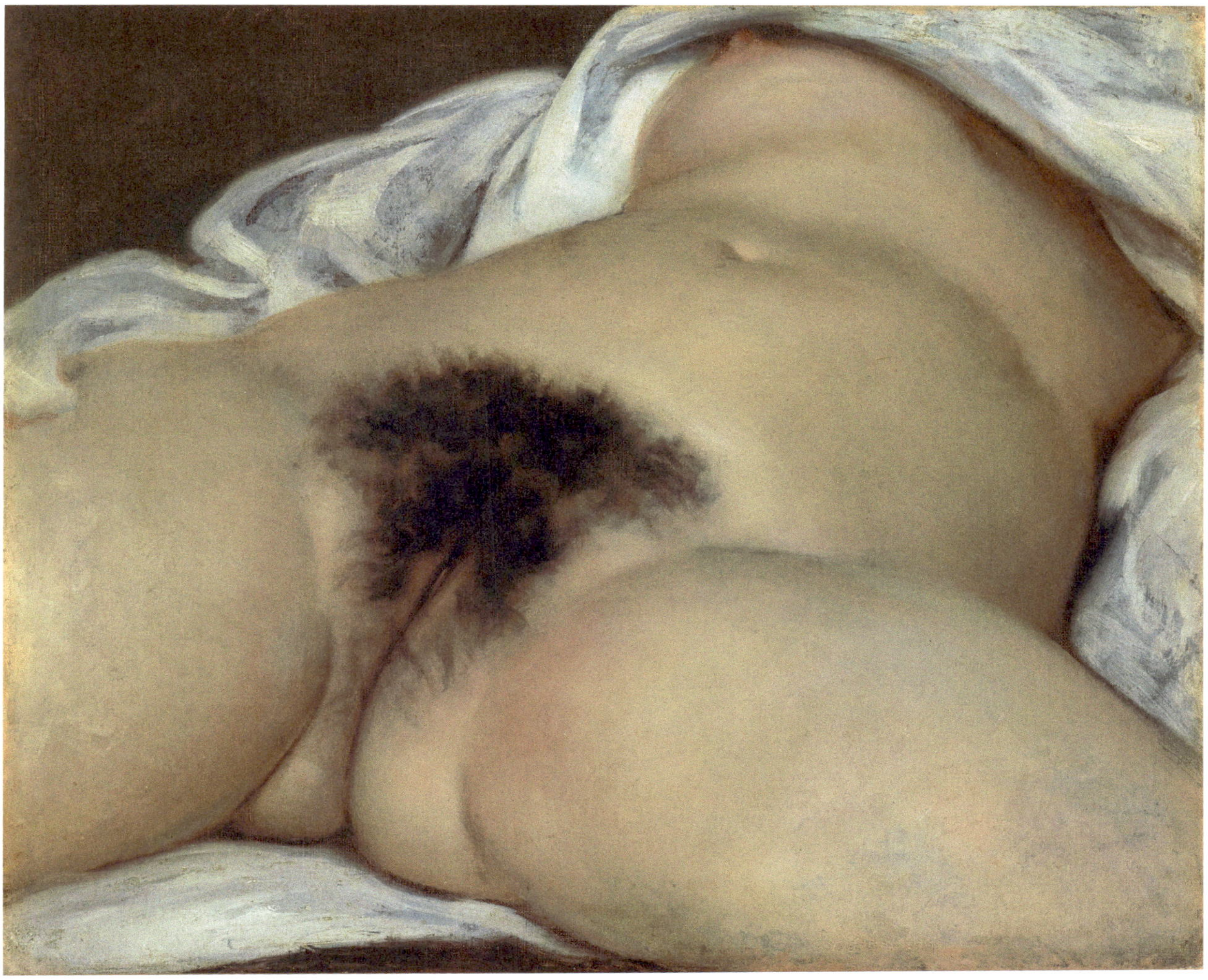

218
Gustave Courbet,
The Origin of the World, 1866
Oil on canvas, 46 × 55 cm
Paris, Musée d'Orsay

WRITING ABOUT THE ORIGINS OF MAN, OR THE EMERGENCE OF ANIMALITY

NICOLAS WANLIN

In the nineteenth century, the new sciences of primitive man (paleontology, comparative anatomy, anthropology, etc.) were rich in hypotheses needing only to be colourfully fleshed out in the style of modern legends. It is thus that we encounter the first popular expressions of Pierre Boitard's transformative theories in the form of fiction (111, P. 159). In one of his works, a time traveller asks a devil who guided him through prehistoric times:

> - Show me, please, beings of my kind, for I am very curious to see the first men. Are they giants or pygmies? Are they white, red or black? Do they walk on two feet or on all fours? Do they live alone in the woods like bears, or do they gather in hordes like gazelles? All these things are extremely important to know, and I will certainly write a fine book in which the reasoning will be based on . . .
> - On dreams, as usual, says the genie.[1]

This story is only a dream for its protagonist, and hypotheses about the animal origin of the human species were considered at the time to be no more than a strange product of the imagination. But while scientists could not hide behind fiction, they did mould narratives to make their hypotheses plausible. The forms that were then invented, blending literature, natural history, philosophy and popularization, had to draw from traditional myths or scientific hypotheses, or try to combine them to achieve the verisimilitude of an acceptable story.

Authors were thus confronted with either the seriousness of a positivist scientific method or the text of the Adamic genesis. In fact, for a long time, it was the Bible that offered the consensual narrative of the birth of man. The poet Jacques Delille echoes this:

> Image of his God, favourite of his king,
> Man was just born; and, subject to his law,
> Animals lived without rebellion or war;
> But all with servile front looked down upon the earth:
> Their sovereign alone, walking among them,
> Raised a sublime forehead and looked up at the heavens.[2]

Walking on two legs from birth, this early biped was as *erectus* as possible and thus distinguished himself from other animals without needing to imagine that he had to apply himself to stand upright. The fixist *doxa* prevailed at the beginning of the century. Cuvier asserted that "there are no human bone fossils,"[3] which seems to imply that man was created suddenly, in the state in which he is today or approximately so, and that it would be wrong to conceive of him emerging from animality. Moreover, cartoons and the use of a satirical tone, even more than scientific arguments, castigated the transformists by accusing them of implausibility and an excess of poetic imagination. Chateaubriand and Lamartine, following the reactionary Bonald, condemned transformism for its godlessness and falseness. Man's animality was repugnant to a religious mind, which believed that man derived his dignity not from ascension or self-perfecting but from divine providence, as he had done since the day of his creation.

Victor Hugo introduced doubt and posed the question in new terms. While he refused to "reduce man to beast" and would never be completely transformist, he was open to scientific daring. In contrast to Cuvier's academic and obtuse positivism, he wanted to believe in the existence of "fossil man," which had in fact been proven by a paleo-anthropological discovery, even if it "gave Genesis the displeasure of confirming the hypothesis of many Adams."[4] Hugo acknowledged that there were once anthropoids other than man and he denounced the blindness of "academic and official science."[5] This romantic view of science challenged short-sighted positivism, and Hugo drew attention to "a whole part of nature that our senses do not recognize and that therefore confounds observation."[6] In his great poem *The Satyr,* he casts this chimera as an intermediary being between the beasts and the gods of Olympus, in order to illustrate the condition of man, with his feet in the mud and his head in the stars (220).[7]

But scientific hypotheses were not limited to fixism. In his *Natural History of Man,* Buffon had already imagined "a man such as we might believe was the first man at the time of creation."[8] He did not venture to speculate on the origin of such a being. Later, in a supplement to *The Epochs of Nature* that was published in 1778, Buffon imagined the first men "trembling on the earth that trembled beneath their feet, naked in spirit and body, exposed to the insults of all the elements, victims of the fury of ferocious animals."[9] This image did not yet depict an emergence from animality but it inaugurated a prehistoric picturesqueness that would achieve great popular success.

Lacépède, in his 1798 course at the Muséum, imagined an original man "very different from what he has become":[10]

> His head bristling with hard, matted hair, his forehead veiled by a sort of bushy mane, his eyes hidden under thick eyebrows, his mouth covered with a very long beard, which falls in disorder on a hairy chest, his whole body covered with hair, his nails long and crooked, such is the image he presents. The majesty of his august face, the features of intelligence, the mark of a superior essence, the seal of genius, everything is, so to speak, still hidden under the cover of a ferocious beast . . . The toes of his feet, frequently exercised, and which no whim has yet deformed, longer and more separated from each other than they will become, make him almost quadrumane; they bring his habits closer to those of the monkey, with which his teeth and almost all parts of his body also present a high degree of conformation.[11]

Assuming that these first men adapted to their environment under the pressure of need, Lacépède suggested the possibility of an evolution of the species, and to Lamarck's credit, he developed this hypothesis into a theory in its own right. He thus invented a new narrative, that of

1 Pierre Boitard, "Études d'histoire naturelle. Paris avant les hommes," *Musée des familles,* vol. 3 (June 1836), p. 277.

2 Jacques Delille, *L'Imagination, poëme* (Paris: Didot, 1816), p. 195.

3 Georges Cuvier, *Recherches sur les ossemens fossiles de quadrupèdes, où l'on rétablit les caractères de plusieurs espèces d'animaux que les révolutions du globe paroissent avoir détruites* (Paris: Déterville, 1812), vol. 1, p. 82.

4 Victor Hugo, "Préface de mes œuvres et Post-scriptum de ma vie" [1863–1864 ?] (formerly the "philosophical preface" to *Les Misérables*), in *Œuvres complètes. Critique,* ed. Jean-Pierre Reynaud (Paris: Robert Laffont, 1985) p. 704.

5 Ibid., p. 703.

6 Ibid.

7 Victor Hugo, "Le satyre," in *La Légende des siècles. Première série. Histoire. Les Petites Épopées* [1859], ed. Claude Millet (Paris: Librairie générale française, "Le Livre de poche" series, 2000).

8 Georges-Louis Leclerc de Buffon, "Des sens en général," in *Histoire naturelle, générale et particulière, avec la description du Cabinet du Roy* (Paris: Imprimerie royale, 1749–1804), vol. 3, p. 364.

9 Georges-Louis Leclerc de Buffon, *Époques de la nature, 7ᵉ et dernière époque,* in *Suppléments à l'Histoire naturelle, générale et particulière* (Paris: Imprimerie royale, 1778), vol. 5, p. 225.

10 Étienne de Lacépède, *Discours d'ouverture et de clôture du Cours d'histoire naturelle des animaux vertébrés et à sang rouge, donné dans le Muséum national d'histoire naturelle* (Paris: Plassan, an VI, 1798), p. 7.

11 Ibid., p. 8.

hominization, which was no longer based on mythological and biblical representations: man descended from a quadrumane who became accustomed to standing upright. But Lamarck was accused by Cuvier of being a maker of fictions, duped by his imagination.[12]

The influence of these scientific debates and various possible narratives can be seen in writers who were well versed in natural history, such as Charles Nodier.[13] But science had always been viewed through the prism of philosophy, religion and contemporary utopias that, from Fourier to Saint-Simon, and through Buchez and Ballanche, reformulated the ideas of progress, evolution or even palingenesis.

A later utopian, Jean Reynaud, reconciled the Bible with the animal origin of man: "Finally, on the fifth day, man rises suddenly from the middle of some astonished tribe of orang-outangs or pongos."[14] Instead of "fairy-tale pictures" of a literal reading of the biblical myth, a religious mind might prefer "the slow and quiet evolutions that suit the majesty of Providence so well."[15] Thus, man's animality was no longer shocking because, "if [man] goes through the phases of animality as he walks, it is like the triumphant one who bursts through the crowd."[16] The alternative to a literal respect for biblical creation consists in duplicating the narrative: the story of evolution is superimposed on the story of creation.

In the second half of the nineteenth century, poets enthusiastically embraced the motif of the emergence from animality. They retained the picturesque images of that moment in time and dealt with the problem of a pre-existing traditional narrative, the Adamic myth, in competition with the transformist vision. The amateur poet Jean Chamard introduced this episode into the epic canto he dedicated to the emergence of humanity:

> The beasts, across the vast span of time,
> With new types of bodies and faces,
> Seemed to look more and more like humans.
> The hour approaches! Soon, perhaps as early as tomorrow,
> The old ancestral couple will finally appear to me![17]

Hominization was underway, but one still awaited the appearance of the "old ancestral couple": we are not renouncing biblical imagery! The poem reconciled the gradualness of evolution with the suddenness of emergence. The act of standing upright of Homo erectus was interpreted as a symbol of his spiritual awakening:

> He will not arch his back like a beast,
> Nor lower his forehead to the ground; but his head,
> Despite so much ugliness, rises to the azure.[18]

In the language of the poet, "the azure" is the ideal to which human efforts, hopes and dreams aspire. Here, he also reminds us that bipedal man's field of vision is no longer that of the simian quadrumane and that his whole being in the world is changed by it. Similarly, in his

12 "Éloge historique du chevalier de Lamarck, par M. G. Cuvier. Lu à l'Académie des sciences de Paris dans la séance publique du 26 novembre 1832 [par M. le baron Silvestre]," *Mémoires de l'Académie royale des sciences de l'Institut de France*, vol. 13 (Paris: Firmin Didot, 1835), p. XIX.

13 Charles Nodier, "De la fin prochaine du genre humain," *Revue de Paris*, vol. 26 (May 1831), p. 224–240, and in particular "De la palingénésie humaine et de la résurrection," *Revue de Paris*, vol. 41 (August 1832), p. 81–106.

14 Jean Reynaud, *Philosophie religieuse. Terre et Ciel* (Paris: Furne, 1854), p. 93.

15 Ibid., p. 199.

16 Ibid., p. 200.

17 Jean Chamard, *L'Épopée des âges. Les Origines, poème* [written between 1870 and 1879] (Paris: Rodstein, 1947), p. 122.

18 Ibid., p. 138.

avant-garde natural epic poem *Le Dire du mieux,* René Ghil describes these early men, the "most-Upright," who hold "high their Head of intelligent anguish": "from the other world / they rise up from the mass of appetites."[19] Man thus elevates and gradually emancipates himself from the physical needs and desires by which other living beings are so narrowly determined. His gaze lifted above the horizon is also at the origin of a continual apprehension, but it gives him the symbolic attitude of a watchman capable of observing the world, that is to say, of dominating it. René Ghil is committed in this epic to merging as much as possible the ideas of evolutionary anthropology and their moral interpretation based on his conception of "poetry-science."

These poets inspired by science synthesized heterogeneous cultural data: the natural sciences contributed their part, but religion and even secular spiritualism, filtered by atheistic philosophy, were involved. Their poems thus bear witness to the coexistence in culture of heterogeneous thoughts and expressions that lend themselves to syncretic designs.

But the science of origins was frightening because it touched on aesthetics as much as on morality. For Joris-Karl Huysmans, fossil man was a "nightmare of science."[20] In *Against the Grain,* Huysmans describes Odilon Redon's drawings inspired by antediluvian times thus (187, P. 271; 188 B, D, P. 273):

> Sometimes the subjects even seemed to have borrowed from the cacodemons of science, reverting to prehistoric times. A monstrous plant on the rocks, queer blocks everywhere, glacial mud, figures whose simian shapes, heavy jaws, beetling eyebrows, retreating foreheads and flat skulls, recalled the ancestral heads of the first quaternary periods, when inarticulate man still devoured fruits and seeds, and was still contemporaneous with the mammoth, the rhinoceros and the big bear.[21]

For the dandy hero of this novel, the animal origin of man is horrifying not only because science appears to confirm it, but because art sets before his eyes the exact opposite of what he aspires to be as a human being. "These designs were beyond anything imaginable; they leaped, for the most part, beyond the limits of painting and introduced a fantasy that was unique, the fantasy of a diseased and delirious mind."[22] The scientific hypothesis produced descriptions that were quasi-fantastic, and by eluding epistemological analysis, they were viewed by the culture from both an aesthetic and a psychological standpoint. Man emerging from animality raises questions, shocks or frightens because he arouses a dread or a complex. This disturbing animality thus demands some form of compensation: the history of the intelligence that raised man above the animals, or at least separated him from them.

19 René Ghil, *Dire du mieux. Le meilleur devenir*, 1889–1926, quoted in René Ghil, *Le Vœu de vivre et autres poèmes,* ed. Jean-Pierre Bobillot (Rennes: Presses universitaires de Rennes, 2004), p. 99.

20 Joris-Karl Huysmans, *À rebours* [1884] (Paris: Crès, 1922), p. 81.

21 Joris-Karl Huysmans, *À rebours*, trans. John Howard, *Against the Grain,* n.p., n.d., n.p. https://www.gutenberg.org/files/12341/12341-h/12341-h.htm.

22 Ibid.

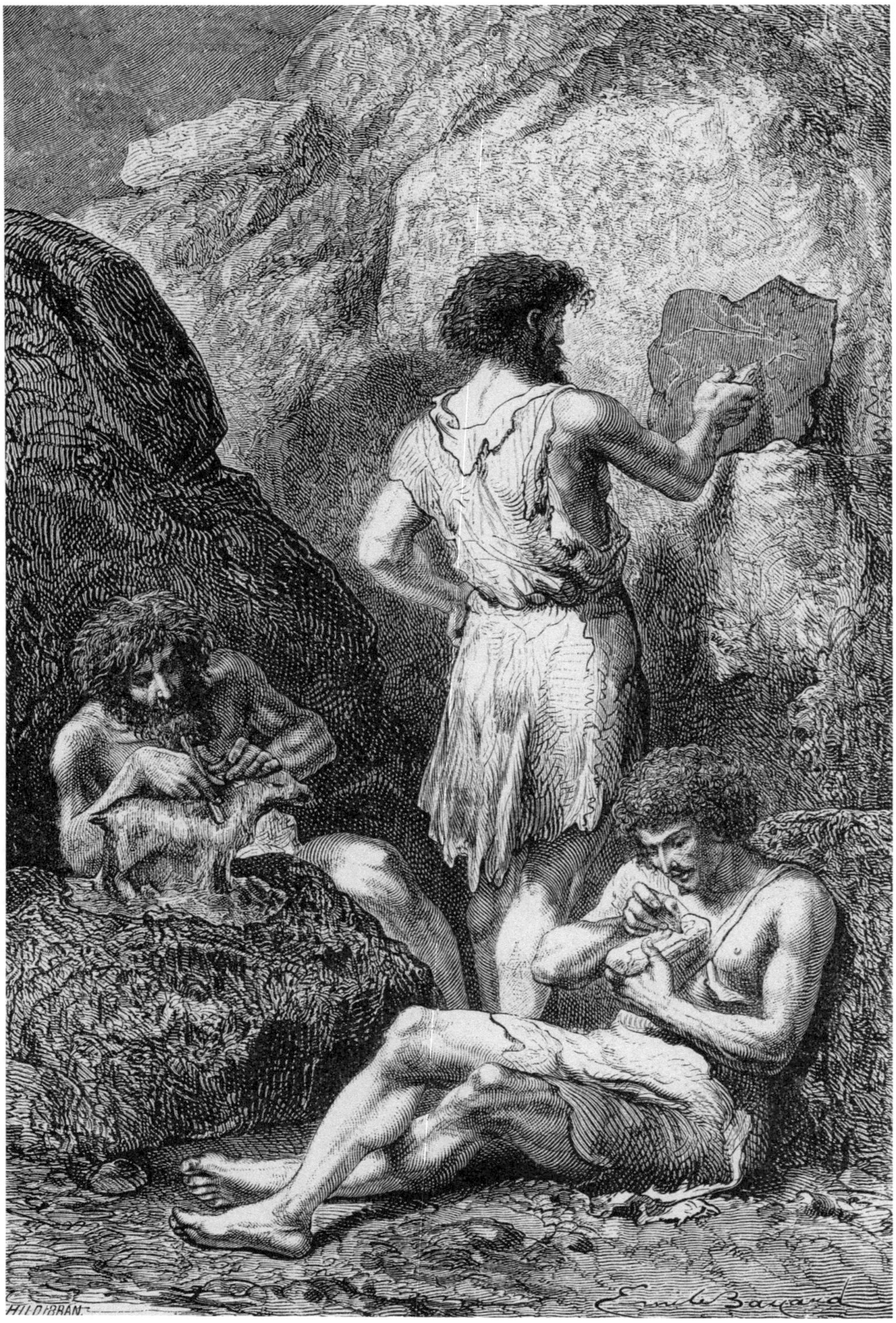

219
Émile Bayard,
"Arts of Drawing and Sculpture
during the Bronze Age,"
in Louis Figuier, *L'Homme
primitif*, Paris: Hachette, 1870
Paris, MNHN

220 FIG.
Fernand Cormon,
The Satyr, 1886–1889
Oil on canvas, 65 × 48 cm
Guernsey, Maison de
Victor Hugo, Hauteville House

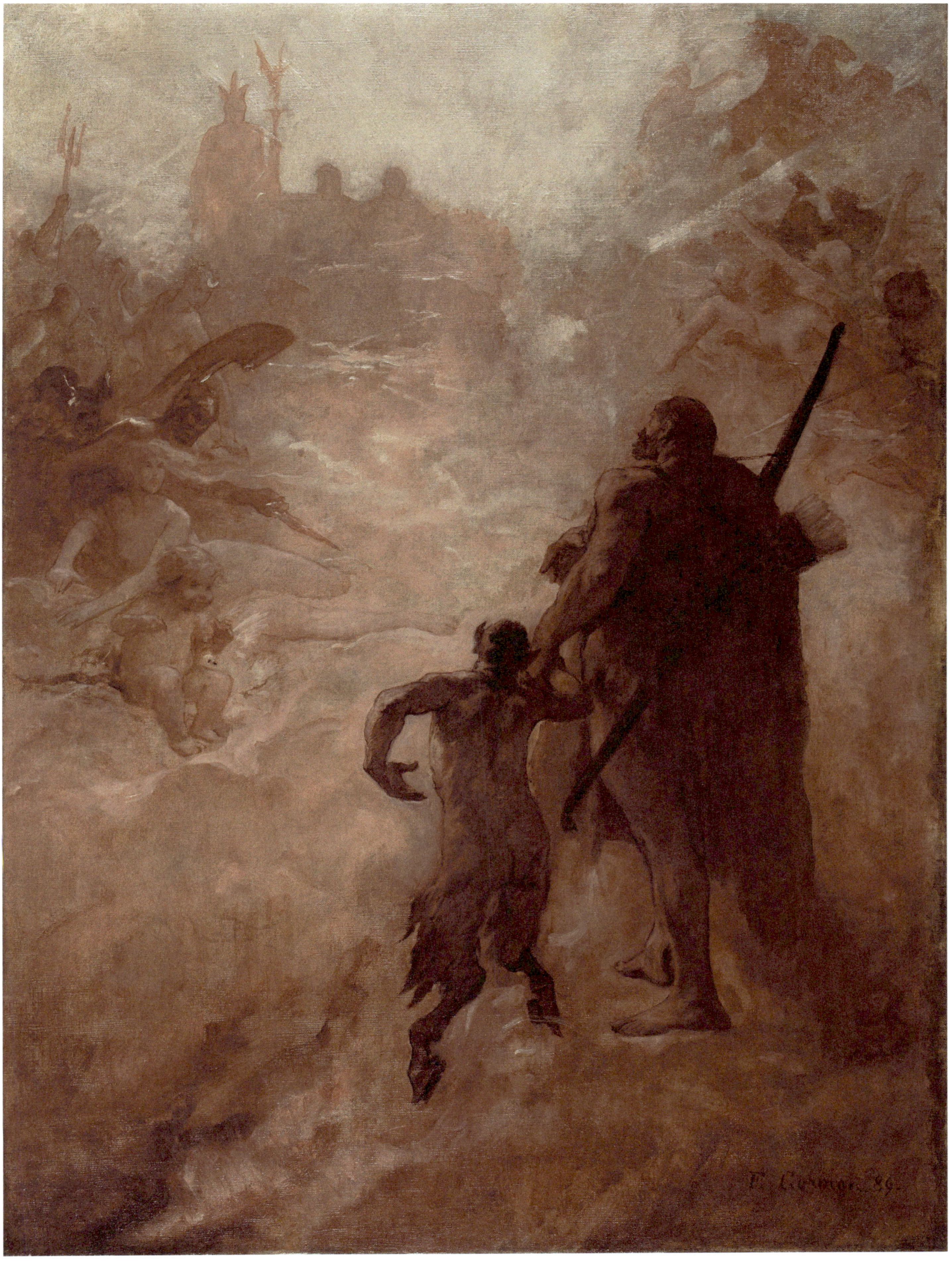

IX

THE SOMBRE FACE OF EVOLUTION: DEGENERATION, DEVOLUTION AND EXTINCTION

FAE BRAUER

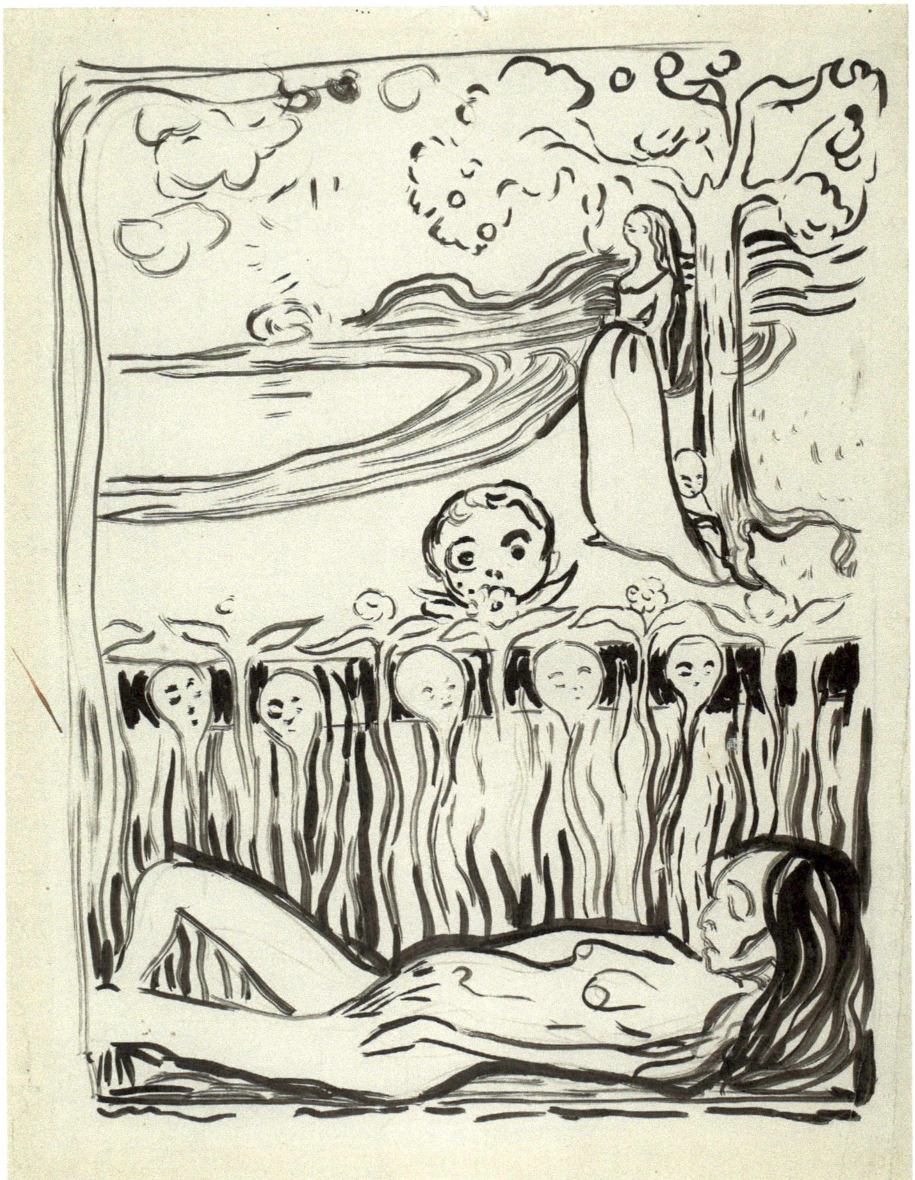

221
Edvard Munch,
Metabolism (Life and Death), 1896
India ink on wove paper,
48 × 36.6 cm
Oslo, Munchmuseet

With the Holocene, or sixth mass extinction, threatening more and more species, the sombre face of evolution has become an everyday experience. Inherent to evolution as conceived by Charles Darwin's theories of "natural selection" and "survival of the fittest," signified by industrial productivity and technological progress, extinction has emerged as the sombre side of its Janus face. As Darwin surmised in *On the Origin of Species*, "The extinction of species and of whole groups of species, which has played so conspicuous a part in the history of the organic world, almost inevitably follows on the principle of natural selection ... extinction and natural selection ... go hand in hand."[1]

Before Darwin, mass extinction had been conceived as catastrophism by Professor of Comparative Anatomy at the Muséum national d'histoire naturelle Georges Cuvier, and as a gradual degenerative spiral of organisms, frequently arising from human territoriality and ecological destruction, by Cuvier's colleague and antagonist Jean-Baptiste Lamarck in his *Philosophie zoologique* (126, P. 190). "One would say that [man] is destined to exterminate himself after having rendered the globe uninhabitable," Lamarck had prophetically concluded.[2] A staunch proponent of gradualism by the uniformitarian geologist Charles Lyell, as well as by Lamarck, Darwin also conceived of the disappearance of species "in the race for life" as a gradual process in *On the Origin of Species*, expounding in *The Descent of Man, and Selection in Relation to Sex* how the intervention of humans into evolution could swiftly lead to mass extinction. "Species and groups of species gradually disappear, one after another," he explained, "first from one spot, then from another, and finally from the world."[3] Once Darwin's theories were correlated with the accumulating medical and scientific evidence of decline and degeneration, they helped mediate a *fin-de-siècle* discourse of devolution, a visual culture of degeneration and a psychopathology of extinction comparable to the *fin-de-siècle* discourses of the sixth mass extinction of the twentieth century.

By the end of the French Commune, the first volume of Darwin's *The Descent of Man, and Selection in Relation to Sex* had been translated into French.[4] In his chapter devoted to extinction, "On the Races of Man," Darwin acknowledged how long it had been speculated that races had endured partial or complete extinction following plagues, famines or such natural catastrophes as hurricanes or volcanic eruptions. In his sly retort to Cuvier catastrophists, Darwin pointed out that extinction did not arise solely from cataclysmic events or environmental change. The survival of the Terai at the foot of the Himalayas and the Fuegians at the southernmost tip of America—whom Darwin had encountered on his *Beagle* voyage—demonstrated a facility among diverse races to withstand environmental conditions hazardous to their existence.[5] In his elaboration of his theory of "natural selection," Darwin deduced that extinction mostly followed from competition between tribes and races, the fittest being the ones that survived. Their fitness did not just depend upon their good health and vigour, but upon their fertility, their numbers, and most of all, upon what Darwin called "the grade of their civilisation."[6] It was due to these factors that Darwin considered those he called "modern civilised nations" had been swiftly able to overcome the Indigenous peoples of Australia and New Zealand. "When civilised nations come into contact with barbarians," he concluded, "the struggle is short."[7] However, the two most potent of all causes of extinction were, he claimed, "lessened fertility" and "ill health."[8] "Decreasing numbers will sooner or later lead to extinction," he warned.[9]

This "doomed race" theory seemed to be corroborated by the impact of Western colonization upon Indigenous peoples, such so-called "vanishing races" as the Australian Aborigine and the New Zealand

1 Charles Darwin, *On the Origin of Species* [1859] (Oxford: Oxford University Press, 1998), p. 384 and p. 141.

2 Jean-Baptiste de Lamarck, "Homme," in *Nouveau dictionnaire d'histoire naturelle, appliquée aux arts, à l'agriculture, à l'économie rurale et domestique, à la médecine, etc.* (Paris: Déterville, 1817), vol. 15, p. 271, as quoted by Jan Sapp, *Genesis: The Evolution of Biology* (Oxford: Oxford University Press, 2003), note 14, p. 274.

3 Charles Darwin, *On the Origin of Species* (London: John Murray, 1859), p. 317–318.

4 Charles Darwin, *La Descendance de l'homme et la sélection sexuelle*, trans. Jean-Jacques Moulinier, 2 vols. (Paris: Reinwald, 1872).

5 Charles Darwin, *The Descent of Man, and Selection in Relation to Sex* [1871], 2nd ed. (London: John Murray, 1874), p. 181–182.

6 Ibid., p. 182–183.

7 Ibid., p. 182.

8 Ibid., p. 183.

9 Ibid., p. 191–192.

10 Anglo-Israel, "The Dying Out of Aboriginal Races before the Anglo-Saxon Race," *New Zealand Herald*, May 10, 1879, p. 6; as quoted by Roger Blackley, "Improper Moves: Maori Haka and Racial Destiny," in *Art, Sex and Eugenics: Corpus Delecti*, ed. Fae Brauer and Anthea Callen (Aldershot: Ashgate, 2008), p. 37.

11 Benedict-Auguste Morel, *Traité des dégénérescences physiques, intellectuelles et morales de l'espèce humaine et des causes qui produisent ces variétés maladives*, 2 vols. (Paris: Baillière, 1857).

12 Cesare Lombroso, *Ricerche sul cretinesimo in Lombardia* (Milan: Chiusi, 1859).

13 Cesare Lombroso, *L'uomo delinquente* (Milan: Hoepli, 1876); five editions were published in Italy. A French edition was published under the title *L'Homme criminel* (Paris: Alcan, 1887).

14 Fae Brauer, "The Transparent Body: Biocultures of Evolution, Eugenics and Scientific Racism," in *A History of Visual Culture: Western Civilization from the 18th to the 21st Century*, ed. Jane Kromm and Susan Benforado Bakewell (Oxford: Berg, 2010), p. 89–103.

15 Fae Brauer, "The Stigmata of Abjection: Degenerate Limbs, Hysterical Skin and the Tattooed Body," in ibid., p. 169–183.

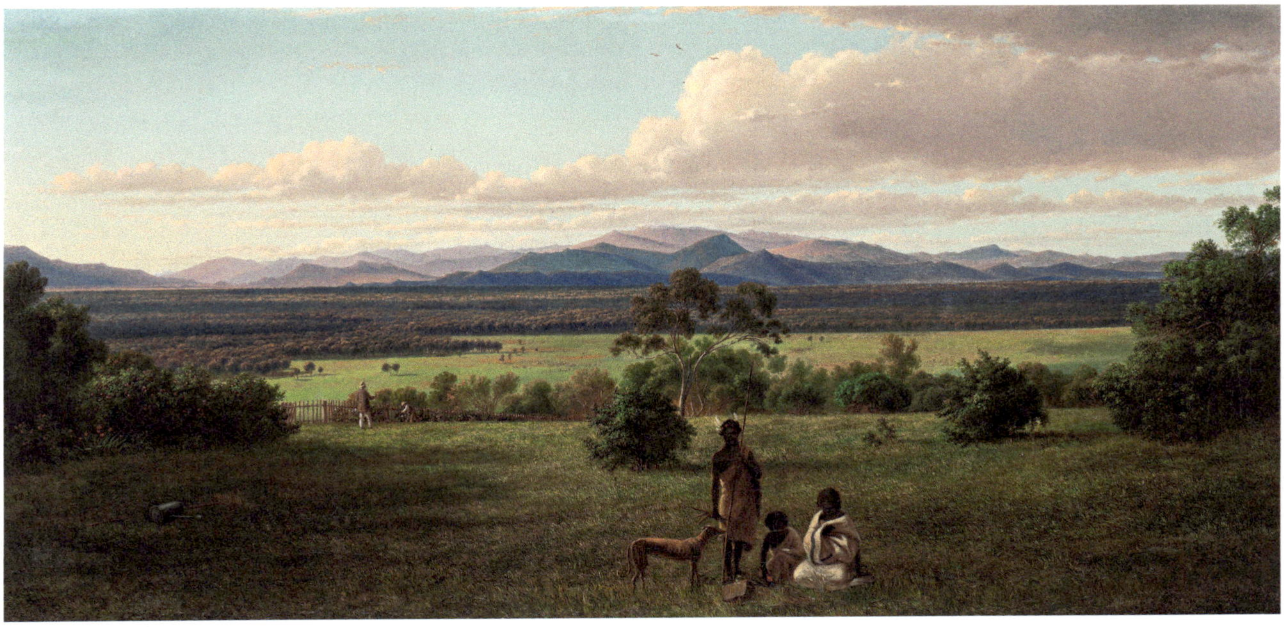

Maori appearing to fulfill Darwin's prophecy that technologically advanced civilizations had a fatal impact upon so-called "inferior" ones. Invariably their disappearance was cloaked in mystery, the Maori population seeming to "mysteriously melt away before the Anglo-Saxons."[10] This denial of the colonizer's agency and disavowal of their complicity seemed to facilitate a mythology of Indigenous people suffering self-extermination as "a dying race" rather than conditions being imposed to induce their extinction, often knowingly so. This image of their mysterious disappearance was invariably reinforced by paintings of Indigenous people before their extinction made by Eugene von Guerard in Australia and by Albert Bierstadt in North America in which they appear submerged within deep melancholic shadow, as if on the verge of disappearing through self-extinction rather than through bloody conflict and colonial agency (222). Yet once degeneration became increasingly documented among the colonizers, the paranoia of extinction was no longer confined to Indigenous people.

Just before the first publication of *On the Origin of Species*, a wide range of European evolutionary scientists, criminal anthropologists and medical psychiatrists had begun to notice a profound paradox: although European economies, industry productivity and scientific discoveries appeared to be progressing with its so-called "civilization" seemingly expanding, more and more illnesses and pathologies were reported. In 1857, Benedict-Auguste Morel provided plentiful documentation in both image and text of cretinism in low-lying areas of France (223).[11] Two years later, Cesare Lombroso did so in Lombardia.[12] Both Morel's images in his *Traité des dégénérescences physiques, intellectuelles et morales de l'espèce humaine* and Lombroso's photographs with measurements of the hollowness in skulls of notorious criminals in Italy, as corroborated by his illustrations for *L'uomo delinquente*, were seized upon as a link between criminality and atavism, and as an explanation for the increase in criminality.[13]

Following the Western concept of the transparent human body in which its exterior was presumed to be a reflection of its interior state, both Morel and Lombroso searched for telltale marks of this devolution upon the outer body.[14] Morel detected them in squint eyes, harelips, webbed fingers, flat palates, twisted limbs, distorted ears, and asymmetries in the two halves of the face and cranium. Since these marks seemed to leave an indelible stain that betrayed their devolution, Morel called them "stigmata" and linked them to the condition he called "degeneration."[15] Published two years before Darwin's *On the Origin of Species*, Morel's *Traité des dégénérescences physiques*

222 FIG.
Eugene von Guerard,
Mr John King's Station, 1861
Oil on canvas, 40 × 84 cm
Brisbane, Taylor Family Collection

defined this condition as "a sickly deviation of a primitive type," as well as "a sickly deviation from the normal type of humanity."[16] At its most extreme, this deviation meant that a particular social group, tribe or race would no longer be able to propagate and was therefore doomed to extinction. To regenerate France, Morel drew upon the evolutionary theories that Lamarck articulated in his *Philosophie zoologique*, postulating the solution of breeding out "dégénérescence héréditaire" through judicious procreation. Nevertheless, after the Franco-Prussian War and the Commune, his successors were far less optimistic.

The first world conflict to be interpreted through Darwin's "survival of the fittest" was the Franco-Prussian War.[17] Although Louis Pasteur believed France's defeat was largely due to the lack of state support for science and technology, French population demographer Jacques Bertillon attributed it to the country's waning fertility and declining birth rate.[18] It was also linked to the unfitness of French soldiers, as well as to their lesser numbers, following *The Descent of Man*, by comparison to the greater numbers, "vigour" and "good health" of German soldiers.[19] In Darwinian terms, Prussia seemed to have proven to possess a higher "grade" of "civilisation."[20] By no means was this anxiety dissipated with the formation of the Third Republic, most French conscripts being found unfit for military service.[21] Not only did this generate a crisis in virile masculinity but it also compounded a Darwinian diagnosis of the French race as a spent force.[22] When France's defeat at the hands of the Germans was diagnosed through Darwin's *The Descent of Man*, some commentators concluded that France showed all the requisite signs of a civilization in decline. This seemed to be confirmed by the Commune.

The analogies often drawn between the violence under the Commune and the bloodshed during the Reign of Terror in the First Republic seemed to reveal a deep-seated fear of a barbaric regression. While the mass culture of the Commune was identified with the seeming degeneracy of native tribes and wild animals, those male Communards deemed responsible for this regression were satirized as savages, as illustrated by the fate of the Realist painter Gustave Courbet who was held solely responsible for dismantling the Vendôme column.[23] Female Communards fared little better, particularly "les pétroleuses" who were derided as incorrigible arsonists and blamed for burning down the Hôtel de Ville and the Palais des Tuileries.[24] Caricatured with hair raised, eyes ablaze, screaming while running, a canister of petrol in one hand, flaming torch in the other, Eugène Girard's "la femme émancipée," as the artist ironically called her, is literally, following the pun of his title, "shining light on the world." Stigmatized as irrational, unnatural and devolved to the state of wild beasts, large numbers of women were admitted as hysterics to the Salpêtrière neurological clinic by Jean-Martin Charcot from the beginning of the Third Republic.[25] Since this visual culture reinforced sensationalist press reports of the Communard's transformation of Paris into a veritable inferno, it was instrumental in spawning paranoia of degeneration and devolution to the condition of savage depravity. Few theorized this fear more articulately than Hippolyte Taine, the philosophical naturalist who lectured on Darwin's "natural selection" at the École des beaux-arts.

In his six-volume history titled *The Origins of Contemporary France*, Taine linked this era of insurrection and barbaric regression to a deep national malaise: self-destruction and racial degeneration.[26] In

16 Morel, *Traité des dégénérescences physiques*, vol. 1, p. 5 and p. 47.
17 John Weiss, *Conservatism in Europe, 1770–1945* (London: Thames & Hudson, 1977), p. 107.
18 Dr. Jacques Bertillon, *La Dépopulation de la France. Ses conséquences, ses causes, mesures à prendre pour la combattre* (Paris: Alcan, 1911).
19 Robert A. Nye, "Degeneration, Neurasthenia and the Culture of Sport in Belle Epoque France," in "Decadence," special issue, *Journal of Contemporary History*, vol. 17, no. 1 (January 1982), p. 51–68.
20 Darwin, *The Descent of Man*, p. 182–183.
21 Nye, "Degeneration, Neurasthenia and the Culture of Sport," p. 51–68.
22 Robert A. Nye, *Masculinity and Male Codes of Honor in Modern France* [1993] (Berkeley: University of California Press, 1998); Christopher E. Forth, "La Civilisation and Its Discontents: Modernity, Manhood and the Body in the Early Third Republic," in *French Masculinities: History, Politics and Culture*, ed. Christopher E. Forth and Bertrand Taithe (Basingstoke: Palgrave Macmillan, 2007); Fay Brauer, "Flaunting Manliness: Republican Masculinity, Virilized Homosexuality and the Desirable Male Body," in "Masculinities," special issue, *Australian and New Zealand Journal of Art*, vol. 6, no. 1 (2005), p. 23–42.
23 Fae Brauer, *Rivals and Conspirators: The Paris Salons and the Modern Art Centre* (Newcastle upon Tyne: Cambridge Scholars Publishing, 2013), p. 21.
24 Jacques Rougerie, *La Commune, 1871* (Paris: Presses Universitaires de France, "Que sais-je ?" series, 1988), p. 117.
25 Georges Didi-Huberman, *Invention of Hysteria: Charcot and the Photographic Iconography of the Salpêtrière*, trans. Alisa Hartz (Cambridge, MA: MIT Press, 2003).
26 Hippolyte-Adolphe Taine, *The Origins of Contemporary France*, trans. John Durand, 6 vols. (New York: Henry Holt and Company, 1876–1894).
27 See Hippolyte-Adolphe Taine, *The Origins of Contemporary France: The Revolution, Volume 1* [1878], trans. John Durand (Indianapolis: Liberty Fund, 2002), chap. 3, p. 62. Darwin refers to Alfred Brehm's *Thierleben* (Hildburghausen: Bibliographisches Institut, 1864).
28 Ibid., Taine acknowledges the source of this quote as Mallet Du Pan, "Revue politique de l'année 1791," *Mercure de France*, January 4, 1792.
29 Hippolyte Taine, *Philosophie de l'Art* [1865] (Paris: Fayard, 1985).
30 Janet Browne, "Darwin in Caricature: A Study of the Popularization and Dissemination of Evolutionary Theory," in *The Art of Evolution: Darwin, Darwinisms, and Visual Culture*, ed. Barbara Larson and Fae Brauer (Hanover, NH: Dartmouth College Press, 2009), p. 30.

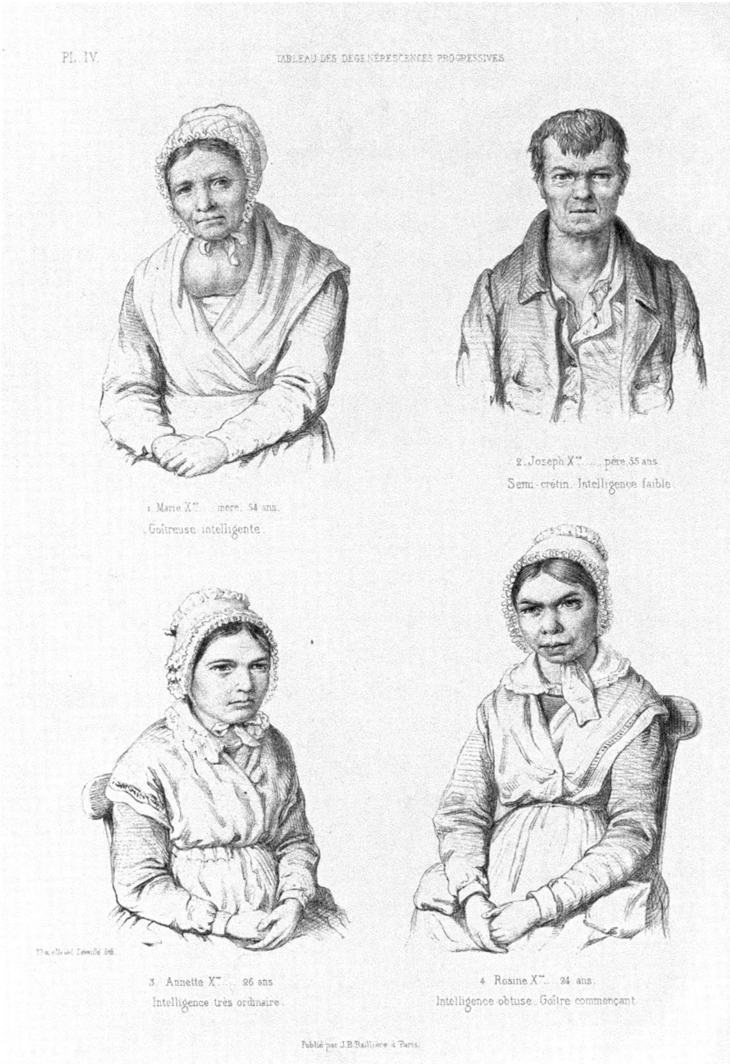

his endeavour to explain this upheaval in Darwinian terms, Taine surmised that those French citizens who had become intoxicated with the fervour of revolution had devolved much further back in the stages of evolution to the most savage state of all "ape-like creatures" notorious for their plundering: the baboon.[27] As he found no difference between them and the invaders of the Roman Empire in the fourth century, he dramatically concluded that "the Huns, the Heruli, the Vandals, and the Goths will come neither from the north nor from the Black Sea; they are in our very midst."[28] Following the legacies of Darwin and Georg Hegel, he insisted that it was time for his generation to judge itself according to race, "milieu and moment."[29] Scrutinizing how France had performed as a race from the 1789 revolution until the Commune was essential for Taine to ascertain whether, through hereditary factors, the French race was destined to devolve to its bloodthirsty Gallic roots and become simian. Singling out Darwin's *On the Origin of Species* and *The Descent of Man* as vital tools to resolve this question, paradoxically the picturing of Darwin during the Third Republic became infected with this psychopathology.

Since first publication of *On the Origin of Species* and the notorious Oxford evolution debate when Thomas Huxley proudly acclaimed a monkey as his ancestor, devolutionary states had been caricatured as monkeys and gorillas. However, it was not until publication of *The Descent of Man* and *The Expression of the Emotions in Man and Animals* that Darwin became caricatured as having devolved to this state, becoming the only Victorian evolutionist to be satirized as an ape.[30] Just as his grizzly beard, bushy eyebrows and vast dome-like skull proved irresistible to caricaturists, so did his corporeal hairiness and its facility for transformation into the animal fur of anthropoid apes, as illustrated in *The Hornet*'s 1871 caricature of Darwin captioned "A Venerable Orang-outang: A Contribution to Unnatural History" (153, P. 224). Three years later, this simian caricature was recaptured by *Figaro's London Sketch Book of Celebrities* in which an ape-like "Prof. Darwin" reveals to a monkey their common ancestry in a mirror (156, P. 225). Once *The Descent of Man* and *The Expression of the Emotions* had finally been published in France after the Franco-Prussian War, these simian caricatures were pursued there, especially given the hostility of the Moral Order regime to Darwinism. Those scientists, botanists and naturalists who supported Darwin also became vulnerable to caricature as apes, especially the lexicographer and positivist philosopher Émile Littré (154, 155).

Upon his election as senator for life in 1874, Littré was caricatured by *L'Éclipse*. Imaged leaning against the "tree of science" with acorns of knowledge scattered around its trunk, Littré is caricatured writing on pages inscribed "Philosophie" and "Positive" with his *Dictionnaire de la langue française* behind him. That it pertains to his Darwinist evolutionary theory of language is signified by the monkey depicted in front of him browsing the alphabet. Despite Littré's support

223 FIG.
Benedict-Auguste Morel, *Traité des dégénérescences physiques, intellectuelles et morales de l'espèce humaine*, pl. 4, Paris: J. B. Baillière, 1857
Waltham, MA, Brandeis University Libraries

for Darwin's membership to the anatomy and zoology section of the Académie des sciences, alongside the eminent zoologists Henri Milne-Edwards and Armand de Quatrefages, Darwin was rejected six times.[31] Only on August 5, 1878, after neo-Lamarckian transformists and radical republicans—many of whom were Darwinists—had gained a slim majority of seats in the Chamber of Deputies, was Darwin finally elected.[32] Yet despite being continuously nominated for the anatomy and zoology section, Darwin was elected to the botany section and only as a foreign correspondent.[33] That he had only been elected to the lowest rung due to his theories exploring the evolution of *Homo sapiens* from simians appears to be satirized by André Gill in his cartoon on the cover of the first edition of *La Petite Lune*, depicting Darwin as a devolutionary ape dangling precariously by one arm from the lowest branch of the "tree of science" (157, P. 225). That his election may have seemed facilitated by Littré appears to be signified by Gill's subsequent cartoon in *La Lune rousse* on August 18, 1878 (155, P. 225), in which both men are caricatured as performing monkeys at the Hippodrome. Entitled "Darwin !," a simian Littré is caricatured as a circus ringmaster beckoning his fellow simian Darwin as the latter bursts through the first hoop, that of credulity, before tackling the next one, that of "Superstitions, Errors and Ignorance"—the very prejudices that seemed to be blocking Darwin's full election to the zoology section at the Académie des sciences.[34]

The growing interest in Darwin's research among neo-Lamarckists in France led to a spate of translations of the British scientist's work beginning in 1877, including *Fertilisation of Orchids, The Structure and Distribution of Coral Reefs, The Effects of Cross and Self Fertilisation in the Vegetable Kingdom, The Different Forms of Flowers on Plants of the Same Species, The Power of Movement in Plants* and *Insectivorous Plants*.[35] For Edmond Perrier, Chair of Natural History of Molluscs, Worms and Corals at the Muséum national d'histoire naturelle, so sure was he of Darwin's theory of the transformation of species and its relationship to Lamarckian transformism that, as President of the Société zoologique de France, in 1879 he took the dangerous step of announcing it publicly to an audience at the Muséum.[36] From the time he did so, neo-Lamarckism, alongside the long-lingering threat of devolutionary atavism, played a dominant role in visual culture, as captured by Fernand Cormon's monumental painting, *Cain*, exhibited at the 1880 Salon (172, P. 251).

Rather than representing the progressive stages of human evolution from the least-evolved Neanderthal to the most evolved *Homo sapiens*, as illustrated by Huxley's *Evidence as to Man's Place in Nature* (118, P. 169), the range of physiognomies captured by Cormon in a jumble of bodies and craniums conveys the infinitely variable process of transmutation and the imminent possibilities of devolution. With *Cain* displayed when full amnesty for Communards was being heatedly debated in the Chamber of Deputies, spectators at the 1880 Salon were faced with the choice of adhering to one of the two evolutionary models inscribed by Cormon.[37] They could degenerate and devolve to the "primitive" bloodthirsty Gallic warrior pictured separated from his tribe who, like Cain, was prepared to destroy his brother and who, like the Communards, would be exiled from his homeland for so doing; alternatively, through the pursuit of neo-Lamarckian collaboration and cooperation, they could evolve and regenerate the French race, as signified by the youngest member of the tribe on the lap of its mother—a pale-skinned, golden-haired child. That collaboration and cooperation may have played a comparable regenerative role even in the Upper Paleolithic period is signified by the mostly dark-skinned nomadic hunter-gatherer family that

---

31 See Robert E. Stebbins, "France," in *The Comparative Reception of Darwinism*, ed. Thomas F. Glick (Chicago: University of Chicago Press, 1988), p. 147–148; see also Joy Harvey, "Darwin in a French Dress: Translating, Publishing and Supporting Darwin in Nineteenth-Century France," in *The Reception of Charles Darwin in Europe*, ed. Eve-Marie Engels and Thomas F. Glick (London: Continuum, 2008), vol. 2, p. 366.

32 Stebbins, "France," p. 149: *Le Moniteur scientifique*, vol. 20 (1878), p. 1027.

33 See Robert Fox and George Weisz, eds., *The Organization of Science and Technology in France 1808–1914* (Cambridge: Cambridge University Press, 1980).

34 On these caricatures, see p. 220.

35 Harvey, "Darwin in a French Dress," p. 371.

36 William Schneider, *Quality and Quantity: The Quest for Biological Regeneration in Twentieth-Century France* (Cambridge: Cambridge University Press, 2002), note 62, p. 302.

37 See Fae Brauer, "'Turquet's Turkey': Ending the Salon," in *The Paris Fine Art Salon/Le Salon, 1791–1881*, selected papers from the first major international conference on the post-1789 Paris Fine Art Salon (University of Exeter, September 4–6, 2013), ed. James Kearns and Alister Mill (Bern: Peter Lang, 2015), p. 399-420.

38 Valentin Magnan and Paul-Maurice Legrain, *Les Dégénérés. État mental et syndromes épisodiques* (Paris: Rueff, 1895), p. 222.

39 Daniel Pick, *Faces of Degeneration: A European Disorder, c. 1848–c. 1918* (Cambridge: Cambridge University Press, 1993).

40 Fae Brauer, "Dégénérescence," in *Dictionnaire du corps*, ed. Michela Marzano (Paris: Presses universitaires de France, "Quadrige" series, 2007), p. 278–285.

41 Max Nordau, *Entartung*, 2 vols. (Berlin: Duncker, 1892–1893).

42 Max Nordau, *Dégénérescence*, trans. from the German by Auguste Dietrich, 2 vols. (Paris: Germer Baillière/Alcan, 1894), dedicated to "Monsieur le Professeur César Lombroso à Turin"; published in English as *Degeneration*, trans. from the second edition of the German work (New York: D. Appleton and Company, 1895).

43 Nordau, "Fin-de-siècle," in *Degeneration*, p. 3–4.

44 Jacques Roubinovitch, *Hystérie mâle et dégénérescence* (Paris: Rueff, 1890).

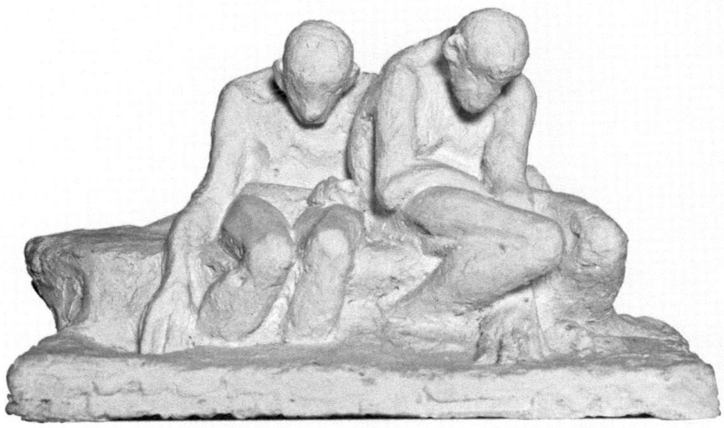

has united to capture a brace of birds upheld on one side of René Rousseau-Decelle's *Prehistoric family* (248, P. 383). At the centre left of this painting, the pale-skinned, golden-haired child cradled by its mother possibly signifies their evolution from *Homo erectus* to *Homo sapiens*. Yet many in French medicine remained sceptical.

In the wake of the social turmoil generated by the Commune, the psychiatrist Valentin Magnan psychopathologized degeneration. In his nosological system of classification, Magnan distinguished far more types of pathological degeneracy than Morel, ranging from the agoraphobic, alcoholic and arithmomanic to the anomatomanic and zoomanic. Far from conceding the possibility of regeneration, let alone the prospect of breeding out degeneration, Magnan considered that the reverse had happened, particularly as he found pathological degenerates generally procreated with one another, leading to "the definite reproduction of more defective types."[38] Capturing the fears of many, the focus of discourses of degeneration shifted from those of individual pathologies to the social types identified by Magnan, revealing that France was suffering devolutionary psychopathologies and rapidly degenerating.

By the last third of the nineteenth century, degeneration became what Daniel Pick calls "the condition of conditions" and was used to explain every physiological, pathological, social and political disaster.[39] A kaleidoscope of new medico-psychiatric taxonomies and sociological paranoia, by the *fin-de-siècle* degeneration purportedly encompassed alcoholism, cretinism, dwarfism, hysteria, neurasthenia and syphilis; agoraphobia, algophobia, arithmomania, claustrophobia, copromania, dipsomania, exhibitionism, kleptomania, necrophilia, nosophilia, nosophobia, nymphomania, onomatomania, pyromania, pyrophobia, philophobia and satyriasis; suicide, poverty, strikes, criminality, anarchism, feminism, decadent literature and Modernist art.[40] With escalating depopulation and rampant degeneration documented in most Western nations, the prospect of their devolving to the point of extinction was regarded as imminent by the *fin de siècle*. This festering paranoia was articulated most dramatically, if not neurotically, by Max Nordau, the Hungarian-born Paris-based German physician, writer and literary critic who had studied cases of hysteria with Charcot at the Salpêtrière alongside Sigmund Freud.

An ardent Darwinist, long inspired by Lombroso's *Genio e follia* and *L'uomo delinquente*, Nordau dedicated his book *Entartung*, published in German in 1892–1893, to Lombroso.[41] Once translated into French in 1894, the book bore a title designated to summon up Morel's investigation of physiological, psychological and moral devolution: *Dégénérescence*.[42] In the first of the book's five parts, titled "Fin de siècle," the first chapter of which is ominously subtitled "Crépuscule des peuples," Nordau identified the French as the most degenerate of all people. Even though the phobia of dehumanization and becoming animal stretched across Europe, particularly in Spain after the Spanish-American War, as illuminated by Carles Mani i Roig's sculpture of simian bodies entitled *Els degenerats* (224), Nordau identified France as the country of origin of degeneration, with Paris as the perfect place to witness its multiple manifestations.[43] Following the neurological research into hysteria, particularly male hysteria, by Charcot and his "charcoterie,"[44] Nordau considered this was highlighted by the increasing numbers of male and female neurasthenic and hysterical patients admitted to the Salpêtrière. "Degenerates, hysterics, and neurasthenics are not capable of adaptation," Nordau declared. "Therefore they are fated to disappear."[45] Nevertheless their rampant degeneracy before extinction explained to Nordau why "it is precisely in France that the craziest fashions in art and literature would necessarily

224 FIG.
Carles Mani i Roig,
Els degenerats, 1891–1904
Plaster, 20.5 × 37 × 22 cm
Barcelona, Museu nacional d'art de Catalunya

arise."[46] Despite connections to be drawn between Darwin and Monet's *Water Lilies*, and between Darwinian paleontology and Paul Cézanne's *In the Quarries of Bibémus*, Nordau concluded that "the curious style of certain recent painters—'impressionists,' 'stipplers,' or 'mosaists,' 'papilloteurs' or 'quiverers,' 'roaring' colourists, dyers in gray and faded tints—becomes at once intelligible to us if we keep in view the researches of the Charcot school into visual derangements in degeneration and hysteria."[47] Paradoxically, what Nordau failed to fathom was why so many French avant-garde artists championed the "primitive" and organicism as an environmental conduit to regeneration, not as a manifestation of their own degeneration. Yet, following the research of Alfred and Edmond Fournier, it was syphilis that appeared to pose a far greater threat of devolution and extinction.

"Everything is syphilis!" exclaimed Joris-Karl Huysmans in *À rebours*.[48] More menacing than onanism, more easily concealed than alcoholism, hysteria, neurasthenia or tuberculosis, more widespread than sexual inversions or perversions, syphilis assumed the whole raft of degeneration paranoia.[49] From the moment that the "Pope of Syphilography," Alfred Fournier, diagnosed congenital syphilis, followed by parasyphilis, almost every kind of malformation, abnormality and morbid form of heredity became attributed to it, from stammering to migraines, ginger hair and hysteria. Due to Fournier's dire revelations of syphilitic hereditary transmission across as many as four generations, syphilis was perceived as sowing the seeds of degeneration and extinction not just for an entire family and its ancestors but for a race or nation. As early as December 1881, this was captured by the publication of Henrik Ibsen's play *Ghosts*. Twenty years later, it continued to be conveyed by Eugène Brieux's play *Les avariés* [*Damaged Goods*], dedicated to Alfred Fournier—which closed down as soon as it opened at the Théâtre Antoine in November 1901 and did not reopen until February 22, 1905.[50] With images identified as the most effective prophylactic instrument, photographs able to demonstrate syphilitic malformations in excruciating detail were extensively commissioned for Alfred Fournier's *La Syphilis héréditaire tardive* in 1886, *L'Hérédité syphilitique* in 1891 and *Les Affections parasyphilitiques* in 1894, as well as for his son Edmond's treatises.[51] Once Alfred Fournier's Musée des Moulages at the Hôpital Saint-Louis opened in 1886, exhibiting 3,500 vividly realistic models of syphilitic afflictions made by Jules Baretta from beeswax and resin, the visual culture of syphilis was resoundingly reinforced.

In picturing the somatic inscriptions of primary, secondary and tertiary syphilis, this visual culture revealed how it was not just manifest in the tertiary lesions of the sexual organs in mature bodies but disclosed by the distorted anatomy and discoloration of infants' bodies. An image proving irresistible for satirical journals, it was caricatured on the cover of the issue of *L'Assiette au beurre* commemorating the reopening of Brieux's play *Les avariés*, with Cupid depicted as an archetypal heredo-syphilitic "runt"—emaciated, gangrenous, stunted and wasted (225). For Edvard Munch, heredo-syphilis became a subject long studied from the time he caught his first live performance of Ibsen's *Ghosts* in Oslo in 1883.[52] It also constituted the contamination that most disrupted Ernst Haeckel's recapitulation theory of evolution.

Drawing upon Haeckel's monism, Munch conveyed the generative energy of nature in his ink drawing entitled *Metabolism (Life and Death)* (221).[53] Created while Munch was working on the illustrations for Baudelaire's *Les Fleurs du mal* commissioned by Alfred Piat, it shows spermatozoa appearing to sprout from a pregnant corpse buried in the earth. As the "fittest" sperm seemingly rises like an umbilical

45 Nordau, "The Twentieth Century," in *Degeneration*, p. 540.

46 Nordau, "Fin-de-siècle," in ibid., p. 43.

47 Ibid., p. 27.

48 Joris-Karl Huysmans, *À rebours* [1884] (Paris: Gallimard, "Folio" series, 1977), p. 193.

49 Alain Corbin, "L'hérédosyphilis ou l'impossible rédemption. Contribution à l'histoire de l'hérédité morbide," *Romanticisme. Revue du dix-neuvième siècle*, no. 31 (1981), p. 131–150.

50 Alfred Fournier, "Sur la diminution de l'accroissement de la population en France," *Bulletin de l'Académie de médecine*, 2nd ser., vol. 14 (1885).

51 See Edmond Fournier, *Stigmates dystrophiques de l'hérédo-syphilis* (Paris: Rueff, 1898) and *Syphilis héréditaire de l'âge adulte* (Paris: Masson, 1912).

52 See Joan Templeton, "The Munch-Ibsen Connection: Exposing a Critical Myth," *Scandinavian Studies*, vol. 72, no. 4 (Winter 2000), p. 445–462.

53 Ernst Haeckel, *Monism as Connecting Religion and Science: A Man of Science*, 1892 Altenburg address, trans. J. Gilchrist (London: Adam and Charles Black, 1894). From Munch's arrival in Berlin, this address was well publicized.

54 Fae Brauer, "Superevolution: Interspecies Modernism and the Naturization of Modernity," in *Superorganism: The Avant-Garde and the Experience of Nature*, exh. cat., ed. Aleksandra Jach and Paulina Kurc-Maj (Lodz, Poland: Muzeum Sztuki, 2017), p. 158–159.

55 While only one fetus appears in the border of the lithograph, several were reportedly depicted in the lost frame of the oil painting as testified by Ragna Stang in *Edvard Munch: The Man and His Art*, trans. Geoffrey Culverwell (New York: Abbeville, 1979), p. 110. Classified as "offensive," this was why, Stang reports, the frame was removed.

56 Shelley Wood Cordulack, *Edvard Munch and the Physiology of Symbolism* (Madison, NJ: Fairleigh Dickinson University Press, 2002), p. 91–92.

57 Ibid., p. 91. See Allison Morehead, *Nature's Experiments and the Search for Symbolist Form* (University Park, PA: Pennsylvania State University Press, 2017), note 70.

58 Rather than living with the "ghosts" of syphilis, Munch considered he lived with the "ghosts" of tuberculosis that had destroyed his sister and his mother; see Lars Roar Langslet, *Henrik Ibsen, Edvard Munch: To genier møtes / Two Geniuses Meet* (Oslo: Cappelen, 1994), and Joan Templeton's review of that publication in *Scandinavian Studies*, vol. 67, no. 3 (Summer 1995), p. 393–395.

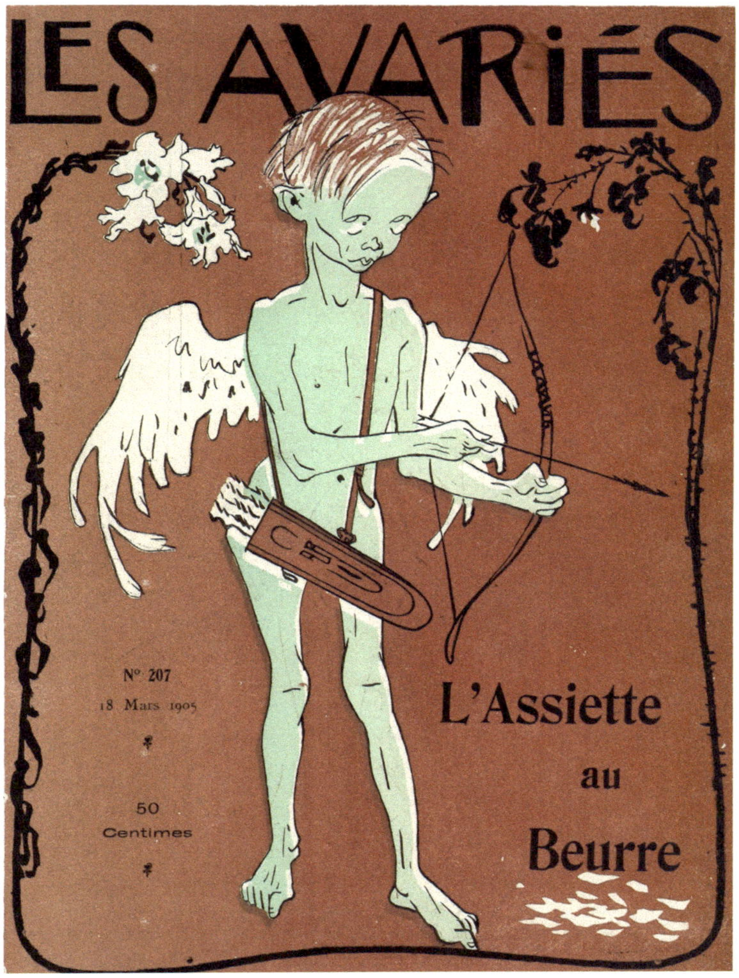

cord from her womb and metamorphoses into a human-faced flower above the earth, it becomes directly aligned with the pregnant woman above the ground and the tree. Following the polyphyletic genealogical trees conceived by Haeckel, which perpetually mutate upwards from the oldest and simplest life forms of monera to increasingly complex organisms, Munch's tree seems to signify this genealogical chain of evolution, most evidently in his *Family Tree* in which *Homo sapiens* appear to be growing (228).[54] However, Haeckel's recapitulation theory in which strong embryos successively evolve to higher organisms seems disrupted by the degeneration and devolution inflicted by heredo-syphilis and its purported transmission by "femme fatales," as signified by Munch's seductively posed Madonna framed by spermatozoa and deformed fetuses (227).[55]

That heredo-syphilis could be traced to such seemingly disconnected anatomical details as the face, skull and skin of newborns was illuminated by Munch in *Woman with Sick Child. Inheritance* (229). By then, Munch had visited Alfred Fournier's Musée des moulages with Dr. Marcel Réja and remained struck by a wax model of a baby with congenital syphilis.[56] Five years earlier, he had made drawings of patients with syphilitic spots in the hospital where August Strindberg was being treated, and where he recalled seeing a woman "crying in the hospital for venereal disease" with a "red mask of syphilis," "contorted face . . . swollen lips . . . bloated crimson chin . . . reddish purplish nose" and the "pallid yellow body" of her baby.[57] This red-faced tearful mother seems to materialize in the figure of the mother in *Inheritance*, crying in grief in the dark background, carrying on her lap the baby who, illuminated in the foreground, commands the focus of the painting, just like the Christ child in Christian iconographies of the Madonna and Child. Yet, with his physiognomy rendered as shrivelled and discoloured as that of the heredo-syphilitic Cupid, together with a death mask face, disproportionately sized skull and pustular red rash dotting the torso like the stain of sin, the baby in Munch's *Inheritance* appears far from promising redemption and regeneration. Instead, this inversion of the Madonna and Child as a Pietà signifies what Munch called "the curse of inheritance" that ends in ruin, as embodied by the tragic fate in Ibsen's *Ghosts* of the heredo-syphilitic artist Oswald Alving, with whom Munch had long identified.[58]

Faced with rampant syphilis infecting populations, together with rising levels of alcoholism, criminality, cholera, hysteria, neurasthenia, tuberculosis and physiological degeneracy (from arthritis to rickets), France, like other modern nation-states, endeavoured to resolve these dire threats of devolution and extinction. While regeneration of the body through "la vie sportive" and "la culture physique" was vigorously pursued, repopulation campaigns and depopulation committees mushroomed, large family legislation was passed, marriage procedures democratized, sexual practices scrutinized, "rational procreation" prescribed, natalism championed, abortion criminalized and eugenic systems explored. To ameliorate France's depopulation and degeneration crises, Professor

225 FIG.
Demetrius Galanis,
"Les avariés," in *L'Assiette au beurre*, March 18, 1905
Paris, bibliothèque du musée d'Orsay

Vacher de Lapouge proposed to replace Eros with Minerva.[59] By dissociating love, sexual pleasure and fertility while implementing a state-run program of "artificial insimination," he envisaged that a few males in perfect health could inseminate all the females worthy of perpetuating the race "to ensure 200,000 births annually."[60] So intent was he on refertilizing the Gallic race that he advertised for sperm to be sent to him at the University of Montpellier. After he received an array of samples through the post, his mission was acclaimed as the first experiment in telegenesis.[61] This scheme proved an irresistible subject for parody, as witnessed by *Les Exploits d'un jeune Don Juan*, in which Guillaume Apollinaire narrates how the healthy sixteen-year-old Don Juan, called Roger, proudly announces, after managing to impregnate his aunt, sister and cousin simultaneously: "I have accomplished my patriotic duty of increasing the population of my country."[62] As bizarre and draconian as these measures may now seem, they were far from isolated.

To produce "robust children," in 1896 Alfred Pichou founded *L'Élite*, which only admitted those who passed a rigorous medical examination of health, hygiene, heredity and morality, and who agreed to marry and procreate with another member.[63] That year, Paul Robin and Eugène Humbert founded the Malthusian League of Human Regeneration to ensure that "quality always precedes quantity" through "prudence procréatrice."[64] Inflamed, the National Alliance against Depopulation, together with large family federations and the staunch Catholic natalist leagues, derided the neo-Malthusian birth-control propaganda as obscene literature and denounced their sexual alternatives to the "procreative imperative" as criminal. "To have quality, you must have quantity!" Jacques Bertillon declared in *La Dépopulation de la France*.[65] Once the Société française d'eugénique was formed in December 1912 and Darwinism became far more influential, the demand for quality increasingly prevailed over quantity, as demonstrated by Charles Richet's treatise drafted that year, *La Sélection humaine*.[66] Yet it was Francis Galton's eugenics that became the dominant model.

Rejoicing in his cousin Charles Darwin's theories of "natural selection" and "survival of the fittest," Galton had been determined to eradicate devolution by accelerating evolution to give what he called "the more suitable races or strains of blood a better chance of prevailing speedily over the less suitable than they otherwise would have had."[67] From his experience of living in Africa—from which he contracted venereal disease—Galton mounted a biocultural model not dissimilar to that of Arthur de Gobineau's *Essai sur L'inégalité des races humaines* to account for the supposed superiority of white races.[68] Yet it was only on the death of Darwin that Galton began to use the term "eugenics," which he defined as "the study of the agencies under social control that may improve or impair the racial qualities of future generations, either physically or mentally."[69] Replacing Darwin's "natural selection" with racial selection and national intervention to secure those he called "the fittest races," Galton explained, "what nature does blindly, slowly, and ruthlessly, man may do providently, quickly and kindly."[70] This was to be facilitated by Galton's composite photography, as illustrated by his composite photographs of Royal Engineers selected by Darwin's son Leonard. Since the faces of these officers, according to Galton, "gives a clue to the direction in which the stock of the English race might most easily be improved," he predicted that his composite portraiture could act as a guide to breeding the English race and restraining those who deviated from it (226).[71] Following the foundation of the Eugenics Education Society, the First International Eugenics Congress and Exhibition was planned.

59 Drawing upon the eugenic research of Alphonse de Candolle and Francis Galton, Vacher de Lapouge published his theory of eugenics in *Les Sélections sociales*, in Paris in 1896, and then in 1899, at the height of the Dreyfus Affair, in *L'Aryen, son rôle social*, which had a profound impact on both American and German race hygiene movements.

60 Georges Vacher de Lapouge, *Les Sélections sociales* (Paris: Fontemoing, 1896), p. 472–473.

61 This was the first experimentation with insemination at a distance. See Jean Rostand, *Biologie et humanisme* (Paris: Gallimard, 1964), p. 161.

62 G. A. [Guillaume Apollinaire], *Les Exploits d'un jeune Don Juan* (Paris, 1911), p. 127. The text was part of a reworked version of *Mémoires d'un jeune Don Juan*, first published in 1907.

63 Alfred Pichou, "L'Élite," *Revue internationale de sociologie*, no. 14 (1896), p. 583–584; see also Dr. Lehmann, "L'Élite," *La Culture physique*, no. 31, April 1, 1906, p. 438–439.

64 Paul Robin, in *Bulletins de la Société d'anthropologie de Paris*, vol. 7 (1896), p. 139–140.

65 Bertillon, *La Dépopulation de la France*, p. 59.

66 Charles Richet, *La Sélection humaine* (Paris: Alcan, 1919); mostly written in 1912. See also G. Papillault, "Galton et la bio-sociologie," *Revue anthropologique*, vol. 21, no. 2 (Feb. 1911), p. 56–65. "… le capital biologique, transmis par l'hérédité, accumulé par la sélection, et qui constitue le trésor où chaque individu puise sa force physique, intellectuelle et morale."

67 Francis Galton, *Inquiries into Human Faculty and Its Development* (London: Macmillan, 1883), p. 24–25.

68 Arthur de Gobineau, *Essai sur L'inégalité des races humaines*, 4 vols. (Paris: Firmin-Didot, 1853–1855).

69 Francis Galton, "The Foundation of Eugenics," *Essays in Eugenics* (London: The Eugenics Education Society/ Macmillan, 1909), p. 81; see also Fae Brauer, "Framing Darwin: A Portrait of Eugenics," in *The Art of Evolution*, p. 136.

70 Francis Galton, "Eugenics: Its Definition, Scope, and Aims," *The American Journal of Sociology*, vol. 10, no. 1 (July 1904), p. 2.

71 Galton, *Inquiries into Human Faculty*, p. 14.

72 Frédéric Houssay, "Eugénique et régimes alimentaires," *Eugénique, organe de la Société française d'eugénique* (January–April 1913), p. 1.

73 Ibid.

74 Ibid., p. 36.

75 "Les infirmités mentales en Angleterre et l'acte de 1913," *Eugénique* (1914), p. 113–118.

76 Ibid, p. 118.

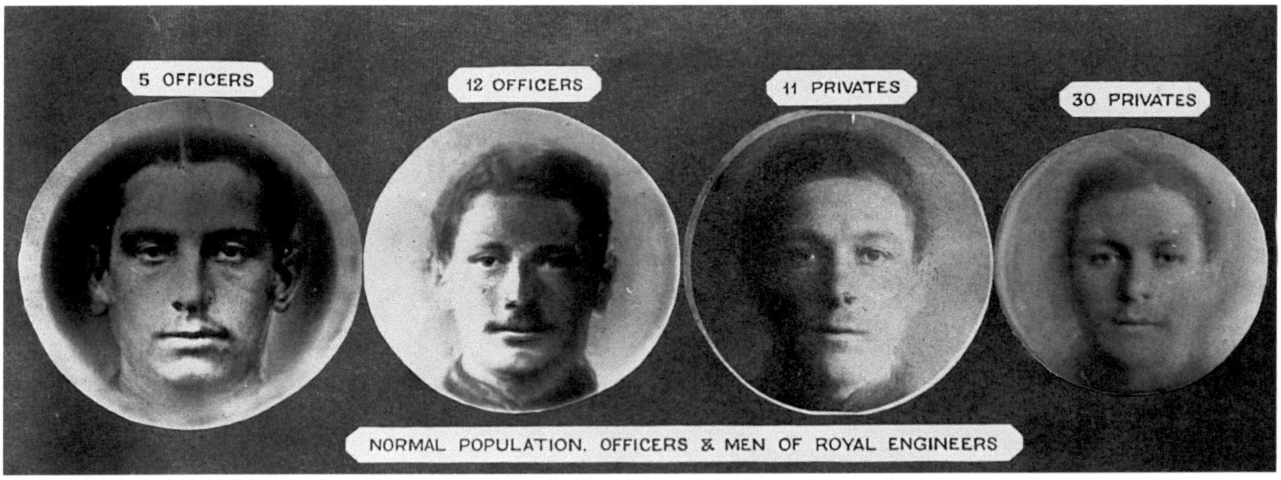

Among the members of the large French Consultative Committee at the First International Eugenics Congress, which opened in London in June 1912, were Lucien March, Director of the Statistique générale de la France; Edmond Perrier, Director of the Muséum national d'histoire naturelle; Louis Landouzy, Dean of the Faculté de médecine de Paris; and the obstetrician Adolphe Pinard. So committed to eugenics did they seem that by the end of the congress, a second was scheduled for Paris in August 1913, but subsequently revised to 1914. The Société française d'eugénique, founded on the Committee's return to France, was by no means short-lived. In 1921 Pinard became its president, and its last meeting was held under the Vichy regime in December 1941. In the first of eight meetings of the Société française d'eugénique, which took place in 1913, Pinard reported on his lecture at the Congrès international de l'éducation physique, after which Frédéric Houssay, Professor of Zoology at the Sorbonne and Vice-President of the Société, declared the dire need for "defectives" to be prevented from reproducing themselves.[72] Like members of the American Breeders Association, he did not flinch from deploying extreme measures, as testified by his recommendation: "The sterilization of these ['defectives'] by vasectomy or vallectomy appeared to many to be the definitive remedy for degeneration."[73] Upon reporting on his visit to the Cold Spring Harbor Laboratory of Eugenics and his correspondence with Karl Pearson at the Francis Galton Laboratory for National Eugenics, March recommended that comparable eugenic institutions be established in France.[74] In the journal *Eugénique*, Galton's *Essays on Eugenics* were reviewed while a long report on the Mental Deficiency Act in Britain appeared.[75] To protect the French race from degeneration and devolution, the journal announced the need for comparable legislation in France: "Another duty imposes itself today: To prepare new legislation that applies to the mentally deficient in general, to those who are called halfwits, and to the backward. As much as those properly called mad, they must be subjected to measures that preserve their fellow citizens and the race, and to measures able to protect their weakness or simply the full use of their diminished faculties."[76]

Disrupted by *La Grande Guerre*, the impetus for this legislation was not revived until Vichy. Yet the fact that leading scientists and physicians, who had long supported Darwin and neo-Lamarckian biopolitics, became so actively engaged in Galtonian eugenics demonstrates how the threat of depopulation and degeneration—as manifest by alcoholism, arthritis, barbarism, cholera, cretinism, criminality, hysteria, insanity, neurasthenia, onanism, sterility, syphilis and tuberculosis—constantly operated as a galvanizing force for evolution in France, even in its most extreme form. Hence, when evolution is located within this dialectical interplay of evolution and extinction, degeneration and regeneration, and viewed from the perspective of eugenics, it is possible to conclude that this became its most sombre face.

226 FIG.
Francis Galton,
Composite Portrait (detail),
"Normal Population. Officers and Men of Royal Engineers,"
1878

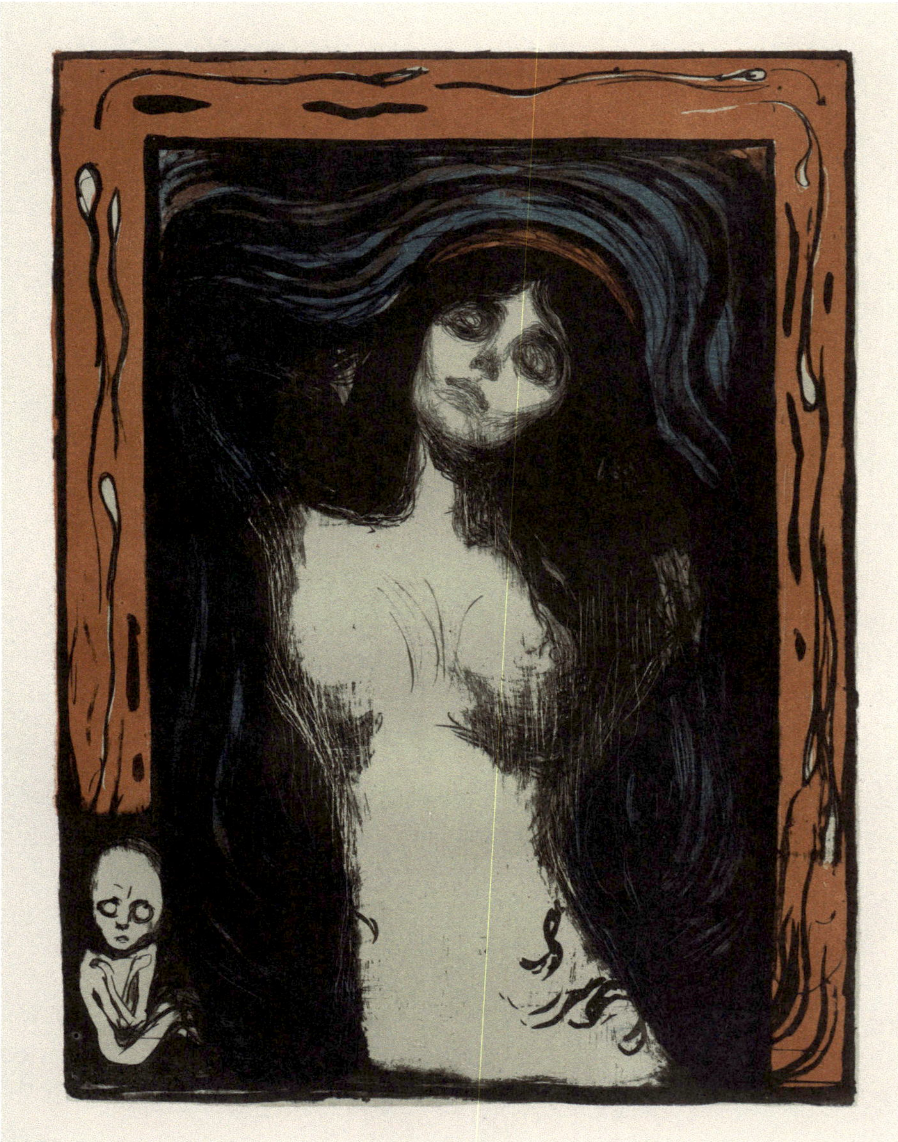

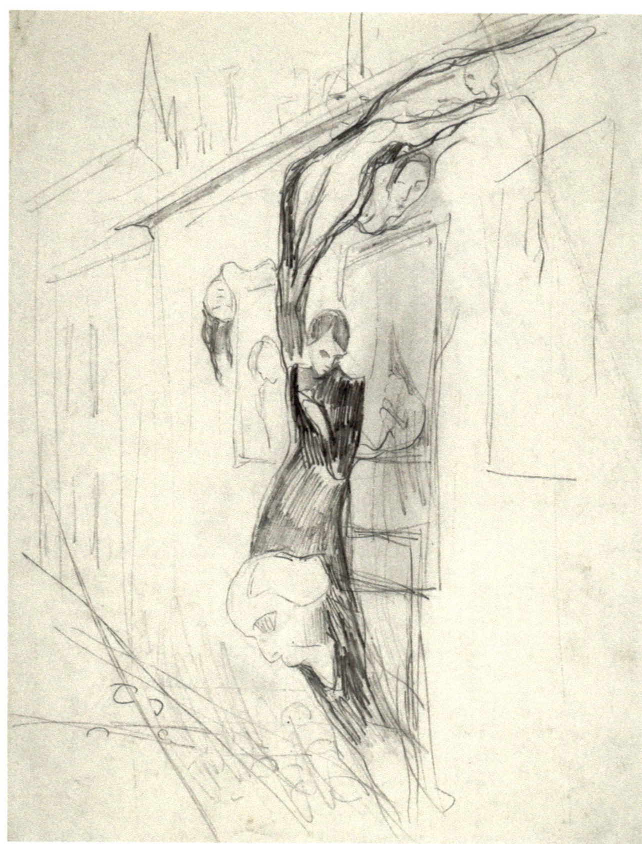

227
Edvard Munch,
Madonna, 1895–1902
Lithograph, 64 × 48 cm
Oslo, Munchmuseet

228
Edvard Munch,
The Family Tree, 1890–1892
Pencil on wove paper,
38.4 × 29 cm
Oslo, Munchmuseet

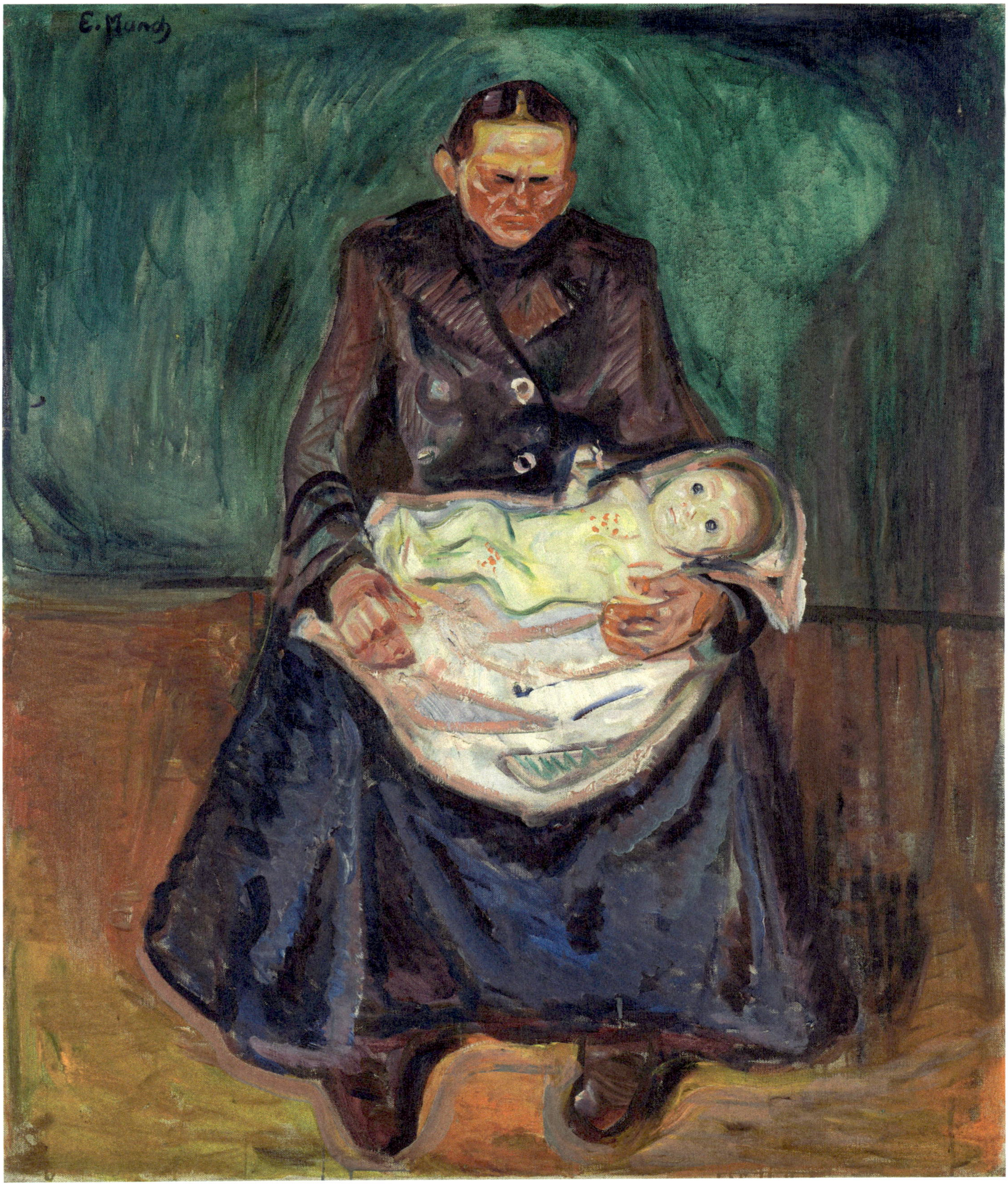

229
Edvard Munch,
Woman with Sick Child.
Inheritance, 1905–1906
Oil on canvas, 120.5 × 100 cm
Oslo, Munchmuseet

X THE ORIGINS ACCORDING TO ABSTRACTION

ARNAULD PIERRE

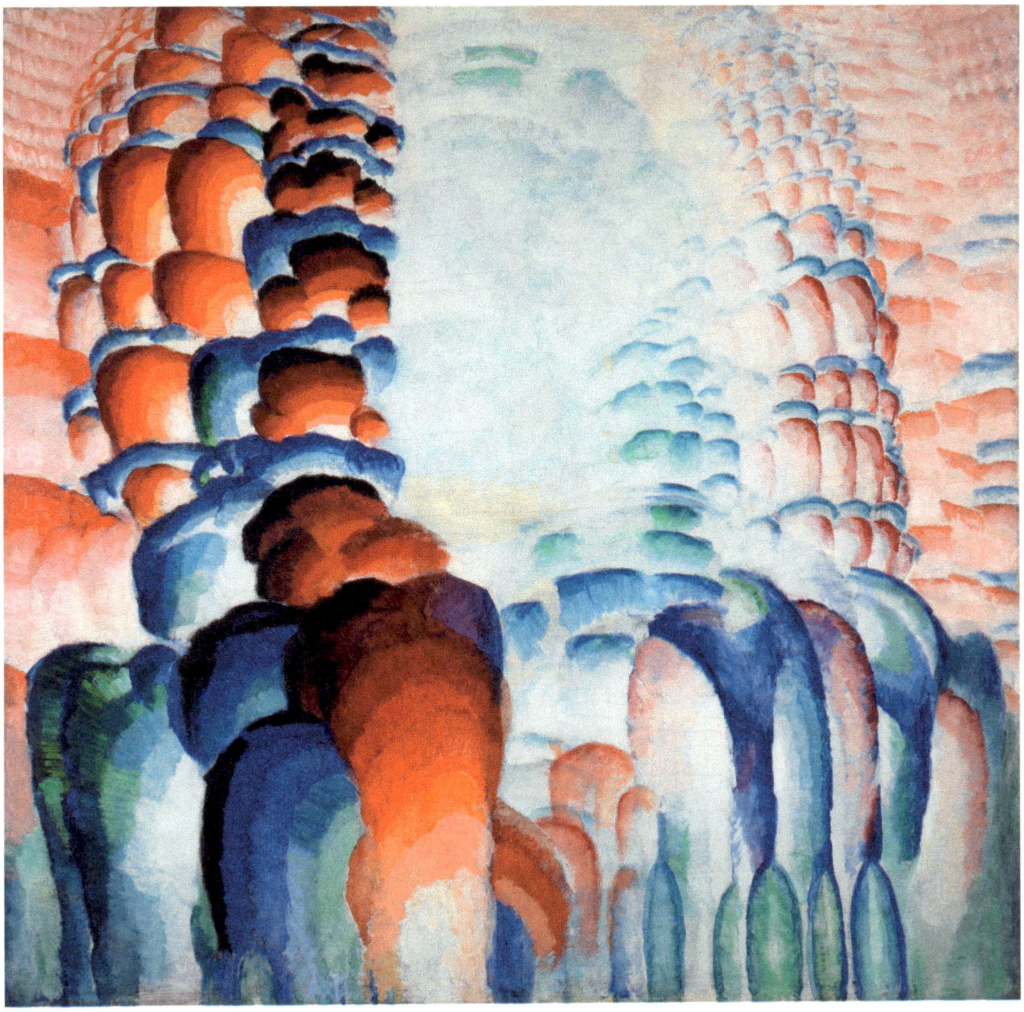

230
František Kupka,
Hindu Motifs
(Graduated Red), 1919
Oil on canvas, 124.5 × 122 cm
Paris, Centre Pompidou –
MNAM, CCI

"Evolution has its own philosophy; why wouldn't it have its own aesthetic?" The question was posed at the turn of the last century by an author named Ernest Marguery in an intriguing essay on aesthetic philosophy, *L'Œuvre d'art et l'évolution*, in which Marguery lays out for the art of his day the path to a renewal that would involve a close concordance between artistic creation and evolutionary biology.[1] This piece of writing has not yet been given any consideration in the historiography of early abstraction. However, it brings a totally new perspective to the mimetic requirements of art with respect to the higher mission it could be invested with: the expression of underlying forces that govern the movements of matter in a nature that is perpetually at work. The nature he refers to is already a broadly abstract entity; he never points to it in its appearances as phenomena but reduces it to a series of essences: matter, motion and rhythm. Rather than the necessarily inaccurate representation of immediate appearances, the path pointed out by Marguery—in terms that many pioneers of abstraction could have endorsed—is that of creating "artificial rhythms" and "imaginary harmonies." And the prospect he outlines is actually that of the "gradual emancipation of the artwork" relative to nature, at the end of which stands an even more breathtaking possibility, that of an art utterly freed from objective reality: "At last, total emancipation arrives. The artist, having transcribed nature, then aspires to control it. He asks no more of it than matter and motion, which he combines in balanced, rhythmic groups according to an artificial order inspired by the harmonies of nature itself, *but independent of any reality*."[2]

AN EVOLUTIONARY AESTHETIC

The argument put forward in Marguery's book is remarkably close to that laid out by František Kupka several years later in his own treatise, *La Création dans les arts plastiques*, an essay, written between 1910 and 1912, that draws upon biological metaphors and evolutionary concepts. At a time when the artist was himself taking his painting in the direction of abstraction, he urged his fellow painters to go beyond the surface of the "objective world" in order to see what actually plays out in "the great theatre of nature," namely, "the drama of the formation of organized matter," where "chemical action" and "motion" play leading roles in the process "that commands matter to take form."[3] Rather than draw samples from nature that are lifeless, because they are cut off from their biotope, "painters and sculptors should try to mimic nature," should borrow "examples of organic connections" and should seek to "create as logically as nature itself."[4] Only then will the work succeed in being what it must be: "a complex whole, an organism with its own, specific properties of existence, living its own life, for itself."[5]

In this way, the life of artworks would be linked to the overall history of life, and abstraction—which, in this context, is no longer the result of a crippling break with visible reality but, on the contrary, a means for delving deeper into it—would establish itself as the preferred idiom by which the story of origins could be told. A story that would no longer take anything from nature apart from matter and motion, to borrow Marguery's words when he asks the artist to "reconstruct or recreate according to his own specific genius, in a plastic, luminous or audible form, the harmonious intertwinings of matter and motion that attracted him."[6] Some of Kupka's early abstract works clearly take on the task of developing this grand narrative of the origins of life and the metamorphoses of matter. A series bearing the programmatic title *Cosmic Spring*, begun in 1911, displays a variety of shapes and structures that are, in turn, vaporous, liquid or crystalline, and that seem to simultaneously summon up the various orders of creation (mineral, vegetable, organic) and the different states (gaseous, liquid, solid) of matter. "Pressed by mysterious forces," Marguery wrote, "the molecules of matter seem to strive to escape from them. They condense, crystallize, join together, combine, organize according to similarly mysterious laws to form groups of varying complexity . . . And each time, the external forces, which have been contained but have never wearied, once again attack these provisional groups, couple with them, disrupt them, split them apart, break them up, liquefy them, volatilize them, leading the matter to undergo other metamorphoses."[7]

---

1 Ernest Marguery, *L'Œuvre d'art et l'évolution* [1899], 2nd ed. (Paris: Alcan, 1904), p. 17; a third edition came out in 1929, a sign of some success.
2 Ibid., p. 56–57 (emphasis added).
3 František Kupka, *La Création dans les arts plastiques* (Paris: Éditions du Cercle d'art, 1989), p. 114. (František Kupka, *Creation in the Plastic Arts*, trans. Edmund Jephcott [Liverpool: Liverpool University Press, 2001] was unfortunately not available at the time of publication.)
4 Ibid., p. 117 and 120.
5 Ibid., p. 262.
6 Marguery, *L'Œuvre d'art*, p. 75–76.
7 Ibid., p. 19.
8 Kupka, *La Création*, p. 114, quoted in Meda Mladek, "Central European Influences," in *Frantisek Kupka, 1871–1957: A Retrospective*, exh. cat. (New York: Solomon R. Guggenheim Foundation, 1975), p. 212.
9 Kupka, *La Création*, p. 254.
10 Ibid., p. 197.
11 Marguery, *L'Œuvre d'art*, p. 22.
12 Arnauld Pierre, *Frank Kupka: In White and Black* (Liverpool: Liverpool University Press/Artists Bookworks, 1998); Arnauld Pierre, "*Quatre histoires de blanc et noir* et l'œuvre en blanc et noir de František Kupka," in *Histoires de blanc et noir*, exh. cat. (Grenoble: Musée de Grenoble; Paris: Réunion des musées nationaux, 1996), p. 27–45.

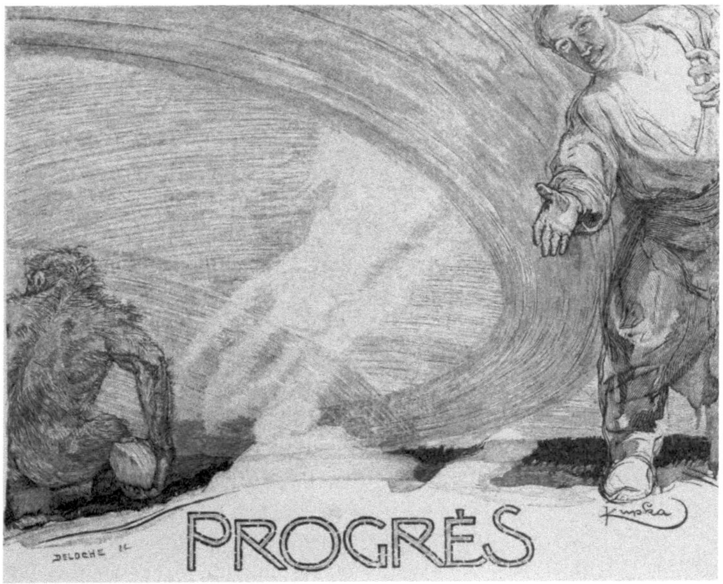

Just after the war, Kupka continued his painting of these metamorphoses in a new cycle of works inspired by the growth of plant organisms. Three paintings, titled *Tale of Pistils and Stamens*, seem to echo this pantheistic evocation by the painter: "The stamens, with their joyous phallic forms fecundate the gracious pistils. It is a real pollen festival in the gynoecium bathed in sunlight, and surrounded by the petals which unfold to protect the event of conception"[8] (233). But here as well, this unfolding falls within an abstract space and is produced from a generative centre whose broad, thick treatment renders palpable the articulation of the lumps of paint, breaks up the ever-widening rhythms of their expansion and so demonstrates the history of their constitution and growth through successive upsurges of coloured impastos: "Modulating the surface, the forms materialize all on their own, open up like the leaves, flowers and fruits of a tree."[9] The painting thus incorporates into its own rhythms each stage in the transformations of matter—"the outcome of enlivened matter," as Kupka puts it[10]—and makes us feel its internal struggles, as described, once again, by Marguery: "Action and reaction collide in a series of rhythmic beats that are like the palpitations of this universal gestation."[11] The series as a whole, set under the narrative heading of "tale," anticipates that of "story" in *Four Stories of White and Black* (1926)—the collection of woodcuts in which the artist summed up the evolution of the four main formal themes in his painting—and so confirms what seems to have been the fundamental project of his evolutionary aesthetic: to produce a natural history of forms, parallel to the official natural history, that of life and of the universe.[12]

"HUMANITY IS NATURE BECOMING SELF-CONSCIOUS"

However, Kupka's keen curiosity in these subjects is evident well before his conversion to abstraction, for which it undeniably paved the way. It dates back at least to the years when he illustrated with dozens of meticulously realistic compositions the final work by the geographer-historian—and major intellectual figure in the anarchist movement—Élisée Reclus. The six volumes of *L'Homme et la Terre*, published between 1905 and 1908, present a vast panorama tracing the universal history of humanity in its interaction with the physical and cosmic environment—a history entirely guided by the evolutionary perspective. The first chapter in the work, "Origines animales de l'homme," amasses the classic proofs of the theory of evolution: anatomical comparativism and the similarity of the human skeleton's organization with those of the large primates, the discoveries of geology, the study of fossils and prehistoric anthropology. This biological Darwinism is extended by a kind of social Darwinism, which takes the opposite view of a Herbert Spencer extolling the free play of the struggle for existence and the elimination of the least fit, and asserts, on the contrary, the existence of a universal solidarity carrying humanity along on the path of progress. This is the idea summed up by Kupka in his heading for the last chapter of the survey, "Progrès" (231). The image contrasts, on the left, an anthropoid with his back turned, gripping a carved flintstone, and on the right, in the final stage of evolution, a man holding out his hand, whose costume and wide fabric sash indicate that he is a revolutionary worker. For Reclus—as well as for the man who was, more than his illustrator, the author of a discourse through images that supported his text in every respect—socialism and Darwinism marched shoulder to shoulder towards the fulfillment of humanity in universal brotherhood, defining together what has every appearance of an evolutionary religion of progress.

231 FIG.
Élisée Reclus,
"Progrès," in *L'Homme et la terre*, Paris: Librairie universelle, 1905–1908
Paris, BnF

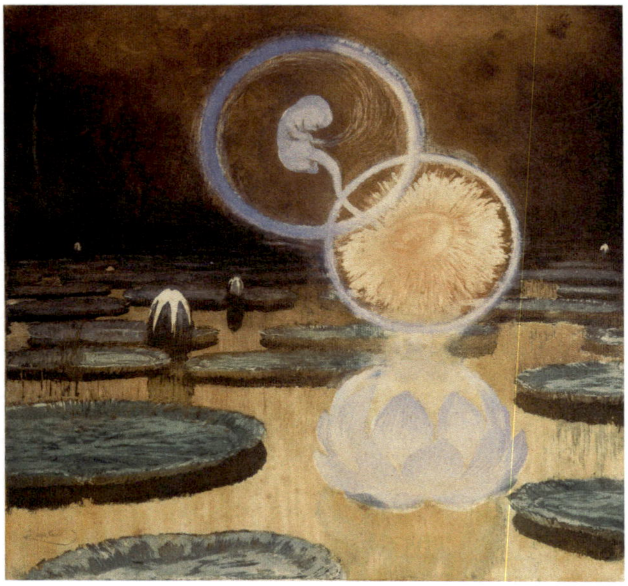

The notes that accompany Kupka's study for this image are unequivocal in this regard: they associate "revolutionary" with "evolve" and evoke the "harmony of individual freedom and the collective will." One of them, "becoming self-conscious," directly echoes the very beginning of the work, where Kupka had singled out a statement from Reclus's introduction: "Humanity is nature becoming self-conscious." This is the pre-eminent evolutionary maxim, which we come across in the writings of other propagators of Darwin's ideas, such as the German naturalist Ludwig Büchner, the author of *L'Homme selon la science*, a well-received work in which he had previously drawn from the new evolutionary concepts the idea that "in man, the world becomes, up to a certain point, conscious of itself."[13] Consciousness, a necessary function of organized matter that has reached a high stage in its evolution, allows nature to see itself in a mirror and look back at its origins. It is in this way, explains Reclus, that man "remembers a whole infinity": "Born of countless generations, other men or anthropoids, animals, plants, primary organisms, man recalls through his structure everything his ancestors lived through over the prodigious duration of the ages. He sums up, in himself, everything that preceded him throughout existence and, during his embryonic life, he successively presents the various forms of structures that are simpler than his."[14] This once again formulated the famous principle of recapitulation, stated by Ernst Haeckel—whose *Anthropogenie* is quoted extensively in the opening pages of *L'Homme et la Terre*—according to which the embryo re-experiences, in accelerated fashion, every stage in the evolution of living beings or, in other words, ontogeny recapitulates phylogeny.[15] Haeckel's synoptic tables comparing the development of the embryo in humans and other species were not hard to find: they circulated in all the popular science works of the day, including those of Camille Flammarion (105, P. 152). Kupka certainly took inspiration from them when he inserted a representation of an embryo floating in the cosmic night of *The Beginning of Life* (1900) (232), one of the most mysterious images ever devoted to the question of origins. Kupka's embryo does not just embody this reference to evolutionary biology; it also takes literally the metaphor which, in the writings of ancient India translated by French philologist James Darmesteter, refers to the primordial light as the "golden embryo" birthed by the lotus, the sacred flower with a calyx in the shape of an egg[16]—the "World-Egg," according to Helena Petrovna Blavatsky, the leading light of theosophy.[17] In a syncretism characteristic of the time, the natural sciences came together with the esoteric concerns of theosophy, of which Kupka was a follower, and its attraction for Eastern philosophies (230).

THE GROWTH OF CONSCIOUSNESS

Was this theosophical imagery incompatible with Haeckel's embryogenic science? Certainly not, if we recall that throughout Haeckel's work the theory of evolution is merged in the substratum of German *Naturphilosophie* and takes up its spiritualist perspective: biological evolution is also the history of the slow, gradual growth of the spirit within matter, and the conclusion of this spiritual evolution is the emergence of a higher being whose consciousness, expanded to encompass the cosmos, would make its way into every corner of the universe. The founding of the Monist League in 1904 set up its doctrine of the

13 Louis Büchner, *L'Homme selon la science. Son passé, son présent, son avenir* (Paris: Reinwald, 1874), p. 307. Originally published in German as *Der Mensch und seine Stellung in der Natur in Vergangenheit, Gegenwart und Zukunft*.

14 Élisée Reclus, *L'Homme et la Terre*, vol. 1: *Les Ancêtres. Histoire ancienne* (Paris: Librairie universelle, 1905), p. 14.

15 Ernst Haeckel, *Histoire de la création des êtres organisés, d'après les lois naturelles* [1868], 3rd ed. (Paris: Schleicher, 1903), p. 225. Published in English as *The History of Creation: Or the Development of the Earth and Its Inhabitants by the Action of Natural Causes* (New York: D. Appleton, 1876).

16 Arnauld Pierre, *Maternités cosmiques. La recherche des origines, de Kupka à Kubrick* (Paris: Hazan, 2010).

17 Helena Petrovna Blavatsky, *The Secret Doctrine: The Synthesis of Science, Religion, and Philosophy*, vol. 1 (London: The Theosophical Publishing House, 1893), p. 58, http://www.gutenberg.org/files/54824/54824-h/54824-h.html.

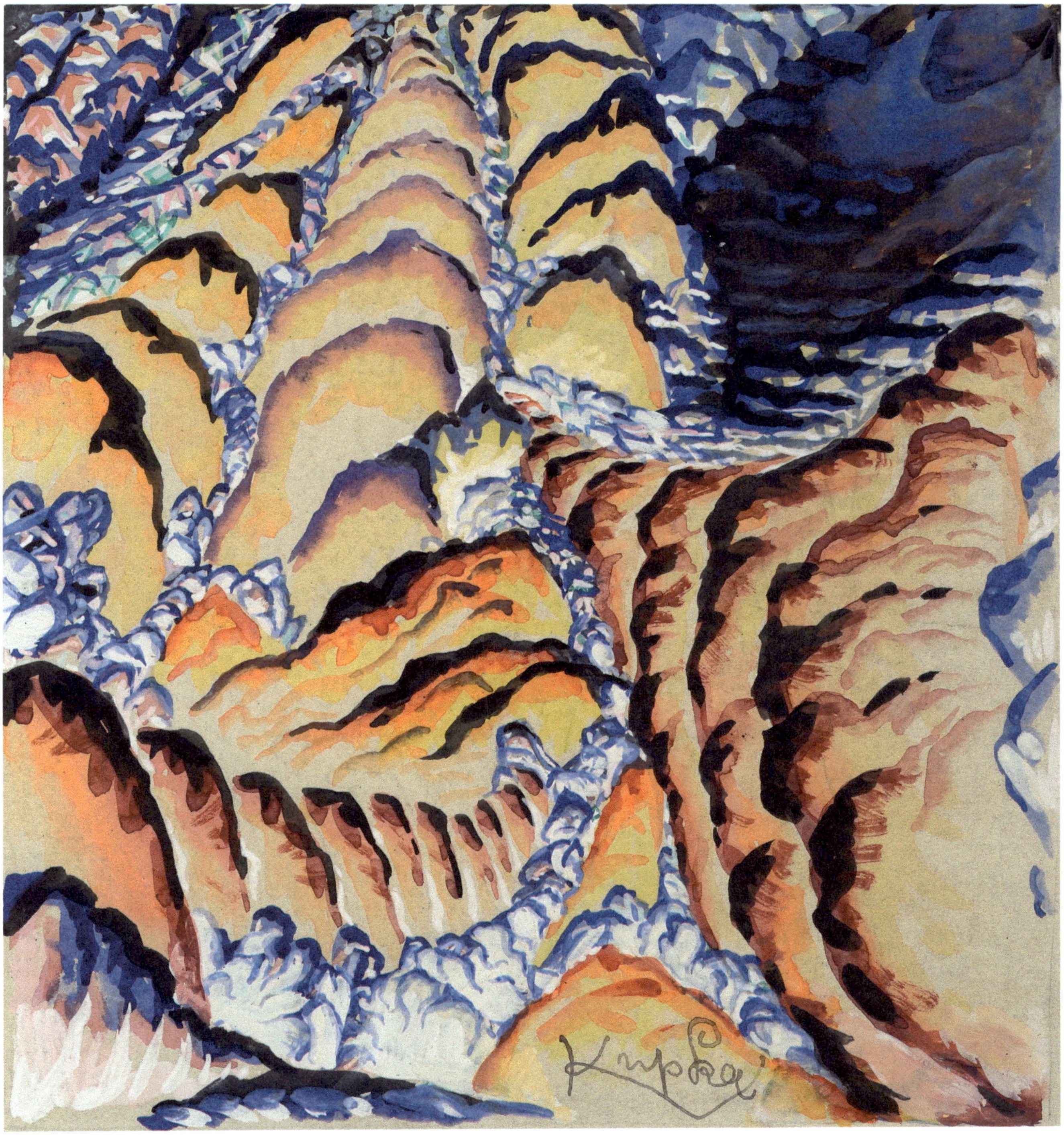

232 FIG.
František Kupka,
The Beginning of Life or Water Lilies, 1900
Coloured aquatint,
34.5 × 34.7 cm (image)
Paris, Centre Pompidou – MNAM, CCI

233
František Kupka,
Study for "Tale of Pistils and Stamens," about 1919
Gouache, 33.3 × 29 cm
Collection de Bueil & Ract-Madoux

cosmic unity of living beings and of all rungs of life as a kind of scientific religion that attracted followers all over Europe. Rudolf Steiner, representative of the Theosophical Society in Germany and a great admirer of Haeckel's panpsychism, brought the two schools of thought together in writings that would nurture the early promoters of abstraction in German-speaking regions and beyond. Among them was Vassily Kandinsky, a reader of the journal *Lucifer Gnosis*, where he may have encountered one of Steiner's major essays on the subject.[18] Before him, Édouard Schuré, in *The Great Initiates*, saw Darwin as having "reverted to the ideas of ancient theosophy," but deplored his "purely external and materialistic conception of nature." This had to be replaced by "a new embryogeny" that also granted "a new intellectual and animistic force," which alone was truly responsible for life's advance along the "degrees of consciousness."[19] The meaning of the *Evolution* triptych painted by Piet Mondrian in 1911 is there in its entirety; its iconography reflects all the implications of this theosophical version of evolutionary theory (234). For indeed, these three women depicted from the front in poses of meditation (those on the sides, eyes closed) and ecstasy (in the middle, eyes open onto a transcendent reality) tell the story of an evolution—that of the spirit gradually freeing itself from matter to ascend to the astral plane, and then the divine plane of the theosophists. This narrative is supported by a geometric symbolism applied to the increasingly luminous coloured auras that frame the faces and that gradually become structured in a play of triangular figures that also affects certain anatomical details, the tips of the breasts and the navels of the figures depicted. This reading has been long established since the pioneering work of Robert Welsh.[20] However, it misses the biological dimension of this evolutionary narrative. Indeed, what has never been noted is the way the blue-tinged chiaroscuro shapes the conical bellies of these "theosophical Graces," as they have been called: they are undeniably pregnant. These three hieratic women are not only pure spirits, they are also heavy with the fruit incarnate of this evolution—no doubt the "future man," endowed with a higher consciousness, to whom Mondrian would dedicate his essay on Neo-Plasticism, his new aesthetic doctrine.[21]

A new art for a new consciousness, a consciousness expanded to encompass the dimensions of the cosmos: that is the justification Malevich would give for his evolution towards Suprematism, his own abstract language. His early works were not impervious to the symbolism of gestation: his *Tree of Life* (1908) shows a broad trunk emblazoned with a stylized embryonic form, while a dozen or so bacchantes in provocative poses hang from its branches, which are laden with heavy, colourful fruit. That same year, the torso of his *Woman in Childbirth* stands out against a background studded with countless fetuses apparently generated by the fertilizing, almost monstrous, power of this kind of mother-goddess.[22] The theme of these works is not unrelated to the early Suprematist compositions, among them the famous *Quadrangle* from 1915, a slightly irregular black square edged in white, which its author had presented as the "embryo of future creation," the "primogenitor" from which would be generated all other forms of Suprematism.[23] It was at that same time that Pavel Filonov formulated the notion of "flowering of the world," which is accompanied by a series of works teeming with cellular forms. In *Flowers of the Universal Flowing* (1915), for example, these forms turn the painting's surface into "the place where the pictorial germinates, grows and blossoms—a pictorial that becomes a grand metonymy for the world and the Universe."[24] But, as Jean-Claude Marcadé points out,[25] in this painting it is not simply a matter of exactly copying natural growth processes; it is also a question of anticipating the evolution of the biosphere to the noosphere, by which the cosmos is able to give birth to a new self-consciousness. In that same circle of artists, who came together in the early 1920s at the State Institute of Artistic Culture in Petrograd, Matyushin saw in the expansion of humanity's sensory capacities, and in particular in the emergence of a new spatial perception, the necessary physiological mutation capable of accompanying and perhaps even generating this new stage of consciousness.

18 Rudolf Steiner, "Haeckel, die Welträtsel und die Theosophie," *Lucifer Gnosis*, no. 31 (1905); see Sixten Ringbom, "Die Steiner-Annotationen Kandinskys," in *Kandinsky und München: Begegnungen und Wandlungen, 1896–1914*, exh. cat., ed. Armin Zweite (Munich: Prestel, 1982), p. 102–105.

19 Édouard Schuré, *The Great Initiates* [1889], trans. Fred Rothwell (London: William Rider & Son, 1913), p. xxv–xxvi.

20 Robert Welsh, "Mondrian and Theosophy," in *Piet Mondrian, 1872–1944: Centennial Exhibition*, exh. cat. (New York: Solomon R. Guggenheim Foundation, 1971), p. 43–48.

21 Piet Mondrian, *Le Néo-Plasticisme* (Paris: L'Effort moderne, 1920).

22 For a more in-depth commentary on these works, see Andréi Nakov, *Kazimir Malewicz, le peintre absolu* (Paris: Thalia, 2006), vol. 1, p. 84–94.

23 Malevich, quoted by Nakov, ibid., vol. 2, p. 54 and 56.

24 Jean-Claude Marcadé, "Filonov, le tableau comme site de l'éclosion universelle," in *Filonov*, exh. cat. (Paris: Centre Pompidou, 1990), p. 68.

25 Ibid., p. 78.

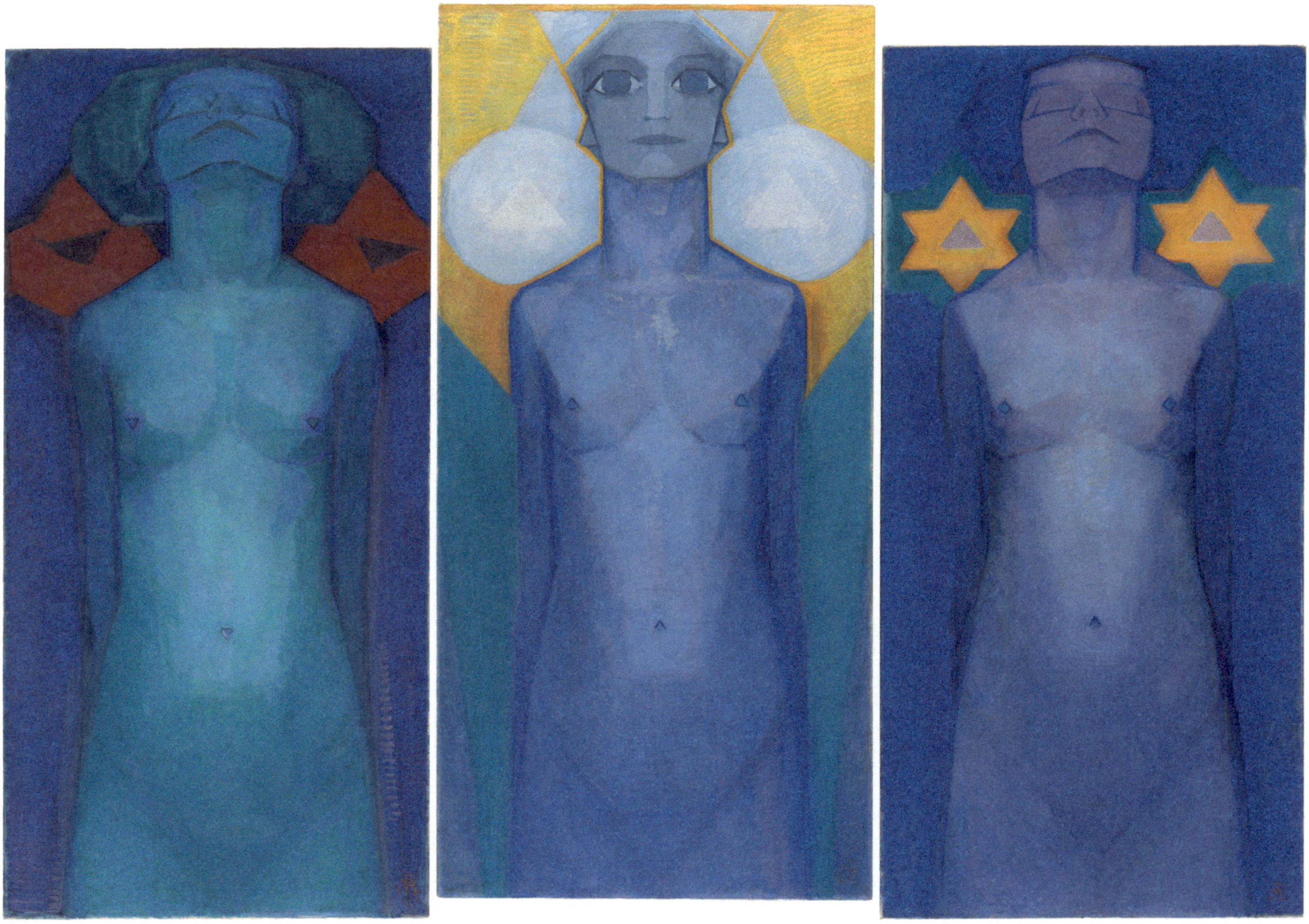

234
Piet Mondrian,
Evolution, 1911
Oil on canvas, 183 × 257.5 cm
The Hague, Stichting
Kunstmuseum Den Haag

EMBRYOGENY

When Kandinsky settled in Munich at the turn of the century, he lived and became formed in an environment totally won over by monism and Haeckelian panpsychism. There, Hermann Obrist, August Endell and Adolf Hölzel were developing an abstract sense of ornamental design that owed much to both the dynamics of biological processes and the psychology of forms posited by Theodor Lipps, who was endeavouring to demonstrate experimentally the capability of visual elements to convey a sensory experience solely through kinesthetic empathy, without prejudging their potential representative function. Obrist, who encountered Haeckel's thinking in his early studies in biology and medicine, and who occasionally lectured for the Monist League, cultivated an approach by which the dynamics of nature, rather than its forms, should be art's true concern. The famous wall hanging known as *Whiplash* or *Cyclamen* (1895) (237) surely stems from a close observation of the plates in Haeckel's *Kunstformen der Natur*: "What can stir our vital emotions more strongly than the graceful, long, sinuous, linear tentacles of a jellyfish swaying in the water?"[26] asks its author, who also collected microphotographs and X-ray photographs of primitive micro-organisms—a type of document that would inspire many more artists, in connection with biomorphic abstraction, for example (235, 236).[27] A former student of Lipps at the Psychological Institute founded by Lipps at the University of Munich in 1894, Endell was also a follower of monism (Haeckel), which venerated nature and unified vital forces, which were translated by the artist into the energy of the curved line (Lipps). As in Obrist's work, Endell's forms rely on the observation of naturalistic sources (the marine world, algae, the flagellates of jellyfish or anemones) which were then driven towards a greater abstraction, so as to create "visual impressions without associations."[28] In Hölzel's case, all of the works he himself called "black" or "abstract ornaments" depend on a conception of creation in which the linear structures "express the inner process of feeling," as he described it in 1905 to the Viennese art critic Arthur Roessler[29]—someone Kupka also spoke to, confiding to him in those same years, through extensive correspondence, his aspirations for a more abstract and spiritual art.[30]

The effect Hölzel's concepts had on Kandinsky—as well as Endell's in his essay *Um die Schönheit* [On beauty] (1896)—has long been a subject of discussion and constant re-evaluation.[31] However that may be, it is practically certain that Munich was the source of Kandinsky's ideas intended to connect "the laws of... art and nature."[32] Ideas that resurface in his treatise summarizing his teaching at the Bauhaus, *Point and Line to Plane* (1926). As in many other treatises by artists in the interwar years, its illustrations draw upon the micro-imagery of the natural sciences. These ideas further link up with those of his colleague Paul Klee, another artist who had adopted Munich as his home in his crucial formative years. For Klee too, his gaze into nature was transformed by the documentation of a fantastical science, as he continued to discover new facets of the invisible and claimed, with respect to nature, a freedom that was not the same as licence but a "freedom which merely demands its rights, the right to develop, as great Nature herself develops." Matter and motion, once more: the process of creation is valued to the detriment of the end result, since the work is a genesis rather than a mere product and continually returns from the model "to the womb," to "Madam Primordial Cell."[33] In this "contribution to the analysis of the pictorial elements" presented by *Point and Line to Plane*, as its subtitle indicates, the primary elements of art are also explicitly compared by Kandinsky to the "modest beginning with protoplasm and cells." Like those forms, and from them, abstract art "gradually advanced to more complex organisms" that promise, for abstraction, perfections that should not be judged according to its current state of development: that would be like considering higher vertebrates according to the evolutionary state of their ancestors, the amphibians, like looking at "not... the end result of creation but, rather, the 'beginning.'"[34] Abstraction is thus also that ability to project towards higher evolutionary states; it is a mere embryo of the future promises of art.

---

26 Quoted by Stephen Escritt, in *Art Nouveau* (London: Phaidon Press, 2000), p. 122.

27 Guitemie Maldonado, *Le Cercle et l'amibe. Le biomorphisme dans l'art des années 1930* (Paris: CTHS/INHA, 2006), p. 274ff.

28 Quoted by Escritt, *Art Nouveau*, p. 124.

29 Lynn Gamwell, *Exploring the Invisible: Art, Science and the Spiritual* (Princeton: Princeton University Press, 2002), p. 102.

30 On the relationship of these three men, see Markéta Theinhardt, "La forme. La voie centre-européenne vers la formalisation," in *Kupka, pionnier de l'abstraction*, exh. cat., ed. Brigitte Léal, Markéta Theinhardt and Pierre Brullé (Paris: Grand Palais/Réunion des musées nationaux, 2018), p. 152–153.

31 Werner Haftmann, *Malerei im 20. Jahrhundert*, rev. 6th ed. (Munich: Prestel, 1979), vol. 1, p. 68, discussed in *Ornament and Abstraction*, exh. cat., ed. Markus Brüderlin (Basel: Fondation Beyeler; Cologne: DuMont, 2001), p. 112.

32 Vassily Kandinsky, *Point and Line to Plane: Contribution to the Analysis of the Pictorial Elements* [1926], trans. Howard Dearstyne and Hilla Rebay (New York: Solomon R. Guggenheim Foundation, 1947), p. 103.

33 Paul Klee, *Théorie de l'art moderne* (Paris: Denoël, 1985), p. 29–30. Quoted in O. K. Werckmeister, *The Making of Paul Klee's Career 1914–1920* (Chicago: University of Chicago Press, 1989), p. 113.

34 Kandinsky, *Point and Line to Plane*, p. 107.

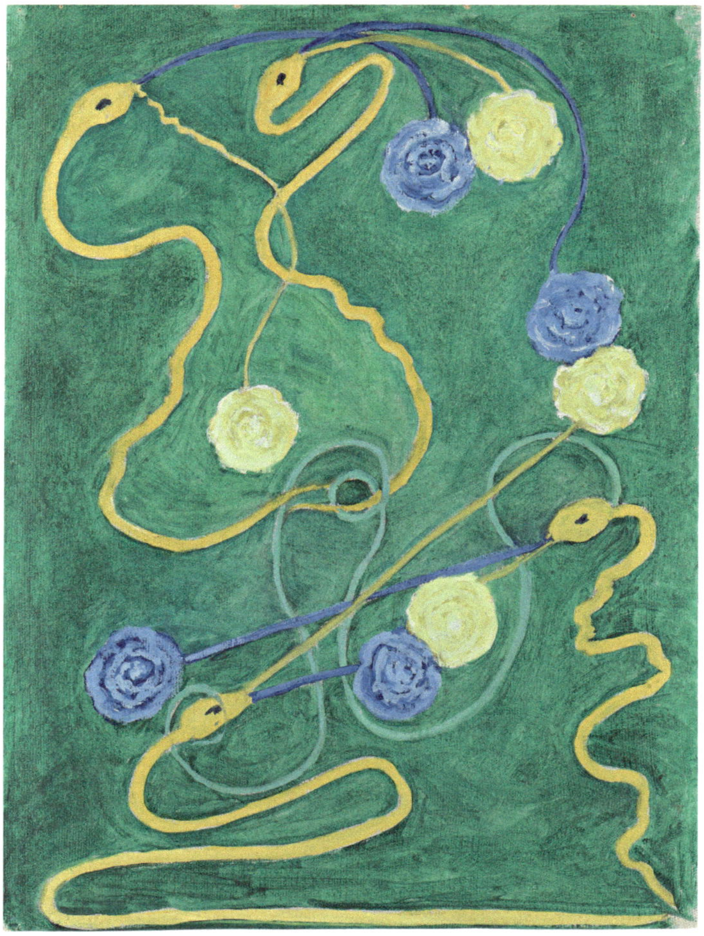

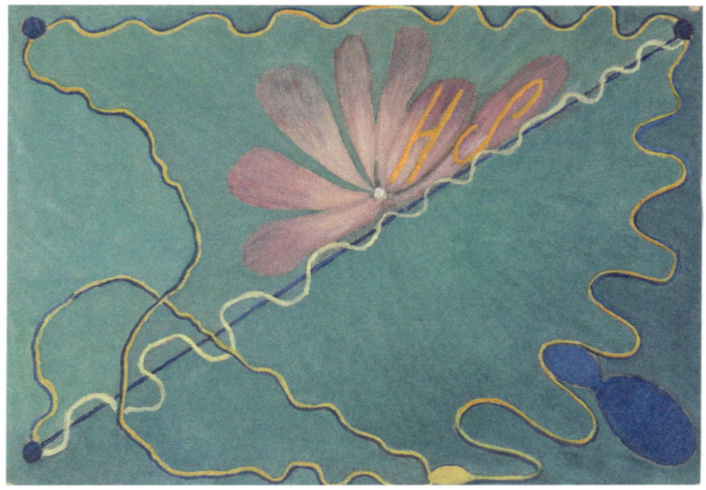

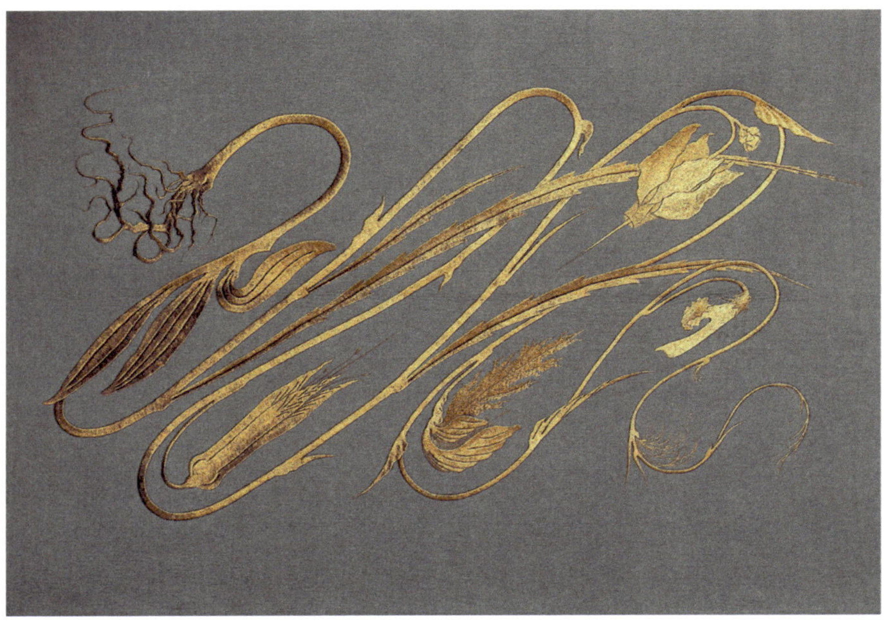

235
Hilma af Klint,
Primordial Chaos,
No. 13, Group 1, Series WU
(The Rose), 1906
Oil on canvas, 53.5 × 37 cm
Stockholm, The Hilma
af Klint Foundation

236
Hilma af Klint,
Eros, No. 1, Group 2,
Series WU (The Rose), 1907
Oil on canvas, 58 × 79 cm
Stockholm, The Hilma
af Klint Foundation

237 FIG.
Hermann Obrist,
Berthe Ruchet,
Cyclamen, 1895
Embroidery in wool and silk,
119 × 183 cm
Munich, Stadtsmuseum

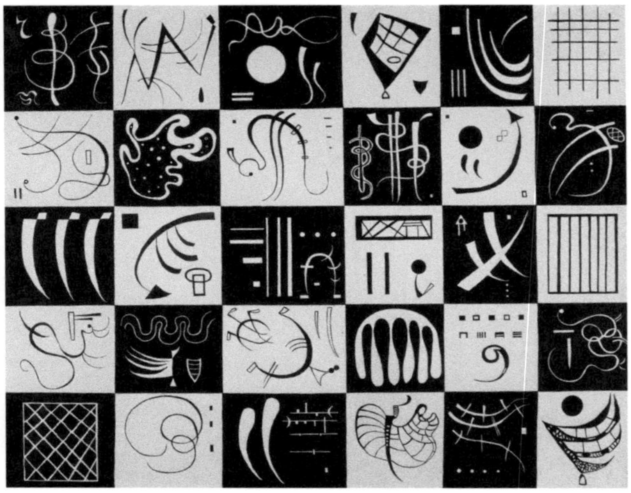

This is not just a metaphor: embryogeny and cellular imagery are *truly* an object of insatiable curiosity and a source of inspiration for Kandinsky, and ultimately dominated the Paris phase of his work, from 1934 on.[35] The volumes of the monumental German encyclopedia *Die Kultur der Gegenwart*, well known by other Bauhaus artists, including Klee, and the origin of many of the scientific images in *Point and Line to Plane*, followed him to Paris. Its plates, with their profusion of different specimens of unicellular life forms, amoebae, larvae and embryos, provided the artist with an endless repertoire of forms and, sometimes, the example of organization in a grid or synoptic table evident in a number of his works. A notable example is *Thirty* (1937) (238), which contains in each of the thirty compartments that cover its surface a zootaxy model "of blurred infusoria, vague flagellata, imprecise monera, bizarre protoplasms"—if we can thus appropriate the enumeration produced by Joris-Karl Huysmans regarding some of the *Noirs* painted by Odilon Redon,[36] whose role as a forerunner of biomorphic abstraction is not sufficiently recognized. The title of one of Kandinsky's first Paris paintings to include this embryological imagery, *Start* (1934), may be an allusion to the new beginning the artist was embarking on at this time in his life or—more certainly—to the undecided form, the first in its line, which would beget all the others: the embryo of all forms. This implies somewhat of a tacit relationship between the iconography of these paintings and their status as abstractions. For, as Vivian Endicott Barnett points out, the indistinct nature of the embryo in the early stages of its development turned it into a "generalized image" already stripped of any figurative power: "At an early stage it is neither ostensibly human nor individual: it is already an abstraction."[37]

THE PRIMORDIAL SUBSTANCE

Along with the embryo, a good candidate for the status of abstract primogenetic form is another non-individualized biological entity: the "bizarre protoplasm" referred to by Huysmans, which Thomas Huxley had named *Bathybius haeckelii*, in homage to the German scientist. In other words, the name of a primitive form of organic life discovered, under the appearance of a fluctuating, amorphous, gelatinous substance, in the sea mud gathered in the laying of transatlantic telegraph cables. And in which Haeckel thought he recognized the "primordial soup" (*Urschleim*), the primeval jelly, the missing link in the transition from inert, inorganic matter to living, organic matter—the very origin of life, as indicated by the caption for the woodcut reproducing the bathybius, that "rudimentary monera," in Flammarion's *Le Monde avant la création de l'homme* (1886): "Their rudimentary, almost undefined forms appear to us . . . as the first faltering steps of the life just born."[38] The emergence of life that can be made out in the fluidity of the contours of the bathybius is clearly also that of the future biomorphic abstraction and its "complicated curves which contain the multiplicity and richness of nature," to describe it in the words of one of its earliest defenders, art critic Anatole Jakovski.[39] Can it be a coincidence that Jakovski speaks of "protoplasm of desire" when looking at Joan Miró's "mother forms" and that Alfred Barr describes

35 As demonstrated by Vivian Endicott Barnett in "Kandinsky and Science: The Introduction of Biological Images in the Paris Period," in *Kandinsky in Paris, 1934–1944*, exh. cat., ed. Vivian Endicott Barnett and Christian Derouet (New York: Solomon R. Guggenheim Museum, 1985), p. 61–87. Reprinted in Oliver A.I. Botar and Isabel Wünsche, eds., *Biocentrism and Modernism* (London: Routledge, 2017). This essay essentially forms the basis for the lines that follow; it sheds considerable new light on our understanding of biomorphism in Kandinsky's work, even if it focuses primarily on what Kandinsky derived from French Surrealism, which seems to be a minor aspect of the discussion now that we have a better sense of the store of knowledge he brought from Munich.

36 Joris-Karl Huysmans, "Le nouvel album d'Odilon Redon," *La Revue indépendante* (February 1885), reprinted in Joris-Karl Huysmans, *Écrits sur l'art*, 1867–1905, ed. Patrice Locmant (Paris: Bartillat, 2006), p. 297. The reference to protoplasm alludes to Haeckel, who is mentioned by Huysmans immediately after the words quoted here. Huysmans, quoted in Dario Gamboni, *The Brush and the Pen: Odilon Redon and Literature*, trans. Mary Whithall (Chicago: University of Chicago Press, 2011), p. 139.

37 Barnett, "Kandinsky and Science," in *Biocentrism and Modernism*, p. 223.

38 Camille Flammarion, *Le Monde avant la création de l'homme. Origines de la Terre, origines de la vie, origines de l'humanité* (Paris: Marpon et Flammarion, 1886), p. 133. This work is actually the French adaptation of *Die Wunder der Urwelt*, by W.F.A. Zimmermann (pseudonym of Carl Gottfried Wilhelm Vollmer).

39 Anatole Jakovski, *Arp, Calder, Hélion, Miró, Pevsner, Seligmann* (Paris: Povolozky, 1933).

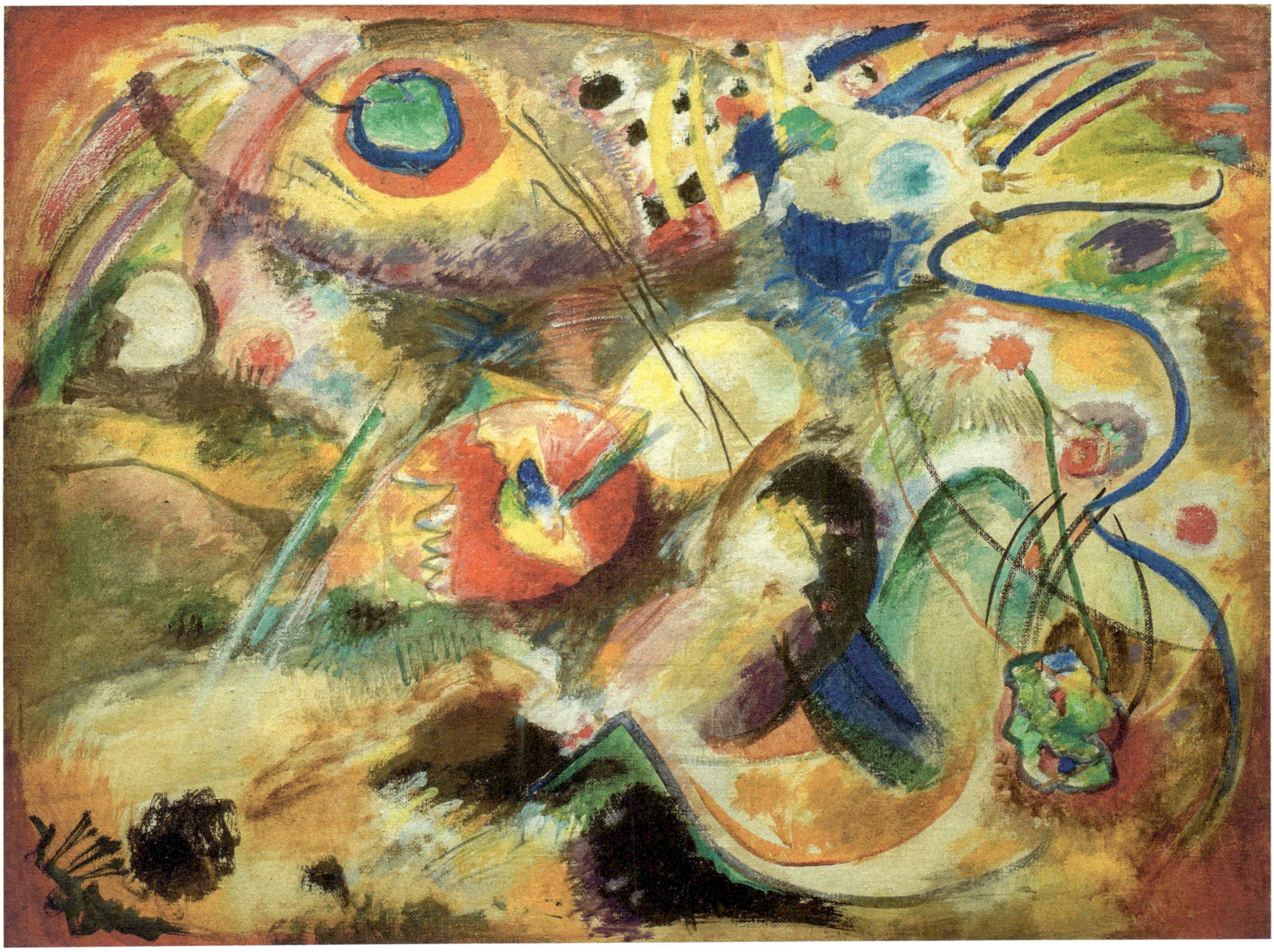

238 FIG.
Vassily Kandinsky,
Thirty, 1936–1937
Oil on canvas, 100 × 81 cm
Paris, Centre Pompidou –
MNAM, CCI

239
Vassily Kandinsky,
Untitled (Deluge), 1914
Oil on canvas, 108 × 139.5 cm
Munich, Städtische Galerie
im Lenbachhaus

a "sculptural protoplasm" with reference to Hans Arp's "concretions" and their budding volumes?[40] More or less consciously, the secret filiation of Haeckel's bathybius asserts itself at the very core of art of the interwar years.

In the origin stories of abstraction, the myth of the protoplasm could be rivalled by that of the protyle, the universal substance which the theosophists held to be the true source of all chemical elements. In *Occult Chemistry* by Annie Besant and Charles Leadbeater, disciples of Blavatsky, the protyle is the primordial environment, the primeval cosmic soup in which the elemental bricks of life are steeped.[41] This "stereochemistry," as the authors also call it, is of Platonic origin: indebted to the cosmogony of the *Timaeus*, it starts from the tetrahedron, the first of Plato's five regular solids, which generates the hexahedron (cube) and the octahedron, with which the dodecahedron and the icosahedron can be built. The polyhedral chain thus gives rise, through growing degrees of complexity, to the diversity of visible forms. The *Timaeus*, theosophy and the occult: these are the familiar sources of the Nabi painter Paul Sérusier who, with his *Tetrahedrons* of about 1910, offered a dramatic glimpse into this mysterious scene of the beginnings of organized matter (240). The origin-based theme of the painting is confirmed with that of a contemporary work: *Origins* (about 1910), a landscape of the beginning of the world showing a range of mountains in the shape of tetrahedrons emerging from the primordial waters, lit by a luminescent triangle dressed in the three primary colours—primary in the sense, naturally, of "primal," "primeval." A third work that came out at the same time, *Golden Cylinder*, completes this triptych of the origins of life and of the world, through which Sérusier provided a contribution to early abstraction that is too seldom considered in the historiography.[42]

When Alfred Barr drew up the famous diagram summarizing the history of modern art from Neo-Impressionism up to the latest developments of abstraction—geometrical and non-geometrical (biomorphic)—for the exhibition *Cubism and Abstract Art* at the Museum of Modern Art in New York in 1936, he applied to the life of forms an obvious evolutionary pattern, exactly copying the phylogenetic trees produced in large quantity by the natural sciences, before and after Darwin. But was the formalist narrative that was deduced from it and that so long dominated historiography in the twentieth century any more accurate? And was it so faithful to its conceptual model? Probably not, if we remember, with Klee, that in fact, "formalism is form without function," that is, the antithesis of "the living form." The living form, that which all evolutionary abstraction sought to produce, as it is the only form that can be linked to the functioning of Life and ensure that "as creation is related to the creator, so is the work of art related to the law inherent in it."[43]

40 Quoted by Maldonado, *Le Cercle et l'amibe*, p. 117.

41 Annie Besant and Charles Leadbeater, *Occult Chemistry: A Series of Clairvoyant Observations on the Chemical Elements* (London: Theosophical Publishing House, 1919). (The work compiles articles published in various theosophical journals, including *Lucifer* and *The Theosophist*, between 1890 and 1909.)

42 Except by Maurice Tuchman and Robert Welsh in *The Spiritual in Art: Abstract Painting, 1890–1985*, exh. cat., ed. Maurice Tuchman and Judi Freeman (Los Angeles: County Museum of Art; New York: Abbeville Press, 1986), p. 20 and 70–71; as well as Pierre, *Maternités cosmiques*, p. 68–79.

43 Klee, *Théorie de l'art moderne*, p. 53–54. Quoted in *Paul Klee Notebooks*, vol. 1: *The Thinking Eye*, ed. Jurg Spiller and trans. Ralph Manheim (London: Humphries, 1973), p. 59–60, https://archive.org/details/PaulKleeNotebooks Vol1The ThinkingEye/.

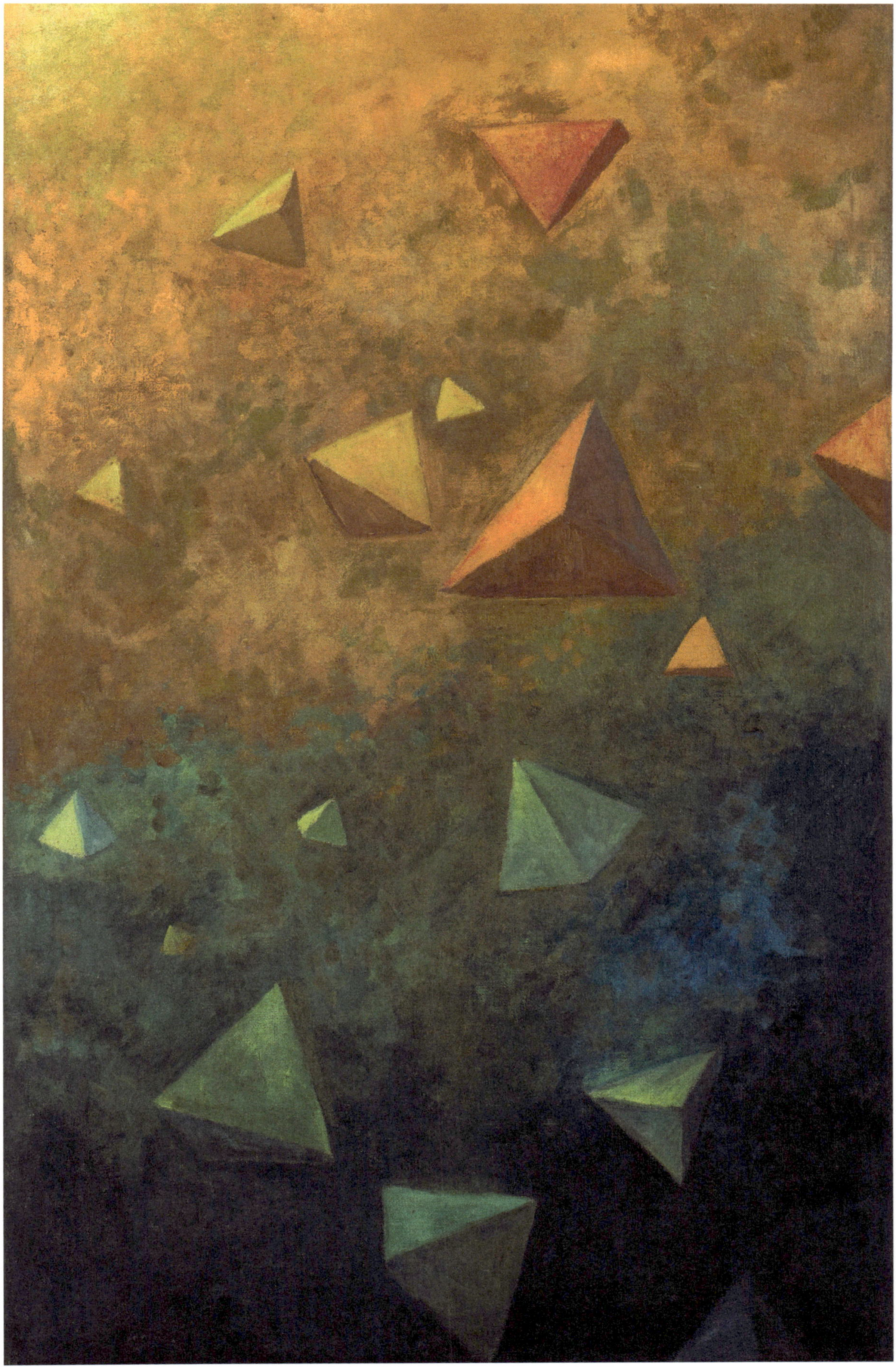

240
Paul Sérusier,
Tetrahedrons, about 1910
Oil on canvas, 91.5 × 57.4 cm
Paris, Musée d'Orsay

NATURE: THE WILD PART OF THE WORLD

VIRGINIE MARIS

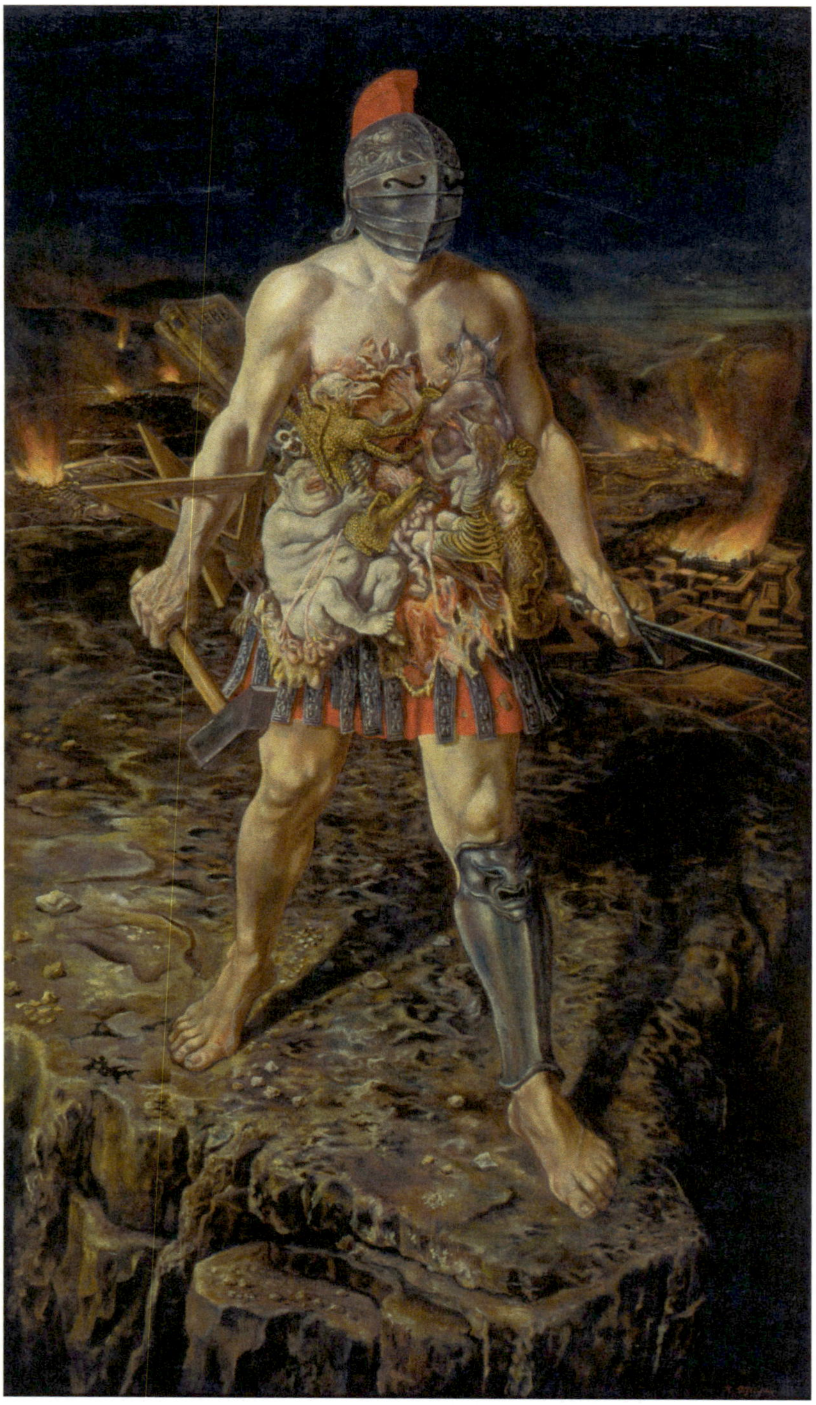

241
Rudolf Schlichter,
Blind Power, 1932–1937
Oil on canvas, 179 × 100 cm
Berlin, Berlinische Galerie

Long an object of admiration, fear, control and scientific investigation, nature became an object of concern and protection in the second half of the twentieth century.

A VULNERABLE NATURE

Although the first voices of alarm were raised in the late 1800s, notably with the publication of George Perkins Marsh's famous *Man and Nature*, in 1864,[1] the real turning point came in the early 1900s. The extension of colonial empires and the accompanying plunder of natural resources, along with the acceleration of the Industrial Revolution, exposed a face of modernity that belied the triumphant mien of material and intellectual progress. Made to seem inexhaustible by colonial-era expansion and culpable blindness, nature was now revealing its limits. Nature was fragile and had to be defended against the assaults of modernity.

During the twentieth century, the vulnerability and degradation of natural environments became a full-fledged scientific and social issue and gradually entered the political arena. From that there emerged three approaches to protecting nature, which, though not entirely dissimilar, gave rise to markedly different environmental practices and policies.

Preservationism, inspired by Romanticism and attuned to the sublime nature of wilderness, emphasizes the need to protect certain natural environments from all human intervention, to preserve wild areas in which humans are merely discreet, temporary visitors. In Europe and the United States, the preferred political response to this concern for the natural world has been to create nature reserves.

Conservationism, in line with modernity and its instrumental conception of nature, promotes management methods that assure the sustainable, if not optimal, use of natural resources. The underlying idea is to improve the human condition but also to achieve equitable access to resources for both individuals and generations. The political implementation of such sound management can take various forms, among them mandating agencies such as France's Office national des forêts, imposing resource-harvesting quotas or moratoria, and introducing best-practice incentives such as agricultural subsidies.

In the 1980s, these two approaches were gradually supplanted by the theme of biodiversity. Repositioning the defense of nature as a scientific issue by focusing on the diversity of the natural world served to go beyond the distinction between humans and other living organisms. The most explicit political form of this approach is the Convention on Biological Diversity, signed in 1992 at the Earth Summit in Rio de Janeiro.

Despite their differences, these three movements remain implicitly or explicitly contingent on the idea that humans bear a certain responsibility vis-à-vis nature.

THE DEATH OF NATURE

If the hallmark of the twentieth century was the protection of nature, that of the early twenty-first century would be the death of nature. Indeed, the very idea of nature as distinct from human culture is under attack today, both as an empirical phenomenon and as a theoretical notion.

From an empirical perspective, we are witnessing what is commonly referred to as the sixth mass extinction, the collapse of life itself, which is far from new but whose catastrophic and largely irreversible impact can no longer be denied.

First, the wild world is collapsing in terms of its diversity. In spring 2019, IPBES[2] reported that a million species had become endangered, among them 40% of amphibians and reptiles, 25% of mammals, 13% of birds and a third of all corals. The 6,000 species deemed critically endangered include some of the most charismatic animals: all gorillas, Sumatran orangutans, all species of rhinoceros and three species of albatross are likely to have vanished from the planet by the end of this century.

But beyond the natural world's diversity, its vigour and abundance are collapsing. It is estimated that, since the late 1970s, close to 60% of wild vertebrates have disappeared.[3] In France, there are nearly a third fewer common birds than there were twenty-five years ago. Wildlife is dying off, so much so that some scientists speak of biotic annihilation in describing the collapse of wild animal populations, including species formerly considered common.

Lastly, the wild part of the world is shrinking in terms of space. The areas not appropriated by human

1 George Perkins Marsh, *Man and Nature: Or, Physical Geography As Modified by Human Action* [1864] (London: Forgotten Books, 2016).

2 The Intergovernmental Science-Policy Platform on Biodiversity and Ecosystem Services (IPBES) is an independent body administered by the United Nations Environment Programme.

3 *Living Planet Report 2018: Aiming Higher*, ed. Monique Grooten and Rosamunde Almond (Gland: World Wildlife Fund, 2018).

activity are being whittled away in favour of urban sprawl, major infrastructures and ever more extensive and intensive farmland.

If wilderness were defined solely by the absence of human influence, we would have to conclude that it has virtually disappeared from the face of the planet. We have entered the Anthropocene, a period of Earth's history characterized by the global impact of human activity on all of Earth's systems: the geosphere, biosphere, atmosphere and the hydrosphere. From continents of plastic in the oceans to the acidification of coral reefs, from phytosanitary products that pollute soils and waterways to the melting of Antarctic glaciers, it appears that Earth is no longer home to autonomous, independent, wild nature.

But such a definition of wilderness seems very restrictive. The fact that the flight path of a golden eagle is accidentally altered by the presence of a wind turbine does not make the eagle an artifact or a domestic animal. The wild world is not that which is untouched by human influence but that which exists on its own, independent of us. This recasts the "wild" in a positive rather than privative perspective: it is the perpetual upwelling of life, uncontrollable, tenacious, incorrigibly resistant to our attempts to control, reason with, manage or pilot it.

A CONCEPT ACCUSED OF ALL ILLS

From a more theoretical standpoint, the concept of nature is also subject to fierce attacks. Deeming it irreparably linked to the modern dualism that separates nature from culture so as to better dominate it, a number of authors strive to deconstruct this concept, to expunge, as it were, our vision of the world from an idea on which human domination is based.

For some, the aim is to show that there is no such thing as unified, homogeneous "Nature" but, rather, a collection of heterogeneous objects and phenomena in which a whole procession of non-human entities and human representations or institutions gather together and hybridize.[4]

Others stress the local character of this concept, the result of a Western world view that constructs its ontology on a distinction that has no meaning in other cultures.[5] Given that the contemporary ecological crisis is largely the consequence of globally imposing exploitation and production rationales derived from Western modernity, we would do well to take inspiration from the non-naturalist ontologies, which imply more substantial, less hierarchical relations between humans and non-humans.

THINKING THE WILD PART OF THE WORLD

Abolishing the very idea of a nature distinct from human affairs breeds a strong temptation to smooth out, once and for all, every alterity and every asperity to manage the natural world as if it were our solely owned factory. The metaphors in this regard are telling. The image of a "spaceship Earth" that requires judicious piloting relates the natural world to an artifact, a mega-machine whose finality is to convey us to greener pastures. Another semantic field that quickly invaded the sphere of nature protection is economics. At this point, the question became less about preserving or respecting a world independent of us than about harnessing a natural capital that generates a flow of ecosystemic services—with a diversity of life forms acting as an insurance policy to deal with changes.

And yet it is possible to consider the separation between human affairs and that which occurs independent of us, unconnected to our aims, representations or institutions, not so as to entrench human domination but so as to recognize and value the radical alterity of natural entities. Rather than validating the self-fulfilling prophecy of nature's death, we could revisit and revitalize the idea of a nature conceived as "the part of the world that we did not create."[6]

For, without this vision of wilderness, we sink into the hell of a world totally closed in on itself, governed only by the instrumental rationality of a narcissistic humanity incapable of perceiving the multitude of existences and finalities that surround it. To think the wildness of nature is to welcome the creativity, wonder and incommensurability of the living world into our trivial human affairs. And this is more than ever necessary at a time when the diversity, vitality and expanse of the wilderness are collapsing before our eyes.

4 Bruno Latour, *Politics of Nature: How to Bring the Sciences into Democracy* [1999], trans. Catherine Porter (Cambridge, MA: Harvard University Press, 2004).

5 Philippe Descola, *Beyond Nature and Culture* [2005], trans. Janet Lloyd (Chicago: University of Chicago Press, 2013).

6 Virginie Maris, *La Part sauvage du monde. Penser la nature dans l'Anthropocène* (Paris: Seuil, 2018).

NATURE: THE IMPOSSIBLE ELSEWHERE

THIERRY HOQUET

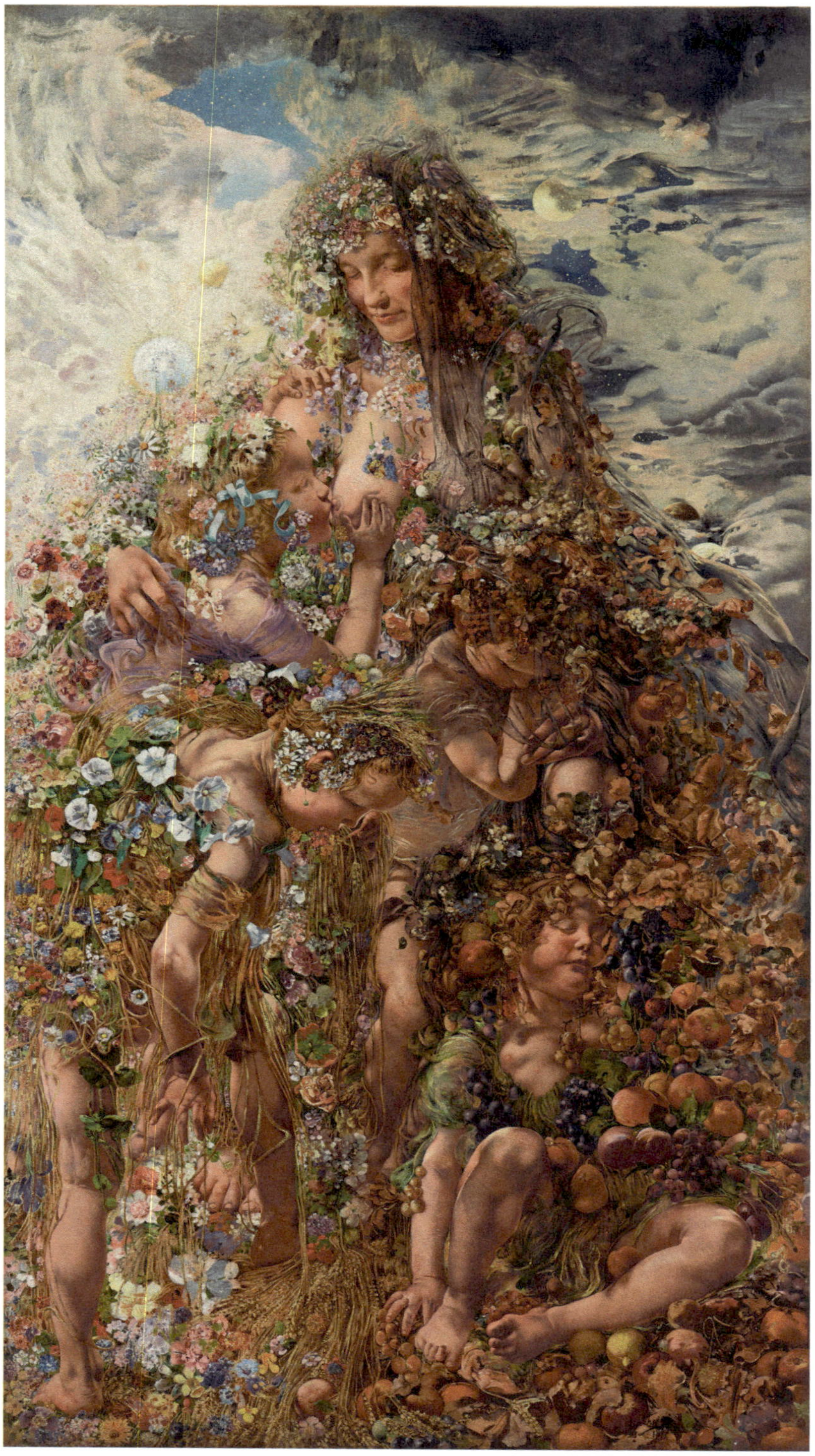

242
Léon Frédéric,
Nature or Abundance, 1897
Oil on canvas, 165 × 90.1 cm
Dallas, Dallas Museum of Art

Let us look to the sky.

Not all that long ago, a hundred years or so, humankind was still bound to the earth. Only a handful of fanciful visionaries were dreaming of flying machines. In 1956, a British scientist could still bluntly declare: "Space travel is utter bilge."[1] Scholars, priests and wise men called us back to reason: humans are not made to fly. The myth of Icarus incarnated the wrenching choice between the mad desire to tear ourselves free from Earth and the price we might pay—our lives—for doing so. The most conservative voices added that, if ever humans managed to make their dream come true, great misfortunes would befall the whole of humanity, for something in the providential order of the world would be permanently fractured. Likewise, if humans were to find the philosopher's stone and be able to turn lead into gold, wouldn't all value collapse? In the past, we could laugh at those fears, but today we face the prospect of disasters such as climate change and repeated pandemics that humans spread more easily in their globalized world, all of which seems to revive and bolster the stances of reactionaries intent on keeping humans in their place, relegated to a well-defined role as users, not "masters and owners," of nature. And woe to anyone, past, present or future, aspiring to step out of their place, to escape their condition.

For eons, humankind was unable to fly and thus limited to gazing at the sky and clouds as a profound source of inspiration. "I love clouds . . . clouds that go by . . . out there . . . over there . . . marvelous clouds!" declares Baudelaire's "puzzling stranger."[2] In the late nineteenth century, the philosophies of Pascal and Kant still held true: "the silence of these infinite spaces" was still frightening, "the starry sky above us" had a transcendent quality, the incarnation of an unyielding moral power. In the world as those thinkers knew it, all human misfortune came from not knowing how to sit quietly in a room, and one could take a stroll each evening at a precise time.

And then we flew. And in an instant everything changed. The age-old fatalism that kept us mired in the mud suddenly shattered. But in realizing our aerial dream of flitting across the sky with the lightness of birds, we mainly acquired an unparalleled destructive power. Italian Futurism embodied the sense of vertigo and the new militarized aesthetic. Surprisingly, that aesthetic continues to find expression, notably in the Japanese films of Hayao Miyazaki, who has made it the theme of a melancholy reflection. "The wind is rising! We must try to live!" he urges, quoting Paul Valéry, in a film haunted by the fanciful flying machines of Gianni Caproni, but also by Stukas and Zeros.

In the novel *The Man Without Qualities*, written in the first third of the twentieth century, Robert Musil points to changes in the way people view clouds in a world where everything has come to be assessed in terms of quantity. Walter (the "man of feeling" character) rails against this: "[There is] no difference anymore between the sun and a kitchen match, or between your mouth at one end of the digestive tract and its other end either."[3] As soon as we learned to fly, there was no more "elsewhere" for us. Even art was no longer eine *Luft von anderen Planeten*, a breath of air from other planets. There was a levelling of values. We no longer recognized greatness.

Like the sky, nature no longer incarnated transcendence in relation to human activity: it was integrated into the scheme of technical life. Descartes, in Chapter 7 of *The World, or Treatise on Light*, explicitly rejects the term "nature" as associated with ancient paganism: "Note, in the first place, that by 'nature' here I do not mean some goddess or any other sort of imaginary power. Rather, I am using this word to signify matter itself, in so far as I am considering it taken together with all the qualities I have attributed to it, and under the condition that God continues to preserve it in the same way that he created it."[4]

Cut off from divine Nature, our world now offered no escape. The world, the animals, the humans: all machines. Having become immanent, there soon was no more Nature outside of us. Everything became technical and technology was everywhere. Everywhere we went, everywhere we looked, we invariably found the human hand. As Herbert Marcuse, disciple and thus successor of Musil, observed: "Obviously, when cities and highways and National Parks replace the villages, valleys, and forests; when motorboats race over the lakes and planes cut through the skies—then these areas lose their character as a qualitatively different reality, as areas of contradiction."[5]

In the face of what we were accustomed to calling "Nature," we were suddenly struck by a strange feeling: it seemed easier to reach things but things seemed to be farther out of reach. As we approached, they retreated. Today, roads and highways make it easier for us to go everywhere, yet it feels as if we are going nowhere: we reach nothing. Hence the melancholy expressed, for example, in *Tristes tropiques* (1955) by Claude Lévi-Strauss. Back when we couldn't go anywhere, we could still dream about elsewhere. But now that we can go here and there as we please, all elsewheres have lost their principal quality: the *aura* that made them the object of an impossible dream.

I can go see some remarkable site, but it is no longer the object of an experience: I reach a scenic viewpoint, I take a picture immortalizing the moment, and I leave, looking for the next place. Venice, Rome, Florence stormed daily by hordes of visitors eager for snapshots. Dubrovnik, a selfie backdrop for *Game of Thrones* fans. Global rationalization and commodification work hand in hand. The concept of disenchantment developed by Max Weber has never seemed more relevant. Hannah Arendt, writing in *The Human Condition* (1958), noted a general contraction: "The decisive shrinkage of the earth was the consequence of the invention of the airplane, that is, of leaving the surface of the earth altogether."[6]

Given this widespread flattening of experience, the wilderness was not far from being deemed a recreational area, a resource available for tourism. Countering that, the American forester Aldo Leopold condemned the consumption of nature by those he derisively labelled "motorized ants": tourists wanting to experience the world quickly but demanding paved roads, spacious parking areas and, at every altitude, well-equipped bathrooms redolent of pine-scented air freshener and, of course, a snack bar with a view where you can get a cold soda—plus, in today's world, a 4G network and WiFi. Since its first publication in 1949, Leopold's *Sand County Almanac* has urged us to reflect on the impact of high-speed trains, on the consequences of building a new highway that opens up the Massif Central or makes Turin closer to Lyon. What happens when railroads begin to carry masses of people to the countryside? When highways shrink the continents? Nature, which is to say the idea of transcendence, disappears.

Should that make us laugh or cry? Nature is sometimes depicted as a woman with generous breasts: the Diane of Ephesus in the garden of the Villa d'Este in Tivoli appeared on the frontispiece of Linnaeus's *Systema naturae* in 1756. But we have long known that the same Nature can behave like an odious stepmother. Pliny the Elder noted this ambiguity at the beginning of Book VII of his *Natural History*: "Our first attention is justly due to Man, for whose sake all other things appear to have been produced by Nature; though, on the other hand, with so great and so severe penalties for the enjoyment of her bounteous gifts, that it is far from easy to determine, whether she has proved to him a kind parent, or a merciless step-mother."[7]

The paradox of our modern condition and our nostalgia for a lost transcendence is that we have long had but one dream: to annihilate nature understood as a "savagery," a faceless entity, brutal and anonymous, whose uncanny countenance would be revealed by Darwinian "natural selection." On the one hand, there is the appearance of beauty and harmony of this face: "We behold the face of nature bright with gladness, we often see superabundance of food; we do not see, or we forget, that the birds which are idly singing round us mostly live on insects or seeds, and are thus constantly destroying life."[8] On the other hand, it hides the cruel reality: "The face of Nature may be compared to a yielding surface, with ten thousand sharp wedges packed close together and driven inwards by incessant blows, sometimes one wedge being struck, and then another with greater force."[9]

Beautiful mask, hideous face! Consider Louis Hémon's *Maria Chapdelaine*, republished in 1967 in the famous Bibliothèque verte children's collection. The back cover blurb declares: "Clear the land! This is the rallying cry of the Canadian woodsman who dares tackle hostile forests to turn them into fertile fields,

1 Remark attributed in 1956 to Sir Richard Woolley, Astronomer Royal and Director of the Royal Greenwich Observatory (1956–1971).

2 Charles Baudelaire, *Paris Spleen: Little Poems in Prose I* [1869], trans. Keith Waldrop (Middletown: Wesleyan University Press, 2012), p. 5.

3 Robert Musil, *The Man Without Qualities* [1930–1943], trans. Sophie Wilkins (London: Pan Macmillan, 2017), p. 65.

4 René Descartes, "The World, or Treatise on Light" [1633], in *The Philosophical Writings of Descartes: Volume I*, trans. John Cottingham, Robert Stoothoff and Dugald Murdoch (Cambridge: Cambridge University Press, 1984), p. 92.

5 Herbert Marcuse, *One-Dimensional Man: Studies in the Ideology of Advanced Industrial Society* [1964] (Boston: Beacon Press, 2012), p. 66.

6 Hannah Arendt, *The Human Condition* (Chicago: University of Chicago Press, 2018), p. 251.

7 Pliny, *The Natural History of Pliny: Volume 2*, trans. John Bostock and H. T. Riley (London: H. G. Bohn, 1855), Book VII, chap. 1, p. 117.

8 Charles Darwin, *On the Origin of Species* (London: John Murray, 1859), p. 62.

9 Ibid., p. 67.

10 Emmanuel Levinas, "Heidegger, Gagarin and Us" [1961], in *Difficult Freedom*, trans. Seán Hand (Baltimore: Johns Hopkins University Press, 1990), p. 231–234.

11 Christophe Bonneuil and Jean-Baptiste Fressoz, *L'Événement anthropocène. La Terre, l'histoire et nous* (Paris: Seuil, 2013), p. 76.

12 Donna J. Haraway, *Staying with the Trouble: Making Kin in the Chthulucene* (Durham: Duke University Press, 2016), p. 101.

13 Virginie Maris, *La Part sauvage du monde. Penser la nature dans l'Anthropocène* (Paris: Seuil, 2018), p. 20–21.

plowed and seeded. In the land of Quebec, life is hard for Samuel Chapdelaine and his daughter Maria."

This takes us back to the pioneer mindset. American settlers saw wilderness primarily as a threat to themselves. It wasn't until after 1890 that their thinking changed: with the frontier line ended and the Pacific Ocean in sight, the world seemed to have reached its limit. This gave rise to the idea of land scarcity. With all wilderness, all actually or supposedly wild land having been conquered, it was time to change course: the need now was to tame the self-destructive tendencies of civilization, to impose self-restraint on human action in order to achieve "sustainability." It was then that emerged the idea that wilderness has rights, an intrinsic value, independent of any use people might hope to make of it.

A comparable "frontier" effect was no doubt born the day Earth was first seen from the sky and famously photographed as the Blue Marble. Some, like the philosopher Emmanuel Levinas, celebrated the "probable opening up of new forms of knowledge and new technological possibilities . . . But what perhaps counts most of all is that he [cosmonaut Yuri Gagarin] left the Place."[10] Today, however, this photograph taken from space is widely criticized for offering a "deterrestrialized" vision that feeds the notion of "Anthropocene": a humanity rejecting Earth as natural otherness and determined to transform it into technonature.[11] The image of Earth seen from afar affords an overview, a dominant, overhanging perspective. It has torn us from our "humility," our primordial "humus," and given us the new intoxication of subjecting the entire Earth to a "geopower": a clique of engineers and mad scientists that would regulate the climate at the global level.

Now, more than ever, as we face the threat of hubris, we need a certain "idea of Nature": a regulatory, moral idea, independent of any human enterprise. We need to remember that "we are all compost": in other words, as Donna Haraway puts it, we need a "compostist" philosophy.[12] More broadly, in addition to nature-totality (the entire world) and nature-normality (the normal functioning of things), we need nature-alterity (reality separate from humans) as an antidote to the project to completely anthropize all that exists.[13]

It matters not here that all of Nature is already anthropized, that it bears the mark of our hand and that we can no longer distinguish falling stars from artificial satellites in the sky above our heads. Just as the inner flow of growth and aging reminds us that our bodies are changed by life forces beyond our control and impervious to our will, we must regain the sense of Nature-transcendence: a power outside of us, independent of us. Does that necessarily entail musing about the mother goddess Gaia, promising to resurrect the animist Pachamama or reviving a providentialist vision of Nature, created by a wise God with the good of humanity in mind?

The answer is No. Humans and their techniques are a necessary part of the earthly adventure and its becoming. They are in Nature and there is nature only for them. But it is not because they are everywhere that they must renounce the concept of Nature. Nature remains a resource for imagining a world that is escaping us, with its own spontaneity: an impossible elsewhere that commands our respect.

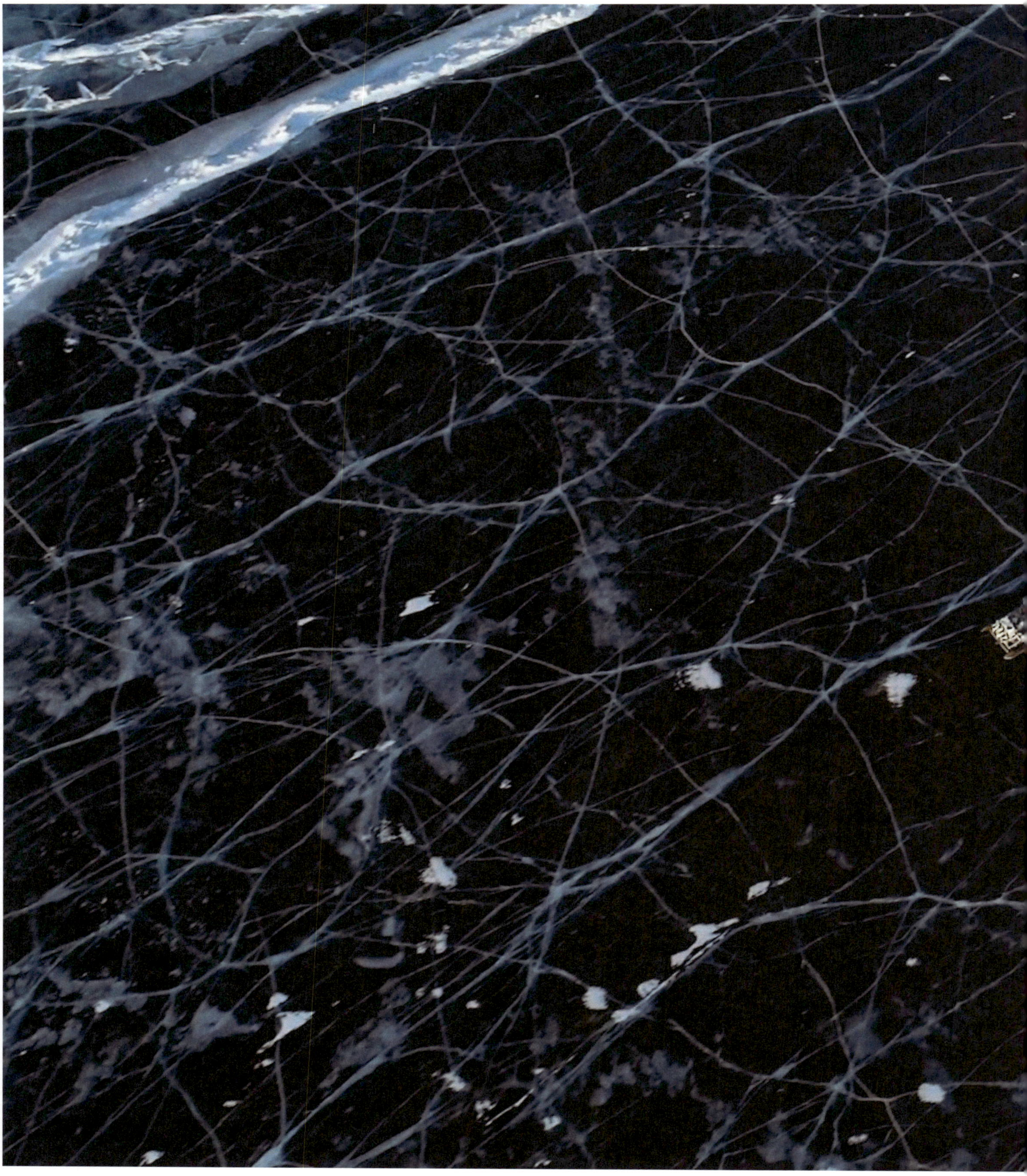

PP. 352-357

243, 244, 245
Laurent Grasso,
Artificialis, 2020
High-resolution film

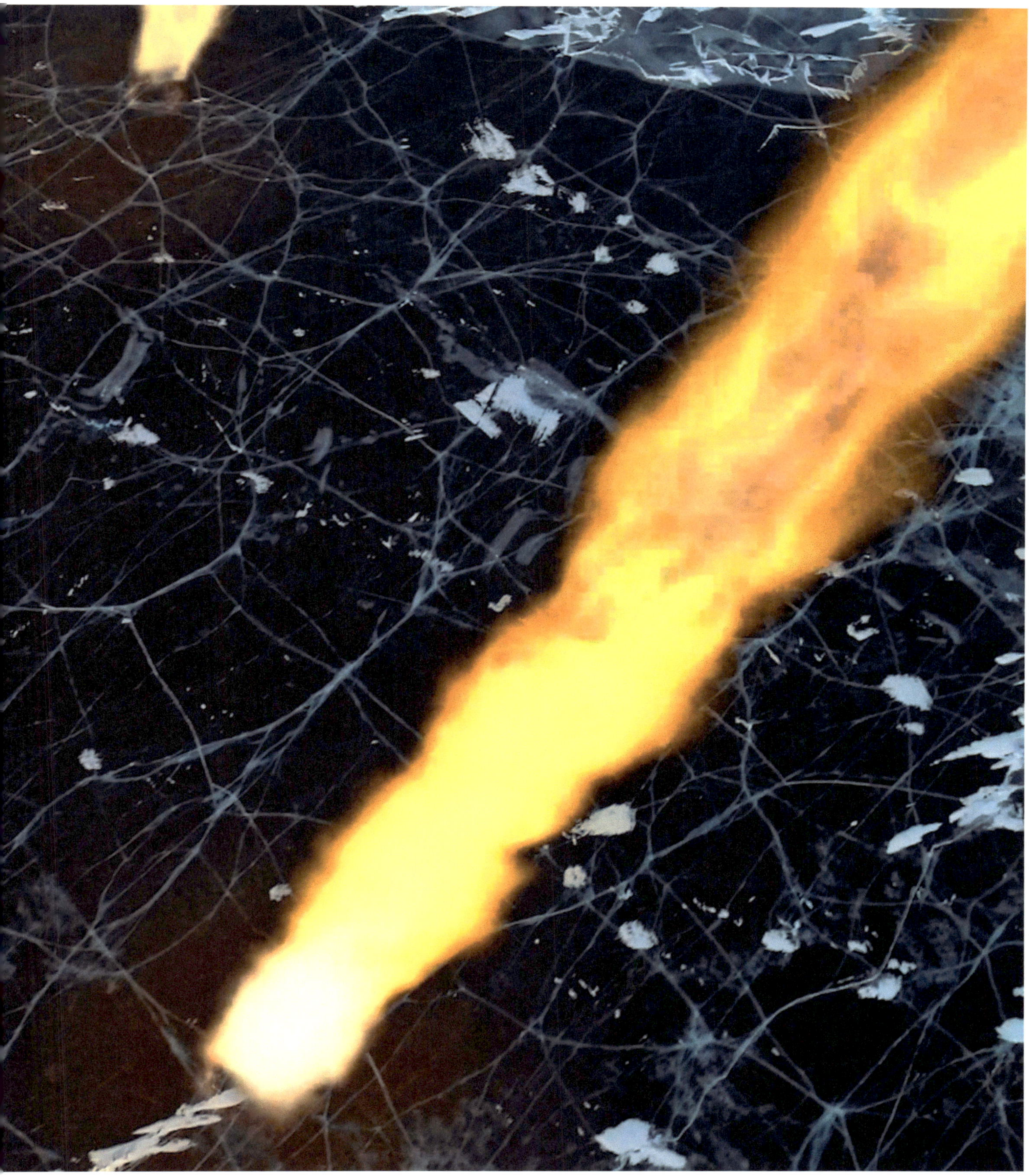

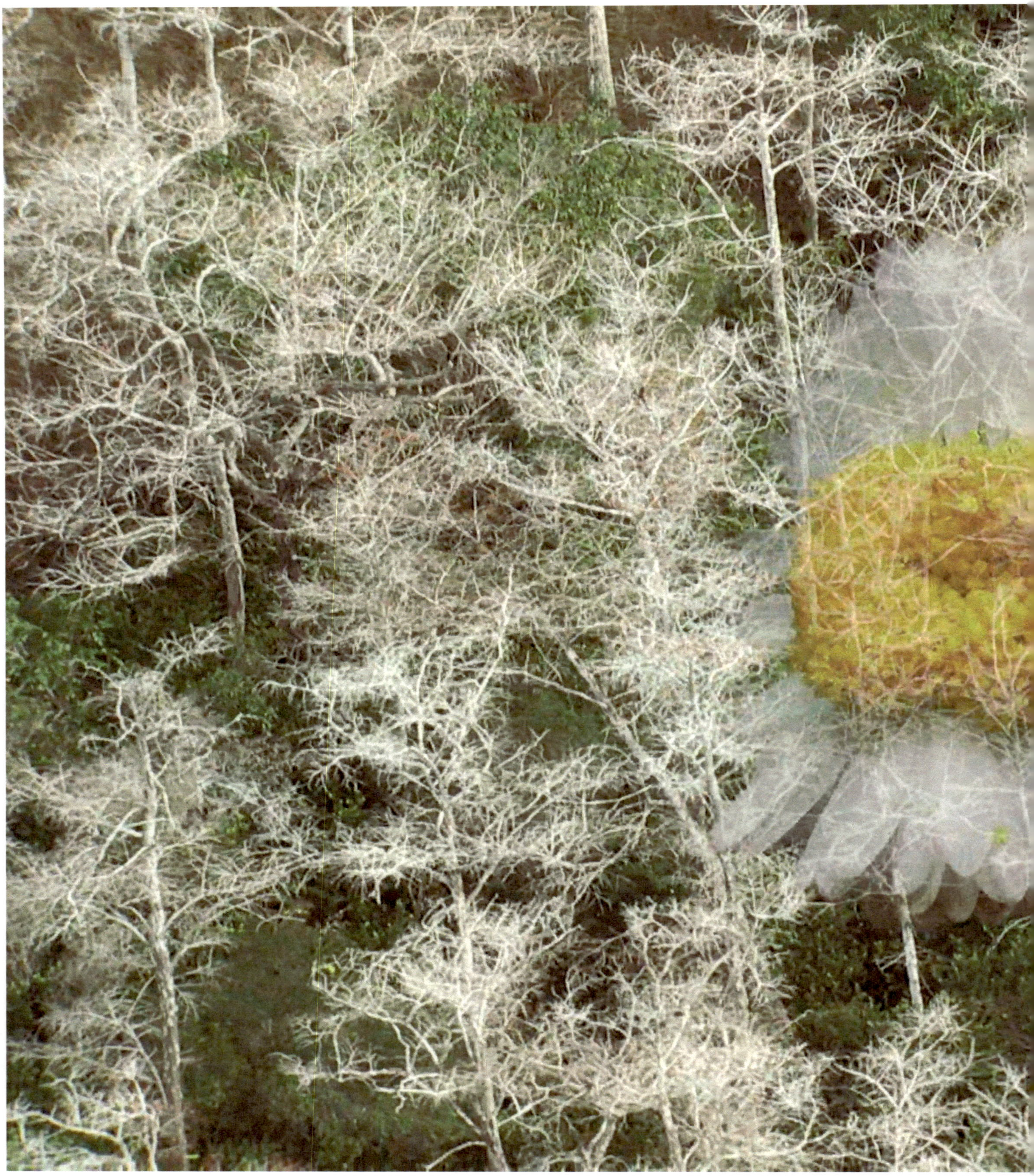

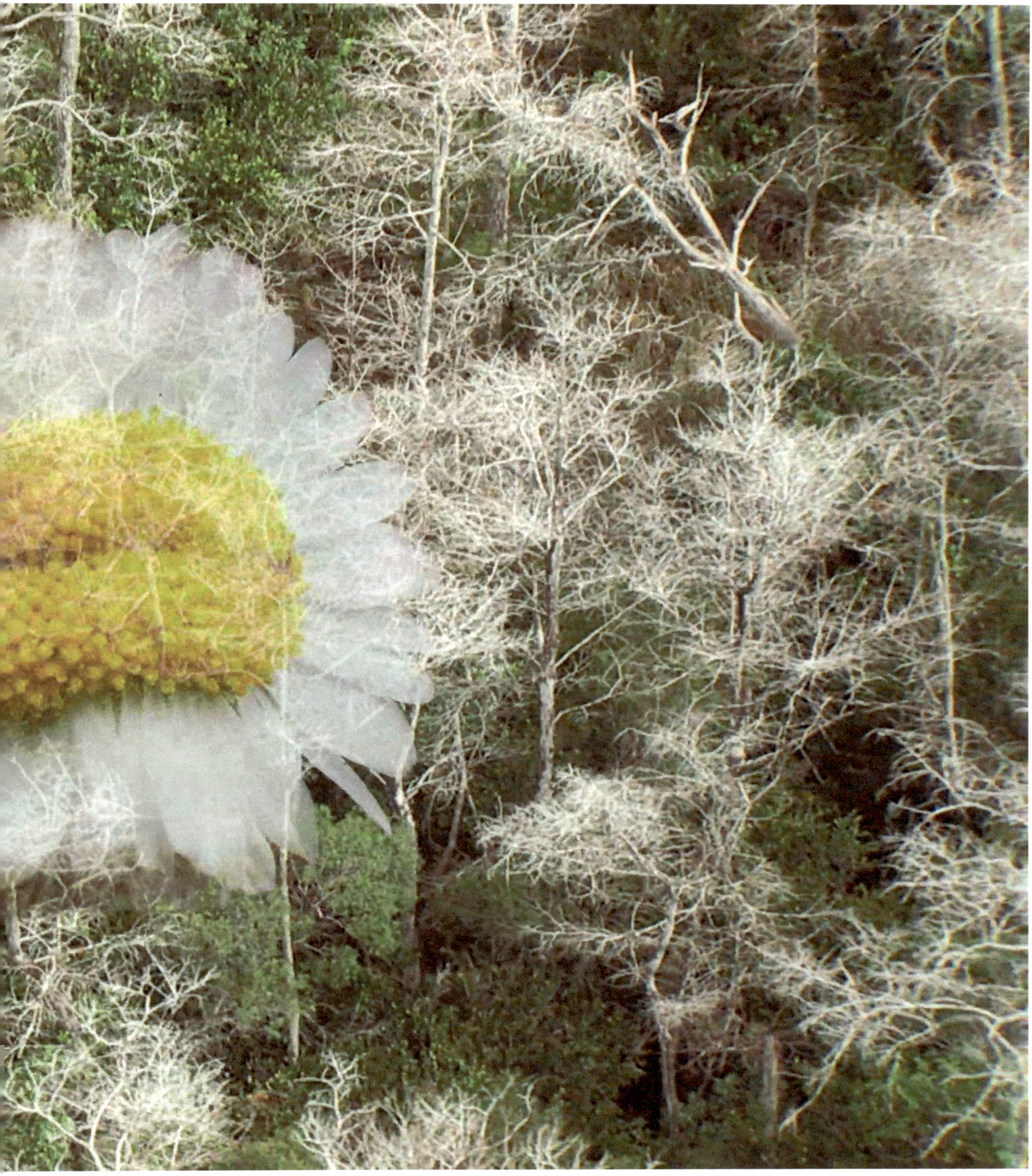

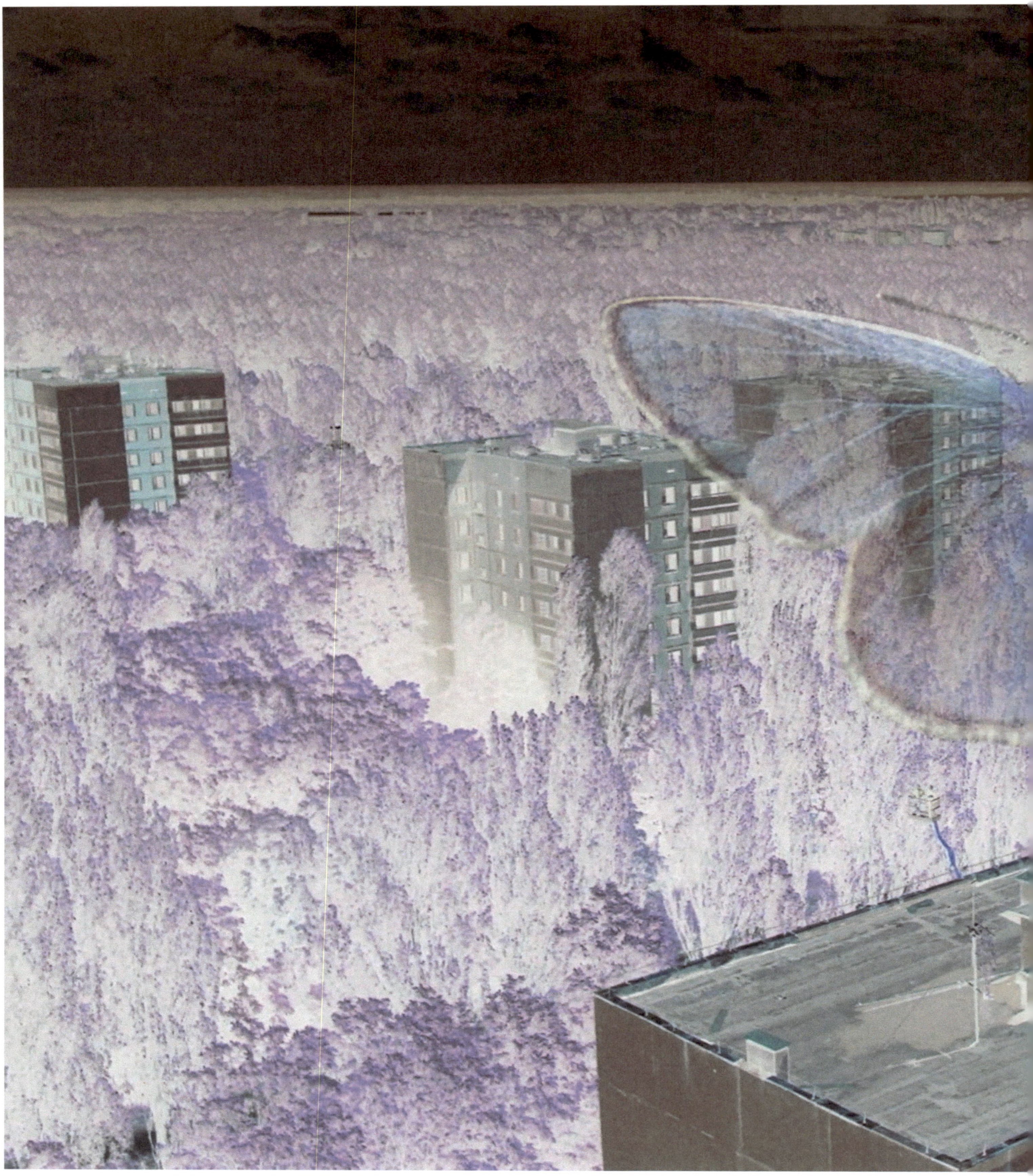

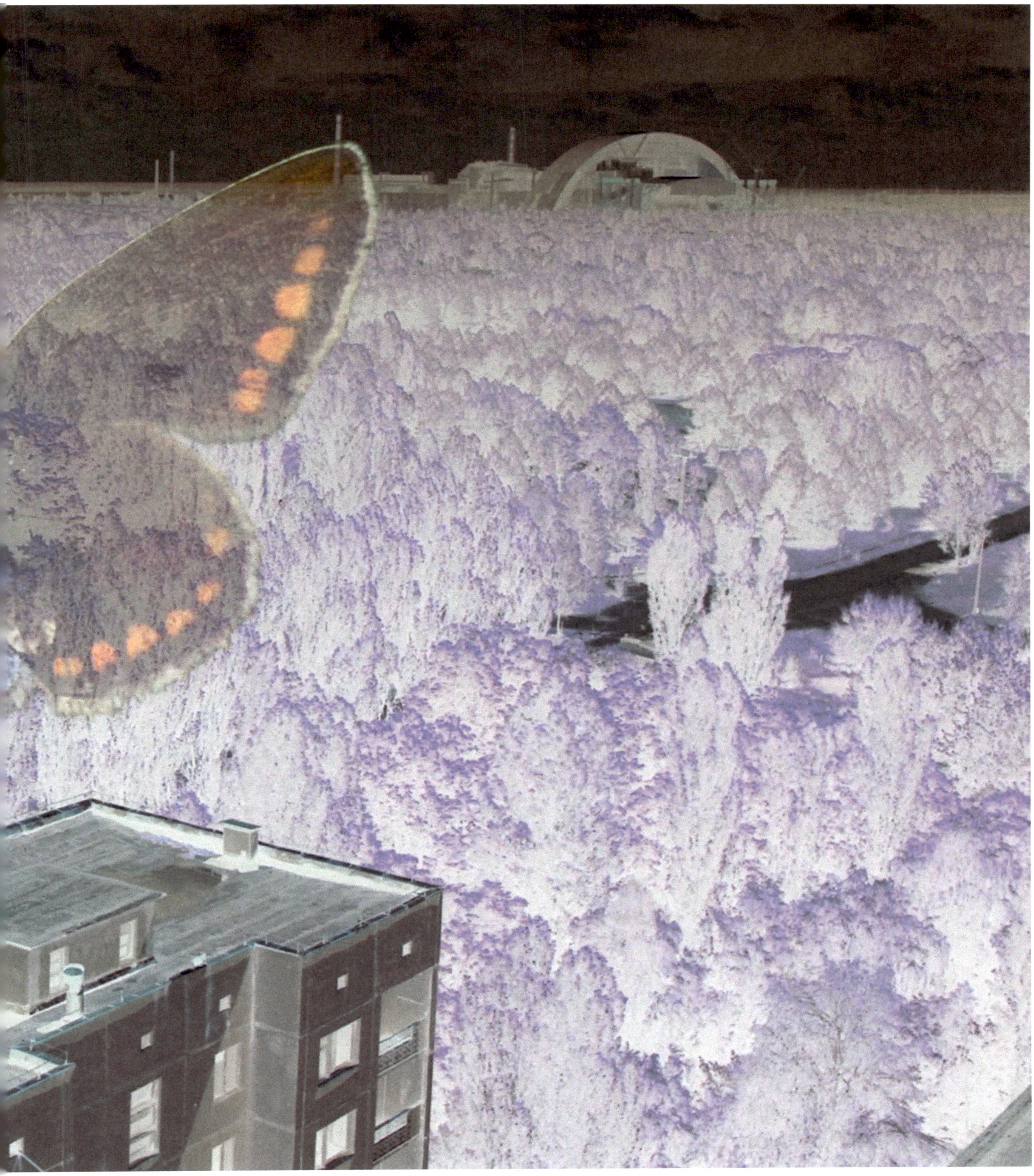

ARTIFICIALIS

LAURENT GRASSO

When Laurence des Cars invited me to create a large-scale work that would dialogue with the space of the Musée d'Orsay, I began thinking about the scale of the building, about the vaulted ceiling and the great hall of the former train station built in the days of industrial boom.[1] My initial research revolved around a scientific fantasy recalling utopian preoccupations and the quite spectacular experiments of Nikola Tesla, and the idea of a film shown on a very large LED screen at the end of the hall quickly came to mind.

Before long, it was suggested that the project interface with a major exhibition on the theory of evolution and its representations in the nineteenth century: *The Origins of the World: The Invention of Nature in the 19th Century*. I then began the research process for the film, focusing on the irreversible transformation of nature by humans and the inseparable intertwining of the natural with the cultural that we are witnessing today. Gradually I assembled a scientific team, including researchers like Grégory Quenet, who teaches environmental history at Université Paris-Saclay and had been an advisor for my 2014 exhibition *Soleil Double*.[2]

Charles Darwin spent close to five years (1832–1836) sailing the seas aboard HMS *Beagle*, voyaging from England to the famous Galapagos Islands and on to Australia. His extraordinary exploring, collecting and recording work, the basis for his entire theory, gave me great inspiration.

In keeping with Darwin's research approach, I attempted to scrutinize the evolutions, mutations and transformations of the living world, as it exists in our 2.0 environment of hyper-accelerated change and phenomenal hybridization of humans and non-humans.

I drew on recent concepts such as those of Philippe Descola, who is renewing the anthropological discipline by rethinking the traditional duality between nature and culture,[3] and that of Eduardo Kohn, who reverses custom by situating his approach not from the human perspective but from the standpoint of thinking forests.[4]

Today, the concept of nature as envisaged in the nineteenth century—pure, virgin, timeless—is an anachronism. Instead, we are seeing the collapse of the nature/culture distinction and entering an era that some are already calling "post-Anthropocene." Although human impact is irreparably imprinted on natural environments, we also see rewilding taking place in certain inhabited areas. With that in mind, I began conceiving a film that would testify to this shift and show the contemporary overlapping of the natural and the artificial.

The idea of probing the erasure of difference between nature and culture raised the question of what means could be used to make it perceptible today. I have always been interested in observation, in how we see, and in the optical tools that change our relationship to perception. This led me to look into new ways of re-envisaging the real—thermal cameras, Lidar scanners, microscopic scanners—that now enable us to plunge into layers of previously inaccessible reality. Consequently, the cinematic techniques used for the project were designed to go beyond their original functions as extensions of the human eye, but also as measuring instruments, survey tools focused on what lies beyond the visible. Lidar (similar to sonar and radar but using laser beams) affords access to the spectral dimension of matter. This new scientific imagery was enormously inspiring and spurred me to add special effects that further accentuate the blurring of borders between reality and fiction by combining images of different sorts. We are witnessing the virtualization of the world today, and these new types of images enable us to access another level of reality.

While looking for places that could embody these ideas, I wondered about exploration in our time, in a world mapped by satellite, a hyperconnected world in which space and time are compressed and discoveries are more closely linked to a conceptual, prospective, experimental geography than to a physical journey such as the one undertaken by Darwin on HMS *Beagle*.

The advent of the current health crisis, and the ensuing unprecedented lockdown, confirmed my intuitions as the project advanced. Instead of opting for live shooting, as for my previous films, I undertook the lengthy task of processing images, facts and narratives already collected, sensing that the challenge of present-day exploration lies more in analyzing big data than in amassing new data to be interpreted. In other words, that the opacity of the world is due more to the accumulation of information to decipher than to geographically unknown areas yet to be made clear.

This was how the idea of a "machine film" gradually emerged, a film that would be written, developed and reassembled like a code, drawing on the world as if on a database and calling up ghosts of places symptomatic of the environmental impact of human activity and engineering. Looking to bioinformatics, I conceived the editing as a living machine, an evolving organism, during which the film would become an autonomous entity.

The quasi autonomous film-object took control of the collected material and rolled out a montage of associations, collisions and micro-narratives whose meaning constantly evolves. A soundtrack specially composed by the Australian musician Warren Ellis completes this apparatus intended as a platform for reflection on a future world.

I designed the monumental LED screen on which the film is shown not just as a projection surface but as an acting object, an emitter of a light signal that acts on the viewer, makes the great hall resonate, questions the museum as a space of freedom and thought. It is a control tower that unfurls a sort of profoundly heterogeneous Warburgian atlas[5] in real time by juxtaposing geographic areas, phenomena and differentiated scales with, as common denominator, the disturbing diagnosis of our times.

1 I want to thank Laurence des Cars, President of the Musée d'Orsay and Musée de l'Orangerie, Donatien Grau, Head of Contemporary Programs, Hélène Flon, Director of Exhibitions, and all of the Musée d'Orsay teams for their commitment to this project.

2 The film *Soleil Double* and the exhibition of the same name were conceived around the notion of disaster and the fictional idea that you might get up one day and find two suns in the sky casting two shadows on every object, every building. *Soleil Double* was presented at the Galerie Perrotin, in Paris, and at the Sean Kelly Gallery, in New York, in September 2014.

3 Philippe Descola, *Beyond Nature and Culture*, trans. Janet Lloyd (Chicago: University of Chicago Press, 2013).

4 Eduardo Kohn, *How Forests Think: Toward an Anthropology Beyond the Human* (Oakland: University of California Press, 2013).

5 Aby Warburg, *Mnemosyne Atlas*, 1924–1929. See Christopher D. Johnson, *Memory, Metaphor, and Aby Warburg's Atlas of Images* (Ithaca: Cornell University Press and Cornell University Library, 2012). See also https://warburg.library.cornell.edu/.

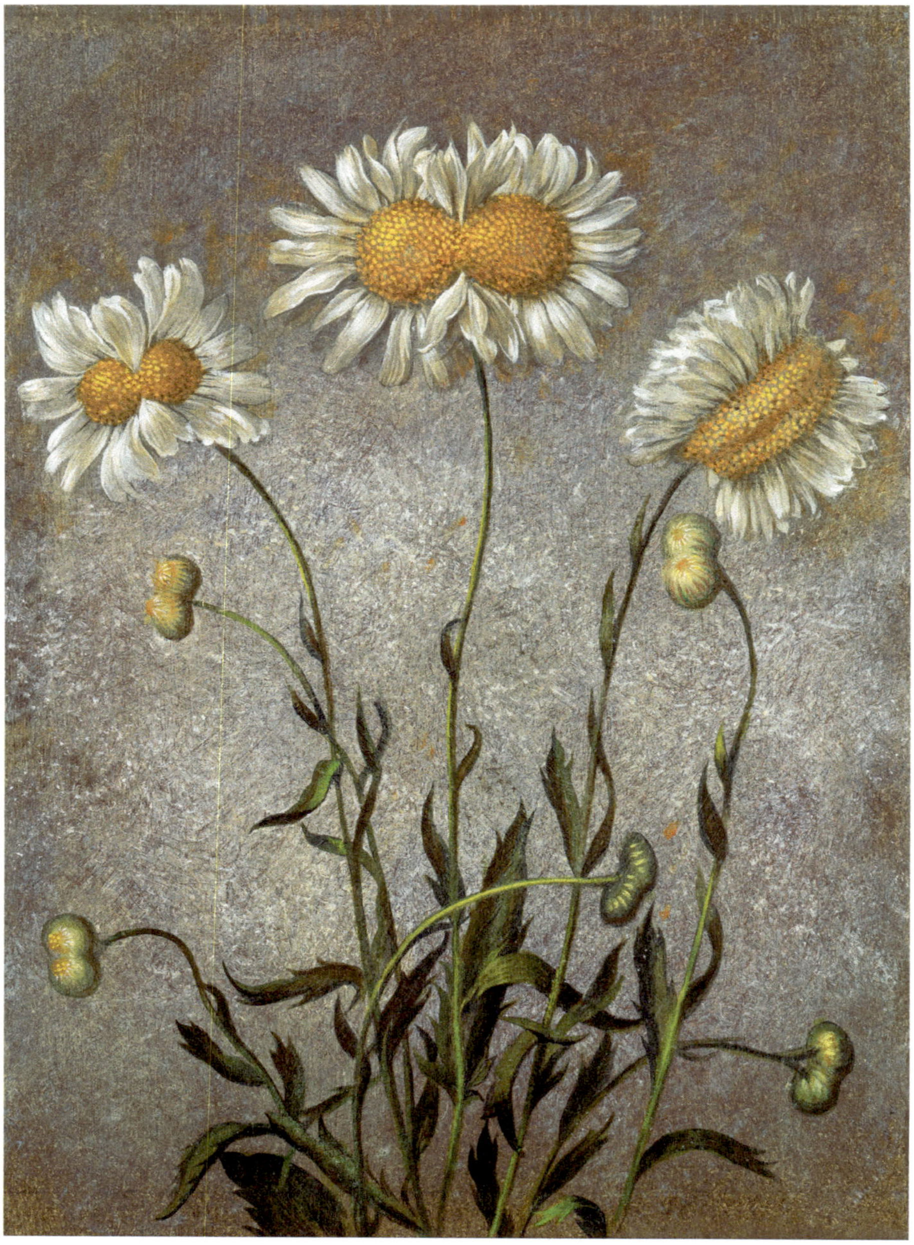

246
Laurent Grasso,
Future Herbarium, n.d.
Oil on wood, 33.5 x 24 cm
Private collection

APPENDICES

As a general rule, if the title of a work has never been translated into English or if the work is known under different titles, the original foreign-language title appears in square brackets. The dimensions of the works are given in centimetres (cm) with height preceding width, except in the case of books, where the order is reversed. "M" (Montreal) and "P" (Paris) indicate at which venue or venues the work was shown. "Not exhibited" indicates that, owing to COVID-19, the work could unfortunately not be presented.
"FIG." in the caption of an illustration indicates that the work was not included in either of the two exhibitions.

PAINTINGS

Jacques-Laurent Agasse (1767–1849), *The Nubian Giraffe*, 1827
Oil on canvas, 127.3 × 101.7 cm
London, The Royal Collection Trust
RCIN 404394
PM
54, p. 77

Nicasius Bernaerts (1620–1678), *Sea Turtle*, 1665–1668
Oil on canvas, 84.5 × 69.5 cm
Paris, Musée du Louvre, département des Peintures
1624
PM
22, p. 44

Nicasius Bernaerts (1620–1678), *Study of an Ostrich*, 1665–1668
Oil on canvas, 91 × 74 cm
Paris, Musée du Louvre, on deposit at the Musée du château des ducs de Wurtemberg, Montbéliard, 1999
337_D.1626
PM
21, p. 43

Carl Blechen (1798–1840), *The Interior of the Palm House*, 1832–1833
Oil on paper mounted on canvas, 64 × 56 cm
Berlin, Staatliche Museen zu Berlin – Preußischer Kulturbesitz
A I 617
P
55, p. 79

Arnold Böcklin (1827–1901), *The Battle of the Centaurs*, 1872–1873
Oil on canvas, 104.2 × 194.3 cm
Basel, Kunstmuseum Basel
Nr.107
P
175, p. 254–255

Arnold Böcklin (1827–1901), *The Calm Sea*, 1887
Oil on wood, 103 × 150 cm
Bern, Kunstmuseum Bern
G0052
Not exhibited
182, p. 262

John Brett (1831–1902), *Glacier of Rosenlaui*, 1856
Oil on canvas, 44.5 × 41.9 cm
London, Tate
N05643
PM
94, p. 138

Jan Brueghel the Elder (1568–1625), *A Paradise Landscape with Animals and Birds in a Wooden Glade by a Pond*, 1617 or 1615
Oil on copper, 19.4 × 15.6 cm
Montreal, The Montreal Museum of Fine Arts, gift of Mr. and Mrs. Michal Hornstein
2014.71
M
Not reproduced

Jan Brueghel the Younger (1601–1678), *The Garden of Eden with the Creation of Eve*, about 1630
Oil on copper, 48.5 × 37 cm
Besançon, Musée des Beaux-Arts et d'Archéologie
855.10.1
P
8, p. 29

Agostino Carracci (1557–1602), *Triple Portrait: Hairy Harry, Mad Peter and Tiny Amon*, 1598–1600
Oil on canvas, 101 × 133 cm
Naples, Museo di Capodimonte
Q 369
P
18, p. 41

Carl Gustav Carus (1789–1869), *Geognostic Landscape*, 1820
Oil on canvas, 91.5 × 133 cm
Stuttgart, Staatsgalerie Stuttgart
L 1309
P
89, p. 134–135

Frederic Edwin Church (1826–1900), *The Icebergs*, 1861
Oil on canvas, 163.8 × 285.7 cm
Dallas, Dallas Museum of Art, gift of Norma and Lamar Hunt
1979.28
Not exhibited
92, p. 137

Frederic Edwin Church (1826–1900), DeWitt Clinton Boutelle (1820–1884), *Cotopaxi, Ecuador*, 1862
Oil on canvas, 87.6 × 142.2 cm
Reading, Reading Public Museum
1929.32.1
P
91, p. 136

Fernand Cormon (1845–1924), *The Human Races*, sketch for the ceiling of the amphitheatre of paleontology in the Muséum d'histoire naturelle de Paris, 1897
Oil on canvas, 76 × 118.5 cm
Paris, Petit Palais, musée des Beaux-Arts de la Ville de Paris
PDUT1805
P
109, p. 159

Gustave Courbet (1819–1877), *The Grotto of the Loue*, 1864
Oil on canvas, 98.4 × 130.4 cm
Washington, National Gallery of Art, gift of Charles L. Lindemann
1957.6.1
P not exhibited
M
217, p. 308

Gustave Courbet (1819–1877), *The Origin of the World*, 1866
Oil on canvas, 46 × 55 cm
Paris, Musée d'Orsay, donation, 1995
RF 1995 10
PM
218, p. 309

Eugène Delacroix (1798–1863), *The Lion Hunt (Sketch)*, 1854
Oil on canvas, 90 × 116.7 cm
Paris, Musée d'Orsay
RF 1984 33
P
Not reproduced

Eugène Delacroix (1798–1863), *The Puma*, 1859
Oil on wood, 41.2 × 30.7 cm
Paris, Musée d'Orsay
RF 1815
P
Not reproduced

Eduard Ender (1822–1883), *Alexander Von Humboldt and Aimé Bonpland in the Jungle*, about 1850
Oil on canvas, 80 × 150 cm
Berlin, Berlin-Brandenburgische Akademie der Wissenschaften
P/BON-1053
PM
49, p. 72

Maxime Faivre (1856–1941), *Two Mothers*, 1888
Oil on canvas, 242.5 × 182 cm
Paris, Musée d'Orsay
RF 590
M
Not reproduced

Robert Farren (1832–1912), *Duria Antiquior, a More Ancient Dorset*, about 1850
Oil on canvas, 166 × 244 cm
Cambridge, Sedgwick Museum of Earth Sciences
CAMSM P.248
PM
99, p. 148

Léon Frédéric (1856–1940), *Nature or Abundance*, 1897
Oil on canvas, 165 × 90.1 cm
Dallas, Dallas Museum of Art, Foundation for the Arts Collection, Mrs. John B. O'Hara Fund
2007.18.FA
PM
242, p. 348

French school, *Buffon's Tree of Apes* [*Arbre des singes de Buffon*], after 1766
Oil on canvas, 106 × 44 cm
Montbard, Musée Buffon
93.20.1
PM
163, p. 237

Michel Garnier (1753–1829), *Date Palm* [*Dattier*], between 1801 and 1810
Oil on paper mounted on canvas, 58.5 × 46.5 cm
Paris, Muséum national d'histoire naturelle
OA 730 (060)
P
79, p. 113

Michel Garnier (1753–1829), *Plum Mango* [*Mangue prune*], between 1801 and 1810
Oil on paper mounted on canvas, 45.5 × 58.4 cm
Paris, Muséum national d'histoire naturelle
OA 730 (059)
P
Not reproduced

Laurent Grasso (born in 1972), *Future Herbarium*, n.d.
Oil on wood, 33.5 × 24 cm
Private collection
P
246, p. 360

John Atkinson Grimshaw (1836–1893), *Il Penseroso*, 1875
Oil on cardboard, 59.7 × 49.5 cm
Private collection
P
2, p. 8

Martin Johnson Heade (1819–1904), *Cattleya Orchid and Three Brazilian Hummingbirds*, 1871
Oil on wood, 34.8 × 45.6 cm
Washington, National Gallery of Art
1982.73.1
P not exhibited
47, p. 69

William Holman Hunt (1827–1910), *The Festival of St. Swithin (The Dovecot)*, 1856–1866
Oil on canvas, 73 × 91 cm
Oxford, Ashmolean Museum, University of Oxford
WA1894.5
PM
120, p. 179

Paul Jamin (1853–1903), *Escaping from the Mammoth*, 1885
Oil on canvas, 125 × 94.5 cm
Paris, Muséum national d'histoire naturelle
OA. 134
PM
117, p. 167

Paul Jamin (1853–1903), *Stone-age Abduction* [*Rapt à l'âge de pierre*], 1888
Oil on canvas, 282 × 200.2 cm
Reims, Musée des Beaux-Arts
892.41.1
P
115, p. 165

Vassily Kandinsky (1866–1944), *Untitled (Deluge)*, 1914
Oil on canvas, 108 × 139.5 cm
Munich, Städtische Galerie im Lenbachhaus
GMS 72
PM
239, p. 341

Gustave Klimt (1862–1918), *Hope I*, 1903
Oil on canvas, 189.2 × 67 cm
Ottawa, National Gallery of Canada
16579
M
208, p. 300

Hilma af Klint (1862–1944), *Eros, No. 1, Group 2, Series WU (The Rose)*, 1907
Oil on canvas, 58 × 79 cm
Stockholm, The Hilma af Klint Foundation
Hak 27
PM
236, p. 339

Hilma af Klint (1862–1944), *Primordial Chaos, No. 13, Group 1, Series WU (The Rose)*, 1906
Oil on canvas, 53.5 × 37 cm
Stockholm, The Hilma af Klint Foundation
Hak 13
PM
235, p. 339

František Kupka (1871–1957), *Hindu Motifs (Graduated Red)*, 1919
Oil on canvas, 124.5 × 122 cm
Paris, Centre Pompidou, musée national d'Art moderne/Centre de création industrielle
AM 4184 P
PM
230, p. 331

Sir Edwin Henry Landseer (1802–1873), *Attachment*, 1829
Oil on canvas, 101.3 × 83.5 cm
St. Louis, Saint Louis Art Museum, gift of Mrs. Eugene A. Perry in memory of her mother, Mrs. Claude Kilpatrick, by exchange
123:1987
M
Not reproduced

Sir Edwin Henry Landseer (1802–1873), *Sleeping Bloodhound*, 1835
Oil on canvas, 101.6 × 127 cm
London, Tate
N00603
P
140, p. 207

Ludger Larose (1868–1915), *Flower Composition with Figure, Logan Greenhouse*, 1899
Oil on canvas, 95.4 × 78.3 cm
Montreal, The Montreal Museum of Fine Arts, purchase, Marcel Boisvert Bequest and Horsley and Annie Townsend Bequest
2007.103
M
Not reproduced

Frederic Leighton (1830–1896), *Pavonia*, 1858–1859
Oil on canvas, 53 × 41.5 cm
Private collection
P
133, p. 199

Bruno Andreas Liljefors (1860–1939), *A Fox Family*, 1886
Oil on canvas, 112 × 218 cm
Stockholm, Nationalmuseum
NM 1376
P
119, p. 174–175

Jacques Linard (1597–1645), *Still Life with Shells and Coral*, 1640
Oil on canvas, 53.3 × 62.2 cm
Montreal, The Montreal Museum of Fine Arts, gift of Mr. and Mrs. Michal Hornstein
1999.149
M
Not reproduced

Pietro Longhi (1701–1785) (attributed to), *The Rhinoceros*, about 1751
Oil on canvas, 55.5 × 72 cm
Vicenza, Gallerie d'Italia – Palazzo Leoni Montanari
Inv. 196
PM
33, p. 59

Pietro Longhi (1701–1785), *The Rhinoceros*, 1751
Oil on canvas, 62 × 50 cm
Venice, Fondazione Musei Civici di Venezia
Cl. ln.1312
P
32, p. 58

John Martin (1789–1854), *The Deluge*, 1834
Oil on canvas, 168.3 × 258.4 cm
New Haven, Yale Center for British Art
B1978.43.11
Not exhibited
84, p. 121

Gabriel von Max (1840–1915), *Abelard and Heloise*, after 1900
Oil on canvas, 41 × 36 cm
The Jack Daulton Collection
PM
166, p. 241 and cover

Gabriel von Max (1840–1915), *Gruss (Monkey with a Bouquet, a Posy of Pansies)*, 1901–1915
Oil on wood, 24 × 16 cm
The Jack Daulton Collection
PM
165, p. 240

Gabriel von Max (1840–1915), *Model Laura*, 1901–1915
Oil on canvas, 34 × 25.4 cm
The Jack Daulton Collection
PM
Not reproduced

Gabriel von Max (1840–1915), *Monkey Before Skeleton*, about 1900
Oil on canvas, 61 × 44.5 cm
Private collection
Not exhibited
164, p. 239

Gabriel von Max (1840–1915), *Pithecanthropus alalus*, 1894
Oil on canvas, 99 × 68.5 cm
Jena, Friedrich-Schiller-Universität, Ernst-Haeckel-Haus
PM
4, p. 23

Paul Meyerheim (1842–1915), *A Darwinian Prehistoric Social Party (The Un-Evolved Club Man of the Period)*, 1865
Oil on canvas, 53.6 × 114.3 cm
Milwaukee, Layton Art Collection, Inc., Purchase, at the Milwaukee Art Museum
L148
PM
Not reproduced

Paul Meyerheim (1842–1915), *The Jealous Lioness*, 1885–1890
Oil on canvas, 49.6 × 69 cm
Frankfurt, Städel Museum
1375
PM
178, p. 259

Piet Mondrian (1872–1944), *Evolution*, 1911
Oil on canvas, 183 × 257.5 cm
The Hague, Stichting Kunstmuseum Den Haag
SCH-1971-0148.1=3
P
234, p. 337

Claude Monet (1840–1926), *Blue Water Lilies*, 1914–1919
Oil on canvas, 200 × 200 cm
Paris, Musée d'Orsay
RF 1981 40
P
197, p. 285

Claude Monet (1840–1926), *Water Lilies*, between 1916 and 1919
Oil on canvas, 130 × 152 cm
Paris, Musée Marmottan Monet, legs Monet Michel
5098
P
198, p. 286

Claude Monet (1840–1926), *Wisteria*, between 1919 and 1920
Oil on canvas, 100 × 300 cm
Paris, Musée Marmottan Monet
5124
M
Not reproduced

Edward Moran (1829–1901), *The Valley in the Sea*, 1862
Oil on canvas, 102.9 × 162.6 cm
Indianapolis, Indianapolis Museum of Art at Newfields
IMA 70.5
P
71, p. 105

Thomas Moran (1837–1926), *Hot Springs of the Yellowstone*, 1872
Oil on canvas, 71.1 × 106.6 cm
Los Angeles, Los Angeles County Museum of Art, gift of Beverly and Herbert M. Gelfand
84.198
M
Not reproduced

Gustave Moreau (1826–1898), *Bertrand and Raton*, 1879–1885
Oil on canvas, 82 × 65 cm
Paris, Musée national Gustave-Moreau
Cat.743
PM
159, p. 229

Edvard Munch (1863–1944), *Woman with Sick Child. Inheritance*, 1905–1906
Oil on canvas, 120.5 × 100 cm
Oslo, Munchmuseet
MM.M.283
PM
229, p. 329

Isaak van Oosten (1613–1661), *The Earthly Paradise [Adam et Ève et les animaux de la Création]*, 1625–1650
Oil on canvas, 58.2 × 84.8 cm
Rennes, Musée des Beaux-Arts
801.2.2
PM
10, p. 30

Filippo Palizzi (1818–1899), *After the Deluge*, 1864
Oil on canvas, 185 × 266 cm
Naples, Museo e Real Bosco di Capodimonte
PS 44
P
12, p. 32–33

Eugen Ransonnet-Villez (1838–1926), *Underwater Landscape*, 1864
Oil on canvas, 50 × 70 cm
Vienna, Naturhistorisches Museum
PM
72, p. 105

Odilon Redon (1840–1916), *Oannes*, 1900–1910
Oil on canvas, 66 × 51.5 cm
Otterlo, Stichting Kröller-Müller Museum
KM 109.934
PM
184, p. 265

Odilon Redon (1840–1916), *Trees on a Yellow Background*, 1901
Oil, tempera, charcoal and pastel on canvas, 249.5 × 185.5 cm
Paris, Musée d'Orsay
RF 1988 30
PM
190, p. 276

Odilon Redon (1840–1916), *Trees on a Yellow Background*, 1901
Oil, tempera, charcoal and pastel on canvas, 247.5 × 173 cm
Paris, Musée d'Orsay, donation, 1988
RF 1988 31
PM
191, p. 277

Briton Riviere (1840–1920), *Beyond Man's Footsteps*, about 1894
Oil on canvas, 119 × 184.5 cm
London, Tate
N01577
P
95, p. 139

Briton Riviere (1840–1920), *Conscience Makes Cowards of Us All*, about 1871–1872
Oil on canvas, 29.7 × 39.8 cm
Manchester, The Whitworth, The University of Manchester
O.1910.9
Not exhibited
139, p. 206

René Rousseau-Decelle (1881–1964), *Prehistoric Family [La Famille préhistorique]*, 1906
Oil on canvas, 115 × 145 cm
La Roche-sur-Yon, Musée de La Roche-sur-Yon
2012.6.18
PM
248, p. 383

Johann Moritz Rugendas (1802–1858), *Giant Tree in Brazil's Tropical Forest*, 1830
Oil on canvas, 47 × 34 cm
Berlin, Die Stiftung Preußische Schlösser und Gärten Berlin-Brandenburg
GKI 4341
P
59, p. 83

Johann Moritz Rugendas (1802–1858), *Indians in the Virgin Forest*, 1830
Oil on canvas, 62 × 49 cm
Berlin, Die Stiftung Preußische Schlösser und Gärten Berlin-Brandenburg
GKI 1271
P
58, p. 82

Roelandt Savery (1576–1639), *The Dodo*, before 1639
Oil on canvas, 96 × 116 cm
London, The Natural History Museum
992066802081
Not exhibited
20, p. 42

Rudolf Schlichter (1890–1955), *Blind Power*, 1932–1937
Oil on canvas, 179 × 100 cm
Berlin, Berlinische Galerie
BG-M 1984/79
P
241, p. 344

Paul Sérusier (1864–1927), *Tetrahedrons*, about 1910
Oil on canvas, 91.5 × 57.4 cm
Paris, Musée d'Orsay
RF MO P 2018 5
PM
240, p. 343

George Stubbs (1724–1806), *A Lion Attacking a Horse*, 1770
Oil on canvas, 102.2 × 127.6 cm
New Haven, Yale University Art Gallery, gift of The Yale University Art Gallery Associates
1955.27.1
M
Not reproduced

Archibald Thorburn (1860–1935), *Peacock and Peacock Butterfly*, 1917
Oil on canvas, 87.5 × 111.5 cm
Private collection
PM
135, p. 201

Joseph Mallord William Turner (1775–1851), *Stormy Sea with Blazing Wreck*, about 1840
Oil on canvas, 99.4 × 141.6 cm
London, Tate
N04658
P
96, p. 140

Joseph Mallord William Turner (1775–1851), *The Evening of the Deluge*, about 1843
Oil on canvas, 76 × 76 cm
Washington, National Gallery of Art, Timken Collection
1960.6.40
M
97, p. 141

Anne Vallayer-Coster (1744–1818), *Sea Plumes, Lithophytes and Shells*, 1769
Oil on canvas, 130 × 97 cm
Paris, Musée du Louvre, département des Peintures
RF 1992-410
P
13, p. 35

Pierre-Jacques Volaire (1729–1799), *The Eruption of Vesuvius*, 1771
Oil on canvas, 73.3 × 162 cm
Le Havre, Musée d'Art moderne André-Malraux
2015.2.11
PM
90, p. 136

George Frederic Watts (1817–1904), *The Minotaur*, 1885
Oil on canvas, 118.1 × 94.5 cm
London, Tate
N01634
PM
171, p. 249

Jacopo Zucchi (1541?–1590?), *Coral Fishing (Allegory of the Discovery of America)*, 1615–1630
Oil on copper, 52 × 42.5 cm
Rome, Galleria Borghese
292
P
14, p. 37

GRAPHIC ARTS

Anonymous, *Study of a Peacock's Feather*, n.d.
Watercolour over graphite on wove paper, 64.3 × 30.7 cm
Oxford, Ashmolean Museum, University of Oxford
WA.RS.REF.116
P
129, p. 196

George Baxter (1804–1867) (after Benjamin Waterhouse Hawkins), *The Crystal Palace and Park in Sydenham*, Day & Son, 1854
Lithograph, 11 × 16 cm
London, The British Library
MS 50150 f.220
Not exhibited
100, p. 149

René Binet (1866–1911), *Design for the Monumental Entrance Gate of the 1900 Universal Exposition [Projet pour la porte monumentale de l'Exposition universelle de 1900]*, 1898
Watercolour on paper, 62 × 95 cm
Sens, Musées de Sens
P
143, p. 214

René Binet (1866–1911), "Electric Lantern [Lanterne électrique]," in *Esquisses décoratives*, Paris, Librairie centrale des beaux-arts, 1903
Portfolio, printed plates, 47 × 69 × 3 cm
Paris, Bibliothèque du musée d'Orsay
Gd FOL A 62
P
Not reproduced

Charles Darwin (1809–1882), *Argus Pheasant Feathers*, 1868
Ink and graphite on paper, 16 × 8 cm
Cambridge, Cambridge University Library, Darwin Papers
DAR 84.1: 149r
Not exhibited
121, p. 181

Henry De La Beche (1796–1855), *Duria Antiquior, a More Ancient Dorset*, 1832
Watercolour on paper, 23 × 32 cm
Cardiff, Amgueddfa Cymru – National Museum Wales
84.20G D368
PM
98, p. 148

Albrecht Dürer (1471–1528), *The Rhinoceros*, 1515
Engraving, woodcut, 21.6 × 30.2 cm
Paris, Musée du Louvre, département des Arts graphiques, collection Edmond de Rothschild
L 37 LR/245 Recto, Folio 109
P
28, p. 55

Ernst Moritz Geyger (1861–1941), *Darwinian Disputation [Darwinistische Disputation]*, 1888
Etching, 24 × 33.5 cm
Frankfurt, Städel Museum
67240
P
158, p. 227

Eugène Grasset (1845–1917), *Peacock Feet, Feathers and Heads, in Profile [Pattes, plumes et têtes de paon de profil]*, between 1890 and 1903
Pencil, 18.7 × 13.9 cm
Paris, Musée d'Orsay, achat avec l'aide de la SAMO, 1993
ARO 1993 9 2 103
P
131, p. 197

Eugène Grasset (1845–1917), *Peacocks in Profile with Decorative Motifs [Paons de profil avec motifs ornementaux]*, between 1890 and 1903
Pencil, 20 × 15.5 cm
Paris, Musée d'Orsay, achat avec l'aide de la SAMO, 1993
ARO 1993 9 2 121
P
Not reproduced

Alfred Kubin (1877–1959), *Fertility (Version I)*, 1901–1902
India ink, pencil, watercolour, spray, on cadastral paper, 37.2 × 27.2 cm
Vienna, Leopold Museum
LM 927
P
205, p. 295

Alfred Kubin (1877–1959), *Sea Floor [Meeresgrund]*, 1906
Tempera on black cardboard, 23.6 × 51.5 cm
Linz, Oberösterreichische Landesmuseum
Ha II 7338
P
214, p. 303

Alfred Kubin (1877–1959), *Suckers [Die Sauger]*, 1905
Watercolour on cadastral paper, 31.6 × 39.2 cm
Linz, Oberösterreichische Landesmuseum
Ha II 7341
P
211, p. 303

Alfred Kubin (1877–1959), *The Ape*, about 1903–1904
Chalk and pencil on paper, 35.7 × 23.8 cm
Vienna, Albertina
INV. 27963
P
207, p. 299

Alfred Kubin (1877–1959), *The Hour of Birth*, 1901–1902
India ink, wash, crachis on paper, 25.5 × 30.6 cm
Vienna, Leopold Museum
LM 877
P
212, p. 303

Alfred Kubin (1877–1959), *The Swamp*, 1903–1905
Pen, ink and crachis on handmade paper, 30.7 × 37.9 cm
Linz, Lentos Kunstmuseum Linz
G817
P
215, p. 304

Alfred Kubin (1877–1959), *Underwater Scene [Unterseestück]*, 1906
Tempera on cardboard, 39.8 × 31.3 cm
Linz, Oberösterreichische Landesmuseum
ZW II 45
P
210, p. 302

Alfred Kubin (1877–1959), *Untitled*, about 1906
Gouache on Bristol board, 25.2 × 29.5 cm
Zug, Kunsthaus Zug
INV.-Nr.779
P
213, p. 303

František Kupka (1871–1957), *Anthropoids: Fight for a Woman*, 1902
Watercolour and gouache on paper, 59.5 × 62 cm
Prague, Galerie Zdeněk Sklenář
P
174, p. 253

František Kupka (1871–1957), *Study for "Tale of Pistils and Stamens,"* about 1919
Gouache on paper, 33.3 × 29 cm
Collection de Bueil & Ract-Madoux
DBRM-112
PM
233, p. 335

Alexandre-Isidore Leroy de Barde (1777–1828), *A Collection of Foreign Birds*, about 1810
Watercolour and gouache on paper, 126 × 90 cm
Paris, Musée du Louvre, département des Arts graphiques
INV. 23692, recto
P
27, p. 51

Alexandre-Isidore Leroy de Barde (1777–1828), *Birds*, 1800–1828
Watercolour and gouache on paper, 124.5 × 88.5 cm
Paris, Musée du Louvre, département des Arts graphiques
INV. 23691, Recto
P
26, p. 50

Charles-Alexandre Lesueur (1778–1846), *Astropecten preissi Müller & Troschel Starfish [Étoile de mer Astropecten preissi Müller & Troschel]*, 1843
Watercolour and gouache on paper, 31.4 × 47 cm
Le Havre, Muséum d'histoire naturelle
INV. 74011
P
Not reproduced

Charles-Alexandre Lesueur (1778–1846), *Terrapene carolina (Linnaeus, 1758) Turtle [Tortue Terrapene Carolina (Linné, 1758)]*, between 1816 and 1837
Watercolour and gouache on paper, 46.7 × 31.1 cm
Le Havre, Muséum d'histoire naturelle
INV. 78027
P
62, p. 89

Charles-Alexandre Lesueur (1778–1846), *Unspecified Starfish [Étoile de mer indéterminée]*, about 1840
Watercolour and gouache on paper, 31.4 × 46.8 cm
Le Havre, Muséum d'histoire naturelle
INV. 74037
P
Not reproduced

Conrad Martens (1801–1878), *Fuegians and the* Beagle *at Portrait Cove, Beagle Channel*, 1835–1836
Graphite, watercolour, heightened with white on paper, 19.4 × 29.5 cm
London, National Maritime Museum
PAF6242
P
52, p. 74

Conrad Martens (1801–1878), *Fuegians in a Canoe off Wollaston Island*, 1835–1836
Graphite, watercolour, heightened with gum arabic on paper, 21.7 × 34.7 cm
London, National Maritime Museum
PAF6241
P
51, p. 74

Edvard Munch (1863–1944), *Madonna*, 1895–1902
Lithograph, 64 × 48 cm
Oslo, Munchmuseet
MM.G 194 - 02
P
227, p. 328

Edvard Munch (1863–1944), *Metabolism (Life and Death)*, 1896
India ink on wove paper, 48 × 36.6 cm
Oslo, Munchmuseet
MM.T 411
P
221, p. 317

Edvard Munch (1863–1944), *The Family Tree*, 1890–1892
Pencil on wove paper, 38.4 × 29 cm
Oslo, Munchmuseet
MM.T 388-recto
P
228, p. 328

Alexis Peyrotte (1699–1769), *An Ape Preaching*, about 1760
Gouache on cardboard, 28.5 × 35.7 cm
Montreal, The Montreal Museum of Fine Arts, gift in memory of Isaac Battat, from his family
2010.95
M
Not reproduced

Odilon Redon (1840–1916), *The Origins*, Paris, Lemercier & Cie, 1883
Lithographs on China paper laid down on wove paper
PM
188, p. 273
(A) Cover-frontispiece, 30.5 × 22.5 cm (image)
Chicago, The Art Institute of Chicago
(B) "The Satyr with the Cynical Smile," pl. 5, 24 × 20.7 cm (image)
(C) "There Was Perhaps a First Vision Attempted in the Flower," pl. 2, 22.3 × 17.2 cm (image)

(D) "The Misshapen Polyp Floated on the Shores, a Sort of Smiling and Hideous Cyclops," pl. 3, 22.3 × 17.2 cm (image)
Washington, National Gallery of Art

Odilon Redon (1840–1916), *The Shell*, 1912
Pastel on paper, 52 × 57.8 cm
Paris, Musée d'Orsay
RF 40494
P
Not reproduced

Odilon Redon (1840–1916), *Underwater Vision*, 1900
Pastel on paper, 66 × 54 cm
Paris, Musée d'Orsay, gift, 1982
RF 40499
P
216, p. 305

Johann Elias Ridinger (1698–1767), *Adam Who Names the Animals* [*Adam qui nomme les animaux*], about 1740
Engraving, 38 × 53 cm
Private collection
P
Not reproduced

Briton Riviere (1840–1920), *Dog "in a humble and affectionate frame of mind," Study for Darwin's* Expression of the Emotions of Man and Animals, about 1871–1872
Pencil, ink and wash, white highlights on grey paper, 20 × 27 cm
Cambridge, Cambridge University Library, Darwin Papers
DAR 53.1: C168r
Not exhibited
Not reproduced

John Ruskin (1819–1900), *Study in Colour of Quartz Rock Weathered*, about 1875
Graphite, watercolour and gouache on wove paper, 13.2 × 24.4 cm
Oxford, Ashmolean Museum, University of Oxford
WA.RS.ED.277
P
249, p. 384

Joseph Mallord William Turner (1775–1851), *The Devil's Bridge and Schöllenen Gorge*, 1802
Graphite, watercolour and gouache on paper, 47.1 × 31.8 cm
London, Tate
D04626
P
Not reproduced

Joseph Mallord William Turner (1775–1851), *The Source of the Arveyron below the Glacier du Bois and Mer de Glace*, 1802
Graphite, watercolour and gouache on paper, 31.3 × 46.8 cm
London, Tate
D04613
P
93, p. 138

Benjamin Waterhouse Hawkins (1807–1894), *Extinct Creatures Including Ichthyosaur*, 1858
Watercolour
London, The Natural History Museum
L&A 428541
Not exhibited
102, p. 150

Benjamin Waterhouse Hawkins (1807–1894), *Ichthyosaur and Other Extinct Creatures*, about 1850
Watercolour and ink on paper, 17.5 × 35.8 cm
London, The Natural History Museum
L&A 428541
Not exhibited
103, p. 150

Benjamin Waterhouse Hawkins (1807–1894), *Invitation to the Iguanodon Dinner at the Crystal Palace*, 1853
Pencil and ink on paper
London, The Natural History Museum
L&A 428541
Not exhibited
1, p. 4

Jean-Charles Werner (1798–1856), *Male Mazikina at 2/3 Life Size* [*Mazikina mâle aux 2/3 de la grandeur naturelle*], 1824
Watercolour on paper, 35 × 49 cm
Paris, Muséum national d'histoire naturelle
Ms 600, fol. 89
P
Not reproduced

Jean-Charles Werner (1798–1856), *Tufted Capuchin at 4/5 of Its Actual Size* [*Sajou cornu, variété à moustaches, au 4/5e de sa taille réelle*], 1842
Watercolour on paper, 35 × 49 cm
Paris, Muséum national d'histoire naturelle
Ms 600, fol. 85
P
Not reproduced

Joseph Wolf (1820–1899), *Cynopithecus niger, in a Placid Condition*; *The Same, when Pleased by Being Caressed*, about 1871–1872
Black chalk, white highlights, 32.5 × 19.5 cm
Cambridge, Cambridge University Library, Darwin Papers
DAR 53.1: C163
Not exhibited
138, p. 205

VELLUMS

Nicolas Huet (1770–1830), *Condor, Head*, 1827
Watercolour on vellum, 46 × 33 cm
Paris, Muséum national d'histoire naturelle
Collection des vélins, portefeuille 77, fol. 10
P
Not reproduced

Charles-Alexandre Lesueur (1778–1846), *Chrysaora Lesueur Jellyfish* [*Méduse Chrysaora Lesueur*], between 1804 and 1810
Watercolour and pencil on vellum, 43.2 × 27.9 cm
Le Havre, Muséum d'histoire naturelle
INV. 70060
P
65, p. 92

Charles-Alexandre Lesueur (1778–1846), *Cyanea lamarckii* (centre), *Chrysaora cyclonota, spiligona, spilhemigona and aspilonota* (corners) *Jellyfish* [*Méduses Cyanea lamarckii* (centre), *Chrysaora cyclonota, spiligona, spilhemigona, aspilonota* (angles)], between 1804 and 1810
Watercolour on vellum, 27.5 × 43.6 cm
Le Havre, Muséum d'histoire naturelle
INV. 70059
P
Not reproduced

Charles-Alexandre Lesueur (1778–1846), *Cyanea lamarckii Jellyfish* [*Méduse Cyanea lamarckii*], between 1804 and 1810
Watercolour on vellum, 44.1 × 28.5 cm
Le Havre, Muséum d'histoire naturelle
INV. 70058
P
66, p. 93

Charles-Alexandre Lesueur (1778–1846), *Macropus fuliginosus (Desmarest, 1817) Kangaroo* [*Kangourou Macropus fuliginosus (Desmarest, 1817)*], between 1804 and 1810
Watercolour, gouache, graphite, gum arabic on vellum, 25.7 × 40.3 cm
Le Havre, Muséum d'histoire naturelle
INV. 80057
P
60, p. 89

Charles-Alexandre Lesueur (1778–1846), *Macropus rufogriseus (Desmarest, 1817) Kangaroo* [*Kangourou Macropus rufogriseus (Desmarest, 1817)*], between 1804 and 1810
Watercolour, gouache, graphite, gum arabic on vellum, 25.7 × 39.7 cm
Le Havre, Muséum d'histoire naturelle
INV. 80063
P
Not reproduced

Charles-Alexandre Lesueur (1778–1846), *Rhizostoma cuvieri Jellyfish* [*Méduse Rhizostoma cuvieri*], between 1804 and 1810
Watercolour on vellum, 28.1 × 42.8 cm
Le Havre, Muséum d'histoire naturelle
INV. 70055-1
P
Not reproduced

Nicolas Maréchal (1753–1802), *Barbary Macaque – Macaca sylvanus Linnaeus (Cercopithecidae): Magot. Simia inuus* [*Macaque magot – Macaca sylvanus* Linné *(Cercopithécidés) : Le Magot. Simia inuus*], 1793–1794
Watercolour on vellum, 46 × 33 cm
Paris, Muséum national d'histoire naturelle
Collection des vélins, portefeuille 69, fol. 31
P
43, p. 66

Nicolas Maréchal (1753–1802), *Male Tufted Capuchin, Simia apella* [*Le Sajou mâle, Simia apella*], late 18th century – early 19th century
Watercolour on vellum, 46 × 33 cm
Paris, Muséum national d'histoire naturelle
Collection des vélins, portefeuille 69, fol. 52
P
Not reproduced

Nicolas Maréchal (1753–1802), *Polar Bear – Ursus maritimus* Phipps *(Ursidae): Polar Bear, One Sixth of the Specimen's Size* [*Ours blanc – Ursus maritimus* Phipps (Ursidés) : *ours polaire, sixième de la grandeur de l'individu*], 1796
Watercolour on vellum, 33 × 46 cm
Paris, Muséum national d'histoire naturelle
Collection des vélins, portefeuille 70, fol. 78
P
45, p. 67

Nicolas Maréchal (1753–1802), *Sighthound – Canis lupus familiaris (Canidae): Sighthound Bitch, One Third of the Specimen's Size* [*Chien lévrier – Canis lupus familiaris (Canidés) : chienne levrette, tiers de la grandeur de l'individu*], late 18th century – early 19th century
Watercolour on vellum, 33 × 46 cm
Paris, Muséum national d'histoire naturelle
Collection des vélins, portefeuille 71, fol. 13
P
Not reproduced

Nicolas Maréchal (1753–1802), *The Green Monkey, Simia sabaea* [*Le Callitriche, Simia sabaea*], 1794–1795
Watercolour on vellum, 46 × 33 cm
Paris, Muséum national d'histoire naturelle
Collection des vélins, portefeuille 69, fol. 10 bis
P
Not reproduced

Pierre-Joseph Redouté (1759–1840), *Iris orientalis* Miller *(Iridaceae): Iris Monnieri De Candolle* [*Iris orientalis* Miller *(Iridacées) : Iris Monnieri Decandolle*], 1808
Watercolour on vellum, 46 × 33 cm
Paris, Muséum national d'histoire naturelle
Collection des vélins, portefeuille 11, fol. 78
P
38, p. 64

Pierre-Joseph Redouté (1759–1840), *Paeonia x suffruticosa* Andrews *(Ranunculaceae): Paeonia moutan Bonpland, Malmaison* [*Paeonia x suffruticosa* Andrews *(Renonculacées) : Paeonia moutan Bonpland, Malmaison*], late 18th century – early 19th century
Watercolour on vellum, 46 × 33 cm
Paris, Muséum national d'histoire naturelle
Collection des vélins, portefeuille 41, fol. 28
P
35, p. 63

Pierre-Joseph Redouté (1759–1840), *Phymosia umbellata (Cavanilles)* Kearney *(Malvaceae): Malva umbellata Cavanilles, Mexico* [*Phymosia umbellata* (Cavanilles) Kearney *(Malvacées) : Malva umbellata Cavanilles, Mexique*], 1810
Watercolour on vellum, 46 × 33 cm
Paris, Muséum national d'histoire naturelle
Collection des vélins, portefeuille 45, fol. 44
P
39, p. 64

Pierre-Joseph Redouté
(1759–1840), *Rhinoceros Underbelly with Penis [Le Ventre du rhinocéros avec la verge]*, late 18th century – early 19th century
Watercolour on vellum,
33 × 46 cm
Paris, Muséum national d'histoire naturelle
Collection des vélins, portefeuille 65, fol. 45
P
46, p. 67

Nicolas Robert
(1614–1685), *Tulipa (Liliaceae): Tulipa lutea, lituris quibusdam viridibus et sanguineis distincta flore maximo, laciniato, H. R. P.*, 17th century
Gouache on vellum,
46 × 33 cm
Paris, Muséum national d'histoire naturelle
Collection des vélins, portefeuille 7, fol. 32
P
Not reproduced

Nicolas Robert
(1614–1685), *Tulipa (Liliaceae): Tulipa munstrosa*, 17th century
Gouache on vellum,
46 × 33 cm
Paris, Muséum national d'histoire naturelle
Collection des vélins, portefeuille 7, fol. 31
P
36, p. 64

Nicolas Robert
(1614–1685), *Tulipae variae. Violet Widow. Passable Variegation but Poor Shapes [Tulipae variae. La veuve violette. Passables de panaches mais de mauvaise forme]*, 17th century
Gouache on vellum,
46 × 33 cm
Paris, Muséum national d'histoire naturelle
Collection des vélins, portefeuille 7, fol. 28
P
37, p. 64

Pierre-François de Wailly
(1775–1852), *Sulphur-crested Cockatoo – Cacatua galerita Latham (Psittacidae): Psittacus sulphureus, Yellow-crested Cockatoo [Cacatoès à huppe jaune – Cacatoe galerita* Latham *(Psittacidés) : Psittacus sulphureus, Kakatoès à huppe jaune]*, 1811
Watercolour on vellum,
46 × 33 cm
Paris, Muséum national d'histoire naturelle
Collection des vélins, portefeuille 80, fol. 92
P
40, p. 65

Pierre-François de Wailly
(1775–1852), *Yellow-tailed Black Cockatoo. Psittacus funereus Shaw, New Holland (Australia) [Le Kakatoès buse. Psittacus funereus Shaw, Nouvelle Hollande]*, 1813
Watercolour on vellum,
46 × 33 cm
Paris, Muséum national d'histoire naturelle
Collection des vélins, portefeuille 80, fol. 95
P
Not reproduced

Jean-Charles Werner
(1798–1856), *Chimpanzee [Chimpanzé]*, 1837
Watercolour on vellum,
46 × 33 cm
Paris, Muséum national d'histoire naturelle
Collection des vélins, portefeuille 69, fol. 6 bis
P
41, p. 66

Jean-Charles Werner
(1798–1856), *Semnopithecus dussumieri*, 1842
Watercolour on vellum,
46 × 33 cm
Paris, Muséum national d'histoire naturelle
Collection des vélins, portefeuille 69, fol. 7 ter
P
42, p. 66

Jean-Charles Werner
(1798–1856), *Young Sumatran Orangutan That Died at the Ménagerie [Jeune orang bicolore mort à la Ménagerie]*, 1836
Watercolour on vellum,
46 × 33 cm
Paris, Muséum national d'histoire naturelle
Collection des vélins, portefeuille 102, fol. 1
P
44, p. 66

SCULPTURES

The Neanderthal Man from Le Moustier, Reconstruction by Mr. Faure and Miss Parvillée [L'Homme du Moustier, reconstitution par M. Faure et Mlle Parvillée], 1923
Plaster, 50 × 60 × 30 cm
Les Eyzies, Musée national de Préhistoire
MNP 1950-12
P
114, p. 161

Louis Ernest Barrias
(1841–1905), *Nature Unveiling Herself Before Science*, 1903
Ivory, gold- and silver-plated bronze, lapis lazuli, marble, h. 58 cm
Paris, Musée des Arts décoratifs
12008
M
3, p. 20

Antoine-Louis Barye
(1795–1875), *Ape Riding a Gnu*, first edition, about 1840
Patinated bronze,
23.2 × 25.8 × 10 cm
Paris, Musée du Louvre, département des Sculptures
OA 5772
PM
161, p. 233

Antoine-Louis Barye
(1795–1875), Honoré Gonon (1780–1850) (founder), *Tiger Devouring a Gavial*, 1832
Brown-red patina, lost-wax casting,
39.7 × 105.6 × 40.5 cm
Paris, Musée du Louvre, département des Sculptures
RF 2991
PM
Not reproduced

Sarah Bernhardt
(1844–1923), Thiébault Frères (founder), *Inkwell: "Self-Portrait as a Sphinx,"* 1880
Bronze, 30 × 30.5 × 34 cm
Montreal, The Montreal Museum of Fine Arts, purchase, Suzanne Caouette Bequest
2019.87.1-4
M
Not reproduced

Jean Carriès
(1855–1894), *Fish-tailed Amphibian [Batracien à queue de poisson]*, 1892
Glazed stoneware,
30 × 84 × 69 cm
Paris, Petit Palais, musée des Beaux-Arts de la Ville de Paris
OGAL33
PM
181, p. 261

Jean Carriès (1855–1894), *Frog with Rabbit Ears*, 1891
Glazed stoneware,
34.4 × 26.8 × 41 cm
Paris, Petit Palais, musée des Beaux-Arts de la Ville de Paris
PPS404
PM
179, p. 261

Jean Carriès
(1855–1894), *Frog-Man*, about 1891
Glazed stoneware,
31.5 × 36 × 38 cm
Paris, Musée d'Orsay
OAO 523
PM
180, p. 261

Jean Carriès
(1855–1894), *Head of a Faun*, 1890–1891
Glazed stoneware,
32 × 28.5 × 21 cm
Paris, Musée des Arts décoratifs
34361
PM
Not reproduced

Jean-Pierre Dantan, known as Dantan the Younger (1800–1869), *Portrait of the Orangutan Called "Jack" [Portrait de l'orang-outang, dit "Jack"]*, 1836
Bronze, 42 × 28 × 23 cm
Paris, Musée Carnavalet – Histoire de Paris
S164
PM
162, p. 235

Emmanuel Fremiet
(1824–1910), *Gorilla Carrying Off a Woman*, 1893
Master model, bronze,
45 × 26 × 40 cm
Dijon, Musée des Beaux-Arts
4197
PM
170, p. 247

Emmanuel Fremiet
(1824–1910), *Gorilla Dragging a Warrior by the Hair*, about 1859
Terracotta bozzetto,
24 × 17.4 × 15.3 cm
Paris, Musée d'Orsay, don de Mme Fauré-Fremiet, 1971
RF 2977
P
167, p. 242

Emmanuel Fremiet
(1824–1910),
Orangutan Strangling a Savage of Borneo, 1893
Polychromed plaster,
58 × 55.5 × 27 cm
Dijon, Musée des Beaux-Arts
4964
P
168, p. 243

Paul Gauguin (1848–1903), *Be Mysterious*, 1890
Low relief, polychromed lime wood, traces of dark-coloured crayon,
73 × 95 × 5 cm
Paris, Musée d'Orsay
RF 3405
P
Not reproduced

Louis Mascré (1871–1929) and Aimé Rutot (1847–1933), *Neanderthal Man*, 1909–1914
Painted plaster,
68 × 55 × 60 cm
Brussels, Institut royal des sciences naturelles de Belgique
P
Not reproduced

Louis Mascré (1871–1929) and Aimé Rutot (1847–1933), *Precursor of the Tertiary Period [Précurseur de l'ère tertiaire]*, 1909–1914
Painted plaster,
70 × 50 × 60 cm
Brussels, Institut royal des sciences naturelles de Belgique
PM
Not reproduced

Louis Mascré (1871–1929) and Aimé Rutot (1847–1933), *Woman of the Neanderthal Race. Descendant of prior generations. Mousterin Age*, 1909–1914
Painted plaster,
84 × 60 × 55 cm
Brussels, Institut royal des sciences naturelles de Belgique
P
Not reproduced

Conrad Meit
(about 1480–about 1550) (entourage of), *Adam and Eve*, 1st half of 16th century
Marble, 48 × 37 × 7 cm
Paris, Musée du Louvre, département des Sculptures
RF 1530
PM
Not reproduced

Alphonse Mucha
(1860–1939), Émile Pinédo (1840–1916) (founder), *Nature*, about 1900
Bronze, silver, gilding, marble, 69.2 × 27.9 × 30.5 cm
Richmond, Virginia Museum of Fine Arts, Sydney and Frances Lewis Art Nouveau Fund
72.13
M
Not reproduced

Paul Richer (1849–1933), *First Artist*, 1890
Plaster, 179 × 67 × 80 cm
Le Puy-en-Velay, Musée Crozatier
2014.4.5.55
PM
206, p. 297

Jean Désiré Ringel d'Illzach (1847–1916), *Batrachian-Chimera*, 1905
Glazed stoneware,
63.5 × 40.6 × 31.7 cm
Los Angeles, Los Angeles County Museum of Art, purchased with funds provided by the Dalzell Hatfield Memorial Fund, the Hearst Foundation, the Mr. and Mrs. Alain C. Balch Collection, the James E. Clark Bequest, Michael Hall, the William Randolph Hearst Collection, Fabrizio Apolloni, and various other donors by exchange
2006.98.2
M
Not reproduced

Jean Désiré Ringel d'Illzach (1847–1916), *Owl-Chimera*, 1905
Enamelled stoneware,
66 × 33 × 38.1 cm
Los Angeles, Los Angeles County Museum of Art, purchased with funds provided by the Dalzell Hatfield Memorial Fund, the Hearst Foundation, the Mr. and Mrs. Alain C. Balch Collection, the James E. Clark Bequest, Michael Hall, the William Randolph Hearst Collection, Fabrizio Apolloni, and various other donors by exchange
2006.98.1
M
Not reproduced

Auguste Rodin
(1840–1917), Émile Godard
(founder), *Centauress*,
about 1887, 1969 (cast)
Bronze, lost-wax casting,
45.8 × 49 × 16.7 cm
Paris, Musée Rodin
S.00479
PM
176, p. 257

Auguste Rodin
(1840–1917), carved
by Victor Peter (1840–1918),
Fish-Woman, 1915–1917
Marble,
34 × 32 × 50 cm
Paris, Musée Rodin
S.01103
PM
183, p. 263

Jules Laurent Terrier
(1853–after 1907), *Dodo, Raphus cucullatus*, 1901
White plaster, moulded,
70 × 48 ×58 cm
Paris, Muséum national
d'histoire naturelle
OA.946
P
Not reproduced

Benjamin Waterhouse
Hawkins (1807–1894),
Iguanodon, 1854
Painted plaster,
31 × 49 × 24 cm
Paris, Muséum national
d'histoire naturelle
P
Not reproduced

Benjamin Waterhouse
Hawkins (1807–1894),
Megalosaurus, 1853
Painted plaster,
28 × 55 × 22 cm
Paris, Muséum national
d'histoire naturelle
PM
Not reproduced

DECORATIVE ARTS

Anonymous,
France's First Giraffe [*La Première Girafe de France*], about 1830–1845
Diorama: wood, oil paint,
earthenware, glass, leather,
85 × 80 × 40 cm
Paris, Muséum national
d'histoire naturelle
OA 1754
P
Not reproduced

Anonymous,
Suspension, about 1900
Bronze, glass,
98.5 × 54 × 15 cm
Montreal, The Montreal
Museum of Fine Arts,
Liliane and David M.
Stewart Collection
D95.142.1a-d
M
Not reproduced

Ferdinand Barbedienne
(1810–1892) (founder),
Édouard Lièvre (1828–1886) (draughtsman),
Aquarium and Pedestal,
about 1875
Glass, silver-plated bronze,
143 cm (total height),
57.7 × 60 cm (aquarium),
85 × 75 × 62 cm (pedestal)
Collection of Ramy Boutros
PM
68, p. 97

Georges Bastard
(1881–1939), *Peacocks* [*Les Paons*], 1913
Broken fan: sculpted
mother-of-pearl, ribbon,
18.5 × 35 cm
Paris, Musée d'Orsay
DO 1977-6
M
Not reproduced

Leopold Blaschka
(1822–1895) and Rudolf
Blaschka (1857–1939),
Argonauta argo, 1866
Glass
London, The Natural
History Museum
BMNH 1883.12.22.1
Not exhibited
148, p. 216

Leopold Blaschka
(1822–1895) and Rudolf
Blaschka (1857–1939),
Octopus vulgaris, 1866
Glass
London, The Natural
History Museum
BMNH 1883.12.22.8
Not exhibited
149, p. 217

Christofle & Cie,
Tray, Decorated with "Natural Impression,"
about 1886
Silver-plated
metal, 46.5 × 2 cm
Paris, Musée d'Orsay
OAO 1178
M
Not reproduced

Auguste Delaherche
(1857–1940), *Vase*, 1889
Enamelled stoneware,
H. 38 cm
Paris, Musée d'Orsay,
don de Mme René Bureau,
en souvenir de son mari,
par l'intermédiaire
de la SAMO, 1993
OAO 1274
PM
Not reproduced

Eugène Feuillâtre
(1870–1916), *"Sea" Vase* [*Vase "La Mer"*], about 1912
Translucent enamel
on copper, 37.6 × 17.6 cm
Paris, Petit Palais, musée
des Beaux-Arts
de la Ville de Paris
OGAL 370
PM
70, p. 104

William Frend De Morgan
(1839–1917), *"Peacocks" Charger*, about 1885
Earthenware, copper and
silver lustre, 36 cm (diam.)
Paris, Musée d'Orsay
OAO 455
M
Not reproduced

Émile Gallé (1846–1904),
Hand with Seaweed and Shells, 1904
Glass modelled under heat,
inclusions, applications,
wheel engraving,
33.4 × 13.4 cm
Paris, Musée d'Orsay, gift,
1990
OAO 1207
P
204, p. 293

Émile Gallé (1846–1904),
Victor Prouvé (1858–1943)
and Louis Hestaux (1856–1919), *Jardinière Flora marina, Flora exotica*, 1889
Sculpted pear wood,
marquetry in different
woods (amaranth, Macassar
ebony, bird's-eye maple,
elm, padauk, pear wood,
plane-tree wood), interior
zinc and lead,
117 × 95 × 56 cm
Nancy, Musée de l'École
de Nancy
NV76
PM
200, p. 291

Émile Gallé (1846–1904),
Jug, between 1884 and 1889
Lead-glazed earthenware,
slip, painted decoration,
29.2 × 24.7 × 16.1 cm
Montreal, The Montreal
Museum of Fine Arts,
purchase, The Montreal
Museum of Fine Arts'
Volunteer Association Fund
2009.56
M
Not reproduced

Émile Gallé (1846–1904),
Rouleau Vase with Marine Decoration, 1902–1903
Blown glass in several
layers, metal inclusions,
wheel-cut decoration,
applications,
28.2 × 9 cm (base)
Paris, Musée d'Orsay, gift of
M. Alain Pierre Bourgogne,
2018
RF MO OAO 2018 12
PM
203 A and B, p. 292

Émile Gallé (1846–1904),
Sea Lily [*Le Lys de mer*],
about 1903
Double-layered glass,
metal inclusions, engraved
decoration,
14.5 × 10.5 × 8 cm
Nancy, Musée de l'École
de Nancy
HH9
PM
202, p. 292

Émile Gallé (1846–1904),
The Bottom of the Sea [*Les Fonds de la mer*], 1889–1903
Multi-layered glass, metal
inclusions and intercalated
bubbles, two wheel-cut
applications, 21.8 × 14.3 cm
Nancy, Musée de l'École
de Nancy
JC7
PM
201, p. 292

Gustav Gurschner
(1873–1970), *Nautilus Lamp*, about 1899–1900
Bronze, nautilus shell,
44.5 × 18.7 × 19.1 cm
Richmond, Virginia
Museum of Fine Arts,
Sydney and Frances Lewis
Art Nouveau Fund
72.19
M
Not reproduced

Henri Husson
(1854–1914), Adrien-Aurélien Hébrard (1865–1937) (founder), *Vessel* [*Coupe*], about 1909
Hammered copper, tooled,
patinated, silver dust,
10 × 44.2 cm (diam.)
Paris, Musée d'Orsay
OAO 65
M
Not reproduced

Sèvres Manufactory,
drawings by Léon Kann
(1859–1925), *"Fenouil" Coffee Service*, 1898–1912
Hard-paste porcelain,
coffee pot: 18.5 × 18 × 9 cm,
sugar bowl: 10.6 × 10.4 × 8.6
cm, cups (2):
6.3 × 7.7 × 5.7 cm, saucers:
13.1 cm (diam.), tray:
42 × 31.2 cm
Montreal, The Montreal
Museum of Fine Arts,
purchase, Liliane and David
M. Stewart Collection
D94.300.1-7
M
Not reproduced

Constant Roux
(1865–1942) (drawing),
for the Maison Baguès,
Radiolarian Chandelier [*Lustre radiolaire*], 1910
Hammered copper
with sheets of opalescent
and red glass, 140 × 80 cm
Monaco, Institut
océanographique,
Fondation Albert Ier, prince
de Monaco
MOB-0077
PM
146, p. 215

Jean-Joseph de Saint-Germain (1719–1791)
(founder), François Viger
(1708–1784) (clockmaker),
Rhinoceros Clock, 1750
Gilt and patinated
bronze with tortoiseshell
and copper on the base,
enamelled face,
83 × 50 × 24 cm
Paris, Musée du Louvre,
département des Objets
d'art
OA 10540
P
31, p. 57

Louis Comfort Tiffany
(1848–1933), *"Jack-in-pulpit" Vase*, manufactured
by Tiffany Studios,
New York, about 1909–1910
Blown glass,
51.8 × 27.3 × 13.3 cm
Montreal, The Montreal
Museum of Fine Arts,
purchase, Deutsche Bank
Fund
2009.28
M
Not reproduced

Louis Comfort Tiffany
(1848–1933), *"Peacock" Bottle* [*Bouteille "Paon"*],
about 1897
Glass, 41.2 × 12.5 cm
Paris, Petit Palais,
musée des Beaux-Arts
de la Ville de Paris
OGAL 61
P
134, p. 200

Louis Comfort Tiffany
(1848–1933), *Peacock-feather Vase* [*Vase plumes de paon*], about 1897
Blown glass, hot-modelled,
veined with iridescent
colours, 32 × 16 cm
(opening)
Paris, Musée des Arts
décoratifs
8554
P
Not reproduced

Louis Comfort Tiffany
(1848–1933), designed
by Clara Driscoll (1861–1944), made by Tiffany
Studios, New York, *Peacock Table Lamp*, about 1905
Leaded glass, bronze,
67 × 47 cm
Montreal, The Montreal
Museum of Fine Arts,
purchase, Claire Gohier
Fund, gift of Gérald-Henri
Vuillien and Christophe
Pilaire in honour of being
granted Canadian
permanent resident status,
Ruth Jackson Bequest,
gift of the International
Friends of the Montreal
Museum of Fine Arts
and gift of Joan and Martin
Goldfarb
2011.373.1-2
M
Not reproduced

Louis Comfort Tiffany
(1848–1933), *Vase*,
1902–1903
"Favrile" glass,
20.6 × 20.6 cm
New York,
The Metropolitan Museum
of Art, gift of Louis Comfort
Tiffany Foundation, 1951
51.121.24
Not exhibited
Not reproduced

Philippe Wolfers
(1858–1929), *Jewel Box "Parure" or "The Peacock,"*
1905
Ivory, silver, gold, enamel,
opal, pearl, 42 × 30 × 20.2 cm
Brussels, Musées royaux
d'art et d'histoire
Sc. 7032
P
Not reproduced

FOSSILS AND MODELS

Aepyornis Egg
[*Œuf d'Aepyornis*]
35 × 27 cm
Paris, Muséum national d'histoire naturelle, don de Nanette et Jean Têtefort, entomologiste acridologue à Madagascar pour le MNHN
MNHN.F.MAD8904
P
Not reproduced

Archaeopteryx lithographica, model of the Eichstätt specimen discovered in 1861
Plaster model, 49 × 39 cm
Paris, Muséum national d'histoire naturelle
MNHN.F.MAD8904
P
Not reproduced

Antoine-Louis Barye (1795–1875), *Deer Muzzle, Mouth Open, Cast from Life* [*Mufle de cerf, gueule ouverte, moulée sur nature*], 19th century
Plaster and model from life, 24.5 × 17 × 30 cm
Paris, Beaux-Arts de Paris
MU 12129
P
Not reproduced

Georges Cuvier (1769–1832), *Cuvier's Opossum (Cast) – Peratherium cuvieri* [*Sarigue de Cuvier (moulage) – Peratherium cuvieri*]
Plaster model in two parts, 9 × 14.5 × 3 cm and 7.5 × 10 × 2.3 cm
Paris, Muséum national d'histoire naturelle
MNHN-F-GY679a and b
P
Not reproduced

Ichthyosaurus tenuirostris Skeleton from Holzmaden, Wurtemberg [*Squelette d'Ichthyosaurus tenuirostris provenant d'Holzmaden, Wurtemberg*], Late Liassic
45 × 115 × 5.3 cm
Paris, Muséum national d'histoire naturelle
MNHN-F-1866-90
P
Not reproduced

Mammut americanum Mastodon Thigh Bone from Ohio [*Fémur de mastodonte Mammut americanum de l'Ohio*]
Resin model, 112 × 51 × 28 cm
Paris, Muséum national d'histoire naturelle
MNHN-F-AC2020
P
Not reproduced

Onchiodon from Autun [*Onchiodon d'Autun*]
Plaster model, 82.5 × 36 × 4 cm
Paris, Muséum national d'histoire naturelle
MNHN-F-AUT-483
P
Not reproduced

Pterodactyl
Fossil, 26 × 25 × 2.5 cm
Paris, Muséum national d'histoire naturelle
MNHN-F-SLN-26
P
86, p. 125

Jules Tramond, *Anatomical Cast of an Ape's Hand* [*Main de singe, moulage anatomique*], 19th century
Plaster and model from life, 29.8 × 11.4 × 15 cm
Paris, Beaux-Arts de Paris
MU 12110
P
Not reproduced

Jules Tramond, *Anatomical Cast of an Ape's Hand* [*Main de singe, moulage anatomique*], 19th century
Plaster and model from life, 30.4 × 11.4 × 15 cm
Paris, Beaux-Arts de Paris
MU 12117
P
Not reproduced

MODELS

Louis-Marc-Antoine Robillard d'Argentelle (1777–1828), *Bullock's Heart* [*Corossol cœur de bœuf*], between 1802 and 1826
Wax, wood, metal, 45 × 30 × 14 cm
Paris, Muséum national d'histoire naturelle
OA 1300 (068)
P
77, p. 112

Louis-Marc-Antoine Robillard d'Argentelle (1777–1828), *Chinese Plum Tree* [*Prunier de Chine*], between 1802 and 1826
Wax, wood, metal, 52 × 45 × 15 cm
Paris, Muséum national d'histoire naturelle
OA 1300 (020)
P
Not reproduced

Louis-Marc-Antoine Robillard d'Argentelle (1777–1828), *Japanese Persimmon* [*Plaqueminier du Japon*], between 1802 and 1826
Wax, wood, metal, 52 × 45 × 16 cm
Paris, Muséum national d'histoire naturelle
OA 1300 (019)
P
74, p. 111

Louis-Marc-Antoine Robillard d'Argentelle (1777–1828), *Two Clusters of Dates* [*Deux grappes de dattes*], between 1802 and 1826
Wax, wood, metal, 45 × 30 × 10 cm
Paris, Muséum national d'histoire naturelle
OA 1300 (086)
P
78, p. 112

BIFACES

Biface, Middle or Late Paleolithic
Cut stone, 22.5 × 10.5 × 4.3 cm
"Antediluvian period, found at about 4(?) metres depth in the diluvium of St Acheul les Amiens. Given by M. Boucher de Perthes."
Paris, Muséum national d'histoire naturelle
MNHN-HP-2127
P
112, p. 160

Biface, Middle or Late Paleolithic
Cut stone, 14.4 × 7.5 × 5.2 cm
"Diluvium; Abbeville, Moulin-Quignon 1825; very large axe, probably one of the oldest, found at 5 metres depth and given by M. Boucher de Perthes."
Paris, Muséum national d'histoire naturelle
MNHN-HP-2143.17
P
113, p. 160

Biface, Middle or Late Paleolithic
Cut stone, 9.4 × 6.5 × 2.6 cm
"Hache roulée [axe], Diluvium 1844, Menchecourt-les-Abbevilles" "Diluvium, Abb., yellow sands at 8 (?) metres depth., found by me in 1844. Given to M. le Marquis de Vibraye. J.B. de Perthes"
Paris, Muséum national d'histoire naturelle
Cote 12683
P
Not reproduced

STUFFED ANIMALS

Marguerite [*Parkie*], *a Female Asian Elephant, a Specimen from the Stathouder's Zoo at Het Loo Taken to the Ménagerie in 1798, Stuffed* [*Marguerite (Parkie), éléphant d'Asie femelle, spécimen historique ramené à la Ménagerie depuis le zoo du Stathouder à Loos en 1798, naturalisé*], 1798
310 × 420 × 130 cm
Paris, Muséum national d'histoire naturelle
MNHN-ZM-1944-296
P
Not reproduced

MANUSCRIPTS

Petrus Comestor (1100?–1179?), *Bible historiale*, leaflet 6, first quarter of 15th century
Book-length manuscript, illuminated on parchment, 35 × 27 × 9 cm
Paris, Bibliothèque Mazarine
Ms 313
P
6, p. 27

Aimé Goujaud, known as Bonpland (1773–1858), *Botanical Field Notes of the Humboldt and Bonpland Expedition* [*Carnet botanique de l'expédition de Humboldt et Bonpland*], 1799–1804
Handwritten notebook, 33 × 40 × 6 cm
Paris, Muséum national d'histoire naturelle
Ms 53
P
48, p. 71

Nicolas Jenson (about 1420–1480), "The Creation of the World," in *Biblia Latina*, Venice, 1476
Book-length manuscript, illuminated on vellum, 29.5 × 21 cm
Paris, Bibliothèque nationale de France
Réserve des livres rares, Rés. Vélins 80
P
5, p. 25

Benjamin Waterhouse Hawkins (1807–1894), *Detail of the Dinner Party Held Inside the Iguanodon, from Hawkins' Letter to Trimmer*, after 1854
Letter handwritten on paper, 20.5 × 13.5 cm
London, The British Library
Add MS 50150 f.215
Not exhibited
101, p. 150

BOOKS

Bernhard Siegfried Albinus (1697–1770), Jan Wandelaar (draughtsman), *Tabulae sceleti et musculorum corporis humani*, pl. 4, Leiden: Joan & Herman Verbeek, 1738–1747
Printed book, 70.5 × 52 × 5 cm
Paris, Muséum national d'histoire naturelle
098 GF
P
30, p. 56

Ulisse Aldrovandi (1522–1605), *Ornithologiae hoc est de avibus historiae, Libri XII*, Bologna: Franciscum de Franciscis, 1599–1603
Printed book, 36.5 × 26 × 7 cm
Paris, Muséum national d'histoire naturelle
Fol RES 295
P
16 A and B, p. 39

John James Audubon (1785–1851), "Virginian Partridge," in *Birds of America*, pl. 76, vol. 1, London: Published by the Author, 1827–1838
Printed book, 70 × 105 × 6 cm
Paris, Muséum national d'histoire naturelle
001 GF – 1
P
247, p. 371

Émile Bayard (1837–1891), "Arts of Drawing and Sculpture during the Bronze Age [Les arts du dessin et de la sculpture à l'époque du bronze]," in Louis Figuier (1819–1894), *L'Homme primitif*, Paris: Hachette, 1870
Printed book, 24.5 × 28.5 × 3 cm
Paris, Muséum national d'histoire naturelle
15 558
P
219, p. 314

Karl Blossfeldt (1865–1932), *Urformen der Kunst: Photographische Pflanzenbilder*, pls. 10 and 11, Berlin: Ernst Wasmuth, 1928
Printed book illustrated with rotogravures, 32 × 50 × 1.5 cm
Paris, Bibliothèque du musée d'Orsay
Rés. B 2 BLO 2
P
Not reproduced

Pierre Boitard (1789–1859), "Fossil Man [L'Homme fossile]," in Études antédiluviennes : Paris *avant les hommes*, Paris: Passard, 1861
Printed book, 24 × 17 × 4 cm
Paris, Muséum national d'histoire naturelle
15 577
P
110, p. 159

Charles Bonnet (1720–1793), "Idea for a Scale of Natural Beings [Idée d'une échelle des êtres naturels," in *Traité d'insectologie ou Observations sur les pucerons*, Paris: Durand, 1745
Printed book, 18 × 53 cm
Paris, Muséum national d'histoire naturelle
P
124, p. 189

William Buckland (1784–1856), *Reliquiae Diluvianae; or, Observations on the Organic Remains Contained in Caves, Fissures, and Diluvial Gravel, and on Other Geological Phenomena, Attesting the Action of an Universal Deluge*, London: J. Murray, 1823
Printed book, 29 × 24 × 4 cm
Paris, Muséum national d'histoire naturelle
14 444
P
Not reproduced

Arthur Conan Doyle (1859–1930), *The Lost World*, 1912
Printed book, 16.5 × 12 × 2.2 cm
Paris, Bibliothèque nationale de France
Tolbiac, rez-de-jardin
Z-32977 (4370)
P
Not reproduced

Charles Darwin (1809–1882), "English Pouter," in *The Variation of Animals and Plants under Domestication*, vol. 1, London: J. Murray, 1875
Printed book, 24 × 14 × 4 cm
Paris, Muséum national d'histoire naturelle
4 740 – 1
P
Not reproduced

Charles Darwin (1809–1882), "Feather of Peacock," in *The Descent of Man and Selection in Relation to Sex*, vol. 2, London: J. Murray, 1871
Printed book, 19.5 × 13.5 × 4 cm
Paris, Muséum national d'histoire naturelle
4741-2
P
Not reproduced

Charles Darwin (1809–1882), *The Expression of the Emotions in Man and Animals*, pl. 11, London: J. Murray, 1872
Printed book, 19.5 × 14 × 5 cm
Paris, Muséum national d'histoire naturelle
4 302
P
Not reproduced

Charles Darwin (1809–1882), "The Probable Effects of the Action of Natural Selection through Divergence of Character and Extinction, on the Descendants of a Common Ancestor," in *On the Origin of Species by Means of Natural Selection, or The Preservation of Favoured Races in the Struggle for Life*, London: J. Murray, 1860
Printed book, 20 × 13.5 × 4.5 cm
Paris, Muséum national d'histoire naturelle
4 734
P
127, p. 190

Louis Figuier (1819–1894), Édouard Riou (1833–1900) (illustrations), "Ideal Scene of the Lias with Ichthyosaurus and Plesiosaurus [L'Ichthyosaure et le Plésiosaure (période du Lias)]," in *La Terre avant le Déluge*, 2nd edition, Paris: Hachette, 1863
Printed book, 25 × 17 × 4 cm
Paris, Muséum national d'histoire naturelle
15 573
P
106, p. 153

Robert Fitzroy (1805–1865), "Fuegian," in *Narrative of the Surveying Voyages of His Majesty's Ships* Adventure *and* Beagle *between the Years 1826 and 1836, Describing Their Examination of the Southern Shores of South America and the* Beagle's *Circumnavigation of the Globe*, vol. 2, London: H. Colburn, 1839
Printed book, 23.5 × 15.5 × 6 cm
Paris, Muséum national d'histoire naturelle
21 432 – 2
P
53, p. 75

Camille Flammarion (1842–1925), *Le Monde avant la création de l'homme : origines de la Terre, origines de la vie, origines de l'humanité*, Paris: Marpon et Flammarion, 1886
Printed book, 28 × 20 × 5.5 cm
Paris, Bibliothèque nationale de France
SCT – Sciences et techniques 4 – S6880
P
Not reproduced

Conrad Gessner (1516–1565), *Historia animalium Liber I De quadripedibus viviparis*, Zurich: Christoph Froschauer, 1554
Printed book, hand-coloured, 39 × 25 × 7.5 cm
Paris, Muséum national d'histoire naturelle
Fol Res 87
P
Not reproduced

Conrad Gessner (1516–1565), *Historia animalium Liber I De quadripedibus viviparis*, Zurich: Christoph Froschauer, 1554
Printed book, 39 × 25 × 9 cm
Paris, Muséum national d'histoire naturelle
Fol Res Bn 96
P
Not reproduced

Philip Henry Gosse (1810–1888), *Actinologia Britannica: A History of the British Sea-Anemones and Corals*, pls. 2 and 6, London: Van Voorst, 1860
Printed book, 23 × 16 × 4 cm
Paris, Muséum national d'histoire naturelle
6 491 (pl. 2): 6 491 (pl. 6)
P
73 A and B, p. 106–107

John Gould (1804–1881), *The Birds of Europe*, pl. 243, vol. 4, London: R. and J. E. Taylor, 1837
Printed book, 57 × 40 × 7 cm
Paris, Muséum national d'histoire naturelle
Y3 331 – IV
P
Not reproduced

Ernst Haeckel (1834–1919), *Anthropogenie, oder, Entwickelungsgeschichte des Menschen*, frontispiece, Leipzig, W. Engelmann, 1874
Printed book, 23.5 × 15 × 4 cm
Paris, Muséum national d'histoire naturelle
4 550
P
Not reproduced

Ernst Haeckel (1834–1919), "Die Familiengruppe der Katarrhinen," in *Natürliche Schöpfungsgeschichte*, frontispiece, Berlin: G. Reimer, 1868
Printed book, 23.5 × 15 × 4 cm
Paris, Muséum national d'histoire naturelle
4 752
P
122, p. 183

Ernst Haeckel (1834–1919), *Die Radiolarien: eine Monographie*, pl. 9, vol. 1, Berlin: G. Reimer, 1862
Printed book, 39 × 30 × 3 cm
Paris, Muséum national d'histoire naturelle
105 454 – 4
P
Not reproduced

Ernst Haeckel (1834–1919), *Generelle Morphologie der Organismen*, pl. 1, vol. 2, Berlin: G. Reimer, 1866
Printed book, 25 × 17 × 5 cm
Paris, Muséum national d'histoire naturelle
172 338 – 2
P
128, p. 191

Ernst Haeckel (1834–1919), *Kunstformen der Natur*, Leipzig, Vienna: Bibliographisches Institut, 1899–1904
Anthology of lithographic plates, 36.1 × 27.4 cm
Paris, private collection
PM
142, p. 214 (pl. 31)
144, p. 215 (pl. 1)
145, p. 215 (pl. 88)
150 A, p. 218 (pl. 8)
150 B, p. 219 (pl. 71)

Alexander von Humboldt (1769–1859), *Atlas zu Alex. v. Humboldt's Kosmos: in zweiundvierzig Tafeln mit erläuterndem Texte*, Stuttgart: Verlag von Krais & Hoffmann, 1851
Printed book, 30 × 36 × 3.5 cm
Paris, Muséum national d'histoire naturelle
Y 2 1270
P
Not reproduced

Ferrante Imperato (1550–1631), *Dell'Historia naturale di Ferrante Imperato Napolitano. Libri XXVIII: Nella quale ordinatamente si tratta della diuersa condition di miniere, e pietre, con alcune historie di piante, & animali; sin'hora non date in luce*, Naples: Costantino Vitale, 1599
Printed book, 31 × 22 × 6 cm
Paris, Muséum national d'histoire naturelle
4 RES 134
P
15, p. 39

Athanasius Kircher (1602–1680), "Cross-section of Noah's Ark," in *Arca Noë, in tres libros digesta, quorum I. de rebus quæ ante diluvium, II. de iis, quæ ipso diluvio ejusque duratione, III. de iis, quae post diluvium a Noëmo gesta sunt, quae omnia novâ methodo, nec non summa argumentorum varietate, explicantur, & demonstrantur*, pls. 122–123, Amsterdam: Johannes Janssonius van Waesberge, 1675
Printed book, 37 × 26 × 5 cm
Paris, Muséum national d'histoire naturelle
Y3 79
P
11, p. 31

Jean-Baptiste Lamarck (1744–1829), "Table Showing the Origin of the Various Animals [Tableau servant à montrer l'origine des différens animaux]," in *Philosophie zoologique ou exposition des considérations relatives à l'histoire naturelle des animaux*, Paris: Dentu, 1809
Printed book, 21 × 14.5 × 4.5 cm
Paris, Muséum national d'histoire naturelle
Geol RES Lamarck
P
126, p. 190

Charles Léopold Laurillard (draughtsman) (1783–1853), Couët (engraver), "*Megatherium,*" in Georges Cuvier (1769–1832), *Recherches sur les ossemens fossiles de quadrupèdes où l'on rétablit les caractères de plusieurs espèces d'animaux que les révolutions du globe paraissent avoir détruites*, vol. 1, Paris: Déterville, 1812
Printed book, 23 × 29.5 cm
Paris, Muséum national d'histoire naturelle
16 234 – 1
P
87, p. 126–127

Georges-Louis Leclerc de Buffon (1707–1788), *Histoire naturelle, générale et particulière par Leclerc de Buffon ; nouvelle édition, accompagnée de notes, et dans laquelle les suppléments sont insérés dans le premier texte, à la place qui leur convient*, vols. 12 and 16, Paris: F. Dufart, an VII [1798]–1808
Printed book, 21 × 14 × 4.5 cm
Paris, Muséum national d'histoire naturelle
10 530 – 12 (t. XII), 10 530 – 16 (t. XVI)
P
Not reproduced

Carl Linnaeus (1707–1778), *Systema naturae, per regna tria naturae, secundum classes, ordines, genera, species, cum charateribus, differentiis, synonymis, locis*, Stockholm: Laurent Salvius, 1758–1759
Printed book, 21.5 × 14 × 5.5 cm
Paris, Muséum national d'histoire naturelle
8 RES 224
P
Not reproduced

Carl Linnaeus (1707–1778), "Troglodita Bontii, Lucifer Aldrovandi, Satyrus Tulpii, Pygmaeus Edwardi," in *Amoenitates academicae, seu Dissertationes variæ physicae, medicae, botanicae… accedit hypothesis nova de febrium intermittentium causa*, vol. 6, Leiden: Cornelius Haak, 1749
Printed book, 20 × 13.5 × 4 cm
Paris, Muséum national d'histoire naturelle
Y1 2459-6
P
160, p. 231

Charles Lyell (1797–1875), *Principles of Geology, Being an Attempt to Explain the Former Changes of the Earth's Surface, by Reference to Causes Now in Operation*, London: J. Murray, 1830–1833
Printed book, 22 × 14.5 × 4.5 cm
Paris, Muséum national d'histoire naturelle
15 716
P
88, p. 129

Edmond Perrier (1844–1921), "Sedentary Annelids [Annélides sédentaires]," in *Les Colonies animales et la formation des organismes*, Paris: Masson, 1881
Printed book, 25 × 16.5 × 5.5 cm
Paris, Muséum national d'histoire naturelle
4 749
P
Not reproduced

Élisée Reclus (1830–1905), "Origine," in *L'Homme et la Terre*, vol. 6, Paris: Librairie universelle, 1905–1908
Printed book, 28 × 22.5 × 5 cm
Paris, Muséum national d'histoire naturelle
B 338
P
Not reproduced

Johannes Scheuchzer (1672–1733), *Physica sacra*, pl. 23, Augsburg and Ulm, 1731–1733
Printed book, 41 × 27 × 9 cm
Paris, Muséum national d'histoire naturelle
7577
P
123, p. 185

Jules Verne (1828–1905), Édouard Riou (1833–1900) (illustrations), *Journey to the Centre of the Earth* [*Voyage au centre de la Terre*], chapter heading 1, Paris: Hetzel, 1867
Printed book, 27.8 × 18.6 × 2 cm
Paris, Bibliothèque nationale de France
Réserve des livres rares, Res M-Y2-1000
P
Not reproduced

Jules Verne (1828–1905), Édouard Riou (1833–1900) and Alphonse Marie de Neuville (1836–1885) (illustrations), *Twenty Thousand Leagues Under the Sea* [*Vingt mille lieues sous les mers*], frontispiece, Paris: Hetzel, 1871
Printed book, 28.3 × 19.5 × 3 cm
Paris, Bibliothèque nationale de France
8-RF-48851
P
69 A, p. 102

Jeanne Villepreux-Power (1794–1871), *Observations physiques sur le poulpe de l'Argonauta Argo, commencées en 1832 et terminées en 1843*, Paris: Charles de Mourgues Frères, 1856
Printed book illustrated with watercolours, 30.5 × 24.5 × 1.5 cm
Paris, Muséum national d'histoire naturelle
Fol RES 130
P
Not reproduced

Benjamin Waterhouse Hawkins (1807–1894), "Skeletons of the Gibbon, Orang, Chimpanzee, Gorilla, Man," in Thomas Henry Huxley (1825–1895), *Evidence as to Man's Place in Nature*, London: Williams and Norgate, 1863
Diagram, 23 × 14.5 cm
Paris, Muséum national d'histoire naturelle
5 066
P
118, p. 169

ILLUSTRATED NEWS MAGAZINES

Anonymous, "Dinner in the Iguanodon Model, at the Crystal Palace, Sydenham," in *The Illustrated London News*, January 7, 1854
Printed magazine, 46 × 51 cm
London, The British Library
Add MS 50150, f. 225
Not reproduced

Pierre Boitard (1789–1859), Johann Theodor Susemihl (1772–1847) (draughtsman), "Fossil Man [L'Homme fossile]," in *Le Magasin universel*, vol. 5, April 1838
Printed magazine, 28.6 × 19 cm
Paris, Bibliothèque nationale de France
Tolbiac, rez-de-jardin Z-5260-5266 Z 5264
P
111, p. 159

Philip Henry Delamotte (1821–1889), "Benjamin Waterhouse Hawkins at Work on Dinosaur Models," in *The Illustrated London News*, December 31, 1853
Printed magazine, 28 × 40.5 cm
London, The British Library
Add MS 50150 f.224
Not exhibited
104, p. 151

DOCUMENTS

Georges Cuvier (1769–1832) and Alexandre Brongniart (1770–1847), *Coupe théorique des divers terrains, roches et minéraux qui entrent dans la composition du sol du bassin de Paris*, Paris: Clerget, 1851
Off-print in watercolour, 24 × 32.5 cm
Paris, Muséum national d'histoire naturelle
Pale RES Cuvier
P
83, p. 119

Johann Wolfgang von Goethe (1749–1832), *Esquisse des principales hauteurs des deux continens dressée par Mr. de Göthe, conseiller intime du Duc de Saxe Weimar d'après l'ouvrage de Mr. de Humboldt, publié en 1807 sous le titre d'« Essai sur la géographie des plantes »*, Paris: n.p., 1813
Printed card, 47 × 55 cm
Paris, Muséum national d'histoire naturelle
CA 3 – FA
P
57, p. 80

Aimé Goujaud, known as Bonpland (1773–1858), *Cordia sebestena Cuba*, Humboldt & Bonpland herbarium, equatorial America, n.d.
44 × 28 cm
Paris, Muséum national d'histoire naturelle
MNHN-P-P00670683
P
50 A, p. 73

Aimé Goujaud, known as Bonpland (1773–1858), *Oreopanax*, Humboldt & Bonpland herbarium, equatorial America, n.d.
44 × 28 cm
Paris, Muséum national d'histoire naturelle
MNHN-P-P00136909
P
50 B, p. 73

Aimé Goujaud, known as Bonpland (1773–1858) and Alexander von Humboldt (1769–1859), Schönberger and Turpin (draughtsmen), Bouquet and Beaublé (engravers), *Géographie des plantes équinoxiales. Tableau physique des Andes et pays voisins*, Paris: Langlois, 1805
Printed card, 61 × 91 cm
Paris, Muséum national d'histoire naturelle
CM 5171-FA
P
56, p. 80

Joseph Papoint (1866–1916), *Menu du dîner du diplodocus*, 1908
Printing on paper, 18 × 12 cm
Paris, Muséum national d'histoire naturelle
MS BOU 8
P
Not reproduced

FILMS

Merian C. Cooper (1893–1973), Ernest B. Schoedsack (1893–1979), *King Kong*
United States, 1933
Full-length silent film in black and white, screenplay: James Ashmore Creelman, Ruth Rose from an idea of Merian C. Cooper and Edgar Wallace; music: Max Steiner; cinematography: Edward Linden, J. O. Taylor, Vernon L. Walker; sets: Thomas Little; special effects: Harry Redmond Jr., Willis O'Brien, Marcel Delgado; editor: Ted Cheesman; with Fay Wray, Robert Amstrong, Bruce Cabot, Frank Reicher, Sam Hardy, Noble Johnson, Steve Clemente, Roscoe Ates. Producers: Merian C. Cooper, Ernest B. Schoedsack, David O. Selznick, RKO Radio Pictures.
P
169, p. 245

Laurent Grasso, *Artificialis*, 2020
High-resolution film
P
243, p. 352–353
244, p. 354–355
245, p. 356–357

Harry O. Hoyt (1885–1961), *The Lost World*
United States, 1925
Full-length silent film in black and white, screenplay: Marion Fairfax from Arthur Conan Doyle's novel *The Lost World* (1912); cinemaphotography: Arthur Edeson; sets: Milton Menasco; special effects: Willis O'Brien assisted by Devereaux Jennings; models: Marcel Delgado; editor: George McGuire; with Wallace Beery, Bessie Love, Lloyd Hughes, Lewis Stone, Arthur Hoyt, Alma Bennett, Bull Montana. Producers: Earl Hudson, Watterson R. Rothacker and Jamie White.
PM
Not reproduced

Winsor McCay (1871–1934), *Gertie the Dinosaur*
France, 1914
Short silent film in black and white, live-action shots and animation
P
Not reproduced

Georges Méliès (1861–1938), *The Conquest of the Pole* [*À la conquête du Pôle Nord*]
France, 1912
Short silent film in black and white, screenplay by Georges Méliès freely adapted from the novel by Jules Verne *The Adventures of Captain Hatteras* [*Les Aventures du capitaine Hatteras*] (1866), with Georges Méliès, Fernande Albany. Producers: Georges Méliès, Charles Pathé.
P
80, p. 115

Georges Méliès (1861–1938), *The Doctor and the Monkey* [*Le Savant et le Chimpanzé*]
France, 1900
Short silent film in black and white
P
Not reproduced

Georges Méliès (1861–1938), *The Mermaid* [*La Sirène*]
France, 1904
Short silent film in black and white
P
82, p. 116

Georges Méliès (1861–1938), *Under the Seas* [*Deux Cents Milles sous les mers ou Le Cauchemar du pêcheur*]
France, 1907
Short silent film in black and white, screenplay by Georges Méliès freely adapted from the novel by Jules Verne (1869–1870)
PM
81, p. 116

Steven Spielberg (born in 1946), *Jurassic Park*
United States, 1993
Feature film in colour, screenplay: David Koepp, Michael Crichton from Michael Crichton's novel; music: John Williams; cinematography: Dean Cundey; sound: Gary Rydstrom; art direction: John Bell and William James Teegarden; sets: Jackie Carr; costumes: Sue Moore and Eric H. Sandberg; editor: Michael Kahn; with Sam Neill, Laura Dern, Jeff Goldblum, Richard Attenborough, Bob Peck, Martin Ferrero, Joseph Mazzello, Ariana Richards, Samuel L. Jackson. Producers: Kathleen Kennedy and Gerald R. Molen, Universal Pictures and Amblin Entertainment.
P
Not reproduced

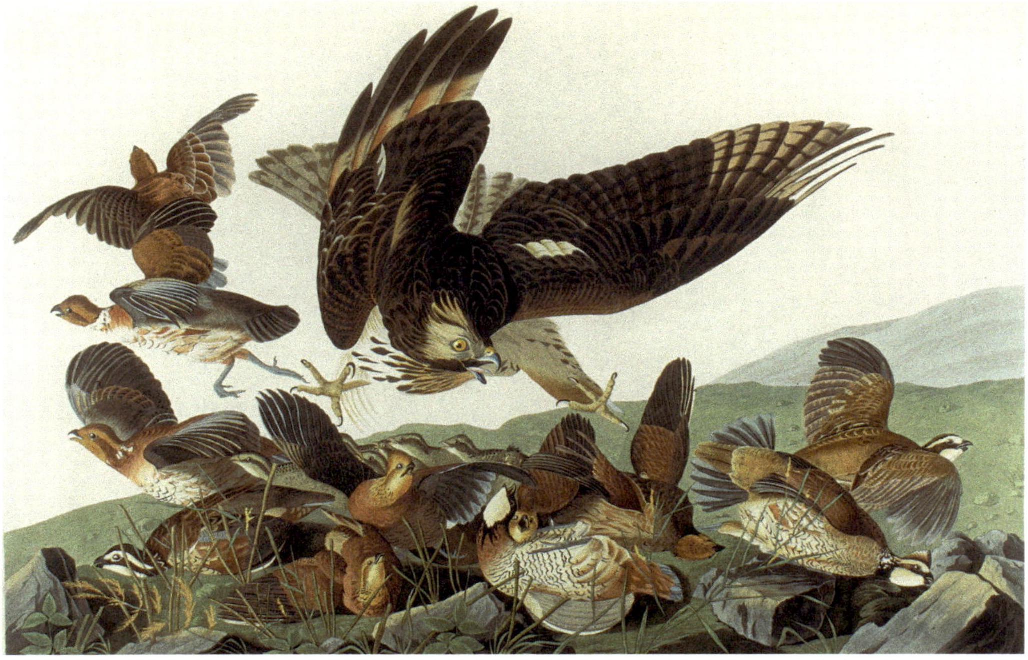

247
John James Audubon,
"Virginian Partridge," in *Birds of America*,
pl. 76, vol. 1, London: Published
by the Author, 1827–1838
Paris, MNHN

A complete list of Charles Darwin's publications can be found at http://darwin-online.org.uk.

Charles Darwin's extensive correspondence is archived at www.darwinproject.ac.uk, along with an abundant body of notes and texts. All of the letters from or to Darwin quoted in this catalogue appear on that site.

A complete collection of Jean-Baptiste Lamarck's theoretical works and the listing of 978 pupils who attended his Muséum lectures from 1795 to 1820 can be found at www.lamarck.cnrs.fr.

BOOKS AND ARTICLES

[Anonymous]. *Catalogue des fruits et des plantes modelés composant le Carporama*. [Paris], n.d.

Albinus, Bernhard Siegfried. *Tabulae sceleti et musculorum corporis humani*. Leiden: Johan & Herman Verbeek, 1747.

Aldrovandi, Ulisse. *Ornithologiae hoc est de avibus historiae, Libri XII*. Bologna: Franciscum de Franciscis, 1599–1603.

Allorge, Lucille and Olivier Ikor. *La Fabuleuse Odyssée des plantes. Les botanistes voyageurs, les jardins des plantes, les herbiers*. Paris: Jean-Claude Lattès, 2003.

Althaus, Karin. *Gabriel von Max. Von ekstatischen Frauen und Affen im Salon*. Munich: Schirmer Mosel Verlag, 2018.

Anselm, Marie-Christine, Anne Forray-Carlier and Nicole Garnier-Pelle. *Singeries et exotisme chez Christophe Huet*. Saint-Rémy-en-l'Eau: Éditions Monelle Hayot, 2010.

Archibald, J. David. *Aristotle's Ladder, Darwin's Tree: The Evolution of Visual Metaphors for Biological Order*. New York: Columbia University Press, 2014.

Argot, Christine and Luc Vivès. *Un jour avec les dinosaures*. Paris: Muséum national d'histoire naturelle/ Flammarion, 2018.

Aristotle. "Parts of Animals." In *The Complete Works of Aristotle: The Revised Oxford Translation*. Edited by Jonathan Barnes. Vol. 1. Princeton: Princeton University Press, 2014.

Audubon, John James. *Birds of America*. London: Published by the Author, 1827–1838.

Aymonin, Gérard. "Les tableaux de végétaux tropicaux de Michel Garnier, 1753–1819." In "Mahé de La Bourdonnais. La Compagnie des Indes dans l'océan Indien," special issue, *Cahiers de la Compagnie des Indes*, no. 4 (1999), p. 103–108.

Bacou, Roseline. "La décoration d'Odilon Redon pour le château de Domecy, 1900–1901." *La Revue du Louvre et des musées de France*, vol. 42, no. 2 (1992), p. 43.

Baglione, Gabrielle and Cédric Cremière. *Charles-Alexandre Lesueur, peintre voyageur. Un trésor oublié*. Paris: Éditions De Conti, 2009.

Barsanti, Giulio. "Storia naturale delle scimmie, 1600–1800." *Nuncius. Annali di storia della scienza*, vol. 5, no. 2 (1990), p. 99–165.

Barsanti, Giulio. *La scala, la mappa, l'albero: immagini e classificazioni della natura fra Sei e Ottocento*. Florence: Sansoni, 1992.

Beaurepaire, Pierre-Yves. *Les Lumières et le Monde. Voyager, explorer, collectionner*. Paris: Belin, 2019.

Bedell, Rebecca. *The Anatomy of Nature: Geology and American Landscape Painting, 1825–1875*. Princeton: Princeton University Press, 2001.

Bellon, Richard. "Inspiration in the Harness of Daily Labor: Darwin, Botany and the Triumph of Evolution, 1859–1868." *Isis*, vol. 102, no. 3 (2011), p. 392–420.

Bendyshe, Thomas. "The History of Anthropology." In *Memoirs Read before the Anthropological Society of London, 1863–1864*. Vol. 1, p. 421–458. London: Trübner, 1865.

Benforado Bakewell, Susan and Jane Kromm, eds. *A History of Visual Culture: Western Civilization from the 18th to the 21st Century*. Oxford: Berg, 2010.

Bianciotto, Gabriel, ed. *Bestiaires du Moyen Âge*. Paris: Stock, 1980.

Binet, René. *Esquisses décoratives*. Preface by Gustave Geffroy. Paris: Librairie centrale des beaux-arts, 1903.

Blair, Ann. *The Theater of Nature: Jean Bodin and Renaissance Science*. Princeton: Princeton University Press, 1997.

Blais, Hélène. *Voyages au Grand Océan. Géographies du Pacifique et colonisation, 1815–1845*. Paris: Comité des travaux historiques et scientifiques, 2005.

Blanckaert, Claude, Claudine Cohen, Pietro Corsi and Jean-Louis Fischert, eds. *Le Muséum au premier siècle de son histoire*. Paris: Muséum national d'histoire naturelle, 1997.

Blanckaert, Claude. "Frontières de l'humanité. Le 'satyre' des Lumières entre science et fiction critique." In "L'homme et la nature," special issue, *Histoire et anthropologie*, no. 25 (2002), p. 13–32.

Blanckaert, Claude. "'Produire l'être singe.' Langage du corps et harmonies spirituelles." *Annales historiques de la Révolution française*, no. 377 (2014), p. 9–35.

Blossfeldt, Karl. *Urformen der Kunst: Photographische Pflanzenbilder*. Berlin: Ernst Wasmuth, 1928. Translated as *Art Forms in the Plant World: 120 Full-Page Photographs*. Introduction by Karl Nierendorf. New York: Dover Publications, 1985.

Blunt, Wilfrid. *The Art of Botanical Illustration: An Illustrated History* [1950]. Updated and revised by William Stearn. Woodbridge: Antique Collectors' Club; London (Kew): The Royal Botanic Gardens, 1994.

Boitard, Pierre. "Études d'histoire naturelle. Paris avant les hommes." *Musée des familles – Lectures du soir*, 1st article in no. 2 (June 1836), p. 257–281; 2nd article in no. 5 (November 1837), p. 45–63.

Boitard, Pierre. "L'homme fossile. Étude paléontologique." *Le Magasin universel*, vol. 5 (1838), p. 240.

Boitard, Pierre. *Paris avant les hommes* [1861]. Paris: Passard, 1961.

Bonnemains, Jacqueline, Elliott Forsyth and Bernard Smith, eds. *Baudin in Australian Waters: The Artwork of the French Voyage of Discovery to the Southern Lands, 1800–1804*. Melbourne: Oxford University Press, 1988.

Bonnet, Charles. *Œuvres d'histoire naturelle et de philosophie*. 8 vols. Neuchâtel: Samuel Fauche, 1779–1783.

Bonneuil, Christophe and Jean-Baptiste Fressoz. *The Shock of the Anthropocene: The Earth, History and Us*. Translated by David Fernbach. London: Verso, 2016.

Bossi, Laura. *Histoire naturelle de l'âme*. Paris: Presses universitaires de France, 2003.

Bossi, Laura. "Fantaisies phylogénétiques." In *Freud. Du regard à l'écoute*, exh. cat. edited by Jean Clair, p. 38–49. Paris: Musée d'art et d'histoire du judaïsme/ Gallimard, 2018.

Bossi, Laura. "Après Darwin, Haeckel." In *Darwin au Collège de France*, edited by Antoine Compagnon and Céline Surprenant, p. 123–147. Paris: Collège de France, "Passage des disciplines" series, 2020. Also available online at https://books.openedition.org/cdf/7356.

Bossi, Laura. *Histoire naturelle de la création. Ernst Haeckel, 1834–1919*. Paris: Les Belles Lettres, forthcoming.

Boucher de Perthes, Jacques. *Antiquités celtiques et antédiluviennes*. Paris: Treuttel et Würtz, 1847–1864.

Bourguet, Marie-Noëlle. *Le Monde dans un carnet. Alexander von Humboldt en Italie* [1805]. Paris: Éditions du Félin, 1999.

Bramwell, Valerie and Robert McCracken Peck. *All in the Bones: A Biography of Benjamin Waterhouse Hawkins*. Philadelphia: Academy of Natural Sciences of Philadelphia, 2008.

Brauer, Fae and Anthea Callen, eds. *Art, Sex and Eugenics: Corpus Delecti*. Aldershot: Ashgate, 2008.

Brauer, Fae and Serena Keshavie, eds. *Picturing Evolution and Extinction. Regeneration and Degeneration in Modern Visual Culture*. Newcastle upon Tyne: Cambridge Scholars Publishing, 2015.

Braun, Emily. "Ornament as Evolution: Gustav Klimt and Berta Zuckerkandl." In *Gustav Klimt: The Ronald S. Lauder and Serge Sabarsky Collections*, exh. cat. edited by Renée Price. New York: Neue Galerie; Munich: Prestel, 2007.

Bredekamp, Horst. *Darwins Korallen. Die frühen Evolutionsdiagramme und die Tradition der Naturgeschichte*. Berlin: Wagenbach, 2005.

Breidbach, Olaf and Robert Proctor. *René Binet: Natur und Kunst*. Munich: Prestel, 2007. Translated as *René Binet: From Nature to Form*. Munich and London: Prestel, 2007.

Breidbach, Olaf and Irenäus Eibl-Eibesfeldt, eds. *Ernst Haeckel. Kunstformen der Natur. Kunstformen aus dem Meer*. Munich: Prestel, 2012.

Browne, Janet. "Darwin in Caricature: A Study in the Popularisation and Dissemination of Evolution." *Proceedings of the American Philosophical Society*, vol. 145, no. 4 (December 2001), p. 496–509.

Browne, Janet. *Charles Darwin: A Biography*. Vol. 1: *Voyaging*. London: Jonathan Cape, 1995; Princeton: Princeton University Press, 1996. Vol. 2: *The Power of Place*. London: Jonathan Cape, 2002; Princeton: Princeton University Press, 2003.

Brunner, Bernd. *The Ocean at Home: An Illustrated History of the Aquarium*. New York: Princeton Architectural Press, 2005.

Buch, Leopold von. *Description physique des îles Canaries* [1825]. Translated by Charles Boulanger. Paris: Levrault, 1836.

Buc'hoz, Pierre-Joseph. *Le Jardin d'Éden. Le paradis terrestre renouvelé dans le jardin de la reine à Trianon*. 2 vols. Paris: Chez l'auteur, 1783–1785.

Buckland, William. *Reliquiae Diluvianae*. London: John Murray, 1823.

Buffon, Georges-Louis Leclerc, comte de. *Histoire naturelle, générale et particulière, avec la description du Cabinet du Roy*. 45 vols. Paris: Imprimerie royale, 1749–1804.

Buffon, Georges-Louis Leclerc, comte de. *Les Époques de la Nature* [1778]. Critical edition with manuscript facsimile, annotated by Jacques Roger [1962]. Paris: Éditions du Muséum, 1988.

Buffon, Georges-Louis Leclerc, comte de. *Œuvres complètes. Histoire naturelle, générale et particulière* [1749–1789]. Edited by Stéphane Schmitt and Cédric Crémière. 14 vols. (38 expected). Paris: Honoré Champion, 2007–2020.

Burkhardt, Richard W. *The Spirit of System: Lamarck and Evolutionary Biology*. Cambridge, MA: Harvard University Press, 1977; new edition 1995.

Burkhardt, Richard W. "Lamarck, Evolution, and the Inheritance of Acquired Characters." *Genetics*, vol. 194, no. 4 (2013), p. 793–805.

Burnet, Thomas. *Telluris theoria sacra*. 2 vols. London: Kettilby, 1681.

Cailleux, André. "Progression du nombre d'espèces de plantes décrites de 1500 à nos jours." *Revue d'histoire des sciences et de leurs applications*, vol. 6, no. 1 (1953), p. 42–49.

Call, Michael J. *Claude Monet, Free Thinker. Radical Republicanism, Darwin's Science, and the Evolution of Impressionist Aesthetics*. New York: Peter Lang, "American University Studies" series, 2015.

Candolle, Augustin-Pyramus de. *Mémoires et souvenirs* [1862]. Edited by Jean-Daniel Candaux and Jean-Marc Drouin. Geneva: Georg Éditeur, 2003.

Carson, Rachel. *Silent Spring*. Boston: Houghton Mifflin, 1962.

Carus, Carl Gustav. *Nine Letters on Landscape Painting* [1831]. Translated by David Britt. Los Angeles: Getty Publications, 2002.

Céard, Jean. *La Nature et les prodiges. L'insolite au XVIe siècle en France* [1977]. Geneva: Droz, 1996.

Chamberlin, J. Edward and Sander Gilman, eds. *Degeneration. The Dark Side of Progress*. New York: Columbia University Press, 1985.

Chambers, Robert. *Vestiges of the Natural History of Creation*. London: John Churchill, 1844.

Ciancio, Luca. *Le colonne del tempo. Il "Tempio di Serapide" a Pozzuoli nella storia della geologia, dell'archeologia e dell'arte, 1750–1900*. Florence: Edifir, Le voci del Museo, 2009.

Clark, Kenneth. *Animals and Men: Their Relationship as Reflected in Western Art from Prehistory to the Present Day*. London: Thames and Hudson, 1977.

Clarke, T. H. *The Rhinoceros from Dürer to Stubbs, 1515–1799*. London: Sotheby's Publications, 1986.

Cohen, Claudine. *L'Homme des origines. Savoirs et fictions en préhistoire*. Paris: Seuil, 1999.

Cohen, Claudine and Jean-Jacques Hublin. *Boucher de Perthes. Les origines romantiques de la préhistoire*. Paris: Belin, 1989; new revised edition, "Alpha" series, 2017.

Comar, Philippe. *Faites à peindre*. Paris: L'Échoppe, 2014.

Comestor, Petrus. *Bible historiale*. [ca. 1400–1425].

Compagnon, Antoine and Céline Surprenant, eds. *Darwin au Collège de France*. Paris: Collège de France, "Passage des disciplines" series, 2020. Also available online at https://books.openedition.org/cdf/7744.

Conry, Yvette. *L'Introduction du darwinisme en France au XIXe siècle*. Paris: Vrin, 1974.

Cook, E. T. and Alexander Wedderburn, eds., *The Works of John Ruskin*. The Library Edition. 39 vols. London: George Allen; New York: Longmans, Green, and Co, 1903–1912.

Corbin, Alain. "L'hérédosyphilis ou l'impossible rédemption. Contribution à l'histoire de l'hérédité morbide." *Romantisme. Revue du dix-neuvième siècle*, no. 31 (1981).

Cordulack, Shelley Wood. *Edvard Munch and the Physiology of Symbolism*. Madison, NJ: Fairleigh Dickinson University Press, 2002.

Corsi, Pietro. *The Age of Lamarck: Evolutionary Theories in France, 1790–1830*. Berkeley: University of California Press, 1988; and the new expanded edition, *Lamarck. Genèse et enjeux du transformisme, 1770–1830*. Paris: CNRS Éditions, 2001.

Corsi, Pietro. "Before Darwin: Transformist Concepts in European Natural History." *Journal of the History of Biology*, vol. 38, no. 1 (2005), p. 67–83.

Corsi, Pietro. "The Revolutions of Evolution: Geoffroy and Lamarck, 1825–1840." *Bulletin du Musée d'Anthropologie préhistorique de Monaco*, no. 51 (2011), p. 97–122.

Corsi, Pietro. "Systèmes de la Nature and Theories of Life: Bridging the Eighteenth and the Nineteenth Centuries." *Republics of Letters*, vol. 6, no. 1 (2018), https://arcade.stanford.edu/rofl/systèmes-de-la-nature-and-theories-life-bridging-eighteenth-and-nineteenth-centuries.

Creed, Barbara. *Darwin's Screens: Evolutionary Aesthetics, Time and Sexual Display in the Cinema*. Melbourne: Melbourne University Press, 2009.

Cronin, Helena. *The Ant and the Peacock: Altruism and Sexual Selection from Darwin to Today*. Cambridge: Cambridge University Press, 1993.

Cronon, William, ed. *Uncommon Ground: Rethinking the Human Place in Nature*. New York: W. W. Norton, 1996.

Crutzen, Paul J. and Eugene F. Stoermer. "The 'Anthropocene.'" *IGBP Newsletter*, no. 41 (2000), p. 17–18.

Cuvier, Georges and Alexandre Brongniart. *Essai sur la géographie minéralogique des environs de Paris*. Paris: Baudouin, 1811.

Cuvier, Georges. *Recherches sur les ossemens fossiles de quadrupèdes, où l'on rétablit les caractères de plusieurs espèces d'animaux que les révolutions du globe paroissent avoir détruites*. 4 vols. Paris: Déterville, 1812.

Cuvier, Georges. *Discours sur les révolutions de la surface du globe et sur les changements qu'elles ont produits dans le règne animal*. Paris: G. Dufour et E. d'Ocagne, 1826. Translated by Ian Johnston as *Discourse on the Revolutionary Upheavals on the Surface of the Earth*. Arlington: Richer Resources, 2009.

Dardaud, Gabriel. *Une girafe pour le roi*. Preface by Georges Poisson. Paris: Dumerchez, 1985.

Darwin, Charles. *Narrative of the Surveying Voyages of HMS* Adventure *and* Beagle. Vol. 3: *Journal and Remarks, 1832–1836*. London: Henry Colburn, 1839.

Darwin, Charles. *On the Origin of Species by Means of Natural Selection, or The Preservation of Favoured Races in the Struggle for Life*. London: John Murray, 1859.

Darwin, Charles. *On the Various Contrivances by Which British and Foreign Orchids Are Fertilised by Insects*. London: John Murray, 1862.

Darwin, Charles. *The Variation of Animals and Plants under Domestication*. London: John Murray, 1868.

Darwin, Charles. *The Descent of Man, and Selection in Relation to Sex*. 2 vols. London: John Murray, 1871.

Darwin, Charles. *The Expression of the Emotions in Man and Animals*. London: John Murray, 1872.

Darwin, Charles. *The Power of Movement in Plants*. London: John Murray, 1880.

Darwin, Charles. *The Formation of Vegetable Mould, Through the Action of Worms*. London: John Murray, 1881.

Darwin, Erasmus. *Zoonomia; Or the Laws of Organic Life*. 2 vols. London: Johnson, 1794–1796.

Daston, Lorraine and Katherine Park. *Wonders and the Order of Nature, 1150–1750*. New York: Zone Books, 2001.

Daugeron, Bertrand. *Collections naturalistes, entre science et empires, 1763–1804*. Paris: Muséum national d'histoire naturelle, 2009.

Debourdieau, Ariane, ed. *Les Grands Textes fondateurs de l'écologie*. Paris: Flammarion, 2013.

De La Beche, Henry Thomas. *Researches in Theoretical Geology*. London: Knight, 1834.

Deléage, Jean-Paul. *Histoire de l'écologie. Une science de l'homme et de la nature*. Paris: La Découverte, 1991.

Delumeau, Jean. *Une histoire du paradis*. Vol. 1: *Le Jardin des délices*. Paris: Fayard, 1992.

des Cars, Laurence. "Gustave Courbet : L'Origine du monde." In *De l'impressionnisme à l'Art nouveau : acquisitions du musée d'Orsay, 1990, 1996*, exh. cat. edited by Henri Loyrette and Marc Bascou, p. 25–35. Paris: Réunion des musées nationaux, 1996.

Descola, Philippe, ed. *Les Natures en question*. Paris: Odile Jacob, 2018.

Desmond, Adrian J. *Archetypes and Ancestors: Palaeontology in Victorian London, 1850–1875*. London, Blond & Briggs, 1982.

Desmond, Adrian and James Moore. *Darwin*. London: Penguin Books, 1992.

Desmond, Adrian and James Moore. *Darwin's Sacred Cause: Race, Slavery and the Quest for Human Origins*. Chicago: University of Chicago Press, 2009.

Di Gregorio, Mario. *From Here to Eternity: Ernst Haeckel and Scientific Faith*. Göttingen: Vandenhoeck & Ruprecht, 2005.

Dobai, Johannes. "Gustav Klimt's 'Hope I.'" *National Gallery of Canada Bulletin*, no. 17 (1971), p. 2–15, 33.

Donald, Diana. *Picturing Animals in Britain 1750–1850*. New Haven: Yale University Press, 2007.

Doyle, Arthur Conan. *The Lost World*. London: Hodder & Stoughton, 1912.

Drahos, Alexis. *Orages et tempêtes, volcans et glaciers. Les peintres et les sciences de la Terre aux XVIIIe et XIXe siècles*. Paris: Hazan, 2014.

Drouin, Jean-Marc. *Réinventer la nature. L'écologie et son histoire*. Paris: Desclée de Brouwer, 1991.

Drouin, Jean-Marc. *L'Herbier des philosophes*. Paris: Seuil, 2008.

Du Chaillu, Paul. *Explorations and Adventures in Equatorial Africa*. London: John Murray, 1861.

du Crest, Sabine and Frank Lestringant, eds. *Le Théâtre de la curiosité, XVIe–XVIIe siècle*. Proceedings of the international colloquium (Paris, Centre V. L. Saulnier, Université de la Sorbonne, March 8, 2007). Paris: Presses de la Sorbonne, 2008.

du Crest, Sabine. *L'Art de vivre ensemble. Objets frontière de la Renaissance au XXIe siècle*. Rome: Gangemi, 2017.

Ducros, Albert and Jacqueline Ducros, eds. *L'Homme préhistorique. Images et imaginaire*. Paris: L'Harmattan, 2000.

Durand, Ariane. "Odilon Redon et Armand Clavaud ou les étroits contacts entre la botanique, l'art et la philosophie." *Revue historique de Bordeaux*, no. 24 (1975), p. 149–182.

Duviols, Jean Paul and Charles Minguet. *Humboldt savant-citoyen du monde*. Paris: Gallimard, "Découvertes Gallimard" series, 1994.

Egerton, Frank N. *A History of the Ecological Sciences*. Parts 1 to 57. *Bulletin of the Ecological Society of America* (2001–2016), http://esapubs.org/bulletin/current/history_links_list.htm.

Engell, James. *The Creative Imagination: Enlightenment to Romanticism*. Cambridge, MA: Harvard University Press, 1981.

Engelmeier, Hanna. *Der Mensch, der Affe. Anthropologie und Darwin-Rezeption in Deutschland 1850–1900*. Cologne: Böhlau, 2016.

Engels, Eve-Marie and Thomas F. Glick, eds. *The Reception of Charles Darwin in Europe*. London: Continuum, 2008.

Figuier, Louis. *La Terre avant le Déluge*. Paris: Hachette, 1863.

Figuier, Louis. *L'Homme primitif*. Paris: Hachette, 1870.

Findlen, Paula. *Possessing Nature: Museums, Collecting, and Scientific Culture in Early Modern Italy*. Berkeley: University of California Press, 1994.

Findlen, Paula, ed. *Athanasius Kircher: The Last Man Who Knew Everything*. New York: Routledge, 2004.

Fisler, Marie and Guillaume Lecointre. "Categorizing Ideas about Trees: A Tree of Trees." *PLoS ONE*, vol. 8, no. 8 (August 7, 2013), https://doi.org/10.1371/journal.pone.0068814.

Fisler, Marie, Cédric Crémière, Pierre Darlu and Guillaume Lecointre. "The Treeness of the Tree of Historical Trees of Life." *PLoS ONE*, vol. 15, no. 1 (January 15, 2020), https://doi.org/10.1371/journal.pone.0226567.

Flammarion, Camille. *Dieu dans la nature*. Vol. 1. Paris: Ernest Flammarion, 1866.

Flammarion, Camille. *Le Monde avant la création de l'homme. Origines de la Terre, origines de la vie, origines de l'humanité*. Paris: Marpon et Flammarion, 1886.

Fonvielle, Wilfrid de. *Les Merveilles du monde invisible*. Paris: Hachette, 1866.

Fournier, Edmond. *Stigmates dystrophiques de l'hérédo-syphilis*. Paris: Rueff, 1898.

Francis, Mark. *Herbert Spencer and the Invention of Modern Life*. Ithaca: Cornell University Press, 2007.

Frängsmyr, Tore, ed. *Linnaeus: The Man and His Work*. Berkeley: University of California Press, 1983.

Freud, Sigmund. *Civilization and Its Discontents* [1930]. Translated by James Strachey. New York: W. W. Norton, 1961.

Freud, Sigmund. *Essais de psychanalyse appliquée*. Translated by Marie Bonaparte and E. Marty. Paris: Gallimard, 1933.

Galton, Francis. *Inquiries into Human Faculty and Its Development*. London: Macmillan, 1883.

Gamboni, Dario. *La Plume et le Pinceau. Odilon Redon et la littérature*. Paris: Minuit, 1989. Translated by Mary Whittall as *The Brush and the Pen: Odilon Redon and Literature*. Chicago: University of Chicago Press, 2011.

Gamwell, Lynn. *Exploring the Invisible: Art, Science and the Spiritual*. Princeton: Princeton University Press, 2002.

Gauchet, Marcel. *Le Désenchantement du monde. Une histoire politique de la religion*. Paris: Gallimard, 1985.

Gayon, Jean, ed. *Buffon 88*. Proceedings of the international symposium for the bicentennial of Buffon's death (Paris, Montbard, Dijon, June 14–22, 1988). Paris: Vrin; Lyon: Institut interdisciplinaire d'études épistémologiques, 1992.

Gayon, Jean. *Darwinism's Struggle for Survival*. Cambridge: Cambridge University Press, 1998.

Gayon, Jean. "Mort ou persistance du darwinisme ? Regard d'un épistémologue." *Comptes rendus Palevol*, no. 8 (2009), p. 321–340.

Geoffroy Saint-Hilaire, Étienne. "Paléontographie. Considérations sur les ossements fossiles la plupart inconnus, trouvés et observés dans les bassins de l'Auvergne." *Revue encyclopédique*, vol. 59 (July–September 1833), p. 76–95.

Geoffroy Saint-Hilaire, Étienne. "Études sur l'orang-outang de la Ménagerie." *Comptes rendus de l'Académie des sciences*, vol. 3 (1836), p. 1–8.

Geoffroy Saint-Hilaire, Isidore. "Description des mammifères nouveaux ou imparfaitement connus de la collection du Muséum d'histoire naturelle. Quatrième mémoire. Famille des singes." *Archives du Muséum d'histoire naturelle*, vol. 10 (1858–1861), p. 78–81.

Geoffroy Saint-Hilaire, Isidore. *Histoire naturelle générale des règnes organiques, principalement étudiée chez l'homme et les animaux*. Vol. 2. Paris: Masson, 1859.

Gere, Charlotte. *Artistic Circles: Design and Decoration in the Aesthetic Movement*. London: V&A Publishing, 2010.

Gessner, Konrad. *Historiae animalium*. 5 vols. Zurich: Christoph Froschauer, 1551–1585.

Glacken, Clarence J. *Traces on the Rhodian Shore: Nature and Culture in Western Thought from Ancient Times to the End of the Eighteenth Century*. Berkeley: University of California Press, 1967.

Glick, Thomas F., ed. *The Comparative Reception of Darwinism*. Chicago: University of Chicago Press, 1988.

Glick, Thomas F. and Elinor Schaffer, eds. *The Literary and Cultural Reception of Charles Darwin in Europe*. London: Bloomsbury, 2014.

Goethe, Johann Wolfgang von. "Fragment über die Natur." [*Tiefurter Journal*, 1783]. In *Goethes Werke, Hamburger Ausgabe*. Vol. 13: *Naturwissenschaftliche Schriften I*, p. 45. Munich: Ch. Beck, 1998.

Goethe, Johann Wolfgang von. *Versuch, die Metamorphose der Pflanzen zu erklären*. Gotha: Ettinger, 1790. Translated by Douglas Miller as *The Metamorphosis of Plants*. Introduction and Photography by Gordon L. Miller. Cambridge, MA: MIT Press, 2009.

Gosse, Philip Henry. *A Naturalist's Rambles on the Devonshire Coast*. London: Van Voorst, 1853.

Gosse, Philip Henry. *Actinologia Britannica: A History of the British Sea-Anemones and Corals*. London: Van Voorst, 1860.

Gould, John. *The Birds of Europe*. 5 vols. London: Richard and John E. Taylor for J. Gould, 1832–1837.

Gould, Stephen Jay. *Ontogeny and Phylogeny*. Cambridge, MA: Harvard University Press, 1977.

Gould, Stephen Jay. *Time's Arrow, Time's Cycle: Myth and Metaphor in the Discovery of Geological Time*. Cambridge, MA: Harvard University Press, 1987.

Gould, Stephen Jay. *Rocks of the Ages: Science and Religion in the Fullness of Life*. New York: Ballantine Books, 1999.

Goy, Jacqueline. *Les Méduses de Péron et Lesueur. Un autre regard sur l'expédition Baudin*. Paris: Comité des Travaux historiques et scientifiques, 1995.

Grandordy, Béatrice. *Charles Darwin et l'évolution dans les arts plastiques de 1859 à 1914*. Paris: L'Harmattan, 2012.

Grandville, Jean-Jacques. *Les Fleurs animées*. Introduction by Alphonse Karr. Text by Taxile Delord. 2 vols. Paris: Gabriel de Gonnet, 1847.

Grigson, Caroline. *Menagerie. The History of Exotic Animals in England*. Oxford: Oxford University Press, 2016.

Guerrini, Anita. *The Courtiers' Anatomists. Animals and Humans in Louis XIV's Paris*. Chicago: Chicago University Press, 2015.

Gusdorf, Georges. *Le Romantisme*. 2 vols. Paris: Payot & Rivages, 1993.

Hadot, Pierre. *Le Voile d'Isis. Essai sur l'histoire de l'idée de nature*. Paris: Gallimard, "Folio Essais" series, 2008.

Haeckel, Ernst. *Die Radiolarien: Eine Monographie*. 2 vols. Berlin: G. Reimer, 1862.

Haeckel, Ernst. *Generelle Morphologie der Organismen*. 2 vols. Berlin: G. Reimer, 1866.

Haeckel, Ernst. *Natürliche Schöpfungsgeschichte*. Berlin: G. Reimer, 1868. Translated as *The History of Creation: Or the Development of the Earth and Its Inhabitants by the Action of Natural Causes*. New York: D. Appleton, 1876.

Haeckel, Ernst. *Anthropogenie*. Leipzig: W. Engelmann, 1872.

Haeckel, Ernst. *Der Monismus als Band zwischen Religion und Wissenschaft. Glaubensbekenntniss eines Naturforschers*. Bonn: Emil Strauss, 1892. Translated by J. Gilchrist as *Monism as Connecting Religion and Science: The Confession of Faith of a Man of Science*. London: Adam and Charles Black, 1895.

Haeckel, Ernst. *Die Welträthsel. Gemeinverständliche Studien über monistische Philosophie*, Bonn, E. Strauß, 1899.

Haeckel, Ernst. *Kunstformen der Natur*. Leipzig and Vienna: Bibliographisches Institut, 1899–1904. Reprint, Munich: Prestel, 1998.

Hamy, Ernest-Théodore. "Les peintures de Michel Garnier au Muséum d'histoire naturelle." *Bulletin du Muséum d'histoire naturelle*, vol. 4, no. 8 (1898), p. 336–344.

Hawkins, Thomas. *The Book of the Great Sea Dragons*. London: William Pickering, 1840.

Heurtel, Pascale and Michelle Lenoir. *Les Vélins du Muséum d'histoire naturelle de Paris*. Paris: Citadelles et Mazenod, 2016.

Hodge, Jonathan and Gregory Radick, eds. *The Cambridge Companion to Darwin*. Cambridge: Cambridge University Press, 2009.

Hogarth, William. *The Analysis of Beauty*. London: J. Reeves, 1753.

Holy Bible: King James Version. Peabody, MA: Hendrickson, 2011.

Hoquet, Thierry, ed. *Les Fondements de la botanique. Linné et la classification des plantes*. Paris: Vuibert, 2005.

Hoquet, Thierry. *Buffon illustré, les gravures de l'Histoire naturelle, 1749–1767*. Paris: Muséum national d'histoire naturelle, 2007.

Hoquet, Thierry. *Buffon/Linné, éternels rivaux de la biologie ?* Paris: Dunot, 2007.

Hoquet, Thierry. *Darwin contre Darwin. Comment lire* L'Origine des espèces. Paris: Seuil, 2009.

Hoquet, Thierry, ed. *Current Perspectives on Sexual Selection: What's Left after Darwin?* Dordrecht, the Netherlands: Springer, 2015.

Houel, Jean-Pierre-Louis-Laurent. *Histoire naturelle des deux éléphans* [sic], *mâle et femelle, du Muséum de Paris, venus de Hollande en France en l'an VI.* Paris: Chez l'auteur, 1803.

Hufnagel, Henning, Frank Jäger and Nicolas Wanlin. "Une approche pluridisciplinaire d'Ernst Haeckel." In "Ernst Haeckel entre science et esthétique," special issue, *Arts et Savoirs*, no. 9 (2018), https://journals.openedition.org/aes/1086.

Humboldt, Alexander von. *Ansichten der Natur mit wissenschaftlichen Erläuterungen*. Stuttgart: J. G. Cotta, 1808. Translated by Elizabeth Sabine as *Aspects of Nature, in Different Lands and Different Climates; with Scientific Elucidations.* 2 vols. London: Longman, Brown, Green, and Longmans, 1849.

Humboldt, Alexander von. *Personal Narrative of Travels to the Equinoctial Regions of the New Continent, During the Years 1799–1804. By Alexander de Humboldt and Aimé Bonpland; with Maps, Plans, &c.* Translated by Helen Maria Williams. 7 vols. London: Longman, Hurst, Rees, Orme, and Brown, 1814–1829.

Humboldt, Alexander von. *Kosmos. Entwurf einer physischen Weltbeschreibung*. 4 vols. Stuttgart and Tübingen: Cotta, 1845–1862. Translated by E. C. Otté and B. H. Paul as *Cosmos: A Sketch of a Physical Description of the Universe.* New York: Harper and Brothers, 1863–1864.

Humboldt, Alexander von and Aimé Bonpland. *Essay on the Geography of Plants* [1807]. Edited with an introduction by Stephen T. Jackson. Translated by Sylvie Romanowski. Chicago: University of Chicago Press, 2009.

Hurel, Arnaud. *La France préhistorienne de 1789 à 1941.* Paris: CNRS Éditions, 2007.

Hurel, Arnaud and Noël Coyé. *Dans l'épaisseur du temps. Archéologues et géologues inventent la préhistoire*. Paris: Muséum national d'histoire naturelle, 2011.

Huret, Jules. *Enquête sur l'évolution littéraire*. Paris: Charpentier, 1891. Reprint, Paris: José Corti, 1999.

Hutton, James. *Theory of the Earth.* Transactions of the Royal Society of Edinburgh. 1788.

Huxley, Thomas Henry. *Evidence as to Man's Place in Nature*. London: Williams and Norgate, 1863.

Huysmans, Joris-Karl. *Écrits sur l'art, 1867–1905.* Edited by Patrice Locmant. Paris: Bartillat, 2006.

Imperato, Ferrante. *Dell'historia naturale di Ferrante Imperato Napolitano. Libri XXVIII: Nella quale ordinatamente si tratta della diversa conditione di miniere, e pietre. Con alcune historie di piante, & animali, sin'hora non date in luce.* Naples: Costantino Vitale [Stamperia Stigliola], 1599.

Impey, Olivier and Arthur MacGregor. *The Origins of Museums: The Cabinet of Curiosities in Sixteenth- and Seventeenth-Century Europe*. Oxford: Ashmolean Museum, 1985. Reprint, 2017.

Janson, Horst Woldemar. *Apes and Ape Lore in the Middle Ages and the Renaissance*. London: Warburg Institute, 1952.

Jaussaud, Philippe and Édouard-Raoul Brygoo, *Du jardin au Muséum en 516 biographies*. Paris: Muséum national d'histoire naturelle, 2004.

Jenson, Nicolas. *Biblia Latina*. Venice, 1476.

Jones, Owen. *The Grammar of Ornament*. London: Day & Son, 1856.

Jussieu, Antoine-Laurent de. *Genera plantarum*. Paris: Herissant & Barrois, 1789.

Kandinsky, Vassily. *Point and Line to Plane: Contribution to the Analysis of the Pictorial Elements* [1926]. Translated by Howard Dearstyne and Hilla Rebay. New York: Solomon R. Guggenheim Foundation, 1947.

Kaselow, H. *Die Schaulust am exotischen Tier. Studien zur Darstellung des zoologischen Gartens in der Malerei des 19. und 20. Jahrhunderts*. Hildesheim: Georg Olms, 1999.

Keraudren-Aymonin, Monique. "Le *Carporama* de L.M.A. Robillard d'Argentelle." *Bulletin du Muséum national d'histoire naturelle*, series no. 4, no. 1 (1979), p. 117–149.

Keynes, Richard Darwin. *Fossils, Finches and Fuegians: Darwin's Adventures and Discoveries on the* Beagle, *1832–1836*. London: Harper Collins, 2002.

Kircher, Athanasius. *Arca Noë, in tres libros digesta, quorum I. de rebus quae ante diluvium, II. de iis, quae ipso diluvio ejusque duratione, III. de iis, quae post diluvium a Noëmo gesta sunt, quae omnia novâ methodo, nec non summa argumentorum varietate, explicantur, & demonstrantur.* Amsterdam: Johannes Janssonius van Waesberge, 1675.

Kolbert, Elizabeth. *The Sixth Extinction: An Unnatural History.* New York: Henry Holt and Company, 2014.

Koyré, Alexandre. *From the Closed World to the Infinite Universe.* Baltimore: Johns Hopkins University Press, 1957.

Kubin, Alfred. *Aus meinem Leben* [1911]. Edited by Ulrich Riemerschmidt. Munich: Ellerman, 1974.

Kupka, František. *La Création dans les arts plastiques*. Paris: Éditions du Cercle d'art, 1989.

Kury, Lorelai. *Histoire naturelle et voyages scientifiques*. Paris: L'Harmattan, 2011.

Lack, Hans Walter. *Un jardin d'Eden. Chefs-d'œuvre de l'illustration botanique*. Cologne: Taschen, 2001.

Lamarck, Jean-Baptiste de. *Hydrogéologie, ou Recherches sur l'influence qu'ont les eaux sur la surface du globe terrestre*. Paris: Agasse, 1802.

Lamarck, Jean-Baptiste de. *Philosophie zoologique ou exposition des considérations relatives à l'histoire naturelle des animaux*. Paris: Dentu, 1809.

Lamarck, Jean-Baptiste de. *Histoire naturelle des animaux sans vertèbres*. Paris: Verdière, 1817.

Lamarck, Jean-Baptiste de. *Système analytique des connaissances positives de l'homme*. Paris: Chez l'auteur and Belin, 1820.

Larrère, Catherine and Raphaël. *Du bon usage de la nature. Pour une philosophie de l'environnement*. Paris: Aubier, 1997. Reprint, Paris: Flammarion, 2009.

Larson, Barbara. *The Dark Side of Nature: Science, Society and the Fantastic in the Work of Odilon Redon*. University Park, PA: Pennsylvania State University Press, 2005.

Larson, Barbara and Fae Brauer, eds. *The Art of Evolution: Darwin, Darwinisms, and Visual Culture*. Hanover, NH: Dartmouth College Press, 2009.

Larson, Barbara and Sabine Flach, eds. *Darwin and Theories of Aesthetics and Cultural History*. Burlington, VT: Ashgate, 2013.

Latour, Bruno. *Face à Gaïa. Huit conférences sur le nouveau régime climatique*. Paris: La Découverte, 2015.

Laubier, Lucien. *Ténèbres océanes, le triomphe de la vie dans les abysses*. Paris: Buchet-Chastel, 2008.

Leakey, Richard E. and Roger Lewin. *The Sixth Extinction: Patterns of Life and the Future of Humankind*. New York: Anchor Books, 1995.

Lecointre, Guillaume and Hervé Le Guyader. *Classification phylogénétique du vivant*. 4th ed. 2 vols. Paris: Belin, 2016–2017.

Lemire, Michel. *Artistes et mortels*. Paris: Chabaud, 1990.

Letissier, Georges and Michel Prum. *Héritage de Charles Darwin dans les cultures européennes*. Paris: L'Harmattan, 2011.

Leys, Simon, ed. *La Mer dans la littérature française*. 2 vols. Paris: Plon, 2003.

Limoges, Camille. *La Sélection naturelle. Étude sur la première constitution d'un concept, 1837–1859*. Paris: Presses universitaires de France, 1970.

Linnaeus, Carl. *Amoenitates academicae, seu Dissertationes variae, physicae, medicae, botanicae, antehac seorsim editae nunc collectae et auctae cum tabulis aeneis*. Vol. 1. Leiden: Cornelium Haak, 1749.

Linnaeus, Carl. *Systema naturae, per regna tria naturae, secundum classes, ordines, genera, species, cum characteribus, differentiis, synonymis, locis*. Stockholm: Laurent Salvius, 1758–1759.

Loisel, Gustave. *Histoire des ménageries de l'Antiquité à nos jours*. 3 vols. Paris: Octave Doin et fils; Henri Laurens, 1912.

Loison, Laurent. *Qu'est-ce que le néo-lamarckisme ? Les biologistes français et la question de l'évolution des espèces*. Paris: Vuibert, 2010.

Lombroso, Cesare. *L'uomo delinquente*. 5th ed. with plates. 4 vols. Turin: Bocca, 1897.

Lorenzi, Camille. "L'Aquarium en France au XIXe siècle, 1855–1870. Diffusion sociale et culturelle d'un objet scientifique." Master's thesis 1, Paris 1 Panthéon-Sorbonne, 2008.

Lovejoy, Arthur Oncken. *The Great Chain of Being: A Study of the History of an Idea*. Cambridge, MA: Harvard University Press, 1976.

Lovelock, James E. *Gaia: A New Look at Life on Earth*. Oxford: Oxford University Press, 1979.

Lucretius. *On the Nature of Things*. Translated by W.H.D. Rouse. Cambridge, MA: Harvard University Press, 1992.

Lugli, Adalgisa. *Naturalia et mirabilia. Les cabinets de curiosités en Europe* [1983]. Paris: Adam Biro, 1998.

Lyell, Charles. *Principles of Geology, Being an Attempt to Explain the Former Changes of the Earth's Surface, by Reference to Causes Now in Operation* [1830–1833]. Reproduced with an introduction by Martin J. S. Rudwick. 3 vols. Chicago: University of Chicago Press, 1990–1991.

MacGregor, Arthur. *Curiosity and Enlightment: Collectors and Collections from the Sixteenth to the Nineteenth Century*. New Haven: Yale University Press, 2007.

Maintenant, Elvire de. "Michel Garnier, peintre de genre sous la Révolution." *L'Estampille – L'Objet d'art*, no. 370 (2002), p. 76–82.

Marchal, Hugues, ed. *Muses et ptérodactyles. La poésie de la science de Chénier à Rimbaud*. Paris: Seuil, 2013.

Marguery, Ernest. *L'Œuvre d'art et l'évolution* [1899]. 2nd ed. Paris: Alcan, 1904.

Maris, Virginie. *La Part sauvage du monde. Penser la nature dans l'Anthropocène*. Paris: Seuil, 2018.

Marret, Bertrand. *Portraits de l'artiste en singe. Les singeries dans la peinture*. Paris: Somogy, 2001.

Maury, L.-F.-Alfred. *Le Sommeil et les rêves*. Paris: Didier, 1861.

Mayr, Ernst. *The Growth of Biological Thought. Diversity, Evolution and Inheritance*. Cambridge, MA: Harvard University Press, 1985.

McGowan, Christopher. *The Dragon Seekers. The Discovery of Dinosaurs during the Prelude to Darwin*. London: Little, Brown, 2002.

Mellerio, André. *Odilon Redon* [1913]. New York: DaCapo Press, 1968.

Meurer, Moritz. *Meurer's Pflanzenbilder. Ornamental verwerthbare Naturstudien für Architekten, Kunsthandwerker, Musterzeichner*. 20 vols. Dresden: Kühtmann, 1896–1903.

Michelet, Jules. *La Mer*. Paris: Hachette, 1861. Translated as *The Sea*. New York: Rudd & Carleton, 1861.

Mills, Eric. *Biological Oceanography: An Early History, 1870–1960*. Ithaca: Cornell University Press, 1989.

Mondrian, Piet. *Le Néo-Plasticisme*. Paris: L'Effort moderne, 1920.

Morel, Benedict-Auguste. *Traité des dégénérescences physiques, intellectuelles et morales de l'espèce humaine et des causes qui produisent ces variétés maladives*. 2 vols. Paris: Baillière, 1857.

Mortillet, Gabriel de. *Le Préhistorique. Antiquité de l'homme*. Paris: Reinwald, 1883.

Morton, Marsha. *Max Klinger and Wilhelmine Culture: On the Threshold of German Modernism*. Burlington, VT: Ashgate, 2014.

Morton, Mary. *Oudry's Painted Menagerie. Portraits of Exotic Animals in Eighteenth-Century Europe*. Los Angeles: J. Paul Getty Museum, 2007.

Moser, Stephanie. *Ancestral Images: The Iconography of Human Origins*. Ithaca: Cornell University Press, 1998.

Mosser, Monique and Georges Teyssot, eds. *The Architecture of Western Gardens: A Design History from the Renaissance to the Present Day*. Cambridge, MA: MIT Press, 1991.

Naish, Darren. "The Dinosaurs of Crystal Palace: Among the Most Accurate Renditions of Prehistoric Life Ever Made." *Scientific American* (August 25, 2016).

Nakov, Andréi. *Kazimir Malewicz, le peintre absolu*. Paris: Thalia, 2006.

Nodier, Charles. "De la fin prochaine du genre humain." *Revue de Paris*, vol. 26 (May 1831), p. 224–240.

Nodier, Charles. "De la palingénésie humaine et de la résurrection." *Revue de Paris*, vol. 41 (August 1832), p. 81–106.

Nordau, Max. *Entartung*. 2 vols. Berlin: Duncker, 1892–1893. Translated from the second edition as *Degeneration*. New York: D. Appleton and Company, 1895.

Olmi, Giuseppe. *Ulisse Aldrovandi. Scienza e natura nel secondo Cinquecento*. Trento: Libera Università degli Studi, 1976.

Olmi, Giuseppe. *L'inventario del mondo. Catalogazione della natura e luoghi del sapere nella prima età moderna*. Bologna: Il Mulino, 1992.

O'Mahony, Claire I. R., ed. *Symbolist Objects: Materiality and Subjectivity at the Fin de Siècle*. High Wycombe: Rivendale Press, 2009.

Owen, Richard. *On the Archetype and Homologies of the Vertebrate Skeleton*. London: Richard and Taylor, 1848.

Paley, William. *Natural Theology: Or, Evidences of the Existence and Attributes of the Deity*. London: J. Faulder, 1802.

Palmer, Alfred H. *The Life of Joseph Wolf, Animal Painter*. London: Longmans, 1895.

Peron, François. *Voyage of Discovery to the Southern Hemisphere, Performed by Order of the Emperor Napoleon during the Years 1801, 1802, 1803, and 1804* [1807]. London: Richard Phillips, 1809.

Perrier, Edmond. *Les Colonies animales et la formation des organismes*. Paris: Masson, 1881.

Pick, Daniel. *Faces of Degeneration: A European Disorder, c. 1848–c. 1918*. Cambridge: Cambridge University Press, 1993.

Pierre, Arnauld. *Frank Kupka: In White and Black*. Liverpool: Liverpool University Press/Artists Bookworks, 1998.

Pierre, Arnauld. *Maternités cosmiques. La recherche des origines, de Kupka à Kubrick*. Paris: Hazan, 2010.

Pietsch, Theodore W. *Trees of Life: A Visual History of Evolution*. Baltimore: Johns Hopkins University Press, 2012.

Pinault, Madeleine. *Le Peintre et l'histoire naturelle*. Paris: Flammarion, 1990.

Pluche, Abbé Antoine. *Le Spectacle de la nature, ou Entretiens sur les particularités de l'histoire naturelle qui ont paru les plus propres à rendre les jeunes gens curieux et à leur former l'esprit*. 9 vols. Paris: Vve Estienne, 1739.

Pomian, Krzysztof. *Collectionneurs, amateurs et curieux (Paris, Venise, XVIe–XVIIIe siècle)*. Paris: Gallimard, 1987.

Prodger, Phillip. *Darwin's Camera: Art and Photography in the Theory of Evolution*. Oxford: Oxford University Press, 2009.

Prum, Richard O. *The Evolution of Beauty: How Darwin's Forgotten Theory of Mate Choice Shapes the Animal World—and Us*. New York: Doubleday, 2017.

Ragan, Mark A. "Trees and networks before and after Darwin." *Biology Direct*, vol. 4, no. 1 (2009), article no. 43, https://biologydirect.biomedcentral.com/articles/10.1186/1745-6150-4-43.

Reclus, Élisée. *L'Homme et la Terre*. Vol. 1: *Les Ancêtres. Histoire ancienne*. Paris: Librairie universelle, 1905.

Redon, Odilon. *À soi-même. Journal, 1867–1915. Notes sur la vie, l'art et les artistes* [1922]. Paris: José Corti, 1979. Translated by Mira Jacob and Jeanne L. Wasserman as *To Myself: Notes on Life, Art, and Artists*. New York: George Braziller, 1986.

Redouté, Pierre-Joseph. *Les roses, peintes par P.J. Redouté . . . décrites et classées selon leur ordre naturel par C. A. Thory*. 3 vols. Paris: printed by Crapelet for P. Dufart, 1828–1829.

Réja, Marcel. *L'Art chez les fous. Le dessin, la prose, la poésie*. Paris: Mercure de France, 1907.

Rice, Anthony. *Voyages of Discovery: Three Centuries of Natural History Exploration*. New York: Clarkson Potter, 1999.

Richard, Nathalie. *L'Invention de la préhistoire. Anthologie*. Paris: Presses Pocket, "Agora, les classiques" series, 1992.

Richard, Nathalie. *Inventer la préhistoire. Les débuts de l'archéologie préhistorique en France*. Paris: Vuibert, 2008.

Richards, Evelleen. *Darwin and the Making of Sexual Selection*. Chicago: University of Chicago Press, 2017.

Richards, Robert J. *The Meaning of Evolution: The Morphological Construction and Ideological Reconstruction of Darwin's Theory*. Chicago: University of Chicago Press, 1992.

Richards, Robert J. *The Tragic Sense of Life: Ernst Haeckel and the Struggle over Evolutionary Thought*. Chicago: University of Chicago Press, 2008.

Ridley, Glynis. *Clara's Grand Tour: Travels with a Rhinoceros in Eighteenth-Century Europe*. New York: Grove Press, 2004.

Ritvo, Harriet. *The Animal Estate: The English and Other Creatures in the Victorian Age*. Cambridge, MA: Harvard University Press, 1989.

Robbins, Louise E. *Elephant Slaves and Pampered Parrots: Exotic Animals in Eighteenth-Century Paris*. Baltimore: Johns Hopkins University Press, 2002.

Roger, Jacques. "Darwin, Haeckel et les Français." In *De Darwin au darwinisme. Science et idéologie*. Proceedings of the international congress for the centenary of Darwin's death (Paris, Chantilly, September 13–16, 1982), edited by Yvette Conry, p. 158–160. Paris: Vrin, 1983.

Roger, Jacques. *Buffon. Un philosophe au Jardin du Roi*. Paris: Fayard, 1989.

Rosny aîné, J.-H. *Les Origines* [1895]. In J.-H. Rosny aîné, *Romans préhistoriques*. Paris: Robert Laffont, "Bouquins" series, 1985.

Rudwick, Martin J. S. *The Meaning of Fossils: Episodes in the History of Palaeontology* [1972]. Chicago: University of Chicago Press, 1985.

Rudwick, Martin J. S. *Scenes from Deep Time: Early Pictorial Representations of the Prehistoric World*. Chicago: University of Chicago Press, 1992.

Rudwick, Martin J. S. *Earth's Deep History: How It Was Discovered and Why It Matters*. Chicago: University of Chicago Press, 2014.

Ruse, Michael. *The Darwinian Revolution. Science Red in Tooth and Claw*. Chicago: University of Chicago Press, 1979.

Ruskin, John. *The Eagle's Nest: Ten Lectures on the Relation of Natural Science to Art*. New York: Maynard, Merrill & Co., 1893.

Sandström, Sven. *Le Monde imaginaire d'Odilon Redon. Étude iconologique*. Lund: Berlingska Boktryckeriet, 1955.

Saussure, Horace-Benedict de. *Voyage dans les Alpes*. Neuchâtel: Samuel Fauche, 1779–1796.

Scheuchzer, Johann Jakob. *Physica sacra*. 4 vols. Augsburg and Ulm: Wagner, 1731–1735.

Schleiden, Matthias Jacob. *Die Pflanze und ihr Leben*. Leipzig: Wilhelm Engelmann, 1848. Translated by Arthur Henfrey as *The Plant; a Biography*. London: Hippolyte Baillière, 1848.

Schnapper, Antoine. *Le Géant, la licorne et la tulipe. Les cabinets de curiosités en France au XVIIe siècle*. Paris: Flammarion, "Champs" series, 2012.

Secord, James. *Victorian Sensation: The Extraordinary Publication, Reception and Secret Authorship of Vestiges of the Natural History of Creation*. Chicago: University of Chicago Press, 2003.

Secord, James A. "Monsters at the Crystal Palace." In *Models: The Third Dimension of Science*, edited by Soraya De Chadarevian and Nick Hopwood, p. 138–169. Stanford: Stanford University Press, 2004.

Séginger, Gisèle. *Naissance et métamorphose d'un écrivain. Flaubert et Les Tentations de saint Antoine*. Paris: Champion, 1997.

Sloan, Phillip R., Gerald McKenny and Kathleen Eggleson, eds. *Darwin in the Twenty-First Century: Nature, Humanity, and God*. Notre Dame, IN: University of Notre Dame Press, 2015.

Stead, Evanghélia. *Le Monstre, le singe et le fœtus. Tératogonie et décadence dans l'Europe fin-de-siècle*. Geneva: Librairie Droz, 2004.

Steigerwald, Joan. *Experimenting at the Boundaries of Life: Organic Vitality in Germany around 1800*. Pittsburgh, PA: University of Pittsburgh Press, 2019.

Taquet, Philippe, ed. *Les Bonnes Feuilles du Jardin des Plantes*. Paris: Artlys/Muséum national d'histoire naturelle, 2013.

Tassy, Pascal.
L'Arbre à remonter le temps.
Paris: Bourgois, 1991.

Tassy, Pascal.
L'Invention du mastodonte.
Paris: Belin pour la science, 2009.

Taylor, Basil.
Stubbs. London: Phaidon, 1971.

Thomas, Keith.
Man and the Natural World: Changing Attitudes in England 1500–1800.
Harmondsworth: Penguin Books, 1983.

Thompson, D'Arcy Wentworth.
On Growth and Form.
Cambridge: Cambridge University Press, 1917.

Tillier, Bertrand.
"Pastorales satiriques."
In *Caricaturer Pasteur*, exh. cat., p. 24–29.
Besançon: Éditions du Sekoya; Dole: Musée Pasteur, 2014.

Tort, Patrick, ed.
Dictionnaire du darwinisme et de l'évolution. 3 vols.
Paris: Presses universitaires de France, 1996.

Tyson, Edward.
Orang-outang, sive Homo Sylvestris: or, The Anatomy of a Pygmie Compared with That of a Monkey, an Ape, and a Man . . .
London: Thomas Bennet & Daniel Brown, 1699.

Venayre, Sylvain.
Panorama du voyage, 1780–1920. Paris: Les Belles Lettres, 2012.

Verne, Jules.
Journey to the Centre of the Earth [1864].
Translated by William Butcher. Oxford: Oxford University Press, 1992.

Verne, Jules.
Twenty Thousand Leagues Under the Sea [1871].
Translated by William Butcher. Oxford: Oxford University Press, 1998.

Villepreux-Power, Jeanne.
Observations physiques sur le poulpe de l'Argonauta Argo, commencées en 1832 et terminées en 1843.
Paris: Charles de Mourgues Frères, 1856.

Viola, Jerome.
"Redon, Darwin and the Ascent of Man."
Marsyas, vol. 11 (1962–1964), p. 42–57.

Virey, Julien-Joseph.
Histoire naturelle du genre humain. Vol. 2. Paris: Crochard, 1824.

Virgil's Eclogues.
Translated by Barbara Hughes Fowler. Chapel Hill: The University of North Carolina Press, 1997.

Voisenet, Jacques.
Bêtes et hommes dans le monde médiéval. Le bestiaire des clercs du v*e au* xii*e siècle.* Turnhout: Brepols, 2000.

Voss, Julia.
Darwin's Pictures: Views of Evolutionary Theory, 1837–1874. Translated by Lori Lantz. New Haven: Yale University Press, 2010.

Voss, Julia.
Hilma af Klint: Die Menschheit in Erstaunen versetzen. Frankfurt am Main: Fischer Verlag, 2020.

Wagner, Rudolph.
"The Numerical Relations of Animals." *The Edinburgh New Philosophical Journal*, vol. 15 (April–October 1833), p. 221–227.

Wallace, Alfred Russel.
Miracles and Modern Spiritualism. London: James Burns, 1875.

Wallace, Alfred Russel.
"The Colours of Animals and Plants: I and II."
Macmillan's Magazine, vol. 36 (September and October 1877), p. 384–408 and p. 464–471.

Wanlin, Nicolas.
"Évolutions. Littérature et sciences au xix*e* siècle."
Habilitation thesis, Université Paris-Diderot, 2018.

Wat, Pierre.
Pérégrinations. Paysages entre nature et histoire.
Paris: Hazan, 2017.

Welsh, Robert.
"Mondrian and Theosophy." In *Piet Mondrian, 1872–1944: Centennial Exhibition*, exh. cat., p. 43–48. New York: Solomon R. Guggenheim Foundation, 1971.

Wichmann, Siegfried.
Jugendstil floral funktional in Deutschland und Österreich und den Einflussgebieten. Zurich: Atlantis, 1984. Translated as *Jugendstil Art Nouveau: Floral and Functional Forms.* Boston: Little, Brown, 1984.

Willmann, Rainer and Julia Voss, eds.
The Art and Science of Ernst Haeckel.
Cologne: Taschen, 2017.

Wulf, Andrea.
The Invention of Nature.
London: John Murray Press, 2015.

Yasui, Hiroo.
"Le baron de Domecy et les décors de la salle à manger du château de Domecy." In *Flore d'Odilon Redon*, exh. cat. edited by Hiroo Yasui, p. 190–198. Tokyo: Mitsubishi Ichigokan Museum, 2018.

Yates, Frances A.
The Art of Memory.
London: Routledge & Kegan Paul, 1966.

Zalasiewicz, Jan, Colin N. Waters, Mark Williams and Colin P. Summerhayes, eds. *The Anthropocene as a Geological Time Unit.* Cambridge: Cambridge University Press, 2019.

EXHIBITION CATALOGUES

Kandinsky und München: Begegnungen und Wandlungen, 1896–1914.
Edited by Armin Zweite. Munich: Prestel, 1982.

Kandinsky in Paris, 1934–1944.
Edited by Vivian Endicott Barnett and Christian Derouet. New York: Solomon R. Guggenheim Museum, 1985.

The Spiritual in Art: Abstract Painting, 1890–1985.
Edited by Maurice Tuchman and Judi Freeman. Los Angeles: County Museum of Art; New York: Abbeville Press, 1986.

Vienne 1880–1938. L'Apocalypse joyeuse.
Edited by Jean Clair. Paris: Centre Pompidou, 1986.

Courbet Reconsidered.
Edited by Sarah Faunce and Linda Nochlin. New Haven: Yale University Press, 1988.

In the Realm of the Wild: The Art of Bruno Liljefors of Sweden. Edited by Bjorn Fredlund. Goteborgs: Konstmuseum, 1988.

L'Âme au corps. Arts et sciences, 1793–1993.
Edited by Jean Clair.
Paris: Réunion des musées nationaux/Gallimard/Electa, 1993.

Darwin und Darwinismus.
Edited by Bodo-Michael Baumunk and Jürgen Riess. Dresden: Deutsches Hygiene-Museum/Akademie Verlag, 1994.

Odilon Redon: Prince of Dreams, 1840–1916.
Edited by Douglas Druick et al. Chicago: The Art Institute of Chicago; New York: Abrams, 1994.

Okkultismus und Avantgarde. Von Munch bis Mondrian, 1900–1915.
Edited by Veit Loers and Pia Witzmann. Frankfurt: Schirn Kunsthalle/Tertium, 1995.

Dessiner la Nature. Dessins et manuscrits des bibliothèques de France xvii*e*–xviii*e*–xix*e* *siècles.*
Edited by Madeleine Pinault-Sorensen. Paris: Fondation Electra, 1996.

The Age of Rossetti, Burne-Jones and Watts: Symbolism in Britain, 1860–1910.
Edited by Andrew Wilton and Robert Upstone. London: Tate Gallery, 1997.

Cosmos: From Romanticism to the Avant-Garde.
Edited by Jean Clair. Montreal: The Montreal Museum of Fine Arts; Munich: Prestel, 1999.

1900.
Edited by Philippe Thiébaut. Paris: Réunion des musées nationaux, 2000.

1900: Art at the Crossroads.
Edited by Robert Rosenblum, Maryanne Stevens and Anne Dumas. London: Royal Academy of Arts Publications, 2000.

Arnold Böcklin.
Edited by Katharina Schmidt, Christian Lenz and Marie-Pierre Salé. Paris: Musée d'Orsay/Réunion des musées nationaux, 2001.

Athanasius Kircher. Il Museo del mondo.
Edited by Eugenio Lo Sardo. Rome: De Luca, 2001.

Ornament and Abstraction.
Edited by Markus Brüderlin. Basel: Beyeler Foundation; Cologne: DuMont, 2001.

Aux origines de l'abstraction, 1800–1914.
Edited by Serge Lemoine and Pascal Rousseau. Paris: Musée d'Orsay/Réunion des musées nationaux, 2003.

Gallé, le testament artistique.
Edited by Philippe Thiébaut. Paris: Musée d'Orsay/Hazan, 2004.

La Mer, terreur et fascination.
Edited by Alain Corbin and Hélène Richard. Paris: Bibliothèque nationale de France/Seuil, 2004.

Fierce Friends: Artists and Animals, 1700–1900.
Edited by Louise Lippincott and Andreas Blühm. London: Merrell, 2005.

Visions du Déluge de la Renaissance au xix*e siècle.*
Edited by Rémi Cariel and Sylvie Wuhrmann. Paris: Réunion des musées nationaux, 2006.

Gustave Courbet.
Edited by Laurence des Cars and Dominique de Font-Reaulx. Paris: Réunion des musées nationaux, 2007.

Jean Carriès, 1855–1894. La matière de l'étrange.
Edited by Amélie Simier. Paris: Paris-Musées/Nicolas Chaudun, 2007.

Darwin and the Search for Origins.
Edited by Pamela Kort and Max Hollein. Cologne: Wienand, 2009.

Endless Forms: Charles Darwin, Natural Science and the Visual Arts.
Edited by Diana Donald and Jane Munro. New Haven and London: Yale University Press, 2009.

Gabriel von Max: Malerstar, Darwinist, Spiritist.
Edited by Karin Althaus. Munich: Hirmer, 2010.

Sciences et curiosités à la cour de Versailles.
Edited by Catherine Arminjon and Béatrix Saule. Paris: Réunion des musées nationaux, 2010.

L'Europe des esprits, ou la fascination de l'occulte, 1750–1950.
Edited by Serge Fauchereau and Joëlle Pijaudier-Cabot. Strasbourg: Musées de Strasbourg, 2011.

Odilon Redon, prince du rêve, 1840–1916.
Edited by Rodolphe Rapetti. Paris: Réunion des musées nationaux, 2011.

1740, un abrégé du monde. Savoirs et collections autour de Dezallier d'Argenville.
Edited by Anne Lafont. Paris: Institut national d'histoire de l'art; Lyon: Fage Éditions, 2012.

Henry van de Velde: Passion, Function, Beauty, 1863–1957.
Edited by Thomas Föhl, Sabine Walter and Werner Adriaenssens. Tielt, Belgium: Lannoo, 2013.

Cet obscur objet de désirs. Autour de L'Origine du monde.
Paris: Lienart, 2014.

Le Monde de Darwin.
Edited by Guillaume Lecointre and Patrick Tort. Paris: La Martinière, 2015.

La Nature silencieuse. Paysages d'Odilon Redon.
Edited by Sophie Barthélémy et al. Bordeaux: Musée des Beaux-Arts; Gand: Snoeck, 2016.

Form Follows Flower: Moritz Meurer, Karl Blossfeldt & Co.
Edited by Angela Nikolai and Sabine Thümmler. Berlin: Deutscher Kunstverlag, 2017.

Hilma af Klint: Paintings for the Future.
Edited by Tracey Bashkoff. New York: Solomon R. Guggenheim Museum, 2018.

Kupka, pionnier de l'abstraction. Edited by Brigitte Léal, Markéta Theinhardt and Pierre Brullé. Paris: Grand Palais/Réunion des musées nationaux, 2018.

L'École de Nancy. Art nouveau et industrie d'art.
Edited by Nicolas Neumann. Nancy: Musée de l'École de Nancy/Paris: Somogy, 2018.

Né(e)s de l'écume et des rêves. Edited by Annette Haudiquet, Denis-Michel Boell and Marc Donnadieu. Oissel-sur Seine: Octopus Éditions, 2018.

Préhistoire, une énigme moderne.
Edited by Cécile Debray, Rémi Labrusse and Maria Stavrinaki. Paris: Centre Pompidou, 2019.

COVER: The Jack Daulton Collection. Don Tuttle, photographer

1: © The Trustees of the Natural History Museum, London / **2**: Private Collection / **3**: © MAD, Paris / Jean Tholance / **4**: akg-images / **5**: Photo BnF / **6**: Cliché Mazarine / **7**: Digital image courtesy of the Getty's Open Content Program / **8**: Bridgeman Images / **10**: Photo © MBA, Rennes, Dist. RMN-Grand Palais / Louis Deschamps / **11**: © Muséum national d'histoire naturelle / **12**: © Christie's Images Ltd - ARTOTHEK / **13**: Photo © RMN-Grand Palais (musée du Louvre) / René-Gabriel Ojéda / **14**: Photo © Archives Alinari, Florence, Dist. RMN-Grand Palais / Mauro Magliani / **15, 16 A AND B**: © Muséum national d'histoire naturelle / **17**: Photo © RMN-Grand Palais / Philippe Fuzeau / **18**: Photo © Archives Alinari, Florence, Dist. RMN-Grand Palais / Luciano Pedicini / **19**: akg-images / Rabatti & Domingie / **20**: © The Trustees of the Natural History Museum, London / **21**: Photo © RMN-Grand Palais (musée du Louvre) / Hervé Lewandowski / **22**: Photo © RMN-Grand Palais (musée du Louvre) / René-Gabriel Ojéda / **23**: Photo © RMN-Grand Palais (musée du Louvre) / Tony Querrec / **25**: Art Collection / Alamy Banque d'Images / **26**: Photo © RMN-Grand Palais (musée du Louvre) / Gérard Blot / **27**: Photo © RMN-Grand Palais (musée du Louvre) / Jean-Gilles Berizzi / **28**: Photo © RMN-Grand Palais (musée du Louvre) / Michel Urtado / **29**: Photo © BnF, Dist. RMN-Grand Palais / image BnF / **30**: © Muséum national d'histoire naturelle / **31**: Photo © RMN-Grand Palais (musée du Louvre) / Jean-Gilles Berizzi / **32**: Photo © Archives Alinari, Florence, Dist. RMN-Grand Palais / Mauro Magliani / **33**: Archive, Art, Culture and Historical Heritage Head Office Department, Intesa Sanpaolo / **34**: Photo © BPK, Berlin, Dist. RMN-Grand Palais / image Staatliches Museum Schwerin / **35 TO 46**: © Muséum national d'histoire naturelle, dist. RMN / Tony Querrec / **47**: Courtesy National Gallery of Art, Washington / **48**: © Muséum national d'histoire naturelle / **49**: Berlin-Brandenburgische Akademie der Wissenschaften, photo: Judith Affolter / **50 A AND B**: © Muséum national d'histoire naturelle / **51, 52**: © National Maritime Museum, Greenwich, London / **53**: © Muséum national d'histoire naturelle / **54**: Royal Collection Trust / © Her Majesty Queen Elizabeth II 2020 / **55**: Photo © BPK, Berlin, Dist. RMN-Grand Palais / Jörg P. Anders / **56, 57**: © Muséum national d'histoire naturelle / **58, 59**: Stiftung Preußische Schlösser und Gärten Berlin-Brandenburg / Photo: Roland Handrick / **60 TO 66**: Le Havre (France), Muséum d'histoire naturelle / **67**: Collection Kharbine-Tapabor / **68**: © Christie's Images, 2020 / **69 A TO 69 E**: Photo BnF / **70**: CC0 Paris Musées / Petit Palais, musée des Beaux-Arts de la Ville de Paris / **71**: Courtesy of the Indianapolis Museum of Art at Newfields / **72**: Natural History Museum Vienna / **73 A AND B TO 79**: © Muséum national d'histoire naturelle / **80**: Photo © Bridgeman Images / **81**: Archives du 7e Art / Star-Film / **82**: Collection ChristopheL / **83**: © Muséum national d'histoire naturelle / **84**: Yale Center for British Art / **85**: Photo BnF / **86 TO 88**: © Muséum national d'histoire naturelle / **89**: Photo © BPK, Berlin, Dist. RMN-Grand Palais / image Staatsgalerie Stuttgart / **90**: © MuMa Le Havre / Florian Kleinefenn / **91**: Courtesy of the Reading Public Museum, Reading, Pennsylvania / **92**: Image courtesy Dallas Museum of Art / **93, 94**: Photo © Tate / **95**: Photo © Tate, London, Dist. RMN-Grand Palais / Tate Photography / **96**: Photo © Tate / **97**: Image courtesy of the National Gallery of Art, Washington / **98**: National Museum of Wales / **99**: © 2019. Sedgwick Museum of Earth Sciences, University of Cambridge. Reproduced with permission / **100**: Wellcome Collection / **101**: © British Library Board. All Rights Reserved / Bridgeman Images / **102, 103**: © The Trustees of the Natural History Museum, London / **104**: Universal History Archive / UIG / Bridgeman Images / **105**: Bridgeman Images / **106**: © Muséum national d'histoire naturelle / **107, 108**: Photo BnF / **109**: Photo © RMN-Grand Palais / Agence Bulloz / **110**: © Muséum national d'histoire naturelle / **111**: Photo BnF / **112, 113**: © Muséum national d'histoire naturelle / **114**: Photo © RMN-Grand Palais (musée de la Préhistoire des Eyzies) / Franck Raux / **115**: Reims, Musée des Beaux-Arts / © Photo: C. Devleeschauwer / **116 A TO 116 D**: Photo BnF / **117, 118**: © Muséum national d'histoire naturelle / **119**: Nationalmuseum / **120**: © Ashmolean Museum, University of Oxford / **121**: Cambridge University Library / **122 TO 124**: © Muséum national d'histoire naturelle / **125**: Image Courtesy the Biodiversity Heritage Library / **126 TO 128**: © Muséum national d'histoire naturelle / **129**: © Ashmolean Museum. University of Oxford / **130**: Photo © Museums Sheffield / Bridgeman Images / **131**: Photo © Musée d'Orsay, Dist. RMN-Grand Palais / Patrice Schmidt / **132**: Bridgeman Images / **133**: © Mallett Gallery, London, UK / Bridgeman Images / **134**: CC0 Paris Musées / Petit Palais, musée des Beaux-Arts de la Ville de Paris / **135**: Bonhams / **136, 137**: © Muséum national d'histoire naturelle / **138**: Cambridge University Library / **139**: Alamy Stock Photo / **140**: Photo © Tate / **141, 142**: Photo © musée d'Orsay / Patrice Schmidt / **143**: Cl. Musées de Sens – E. Berry / **144, 145**: Photo © musée d'Orsay / Patrice Schmidt / **146, 147**: Photo courtesy Musée océanographique de Monaco / Michel Dagnino / **148, 149**: © The Trustees of the Natural History Museum, London / **150 A AND B**: Photo © musée d'Orsay / Patrice Schmidt / **151**: Photo © Gusman / Bridgeman Images / **152**: Granger / Bridgeman Images / **153**: Bridgeman Images / **154**: akg-images / **155**: CC0 Paris Musées / Musée Carnavalet / **156**: Wellcome Collection / **157**: CC0 Paris Musées / Musée Carnavalet / **158**: © Städel Museum – ARTOTHEK / **159**: Photo © RMN-Grand Palais / René-Gabriel Ojéda / **160**: © Muséum national d'histoire naturelle / **161**: Photo © RMN-Grand Palais / René-Gabriel Ojéda / **162**: Paris Musées / Musée Carnavalet / **163**: Photo © Xavier Spertini / Musée Buffon, ville de Montbard / **164**: akg-images / **165, 166**: The Jack Daulton Collection. Don Tuttle, photographer / **167**: Photo © RMN-Grand Palais (musée d'Orsay) / René-Gabriel Ojéda / **168**: © Musée des Beaux-Arts de Dijon / François Jay / **169**: Christie's Images, London / Scala, Florence / **170**: © Musée des Beaux-Arts de Dijon / François Jay / **171**: Photo © Tate / **172**: Photo © RMN-Grand Palais (musée d'Orsay) / Hervé Lewandowski / **173**: The Jack Daulton Collection / **174**: Photo © Galerie Zdenek Sklenár Archive, Martin Polák / **175**: Kunstmuseum Basel, Martin P. Bühler / **176**: Photo © The Philadelphia Museum of Art, Dist. RMN-Grand Palais / image Philadelphia Museum of Art / **177**: Photo © BPK, Berlin, Dist. RMN-Grand Palais / Dietmar Kat / **178**: © Städel Museum - ARTOTHEK / **179**: CC0 Paris Musées / Petit Palais, musée des Beaux-Arts de la Ville de Paris / **180**: Photo © Musée d'Orsay, Dist. RMN-Grand Palais / Patrice Schmidt / **181**: Photo © RMN-Grand Palais / Agence Bulloz / **182**: akg-images / **183**: Photo © The Philadelphia Museum of Art, Dist. RMN-Grand Palais / image Philadelphia Museum of Art / **184**: © 2019 Collection Kröller-Müller Museum, Otterlo, the Netherlands / Photography Rik Klein Gotink, Harderwijk / **185**: Image Courtesy of the Art Institute of Chicago / **186**: © Muséum national d'histoire naturelle / **187**: Image Courtesy of the National Gallery of Art, Washington / **188 A**: Image Courtesy of the Art Institute of Chicago / **188 B TO 188 D**: Image Courtesy of the National Gallery of Art, Washington / **189**: Image Courtesy of the Art Institute of Chicago / **190**: Photo © Musée d'Orsay, Dist. RMN-Grand Palais / Patrice Schmidt / **191 TO 195**: Photo © Musée d'Orsay, Dist. RMN-Grand Palais / Hervé Lewandowski / **196**: Photo © Musée d'Orsay, Dist. RMN-Grand Palais / Patrice Schmidt / **197**: Photo © RMN-Grand Palais (musée d'Orsay) / Hervé Lewandowski / **198**: Bridgeman Images / **199**: Photo © RMN-Grand Palais (musée d'Orsay) / Adrien Didierjean / **200**: Nancy, Musée de l'École de Nancy, cliché Studio Image / **201, 202**: Nancy, Musée de l'École de Nancy, cliché Studio Image / **203, 204**: Photo © Musée d'Orsay, Dist. RMN-Grand Palais / Patrice Schmidt / **205**: Leopold Museum, Vienna / **206**: Musée Crozatier © Luc Olivier / **207**: © Albertina, Wien / **208**: Photo: NGC / **209**: akg-images / **210, 211**: OÖ Landes-Kultur GmbH, Land Oberösterreich, Grafische Sammlung / **212**: Leopold Museum, Vienna / **213**: Kunsthaus Zug, Alois Ottiger / **214**: OÖ Landes-Kultur GmbH, Land Oberösterreich, Grafische Sammlung / **215**: LENTOS Kunstmuseum Linz / **216**: Photo © RMN-Grand Palais (musée d'Orsay) / Hervé Lewandowski / **217**: Courtesy National Gallery of Art, Washington / **218**: Photo © RMN-Grand Palais (musée d'Orsay) / Hervé Lewandowski / **219**: © Muséum national d'histoire naturelle / **220**: CC0 Paris Musées / Maisons de Victor Hugo Paris-Guernesey / **221**: Photo: Munchmuseet / **222**: Image courtesy of Deutscher and Hackett / **224**: Courtesy Museu Nacional d'Art de Catalunya of Barcelona / **225**: Collection Kharbine-Tapabor / **226**: Historical Views / agefotostock / **227 TO 229**: Photo: Munchmuseet / **230**: Photo © Centre Pompidou, MNAM-CCI, Dist. RMN-Grand Palais / Philippe Migeat / **231**: Photo BnF / **232**: Photo © Centre Pompidou, MNAM-CCI, Dist. RMN-Grand Palais / Philippe Migeat / **233**: Photo Jean-Louis Losi, Paris / **234**: Photo Kunstmuseeum den Haag / **235, 236**: By courtesy of the Hilma af Klint Foundation. Photo: Moderna Museet, Stockholm, Sweden / **237**: Interfoto / Alamy Banque d'Images / **238**: Photo © Centre Pompidou, MNAM-CCI, Dist. RMN-Grand Palais / Philippe Migeat / **239**: akg-images / **240**: Photo © Musée d'Orsay, Dist. RMN-Grand Palais / Patrice Schmidt / **241**: Photo © BPK, Berlin, Dist. RMN-Grand Palais / image BPK / **242**: Image courtesy Dallas Museum of Art / **243 TO 245**: Photo © Studio Laurent Grasso, Courtesy Perrotin / **246**: Photo: Claire Dorn, Courtesy Perrotin / **247**: © Muséum national d'histoire naturelle / **248**: Photo Jacques Boulissière 2019 / **249**: © Ashmolean Museum. University of Oxford

RIGHTS HOLDERS

MUSÉES D'ORSAY
ET DE L'ORANGERIE

Head of Publications
Annie Dufour

Editor
Jean-Benoit Ormal-Grenon
assisted by Sonia Georget

Picture research
Sonia Georget

THE MONTREAL
MUSEUM OF FINE ARTS

Head of Publications
Sébastien Hart

Managing Editor
Louise Gauthier

Translators
Valentina Baslyk
Jill Corner
Marcia Couëlle
Alexis Diamond
Susan Le Pan
Donald McGrath
Shelley Pomerance
Kathleen Putnam

Proofreader
Jane Jackel

GALLIMARD

Publications Manager, Illustrated Books
Nathalie Bailleux

Project Manager
Céline Guichard

Production Manager
Caroline Artémon

Production
Amélie Airiau

Co-publications Manager
Mathilde Barrois

Press Attaché
Béatrice Foti

Design and Layout
Line Célo

Colour Separation
Les Artisans du Regard

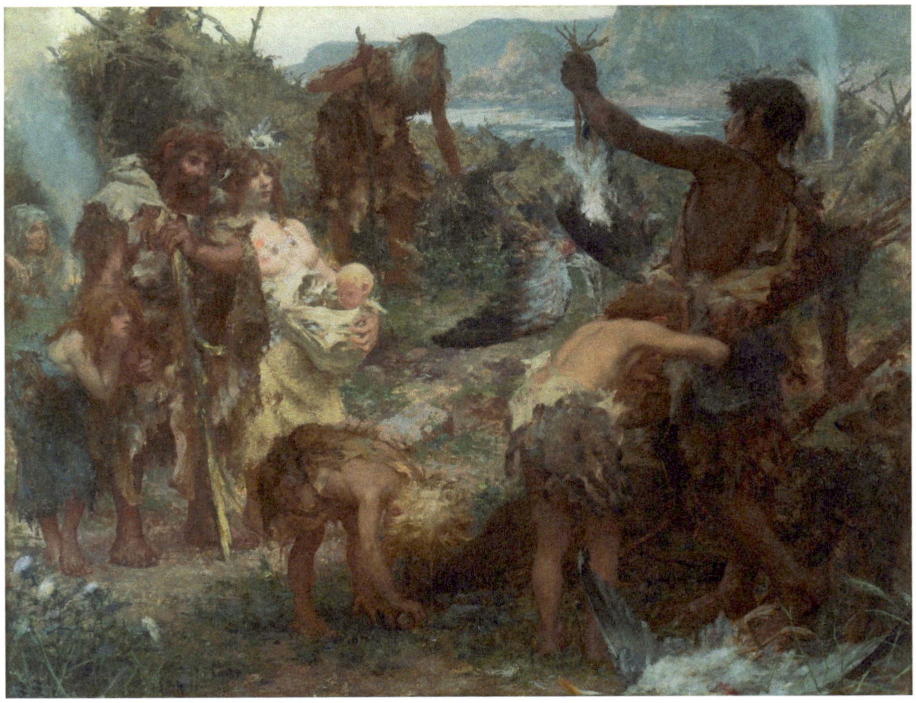

248
René Rousseau-Decelle,
Prehistoric Family, 1906
Oil on canvas, 115 × 145 cm
La Roche-sur-Yon,
Musée de La Roche-sur-Yon

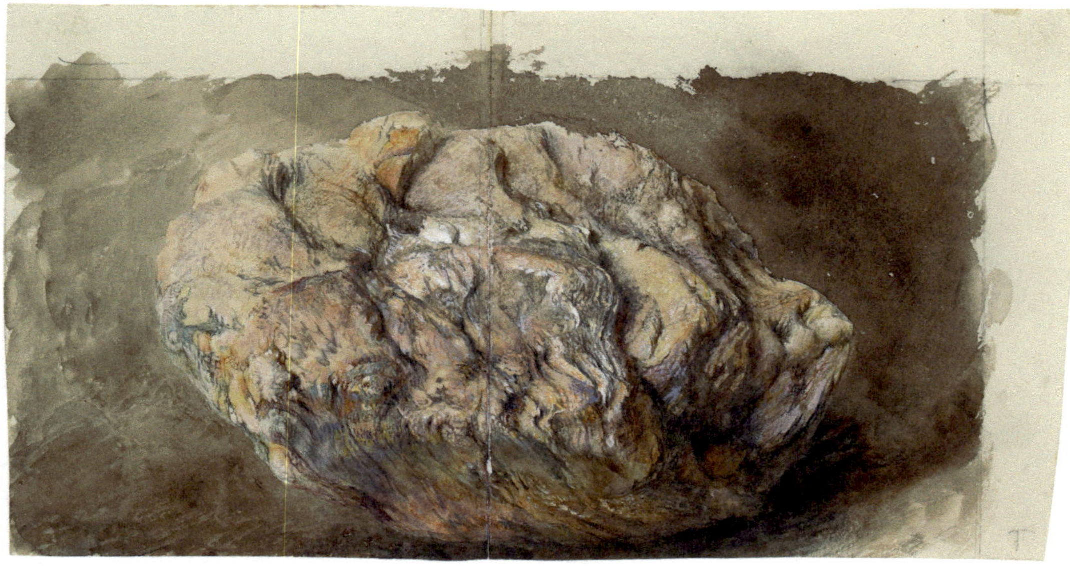

62, rue de Lille – 75007 Paris
ISBN 978-2-35433-324-9

5, rue Gaston-Gallimard –
75007 Paris
www.gallimard.fr
ISBN 978-2-07-292700-3
Edition Number 374856

Printed and bound
by Graphius, Ghent (Belgium)
in February 2021
Legal deposit: February 2021

ON THE COVER
Gabriel von Max,
Abelard and Heloise,
after 1900
Oil on canvas, 41 × 36 cm
The Jack Daulton Collection

249
John Ruskin,
Study in Colour of Quartz Rock Weathered, about 1875
Graphite, watercolour and gouache on wove paper, 13.2 × 24.4 cm
Oxford, Ashmolean Museum, University of Oxford